S0-ADF-929

Mechanical Engineering Series

Frederick F. Ling
Series Editor

Mechanical Engineering Series

Introductory Attitude Dynamics
F.P. Rimrott

Balancing of High-Speed Machinery
M.S. Darlow

Theory of Wire Rope
G.A. Costello

Theory of Vibration
  Vol. I An Introduction
  Vol. II Discrete and Continuous Systems
A.A. Shabana

Laser Machining: Theory and Practice
G. Chryssolouris

Underconstrained Structural Systems
E.N. Kuznetsov

Principles of Heat Transfer in Porous Media
M. Kaviany

Mechatronics: Electromechanics and Contromechanics
D.K. Miu

Structural Analysis of Printed Circuit Board Systems
P.A. Engel

Kinematic and Dynamic Simulation of Multibody Systems: The Real-Time Challenge
J. García de Jalón and E. Bayo

High Sensitivity Moiré: Experimental Analysis for Mechanics and Materials
D. Post, B. Han, and P. Ifju

Principles of Convective Heat Transfer
M. Kaviany

M. Kaviany

Principles of Convective Heat Transfer

With 196 Figures and Charts

Springer-Verlag
New York Berlin Heidelberg London Paris
Tokyo Hong Kong Barcelona Budapest

M. Kaviany
Department of Mechanical Engineering
and Applied Mechanics
The University of Michigan
Ann Arbor, MI 48109-2125
USA

Series Editor
Frederick F. Ling
Ernest F. Gloyna Regents Chair in Engineering
Department of Mechanical Engineering
The University of Texas at Austin
Austin, TX 78712-1063
USA
and
Distinguished William Howard Hart Professor Emeritus
Department of Mechanical Engineering,
Aeronautical Engineering, and Mechanics
Rensselaer Polytechnic Institute
Troy, NY 12180-3590
USA

Library of Congress Cataloging-in-Publication Data
Kaviany, M. (Massoud)
Principles of convective heat transfer/Massoud Kaviany.
p. cm.
Includes bibliographical references and index.
ISBN 0-387-94271-8
1. Heat–Convection. I. Title.
TJ260.K288 1994
536′.25–dc20 94-1483

Printed on acid-free paper.

Production managed by Francine McNeill; manufacturing supervised by Vincent Scelta.
Camera-ready copy supplied by the author using TeX.
Printed and bound by Edwards Brothers, Inc., Ann Arbor, MI.
Printed in the United States of America.

9 8 7 6 5 4 3 2 1

ISBN 0-387-94271-8 Springer-Verlag New York Berlin Heidelberg
ISBN 3-540-94271-8 Springer-Verlag Berlin Heidelberg New York

To my wife Mitra and my daughters Saara and Parisa.

Series Preface

Mechanical engineering, an engineering discipline born of the needs of the industrial revolution, is once again asked to do its substantial share in the call for industrial renewal. The general call is urgent as we face profound issues of productivity and competitiveness that require engineering solutions, among others. The Mechanical Engineering Series is a new series, featuring graduate texts and research monographs, intended to address the need for information in contemporary areas of mechanical engineering.

The series is conceived as a comprehensive one that will cover a broad range of concentrations important to mechanical engineering graduate education and research. We are fortunate to have a distinguished roster of consulting editors, each an expert in one of the areas of concentration. The names of the consulting editors are listed on the first page of the volume. The areas of concentration are applied mechanics, biomechanics, computational mechanics, dynamic systems and control, energetics, mechanics of materials, processing, thermal science, and tribology.

Professor Bergles, the consulting editor for thermal science, and I are pleased to present this volume of the series: *Principles of Convective Heat Transfer* by Professor Kaviany. The selection of this volume underscores again the interest of the Mechanical Engineering Series to provide our readers with topical monographs as well as graduate texts.

Austin, Texas — Frederick F. Ling

Preface

Convective heat transfer is a result of the simultaneous presence of a fluid flow and a temperature gradient. The velocity field and the temperature gradient may be present within a *single phase*, as in exo- or endothermically *reacting flows*, or may be present within two or more neighboring phases, as in nonisothermal *multiphase* flows (which may include *phase change*). Therefore, the convective heat transfer can be *intramedium* (or *intraphasic*) or *intermedium* (or *interfacial*). In most cases the heat transfer also *causes* or *influences* the fluid motion through the changes in the thermophysical properties.

The *objective* of this *monograph* is to review, in a *concise* and *unified manner*, the contributions made to the *principles* of the convective heat transfer for *single*- and *multiphase* flow systems (including *phase change* and *chemical reaction*). The *aim* is to summarize the role of the various mechanisms, to discuss the *governing differential equations* (including the practical needs for their *simplification*, *approximation*, and *phenomenological modeling*), and to examine some of their solutions in order to demonstrate their *applications* and to gain further *insight*. The available *semi-empirical* treatments and comparison of the predicted and *experimental results* are also discussed.

The *approach* used is to first review (Chapter 1) some of the preliminaries, such as the molecular approach to the prediction of the thermophysical properties and phase transitions, the heat transfer aspects and the flow structure of multiphase flows, *pointwise* conservation equations, the principles of time and space *averaging*, the local thermal equilibrium or lack of it among the phases, and the various effects of heat transfer on the fluid motion. The need in practice for simplification, approximation, and modeling is addressed, and the development and the *judicial* use of the phenomenological models (with some *empiricisms*) are discussed. Then, based on the number of phases and the presence or lack of the local thermal equilibrium, the discussion is divided into *three parts*. Part I deals with the *single-medium treatments*, addressing the *intraphase* heat transfer in single-phase flows (Chapter 2) and the *intramedium* heat transfer in two-phase flow systems with local thermal equilibrium (Chapter 3). Part II deals with the *two-medium treatments* of two-phase flows. The *interphasic* heat transfer is due to the *lack* of local thermal equilibrium between the phases, where in Chapter 4 the *solid* and *fluid* phases are both *continuous* and with small volumetric interfacial areas. In Chapter 5 the *solid-fluid* volumetric interfacial area is *large*, and the solid phase is *continuous* or *dispersed*; in Chapter 6 the heat transfer is between *two fluids*. Part III deals with the *three-medium treatments* of three-phase systems with Chapter 7 addressing the *solid-solid-fluid* systems and with Chapter 8 addressing the *solid-liquid-gas* systems. A description of each chapter is given below.

In Chapter 1, after a short review of the *applications* and the *history* of convective heat transfer, the *scope* of the monograph is outlined. The *thermophysical properties* of the *three* states of the matter are discussed, using the results of *equilibrium classical* and *statistical* (i.e., *molecular*) thermodynamics and from the *nonequilibrium molecular theory* (i.e., the *kinetic theory*). The *phase transitions*, including *metastable* states and *multicomponent* systems, are discussed next. The flow and heat transfer in *multiphase* (*two*- and *three-phase*) *flows* are discussed, and the role of the *confining* and *submerged* solid surfaces is addressed. Then the *conservation equations* for overall mass, species, momentum, and energy (thermal and mechanical), written for the *most elementary volume* (i.e., a single-phase continuum), are reviewed. The *time* and *spatial averaging* of these differential conservation equations *required* for practical treatment of transient, multiphase, and multidimensional problems are discussed next. In multiphase flow and heat transfer, the assumption of *local thermal equilibrium* among the phases, which, when valid, simplifies the analysis significantly, is examined. The need for both *rigorous* and *approximate* (i.e., *empiricism* and *modeling*) treatments of multiphase systems, for gaining insight into the transport phenomena and for obtaining solutions with reasonable effort, is addressed, and an example is given. *Intramedium* versus *interfacial* heat transfer is addressed next by pointing to the practical *intramedium* convective heat transfer problems occurring in a single-phase system and in a multiphase system treated as a single medium. The fluid motion *caused* or *influenced* by the temperature gradient resulting from heat transfer is discussed. This gradient causes gradients in *density* (resulting in *thermobuoyant* flows or in *thermoacoustics*), surface tension (resulting in *thermocapillary* flows), or electrical conductivity (resulting in *electrothermal* convection). These and the changes in other thermophysical properties are reviewed.

Chapter 2 addresses *intraphasic* convective heat transfer in *single-phase flows*. The flow and heat transfer in *compressible* flows, with examples of the temperature structure of a *shock wave* and the attenuation of a *thermoacoustic wave*, are discussed. *Reacting flows*, that is, *premixed* and *diffusion flames*, are reviewed as examples of intraphasic convective heat transfer with *exothermic* reactions. The role of flow *turbulence* in intraphasic heat transfer is addressed by first reviewing the existing treatments of the turbulent flow and heat transfer and then applying a simple turbulence model to an isothermal turbulent jet. These provide the needed materials for analyzing heated jets in the later sections. The role of *radiative* heat transfer and the semi-transparency of fluids are discussed, and the material needed for the analysis of the role of radiation in exothermic reactions is discussed. The effect of an imposed *electric* or *magnetic* field on the motion of electrically *conducting* fluids is discussed next. This is followed by a discussion of the thermal features of flowing *plasmas*. Plasmas in addition to being at high temperatures, where dissociation and ionization occur, are generally subject to an electromagnetic field. *Thermobuoyant-* and *thermobuoyancy-influenced* flows are discussed next. Examples of turbulent plumes and turbulent jets are reviewed. In discussing the various mechanisms causing the flow or influencing the convective heat transfer, the governing equations are examined, and whenever simple analytical solutions are available, they are also reviewed along with available numerical and/or experimental results.

Chapter 3 considers the treatment of the *two-phase systems* as a *single medium* by assuming that local thermal equilibrium exists between the phase pairs. After discussing the requirement for this equilibrium, *solid-fluid systems* with both phases being *continuous* and the solid being stationary (i.e., *porous media*) are discussed first. The result of the *local volume averaging* of the energy equation over *both* phases and the resulting *Taylor dispersion* are discussed along with the details of the *closure conditions*. The effective thermal conductivity and the dispersion tensors for *periodic structures* are evaluated, and examples are given. The inclusion of *radiative heat transfer*, *exothermic reaction*, and *phase change* (i.e., melting or solidification, including those for *multicomponent systems*) are addressed. The role of judicial *approximations* in the development of local volume-averaged conservation equations is emphasized. The two-phase flows (with *both* phases moving), with the possibility of one of the phases being *dispersed*, and the time and space averaging of such systems, are discussed. The need for a rigorous treatment and the need for reducing the large number of variables emerging from the averaging are discussed. Some approximate forms of the averaged equations are reviewed along with the available results for the *effective medium properties*. An example of *exothermic reaction* in a particulate flow (solid particles flowing with a gas) is discussed. In Chapter 3, in examining the single-medium treatment of two-phase systems, the *principles* of local volume averaging and the emergence of the effective medium properties and the need for the *modeling* of these properties are emphasized. Some examples of solutions to these local volume-averaged equations are examined, and aspects of the intramedium convective heat transfer are discussed.

In Chapter 4 the *interfacial* heat transfer between a *solid surface* and the *fluid flowing* around it is discussed. Because of the interest in heat exchangers and the possibilities of analytical treatments of some of the simpler flows and geometries (especially by using the boundary-layer theory), this solid-fluid *nonequilibrium* convective heat transfer problem has been studied most extensively in the past. The materials for this chapter are selected to show the effect of the various mechanisms on the *interfacial heat transfer*. These include *viscosity*, *oscillation*, *turbulence*, *surface roughness*, *compressibility*, *rarefaction*, *non-Newtonian fluid behavior*, *surface injection*, *suction*, *impinging jets*, *chemical reaction*, *buoyancy*, *electro- and magneto-hydrodynamics*, *particle suspension*, *porous media*, and *solidification*. The emerging dimensionless parameters expressing the influence of these mechanisms on the interfacial heat transfer are identified, and some specific results (closed form, numerical, and experimental) are presented. The fundamentals treated rigorously in this chapter, because of the advantage of simplicity of flow and geometry, are used in the following chapters where the complexity of the phase distributions and the addition of a third phase will make a rigorous treatment rather involved, if not impossible.

Chapter 5 examines the *intraphasic* and *interfacial heat transfer* in *solid–fluid* systems with large, specific interfacial areas, and begins by considering an *isolated element* of a *dispersed solid phase*. Because of its simplicity, the flow and heat transfer around a *sphere* allow for a review of the hydrodynamics and heat transfer and an examination of the parameters affecting this *interfacial heat transfer*. After examining the flow and heat transfer about an isolated sphere, the *convective interaction among elements* is addressed by reviewing and summarizing the results for the interfacial heat transfer for *two* or *more* elements placed next to each other. Then a treatment of the *dense element-concentration* is given, and the phasic, local volume-averaged equations are discussed. When the solid is *stationary* and *continuous*, that is, a *porous medium*, the two-medium treatment is made by starting from the principles of the *local volume averaging* and is carried out with relative rigor. This is done for *periodic solid structures*, and the practical simplifications of these averaged equations are discussed. An example demonstrates the *need* for the two-medium treatment in *transient* heat transfer in porous media. Another example dealing with exothermic reactions in the fluid (i.e., gas) phase is given, and the solution and the physical features are discussed. *Particulate flows*, with each flowing solid particle treated as having a *uniform* or a *nonuniform* (transient behavior only) temperature, are discussed. The *force* acting on a particle moving relative to the fluid is first generalized to include the viscous, pressure, and buoyancy forces and the transient effects. The recent developments in the treatment of dense particulate flows, using the result of the *nonequilibrium molecular theory*, are discussed. These treatments use the particle collisions to arrive at *modified continuum treatments* of the fluid and the particle flow and separate continuum energy equations for the fluid and the particles. As an example, *exothermic* reactions in a particulate flow are considered, and specific references are made to the burning of coal particles. Another example given, which also allows for the examination of the gas rarefaction, is the *rarefied plasma-particle* flow and heat transfer. Chapter 5 addresses the local volume-averaged treatment of the solid–fluid systems and examines the averaged equations through some examples.

Chapter 6 is on *two-medium* treatment of *fluid-fluid* systems and first considers *immiscible liquid-liquid* flow and heat transfer, followed by the most encountered *gas-liquid* systems, and the special case of *immiscible gases*, as in electrons and heavier species in plasmas. In the discussion of liquid-liquid flow and *interliquid* and *interfacial* heat transfer, the conditions for the *vapor bubble nucleation* at the interface of two immiscible liquids are addressed. Then the interfacial heat transfer for the case of isolated liquid *drops* flowing in relative motion in a host liquid is examined, and the internal *drop motion* and the *interdrop* heat transfer are discussed. For the case of a liquid dispersed in another liquid, the *energy equation* for each liquid is evaluated. The *gas-liquid* systems are first examined for a continuous interface, and the *condensation* (or *evaporation*) rate and the interfacial heat transfer for the *nonwavy* interfaces are examined. Then the discussion of gas-liquid flows with one phase dispersed begins with the *bulk nucleation* (*bubbles* and *drops*) using the results of classical thermodynamics and the kinetic theory. Then the phase dynamics of the *isolated bubbles* and *droplets* are addressed. The discussion begins with the bubbles. The *bubble growth or decay* due to heat transfer is examined, and the effect of *bubble motion*, *liquid-phase turbulence*, and the *noncondensables* (i.e., multicomponent systems) are discussed. Then a discussion of the *interbubble* interactions begins with the study of a train of bubbles, and general *conservation equations* for *multibubble systems* are reviewed. Similar discussions follow for *droplet growth or decay* for which the *internal droplet motion*, the *transient droplet behavior*, and the *variations* in the *thermophysical properties* in vaporation to high-temperature gases have been extensively studied because of interest in liquid spray combustion. Then the *interdroplet* interaction is examined using various *unit-cell models* and the results of *direct formulation* of multidroplet evaporations. The *conservation equations* for the multidroplet systems are examined next. The chapter ends with a discussion of the two-medium treatment of *gases*. This chapter deals with the internal motion in the dispersed phase and the interfacial condensation/evaporation which requires a treatment different from the solid dispersed particles discussed in Chapter 5.

Chapter 7 is on *solid-solid-fluid three-medium* treatment, for example, *internal particulate flow* and heat transfer, heat exchange between a *fluidized bed* and the *surfaces bounding* the bed or *submerged* in the bed, and heat transfer resulting in *phase change* occurring in *solidification*, *melting*, and *frosting*. In these examples, one *solid phase* is the *bounding* or the *submerged* surface, and the other is the *dispersed* (or *packed* in the case of frosting) *phase*. The *fluid phase* is continuous. Lack of the local thermal equilibrium results in interfacial heat transfer between *any* of these *phase pairs*. The hydrodynamic and heat transfer aspects of internal particulate flows, both *dilute* and *dense* particle concentrations, are reviewed. The heat transfer to surfaces submerged in internal particulate flows is also examined. The *hydrodynamics* of the fluidized beds and the heat transfer to surfaces submerged in these beds, including the effect of flow instabilities, are discussed. The solidification of the multicomponent systems, which may result in the formation of a fragment of solid traveling in the melt, is examined. The formation of frost on the solid surfaces and the structure of the matrix and the local thermal nonequilibrium are addressed. This chapter aims at using the *simplified* forms of the *time-* and *local phase-volume-averaged* conservation equations in the analysis of some practical problems. These problems generally involve a third phase, which is a solid surface, and the heat transfer through this solid surface either is of prime interest or is in part controlling the heat transfer between the other two phases. The need for practical approximations and modeling is emphasized through examples.

Chapter 8 is on *solid-liquid-gas*, *three-medium* treatment and concentrates on the *gas-liquid phase change* (*evaporation* and *condensation*) in the presence of a *solid surface*. The major classification is based on the *continuity* or *discontinuity* of the *gas-liquid interface*. The *rate* of phase change depends on the solid-liquid-gas interfacial geometry, the flow directions, and the velocity and the temperature distributions within each phase. The *continuous* gas-liquid interface is considered first with the analyses of *vapor films* and *liquid films*. Both single- and multicomponent systems are considered. Then the case of *discrete* gas-liquid interfaces is examined starting with a discussion of the solid-liquid-gas *common-* or *interline*. The evaporation and condensation rates are generally limited by the resistance to heat transfer to the liquid-vapor interface and the resistance to vapor or condensate removal from it. The upper bound on these rates, that is, when these resistances vanish, is discussed. The *surface bubble nucleation* and *dynamics* and *surface droplet nucleation* and *dynamics* are reviewed. The *impingement* of *drops* on heated *solid surfaces* is examined. This chapter illustrates that these complex three-phase systems involve variations in temperature and velocity *occurring* over *very short distances*, with *large* variations of the temperature *between* the phases, such that the local phase-volume averaging is *not* valid and a direct, pointwise treatment is needed.

The monograph assumes an *introductory* understanding of *thermodynamics*, *fluid mechanics*, and *heat transfer*. Since the aim has been to give a *broad* coverage of the field, some of the intermediate derivations have been intentionally omitted. However, complete citations are given regarding every derivation, where further discussions (including derivations) can be found. A *nomenclature* defining the symbols is given at the end of the monograph.

Ann Arbor, Michigan — Massoud Kaviany

Acknowledgments

I am most grateful to those whose contributions to the principles of convective heat transfer have made this monograph possible. I am indebted to the late Professor Ralph Seban, who has influenced many of us by his insight, contributions, and dedication to this field. My main source of inspiration and my perpetual teachers have been the students, in the classroom and in the research laboratory, who have wanted to know and to be relevant. I am also indebted to them.

The illustrations were made by Ms. Sandra Ackerman, and the typing was done by several of my associates, with the completion by Mr. Sander Foster. I would like to thank them all. A special thanks goes to Dr. Melik Sahraoui, whose research and support facilitated this project.

Contents

1

Introduction

In order to discuss the convective heat transfer in the context of its applications in multiphase systems with or without phase change and reaction, and in order to allow for the *analysis* of the convective heat transfer at the *molecular*, *phase*, and *multiphase* scales, this chapter reviews some of the preliminaries needed for the discussions in the following chapters. It begins with the spectrum of applications and length scales and after a short review of the historical contributions to the field, gives the scope of the monograph. The various molecular and continuum aspects of the thermphysical properties, phase transitions, and multiphase flows are discussed. The analysis of the convective heat transfer in multiphase systems is addressed beginning with the conservation equations for the *most elementary*, *continuum* volume and then moving to the volume averaging over phases. The extent of the *rigor* in the treatment of the multiphase flow and heat transfer, the applied *approximations* and *simplifications*, and the practical, *semi-empirical models*, are addressed. The objective of this treatise, which is an examination of both the convective heat transfer *within* a medium and that *across* the interfaces of the various phases, as well as the *influence* of the heat transfer on the fluid motion, is discussed at the end of this chapter.

1.1 Applications and Length Scales

Convective heat transfer occurs in many engineering and natural systems. In the following, some examples are given with a rather loose definition of the *application areas*. As evident, various *length*, *velocity*, and *temperature* scales are encountered. Also, an extensive range of *phases* and *phase changes*, as well as *reactions*, are involved.

Atmospheric: Influence of cyclic heat and moisture addition/removal on the earth's climate. Thermally effected motions in the atmosphere of the earth and other planets. Evaporative cooling and increase in the salinity of sea water (leading to the formation of dense bottom waters in oceans and affecting climate).

Biological and Physiological: Heat transfer of gas and liquid flow in plants and animals. Hypo- and hyperthermia (local and whole-body). Cryosurgery and cell freezing. Cell reactors, incubators, and biomass systems.

Chemical Processes: Catalytic and noncatalytic reactions in exothermic and endothermic reacting flows. Polymer processing. Catalytic converters. Separation. Extraction.

Climate Control: Comfort control of habitats and vehicles. Refrigeration. Water heaters. Influence of thermal convection on life in large bodies of water.

Energy Conversion: Fossil fuel to electrical. Solar to electrical and also in low-temperature process applications. Thermal energy in hydrogen and synfuel production. Geothermal brine heat content to electrical and also in low-temperature process applications. Nuclear to electrical (thermal hydraulics of the reactor). Thermoacoustic engines.

Energy Harvesting: Thermal-enhanced oil recovery. Geothermal energy recovery. In-situ coal gasification.

Energy Storage and Transport: Sensible and phase-change thermal energy storage and/or transport. Thermochemical and thermoelectrical energy storage and/or transport.

Environmental: Discharge of plumes and jets into air and water. Catalytic converters for automobile exhaust. Thermal venting and flooding of contaminated soil. Thermal energy use in recycling. Underground heat-generating waste deposit.

Extraterrestrial: Microgravity convection and thermal processes (including crystal growth). Two-phase flow and heat transfer in process and fluid transfer in microgravity.

Geological: Thermoconvective motion of magmas. Melt solidification in magma chambers and during volcanic eruptions. Motion of heated liquids and gases in geothermal reservoirs. Petroleum (oil and gas) formation and motion.

Heat Dissipation: Cooling of electronic equipment. Cooling of engines and turbine blades. Heat dissipation in biological tissues. Nuclear and chemical reactors. Meltdown prevention. Cryogenic thin-film detectors. Cooling of thin films used in lasers. Cooling of particle-beam dumps.

Heat Exchangers: Indirect (i.e., without fluid-fluid contact) heat exchange between fluids with and without phase change. Direct contact heat exchange between fluids with and without phase change. Wicked and non-wicked heat pipes. Regenerators.

Heat Generation: Exothermic chemical reaction involving combinations of gas, liquid, and solid phases. Thermoacoustic heat generation. Viscous dissipation.

Manufacturing: Forming. Cutting. Drilling. Grinding. Welding. Joining. Melting. Casting. Rapid solidification. Coating (including painting). Materials removal.

Materials Processing: Crystal growth. Thin-film coating. Combustion synthesis. Powder forming. Drying. Curing. Heat treating. Composite forming. Ceremic debindering. Containerless processing.

Transportation: Internal combustion engines. Stirling engines. Propulsions (gas, solid, and liquid fuels). Exhaust gas reuse.

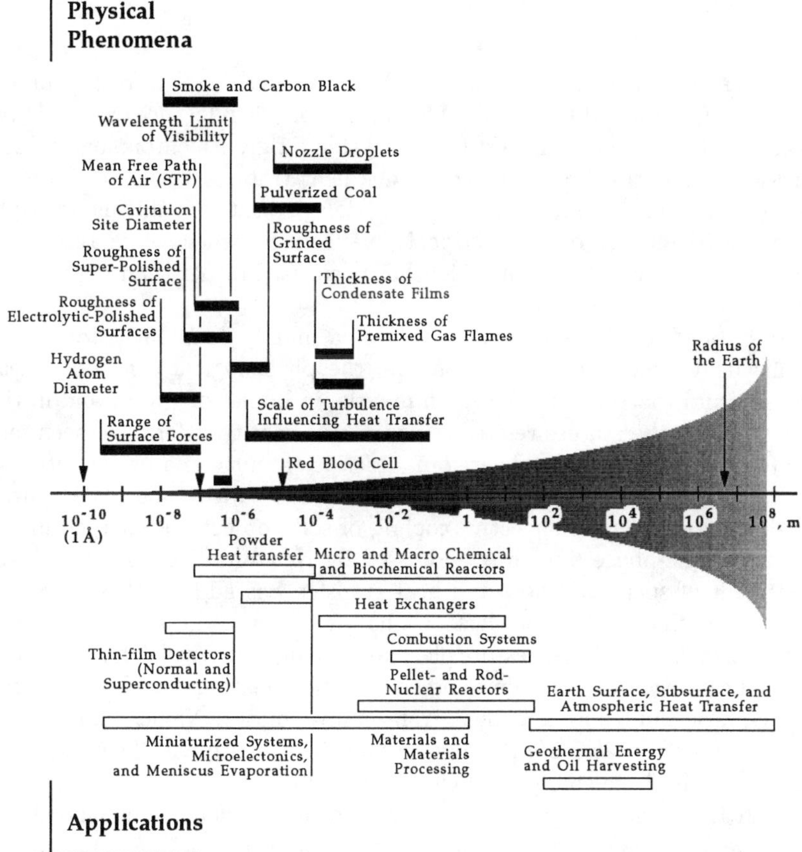

Figure 1.1. The linear dimensions or length scales of interest to convection heat transfer. The physical phenomena are given on the top and some applications are given at the bottom.

Figure 1.1 shows the range of length scales, i.e., the linear dimensions, of importance to convective heat transfer. Some applications are also shown at the bottom. While the cellular motion within the earth mantle and the motion in its atmosphere involve cell sizes reaching the magnitude of the radius of the earth, the coolant flow in miniaturized solid-state devices involve length scales of one micron and smaller. The drastic difference in behavior of gases, liquid, and solids, originates from their intermolecular forces and their average intermolecular separations. For a phase change to occur at the site of a surface cavity, the cavity size must be smaller than the surface roughness characteristic of a superpolished surface.

1.2 Historical Background

An extensive review of the history of heat transfer is given in the volume edited by Layton and Lienhard (1988) and a specific reference to convective heat transfer is given by Eckert (1981). Table 1.1 gives a chronological brief on the contributions made to *convective* heat transfer with the emphasis being on the studies influencing the materials presented in this monograph. In order to include other historical perspectives, whenever a recent review has been available, that review has been cited rather than the original work.

As part of the *thermal science* and as a mainly *technology-driven* scientific field (among the exceptions are the physiological convective heat transfer and those within the earth mantle, in bodies of water, and in the atmosphere), the studies related to convective heat transfer have been motivated by existing needs. Therefore, description of the major constituents of convective heat transfer ought to be modified with time. The heuristic description of the convective cooling of solid objects submerged into a moving single-phase fluid was addressed by Newton. The boundary-layer treatment of such fluid flow and heat transfer was addressed by Prandtl. The heat exchange between electrons, heavier species, and suspended small particles in flowing, low-pressure plasmas is being addressed. The rigorously derived conservation equations and constitutive relations developed for the single-phase fluid flow and heat transfer by Fourier, Navier, and Stokes are being used in arriving at solvable *models* of multiphase flow and heat transfer with phase change and chemical reaction.

With time, the emphasis of convective heat transfer has expanded beyond the single- and two-phase heat exchangers to applications listed above, which include materials and manufacturing processes such as the solidification of multicomponent liquids and the interaction of the weld pool and its surrounding plasma. Also, among these expansions are the reacting flows such as in the liquid and solid spray combustion.

The accounts of the recent advances in the convective heat transfer are given in the main text. The contributions listed in Table 1.1 demonstrate the past *fundamental contributions* to the transport phenomena (fluid, heat, and mass transfer, including phase change and reaction) as well as attempts in the *integration* of these in relation to the convective heat transfer. Since the technological needs are generally first met with empiricism, there have also been many contributions which are *yet* in *semiempirical* forms. These *recent* empirical results, which are mostly in the multiphase flow and heat transfer, are also discussed in the text. The time and local phase-volume averaging of the conservation equations allows for the formulation of the multiphase transport; however, the needed constitutive equations are *yet* being developed.

Table 1.1. Contributions to principles of convective heat transfer.

1701	*Newton*'s law of convective cooling and introduction of the heat transfer coefficient (Layton and Lienhard, 1988).
1811	*Fourier*'s law of heat conduction and introduction of the thermal conductivity (Layton and Lienhard, 1988).
1827	*Navier*'s and *Stokes*' (1845) equations of conservation of fluid mass and momentum (Schlichting, 1979).
1847	*Helmholz*'s formulation of principle of conservation of energy (Layton and Lienhard, 1988).
1861	*Joule*'s film condensation heat transfer experiment establishing the thermal resistance across the film (Fujii, 1991).
1871	*Marangoni*'s description of the liquid motion caused by a gradient in the surface tension (Probstein, 1989).
1873	*van der Waals*' theoretical description of the region around the liquid-vapor critical point (Stanley, 1987).
1875	*Grashof*'s introduction of the log-mean-temperature-difference equation for heat exchangers (Layton and Lienhard, 1988).
	Maxwell and *Boltzmann*'s molecular theory of the specific heat (Layton and Lienhard, 1988).
1877	*Boussinesq*'s introduction of the turbulent eddy viscosity (Bird, Stewart, and Lightfoot, 1960).
1879	*Oberbeck*'s application of the Stokes equation to thermobuoyant convection (Layton and Lienhard, 1988).
	Reynolds' introduction of the Reynolds number (Layton and Lienhard, 1988).
1881	*Lorenz*'s analysis of natural convection (similar to *Oberbeck*'s) adjacent to a heated vertical surface (Gebhart et al., 1988).
1885	*Graetz*'s analysis of the thermal entrance region in tubes (Layton and Lienhard, 1988).
	Mallard and *LeChatelier*'s thermal theory for premixed gas flame speed based on the ignition temperature (Liñán and Williams, 1993).
1889	*Arrhenius*' postulate that only molecules possessing energies greater than an activation energy can react (Liñán and Williams, 1993).
1896	*Rayleigh*'s description of thermoacoustics in his treatment of the theory of sound (Rott, 1980).
1901	*Gibbs*' monograph on the elementary principles of the statistical mechanics (Domb and Green, 1972).
1903	*Boussinesq*'s approximations to the equations governing the thermobuoyant motion (Gray and Giorgini, 1976).
1904	*Prandtl*'s introduction of the viscous boundary layer (Schlichting, 1979).
1907	*Blasius*' calculation of the viscous boundary layer along a flat plate (Oswatitsch and Wieghardt, 1987).

continued

Table 1.1 (Continued).

1909	*Nusselt*'s boundary-layer analysis of solid-fluid interfacial heat transfer in flow over a semi-infinite solid and introduction of the Prandtl number (Oswatitsch and Wieghardt, 1987).
1911	*Hadamard Rybizynski*'s analytical solution of the Stokes flow around a fluid particle, showing the internal vortical motion (Clift et al., 1978).
1916	*Nusselt*'s boundary-layer analysis of the laminar, phase-density buoyant flow film-condensation heat transfer (Fujii, 1991).
1921	*von Kármán* and *Pohlhausen*'s approximate calculation of the laminar boundary layer (Oswatitsch and Wieghardt, 1987).
1925	*Prandtl*'s mixing-length theory of the turbulent flow (Oswatitsch and Wieghardt, 1987).
1929	*Tollmien* and *Schlichting*'s analysis of instability of the laminar boundary layer along a flat plate (Oswatitsch and Wieghardt, 1987).
1930	*Schmidt*'s experimental demonstration of existance of natural convection boundary layer adjacent to a heated vertical surface (Gebhart et al., 1988).
	von Kármán's introduction of the similarity hypothesis for the turbulent shear stresses (Bird et al., 1960).
1933	*McAdams*' introduction of the *Heat Transmission* as the first heat transfer book (Layton and Lienhard, 1988).
1934	*Nukiyama*'s introduction of the boiling curve (Layton and Lienhard, 1988).
	Colborn's boundary-layer analysis of the turbulent, phase-density buoyant flow, film-condensation heat transfer (Layton and Lienhard, 1988).
1935	*Taylor*'s statistical theory of turbulence (Oswatitsch and Wieghardt, 1987).
1936	*Senftleben-Braun*'s and *Kronig-Ashmann*'s (1949) study of the effect of strong electrical fields on convective heat transfer, first in polar gases and next in dielectric liquids (Jones, 1978).
1938	*Zel'dovich* and *Frank-Kamentesky*'s thermal-species theory for the premixed-gas flame speed using the adiabatic flame temperature (Liñán and Williams, 1993).
1941	*Jakob*'s measurement of the superheat required for incipient surface bubble nucleation in pool boiling (Jakob and Hawkins, 1957).
	Kolmogorov's hypothesis that the small-scale components of turbulence are approximatly in statistical equilibrium (Turner, 1979).
1942	*Eckert*'s analysis of the wedge boundary-layer flow and heat (Eckert and Drake, 1972).
1950	*Bromley*'s boundary-layer analysis of the laminar, phase-density buoyant flow, film-boiling heat transfer (Sakurai and Shiotsu, 1992).
1951	*Seban* and *Shimazaki*'s analysis of the turbulent flow and heat transfer in tubes (Kays, 1966).

continued

Table 1.1 (Continued).

1953	*Spalding*'s boundary layer-integral analysis of the diffusion flame around a liquid film (Emmons, 1956).
	Batchelor's treatise on the theory of homogeneous turbulence (Turner, 1979).
	Taylor's treatment of the hydrodynamic dispersion in tubes (Taylor, 1953).
	Tribus and *Klein*'s heat sink model of the film cooling in boundary layers (Goldstein, 1971).
1955	*Deissler*'s empirical relation for the turbulent shear stresses near a solid surface (Bird et al., 1960).
1956	*Rohsenow* and *Griffith*'s model for the hydrodynamic and heat transfer aspects of maximum surface heat flux in evaporation with surface bubble nucleation (Layton and Lienhard, 1988).
1959	*Goulard* and *Goulard*'s analysis of coupled convective-radiative heat transfer in tubes (Özisik, 1985).
	Sparrow and *Gregg*'s analysis of the effect of thermobuoyancy on the boundary-layer forced flow and heat transfer (Gebhart et al., 1988).
	Zuber's analysis of the hydrodynamic instabilities in boiling heat transfer from horizontal surfaces (Sakurai and Shiotsu, 1992).
1960	*Bird*, *Stewart*, and *Lightfoot*'s unified and comprehensive treatment of the transport phenomena (Bird et al., 1960).
	Hsu and *Westwater*'s boundary-layer analysis of the turbulent, phase-density buoyant flow, film-boiling heat transfer (Sakurai and Shiotsu, 1992).
1961	*Tien*'s analysis of the internal, particulate flow and heat transfer with an allowance for thermal nonequilibrium between fluid and particles (Tien, 1961).
1962	*Kraichnan*'s turbulent thermal convection analysis using a three-layer model and the assumption of an explicit Prandtl number dependence (Turner, 1979).
1963	*Kutateladze*'s monograph on heat transfer summarizing the contributions in eastern Europe (Kutateladze, 1963).
1964	*Lienhard* and *Wong*'s hydrodynamic analysis of the minimum heat flux for phase-density buoyant flow, film-boiling heat transfer (Sakurai and Shiotsu, 1992).
	Mullins and *Sekerka*'s analysis of the stability of a planar interface during the solidification of a dilute, binary alloy (Mullins and Sekerka, 1964).
1965	*Boelter's* lecture notes published as *Heat Transfer Notes* (Boelter, 1965).
1966	*Kays*' introduction of *Convective Heat and Mass Transfer* as the first textbook solely dedicated to the subject (Kays, 1966).

1.3 Scope

Description of convective heat transfer begins with the conservation equations describing the conservation of mass, momentum, and energy (discussed in Section 1.7). These *volumetric* equations describe the rate of change of a quantity assuming a *continuum* and using the *most elementary* and *infinitesimally* small volume. For multiphase systems (discussed in Section 1.6), when describing an *average* behavior for a *local* collection of phases, the word infinitesimal would then mean *large enough* to include a sufficient volume of all the phases involved, *while* small enough to describe a *local* behavior (this is further discussed in Section 1.8). When we choose to *not* describe an *average* behavior, then the heat transfer *within* each phase as well as across the *interfaces* of these phases, is described. In phase-change problems, these interfaces can move, thus altering the heat transfer within the phases as well as across the interface. Chart 1.1 gives a classification of the methods of *analysis* of the convective heat transfer. The classification is further discussed below. The mathematical treatments of the convective heat transfer are simplified when valid assumptions are made about the *dominant mechanisms* and the *prevailing directions* over which the most significant changes occur. This is also discussed below.

1.3.1 Transport at Elementary Differential Level

In multiphase flow and heat transfer applications such as in spray combustion, in particle-plasma-surface interactions, in liquid-gas two-phase flow over surfaces, and in single- and two-phase flow in porous media, it might be justifiable to assume that two or more phases are in *local thermal equilibrium*, i.e., that the difference in the local temperature of the phases is small compared to the main temperature difference of interest (this is further discussed in Section 1.9). Then a local volume containing the two or more phases is chosen over which an average behavior is found. Thus a multiphase system can be treated as a *single medium*, if all the phases are assumed to be in the local thermal equilibrium. Chart 1.1 considers the *single-phase* flow and heat transfer, where the interest is in the heat transfer *within* this *medium*. Also shown in the chart are *two-* and *three-phase* systems, and depending on the imposition of the local thermal equilibrium, a *single-* or a *multimedium* treatment of the convective heat transfer is made. In two-phase systems, any phase pairs can exist, including two *immiscible* liquids. In the three-phase systems, fluid-solid-solid systems of interest include particulate flows over a solid surface and porous media bounded by a solid surface.

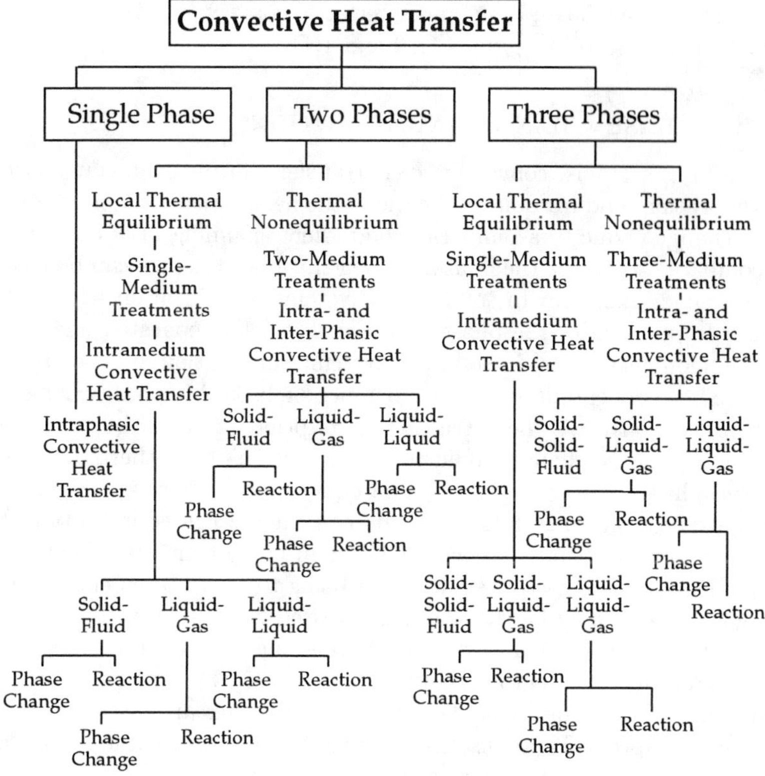

Chart 1.1. A classification of the convective heat transfer and its treatment based on the number of phases, the presence or lack of local thermal equilibrium, and the presence or lack of phase change.

1.3.2 Intramedium and Interfacial Heat Transfer

As was discussed above, the convective heat transfer of interest can be that *within* a medium, and this medium can be a single phase or a collection of phases treated as a medium by assuming a local thermal equilibrium. The convective heat transfer within the medium, or *intraphasic* or *intramedium* convective heat transfer, of interest includes jet- and shear-flow thermal mixing, single- and multiphase combustion, heat transfer in porous media, and heat transfer in solidification.

The convective heat transfer of interest can also be that *across* the phases. The *interfacial* convective heat transfer occurs across the fluid-solid or fluid-fluid interface in two-phase nonequilibrium systems and across the *common line* of gas-liquid-solid or the interface of fluid-solid-solid three-phase nonequilibrium systems. Each phase may be treated as *lumped* (i.e., uniform temperature) or *distributed*. The distributed treatment may use *local, phase-volume* averaged quantities or *pointwise* quantities in each phase. The spatial distribution may be in one, two, or three dimension. The intramedium versus intraphasic and interfacial convective heat transfer will be further discussed in Sections 1.10 and 1.11.

1.3.3 Phase Change and Reaction

In liquid-gas systems, convective heat transfer can result in condensation or evaporation, and in some applications the heat transfer required for this phase change is due to a temperature gradient spanning *over* both phases (as compared to *across* the phases). Then this phase change can be treated using a single medium. In solid-liquid systems, solidification and melting occurs. In some cases the phases are *not* distinctly separated and a two-phase region also exists. In solid-gas systems sublimation and frosting can occur. Then, depending on the occurrence or lack of a phase change, an appropriate convective heat transfer description needs to be made. Addressing the phase change would require inclusion of the thermodynamics and the phases may *not* be in stable equilibrium states (i.e., *metastable* states are possible, as in liquid-gas phase change discussed in Section 1.5). Also, exothermic chemical reactions occur in many convective heat transfer applications, the gas-phase reactions being the most often encountered. The gaseous fuel may be the volatiles released from solid or liquid phases as a result of the heat transfer from the reacting gaseous flow.

Following the introductory chapter, the single-, two-, and three-medium treatments of the multiphase flow and heat transfer are discussed in the remaining chapters. These discussions include the *role* of *phase change* and *reaction* in convective heat transfer.

1.3.4 Multidimensionality, Asymptotics, and Similarity

In description of the convective heat transfer by the differential governing equations, variable properties, time-dependence, and three-dimensionality are assumed. Whenever justifiable, *time* and *space averaging* are performed. Space averaging can be taken over one or more dimensions and a *phasic local volume average* is also taken when appropriate. In order to gain further insight, whenever possible the governing equations are solved with the goal of obtaining closed-form solutions. Since the momentum equation is nonlinear and in some cases coupled with the energy equation, this would restrict the solution to rather simple geometries and to asymptotic or to self-similar temperature and velocity field behaviors (this is further discussed in Section 1.10.4). Examples given in Chapters 2 to 8 are chosen among the available results to *demonstrate* the role of various mechanisms. Some of these examples are for relatively *idealized* cases where *closed-form* solutions exist. For some other problems, the solutions are found *numerically*. For some problems *experimental* results are presented, but a complete prediction may or may not exist.

1.4 Three Phases of Matter

Descriptions of the three phases of matter from the *molecular* and *continuum* points of view have been reviewed by Hirschfelder et al. (1959), Moelwyn-Hughes (1961), Tien and Lienhard (1971), Goodstein (1975), Gray and Gubbins (1984), Walton (1989), Barber and Loudon (1989), and Hockney and Eastwood (1989), among others. The followings are excerpts from these reviews on the relationship between the molecular and continuum *properties* of the matter. The thermophysical properties of three phases, and their variations with respect to temperature and other state variables, are given in tabular form by Liley (1985).

1.4.1 Gas, Liquid, and Solid Phases

The constituents of matter are *atoms* and *molecules* which are generally treated as *particles*. Each particle has a set of characteristics and in classical mechanics these are mass, charge, vorticity, position, momentum, etc. The state of the matter is defined by the attributes of a finite ensemble of particles, and the evolution of the system is determined by the laws of interactions of the particles. In a collection of particles the *interparticle attractive* and *repulsive* forces are represented by a potential. These particles will have an average *kinetic* energy and an average *intermolecular potential* (or *dissociation*) *energy*. The *association* or *bounding forces* are related to the electrical structure of the particles. The kinetic energy of a particle of mass m † and velocity u_m is $mu_m^2/2{=}3k_BT/2$, where k_B is the *Boltzmann constant* and is equal to $1.381{\times}10^{-23}$ J/K, and T is the *absolute temperature.* In a *single-pair* model (as compared to *many-body models*), to separate two particles a *dissociation energy* ΔE_d is required. A rather intuitive description of the three phases of the matter, in terms of the kinetic energy and the dissociation energy, is given by Walton (1989) and is repeated below.

When a particle of kinetic energy $3k_BT/2$ collides with an *associated* (or *bound*) *pair* and when this energy is larger than ΔE_d, the pair will dissociate. In a system of particles where the *average* kinetic energy is $3k_BT/2$, this energy must be much larger than ΔE_d in order to have *nearly* no bound pairs existing. This system of individual particles will occupy any given volume. This *unconstrained* particle population is a property of the ideal *gaseous phase.*

When in a *system* of particles the *average* kinetic energy $m\overline{u}_m^2/2$, where $(\overline{u}_m^2)^{1/2}$ is the *molecular root-mean-square speed*, is of the same order of magnitude as ΔE_d, then some *clustering* will occur. Those particles that

---

†The units for the symbols are given in the *nomenclature* at the end of the monograph.

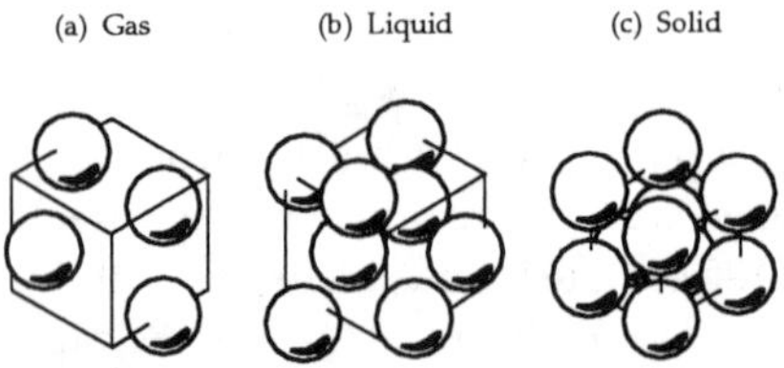

Figure 1.2. Molecular arrangements depicted for (a) gases, (b) liquids, and (c) solids.

possess a larger kinetic energy larger than the average will tend to dissociate the clusters. This is a *fluctuating* system with partly bound and partly free particles. This *semiconstrained structure* is the property of the *liquid phase.*

When the average kinetic energy of the colliding particles is very much less than ΔE_d, no dissociation will occur. The clusters become connected and larger conglomerate will form in which the particles are tightly connected. This is a *constrained* (i.e.,*rigid*) *structure* and is a property of the *solid phase.*

In order to examine other properties of the three phases, we consider a *simple compressible substance* (Sonntag and Van Wylen, 1982; Howell and Buckius, 1992) and use the *pressure p*, the *specific volume v*, the *specific entropy s*, the *specific internal energy e*, and the *temperature T*. From these we also define the *specific enthalpy* as $i=e+pv$, the *specific Gibbs energy* as $g=i-Ts$, and the *specific Helmholtz energy* as $h=e-Ts$. Some of the properties related to the derivatives of these properties are the *isothermal compressibility* κ and the *coefficient of thermal expansion* β

$$\kappa = -\frac{1}{v}\frac{\partial v}{\partial p}\bigg|_T \ , \quad \beta = \frac{1}{v}\frac{\partial v}{\partial T}\bigg|_p = -\frac{1}{\rho}\frac{\partial \rho}{\partial T}\bigg|_p \tag{1.1}$$

and the *specific heats* at *constant pressure* and *constant volume*

$$c_p = \frac{\partial i}{\partial T}\bigg|_p \ , \quad c_v = \frac{\partial e}{\partial T}\bigg|_v \ . \tag{1.2}$$

The *Joule-Thompson coefficient* is defined as

$$\mu_{JT} = \frac{\partial T}{\partial p}\bigg|_i \ . \tag{1.3}$$

For cooling purposes, using the gas expansion (i.e., drop in the pressure), a positive and relatively large μ_{JT} is required. Recent developments in the Joule-Thompson cooling (to as low as 68 K) involve using mixtures of hydrocarbon gases.

The isothermal compressibility decreases as the kinetic energy decreases and as more clustered and ordered structures are formed. The gas compressibility is several orders of magnitude larger than that of the liquid which in turn is a few orders of magnitude larger than the solid. This is depicted in Figure 1.2, where the molecular arrangements for the three phases

Table 1.2. Some molecular features of gas, liquid and solid phases.

Phase	Intermolecular Forces	$\frac{3k_BT/2}{\Delta E_d}$	$\frac{\lambda}{d_m}$	Molecular Arrangement	Statistics
Gas	Weak	$\gg 1$	$\gg 1$	Disordered	Classical
Liquid	Intermediate	$\simeq 1$	$\simeq 1$	Partially-Ordered	Classical-Quantal
Solid	Strong	$\ll 1$	$\ll 1$	Ordered	Quantal

are illustrated in terms of the intermolecular spacing and the regularity of the arrangement.

For particles possessing a kinetic energy, the compressibility of a collection of them is a manifestation of the spacing between particles. Then the ratio of the *mean free path* λ, i.e., the average distance a particle travels between successive collisions, to the particle diameter d_m (which is of the order one angstrom for short-chain molecules), is indicative of the compressibility. When this ratio is very large, the compressibility is characteristic of the gaseous phase; when it is of the order of unity it is characteristic of the liquids; and when it is much smaller than unity it characterizes the solid. Table 1.2 gives some of the molecular features of the three phases (Moelwyn-Hughes, 1961). These features are for the conditions in which each phase is in an *equilibrium state*, i.e., in a *uniform* temperature and pressure field, and *far* from *phase transition states*. The *classical dynamics* referred to in the table, implies *continuity* of the momentum and energy, while the *quantum theory* show that atomic (and molecular) energy levels are *discrete*. As the intermolecular potential energy becomes more significant, the quantal description of the average energy becomes necessary (Incropera, 1974).

More complete *molecular theories* exist for the description of the molecular structure of the three phases (and to some extent for the *transitions* among these phases). Some of these theories and their predictions of the properties of the phases are considered next.

1.4.2 Equilibrium Molecular Theory

The magnitude of the two intermolecular forces, namely, the *attractive* and the *repulsive forces*, determines if molecules *condense* into a minimum volume or completely fill the volume of their containers (gas phase). These magnitudes depend on the *temperature*, the *pressure*, and the *composition* (i.e., the specific atoms or molecules). The numerous materials with variety of properties are the results of these intermolecular boundings. The intermolecular forces are modeled and a commonly used approximation for the force (or potential energy) distribution around a molecule is the *Lennard-Jones* model. In this model, for a radial distance r measured from the center of a molecule (or particle), the *total force* $f(r)$ has a r^{-a_1} dependent force of *mutual repulsion* between the molecule pair and a force of *mutual attraction* which is proportional to r^{-a_2}, with $a_1 > a_2$. This force is the basic element of the *mean field theory* of molecular interactions as laid out in 1873 by van der Waals (Walton, 1989). The *differential potential energy* $d\varphi$, i.e., the energy required for a displacement dr, is

$$d\varphi = -f(r)\,dr \tag{1.4}$$

and the potential energy for bringing a particle from $r \to \infty$ to r is

$$\varphi(r) = -\int_r^\infty f(r)\,dr. \tag{1.5}$$

The *minimum* of $\varphi(r)$ occurs at r_o, where $f(r_o)$=0, and this minimum is the *negative* of the dissociation energy, i.e.,

$$\Delta E_d = -\varphi(r_o). \tag{1.6}$$

The modeling of $f(r)$ for gases, liquids and solids, is done using the classical mechanics whenever the intermolecular interactions are due to the ionic bounding, various van der Waals forces, and the covalent bounding. The quantum mechanics is used for other forces.

(A) Gases

The *ideal* (or *perfect*) *gas* is made of molecules with *zero* intermolecular dissociation energy, i.e., ΔE_d=0. The pressure required to contain a perfect gas is

$$p = \frac{1}{3} m n \overline{u_m^2}\,, \tag{1.7}$$

where as before $(\overline{u_m^2})^{1/2}$ is the *molecular root-mean-square speed*, m is the *mass* of the *molecules*, and n is the *number density* of molecules (related to *density* ρ by n=ρ/m and to the volume V by n=N/V, where $m = M/N_A$, M is the *molecular weight*, N_A is the Avogadro number, and N is the *total* number of molecules). This equation can be written as

$$pV = \frac{1}{3} m N \overline{u_m^2}\,. \tag{1.8}$$

Now using the temperature from

$$\frac{m\overline{u_m^2}}{2} = \frac{3k_BT}{2} \tag{1.9}$$

we have

$$p = nk_BT \quad \text{or} \quad pV = Nk_BT = \frac{N}{N_A}R_gT\,, \quad R_g \equiv N_A k_B\,, \quad \frac{R_g}{M} = \frac{k_B}{m} \tag{1.10}$$

or

$$p = \frac{\rho R_g T}{M}, \tag{1.11}$$

where R_g is called the *universal gas constant*. The specific heat capacities at constant pressure and constant volume are related through

$$c_p - c_v = \frac{R_g}{M}. \tag{1.12}$$

The *speed* of *sound* is (Tien and Lienhard, 1971)

$$\begin{aligned} a_s &= (\frac{\partial p}{\partial \rho}|_s)^{1/2} = (\frac{c_p}{c_v}\frac{\partial p}{\partial \rho}|_T)^{1/2} = (\frac{c_p}{3c_v})^{1/2}(\overline{u_m^2})^{1/2} \\ &= (\frac{c_p}{c_v}\frac{R_g}{M}T)^{1/2} = (\frac{c_p}{3c_v})^{1/2}\overline{u_m^2}\,. \end{aligned} \tag{1.13}$$

The mean free path is given by (Walton, 1989)

$$\lambda = \frac{1}{2^{1/2}}\frac{1}{\pi d_m^2 n} = \frac{1}{2^{1/2}}\frac{k_BT}{\pi d_m^2 p}\,. \tag{1.14}$$

As was mentioned, for short-chain molecules the molecular diameter is of the order of one angstrom. For example it is 2.70 Å for hydrogen, 3.68 Å for nitrogen, 4.60 Å for water, and 4.00 Å for carbon dioxide (Tien and Lienhard, 1971).

The *collision frequency* (or inverse of the *intercollision period* t_c) is given through (Walton, 1989)

$$\frac{1}{t_c} = \pi d_m^2 (\frac{4k_BT}{\pi m})^{1/2} n^2. \tag{1.15}$$

For a *binary* mixture, the *reaction rate* of *species* A with *species* B (assuming the *same* molecular size and mass) $\dot{n}_A$ (numbers per unit volume per unit time) is given by (Walton, 1989)

$$\dot{n}_A = \pi d_m^2 n_A n_B (\frac{4k_BT}{\pi m})^{1/2} e^{-\Delta E_a/k_BT}, \tag{1.16}$$

where ΔE_a is the *molecular activation energy* for the *reaction* and is generally substantially *less* than the dissociation energy of species A or B. The *reaction rate equation* (1.16) will be further discussed in Section 2.3, and examples for reacting flows will be given in Sections 2.3, 3.2.6, 3.35, 4.11, 4.12.2, 5.3.4, 5.3.5, and 5.4.4.

For a *real* (or *imperfect*) gas, $\Delta E_d > 0$ and the intermolecular interactions need to be included. In one of the models (the *van der Waals* model), (1.10) is modified to

$$(p + \frac{aN^2}{N_A^2V^2})(V - \frac{bN}{N_A}) = \frac{N}{N_A}R_gT, \tag{1.17}$$

where

$$a = \frac{4}{9}\pi N_A^2 d_m^3 \Delta E_d \ , \quad b = \frac{2}{3}\pi N_A d_m^3. \tag{1.18}$$

The *compression factor* (i.e., the *deviation* from an ideal behavior) is defined as

$$Z \equiv \frac{pV}{Nk_B T} \tag{1.19}$$

and (1.17) predicts the compression factor at the *critical state* to be given by

$$Z_c = \frac{p_c V_c}{Nk_B T_c} = \frac{3}{8}. \tag{1.20}$$

The experimental results for Z_c are between 0.2 and 0.3. Also, for a real gas the two heat capacities are related through

$$c_p - c_v = \frac{R_g}{M}(1 + \frac{2aN^2}{N_A^2 pV^2}). \tag{1.21}$$

The *specific heat of evaporation* $\Delta i_{\ell g}$, estimated using a *coordination number* (number of nearest neighboring molecules) of ten, is given through (Walton, 1989)

$$\frac{R_g T_c}{M\Delta i_{\ell g}} = \frac{16}{405}. \tag{1.22}$$

The experimental results for this ratio is *larger* (e.g., 0.13 for water).

(B) Liquids

For *simple liquids*, i.e., those that can be modeled as spherical molecules with a rather simple intermolecular potential energy φ, the equation of state can be written as

$$pV = \frac{R_g}{M}T - \frac{2\pi N_A}{3M}\int_0^\infty \frac{d\varphi}{dr} n_m(r) r^3 \, dr \ , \tag{1.23}$$

where $n_m = n_m(r, p, T)$ is the *mean number* of *surrounding molecules per unit volume* at a location r from the center of a molecule. The last term in (1.23) is the so-called *negative pressure* which allows for the *high* density of the liquid phase. The *specific internal energy* is given as

$$e = \frac{3}{2}\frac{R_g}{M}T + \frac{2\pi N_A}{M}\int_0^\infty \varphi(r) n(r) r^2 \, dr. \tag{1.24}$$

For some single- and multicomponent liquids, $\varphi(r)$ has been modeled and used in the statistical molecular theory or in the *molecular dynamic simulations* (Hirschfelder et al., 1954; Gray and Gubbins, 1984; Lee, 1988; Hockney and Eastwood, 1989).

(C) Solids

For stationary molecules, as occurring in the *bulk* (as compared to the surface) of a solid phase, a zero net resultant force is applied to each molecule. This requires a local *periodic arrangement* of molecules (or *lattice structure*). For amorphous materials, like glass and some plastics, the structure is *nonperiodic* (i.e., *disordered*) and these molecules are *not* in *equilibrium*.

The *lattice energy* of a *crystal* made of single-constituent or multiconstituent molecules can be determined by identifying the lattice structure (*arrangements and spacings*) and by using the appropriate intermolecular potential (*crystal bonds* can be covalent, ionic, metallic or molecular). For example, for alkali halides (such as sodium chloride) the *heat of sublimation* Δi_{sg} (i.e., the negative of the *internal energy* of the crystal) is given by (Walton, 1989)

$$\Delta i_{sg} = \frac{N_A e^2 a_2}{M \epsilon_1 r_o}(1 - \frac{1}{a_1}) \ , \tag{1.25}$$

where constants ϵ_1 and a_1 are prescribed as part of the potential energy φ, a_2 is a geometrical factor, and r_o is again the location where the internal energy is a minimum. The intermolecular potential energy distribution used for ionic crystals is

$$-\varphi = \pm \frac{e^2}{4\pi\epsilon_1 r} + \frac{B}{r^{a_1}}, \tag{1.26}$$

where + is used for the interaction between similar pairs and − is used for the opposite charges. The constant B is called the *interaction strength* or the *Hamaker constant*.

The heat capacity of *nonmetallic monatomic* solids is given by (Walton, 1989)

$$c_v = \frac{9R_g}{M}(\frac{T}{\theta_D})^3 \int_0^{\theta_D/T} \frac{x^4 e^x}{(e^x - 1)^2} \, dx \ , \tag{1.27}$$

where θ_D is the *Debye temperature* and depends on the lattice spacing and other molecular-structural properties. The specific heat capacity of the *metallic* solids has a temperature dependence given by

$$c_v = a_1 T^3 + a_2 T \ , \tag{1.28}$$

where a_2 and a_3 are constants.

1.4.3 Nonequilibrium Molecular Theory

The existence of *nonuniformity* in the *temperature*, the *velocity*, or the *concentration* results in a *departure* from equilibrium and the flow of *heat*, *momentum*, or *mass*. Concerning the *rate* of flow of these quantities, the *nonequilibrium molecular theory* predicts how the *transport properties* of a medium (i.e., *thermal conductivity* k, *dynamic viscosity* μ, and *mass diffusivity* D) influencing these rates are related to its more fundamental and molecular properties.

(A) Gases

In the presence of a temperature gradient ∇T, the *continuum conduction heat flux vector* $\mathbf{q}$ [†] is given by

$$\mathbf{q} = -k\nabla T, \tag{1.29}$$

where k is the thermal conductivity (or *molecular conductivity*). This heat flux is carried by the molecules as they travel through the nonuniform temperature field, and their kinetic energy changes, i.e., this continuum conduction heat flux, is carried by an instantaneous local *molecular flux* of gas molecules (with no net motion). The molecular theory predicts the relationship between k and the molecular flux as (Tien and Lienhard, 1971)

$$k = \frac{5}{2^{5/2}} m c_v \frac{\overline{u}_m}{\pi d_m^2} = \frac{5}{2} c_v \frac{(m k_B T)^{1/2}}{\pi^{3/2} d_m^2}, \tag{1.30}$$

$$\overline{u}_m = (\frac{8}{3\pi})^{1/2} (\frac{3 k_B T}{m})^{1/2} = (\frac{8 k_B T}{\pi m})^{1/2} = (\frac{8 R_g T}{\pi M})^{1/2}, \tag{1.31}$$

where the *difference* between the *root-mean-square speed* $(\overline{u_m^2})^{1/2}$ and the *mean speed* $\overline{u}_m$, which is equal to $(3\pi/8)^{1/2}$, has been accounted for (Walton, 1989). The mean speed is taken with the probability distribution function for u_m as the weighting function. The magnitude and the temperature dependence of the specific heat capacities for monatomic and polyatomic gases are given by Liepmann and Roshko (1967), Tien and Lienhard (1971), and Walton (1989).

The presence of a nonuniform *continuum* velocity vector $\mathbf{u}$ results in a continuum *viscous stress tensor* $\mathbf{S}$. For an *incompressible* and *Newtonian* fluid the elements of the viscous stress tensor τ_{ij} are given by (this will be further discussed in Section 1.6.2)

$$\tau_{ij} = \mu(\frac{\partial u_i}{\partial x_j} + \frac{\partial u_j}{\partial x_i}) . \tag{1.32}$$

This can be written as $\tau_{ij} = \mu d_{ij}$, where d_{ij} are the elements of the deformation rate tensor. The dynamic viscosity μ is also called the *coeffiecient of viscosity*. The nonuniformity in the continuum velocity corresponds to

---

[†] A lowercase *bold* letter indicates that the quantity is a *vector* and an uppercase *bold* character indicates a *tensor*. Reviews of vectors and tensors used in transport phenomena are given by Bird et al. (1960), Lai et al. (1970), Slattery (1981), Aris (1989), and Truesdell and Noll (1992).

a uniform molecular momentum flux and from the nonequilibrium molecular theory the viscosity is given by

$$\mu = \frac{1}{3} n m \overline{u}_m \lambda \tag{1.33}$$

or

$$\mu = \frac{m\overline{u}_m}{2^{1/2} 3\pi d_m^2} = \frac{2}{3\pi^{3/2} d_m^2}(m k_B T)^{1/2} . \tag{1.34}$$

For a binary gas mixture of species A and B and when the continuum density of species A is nonuniform, a *mass flux* vector $\dot{\mathbf{m}}_A$ of species A occurs as given by the *continuum mass diffusion* equation

$$\dot{\mathbf{m}}_A = -D_{AB}\ \rho \nabla \frac{\rho_A}{\rho} \quad \text{at constant } T \text{ and } p, \tag{1.35}$$

where D_{AB} (or D) is the *binary mass diffusivity* or *diffusion coefficient* (species A diffusing in species B), and $\rho = \rho_A + \rho_B$ is the mixture density. The presence of species of different diameters affects the mean free path as given by

$$\lambda_A^{-1} = n_B \frac{\pi}{4}(d_{m,A} + d_{m,B})^2 (1 + \frac{M_A}{M_B})^{1/2}. \tag{1.36}$$

From the nonequilibrium molecular theory, D_{AB} is found as

$$\begin{aligned} D_{AB} &= \frac{4.2}{\pi^{3/2} n (m_A m_B)^{1/2} (d_{m,A} + d_{m,B})^2} [(m_A + m_B) k_B T]^{1/2} \\ &= \frac{4.2 (R_g T)^{3/2}}{\pi^{3/2} (d_{m,A} + d_{m,B})^2 p} (\frac{1}{M_A} + \frac{1}{M_B})^{1/2}. \end{aligned} \tag{1.37}$$

A more complete discussion of the kinetic (i.e., molecular) theory of gases is given by Woods (1993). He discusses the approximations in the first-order kinetic theory and introduces some higher-order theories and models.

(B) Liquids

The molecular theory of the *rheological* properties of *polymeric* liquids, using the mechanical models for *macromolecules* and the principles of nonequilibrium statistical mechanics, is reviewed by Bird et al. (1987). The *cell* model of a *simple* liquid, which allows for the movement of a molecule within a unit cell made of other stationary particles (Walton, 1989), can predict some of the transport properties of the simple liquids. The models based on the gas molecular theory are also used for the liquid by replacing the *speed of sound* $\overline{u}_m$ by the measured speed of sound $\overline{u}_m = \overline{u}_m(\mathrm{T})$ for the liquids. This gives

$$k = \frac{1}{2} n k_B (\frac{vM}{N_A})^{1/3} \overline{u}_m , \tag{1.38}$$

where v is the specific volume. The speed of sound given in terms of *isothermal bulk modulus* E_p is given as

$$a_s = (\frac{c_p}{c_v}\frac{E_p}{\rho})^{1/2}, \quad E_p = \rho(\frac{\partial p}{\partial \rho})_T . \tag{1.39}$$

For liquids $c_p/c_v = 1$ and for water at room conditions, $E_p = 2.2 \times 10^9$ Pa (White, 1991).

An applied shear force causes a nonuniformity in the kinetic energy of the particles in the cell, and if for a particle the increase in the kinetic energy is larger than the *escape activation energy* ΔE_μ, the particle can escape from the cell. Using this model the predicted viscosity is given by (Walton, 1989)

$$\mu = \frac{1}{\ell^2}(8\pi^2 m \Delta E_\mu)^{1/2} e^{\Delta E_\mu / k_B T}, \tag{1.40}$$

where ℓ is the linear dimension of the unit cell.

For diffusion of *similar* molecules (*self-diffusion*), the cell model predicts the diffusion coefficient as

$$D = \frac{a}{2}(\frac{\pi k_B T}{2m})^{1/2} e^{-(\Delta E_D + \ell^3 p)/k_B T}, \tag{1.41}$$

where p is the pressure and ΔE_D is the *activation energy* for *diffusion* which is nearly the same as ΔE_μ.

(C) Solids

The solids may be divided into *metals, ceramics,* and *polymers* and combinations of these, i.e., *composites* (Callister, 1991). In *metallic* crystals, the heat conduction is by *electrons* and in *nonmetals* by the *coupled vibration* of the molecules. Around room temperature a nonmetal such as diamond (carbon) has a higher conductivity than copper. The thermal conductivity of both metallic and nonmetallic (i.e., ceramics and polymers) solids depends substantially on temperature.

For *nonmetals*, the frequency of this vibration and the probability of having a given frequency is combined to give a *particle character* (and the name *phonon*) to this mode of energy. The probability (or population) depends on the temperature and when there is a temperature gradient, the transfer of heat by conduction is modeled by the flow (at the speed of sound) of energy of the phonons (the temperature gradient is equivalent to the phonon energy gradient). The results of the analysis show that at *low temperatures*, i.e., $T < \theta_D$ [where θ_D is the *Debye temperature* and its value is available for most solids (Tien and Lienhard, 1971)], the thermal conductivity *increases* rapidly with temperature (following k is proportional to T^3). For $T > \theta_D$ it *decreases* (following k is proportional to $1/T$).

For metals, the free electron translation (and the impeding impurity and phonon scattering) is responsible for the heat conduction, and the conductivity is given through a relation such as

$$\frac{T}{k} = a_1 T^3 + a_2 \quad \text{for} \quad T < \theta_D. \tag{1.42}$$

As with the nonmetals, this shows a maximum with respect to the temperature.

For metals k varies between 20 and 400 W/m-K, for ceramics between 2 and 50 W/m-K, and for polymers it is about 0.3 W/m-K. In discussing the *equilibrium* properties such as $\rho, c_v, \beta, \kappa, \sigma$ of solids, liquids and gases, the *temperature dependence* (and in general, also the *concentration dependence*) of these properties have become evident. Further discussions and tabulations of these properties are given by Liley (1985). In heat transfer applications, the inherent nonuniformity in temperature results in the spatial variations of these properties, and these can in turn significantly affect the fluid flow and the heat transfer rate. This will be discussed in Section 1.12.

1.4.4 Bulk Versus Surface State

The *bulk* (far from the interface) equilibrium and nonequilibrium properties of materials are determined from the knowledge about the constituent molecules and their arrangements in an *elementary representative volumes* (i.e., the *smallest* volume which contains a *sufficient* representation for the description of the local equilibrium and nonequilibrium properties) located *far* from any *interfaces*. The presence of an interface, e.g., a liquid-gas interface, indicates a *lack* of symmetry of the intermolecular forces for the surface molecular layers of each phase. This *asymmetry* ensues for several molecular layers into each phase and results in many interesting *interfacial* equilibrium and nonequilibrium phenomena. A general discussion of surfaces is given by Somorjai (1994). For liquid-gas interfacial properties, a review is given by Defay and Prigogine (1966) and for the solid surface a review is given by Prutton (1987). A review of the solid-liquid-gas interface (called the *common line*) is given by de Gennes et al. (1990) and a general treatment of the interfacial transport phenomena is given by Slattery (1990). A discussion of the liquid-gas, solid-gas, and solid-liquid-gas interfaces is given below.

(A) Liquid–Gas Interface

When the surface area of a liquid-gas interface is increased, some molecules in the liquid phase move to this surface. Because the bulk (far from the surface) interatomic forces are different from the surface interatomic surfaces, work is required to move these molecules to the surface, i.e., work must be applied along the surface in order to increase the surface area. This *force* appears on the surface, and when expressed as tension per unit length along the surface it is called the *surface tension* of the *liquid-gas interface* $\sigma_{\ell g}$ (or σ). The liquid-gas interfacial *layer* is a region of large density gradient (several molecular layers thick). The stress in this layer is *anisotropic* (unlike the bulk liquid and gas phases). The molecular theory has predicted $\sigma_{\ell g}$ successfully for simple molecules such as argon (Rowlinson and Widom, 1989; Walton, 1989).

The magnitude of $\sigma_{\ell g}$ approaches zero as the *critical point* is reached. Correlation for the estimation of the surface tension of the single-component liquids are reviewed by Reid et al. (1987). In one of these correlations given by

$$\sigma_{\ell g}^{1/4} = P(\rho_\ell - \rho_g) \,, \tag{1.43}$$

the density difference between the two phases is in turn related to the *reduced temperature* T_r (absolute temperature divided by the critical temperature) and the *Parachor* P is related to the atomic bounds. A *simple* correlation, given by Hsieh (1975), is

$$\sigma_{\ell g} = \sigma_o \left(1 - \frac{T - 273.16}{T_c - \Delta T_m - 273.16}\right)^n , \tag{1.44}$$

where ΔT_m indicates that by using this simple equation the interfacial tension diminishes *before* the critical state is reached. For water, σ_o=0.0755 N/m, T_c=647.3 K, ΔT_m=6.14 K and n=1.2.

At the single-component liquid-gas interface, continuous escape of the molecules from the liquid phase to the gas phase occurs (i.e., evaporation). The liquid molecule must have an energy equal to or larger than $\Delta E_{\ell g}$ (evaporation energy) to escape. For an intermolecular spacing ℓ_ℓ at the liquid surface, the *number* of molecules evaporating per unit *area* and time (i.e., *molecular flux*) is given by (Walton, 1989)

$$\frac{\dot{m}_{\ell g}}{m} = \frac{1}{\ell_\ell^3}\left(\frac{\pi k_B T}{2m}\right)^{1/2} e^{-\Delta E_{\ell g}/k_B T} . \tag{1.45}$$

In *dynamic equilibrium*, the evaporation rate is exactly balanced by the *condensation* rate (i.e., rate of vapor molecules *sticking*, with a *probability* of θ, to the liquid surface). The number density of the vapor molecules near the surface, n_g=$1/\ell_g^3$ (ℓ_g is the gas intermolecular spacing), needed to keep this *equilibrium* can then be found from the vapor pressure $p_g = n_g k_B T$. The result is

$$p_g = \frac{\pi k_B T}{\theta \ell_\ell^3} e^{-\Delta E_{\ell g}/k_B T} = \frac{\pi k_B T}{\theta r_\ell^3} e^{-\Delta i_{\ell g}/R_g T} \tag{1.46}$$

at liquid-vapor equilibrium, where $\Delta i_{\ell g}$ is the *molar enthalpy* or *heat of evaporation*. For liquid argon, the computed value of θ is 0.09. The phase equilibrium condition (1.46) will also be discussed in Section 1.5.1(A).

(B) Solid–Gas Interface

When a solid surface is in contact with a gaseous phase, the intermolecular attractive forces that exist between a gas molecule (in the vicinity of a solid surface) and the solid surface molecules result in the *physical adsorption* of the gas molecule to the solid surface. The adsorption of other gas molecules to the surface will also occur until the *adsorbed layers* of the gas molecules reduce these attractive forces substantially so that these layers will not grow any further (at *equilibrium* the rate of *adsorption* and *desorption* are equal). The types of intermolecular forces and their potential energy models, as well as the prediction of the amount of gas adsorbed per unit solid surface area (for porous and nonporous surfaces) as a function of the gas pressure, are discussed by Gregg and Sing (1982). For nonporous solid surfaces and when the intermolecular forces are weak such that they can only form monolayers, the equilibrium number of gas adsorbed per unit area normalized with respect to the maximum *monolayer capacity* of the surface $N/N_{\rm max}$ can be predicted. This is a function of the molecular properties of the gas and the solid, the structural properties of the solid, and the ratio of the gas actual pressure and its *saturation* pressure (at the same temperature) p/p_s. The phenomenon of adsorption is central in the *dehumidification* and in the *catalytic* (*heterogeneous*) *chemical reaction* and plays an important role in how liquid *wets* a surface. When a monolayer of the liquid constituents is already adsorbed to the solid surface, the wetting is more complete. This is discussed next.

(C) Three-Phase Common Line

When any two *immiscible* fluids separated by an interface become in contact with a solid surface, the intersection of this interface and the solid forms a *common* (or *contact*) *line*. One of these fluids preferentially wets the solid surface and this fluid is called the *wetting fluid*. The extent of the wettability is expressed in terms of the angle measured *within* this wetting fluid between the tangents to the solid surface and the interface. This angle is called the *contact angle* θ_c. The contact angle is zero for a *perfectly* wetting fluid. When the *wetting fluid* is a *liquid* and the *nonwetting* fluid is a gas, this common line is called the *solid-liquid-gas* or the *three-phase common line*. The *projected* (shown as a point) common line and the contact angle are shown in Figure 1.3. The contact angle is *defined* for the common line; however, in practice it is observed and measured at a distance from the common line. The various *static* and *dynamical* forces make the *observed* (or *apparent*) contact angle different from the *common-line* (or *intrinsic*) contact angle (de Gennes et al., 1990).

In 1805 Young obtained an expression for the *static contact angle* in terms of the various interfacial tensions. Under *static equilibria* conditions, the *tangential* components of the surface tensions at the interface of the

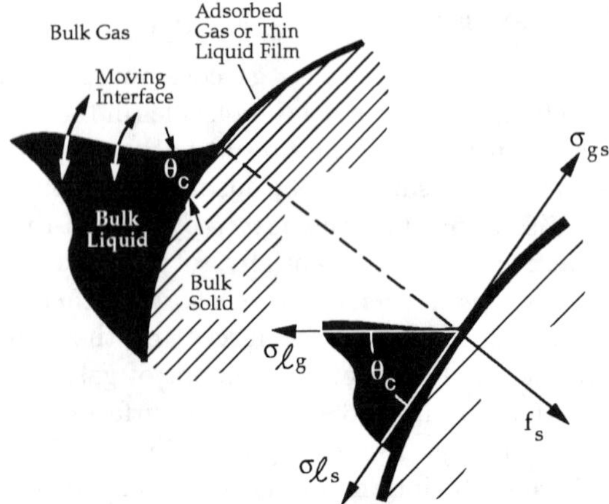

Figure 1.3. The three-phase common line, the contact angle, the three interfacial forces, and the reaction force.

liquid and gas $\sigma_{\ell g}$, at the interface of the gas and solid σ_{gs}, and at the interface of the liquid and solid $\sigma_{\ell s}$ are in *balance*. The *normal* component of the liquid-gas surface tension is also in balance with the *solid reaction force* f_s. The presence of f_s has been demonstrated by the deformation of the solids, by using a very thin solid layer (Chappuis, 1982). These *static equilibrium* conditions and the definition of the static contact angle (called the *Young* equation) at the common line are

$$\sigma_{gs} - \sigma_{\ell s} = \sigma_{\ell g} \cos\theta_c \quad \text{or} \quad \cos\theta_c = \frac{\sigma_{gs} - \sigma_{\ell s}}{\sigma_{\ell g}} , \tag{1.47}$$

$$f_s = \sigma_{\ell g} \sin\theta_c \, . \tag{1.48}$$

In an alterative formulation, the treatment of the liquid-gas interface is extended to include the *interaction* with the solid surface as the common line is approached. In this formulation the contact angle is directly related to the strength of the various fluid-fluid and fluid-solid intermolecular forces (Miller and Ruckenstein, 1974; Hocking, 1993). This will be discussed in Section 8.3.1.

In practice, θ_c and $\sigma_{\ell g}$ are *measured* (instead of σ_{gs}, $\sigma_{\ell s}$ and $\sigma_{\ell g}$). These measurements are made by either *initiating* the contact line by placing a droplet on a solid surface, or by immersing or withdrawing a solid surface through a liquid-gas interface. In these experiments the common line is either *advancing* (i.e., it is moving *away* from the liquid) or *receding*. For a receding common line, in the immediate region of the line a *thin liquid layer* remains on the surface. In the advancing common lines, the vapor molecules which have evaporated from the liquid form an adsorbed layer on the solid surface. This layer moves along the surface and away from the common line at a surface diffusion speed. In addition, the liquid and gas phase *hydrodynamics* also influence the *advancing* and *receding contact lines*. Chen and Wada (1989), Slattery (1990), Dussan V et al. (1991), and Hocking (1992) consider these dynamic effects. For a common line *advancing* at a *speed* u_c and when the *adsorption* is *absent* (i.e., *nonvolatile* liquids) and $\mu_g/\mu_\ell \to 0$, Slattery (1990) suggests the correlation

$$\frac{\cos\theta_c(u_c=0)-\cos\theta_c}{\cos\theta_c(u_c=0)-1} = \tanh(4.96\, Ca^{0.72}) ,$$
$$We_L < 10^{-3}, \quad Bo_L < 5\times 10^{-2} , \tag{1.49}$$

where the *capillary number* Ca (ratio of viscous to surface tension forces), *Weber number* We_L (ratio of inertial to surface tension forces), and *Bond number* Bo_L (ratio of gravitational to surface tension forces) are defined as

$$Ca = \frac{\mu_\ell u_c}{\sigma_{\ell g}} , \quad We_L = \frac{\rho_\ell u_c^2 L}{\sigma_{\ell g}} , \quad Bo_L = \frac{\rho_\ell g L^2}{\sigma_{\ell g}} \tag{1.50}$$

and L is the *system* linear length scale chosen to be the *static meniscus height.* The correlation predicts an *increase* in the contact angle as the common-line velocity *increases.*

For a *receding* common line, the contact angle *decreases* (indicating an improved wetting when a liquid film is present on the solid surface). This *dependence* of θ_c on *wetting* (i.e., advancing) or *dewetting* (i.e., receding) is called the *the contact angle hysteresis.* The prediction of this hysteresis, based on the fluid dynamics and the solid surface adsorption dynamics, is not yet available and is being examined (Slattery, 1990; Hocking, 1992). The *nonisothermal* common line, with evaporation/condensation, will be examined in Section 8.3. There the simultaneous analysis of thehydrodynamics (including the surface forces) and heat transfer will be made.

1.5 Phase Transitions

Phase transitions of the single-component systems (or *pure substances*), i.e., those which have the *same chemical composition* in any phase, have been extensively studied. Although substances which undergo changes in the chemical composition of given phases in response to pressure and temperature changes are in common usage, only a few phase transitions of such substances have received the same attention (e.g., metal alloys in the solid-liquid phase change). For single-component systems, the liquid-vapor (evaporation and condensation), solid-liquid (melting and solidification), and solid-gas (sublimation and crystallization or frosting) are *first-order phase-change transitions* (i.e., discontinuity in the first derivative of the Gibbs energy g occurs as a result of the transition). In these transitions T and p remain constant, while the derivatives of g, such as the entropy and the specific volume undergo finite changes (Callen, 1985). Examples of phase change will be given in Sections 3.2.7, 3.3.5, 4.11.1, 4.16, 6.2, 6.3, 6.4, and 7.3, and throughout Chapter 8.

1.5.1 Single-Component Systems

(A) Transitions

For pure substances all the possible *equilibrium phases* and *phase changes* are shown in the *p-v-T surface* (the remainder of the *p-v-T space* is for *nonequilibrium* states). Figure 1.4(a) shows the surface for a substance that *contracts* upon melting and Figure 1.4(b) is for a substance that *expands* upon melting. The projections of the *p-v-T* surface into the *p-T* and *p-v* planes are also shown in Figure 1.4(b), and the slope $\partial p/\partial T$ of all phase equilibrium lines obeys the *Clapeyron equation* (Desloge, 1968)

$$\frac{\partial T}{\partial p} \simeq \frac{\Delta v_{12}}{\Delta s_{12}} = \frac{T_s \Delta v_{12}}{\Delta i_{12}} \tag{1.51}$$

at phase equilibrium, where Δi_{12} is the specific *heat of the phase change* (i.e., heat absorbed in the phase change), Δv_{12} is the change in the specific volume for transition between phases 1 and 2, and $T_s = T_{12}$ is the *saturation* (i.e., the *coexistance* or *equilibrium*) *temperature*. A line separating a single-phase region from a two-phase region is the loci of the *saturation states*. For a liquid-vapor phase change, when (1.51) is integrated, the resulting equation is similar to (1.46) which is based on the kinetic (i.e., molecular dynamic) theory.

(B) Critical State

The critical point is the *terminal* point of a line representing first-order transitions. At the critical point, the transitions occur continuously and are called the *second-order* transitions (i.e., the second *derivative* of the Gibbs energy undergoes *discontinuity*). In these transitions, while T, p, v, and s remain constant, the second derivatives of g such as the specific heat at constant pressure c_p, the isothermal compressibility κ, and the coefficient of thermal expansion β, undergo finite changes. Discussions of the classical and modern theories of the critical phenomena are given by Stanley (1987).

The thermal conductivity, viscosity, and specific heat at constant volume undergo substantial changes near the critical point, and therefore, influence the motion of the fluid and the heat transfer rate. An example is the thermal disturbance initiated at the solid surfaces bounding a *nearly supercritical* single-component fluid, where the substantial compressibility of the fluid results in the generation of acoustic waves and an increase in the heat transfer rate (Zappoli, 1992).

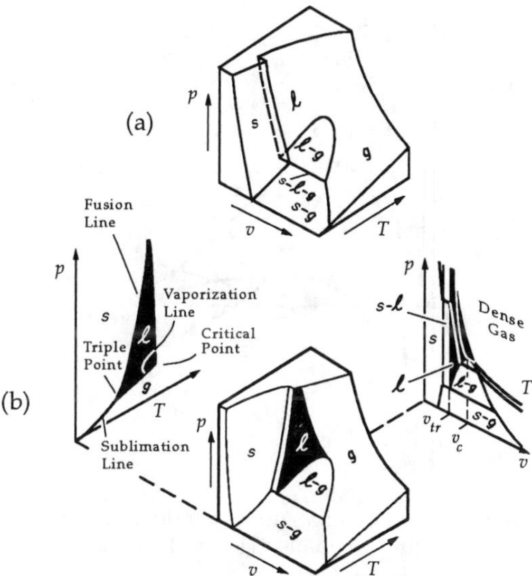

Figure 1.4. (a) Phase diagram for a pure substance that contracts upon melting and (b) for a substance that expands upon melting (also shown are the projections).

(C) Metastable States and Nucleation

Upon changing the state variables such as the temperature and the pressure, a new equilibrium state is reached at the *speed* associated with the molecular kinetics of the system. When the pressure of a *gas* held at a *constant* temperature T below the critical temperature is *raised* to the saturation pressure $p(T)$, condensation may *not* immediately begin. If a liquid droplet or *impurity nucleation site* are not initially available (i.e., in the absence of any *heterogeneous nucleation* sites), the *bulk condensation* (i.e., *homogeneous droplet nucleation*) may *not* occur when the pressure is raised substantially above the saturation pressure. The gas is then in a *metastable state* and is called a *supercooled* or *supersaturated vapor*. Other than the *isothermal* compression, supersaturation can be achieved by *isobaric cooling*, *isometric cooling*, or *isentropic expansion*. A discussion of supersaturation, and the formation of aggregates which grow to be one droplet, is given by Springer (1978).

When the pressure of a *liquid* held at a temperature below the critical temperature is *reduced* below the saturation pressure and when gas-filled cavities and other impurities are not present (i.e., in the absence of *heterogeneous bubble nucleation sites*), *bulk evaporation* (or *homogeneous bubble nucleation*) may *not* begin immediately and the liquid is in a metastable state and is called a *superheated liquid*. Discussions on homogeneous bubble nucleation are given by Skripov (1974), Avedisian (1992), and

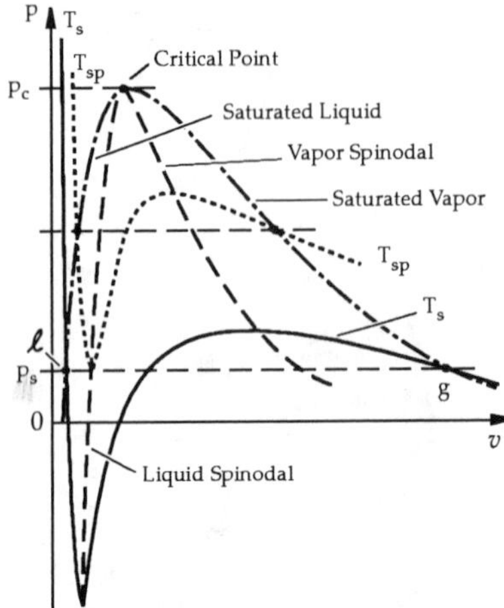

Figure 1.5. The saturated liquid and vapor lines and the liquid and vapor spinodal lines shown in the p-v plane. (From Lienhard et al., reproduced by permission. ©1986 North-Holland.)

Carey (1992), and on heterogeneous bubble nucleation by Nghiem et al. (1981) and Carey (1992).

In *solid-liquid* phase change similar behaviors, i.e., *supercooled liquid* states and *superheated solid* states, are found. Supercooled water and the limits of homogeneous and heterogeneous supercooling have been studied experimentally by Saito et al. (1992). They suggest that the growth of the ice embryo is induced by molecular fluctuations which are in turn affected by the external oscillating or pulsed motions. Metastable phase diagrams for liquid metals are disucssed by Boeltinger and Perepezko (1993) and during the *rapid solidification* there are competing product phases which are selected based on the cooling rate. In Section 7.3.1 homogeneous nucleation of crystals in supercooled melts will be examined.

In Figure 1.4(b) the projection into the $p - v$ plane, the *isotherms* in the *saturation dome* (i.e., the area under the liquid and vapor saturation curves) were kept at their saturation pressure. As we discussed above, superheated liquid and a supercooled vapor can also exist and Figure 1.5 shows these extensions *under* the saturation dome. The isotherms given in this figure can be represented by the van der Waals equation given by (1.17) and by modifications of it as discussed by Skripov (1974), Lienhard and Karimi (1981), Lienhard et al. (1986), Avedisian (1992), and Carey (1992). The *thermodynamic limit* of this liquid superheat is discussed below, and the *kinetic limit* of the superheat, which uses the kinetic or molecular theory, will be discussed in Sections 6.1.1 and 6.2.2. Other models for the equation of state will be discussed in Section 6.2.5(A).

When the slope of the isotherm, i.e., $\partial p/\partial v|_T$ is *positive*, the fluid state is *stable* and when the slope is negative the state is *unstable*. In between these stable and unstable states, the isotherm undergoes local maxima and minima or *spines* (where $\partial p/\partial v|_T = 0$) and the loci of these spines is the *spinodal lines* (one for the liquid and one for the vapor). The region between the liquid and vapor spinodal lines is unstable and as suggested by Lienhard and Karimi (1981) the spinodal lines can be used as *limits* on the liquid superheat and vapor supersaturation, i.e., the limits for *homogeneous nucleation*.

The van der Waals equation (1.17) can be written in terms of the specific volume as

$$p = \frac{R_g T/M}{v - b_1} - \frac{a_1}{v^2} \tag{1.52}$$

with the constants a_1 and b_1 found through the fundamental molecular properties as given in (1.18).

By normalizing the variables with respect to their values at the critical state, i.e., defining the *reduced temperature* $T_r^* = T/T_c$, *reduced pressure* $p_r^* = p/p_c$, *reduced specific volume* $v_r^* = v/v_c$, and using (1.20) and other conditions at the critical point, we have

$$p_r^* = \frac{8T_r^*}{3v_r^* - 1} - \frac{3}{v_r^{*2}} \qquad v_r^* = \begin{cases} v_{r,\ell}^* < 1 & \text{for liquid} \\ v_{r,g}^* > 1 & \text{for vapor.} \end{cases} \tag{1.53}$$

This *dimensionless* van der Waals equation containing *no* fluid-specific constants is called the *law of corresponding states*.

By using $\partial p/\partial v\,|_{T_c} = 0$ and substituting the results in the above, we have the spinodal line *conditions* given by (Skripov, 1974)

$$p_{r_{sp}}^* = \frac{3}{v_r^{*2}} - \frac{2}{v_r^{*3}}, \tag{1.54}$$

or

$$T_{r_{sp}}^* = \frac{(3v_r^* - 1)^2}{4v_r^{*3}}. \tag{1.55}$$

Note that the van der Waals equation when applied to the liquid-vapor saturation, i.e., using $\int_\ell^g p\,dv = 0$, gives the relationship between the *reduced saturation pressure* $p_{r_s}^*$ and *reduced saturation temperature* $T_{r_s}^*$ as (Lienhard and Karimi, 1978)

$$p_{r_s}^* = \frac{8T_{r_s}^*}{v_{r,g}^* - v_{r,\ell}^*} \ln \frac{3v_{r,g}^* - 1}{3v_{r,\ell}^* - 1} - \frac{3}{v_{r,\ell}^* v_{r,g}^*}. \tag{1.56}$$

Equations (1.53) and (1.55) are used for the liquid and vapor spinodal temperatures $T_{r_{sp,\ell}}^*$ and $T_{r_{sp,g}}^*$, respectively, in terms of $T_{r_s}^*$, which in turn is found through (1.53) and (1.56), all iteratively (Lienhard and Karimi, 1978).

Lienhard et al. (1986) suggest a correlation for the *liquid* spinodal line as

$$\frac{T_{sp}}{T_c} = 0.923 - 0.077(\frac{T_s}{T_c})^9 \tag{1.57}$$

and discuss more accurate equations of state for the prediction of the properties of the metastable states. For many liquids, the *superheat* $T_{sp} - T_s$ is over 100°C, and therefore within this range of temperature (over the saturation temperature) the liquid phase may exist without any phase change.

The extension of the application of the spinodal line to the prediction of the maximum superheat, to cases where the intermolecular forces are influenced by the presence of a *solid* surface (i.e., at the common line) is done by Gerwick and Yadigaroglu (1992).

The molecular (i.e., kinetic) theory also predicts that the volumetric rate of formation of *vapor droplets* increases with the superheat and becomes very large (but generally prescribed) as a metastable (superheated) liquid temperature is reached, and this temperature is called the *kinetic-theory limit of superheat*. This and other aspects of the vapor nucleation will be discussed in connection with the nucleation at the liquid-liquid interface in Section 6.1.1, in connection with the bulk liquid nucleation in Section 6.2.2, and in connection with the nucleation on solid surfaces in Section 8.5.1.

1.5.2 Multicomponent Systems

In the above the phase transitions and the equilibrium states were discussed for the single-component systems, while many convective heat transfer processes involve fluids which have multiple constituents. For example, in distillation (a separation process involving evaporation of one or more of the constituents), the *equilibrium concentration* of the various species in the liquid and the gaseous phases depends on the equilibrium temperature and pressure (in the gaseous phase, the *total pressure*). The combination of the *Gibbs phase rule*, which states that the number of independent *intensive* properties is equal to the number of components minus the phases plus two, and the phase diagram of the mixtures are used to determine the equilibrium states for the mixtures. The molecular theory of *fluid-phase* equilibria are reviewed by Prausnitz et al. (1986) and a general background for *all* the three phases is given by Hsieh (1975) and Levine (1988).

Consider phase equilibrium in a *liquid-gas* system. Addition of a *miscible liquid* (e.g., species A) to a liquid which is initially made of species B only results in a new equilibrium state, where in addition of having the same pressure and temperature (*mechanical* and *thermal* equilibrium) in each phase, it is required that

$$\mu_{A,\ell} = \mu_{A,g}, \quad \mu_{B,\ell} = \mu_{B,g}, \tag{1.58}$$

that is, the *chemical potential* μ (which is the driving force for mass transfer) of each species must be the same in each phase (*chemical* equilibrium). The *mole fraction* of species A (and B) in the liquid and gas phases, $Y_{A,\ell}$, $Y_{A,g}$, are *not* equal. Figure 1.6(a) is a schematic of the $p - T$ phase diagram of a binary liquid-gas mixture of species A and B with *miscible*

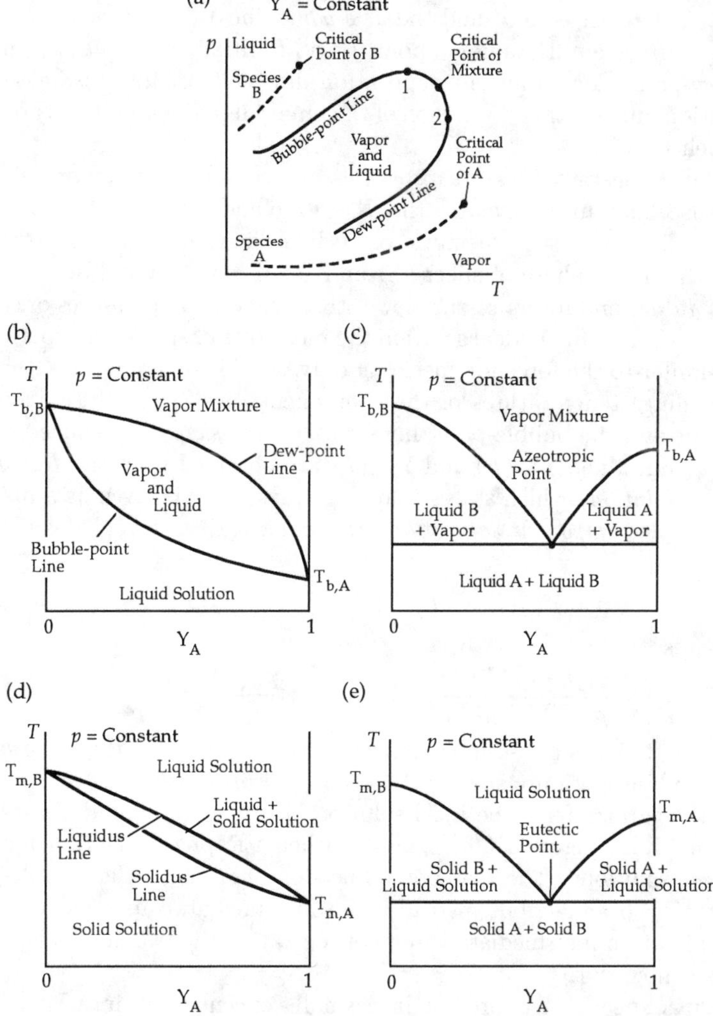

Figure 1.6. Typical phase diagrams for binary mixtures. (a) A liquid-vapor phase diagram shown in the $p - T$ plane for pure species A and pure species B and also for equilbrium states of a given mole fraction of A in the mixture (miscible liquids). (b) A liquid-vapor phase diagram with miscible liquids. (c) A liquid-vapor phase diagram with immiscible liquids. (d) A liquid-solid phase diagram with miscible solids. (e) A liquid-solid phase diagram with immiscible solids.

liquids. For this particular concentration, species B has a critical pressure higher than that of species A and the mixture has a *critical pressure* which is *lower* than the maximum pressure (at point 1) and a *critical temperature* lower than the maximum temperature (at point 2) on the saturation line (i.e., the *bubble-point* and the *dew-point* lines). Note that for a *single-component* system, the bubble-point (loci of initial *condensation*) and the the dew-point (loci of initial evaporation) line collapse into a *single-valued* saturation line. Further discussion of the binary liquid-gas systems are given by Hsieh (1975).

Because constant pressure processes are common, for binary systems the phase diagram is given in the $T - Y_A$ plane for a *constant pressure*. Figure 1.6(b) shows a schematic of such a diagram for a binary system where the liquid phase of species A and B are *miscible* and behave as an *ideal solution* and for a pressure lower than the critical pressures of each of the components. In an ideal solution the molecules of species A experience a force similar to the force for species A only. At this pressure the saturation (i.e., boiling) temperatures of the components are $T_{b,A}$ and $T_{b,B}$ and the dewpoint and the bubble-point lines form a *lens-shaped* two-phase region. The *two* lines, i.e., $Y_{A,\ell}(T)$ and $Y_{A,g}(T)$ are obtained from the *Raoult law* for *ideal solutions* which states that total pressure p is given as a function of the *single-component saturation pressures* $p_A(T)$ and $p_B(T)$ as (Levine, 1988)

$$p = Y_{A,\ell}(T)p_A(T) + [1 - Y_{A,\ell}(T)]\, p_B(T). \tag{1.59}$$

This leads to

$$Y_{A,\ell}(T) = \frac{p - p_B(T)}{p_A(T) - p_B(T)}\,, \quad Y_{A,g}(T) = \frac{p_A(T)}{p} = \frac{p_A(T)}{p} Y_{A,\ell}\,. \tag{1.60}$$

Note that the gas-phase mole concentration is equal to the ratio of the partial pressure of pure species A to the total pressure.

As a *departure* from the ideal solution behavior, the liquid phases may be immiscible. Figure 1.6(c) shows a schematic phase diagram for such *nonideal* solutions. The dew-point line does *not* reach the bubble-point line at $Y_A = 0$ and 1, but instead at an intermediate vapor mole fraction (i.e., only at an intermediate vapor concentration the two liquids and their vapors can coexist).

When a species A is present in a small concentration in a host liquid (i.e., *dilute solution*), the proportionality between $Y_{A,g}$ and $Y_{A,\ell}$ will not be $p_A(T)/p$. This is because the molecules of species A in an *ideally dilute solution* experience a different force compared to the force for species A only. For ideally dilute solutions, the relation between $Y_{A,g}$ and $Y_{A,\ell}$ is given by the *Henry law*, and instead of $p_A(T)$ a proportionality constant called the *Henry law constant* is used. This will be discussed in Section 6.2.4(D) in relation to gas–liquid phase change in multicomponent systems.

For a *binary liquid–solid* phase equilibrium, the miscibility of the components in the *solid* as well as the liquid phase influences the phase diagram.

For a complete miscibility of *both* phases (such as Cu-Ni, Sb-Ni, Pd-Ni and KNO_3-$NaNO_3$) a phase diagram such as the one shown in Figure 1.6(d) is obtained. This is a lens-shaped solid–liquid two-phase region bounded by the melting point of components A and B, $T_{m,A}$ and $T_{m,B}$ at $Y_A=0$, and 1. For the miscible liquid phase and the *immiscible* solid phase, the phase diagram will be similar to that for the nonideal liquid phase in a liquid-gas system. The nonideal solid–liquid system is depicted in Figure 1.6(e). As was the case in Figure 1.6(c) for the liquid-gas systems, the *liquid* concentration at the *eutectic point* is the only liquid phase mole fraction at which both solid phases and the liquid can coexist at equilibrium. At a given pressure, when a liquid phase with the eutectic-point mole fraction is cooled, *both* components solidify (an advantage in alloy casting). For the particular example shown in Figure 1.6(e), which is representative of most *aqueous solutions*, the composition of the solid is *independent* of the temperature. In Section 3.2.7 a more general form of this phase diagram will be considered.

In the above discussion of multicomponent systems, no *chemical reaction* was allowed between the species, and therefore, the physical interactions such as miscibility and ideal-solution behavior were addressed. The extension of the discussion to *ternary* and higher component systems requires a more elaborate displaying of the phase equilibria diagram (examples are given by Prausnitz et al., 1986). In some heat transfer applications, the condensation temperature of one or more components of a multicomponent system is very low, and therefore, in high-temperature applications these components are considered as *noncondensables* (an example is the presence of air, which is a mixture of oxygen and nitrogen and trace of other gases, in condensation of water vapor).

1.6 Multiphase Flows

The *simultaneous* flow of phases occurs when as a result of the heat transfer with a flowing fluid, a phase change occurs, or when various phases are flown together for interfacial heat transfer among them, or when heat transfer occurs between a stream containing several phases and the solid surfaces *containing* the stream or the solid surfaces *submerged* in the stream. We consider the simultaneous flows of *two* and *threephases*. A stream containing two *immiscible* liquid is also considered a two-phase flow. In some *batch* processes, one of the fluid phases recirculates (instead of having a continuous through flow) while the other phases have continuous flows. In Chart 1.2, a classification of the two- and three-phase flows is given with the emphasis on whether the main (i.e., the largest) heat transfer is occurring *among* the flowing phases or between them and the *bounding* (or *confining*) or *submerged surfaces*. Examples of multiphase flow will be given in Sections 3.3, 4.14, and 5.4 and throughout Chapters 6, 7, and 8.

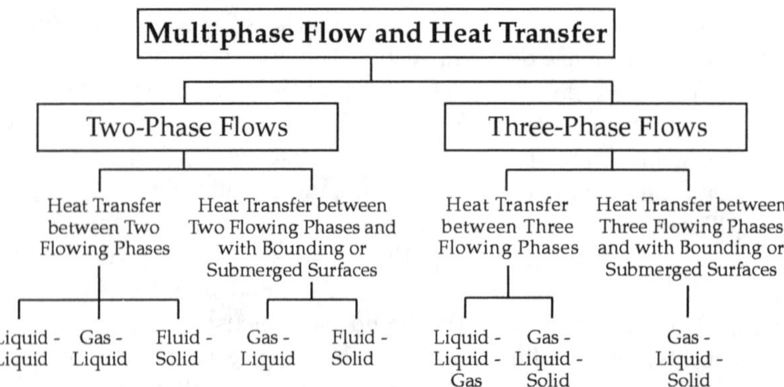

Chart 1.2. A classification of multiphase flow and heat transfer based on the number of phases and the interface through which the main heat transfer of interest occurs.

1.6.1 Two-Phase Flows

In a stream made of two phases, either *one* of the phases is dispersed or *both* phases are *continuous.* While one of the phases is a *fluid* (liquid or gas) the other phase can be liquid, gas, or solid. For the two-fluid flows, interfacial *deformation* and *intraphase motion* can occur, and these influence the heat transfer. This is the reason for the distinction made with a fluid-solid system. In two-phase flows, a *phase change* (i.e., solid-gas, solid-liquid, or liquid-gas) can also occur.

In the following, various relevant two-phase flows, with the main heat transfer of interest occurring between the two phases, are considered first. Then discussed are the practical systems in which the main heat transfer is between the bounding or submerged surfaces and the two-phase stream.

(A) Heat Transfer Between Two Phases

The heat transfer between any two phases through their *common* interface is called the *direct-contact heat transfer* (in contrast to the *noncontact* heat transfer which, for example, occurs between two fluid phases flowing in a heat exchangers with a solid surface *separating* the two phases). Various forms of the direct-contact heat exchange are reviewed by Kreith and Boehm (1988) and Jacobs (1988b).

(i) Liquid–Liquid

A pair of immiscible liquids, such as the organic-aqueous pairs (Ishikubo et al., 1992) at different temperatures are brought together in various processes. For example, when the liquids tend to leave a significant deposit on the bounding solid surfaces (i.e., *scaling* or *fouling* of the solid surface), the direct-contact heat transfer becomes attractive. Any *miscibility* (*dissolution*) or inability to later completely separate the liquids (entrainment), makes this method less attractive (Letan, 1988). In order to increase the interfacial area, one of the phases is *dispersed*. The *linear dimension* of the dispersed element (e.g., drop) can range from 1 μm to 10 mm and the elements can have small or large *interelement clearance*. The motion within the elements, as well as in the continuous phase, influences the interfacial heat transfer rate. In general, for an increase in the overall heat transfer, dispersed elements should be small in size (any permanent coalescing is avoided by staged sieving); the *relative* (i.e., *slip*) *velocity* of the elements should be large; and the number of elements per unit volume should also be large. However, the heat transfer does *not monotonically* increase with any of these changes, i.e., the *interelemental* convective heat transfer *interaction* can be so large that the heat transfer to the continuous phase becomes hindered. The intraelement motion also influences the overall heat transfer significantly (Jacobs and Golafshani, 1989).

Central to the analysis of the interfacial heat transfer is the knowledge of the *interfacial area concentration* (i.e., *specific area*) which is determined from the fluid dynamics. In the liquid-liquid heat transfer, one of the liquids is *dispersed* and the motion of the dispersed liquid is by the *gravity* through a vertical tube (i.e., *column*). A *shape regime diagram* for the *buoyant* motion of drops in an immiscible liquid, in an otherwise *stationary* host liquid and far from bounding surfaces, is given by Clift et al. (1978). The dimensionless governing parameters are the *Bond* (or *Eötvös*) number Bo, the *Reynolds* number Re, and the *Morton* number Mo, defined as

$$Bo_d = \frac{g\Delta\rho\, d^2}{\sigma}\,, \qquad Re_d = \frac{\rho u d}{\mu}\,, \qquad Mo = \frac{g\mu^4\Delta\rho}{\rho^2\sigma^3}\,, \tag{1.61}$$

where ρ and μ are the properties of the continuous liquid and d is the droplet diameter. When the densities and the viscosities of the two liquids are substantially different, the *density* and *viscosity ratios* must also be used for the classification. Three distinct regimes, namely, spherical, ellipsoidal, and spherical cap regimes, are identified. Figure 1.7 is from Clift et al. (1978) and is based on the *normal gravity* (as compared to the *reduced gravity* encountered at high elevations from the earth surface, or the *increased gravity* produced by, for example, centrifuging). For liquid-gas systems, drops *falling under gravity* in gases remain fairly spherical for $Bo_d \leq 0.4$. A further discussion of the drop shape will be given in Section 6.1.2.

(ii) Gas–Liquid

The gas and liquid phases can consist of the same *single component* or can be made up of a *mixture* of components. The single-component systems are common in the gas-liquid phase-change heat transfer processes

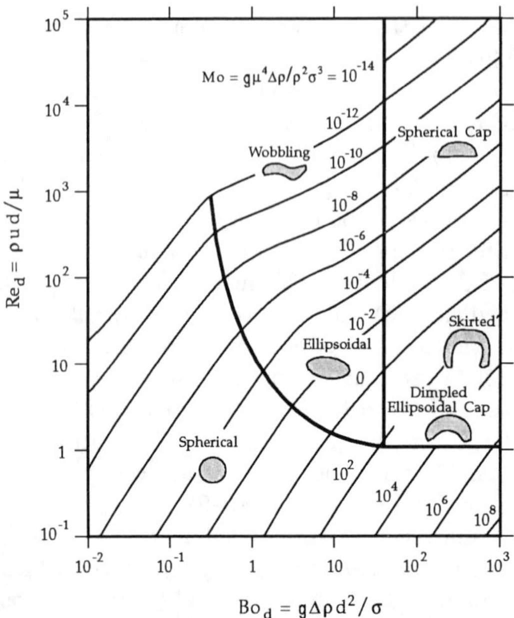

Figure 1.7. A shape diagram for the isothermal buoyant motion of a drop in an otherwise quiescent, immiscible liquid. (From Clift et al., reproduced by permission ©1978 Academic Press.)

and the interface temperature is determined only by the local pressure. In the multicomponent systems, the interface temperature *changes* with the concentration.

Because of the substantial difference in the densities of the two phases, the phase distributions are significantly influenced by the magnitude and the direction of the gravity (with respect to the flow direction). For applications such as in *film condensation* and *evaporation*, the interface is *continuous*. In forced flows, depending on the relative volumetric flow rates of phases, one of the phases can be *dispersed*. A review of the *flow pattern diagrams* for the *upward* and the *downward* gas-liquid flows in *horizontal* and *vertical* tubes under *normal gravity* is given by Whalley (1987). The flow patterns and their transitions under *microgravity* are discussed by Duckler et al. (1988). References to the fluid dynamics of this two-phase flow and its modeling are given by Lahey and Drew (1990) and Lahey and Lopez de Bertodano (1991).

As an example, the flow pattern diagram for the *upward*, isothermal gas-liquid flow in a tube as given by Hewitt (1982) is repeated in Figure 1.8. Note that the *fluid* properties such as the surface tension, the density difference, the ratio of the densities and the viscosities, the *flow* properties such as the the relative direction of the flows, and the volumetric flow rates, and the *magnitude* of the gravitational acceleration, all influence the flow

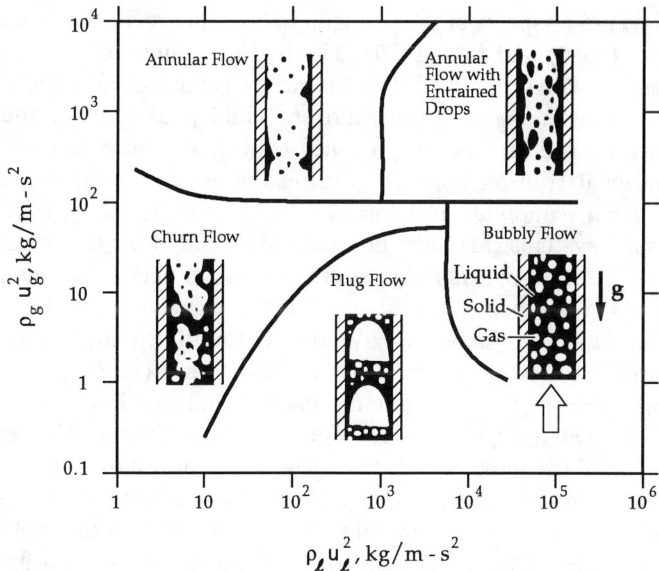

Figure 1.8. Flow pattern diagram for isothermal upward gas-liquid flow in a tube. (From Whalley, reproduced by permission ©1987 Oxford University Press.)

pattern. In this figure the *dimensional phase momentum fluxes*, $\rho_\ell u_\ell^2$ and $\rho_g u_g^2$, are used for the flow pattern mapping. In the *bubbly-flow* regime, the gas bubbles are approximately uniform in size. In the *plug-* (or *slug-*) *flow* regime, the gas phase is distributed in large caplike bubbles as well as smaller spherical bubbles. The *churn-flow* regime is oscillatory and generally unstable. In the *annular-flow* regime, the liquid is present as a film on the tube surface and at high liquid flow rates some liquid drops are also entrained in the gas phase.

(iii) Fluid–Solid

In fluid-solid systems, either *both* phases are moving or the solid is *stationary*. The stationary solid with the fluid moving through it is considered as a *porous medium* and the solid phase is envisioned as being made of particles. These particles can be *consolidated* (as in rocks, foams, sintered powders) or *nonconsolidated* as in *packed beds* of *pellets* (i.e., *granular materials*). Granular materials can also be made to flow while the particles remain in contact with each other (called *liquification*) or they can be made to flow in pipes (*particulate pipe flow* or *slurry flow*), allowed to *settle* under gravity (gravity *sedimentation*), or *fluidized* in *upward* moving fluid flows. The micromechanics of granular materials, from a particle and a continuum view point, are reviewed by Satake and Jenkins (1988). The fluid mechanics and the heat transfer aspects of the particulate flows are reviewed by van Swaaij and Afgan (1986) and Soo (1989). The hydrodynamics of the fluidization is reviewed by Davidson et al. (1985) and by Homsy et al. (1992).

In heat transfer between the solid and fluid phases, for porous media a *two-medium* (i.e., a fluid and a solid) continuum *model* is used and the interfacial heat transfer is *modeled* using the *specific interfacial area* and the *interfacial* (coupling) *heat transfer coefficients* (Kaviany, 1991). In the moving solid systems this must be expanded to allow for the mixing of the particles. This mixing generally results in a rather very *small* temperature gradient along the bed. For nonconsolidated particles and when the flow of the fluid is against the gravity vector, as the fluid velocity increases an incipient fluidization of the particles occurs. Grace (1986) has assembled a *gas-solid flow regime diagram* for packed and fluidized beds and this diagram is shown in Figure 1.9. The figure uses the *dimensionless* gas-phase velocity u^* and the dimensionless particle size d^* and shows the boundaries of the *minimum fluidization velocity* (*incipient* fluidization). Also shown is the *terminal velocity* (i.e., terminal settling velocity of a single sphere in an unbounded fluid). The dimensionless terminal velocity u_t^* is a function of d^* as given by Clift et al. (1978). In a *spouted bed*, the fluid enters as a *jet* through a single centrally located orifice (as compared to the *conventional fluidized bed* where the fluid is distributed *uniformly* across the inlet). The *internal particulate flows* are called the *circulating beds* and are different from the fluidized beds in that the particles are circulating through a loop. As discussed by Grace (1986) and Soo (1989), the various *particulate flow regimes*, identifying the particle concentration distribution along and across the tubes (including instabilities), have not yet been clearly identified for circulating fluidized beds. The upward moving particulate flows will be discussed in Sections 4.14.1 and 7.1.1.

The *liquid-solid*, two-phase flow can occur in the *solidification* or in the *melting* processes. The *sublimation* of a dispersed solid phase is an example of the *gas-solid* two-phase flow and heat transfer. Some of these processes are discussed by Vere (1987) and Szekely et al. (1988).

(B) Heat Transfer with Bounding or Submerged Surfaces

In the two-phase flows mentioned above, the heat transfer through the interfacial between the two phases was the primary focus. However, in many applications the primary heat transfer is between the two-phase stream and the *surfaces bounding* it (such as the tube walls and plates), or the *surfaces submerged* in it (such as tubes). In these cases the presence of these surfaces alter the hydrodynamics and the phase distributions adjacent to them. Examples of the *gas-liquid* systems are the boiling inside and outside heated tubes and the impingement of evaporating drops and jets on surfaces (Rohsenow, 1982; Griffith, 1982; Lienhard V et al., 1992; Yao, 1993).

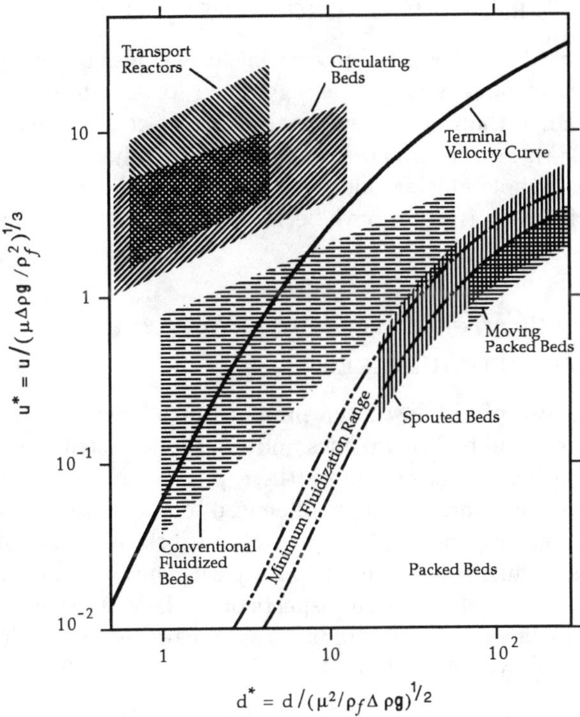

Figure 1.9. Flow regime diagram for the solid-gas two-phase flow systems, including the packed bed regime and the fluidized bed regime. (From Grace, reproduced by permission ©1986 Can. J. Chem. Eng.)

Examples of *fluid-solid* systems are the heat transfer between particulate flows (such as a slurry) and their confining tube walls and between the tubes submerged in fluidized beds (Grace, 1982a; Saxena, 1989; Wu et al., 1991).

1.6.2 Three-Phase Flows

An example of a *three-phase* flow is the gas injection into a liquid-solid flow to assist the flow through the *pumping action* of the buoyancy and through the *drag reduction* at the bounding surfaces. The heat transfer among the phases and between them and the bounding or submerged surfaces is also enhanced. The hydrodynamics of the three-phase flow is discussed by Grace (1982b), Giot (1982), and Fan (1989). The three phases consist of two immiscible liquids and a gas phase in a liquid-liquid direct contact heat transfer where evaporation of one of the liquids occurs.

(A) Heat Transfer Among Three Phases

For *liquid-liquid-gas* flows, the interfacial heat transfer among the phases and the hydrodynamics have been examined for the direct contact *evaporation* by Bharathan (1988) and for the direct-contact *condensation* by Jacobs (1988a). For gas-liquid-solid flows, especially in the case of a single large gas bubble moving due to the buoyancy, the hydrodynamics and the interfacial mass transfer have been examined in detail (Jean and Fan, 1990).

(B) Heat Transfer with Bounding or Submerged Surfaces

As with the two-phase flows, the presence of a bounding or submerged surface changes the hydrodynamics and the phase distribution near these surfaces. The heat transfer among these phases and also between them and these surfaces is also strongly influenced by the phase distribution. For example, in a gas-liquid-solid flow over a solid surface the wettability of the solid surfaces influences the liquid contact with surface, and therefore, the convection heat transfer. Various aspects of the heat transfer to submerged surfaces in *gas-liquid-solid* systems such as three-phase slurry-fluidized beds are discussed by Kim et al. (1986) and Saxena et al. (1992).

1.7 Heat Flux Vector and Conservation Equations

The *heat flux vector* $\mathbf{q}$ through the *most elementary area* and the conservation equations for the *overall mass*, *species*, *momentum*, and *energy*, for the *most elementary volume*, are discussed below.

1.7.1 Heat Flux Vector

In the presence of a velocity field and a nonuniform temperature field, the heat *flow* is by *conduction*, *convection*, and *radiation* with the *heat flux* (i.e., heat flow per unit area and time) *vector* for any point in a *multicomponent*, *simple compressible* fluid, is given by

$$\mathbf{q} = -k\nabla T + \rho e\mathbf{u} + \sum_{i=1}^{n} \rho_i e_i(\mathbf{u}_i - \mathbf{u}) + \mathbf{q}_r \ , \tag{1.62}$$

where

$$de = c_v dT + (T\frac{\partial p}{\partial T}|_v - p)dv \ . \tag{1.63}$$

The *mass-averaged velocity* is defined as

$$\mathbf{u} = \frac{\sum_{i=1}^{n} \rho_i \mathbf{u}_i}{\rho} , \quad \rho_i(\mathbf{u}_i - \mathbf{u}) = -D_i \rho \nabla \frac{\rho_i}{\rho} , \quad \rho = \sum_{i=1}^{n} \rho_i , \tag{1.64}$$

where n is the total number of species, ρ_i is the density, $\mathbf{u}_i$ is the velocity, and D_i is the mass diffusivity of species i. Note that the *mass flux* vector of species i is the sum of the mass-averaged convective flux and the diffusive flux (i.e., both per unit area) and is given by

$$\dot{\mathbf{m}}_i = \rho_i \mathbf{u} - D_i \rho \nabla \frac{\rho_i}{\rho} = \rho_i \mathbf{u}_i . \tag{1.65}$$

The *mixture*-specific internal energy appearing above is also mass averaged, but the mixture thermal conductivity is generally determined experimentally.

The radiation heat flow is by *photons* and when the medium is *semi-transparent* but *optically thick* (i.e., the photons attenuate significantly as they travel) the radiative heat flux is described similar to conduction (i.e., radiative transport is approximated as diffusive). For semi-transparent but *optically thin* media the flux not only depends on the local radiative sources, sinks and resistances, but also on those at distant locations (i.e., radiative transport is *diffuso-integral*). The general treatment of the radiative heat flux vector will be discussed in Section 2.5, and particular applications are discussed throughout Chapters 3 to 8.

For a *stationary solid*, in the heat flux vector $\mathbf{u} = 0$, the heat transfer is by diffusion (heat conduction and internal energy flow by mass diffusion), and radiation and the thermal conductivity may be *anisotropic* (and described by a *tensor*). The incompressibility of solids and liquids results in a single specific heat capacity under constant pressure or volume condition.

While the conductive and radiative components of the heat flux vector are directed towards the lower temperature surrounding, the convective components are pointing in the directions of $\mathbf{u}$ and $\mathbf{u}_i - \mathbf{u}$.

1.7.2 Conservation Equations for Elementary Volumes

Conservation equations for infinitesimal (or elementary) volumes are derived using the principles of conservation of mass (and species), momentum, and energy (thermal and mechanical) in the presence of a fluid flow. A *simple-differential* (or *elementary*) *volume* or a *pointwise* differential volume is an infinitesimal volume taken within a *single phase* and when a *continuum treatment* is allowable (i.e., when the *Knudsen number* $Kn = \lambda/C$, where λ is the molecular mean free path and C is the average intermolecular distance, is smaller than 0.1). This is in contrast to the representative elementary volumes containing multiple phases where the monotonic decrease eventually causes the disappearance of one or more phases from the differential volume.

The balance of mass, species, momentum, and energy is taken over a fixed, simple differential volume as the control volume and then the Taylor series approximation are applied and the limit of this volume approaching zero in imposed. Conservation equations are derived in Bird et al. (1960), Landau and Lifshitz (1968), Eckert and Drake (1972), Schlichting (1979), Slattery (1981), Burmeister (1983), Arpaci and Larsen (1984), Bejan (1984), Cebeci and Bradshaw (1984), Platten and Legros (1984) Aris (1989), Incropera and DeWitt (1990), Sherman (1990), White (1991), and Kays and Crawford (1993), among others.

(A) Conservation of Mass and Species

Using the principle of mass conservation, the *overall* (i.e., including all the species) mass conservation for a fixed (Eulerian) infinitesimal control volume is found by allowing for the *mass flux vector* $\rho\mathbf{u}$ (inflow and outflow through the boundaries) and the *volumetric storage rate* $\partial\rho/\partial t$, within this volume. Using the mass-averaged velocity and allowing for the variation of the density (i.e., *compressibility*), we have the *mass conservation* or the *continuity equation* given by

$$\frac{\partial\rho}{\partial t}+\nabla\cdot\rho\mathbf{u}=0\ , \tag{1.66}$$

or

$$\frac{\partial\rho}{\partial t}+\mathbf{u}\cdot\nabla\rho+\rho\nabla\cdot\mathbf{u}=0\ , \tag{1.67}$$

or using the *total* or *substantial derivative* D/Dt we have

$$\frac{D\rho}{Dt}+\rho\nabla\cdot\mathbf{u}=0\ . \tag{1.68}$$

For *incompressible* (*constant* and *uniform* density) flows we have

$$\frac{\partial\rho}{\partial t}+\mathbf{u}\cdot\nabla\rho=0 \qquad \text{incompressible flow.} \tag{1.69}$$

Then from (1.67) we also have

$$\nabla\cdot\mathbf{u}=0 \qquad \text{incompressible flow.} \tag{1.70}$$

The continuity equation can be written in the *indicial notation* as

$$\frac{\partial\rho}{\partial t}+u_j\frac{\partial\rho}{\partial x_j}+\rho\frac{\partial u_j}{\partial x_j}=0\ . \tag{1.71}$$

The *conservation equation* for the *species* i is found by allowing for its *inflow* and *outflow* using the net mass flux vector for species i, given by (1.65), its *volumetric storage rate* $\partial\rho_i/\partial t$, and its *volumetric production rate* $\dot{n}_i$. Then the *species conservation equation* is given by

$$\frac{\partial\rho_i}{\partial t}+\nabla\cdot\rho_i\mathbf{u}=\nabla\cdot\rho D_i\nabla\frac{\rho_i}{\rho}+\dot{n}_i\ , \qquad \rho_i=\rho_i(\mathbf{x},t). \tag{1.72}$$

Note that both the *temporal* and *spatial* variations of the mixture and the species densities are allowed for in the above equations. For example, low- and high-frequency time- and space-periodic variations (e.g., oscillating and turbulent flows) are described by these equations.

(B) Conservation of Momentum

Using the principle of motion (i.e., the rate of change of the fluid momentum in a volume is equal to the change in the external forces across this volume), the fluid *momentum* (or mass flux) $\rho\mathbf{u}$ *within* a fixed (Eulerian) infinitesimal control volume can change with time $\partial\rho\mathbf{u}/\partial t$ (*local average* or the *first part* of the *acceleration*), or a change can occur in the *momentum flux tensor* $\rho\mathbf{uu}$ *across* the control volume (i.e., the *second part* of the *acceleration*). The changes in the external forces across the volume are the *pressure gradient* $-\nabla p$, the *volumetric gravitational body force* $\rho\mathbf{g}$, the *volumetric electromagnetic force* $\mathbf{f}_e$, and the *volumetric retarding* forces such as the gradient of the viscous *stress* tensor $\nabla\cdot\mathbf{S}$. Then the *momentum conservation equation* can be written as (Bird et al., 1960)

$$\frac{\partial\rho\mathbf{u}}{\partial t}+\nabla\cdot\rho\mathbf{uu}=-\nabla p+\nabla\cdot\mathbf{S}+\rho\mathbf{g}+\mathbf{f}_e\,. \tag{1.73}$$

By using the continuity equation (1.65), this becomes

$$\rho\frac{D\mathbf{u}}{Dt}\equiv\rho\frac{\partial\mathbf{u}}{\partial t}+\rho\mathbf{u}\cdot\nabla\mathbf{u}=-\nabla p+\nabla\cdot\mathbf{S}+\rho\mathbf{g}+\mathbf{f}_e\,, \tag{1.74}$$

where again D/Dt indicates the substantial derivative. This is valid for compressible fluids with any shear stress-strain rate relation. Further assumptions of an *incompressible* and a *Newtonian* fluid, (1.32), will allow the use of $\nabla\cdot\mathbf{S}=\nabla\cdot\mu\nabla\mathbf{u}$.

The *total stress tensor* $\mathbf{T}$, is related to the viscous stress tensor $\mathbf{S}$ through

$$\mathbf{T}=\mathbf{S}-p\mathbf{I}\,, \tag{1.75}$$

where $\mathbf{I}$ is the identity tensor.

The *distant* force $\mathbf{f}_e$, due to electric or magnetic fields applied to charged or electrically conducting fluids, will be discussed in connection with the electro- and magnetohydrodynamics in Sections 2.6 and 4.13. In *very thin* liquid films (film thickness δ of order of 100 Å), the van der Waals *interfacial* forces which penetrate into the film must be included as a volumetric force. This force designated as $\mathbf{f}_\delta=-\nabla\varphi$, where φ is the molecular potential energy, will be discussed in Section 8.3.

The momentum equation can be written in indicial notation as

$$\rho\frac{Du_i}{Dt}=\rho\frac{\partial u_i}{\partial t}+\rho u_j\frac{\partial u_i}{\partial x_j}=-\frac{\partial p}{\partial x_i}+\frac{\partial}{\partial x_j}\tau_{ij}+\rho g_i+f_{e_i}\,, \tag{1.76}$$

where in general the relation between the elements of the viscous stress tensor τ_{ij} and the *elements* of the *deformation rate tensor* d_{ij} is given by a *constitutive* relation. The momentum equation (1.76) is called the

Navier-Stokes momentum equation after its originators. For a *Newtonian, compressible* fluid we have

$$\tau_{ij} = \mu(\frac{\partial u_i}{\partial x_j} + \frac{\partial u_j}{\partial x_i}) + \delta_{ij}\lambda_\mu \frac{\partial u_k}{\partial x_k} , \tag{1.77}$$

where λ_μ is called the *second* (or *bulk* or *expansion*) *viscosity* and δ_{ij} is the *Kronecker delta*. The value for λ_μ for various fluids is not yet measured but is generally taken as suggested by the *Stokes' hypothesis* as $\lambda_\mu = -2\mu/3$ (White, 1991). Non-Newtonian fluids will be discussed in Section 4.8 and the magnitude of λ_μ will be further discussed in Section 2.2.1.

(C) Conservation of Energy

When in addition to the heat flux vector given by (1.62) the *kinetic energy flux* $(\rho\mathbf{u}\cdot\mathbf{u}/2)\mathbf{u}$ *into* and *out* of a fixed (Eulerian) infinitesimal control volume is also added and when the *rates* of the *pressure*, the *viscous stress*, and the *body force work* done or the control surfaces are included and the volumetric *storage* of the energies, and the *volumetric* heat *generation* $\dot{s}$ are also allowed for, we have the *total* (i.e., *thermal* and *mechanical*) *energy conservation equation* for a single-component, simple compressible fluid given by (Bird et al., 1960)

$$\begin{aligned} \frac{\partial}{\partial t}(\rho e + \frac{\rho}{2}\mathbf{u}\cdot\mathbf{u}) + \nabla\cdot(-\mathbf{K}\cdot\nabla T + \rho e\mathbf{u} + \mathbf{q}_r + \frac{\rho}{2}\mathbf{u}\cdot\mathbf{u}) = \\ -\nabla\cdot p\mathbf{u} + \nabla\cdot(\mathbf{S}\cdot\mathbf{u}) + \rho\mathbf{u}\cdot\mathbf{g} + \mathbf{u}\cdot\mathbf{f}_e + \dot{s} , \end{aligned} \tag{1.78}$$

or

$$\begin{aligned} &\rho\frac{D}{Dt}(e + \frac{1}{2}\mathbf{u}\cdot\mathbf{u}) = \nabla\cdot\mathbf{K}\cdot\nabla T - \nabla\cdot\mathbf{q}_r - \\ &\nabla\cdot p\mathbf{u} + \nabla\cdot(\mathbf{S}\cdot\mathbf{u}) + \rho\mathbf{u}\cdot\mathbf{g} + \mathbf{u}\cdot\mathbf{f}_e + \dot{s} , \end{aligned} \tag{1.79}$$

where $\mathbf{K}$ is the thermal conductivity tensor. The conservation of the *mechanical* energy is obtained by taking the dot product of the velocity vector and the vectorial momentum equation, i.e.,

$$\begin{aligned} &\frac{\rho}{2}\frac{D}{Dt}\mathbf{u}\cdot\mathbf{u} = -\mathbf{u}\cdot\nabla p + \mathbf{u}\cdot(\nabla\cdot\mathbf{S}) + \rho\mathbf{u}\cdot\mathbf{g} + \mathbf{u}\cdot\mathbf{f}_e = \\ &p\nabla\cdot\mathbf{u} - \nabla\cdot p\mathbf{u} + \nabla\cdot(\mathbf{S}\cdot\mathbf{u}) - \mathbf{S}:\nabla\mathbf{u} + \rho\mathbf{u}\cdot\mathbf{g} + \mathbf{u}\cdot\mathbf{f}_e . \end{aligned} \tag{1.80}$$

Using this mechanical energy equation, (1.79) can be written as

$$\rho\frac{De}{Dt} = \nabla\cdot\mathbf{K}\cdot\nabla T - \nabla\cdot\mathbf{q}_r - p\nabla\cdot\mathbf{u} + \mathbf{S}:\nabla\mathbf{u} + \dot{s}, \tag{1.81}$$

which is called the *thermal* energy equation. Further, using $de = c_v dT + (-p + T\partial p/\partial T|_v)dv$, we have

$$\begin{aligned} \rho c_v\frac{DT}{Dt} = \nabla\cdot\mathbf{K}\cdot\nabla T - \nabla\cdot\mathbf{q}_r - T\frac{\partial p}{\partial T}\bigg|_v \nabla\cdot\mathbf{u} + \\ \mathbf{S}:\nabla\mathbf{u} + \dot{s} . \end{aligned} \tag{1.82}$$

It is customary to use $\mathbf{S} : \nabla\mathbf{u} = \Phi_\mu$, where Φ_μ is a *positive* definite and states the *conversion* of *mechanical* to *thermal* energy and is called the *viscous dissipation rate*.

Using the specific *enthalpy* $i = e + p/\rho$ and $di = c_p\, dT + (1 - \beta T)\, dp/\rho$, where β is defined by (1.1), we can write the *thermal* energy equation (1.81) as (Burmeister, 1983)

$$\rho\frac{Di}{Dt} = \nabla\cdot\mathbf{K}\cdot\nabla T - \nabla\cdot\mathbf{q}_r + \frac{Dp}{Dt} + \Phi_\mu + \dot{s} \tag{1.83}$$

or

$$\rho c_p\frac{DT}{Dt} = \nabla\cdot\mathbf{K}\cdot\nabla T - \nabla\cdot\mathbf{q}_r + \beta T\frac{Dp}{Dt} + \Phi_\mu + \dot{s}\,. \tag{1.84}$$

The term involving the pressure is generally called the *compressibility work*. While Φ_μ is positive (i.e., *viscous dissipation heating*), $\beta T Dp/Dt$ is generally *negative* (i.e., *expansion cooling*) and in part *nullifies* the local temperature rise due Φ_μ. In the indicial notation, (1.84) can be written as

$$\begin{aligned}\rho c_p\frac{\partial T}{\partial t} + \rho c_p u_j\frac{\partial T}{\partial x_j} \;&=\; \frac{\partial}{\partial x_j}k_{jj}\frac{\partial T}{\partial x_j} + \frac{\partial}{\partial x_j}q_{r_j} + \beta T\frac{\partial p}{\partial t} + \\ &\beta T u_j\frac{\partial p}{\partial x_j} + \tau_{ij}\frac{\partial u_i}{\partial x_j} + \dot{s}\,,\end{aligned} \tag{1.85}$$

where the components of the thermal conductivity tensor k_{ij} have been used.

For *multicomponent* fluids, the term $\sum_i \rho_i c_{p_i}(\mathbf{u}_i - \mathbf{u})\cdot\nabla T$, where $\mathbf{u}_i - \mathbf{u}$ is given by (1.64), must be *added* to the left-hand side of (1.84) to allow for the enthalpy transport due to the species diffusion.

The *total* energy equation can also be written in terms of the *total specific enthalpy* $i + \frac{1}{2}\mathbf{u}\cdot\mathbf{u}$. Using $i = e + p/\rho$ and the continuity equation (1.68), we have

$$\begin{aligned}\rho\frac{D}{Dt}(e + \frac{1}{2}\mathbf{u}\cdot\mathbf{u}) = \rho\frac{D}{Dt}(i + \frac{1}{2}\mathbf{u}\cdot\mathbf{u}) - \frac{Dp}{Dt} + \frac{p}{\rho}\frac{D\rho}{Dt} &= \\ \rho\frac{D}{Dt}(i + \frac{1}{2}\mathbf{u}\cdot\mathbf{u}) - \frac{\partial p}{\partial t} - \mathbf{u}\cdot\nabla p - p\nabla\cdot\mathbf{u} &= \\ \rho\frac{D}{Dt}(i + \frac{1}{2}\mathbf{u}\cdot\mathbf{u}) - \frac{\partial p}{\partial t} - \nabla\cdot p\mathbf{u}\,.\end{aligned} \tag{1.86}$$

Then the *total* energy equation (1.79), in terms of the total specific enthalpy, becomes

$$\begin{aligned}\rho\frac{D}{Dt}(i + \frac{1}{2}\mathbf{u}\cdot\mathbf{u}) \;&=\; \nabla\cdot\mathbf{K}\cdot\nabla T + \nabla\mathbf{q}_r + \frac{\partial p}{\partial t} + \\ &\nabla\cdot(\mathbf{S}\cdot\mathbf{u}) + \rho\mathbf{u}\cdot\mathbf{g} + \mathbf{u}\cdot\mathbf{f}_e + \dot{s}\,.\end{aligned} \tag{1.87}$$

1.8 Temporal and Spatial Averaging

The detailed temporal and spatial variations of the species-density, velocity, and temperature, i.e., resolution to the smallest time and length scales (the limits of the continuum treatments which are the molecular relaxation time and the mean free path), is not required in every heat transfer phenomenon. In some problems the *time-averaged* (over a finite or infinite period) and/or the *spatial-averaged* (over a finite, semi-finite, or infinite spatial domain) behavior is of interest. Also, in multiphase flows and under certain conditions, the *phase-averaged* or *multiphase-averaged local* variations are sought.

In addition, for *stochastic behaviors* an *ensemble average* is made over collection of realizations (in time and/or space). The probability of the occurrence of the realizations is given by a distribution and used as the weighting function in the averaging.

Among the various averages taken over time and space, *two* classes are identified and discussed below. The *first* one is the average taken over a scale which is smallest yet large enough to *represent* the *averaged local* quantity. For example, in a solid-fluid system a two-phase-averaged local density is a density taken over a volume which contains both phases but is small enough to represent the local density. When this average density is used for the continuum treatment of this two-phase system as a single medium, then this local density is assigned to a *point* in space (instead of a finite volume containing representatives of both phases). The *second* average is taken over a *finite* and *prescribed* domain, such as across the flow cross-section of a tube.

1.8.1 Averaging over Representative Elementary Scales

(A) Time Averaging

The temporal fluctuations in the fluid motion can significantly influence the convective heat transfer. These fluctuations can be *ordered* as in *oscillatory* or *periodic flows*, or can be *stochastic* as in *turbulent flows* (there are also some ordered or *coherent* flow structures in some turbulent flows).

The smallest time scale considered in the continuum treatment is the one below which the *viscous flow* treatment of the fluid is not accurate and rather an *elastic* behavior is found. This is the *molecular relaxation time* and is of the order of 10^{-10} s for liquids of moderate molecular weights (Walton, 1989). The temporal fluctuations of the temperature is also generally studied for time scales larger than this relaxation time.

The temporal fluctuation in the species density, velocity and temperature fields, in a single-phase turbulent flow or in a laminar or turbulent multiphase flow, can be *rapid*. Then the *time resolution* of the solution to the conservation equations, or resolution in measurement, should be high enough to accurately describe these fluctuations. Since the time-averaged behavior is of interest in many problems, it is assumed that these fluctuations have a time-averaged behavior, and then an average is taken over the

(a) Continuous Phases in One Direction in a Two-Dimensional Rendering

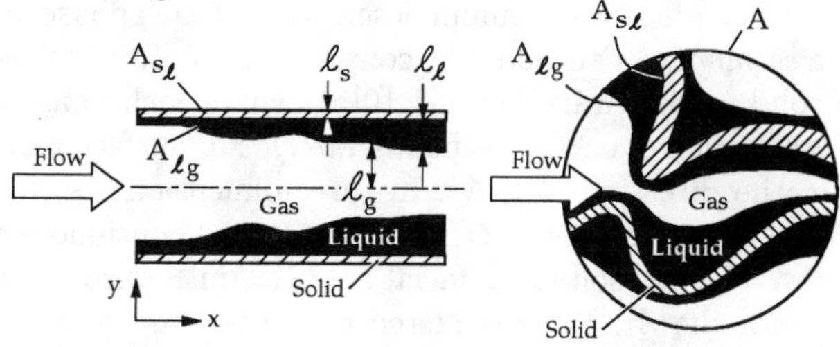

(b) One Phase Discontinuous in Both Directions in a Two-Dimensional Rendering of Gravity Affected Flows

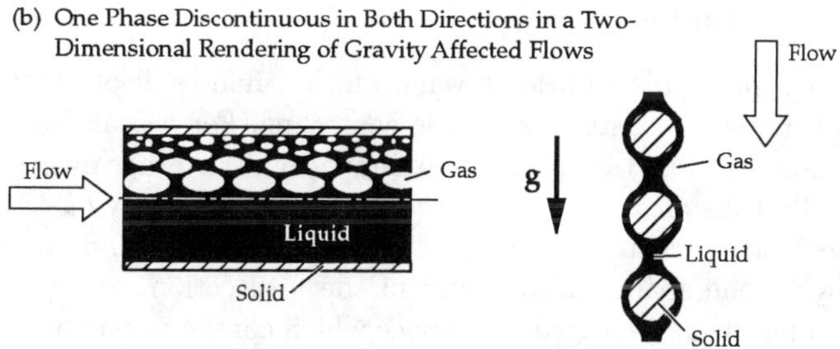

Figure 1.10. Examples of phase distributions. (a) Continuous phases, and (b) one phase is discontinuous.

time period of the slowest fluctuation. Since in some cases an exact periodic behavior is not found, then a *finite*, *representative averaging time* (which should in principle be infinite for nonperiodic fluctuations) is *assumed*. The time-averaged quantities are shown with a bar, e.g., $\overline{\phi}$. As examples, the *time* averaging of turbulent transport equations is discussed in Section 2.4.1, the *time* averaging of oscillatory flows is discussed in Section 4.3, and the *time and space* averaging is discussed in Section 3.3.1.

(B) Space Averaging

The spatial resolutions of interest to heat transfer are at the *lower* bound of the order of the mean free path and at the *upper* bound the system linear dimension. In *practical* descriptions of the spatial distributions, the smallest spatial resolution is taken as that beyond which the variation in the fields are not of any significant consequence. In prediction of the turbulent single-phase flows using temporal and spatial discretizations, lower bounds resolutions are used. Since in general the required spatial resolution is not known and in some cases not as distinct, a *representative spatial averaging* scale is used which is expected to represent the spatial variations of the quantities of interest without any further reduction in the spatial scale.

In multiphase flows, a smallest and yet representative space can be used for each phase and this would lead to a local, representative-volume-averaged description of the various fields for *each phase*. This approach leads to the *phase-averaged* continuum description of each phase or the *multicontinuum treatment*. As an example, consider the liquid-gas flow through a stationary solid as shown in Figure 1.10(a), where each phase is continuous in the y-direction (i.e., a continuous *interfacial* surface exit between phase pairs in the direction of flow). In three-dimensional structures, all interfaces, i.e., $A_{\ell g}$, $A_{\ell s}$ and A_{gs}, can be present and continuous in one or more directions. The gas-solid interfacial area A_{gs} in Figure 1.10(a) is not shown in order for the $A_{\ell s}$ to be rendered as continuous.

(i) Line Averaging

Consider the phase distribution given in the left-hand side of Figure 1.10(a), where a symmetry exists around the centerline. For a scalar $\phi = \phi(x, y)$, we can make a y-direction (i.e., a one-dimensional) average over each of the phases to find the *local y-direction phase-averaged* quantity $\langle\phi\rangle_y^j = \langle\phi\rangle_y^j(x)$, where the *phase indicator* j is $j = s, \ell, g$, the *space-dimension indicator* here is y showing one space variable and in the y direction, and $\langle\ \rangle$ indicates that the quantity is averaged over space which can be in one or more space directions. This gives

$$\langle\phi\rangle_y^s = \frac{1}{\ell_s}\int_{\ell_s}\phi\, dy\ , \quad \langle\phi\rangle_y^\ell = \frac{1}{\ell_\ell}\int_{\ell_\ell}\phi\, dy\ , \quad \langle\phi\rangle_y^g = \frac{1}{\ell_g}\int_{\ell_g}\phi\, dy\ . \qquad (1.88)$$

The qualifier *local* is used to indicate that the spatial variations in at least one space variable is maintained. If the averaging is simultaneously taken over all the phases, we then have the *y-direction three-phase*-averaged or y-direction-averaged *local* quantity $\langle\phi\rangle_y = \langle\phi\rangle_y(x)$ as

$$\begin{aligned}\langle\phi\rangle_y &= \frac{1}{\ell}\int_{\ell_s,\ell_\ell,\ell_g}\phi\, dy \\ &= \frac{\ell_s}{\ell}\langle\phi\rangle_y^s + \frac{\ell_\ell}{\ell}\langle\phi\rangle_y^\ell + \frac{\ell_g}{\ell}\langle\phi\rangle_y^g\ , \ell = \ell_s + \ell_\ell + \ell_g\ . \qquad (1.89)\end{aligned}$$

(ii) Area Averaging

In general, the phase distributions are not as geometrically simple as depicted on the left-hand side of Figure 1.10(a) and may be as depicted on the right-hand side of Figure 1.10(a), and because of the complex phase distributions the spatial variation over the length scales ℓ_j may not be of interest. In this case a *local* area average over each phase is taken and the *local area phase-averaged* quantity $\langle\phi\rangle^j = \langle\phi\rangle^j(x, y)$ is defined as

$$\langle\phi\rangle_A^s = \frac{1}{A_s}\int_{A_s}\phi\, dA_s\ , \langle\phi\rangle_A^\ell = \frac{1}{A_\ell}\int_{A_\ell}\phi\, dA_\ell\ , \langle\phi\rangle_A^g = \frac{1}{A_g}\int_{A_g}\phi\, dA_g\ , \qquad (1.90)$$

where $A = A_s + A_\ell + A_g$ is the sum of the *representative elementary areas*.

Then a *local area-averaged* quantity $\langle\phi\rangle = \langle\phi\rangle(x,y)$ is defined as

$$\langle\phi\rangle_A = \frac{1}{A}\int \phi\, dA = \frac{A_s}{A}\langle\phi\rangle_A^s + \frac{A_\ell}{A}\langle\phi\rangle_A^\ell + \frac{A_g}{A}\langle\phi\rangle_A^g\,. \tag{1.91}$$

(iii) Volume Averaging

A *local volume-averaged* quantity $\langle\phi\rangle = \langle\phi\rangle(\mathbf{x})$ is also defined as

$$\langle\phi\rangle_V = \langle\phi\rangle = \frac{1}{V}\int_V \phi\, dV = \frac{V_s}{V}\langle\phi\rangle^s + \frac{V_\ell}{V}\langle\phi\rangle^\ell + \frac{V_g}{V}\langle\phi\rangle^g\,, \tag{1.92}$$

where the volumes and the local *phase-volume-averaged* quantities are

$$V = V_s + V_\ell + V_g\,,$$

$$\langle\phi\rangle^s = \frac{1}{V_s}\int_{V_s}\phi\, dV\,, \langle\phi\rangle^\ell = \frac{1}{V_\ell}\int_{V_\ell}\phi\, dV\,, \langle\phi\rangle^g = \frac{1}{V_g}\int_{V_g}\phi\, dV\,. \tag{1.93}$$

The *representative elementary volume* is chosen to be the smallest volume that can accurately describe the *local* averaged property. As an example, consider the local volume-averaged *density* in a solid-liquid-gas system. The selected representative elementary volume can originate in any of the phases; then the volume-averaged density will have the value of the density of that phase. This is shown in Figure 1.11. As the representative elementary volume is made larger and encloses other phases, a fluctuation in the averaged density appears up to a critical volume where any further increase in the volume does not change $\langle\rho\rangle$. This is the *critical* representative elementary volume V_{cr}. In general, since the local volume fraction of the phases, i.e., V_s/V, V_ℓ/V, and V_g/V, or the phase densities (and their local phase averages) ρ_s, ρ_ℓ, and ρ_g (or $\langle\rho\rangle^s$, $\langle\rho\rangle^\ell$ and $\langle\rho\rangle^g$) change in space, the representative elementary volume is not taken beyond the critical volume in order to be able to represent these variations, i.e., in order for $\langle\phi\rangle$ to be a *local* value.

In Figure 1.10(b) the cases of discontinuous gas (left) and solid (right) phases in both directions are shown. In these cases, a *continuum phase-averaged* description of the fields in these dispersed phases in any direction cannot be made (however, in some phase-averaged *models*, *heuristic* continuum descriptions are used even for dispersed phases such as the gas phase in the gas-liquid bubbly flows).

1.8.2 Averaging over Finite Scales

Examples of the usage of a finite-scale (known a priori) averaging are in the time- and/or space-periodic species-density, velocity, and temperature fields (e.g., in oscillations or pulses in the species-density, flow, or temperature at the boundaries and in the flow, and heat transfer through periodic structures). Averaging is also used on heat flux over various interfaces, e.g., over submerged and bounding surfaces, and in the averages taken in the

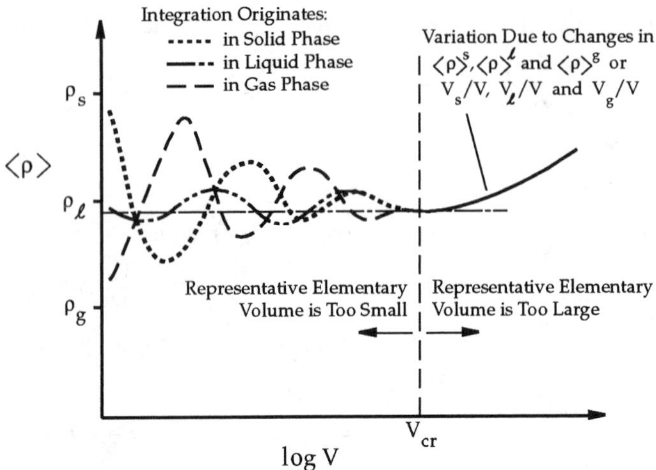

Figure 1.11. A rendering of the variation of the local volume-averaged density with respect to the magnitude of the local representative averaging volume. The density of the three phases, over which the average is made, are also shown.

directions of less significant field variations in order to reduce the number of the space variables. Often the elimination of one or two space variables, by line or area averaging, results in the inclusion of the effect of the transport in these directions through added terms. An example is the *Taylor dispersion* in tube flows which is included as an added axial diffusion and in part accounts for the lateral diffusion which is eliminated by space integration in that direction. This is discussed in Section 3.2.2.

As before, depending on the heat transfer phenomenon, the space averages are taken in one direction (line average), two directions (planar or area average), or three directions (volume average).

1.9 Local Thermal Equilibrium Among Phases

In multiphase flow and heat transfer—for example, in a solid-liquid-gas flow with local *phase-volume-averaged* temperatures $\langle T\rangle^s$, $\langle T\rangle^\ell$ and $\langle T\rangle^g$—the local phase-averaged temperatures may not be equal, and therefore, a *thermal nonequilibrium* may exist among the phases. The lack of the thermal equilibrium (i.e., nonvanishing of *interphasic temperature difference* $\langle T\rangle^s - \langle T\rangle^\ell$, $\langle T\rangle^s - \langle T\rangle^g$, and $\langle T\rangle^\ell - \langle T\rangle^g$, results in the local heat transfer among the phases, and these temperature differences are either sustained through the appropriate *phasic* heat inflow/outflow or sink/source/ or they will decay with time.

The statement of the *presence* of *local thermal equilibrium* in a solid-liquid-gas system is

$$\langle T\rangle = \langle T\rangle^s(\mathbf{x}) = \langle T\rangle^\ell(\mathbf{x}) = \langle T\rangle^g(\mathbf{x}) \ , \tag{1.94}$$

where

$$\langle T\rangle = \frac{V_s}{V}\langle T\rangle^s + \frac{V_\ell}{V}\langle T\rangle^\ell + \frac{V_g}{V}\langle T\rangle^g \ . \tag{1.95}$$

In practice, the local interphasic temperature difference can be *neglected* and a local thermal equilibrium can be assumed, if the temperature difference *across* the *entire system* (over linear dimension L) ΔT_L is much *larger* than those local interphasic differences (occurring over a linear dimension ℓ), i.e.,

$$\begin{aligned} |\langle T\rangle^s - \langle T\rangle^\ell| &\ll |\Delta T_L| \ , \\ |\langle T\rangle^s - \langle T\rangle^g| &\ll |\Delta T_L| \ , \\ |\langle T\rangle^\ell - \langle T\rangle^g| &\ll |\Delta T_L| \ . \end{aligned} \tag{1.96}$$

This may be valid for most cases involving heat flow through the system *boundaries*, but may *not* be so for *volumetric* heat generation/removal within one or more phases. Examination of the conditions for the validity of the assumption of the local thermal equilibrium for the heat flow through the *system boundaries* has been addressed by Whitaker (1991). The various *flow* and *geometric* parameters, as well as *thermophysical properties*, will appear in the expressions for the specific constraints (which need to be developed), and except for very simple systems these constraints are not readily usable.

When the local thermal equilibrium assumption is valid, i.e., when the main temperature change is over the system rather than across the phases, then a *multiphase* system can be treated as a *single* continuum. In the following chapters, Chapters 2 and 3 (i.e., Part I) are on the discussion of such systems. Chapters 4, 5 and 6 (i.e., Part II) are on two-phase systems under the conditions that the main temperature difference is across the phases and the local thermal equilibrium *cannot* be assumed. Chapters 7 and 8 (i.e., Part III) are on three-phase systems with the main temperature difference across the phases.

1.10 Approximation and Modeling

In convective heat transfer involving two or more phases, the detailed space and time *descriptions* (and *solutions*) given by the point (i.e., applied to the *most* elementary continuum volume) conservation equations may require unrealistic efforts and computing resources. Then some *approximations* are introduced into the *averaged* equations (with the space and time scales chosen based on the prior knowledge of the variations in the velocity, the temperature and the species-density fields). In some cases instead of a detailed or approximate field descriptions, some fundamental parameters are defined and used in developing *empirical* (i.e., *experimental*) descriptions.

The complexities which generally render the detailed description impractical are in the *phase distributions* (i.e., *heterogeneity* distribution), in the *interfacial heat transfer* (i.e., thermal nonequilibrium), and in the *multidimensionality* of the fields. In the following, the *general* approaches to these complexities are discussed further.

1.10.1 Rigor, Approximation, and Empiricism

The *spatial* and *temporal* variations of $\mathbf{u}$, T, and ρ_i in turbulent single- and multiphase flows, when added to the spatial and temporal variations of the phase distributions, result in rather extensive and difficult-to-use solution sets. In addition, the quantities of interest are general sought as averaged over a given set of scales. The averaged equations, in principle, are strictly accurate (i.e., are most *rigorous*) if they are based on the detailed *subaverage* fields used for deriving and evaluating them. However, in general, some *approximations* are introduced through some *closure* conditions such that the subaverage fields are not exactly evaluated and used, or used in an approximate manner. In some cases these *closure-based* descriptions are still too elaborate and *further approximations* are used.

Consider the following three examples, i.e., a closure-based description, an approximate description originating from the closure-based description, and a heuristic (empirical) description. The problem is that of heat transfer (with *no* phase change or heat generation) in a solid-fluid system with the solid phase remaining stationary while the fluid flows (i.e., a single-phase flow and heat transfer in porous media). Consider the case where the assumption of local thermal equilibrium is not valid (i.e., *thermal nonequilibrium*). The description of the energy flow (i.e., the energy equation) for this system has been evolving over nearly a century with a gradual evolution from empiricism to the closure-based description.

(A) Closure-Based Treatment

When each phase is continuous, i.e., both phases can be *treated* as a continuum, but the phases are not in a *local* thermal equilibrium, i.e., $\langle T\rangle^f \neq \langle T\rangle^s$, then the local temperature in each phase and the local velocity in the fluid phase are given by (Carbonell and Whitaker, 1984)

$$T_f(\mathbf{x}) = \langle T\rangle^f + T_f^{'} \,, \tag{1.97}$$

$$T_s(\mathbf{x}) = \langle T\rangle^s + T_s^{'} \,, \tag{1.98}$$

$$\mathbf{u}_f(\mathbf{x}) = \langle \mathbf{u}\rangle^f + \mathbf{u}_f^{'} \,, \tag{1.99}$$

where $T_f^{'}$, $T_s^{'}$, and $\mathbf{u}_f^{'}$ are the local *spatial deviations*. The point energy equation (1.84), with no radiation heat flux, no compressibility work, no viscous dissipation, and no heat generation, and for a single-component fluid, can be local volume-averaged for each phase as described in Section 1.8.1 (details will be given in Section 5.3) to give, for example, for the fluid phase

$$\frac{\partial \langle T\rangle^f}{\partial t} + \langle \mathbf{u}\rangle^f \cdot \nabla \langle T\rangle^f = \alpha_f \nabla^2 \langle T\rangle^f - \nabla \cdot \langle \mathbf{u}'T'\rangle^f + \frac{\alpha_f}{V_f}\nabla \cdot \int_{A_{fs}} \mathbf{n}_{fs} T_f' dA + \frac{\alpha_f}{V_f}\int_{A_{fs}} \mathbf{n}_{fs} \cdot \nabla T_f' dA\,, \tag{1.100}$$

where $\mathbf{n}_{fs}$ is the local normal to the f-s interface, which is known a priori through the phase distribution prescription. A similar equation is derived for the solid phase and then a *closure* condition (a first-order approximation-modeling) is introduced which relates T_f' (and T_s') to $\nabla\langle T\rangle^f$, $\nabla\langle T\rangle^s$ and $\langle T\rangle^s - \langle T\rangle^f$ using *transformation vectors* and *scalars* defined through

$$T_f' = \mathbf{b}_{ff} \cdot \nabla\langle T\rangle^f + \mathbf{b}_{fs} \cdot \nabla\langle T\rangle^s + f_f(\langle T\rangle^s - \langle T\rangle^f). \tag{1.101}$$

Closure conditions can be viewed as expansions of the unknown in terms of the *first* and *higher* derivatives of known quantities or as direct relations with the known quantities. Most closure conditions are *first-order* approximations. Using these, the local volume-averaged energy equation for the fluid phase becomes

$$\begin{aligned}
&\frac{\partial \langle T\rangle^f}{\partial t} + \\
&[\langle \mathbf{u}\rangle^f + \alpha_f \frac{A_{fs}}{V_f}(2\langle \mathbf{n}_{fs} f_f\rangle_{A_{fs}} - \langle \mathbf{n}_{fs}\cdot\nabla \mathbf{b}_{ff}\rangle_{A_{fs}} - \langle \mathbf{u}' f_f\rangle^f]\cdot\nabla\langle T\rangle^f + \\
&\quad [\alpha_f \frac{A_{fs}}{V_f}(-2\langle \mathbf{n}_{fs} f_f\rangle_{A_{fs}} - \langle \mathbf{n}_{fs}\cdot\nabla \mathbf{b}_{fs}\rangle_{A_{fs}} + \langle \mathbf{u}' f_f\rangle^f]\cdot\nabla\langle T\rangle^s = \\
&\qquad \nabla\cdot[\alpha_f(\mathbf{I} + 2\frac{A_{fs}}{V_f}\langle \mathbf{n}_{fs}\mathbf{b}_{ff}\rangle_{A_{fs}} - \langle \mathbf{u}'\mathbf{b}_{ff}\rangle^f)]\cdot\nabla\langle T\rangle^f + \\
&\qquad \nabla\cdot(2\alpha_f \frac{A_{fs}}{V_f}\langle \mathbf{n}_{fs}\mathbf{b}_{fs}\rangle_{A_{fs}} - \langle \mathbf{u}'\mathbf{b}_{fs}\rangle^f)\cdot\nabla\langle T\rangle^s + \\
&\qquad \alpha_f \frac{A_{fs}}{V_f}\langle \mathbf{n}_{fs}\cdot\nabla f_f\rangle_{A_{fs}}(\langle T\rangle^s - \langle T\rangle^f)\,,
\end{aligned} \tag{1.102}$$

where subscript A_{fs} indicates area integration on the f-s interface and $\mathbf{I}$ is the identity tensor.

The transformation vectors and scalars need to be determined using the equations for T_f', T_s' and $\mathbf{u}'$ and the procedure is outlined in Section 5.2. An equation similar to the above is written for the solid phase. By defining the *total diffusivity tensors* $\mathbf{D}_{ff}$, $\mathbf{D}_{ss}$, $\mathbf{D}_{fs}$, and $\mathbf{D}_{sf}$ (which are defined from the tensors containing $\nabla\cdot\nabla\langle T\rangle^f$ and $\nabla\cdot\nabla\langle T\rangle^s$), the *interfacial conduction heat transfer coefficient* h_c (which is defined from the coefficient of $\langle T\rangle^s - \langle T\rangle^f$) and the *convective velocities* $\mathbf{v}_{ff}$, $\mathbf{v}_{fs}$, $\mathbf{v}_{sf}$, and $\mathbf{v}_{ss}$ (defined from the coefficients of $\nabla\langle T\rangle^f$ and $\nabla\langle T\rangle^s$), (1.102) becomes

$$\frac{\partial\langle T\rangle^f}{\partial t} + \mathbf{v}_{ff}\cdot\nabla\langle T\rangle^f + \mathbf{v}_{fs}\cdot\nabla\langle T\rangle^s = \nabla\cdot\mathbf{D}_{ff}\cdot\nabla\langle T\rangle^f + \nabla\cdot\mathbf{D}_{fs}\cdot\nabla\langle T\rangle^s + \frac{A_{fs}}{V_f}\frac{h_c}{(\rho c_p)_f}(\langle T\rangle^s - \langle T\rangle^f)\,. \quad (1.103)$$

For the solid phase we have

$$\frac{\partial\langle T\rangle^s}{\partial t} + \mathbf{v}_{sf}\cdot\nabla\langle T\rangle^f + \mathbf{v}_{ss}\cdot\nabla\langle T\rangle^s = \nabla\cdot\mathbf{D}_{sf}\cdot\nabla\langle T\rangle^f + \nabla\cdot\mathbf{D}_{ss}\cdot\nabla\langle T\rangle^s + \frac{A_{fs}}{V_s}\frac{h_c}{(\rho c_p)_s}(\langle T\rangle^f - \langle T\rangle^s)\,, \quad (1.104)$$

where A_{fs}/V_f and A_{fs}/V_s are the *specific interfacial areas* based on the *fluid* and *solid* volumes, respectively. The coefficient h_c is independent of the flow field and depends on the pore geometry and size and the phase-pair properties. This will be further discussed in Section 5.3.

These equations are the *two-medium* continuum descriptions of the energy conservation with intermedium heat transfer (i.e., interaction). The tensor, vector and scalar coefficients in the above equations are evaluated once the transformation quantities are determined. However, this is rather very involved even for the *simplest periodic* phase distributions with periodic, steady-state laminar flow fields.

(B) Approximation

The coupling (i.e., interaction) among the solid and the fluid phases has been described in (1.103) and (1.104) through the difference in local volume-averaged temperatures and through the first and second derivatives of the local volume-averaged temperatures. If the fluid *velocity* is *large* and if the fluid is a *gas*, the second derivative terms (and in general the diffusion tensors) can be neglected within the fluid. The convective term for the solid phase is negligible if the *solid conductivity* is rather *large*. Then the approximate forms of (1.103) and (1.104) are

$$\frac{\partial\langle T\rangle^f}{\partial t} + \mathbf{v}_{ff}\cdot\nabla\langle T\rangle^f + \mathbf{v}_{fs}\cdot\nabla\langle T\rangle^s = \frac{A_{fs}}{V_f}\frac{h_{fs}}{(\rho c_p)_f}(\langle T\rangle^s - \langle T\rangle^f) \quad (1.105)$$

$$\frac{\partial\langle T\rangle^s}{\partial t} = \nabla\cdot\mathbf{D}_{sf}\cdot\nabla\langle T\rangle^f + \nabla\cdot\mathbf{D}_{ss}\cdot\nabla\langle T\rangle^s + \frac{A_{fs}}{V_s}\frac{h_{fs}}{(\rho c_p)_s}(\langle T\rangle^f - \langle T\rangle^s)\,, \quad (1.106)$$

where the *interfacial convective heat transfer coefficient* h_{fs} is introduced and *depends* on the flow field. Further approximation can be made depending on the specific fluid-solid thermophysical properties, phase distributions, phase characteristic dimensions, and the velocity. Various approximations are discussed by Wakao and Kaguei (1982).

(C) Empiricism

In the 1920's Anzelius and later Schumann (e.g., Vortmeyer and Schaefer, 1974) developed an empirical, *one-dimensional transient* model for a gas–solid system given by

$$\frac{\partial \langle T\rangle^f}{\partial t} + \langle \mathbf{u}\rangle^f \frac{\partial \langle T\rangle^f}{\partial x} = \frac{A_{fs}}{V_f}\frac{h_{fs}}{(\rho c_p)_f}(\langle T\rangle^s - \langle T\rangle^f) \,, \qquad (1.107)$$

$$\frac{\partial \langle T\rangle^s}{\partial t} = \frac{A_{fs}}{V_s}\frac{h_{fs}}{(\rho c_p)_s}(\langle T\rangle^f - \langle T\rangle^s) \,. \qquad (1.108)$$

In these equations, the effect of $\mathbf{v}_{fs}$, $\mathbf{D}_{fs}$ and $\mathbf{D}_{ss}$ are also included in h_{fs}, the *interfacial convective heat transfer coefficient*, which is determined experimentally using these equations for transient temperature fields. Then h_{fs} found from these equations will depend on the flow field, the *specific* fluid-solid pair (and their thermophysical), and the geometric properties. Therefore, h_{fs} is not the same as h_c in (1.104) and (1.105). This will be further discussed in Section 5.3.

The above examples of a *rigorous*, an *approximate*, and an *empirical treatment* of a fluid flow and heat transfer in a multiphase system illustrate the need for rigorous treatments and the subsequent approximations for arriving at accurate and yet useful descriptions. In liquid-gas systems, the phase distributions are *not* a priori known and complex flow structures are present in both phases and all of these depend on the phase-flow rates and the magnitude and the direction of gravity (as discussed in Section 1.6.1). Then the need for the approximation of the rigorously developed equations will be even greater (this will be discussed in Chapter 6).

1.10.2 Modeling of Medium Heterogeneity

Heterogeneity is generally referred to as discontinuity in the density (and other thermophysical properties) and the focus here is on the *phasic heterogeneities* (as opposed to *intraphase heterogeneities*; for example, in composite solids where there are material heterogeneities), i.e., where the density discontinuity is due to the simultaneous presence of two or more phases. *Modeling* is referred to as a mathematical description which is based on *convincing simplification* of the reality. In multiphase flows, the heterogeneities are generally modeled using *geometrical simplifications* along with *regime diagrams*.

(A) Hydrodynamics

As discussed in Section 1.5, in multiphase systems with all the phases moving, the local *phase-averaged velocities* may not be the same, and therefore, a velocity *slips* may occur between any of the *phase pairs*. This slip is due to the magnitude and directional differences in the *inertial*, *viscous interfacial drag* (over all interfaces), and *gravitational* force acting on each phase. The local phase *pressures* and *pressure gradients* may also be different, due to the combined effects of the *interfacial curvature* and *surface tension*.

Based on the allowance (i.e., recognized significance) for the interfacial slips, some *multimedium* momentum equations are developed (i.e., models representing the momentum conservation for each phase). Then these momentum equations are used along with the continuity equation to determine the local phase-volume concentration and the local phase velocities. If the slip is negligible, then a *single-medium* momentum equations is developed, which along with the continuity equations, will allow for the determination of the local concentration of each phase (e.g., *void fraction* in liquid-gas flows through porous media; solid, liquid and gas volume fractions in solid-liquid-gas flows) as well as a local velocity for all the phases.

The *geometrical modeling* of the phase distributions (and the modeling of the detail *interfacial* and *intraphasic* motions) have been advanced for *isothermal* and *nonisothermal* multiphase flows. This includes *turbulence* adjacent to liquid-gas interfaces in the presence or absence of a solid surface near this interface (Banerjee, 1981). The modeling of the heterogeneities must be made such that the *overall* features (e.g., the pressure gradient and phase velocities) as well as *detail* features such as the intraphasic mixings (fine and course structures) can be accurately included. These various *hydrodynamic aspects*, as related to convective heat transfer, will be discussed in the relevant specific multiphase systems in the following chapters.

(B) Heat and Mass Transfer

Under local thermal equilibrium, i.e., the absence of any interfacial heat and mass transfer, the *intermedium* (i.e., within a single medium representing all the phases) *effective diffusion* (which is influenced by the fluid motion), and other effects of the heterogeneity on the intermedium transport, becomes significant. The effective diffusion (which has a *molecular* component and a *dispersive* or hydrodynamic-affected component), and other effects on the transport, is analyzed using modeled phase distributions. For example in a solid-liquid-gas three-phase flow, the liquid- and gas-phase topologies are generally modeled using simple geometries identified in the flow regime diagrams. The *motion* within droplets and bubbles, i.e., within the *dispersed* phase, also strongly influences the effective properties of the medium.

In local thermal and chemical nonequilibriums, the interfacial heat and mass depend on the *specific interfacial area* (i.e., interfacial area per unit volume) as well as the flow fields adjacent to the interface. The accurate modeling of the phase distributions and the interfacial geometry (i.e., phase *topology*) is central in the accurate prediction of the interfacial transfer rates.

1.10.3 Modeling of Thermal Nonequilibrium

When no phase change occurs, i.e., *no thermodynamics phase-coexistence constraint* exists on the pressure and temperature (and for multicomponent systems also on the species density), the *interfacial* temperature is *not* directly addressed. The local phase-averaged temperatures, such as $\langle T \rangle^\ell$, are the average local phase temperature, and therefore, the interfacial temperature is not defined (but can be assumed to be, for example, the arithmetic mean of the phase temperatures).

In phase change, the thermodynamic constraints (e.g., saturated states) are also used in modeling the local thermal nonequilibrium. These constraints are applied to the interfaces (where two or more phases *coexist* in thermodynamic equilibrium), but within each phase and away from the interfaces the temperature, pressure and species-density can be *far* from these conditions. Since the local interfacial temperature is defined and used, its *difference* with the local phase temperatures is the *driving* force for the interphasic heat transfer.

Then depending on the phase pairs and the velocity fields, a *single* local interfacial heat transfer coefficient *or two different* heat transfer coefficients (one for each side) must be used. The use of the latter—for example, for a liquid-gas interface with a volumetric interfacial gas phase *production rate* of $\langle \dot{n}_g \rangle$—leads to

$$\frac{A_{\ell g}}{V} h_{\ell g}(\langle T \rangle^\ell - T_{\ell g}) - \frac{A_{g\ell}}{V} h_{g\ell}(T_{\ell g} - \langle T \rangle^g) = \langle \dot{n}_g \rangle \Delta i_{\ell g} \, , \tag{1.109}$$

where $T_{\ell g}$ is the *interfacial temperature*, which is generally taken as the *equilibrium* (i.e., *saturation*) temperature, and $A_{\ell g} = A_{g\ell}$. Note that $h_{\ell g}$ may *not* be equal to $h_{g\ell}$, because of the phasic thermophysical properties, geometry and motion. Then the product of volumetric production rate of phase g and the heat of its formation, $\langle \dot{n}_g \rangle \Delta i_{\ell,g}$, are equal to the *net* heat flowing toward the interface.

If a phase is dispersed (i.e., a phase discontinuity is present), the *continuum* modeling will not be strictly accurate. For example, in direct contact heat transfer between a dispersed liquid phase and a continuous liquid phase (a batch or a continuous flow process), the motion within each element (i.e., droplet) of the dispersed phase causes significant *interdroplet* mixing. Significant interactions may also exist between the droplets. Therefore, modeling of the *effective diffusivity* based *only* on the interdroplet mixing (i.e., *enhanced diffusion*, since no *net* motion occurs within the droplets) leads to significant errors. As an alternative, a *discrete-phase model*, which allows for the inclusion of the behavior of the individual elements and is based on the ensemble averaging over a large number of element *trackings*, has been proposed by Andreani and Yadigaroglu (1991). This will be discussed further in Chapter 8.1.3.

The above examples show that depending on the applications, the thermal nonequilibrium is modeled by keeping the significant transport mechanisms at the small length scale, while arriving at *solvable, local* averaged descriptions.

1.10.4 Modeling of Multidimensional Fields

The temperature gradient and the velocity field resulting in a convective heat transfer are generally time dependent and multidimensional. However, for practical reasons the solutions to the conservation equations are sought in as few space dimensions as reasonable. The *effective* properties (thermal and hydrodynamics) of multiphase systems are *not* generally *isotropic*, and significant order-of-magnitude *differences* exist between the components. In addition, when the problem has a dominant flow and heat transfer direction, then *scaling* and the *order-of-magnitude* analysis are used to reduce the space variables. Also, in some cases, while keeping a space variable, it may be feasible to neglect the higher-order derivatives with respect to this variable.

(A) Reduction in Space Variables

Convective heat transfer is the largest where *both* $\mathbf{u}$ and ∇T have their largest components. However, this dominant heat convection may yet be controlled by *diffusion* and *convection* from the *other* directions, and therefore, their inclusion becomes necessary. The *spatial extent* or *boundedness* (i.e., *infinite, semi-infinite* or *finite*) generally determines if an *asymptotic* field behavior (and thus the elimination of the space variable tending to infinity) is possible.

In forced flows when the gravity effect also becomes important (the flow is then called a *mixed flow*) and in both single- and multiphase flows, the space variable reduction may be possible when the direction of the forced flow and the gravity vector coincide (i.e., *gravity assisted* or *opposed* flows). For example, in the case where the heat transfer is controlled by *axial* diffusion (as opposed to the *lateral* diffusion). However, when the forced flow direction is at an angle to the gravity vector, multidimensional descriptions and solutions become necessary.

(B) Reduction in Higher Derivatives

When the Péclet number $Pe = uL/\alpha$ in a given direction is large, then the diffusion in that direction can be neglected. An example is in the two-medium treatment of gas flow in porous media, where $Pe_{g,x} = \langle u \rangle^g L/\alpha_{f,xx}$ can be much larger than unity and then $\partial/\partial x(\alpha_{g,xx}\partial\langle T\rangle^g/\partial x)$ can be neglected compared to $\partial/\partial x(\langle u\rangle^g\langle T\rangle^g)$, where $\alpha_{g,xx}$ is the *x-principal axis* component of the effective diffusivity tensor $\mathbf{D}_g$ and L and $\langle u\rangle^g$ are the characteristic dimension and velocity in the x-direction.

This approximation, i.e, neglecting of the *axial* (i.e., along the direction of *dominant* flow) diffusion, is also applied in the *single-phase* fluid flow over *stationary planar* solid surfaces and is termed the *boundary-layer approximations* (originated by Prandtl and discussed by Schlichting, 1979). This boundary-layer approximation when applied to heat transfer allows for the *lateral* diffusion and the *axial* convection (in transferring heat between the solid surface and the fluid).

1.11 Single-, Two-, and Three-Medium Treatments

With regard to heat transfer, a *medium* or a *single-medium treatment* is referred to as either a *single-phase* medium or a *multiphase* (i.e., heterogeneous) medium where the assumption of *local thermal equilibrium* among the phases is imposed. Convective heat transfer, which is a result of the simultaneous presence of a flow field and a temperature gradient, can occur *within* the medium (i.e., *intramedium*). An example is the exothermic reaction (as in combustion) within a medium. This can be a single-phase (i.e., liquid or gas flows) or a multiphase medium (such as droplet or particle sprays, particulate flows, emulsion flows, flows through porous media). Figure 1.12(a) shows this intramedium convective heat transfer, where, for example, the reaction zone (e.g., the flame) can be stationary if the reaction front velocity $\mathbf{u}_F$ is equal in magnitude and opposite in sign to the flow velocity $\mathbf{u}_f$. The heat source distribution $\dot{s}(\mathbf{x})$ depends on the distribution of the species-density, temperature and pressure. The intramedium heat flux vector $\mathbf{q}_f$ varies substantially in the reaction zone. In the following chapters, in Part I, i.e., Chapters 2 and 3, the single-medium treatment of the intramedium convective heat transfer in single- and multiphase systems, is considered.

When the assumption of local thermal equilibrium in *not* valid, both the *intraphasic* and the *interphasic* heat transfers are analyzed (Parts II and III) by determining the velocity, temperature, and species-density fields in each phase along with the interfacial conditions. In the discussion of the *two-medium treatment* of *two phases* in *thermal nonequilibrium*, the phase pair can be solid-fluid (Chapter 4 for both phases being continuous and

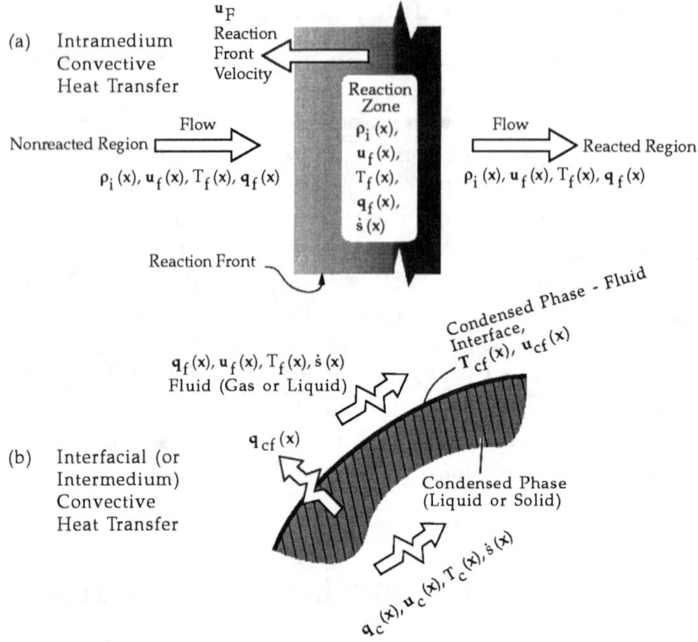

Figure 1.12. (a) Intramedium convective heat transfer. (b) Interfacial convective heat transfer.

having a simple interface and Chapter 5 for large specific interfacial area or where the solid phase being dispersed), or fluid-fluid (Chapter 6). The interfacial heat transfer is depicted in Figure 1.12(b), where the heat exchange is between a fluid phase and a *condensed phase* (liquid or solid) in addition to the heat flow within each phase. Both phases (including the interface) can be moving with respect to a fixed coordinate and the volumetric heat generation can be present in both phases. The interfacial heat flux vector $\mathbf{q}_{cf}$, and the interfacial temperature and velocity distributions are also referred to. Note that through the appropriate line, area or volume averaging over each phase, various models for the determination of the interfacial heat flow can be constructed. Examples range from convective heat transfer in flow of *fluids* over continuous *solid* surfaces to *liquid* to *liquid* heat transfer with one of the liquids dispersed as droplets.

When heat transfer occurs among *three phases* (i.e., *three-phase thermal nonequilibrium*), there are three local interfacial heat flux vectors (one for each phase pair) to be considered. Examples of the *three-medium treatments* include heat transfer between a particulate-gas flow and a submerged solid surface, or when solid, liquid, and gas phases are in contact, as in surface nucleate boiling (Part III, Chapters 7 and 8).

1.12 Fluid Motion Caused or Influenced by Heat Transfer

Heat transfer to or from fluids, i.e., nonuniformity in the fluid temperature distribution, causes changes in the temperature-dependent *thermophysical properties*, and therefore, can *cause* motion in an otherwise quiescent fluid or *influence* the motion *already* present. This occurs in *single-phase* flows of gases and liquids as well as in *multiphase* flows such as in liquid-gas and in fluid-solid systems. In *multicomponent* systems, the heat transfer also influences the species diffusion.

Among the changes in the thermophysical properties considered are the variation of *density* with respect to *temperature* and *species-concentration* (which causes the *thermo- and diffusobuoyancy* effects in liquids and gases and the *thermoacoustic* waves in gases), the variation of the *surface tension* with respect to *temperature* and species-concentration (which causes the *thermo- and diffusocapillarity*), changes in the *density* (or specific volume) resulting from the *phase change* (which causes the *phase-density buoyancy* in phase change), influence of *temperature gradient* on the *species* mass *diffusion flux* (called the *thermodiffusion* or the *Soret* effect), influence of a *temperature gradient* on the particle motion (called the *thermophoresis*), change in the electrical conductivity with respect to temperature (which causes the *electrothermal convection* in dielectric liquids), and variation in the *viscosity* with respect to temperature within the *viscous boundary layer*. These are discussed below and then are used in Chapters 2 through 8, as the inter- and intraphasic heat transfers in moving fluids are examined.

1.12.1 Thermo- and Diffusobuoyancy

The variation of the fluid density with respect to temperature and species concentration can result in a noticeable alteration of the fluid motion. We begin with a general description where the *density* of the fluid is a function of temperature T, pressure p, and species mass concentration ρ_i/ρ. This can be written as

$$\rho(\mathbf{x}) = \rho(\mathbf{x}, T, p, \frac{\rho_i}{\rho}) \, . \tag{1.110}$$

In order to evaluate the significance of the variations in ρ, as compared to that in the other thermophysical properties, first consider the variation in ρ with respect to p and T only. The effect of the species concentration will be discussed next.

(A) Variation with Temperature and Pressure

In *quiescent* fluids, the presence of a temperature gradient *perpendicular* to the gravity vector will result in a *horizontal* density stratification and then the lighter fluid moving upward and the denser fluid moving downward. This motion is called the *thermobuoyancy* or the *natural* (or *free*) *convection* and occurs with the slightest *horizontal* temperature gradient. On the other hand, in the presence of a *vertical* (*along* the gravity vector) temperature gradient (and the resulting density gradient) only when the higher density fluid is on the *top* and when the gradient of density is larger than a *critical* value, can a fluid motion begin (i.e., the *onset* of *thermobuoyant motion*).

The conservation equations governing *thermobuoyant flows* has been examined, for the significance of the variations in the thermophysical properties, by Gray and Giorgini (1976). The properties appearing in the *continuity*, *momentum*, and *energy* equations for a single component, Newtonian fluids in a single-phase flow are ρ, μ, c_p, k, and β. Assuming *linear* relationships with temperature and pressure, we have for any thermophysical property ϕ_j a T- and p-dependence given by

$$\phi_j = \phi_{o_j} (1 + a_j \Delta T + b_j \Delta p) \,, \tag{1.111}$$

where

$$a_j = \frac{1}{\phi_j} \frac{\partial \phi_j}{\partial T}\bigg|_p \,, \quad b_j = \frac{1}{\phi_j} \frac{\partial \phi_j}{\partial p}\bigg|_T \tag{1.112}$$

and $j = \rho, \mu, c_p, k$ and β, and $\Delta T = T - T_o$ and $\Delta p = p - p_o$ are deviations from the *reference* state T_o and p_o. Note that based on the notations used in (1.1) we have $a_\rho = -\beta$ and $b_\rho = \kappa$.

For *thermobuoyant* flows and far from the temperature disturbances, the momentum equation (1.74) reduces to the *hydrostatic* (i.e., *stable stratification*) equation

$$\frac{\partial p_\infty}{\partial x_i} = -\rho_\infty g_i \,. \tag{1.113}$$

This is generally subtracted from the momentum equation (1.76) and results in the elimination of the pressure if $\partial(p - p_\infty)/\partial x_i = 0$ can be justified. Now, scaling the variables using the *buoyancy* parameters, we have a *length* scale L across which the maximum *temperature difference* in the fluid is ΔT_∞, and a *velocity scale* $(g\beta_o \Delta T_\infty L)^{1/2}$. The *time scale* is then based on the length and the velocity scales and becomes $(L/g\beta_o\Delta T_\infty)^{1/2}$. Also the *pressure scale* becomes $\rho_o g \beta_o \Delta T_\infty L$. Using these in (1.68), (1.72), (1.76), and (1.84), the *dimensionless* conservation equations (for a single-component Newtonian fluid and without the viscous dissipation, heat generation, and no radiative heat flux) become (Gray and Giorgini, 1976)

$$a_\rho^* \frac{DT^*}{Dt^*} + b_\rho^* \frac{Dp^*}{Dt^*} + (1 + a_\rho^* T^* + b_\rho^* p^*) \frac{\partial u_i^*}{\partial x_i^*} = 0 \tag{1.114}$$

$$(1 + a_\rho^* T^* + b_\rho^* p^*) \frac{Du_i^*}{Dt^*} = -\frac{\partial (p^* - p_\infty^*)}{\partial x_i^*} +$$

$$(\frac{Pr}{Ra})^{1/2}(1 + a_\mu^* T^* + b_\mu^* p^*)\frac{\partial d_{ij}^*}{\partial x_j^*} + (\frac{Pr}{Ra})^{1/2}(a_\mu^* \frac{\partial T^*}{\partial x_j^*} + b_\mu^* \frac{\partial p^*}{\partial x_j^*})d_{ij}^* -$$
$$(T^* - T_\infty^*)\frac{g_i}{g} + b_\rho^*(p^* - p_\infty^*)\frac{g_i}{g} \tag{1.115}$$

$$(1 + a_\rho^* T^* + b_\rho^* p^*)(1 + a_{c_p}^* T^* + b_{c_p}^* p^*)\frac{DT^*}{Dt^*} =$$
$$(PrRa)^{-1/2}[(1 + a_k^* T^* + b_k^* p^*)\frac{\partial^2 T^*}{\partial x_j^{*2}} + (a_k^* \frac{\partial T^*}{\partial x_j^*} + b_k^* \frac{\partial p^*}{\partial x_j^*})\frac{\partial T^*}{\partial x_j^*}] -$$
$$a^*(1 + a_\beta^* T^* + b_\beta^* p^*)[a_\rho^*(T^* + T_o^*)\frac{D(p^* - p_\infty^*)}{Dt^*} +$$
$$(T^* + T_o^*)\frac{\mathbf{u}^* \cdot \mathbf{g}}{g}(1 + a_\rho^* T_\infty^* + b_\rho^* p_\infty^*)] \,. \tag{1.116}$$

For a *Newtonian* fluid (1.77) can be used with the Stokes hypothesis. Then the viscous stress can be written as $\tau_{ij} = \mu d_{ij}$, where d_{ij} are the elements of deformation rate tensor. The dimensionless coefficients in (1.114) to (1.116) are

$$a_i^* = a_i \Delta T_\infty \,, \quad b_i^* = b_i \rho_o g L \,, \quad a^* = \frac{\beta_o g L}{c_{p_o}} \,,$$
$$Ra = \frac{\beta_o g \Delta T_\infty L^3}{\nu_o \alpha_o} \,, \quad T^* = \frac{T - T_o}{\Delta T_\infty} \,,$$
$$Pr = \frac{\nu_o}{\alpha_o} \,, \quad p^* = \frac{p - p_o}{\rho_o g L} \,, \quad T_o^* = \frac{T_o}{\Delta T_\infty} \,. \tag{1.117}$$

Note that $p - p_o$ is scaled differently from $p - p_\infty$ (the latter is scaled using $\rho_o \beta_o g \Delta T_\infty L$). The stable stratification equation is used to rewrite TDp/Dt as

$$T\frac{Dp}{Dt} = (T - T_o)\frac{D(p - p_\infty)}{Dt} + T_o\frac{D(p - p_\infty)}{Dt} -$$
$$(T - T_o)\rho_\infty \mathbf{u} \cdot \mathbf{g} - T_o \rho_\infty \mathbf{u} \cdot \mathbf{g} \,. \tag{1.118}$$

If the reference temperature T_o is taken as the arithmetic average between the maximum and minimum temperatures occurring in the fluid (*one* of these extremes is generally T_∞), then $|T^*| \leq 0.5$. Gray and Giorgini (1976) have calculated a^*, a_i^*, b_i^*, and Pr for *air* and *water* at $T_o = 15°\mathrm{C}$ and $p_o = 10^5$ N/m^2, and their results are given in Table 1.3.

In 1903 Boussinesq introduced the assumption that the temperature dependence of the thermophysical properties can be reasonably approximated as constants evaluated at T_o (taken as the arithmetic average between the maximum and the minimum temperatures), *except* for the density variation in the body force. Then the *dimensional, Boussinesq-approximate conservation* equations for *thermobuoyant* motion in an otherwise incompressible, Newtonian fluid, become

$$\frac{\partial u_i}{\partial x_i} = 0 \,, \tag{1.119}$$

Table 1.3. Parameters appearing in the dimensionless conservation equations.

	Air	Water
a^*	$3.6 \times 10^{-5} L$	$3.5 \times 10^{-7} L$
a^*_ρ , b^*_ρ	$-3.5 \times 10^{-3} \Delta T_o$, $1.2 \times 10^{-4} L$	$-1.5 \times 10^{-4} \Delta T_o$, $4.8 \times 10^{-6} L$
a^*_μ , b^*_μ	$2.8 \times 10^{-3} \Delta T_o$, 0.0	$-2.7 \times 10^{-2} \Delta T_o$, $-2.7 \times 10^{-6} L$
$a^*_{c_p}$, $b^*_{c_p}$	$4.5 \times 10^{-5} \Delta T_o$, $2.3 \times 10^{-7} L$	$-2.4 \times 10^{-4} \Delta T_o$, $-2.4 \times 10^{-5} L$
a^*_k , b^*_k	$2.4 \times 10^{-3} \Delta T_o$, 0.0	$1.7 \times 10^{-3} \Delta T_o$, $4.2 \times 10^{-5} L$
a^*_β , b^*_β	$-3.6 \times 10^{-3} \Delta T_o$, 0.0	$8.0 \times 10^{-2} \Delta T_o$, 0.0
Pr	0.72	8.1

At atmospheric pressure and $T_o = 15°\text{C}$, L is in m and ΔT_o is in °C.

$$\rho_o \frac{Du_i}{Dt} = -\frac{\partial (p - p_\infty)}{\partial x_i} + \mu_o \frac{\partial^2 u_i}{\partial x_j{}^2} - \rho_o \beta_o (T - T_\infty) g_i , \tag{1.120}$$

$$\rho_o c_{p_o} \frac{DT}{Dt} = k_o \frac{\partial^2 T}{\partial x_j{}^2} . \tag{1.121}$$

The Boussinesq approximations are *valid*, for the absolute value of the magnitude of the scaled variables u^*, T^*, p^*, x^* and t^* varying between zero and unity, when in (1.114) to (1.116) the parameters are bound by

$$a^* \frac{T_o}{\Delta T_\infty} \leq 0.1 \,, \quad a^*_j, b^*_j \leq 0.1 \,, \quad |a^*| (\frac{Pr}{Ra})^{1/2} \leq \frac{0.1}{(PrRa)^{1/2}} . \tag{1.122}$$

These *constraints* are generally satisfied for the range of L and ΔT_∞ encountered in practice (when *only* Ra is used, the constraint becomes $Ra < 10^{17}$ for air and $Ra < 10^{19}$ for water); otherwise, more extended equations must be used.

There are many atmospheric, geological, and engineering occurrences of the *thermobuoyancy* effects in fluids. A summary is given by Turner (1979) and Gebhart (1979) and the effect of thermobuoyancy on the heat transfer will be discussed in Chapters 2 through 8 in the single-, two-, and three-medium treatments.

(B) Variation with Temperature and Concentration

As expressed by (1.110), the density variation can be due to the species mass concentration variation, and again a *linear* approximation to the dependence leads to (assuming no pressure dependence)

$$\rho = \rho_o (1 - \beta_o \Delta T - \sum_i \beta_{m_i} \Delta \frac{\rho_i}{\rho}) ,$$

$$\beta_{m_i} = -\frac{1}{\rho_o}\frac{\partial \rho}{\partial \rho_i}\bigg|_{T,p,\rho_j} \simeq -\frac{1}{\rho_o}\frac{\Delta \rho}{\Delta \rho_i}, \tag{1.123}$$

where β_{m_i} is the *volumetric species-concentration expansion coefficient* (with the dimension of the inverse of the mass fraction of species i, i.e., kg-mixture/kg-species i) for species i, and $\Delta\rho_i/\Delta\rho$ is the mass fraction deviation from the reference state $(\rho_i/\rho)_o$.

By using (1.123) and assuming that the *Boussinesq approximations* are valid, the momentum equation (1.120) will include the influence of the concentration variation in the body force and the species conservation equation (1.72) will also be required. The *thermo-* and *diffusobuoyancy* included momentum equation and the species conservation equation are

$$\rho_o \frac{Du_j}{Dt} = \frac{\partial (p - p_\infty)}{\partial x_j} + \mu_o \frac{\partial^2 u_i}{\partial {x_j}^2} + \rho_o[-\beta_o(T - T_\infty) - \sum_i \beta_{m_{io}}(\frac{\rho_i}{\rho} - \frac{\rho_{i\infty}}{\rho_\infty})]g_i \,, \tag{1.124}$$

$$\frac{D}{Dt}\frac{\rho_i}{\rho} = D_o \frac{\partial^2}{\partial {x_j}^2}\frac{\rho_i}{\rho} + \dot{n}_i \,. \tag{1.125}$$

There is a great interest in the motions solely caused by the thermo- and diffusobuoyancies (Hupport and Turner, 1981; Maxworthy, 1983). For example, in an otherwise *quiescent horizontal* fluid layer, these motions can only begin when the *unstable overall density* gradient has a magnitude larger than a critical value. The term *double-diffusive* (heat and mass) *convection* has been used to described the *onset* of *motion* from this quiescent base state (i.e., in the base-state *diffusion transport* is the *only* transport mechanism).

The applications mostly involve liquids and one such application is in the *salt-gradient* solar pond, where by keeping a stabilizing salt concentration distribution the effect of an unstable temperature distribution can be nullified and the fluid motion prevented (e.g, Kaviany and Vogel, 1986; Mullett, 1993). Examples will be given in Sections 4.12.3, 4.16.1, and 8.3.1(A).

1.12.2 Thermoacoustics

When a *compressible* fluid is subjected to a sudden or a sinusoidal local temperature fluctuations (as a result of local heating or cooling), the corresponding local pressure fluctuations travel through the fluid at the *speed of sound* (thus the name *thermoacoustics*). The fluid velocity also undergoes fluctuations and the relation between T, ρ, p, and $\mathbf{u}$ is determined from the conservation equations for a compressible fluid. The constitutive equations for a viscous-stress tensor (or the viscosity related stresses) of a *compressible, Newtonian* fluid is given by (1.77) and is repeated here

$$\tau_{ij} = \mu(\frac{\partial u_i}{\partial x_j} + \frac{\partial u_j}{\partial x_i}) + \delta_{ij}\lambda_\mu \frac{\partial u_k}{\partial x_k} , \tag{1.126}$$

where according to the *Stokes' hypothesis* $\lambda_\mu = -2\mu/3$, leading to

$$\tau_{ij} = \mu(\frac{\partial u_i}{\partial x_j} + \frac{\partial u_j}{\partial x_i} - \frac{2}{3}\delta_{ij}\frac{\partial u_k}{\partial x_k}) . \tag{1.127}$$

Using this in the momentum equation (1.74), the *Navier-Stokes* equation for the fluid momentum conservation is

$$\frac{D\rho u_i}{Dt} = -\frac{\partial p}{\partial x_i} + \frac{\partial}{\partial x_j}\mu(\frac{\partial u_i}{\partial x_j} + \frac{\partial u_j}{\partial x_i} - \frac{2}{3}\delta_{ij}\frac{\partial u_k}{\partial x_k}) + \rho g_i \tag{1.128}$$

where δ_{ij} is the Kronecker delta, $\delta_{ij} = 0$ for $i \neq j$ and $\delta_{ij} = 1$ for $i = j$.

Since thermoacoustics is a phenomenon observed generally in gases, the p-T-v relationship for the gas is also used. The continuity and the energy equations are also written in the variable density forms, i.e., (1.67), (1.81) or (1.83). For a perfect gas $p = R_g\rho T/M$, $c_p - c_v = R_g/M$ and $\beta = 1/T$ and the *isentropic speed* of *sound* is given by (1.13) as

$$a_s = (\frac{\partial p}{\partial \rho})_s^{1/2} = (\frac{c_p}{c_v}\frac{R_g}{M}T)^{1/2} . \tag{1.129}$$

The thermoacoustics phenomenon can result in an *enhanced heat transfer* without a net (i.e., time-averaged) fluid motion (thus the name *thermoacoustic diffusion* or *thermoacoustic dispersion*) and has been studied for superheated gases by Spradley and Churchill (1975), Ozoe et al. (1981), Swift (1988), and Brown and Churchill (1993), and for fluids near the critical point by Zappoli (1992). This enhanced heat transfer becomes especially significant when the buoyant motion is suppressed (as in a microgravity environment).

The convective heat transfer from an isolated sphere to a surrounding compressible gas with standing acoustic waves has been analyzed by Gopinath and Mills (1993). An oscillating compressible fluid in a one-end closed tube can generate heat due to the *viscous dissipation* in the regions near the tube surface (*boundary-layer dissipation*) as well as in the central core. This is called the *resonance tube* and some of its features have been studied by Rott (1974) and Merkli and Thomann (1975) for one-sided piston-driven *resonance tubes*. Swift (1988) gives a review of the principles and applications of thermoacoustic engines (prime-movers and heat-pumps and refrigerators).

The sudden local heating which generates thermoacoustic waves can be due to absorption of the irradation, as in absorption of visible monochromatic radiation (i.e, a visible laser used as the source). This is then referred to as *photoacoustics* and when the heating is periodic, the pressure can be detected using a sensitive microphone. This method has been used for the measurement of the gas thermal diffusivity by Stephan and Biermann (1992). The photoacoustics in fluids can be coupled to that of the bounding solids when the method of *photoacoustic microscopy* is used to evaluate various geometrical and physical properties of thin and thick films. A review of application to semiconductors is given by Mandelis (1987).

Since for perfect gases the speed of sound only varies with the temperature, this is used for temperature distribution measurement (called *acoustic temperature tomography*). This *intrusive* technique has been applied by Venkatesan et al. (1989) by measuring the time of flight of low-frequency sound waves passed through a local volume of a gas.

In Section 4.11 the thermoacoustic diffusion and resonance tube will be discussed in detail.

1.12.3 Thermo- and Diffuso-Capillarity

The surface tension of a *liquid-gas* (and for immiscible liquid-liquid) interface σ can be described as a function of temperature and species concentrations through a linear relation as

$$\sigma = \sigma_o(1 + \left.\frac{\partial \sigma}{\partial T}\right|_{p,\rho_i/\rho} \Delta T + \sum_i \left.\frac{\partial \sigma}{\partial \rho_i/\rho}\right|_{p,T} \Delta \frac{\rho_i}{\rho}) . \tag{1.130}$$

The force balance on the liquid-gas (Newtonian fluids) interface yields (Levich and Krylov, 1969)

$$\begin{aligned}[p_\ell - p_g + \sigma(\frac{1}{R_1} + \frac{1}{R_2})]n_i &= [\mu_\ell(\frac{\partial u_{\ell i}}{\partial x_j} + \frac{\partial u_{\ell j}}{\partial x_i}) \\ -\mu_g(\frac{\partial u_{gi}}{\partial x_j} + \frac{\partial u_{gj}}{\partial x_i})]n_j &- \frac{\partial \sigma}{\partial x_i} - \mu_s \delta_{ij} \frac{\partial^2 u_{sj}}{\partial x_i \partial x_j} ,\end{aligned} \tag{1.131}$$

where R_1 and R_2 are the *principal radii* of the interfacial curvature, n_i is the *component* of the *normal* unit vector directed towards the gas phase and μ_s is a *phenomenological constant* called the *surface viscosity* which is used when *surfactants* (or *surface-active* species) are present *in* the *surface layer* and u_s is the velocity in the surface layer. The *average surface curvature* is defined as $2H = 1/R_1 + 1/R_2$. For two-dimensional menisci, one of the principal radii is zero, and for an axisymmetric meniscus the two principal radii are equal.

The presence of temperature and species concentration gradients on the interface results in the surface force given by

$$\frac{\partial \sigma}{\partial x_j} = \left.\frac{\partial \sigma}{\partial T}\right|_{p,\rho_i/\rho} \frac{\partial T}{\partial x_j} + \sum_i \left.\frac{\partial \sigma}{\partial \rho_i/\rho}\right|_{p,T} \frac{\partial \rho_i/\rho}{\partial x_j} . \tag{1.132}$$

From (1.131) these gradients can influence the interfacial force balance through *thermo-* and *diffusocapillarity*, and therefore, they influence the fluid motion. As present in (1.131), the surface tension also influences the motion whenever there is a *curvature* on the interface. Thus is called the *curvature capillarity*.

(A) Surface-Tension Variation with Temperature

Thermocapillary fluid motion occurs whenever a temperature nonuniformity exists along the interface of any two *immiscible fluids*. The *characteristic velocity* for a temperature difference ΔT along a length can be found from approximating (1.122) and is (Rivas and Ostrach, 1992)

$$u_\sigma = -\left.\frac{\partial \sigma}{\partial T}\right|_{p,\rho_i/\rho} \frac{\Delta T\, x}{\mu_\ell} . \tag{1.133}$$

The *Reynolds number* (which can be positive or negative depending on the sign of ΔT) can be defined as

$$Re_x = \frac{\rho_\ell u_\sigma x}{\mu_\ell} = -\left.\frac{\partial \sigma}{\partial T}\right|_{p,\rho_i/\rho} \frac{\rho_\ell \Delta T\, x}{\mu_\ell^2} \tag{1.134}$$

and the *Marangoni number* (which can be also positive or negative) is defined as $Ma_x = PrRe_x$, and is

$$Ma_x = -\left.\frac{\partial \sigma}{\partial T}\right|_{p,\rho_i/\rho} \frac{\Delta T x}{\alpha_\ell \mu_\ell} . \tag{1.135}$$

The surface tension of fluids decreases with the temperature and vanishes at the critical point. The relation σ-T for water is approximated by (1.44). In addition to the applications such as welding pools and crystal growth melts, the motion of bubbles in nonisothermal liquids (Satrape, 1992) and the bubble nucleation and fluid motion in the nucleate boiling are also influenced by the thermocapillarity. In Section 6.1.2 the effect on the motion of drops forming in an immiscible nonvolatile host liquid will be discussed.

The *combined* effects of the thermocapillarity and buoyancy (under normal and microgravity environments) has been studied for *cavities* (Ben Hadid and Roux, 1992; Keanini and Rubinsky, 1993) in order to determine the regimes of the dominance of each effect. The liquid is *pulled* towards the *interfacial* locations with a lower temperature and *falls* along the cooled *bounding* surfaces. The effect of the thermocapillarity will be discussed in connection with the solidification of melts in Section 7.3.1 and in relation to the nonisothermal menisci in Section 8.3.

(B) Surface Tension Variation with Temperature and Concentration

The presence of a gradient in *surface active* species (i.e., those which can change the surface tension) along the interface can cause interfacial and bulk motion in the fluids joined at the interface. The concentration distributions in the bulk and the interfacial layer are determined from the *bulk* and the *surface* mass diffusivities and velocities. Probstein (1989) and Slattery (1990) review the concentration distributions of the surface active species on the interlayer as well as in the bulk of the fluids.

1.12.4 Changes in Other Thermophysical Properties

As was reviewed in Section 1.4.3, the *viscosity* of gases *increases* as the temperature increases, and the *reverse* occurs in the liquids. This results in an increase (i.e., *thickening*) in the *viscous boundary-layer thickness* as the gases are heated and the *thinning* of it when a liquid is heated. The reverse happens during cooling. The *thermal conductivity*, the *heat capacity*, and the *mass diffusion coefficient* are also temperature dependent and can influence the motion when a temperature gradient is present. The fluid motion is also influenced by the following phenomena occurring as a result of the heat transfer.

(A) Volume Change Resulting from Phase Change

The *shrinkage* occurring during the *solidification* and in *condensation* (except for the solidification of fluids such as water where the solid density is smaller than the liquid) and the *expansion* during the *melting* and in *evaporation*, can influence or cause a fluid motion by the *phase-density* buoyancy. The ratio $(\rho_s - \rho_\ell)/\rho_\ell$ can be as large as 1.2 during solidification, as discussed in the negative phase-density bouyancy or shrinkage-induced fluid motion during solidification by Chiang and Tsai (1992). The expansion resulting from melting, and its rather minor phase-density buoyancy effect compared to thermobuoyancy, are considered by Kim and Kaviany (1992). In a *cyclic* solid-liquid phase change, a *hysteresis* is found and is partly due to this volume change.

During *bubble nucleation* on heated solid surfaces, the phase-density buoyancy-induced motion or expansion-induced fluid motion in the liquid can be very significant, i.e., the ratio $(\rho_\ell - \rho_v)/\rho_v$ is very large, and this *phase-density buoyant motion* tends to *destratify* the temperature field near the heated surface. This will be discussed in Chapter 8.

(B) Thermodiffusion

The presence of a temperature gradient influences the species concentration in a mixture (gas or liquid) by a molecular diffusive mass flux $\dot{\mathbf{m}}_T$ (called the *Soret diffusion*). The relation for *binary* mixtures is

$$\dot{\mathbf{m}}_T = -[\rho S_T \frac{\rho_1}{\rho}(1 - \frac{\rho_1}{\rho})] D \nabla T \ , \tag{1.136}$$

where S_T is the *thermodiffusion* or the *Soret coefficient* (Zimmermann et al., 1992), and ρ_1 is the species-density for any of the two components. When the convective mass flux term in the species conservation equation (1.72) is of the order of the species diffusive mass flux, then the thermodiffusion can noticeably influence the species flux.

A *diffusothermal* flux, which is a molecular diffusive heat flux $\mathbf{q}_m$ (called the *Dufour diffusion*) due to the presence of a species concentration gradient, for a binary mixture is also defined as

$$\mathbf{q}_m = -[T\rho_1(\frac{\partial \mu_1}{\partial \rho_1/\rho})]D_m \nabla \frac{\rho_1}{\rho} , \tag{1.137}$$

where D_m is the *diffusothermal* or the *Dufour coefficient* and μ_1 is generally taken as the chemical potential of the *lower* concentration species (which in the liquid phase, where the Dufour effect can be important, is the *solute*).

The relationship between D_m and S_T is (Hurle and Jakeman, 1971)

$$D_m = S_T D \tag{1.138}$$

and in liquids D_m is orders of magnitude smaller than S_T.

The thermodiffusion can *cause* a fluid motion (as in onset of convection caused by a vertical unstable gradient) and aspects of this motion have been studied by Hurle and Jakeman (1971) and Zimmermann and Müller (1992). Note that both the Soret and the Dufour diffusive fluxes are towards the lower potential, i.e., the mass flux is towards the *lower* temperature locations and the heat flux is towards the lower species-concentration. Effects of the Soret and Dufour diffusive fluxes in the chemical vapor deposition process have been examined by Mahajan and Wei (1991).

(C) Thermophoresis

In the thermodiffusion, the particles were *molecules* in a multicomponent fluid and the imbalance of the kinetic energy (and therefore the collision) of the surrounding *host* molecules results in their motion. When *small particles* are *suspended* in a nonisothermal gas, a similar and more noticeable phenomenon occurs and is named the *thermophresis*. An example of this is the soot collection is the glass globe of a kerosene lantern (Talbot et al. 1980). The *migration* (i.e., diffusion) of small particles in a *direction opposite* to ∇T_g has been studied using the *continuum* approach of hydrodynamics and the *molecular* theory. The molecular theory has been reviewed and expanded by Yamamoto and Ishihara (1988) to cover the continuous, transition and rarefied gas regimes, and by Williams (1988) to include the effect of the presence of wall (and therefore a mean velocity nonuniformity).

The approach of Talbot et al. (1980) combines the hydrodynamic treatment with an empirical extension to cover the large Knudsen number regime (i.e., rarefied gas regime). In the hydrodynamics treatment, the gas temperature stratification ∇T_g is imposed far from a particle of radius R and the temperature field within the particle as well as in the surrounding continuum gas is determined by using the conduction energy equation and by matching the heat flux at the particle-gas interfaces using the particle and gas thermal conductivities k_p and k_g. However, a *temperature slip* is allowed at this interface according to

$$T_g - T_p = \alpha_T \lambda \frac{\partial T_g}{\partial r} \quad \text{at } r = R\,, \tag{1.139}$$

where T_p is the particle temperature, the *temperature slip coefficient* α_T is a constant, and λ is the mean free path of the gas molecules. A similar *velocity slip* is also used for the tangential (or radial) component of the velocity, in solving for the Stokes (i.e., creeping) flow. This boundary condition gives the magnitude of the *slip*, polar-component of the velocity u_{θ_i} as

$$u_{\theta_i} = \alpha_u \lambda \left(r \frac{\partial}{\partial r}\frac{u_\theta}{r} + \frac{1}{r}\frac{\partial u_r}{\partial \theta}\right) + \frac{\alpha_d \nu}{T_g R}\frac{\partial T_g}{\partial \theta} \quad \text{at } r = R\,, \tag{1.140}$$

where α_u is the *momentum exchange coefficient* and α_d is the *thermodiffusive slip coefficient.* The radial component of the velocity vanishes at the surface, i.e., $u_{r_i} = 0$. The resulting *thermophoretic force* on the particle is given by

$$\mathbf{f}_T = -\frac{12\pi\mu\nu\alpha_d R\left(\frac{k_g}{k_p} + \alpha_T \frac{\lambda}{R}\right)\frac{\nabla T_g}{T_g}}{\left(1 + 3\alpha_u \frac{\lambda}{R}\right)\left(1 + 2\frac{k_g}{k_p} + 2\alpha_T \frac{\lambda}{R}\right)}\,, \tag{1.141}$$

where they suggest $\alpha_u = \alpha_d = 1.14$ and $\alpha_T = 2.18$.

The *thermophoretic velocity* is given by

$$\mathbf{u}_T = -\frac{2\alpha_d \nu \left(\frac{k_g}{k_p} + \alpha_T \frac{\lambda}{R}\right)\left[1 + \frac{\lambda}{R}\left(a_1 + a_2 e^{-a_3 R/\lambda}\right)\right]\frac{\nabla T_g}{T_g}}{\left(1 + 3\alpha_u \frac{\lambda}{R}\right)\left(1 + 2\frac{k_g}{k_p} + 2\alpha_T \frac{\lambda}{R}\right)}\,, \tag{1.142}$$

where a_1=1.20, a_2=0.41, and a_3=0.88. Talbot et al. (1980) show that (1.141) and (1.142) reasonably predict the experimental results for a large range of *Knudsen number* $Kn = \lambda/R$.

The motion of particles, as influenced by the presence of a temperature gradient, in a gas has applications in the chemical vapor deposition (Simpkins et al., 1979; Lin et al., 1992) and in the aerosol transport (Walker et al., 1979). The particles of interest are generally of the order of 1 μm and smaller. Particles as large as 10 μm in temperature gradients of about 50°C/cm can be significantly influenced.

(D) Electrothermal Convection

The heat transfer in dielectric liquids can be increased substantially through motion induced by applied electric fields (Martin and Richardson, 1984; Fujino et al., 1989). The *electric conductivity* σ_e is a function of temperature, and even when a stabilizing vertical temperature gradient is present, application of direct-current voltage (electric field) can result in the onset of motion in a horizontal fluid layer. This is called the *electrothermal convection* (Jones, 1978; Martin and Richardson, 1984). This is due to the volumetric electrostatic force on the fluid containing free charge. This force will be discussed in Section 2.6, and by neglecting the variation in the *dielectric constant* ϵ_e, as discussed by Turnbull (1968), we have

$$\mathbf{f}_e = \rho_e \mathbf{e} \, , \quad \frac{D\rho_e}{Dt} = -\nabla \cdot \sigma_e \mathbf{e} \, , \tag{1.143}$$

where $\mathbf{f}_e$ is the *coulombic* (or *electrostatic*) force resulting from the interaction of an electric field $\mathbf{e}$ with the free space-charge of *charge density* ρ_e which in turn depends on the fluid electrical conductivity σ_e and on $\mathbf{e}$. For an *ideal* dielectric, $\rho_e = 0$. The electrohydrodynamics will be discussed in Sections 2.6 and 4.13.

The electrical conductivity of fluids σ_e *increases* with temperature and the experimental results are *correlated* using quadratic or Arrhenius-type functions (e.g., Martin and Richardson, 1984), i.e.,

$$\frac{\sigma_e}{\sigma_{e_o}} = 1 + a_1(T - T_o)[1 + a_2(T - T_o)] \tag{1.144}$$

or

$$\frac{\sigma_e}{\sigma_{e_o}} = e^{\Delta E_a^*(1 - T_o/T)} \, , \tag{1.145}$$

where a_1 and a_2 are fluid constants, ΔE_a^* is a *dimensionless activation energy* which is also a fluid property, and T_o and σ_{e_o} are the reference values.

The volumetric force $\mathbf{f}_e$ will be discussed in Sections 2.6 and 4.11, along with the discussion of the Maxwell equations governing the electromagnetism. The *ohmic heating*, i.e., the volumetric heat source $\sigma_e \, \mathbf{e} \cdot \mathbf{e}$, causes a local change in the density and can also influence the fluid motion (Oued El Moctar et al., 1993).

1.13 References

Andreani, M., and Yadigaroglu, G., 1991, "A Mechanistic Eulerian-Lagrangian Model for Dispersed Flow Film Boiling," in *Phase-Interface Phenomena in Multiphase flow*, Hewitt, G.F., et al., Editors, Hemisphere Publishing Corporation, Washington, D.C.

Aris, R. 1989, *Vectors, Tensors, and the Basic Equations of Fluid Mechanics*, Dover, New York.

Arpaci, V.S., and Larsen, P.S., 1984, *Convective Heat Transfer*, Prentice-Hall, Englewood Cliffs, NJ.

Avedisian, C.T., 1992, "Homogeneous Bubble Nucleation within Liquids: A Review," in *Two-Phase Flow and Heat Transfer*, Kim, J.-H., et al., Editors, *ASME HTD-Vol.*197, American Society of Mechanical Engineers, New York.

Banerjee, S., 1991, "Turbulence/Interface Interactions," in *Phase-Interface Phenomena in Multiphase Flow*, Hewitt, G.F., et al., Editors, Hemisphere Publishing Corporation, Washington, D.C.

Barber, D.J., and Loudon, R., 1989, *An Introduction to the Properties of Condensed Matter*, Cambridge University Press, Cambridge.

Batchelor, G.K., 1953, *The Theory of Homogeneous Turbulence*, Cambridge University Press, Cambridge.

Bejan, A., 1984, *Convection Heat Transfer*, John Wiley and Sons, New York.

Ben Hadid, H., and Roux, B., 1992, "Buoyancy- and Thermocapillary-Driven Flows in Differentially Heated Cavities for Low Prandtl Number Fluids," *J. Fluid Mech.*, 235, 1–36.

Beysens, D., Straub, J., and Turner, D.J., 1987, "Phase Transitions and Near-Critical Phenomena," in *Fluid Sciences and Materials Science in Space: A European Perspective*, Walter, H.U., Editor, Springer-Verlag, Berlin.

Bharathan, D., 1988, "Direct Contact Evaporation," in *Direct Contact Heat Transfer*, Kreith, F., and Boehm, R.F., Editors, Hemisphere Publishing Corporation, Washington, D.C.

Bird, R.B., Curtiss, C.F., Armstrong, R.C., and Hassager, O., 1987, *Dynamics of Polymeric Liquids, Volume 2, Kinetic Theory*, Second Edition, John Wiley and Sons, New York.

Bird, R.B., Stewart, W.E., and Lightfoot, E.N., 1960, *Transport Phenomena*, John Wiley and Sons, New York.

Boelter, L.M.K., 1965, *Heat Transfer Notes*, McGraw-Hill, New York.

Boeltinger, W.J., and Perepezko, J.H., 1993, "Fundamentals of Solidification at High Rates," in *Rapidly Solidified Alloys*, Liebermann, H.H., Editor, Marcel Dekker, New York.

Brout, R., 1965, *Phase Transitions*, W.A. Benjamin, New York.

Brown, M.A., and Churchill, S.W., 1993, "Transient Behavior of Impulsively Heated Fluid," *Chem. Eng. Technol.*, 16, 82-88.

Burmeister, L.C., 1983, *Convective Heat Transfer*, John Wiley and Sons, New York.

Callen, H.B., 1985, *Thermodynamics and Introduction to Thermostatics*, John Wiley and Sons, New York.

Callister, Jr., W.D., 1991, *Materials Science and Engineering*, Second Edition, John Wiley and Sons, New York.

Carbonell, R.G., and Whitaker, S., 1984, "Heat and Mass Transfer in Porous Media," in *Fundamentals of Transport Phenomena in Porous Medis*, Bear, J., and Corapcioglu, M.Y., Editors, Martinus Nijhoff Publishers, Dordrecht.

Carey, V.P., 1992, *Liquid-Vapor Phase-Change Phenomena*, Hemisphere Publishing Corporation, Washington, D.C.

Cebeci, T., and Bradshaw, P., 1984, *Physical and Computational Aspects of Convection Heat Transfer*, Springer-Verlag, New York.

Chappuis, J., 1982, "Contact Angle," in *Multiphase Science and Technology*, 1, 387–505, Hemisphere Publishing Company, Washington, D.C.

Chen, J.D., and Wada, N., 1982, "Wetting Dynamics of the Edge of a Spreading Drop," *Phys. Rev. Letters*, 62, 3050–3053.

Chiang, K.-C., and Tsai, H.-L., 1992, "Shrinkage Induced Fluid Flow and Domain Change in Two-Dimensional Alloy Solidification," *Int. J. Heat Mass Transfer*, 35, 1763–1770.

Clift, R., Grace, J.R., and Weber, M.E., 1978, *Bubbles, Drops and Particles*, Academic, New York.

Davidson, J.F., Clift, R., and Harrison, D., Editors, 1985, *Fluidization*, Second Edition, Academic, London.

de Gennes, P.G., Hua, X., and Levinson, P., 1990 "Dynamics of Wetting: Local Contact Angles," *J. Fluid Mech.*, 212, 55–63.

Defay, R., and Prigogine, I., 1966, *Surface Tension and Absorption*, John Wiley and Sons, New York.

Desloge, E.A., 1968, *Thermal Physics*, Holt, Rinehart and Winston, New York.

Domb, C., and Green, M.S., 1972, *Phase Transitions and Critical Phenomena, Volumes 1 and 2*, Academic, London.

Duckler, A.E., Fabre, J.A., McQuillen, J.B., and Vernon, R., 1988, "Gas-Liquid Flow at Microgravity Conditions: Flow Pattern and Their Transitions," *Int. J. Multiphase Flow*, 14, 389–400.

Dussan V, E.B., Ramé, E., and Garoff, S., 1991, "On Identifying the Appropriate Boundary Conditions at a Moving Contact Line: An Experimental Investigation," *J. Fluid Mech.*, 230, 97–116.

Eckert, E.R.G., 1981, "Pioneering Contributions to our Knowledge of Convective Heat Transfer," *ASME J. Heat Transfer*, 103, 409–414.

Eckert, E.R.G., and Drake, R.M., 1972, *Analysis of Heat and Mass Transfer*, McGraw-Hill, New York.

Emmons, H.W., 1956, "The Film Combustion of Liquid Fuel," *Z. Angew. Math. Mech.*, 36, 60–71.

Fan, L.-S., 1989, *Gas-Liquid-Solid Fluidization Engineering*, Butterworths, Boston.

Fujii, T., 1991, *Theory of Laminar Film Condensation*, Springer-Verlag, New York.

Fujino, T., Yokoyama, Y., and Mori, Y.H., "Augmentation of Laminar Forced-Convective Heat Transfer by the Application of a Transverse Electric Field," *ASME J. Heat Transfer*, 111, 345–351.

Gebhart, B., 1979, "Buoyancy-Induced Fluid Motions Characteristic of Application in Technology — 1978 Freeman Scholar Lecture," *ASME J. Fluid Eng.*, 101, 5-28.

Gebhart, B., Jaluria, Y., Mahajan, R.L., and Sammakia, B., 1988, *Buoyancy-Induced Flows and Transport*, Hemisphere Publishing Corporation, Washington, D.C.

Gerwick, V., and Yadigaroglu, G., 1992, "A Local Equation of State for a Fluid in the Presence of a Wall and its Application to Rewetting," *Int. J. Heat Mass Transfer*, 35, 1823–1832.

Giot, M., 1982, "Three-Phase Flow," in *Handbook of Multiphase Systems*, Hestroni, G., Editor, Hemisphere Publishing Corporation, Washington, D.C.

Goldstein, R.J., 1971, "Film Cooling," *Advan. Heat Transfer*, 7, 321–379.

Goodstein, D.L., 1975, *States of Matter*, Prentice-Hall, Englewood Cliffs.

Gopinath, A., and Mills, A.F., 1993, "Convective Heat Transfer from a Sphere due to Acoustic Streaming," *ASME J. Heat Transfer*, 115, 332–341.

Grace, J.R., 1982a, "Fluidized Bed Heat Transfer," in *Handbook of Multiphase Systems*, Hestroni, G., Editor, Hemisphere Publishing Corporation, Washington, D.C.

Grace, J.R., 1982b, "Fluidized Bed Hydrodynamics," in *Handbook of Multiphase Systems*, Hestroni, G., Editor, Hemisphere Publishing Corporation, Washington, D.C.

Grace, J.R., 1986, "Contacting Modes and Behavior Classification of Gas-Solid and Other Two-Phase Suspensions," *Can. J. Chem. Eng.*, 64, 353–363.

Gray, C.G., and Gubbins, K.E., 1984, *Theory of Molecular Fluids*, Oxford University Press, Oxford.

Gray, D.D., and Giorgini, A., 1976, "The Validity of the Boussinesq Approximation for Liquids and Gases," *Int. J. Heat Mass Transfer*, 19, 545–551.

Gregg, S.J., and Sing, K.S.W., 1982, *Adsorption, Surface Area and Porosity*, Second Edition, Academic, London.

Griffith, P., 1982, "Condensation," in *Handbook of Multiphase Systems*, Hestroni, G., Editor, Hemisphere Publishing Corporation, Washington, D.C.

Hewitt, G.E., 1982, "Flow Regimes," in *Handbook of Multiphase Systems*, Hestroni, G., Editor, Hemisphere Publishing Corporation, Washington, D.C.

Hirschfelder, J.O., Curtiss, C.F., and Bird, R.B., 1954, *Molecular Theory of Gases and Liquids*, John Wiley and Sons, New York.

Hocking, L.M., 1992, "Rival Contact-Angle Models of the Spreading of Drops," *J. Fluid Mech.*, 239, 671–681.

Hocking, L.M., 1993, "The Influence of Intermolecular Forces on Thin Fluid Layers," *Phys. Fluids*, A5, 793-799.

Hockney, R.W., and Eastwood, J.W., 1989, *Computer Simulation Using Particles*, Adam Hilger, Bristol.

Homsy, G.M., Jackson, R.J., and Grace, J.R., 1992, "Report of a Symposium on Mechanics of Fluidized Beds," *J. Fluid Mechanics*, 230, 477–495.

Howell, J.R., and Buckius, R.O., 1992, *Fundamentals of Engineering Thermodynamics*, McGraw-Hill, New York.

Hsieh, J.-S., 1975, *Principles of Thermodynamics*, McGraw-Hill, New York.

Huppert, H.E., and Turner, J.S., 1981, "Double Diffusive Convection," *J. Fluid Mech.*, 106, 299–329.

Hurle, D.T.J., and Jakeman, E., 1971, "Soret-Driven Thermosolutal Convection," *J. Fluid Mech.*, 47, 667–687.

Incropera, F.P., 1974, *Introduction to Molecular-Structure and Thermodynamics*, John Wiley and Sons, New York.

Incropera, F.P., and DeWitt, D.P., 1990, *Introduction to Heat Transfer*, Second Edition, John Wiley and Sons, New York.

Ishikubo, Y., Ishida, K., and Mori, Y.H., 1992, "Behavior of Hydrocarbon Liquid Dipped onto the Surface of a Flowing Water Layer," *Int. J. Heat Mass Transfer*, 35, 1559–1604.

Jacobs, H.R., 1988a, "Direct Contact Condensation," in *Direct Contact Heat Transfer*, Kreith, F., and Boehm, R.F., Editors, Hemisphere Publishing Corporation, Washington, D.C.

Jacobs, H.R., 1988b, "Direct-Contact Heat Transfer for Process Technologies," *ASME J. Heat Transfer*, 110, 1259–1270.

Jacobs, H.R., and Golafshani, M., 1989, "A Heuristic Evaluation of the Governing Mode of Heat Transfer in a Liquid-Liquid Spray Column," *ASME J. Heat Transfer*, 111, 773–779.

Jakob, M., and Hawkins, G.A., 1957, *Elements of Heat Transfer*, John Wiley and Sons, New York.

Jean, R.-H., and Fan, L.-S., 1990, "Rise Velocity and Gas-Liquid Mass Transfer of a Single Large Bubble in Liquids and Liquid-Solid Fluidized Beds," *Chem. Engng. Sci.*, 45, 1057–1070.

Jones, T.B., 1978, "Electrohydrodynamically Enhanced Heat Transfer in Liquids," *Advan. Heat Transfer*, 14, 107–148.

Kaviany, M., 1991, *Principles of Heat Transfer in Porous Media*, Springer-Verlag, New York.

Kaviany, M., and Vogel, M., 1986, "Effects of Solute Concentration Gradient on the Onset of Convection: Uniform and Nonuniform Initial Gradients," *ASME J. Heat Transfer*, 108, 776–782.

Kays, W.H., 1966, *Convective Heat and Mass Transfer*, McGraw-Hill, New York.

Kays, W.M., and Crawford, M.E., 1993, *Convective Heat and Mass Transfer*, Third Edition, McGraw-Hill, New York.

Keanini, R.G., and Rubinsky, B., 1993, "Three Dimensional Simulation of the Plasma Arc Welding Process," *Int. J. Heat Mass Transfer*, 36, 3283–3298.

Kim, C.-J., and Kaviany, M., 1992, "A Numerical Method for Phase-Change Problems with Convection and Diffusion," *Int. J. Heat Mass Transfer*, 35, 457–467.

Kim, S.D., Kang, Y., and Kwon, H.K., 1986, "Heat Transfer Characteristics in Two- and Three-Phase Slurry-Fluidized Beds," *AIChE J.*, 32, 1397–1400.

Kreith, F., and Boehm, R.F., Editors, 1988, *Direct Contact Heat Transfer*, Hemisphere Publishing Corporation, Washington, D.C.

Kutateladze, S.S., 1963, *Fundamentals of Heat Transfer*, English Edition, Academic, New York.

Lahey, Jr., R.T., and Drew, D.A., 1990, "The Current State-of-the-Art in the Modeling of Vapor/Liquid Two-Phase Flows," *ASME Paper no.* 90-WA/HT-13, American Society of Mechanical Engineers, New York.

Lahey, Jr., R.T., and Lopez de Bertodano, M., 1991, "The Prediction of Phase Distribution Using Two-Fluid Models," *Proceedings* of the ASME/JSME Joint Thermal Engineering Conference, Vol. 2, 193-200, American Society of Mechanical Engineers, New York.

Lai, W.M., Rubin, D., and Kreaple, E., 1978, *Introduction to Continuum Mechanics*, Pergamon, Oxford.

Landau, L.D., and Lifshitz, E.M., 1968, *Fluid Mechanics*, Addison-Wesley, Reading, MA.

Layton, E.T., Jr., and Lienhard, J.H., Editors, 1988, *History of Heat Transfer*, American Society of Mechanical Engineers, New York.

Lee, L.L., 1988, *Molecular Thermodynamics of Nonideal Fluids*, Butterworths, Boston.

Letan, R., 1988, "Liquid-Liquid Processes," in *Direct Contact Heat Transfer*, Kreith, F., and Boehm, R.F., Editors, Hemisphere Publishing Corporation, Washington, D.C.

Levich, V.G., and Krylov, V.S., 1969, "Surface-Tension-Driven Phenomena," *Ann. Rev. Fluid Mech.*, 1, 293–317.

Levine, I.N., 1988, *Physical Chemistry*, Third Edition, McGraw-Hill, New York.

Lienhard, J.H., and Karimi, A., 1978, "Corresponding State Correlations of the Extreme Liquid Superheat and Vapor Subcooling," *ASME J. Heat Transfer*, 100, 492–495.

Lienhard, J.H., and Karimi, A., 1981, "Homogeneous Nucleation and the Spinodal Line," *ASME J. Heat Transfer*, 103, 61–64.

Lienhard, J.H., Shamsundar, N., and Biney, D.O., 1986, "Spinodal Lines and Equation of State: A Review," *Nuc. Eng. Des.*, 95, 297–314.

Lienhard V, J.H., Liu, X., and Gabour, L.A., 1992, "Splattering and Heat Transfer During Impingement of a Turbulent Liquid Jet," *ASME J. Heat Transfer*, 114, 362–372.

Liepmann, H.W., and Roshko, A., 1967, *Elements of Gasdynamics*, John Wiley and Sons, New York.

Liley, P.E., 1985, "Thermophysical Properties," in *Handbook of Heat Transfer: Fundamentals*, Chapter 3, Rosenow, W.M., et al., Editors, McGraw-Hill, New York.

Lin, Y.-T., Choi, M., and Grief, R., 1992, "A Three-Dimensional Analysis of Particle Deposition for the Modified Chemical Vapor Deposition," *J. Heat Transfer*, 114, 735–742.

Liñán, A., and Williams, F.A., 1993, *Fundamental Aspects of Combustion*, Oxford University Press, New York.

Luikov, A.V., and Berkovsky, B.M., 1970, "Thermoconvective Waves," *Int. J. Heat Mass Transfer*, 13, 741–747.

Mahajan, R.L., and Wei, C., 1991, "Buoyancy, Soret, Dufour and Variable Property Effects in Silicon Epitaxy," *ASME J. Heat Transfer*, 113, 688–695.

Mandelis, A., Editor, 1987, *Photoacoustic and Thermal Wave Phenomena in Semiconductors*, North Holland, New York.

Martin, P.J., and Richardson, A.T., 1984, "Conductivity Models of Electrothermal Convection in a Plane Layer of Dielectric Liquid," *ASME J. Heat Transfer*, 106, 131–142.

Maxworthy, T., 1983, "The Dynamics of Double-Diffusive Gravity Currents," *J. Fluid Mech.*, 128, 259–282.

Merkli, P., and Thomann, H., 1975, "Thermoacoustic Effects in a Resonance Tube," *J. Fluid Mech.*, 1970, 161–177.

Miller, C.A., and Ruckenstein, E., 1974, "The Origin of Flow During Wetting of Solids," *J. Colloid Interface Sci.*, 48, 368–373.

Moelwyn-Hughes, E.A., 1961, *States of Matter*, Oliver and Boyd, Edinburgh.

Mullett, L., 1993, "The Role of Buoyant Thermals in Salt Gradient Solar Ponds and in Convection More Generally," *Int. J. Heat Mass Transfer*, 36,1923-1941.

Mullins, W.W., and Sekerka, R.F., 1964, "Stability of a Planar Interface During Solidification of a Dilute Binary Alloy," *J. Appl. Phys.*, 35, 444–451.

Nghiem, L., Merte, H., Winter, E.R.F., and Beer, H., 1981, "Prediction of Transient Inception of Boiling in Terms of Heterogeneous Nucleation Theory," *ASME J. Heat Transfer*, 103, 69-73.

Oswatitsch, K., and Wieghardt, K., 1987, "Ludwig Prandtl and His Kaiser-Wilhelm-Institute," *Ann. Rev. Fluid Mech.*, 19, 1–27.

Oued El Moctar, A., Peerhussaini, H., Le Peurian, P., and Bardon, J.P., 1993, "Ohmic Heating of Complex Fluids," *Int. J. Heat Mass Transfer*, 36, 3143–3152.

Ožisik, M.N., 1985, *Radiative Transfer and Interactions with Conduction and Convection*, Werbel and Peck, New York.

Ozoe, H., Sato, N., and Churchill, S.W., 1980, "The Effect of Various Parameters on Thermoacoustic Convection," *Chem. Eng. Commun.*, 5, 203–221.

Platten, J.K., and Legros, J.C., 1984, *Convection in Liquids*, Springer-Verlag, Berlin.

Prausnitz, J.M., Lichtenther, R.N., and de Azevedo, E.G., 1986, *Molecular Thermodynamics of Fluid-Phase Equilibria*, Second Edition, Prentice-Hall, Englewood Cliffs, NJ.

Probstein, R.F., 1989, *Physiochemical Hydrodynamics*, Butterworths, Boston.

Prutton, M., 1987, *Surface Physics*, Oxford University Press, Oxford.

Ried, R.C., Prausnitz, J.M., and Poling, B.E., 1987, *The Properties of Gases and Liquids*, McGraw-Hill, New York.

Rivas, D., and Ostrach, S., 1992, "Scaling of Low-Prandtl-Number Thermocapillary Flows," *Int. J. Heat Mass Transfer*, 35, 1469–1479.

Rohsenow, W.H., 1982, "General Boiling," in *Handbook of Multiphase Systems*, Hestroni, G., Editor, Hemisphere Publishing Corporation, Washington, D.C.

Rott, N., 1974, "The Heating Effect Connected with Non-Linear Oscillation in a Resonance Tube," *J. Appl. Math. Phys. (ZAMP)*, 25, 619–634.

Rott, N., 1980, "Thermoacoustics," *Advan. Appl. Mech.*, 20, 135–175.

Rowlinson, J.S., and Widom, B., 1989, *Molecular Theory of Capillarity*, Oxford University Press, Oxford.

Saito, A., Okawa, S., Tojiki, A., Une, H., and Tanogashira, K., 1992, "Fundamental Research on External Factors Affecting the Freezing of Supercooled Water," *Int. J. Heat Mass Transfer*, 35, 2527–2536.

Sakurai, A., and Shiotsu, M., 1992, "Pool Film Boiling Heat Transfer and Minimum Film Boiling Temperatures," in *Pool and External Flow Boiling*, Dhir, V. K., and Bergels, A.E., Editors, American Society of Mechanical Engineers, New York.

Satake, M., and Jenkins, J.T., Editors, 1988, *Micromechanics of Granular Materials*, Elsevier, Amsterdam.

Satrape, J.V., 1992, "Interactions and Collisions of Bubbles in Thermocapillary Motion," *Phys. Fluids*, A4, 1883–1900.

Saxena, S.C., 1988, "Heat Transfer Between Immersed Surfaces and Gas-Fluidized Beds," *Adv. Heat Transfer*, 19, 97–190.

Saxena, S.C., 1989, "Heat Transfer Between Immersed Surfaces and Gas-Fluidized Beds," in *Adv. Heat Transfer*, 19, 97–190.

Saxena, S.C., Rao, N.S., and Saxena, A.C., 1992, "Heat Transfer and Gas Holdup Studies in a Bubble Column: Air-Water-Sand System," *Can. J. Chem. Eng.*, 70, 33–41.

Schlichting, H., 1979, *Boundary-Layer Theory*, Seventh Edition, McGraw-Hill, New York.

Sherman, F.S., 1990, *Viscous Flow*, McGraw-Hill, New York.

Simpkins, P.G., Greenberg-Kosinski, S., and MacChesney, J.B., 1979, "Thermophresis: The Mass Transfer Mechanism in Modified Chemical Vapor Deposition," *J. Appl. Phys.*, 50, 5676–5681.

Skripov, V.P., 1974, *Metastable Liquids*, John Wiley and Sons, New York.

Slattery, J.C., 1981, *Momentum, Energy, and Mass Transfer in Continua*, Second Edition, R.E. Krieger Publishing Company, Huntington, NY.

Slattery, J.C., 1990, *Interfacial Transport Phenomena*, Springer-Verlag, New York.

Somorjai, G.A., 1994, *Introduction to Surface Chemistry and Catalysis*, John Wiley and Sons, New York.

Sonntag, R.E., and Van Wylen, G.J., 1982, *Introduction to Thermodynamics, Classical and Statistical*, Second Edition, John Wiley and Sons, New York.

Soo, S.-L., 1989, *Particulates and Continuum*, Hemisphere Publishing Corporation, Washington, D.C.

Spradley, L.W., and Churchill, S.W., 1975, "Pressure and Buoyancy-Driven Thermal Convection in a Rectangular Enclosure," *J. Fluid Mech.*, 70, 705–720.

Springer, G.S., 1978, "Homogeneous Nucleation," *Advan. Heat Transfer*, 14, 281–346.

Stanley, H.E., 1987, *Introduction to Phase Transitions and Critical Phenomena*, Oxford University Press, Oxford.

Stephan, K., and Biermann, J., 1992, "The Photoacoustic Technique as Convenient Instrument to Determine Thermal Diffusivity of Gases," *Int. J. Heat Mass Transfer*, 35, 605–612.

Swift, G.W., 1988, "Thermoacoustic Engines," *J. Acoust. Soc. Am.*, 84, 1145–1180.

Szekely, J., Evans, J.W., and Brimacombe, J.K., 1988, *The Mathematical and Physical Modeling of Primary Metals Processing Operations*, John Wiley and Sons, New York.

Talbot, L., Cheng, R.K., Schefer, R.W., and Willis, D.R., 1980, "Thermophresis of Particles in a Heated Boundary Layer," *J. Fluid Mech.*, 101, 737–758.

Taylor, G.I., 1953, "Dispersion of Soluble Matter in Solvent Flowing Slowly Through a Tube," *Proc. Roy. Soc.* (*London*), A219, 186–203.

Tien, C.-L., 1961, "Heat Transfer by a Turbulently Flowing Fluid-Solid Mixture in a Pipe " *ASME J. Heat Transfer*, 83, 183–188.

Tien, C.-L., and Lienhard, J.H., 1971, *Statistical Thermodynamics*, Holt, Rinehart and Winston, New York.

Truesdell, C., and Noll, W., 1992, *The Non-Linear Field Theories of Mechanics*, Second Edition, Springer-Verlag, Berlin.

Turnbull, R.J., "Electro-Convective Instability with a Stabilizing Temperature Gradient. I. Theory," *Phys. Fluids*, 11, 2588–2596.

Turner, J.S., 1979, *Buoyancy Effects in Fluids*, Cambridge University Press, Cambridge.

van Swaaij, W.P.M., and Afgan, N.H., Editors, 1986, *Heat and Mass Transfer in Fixed and Fluidized Beds*, Hemisphere Publishing Company, Washington, D.C.

Venkatesan, S.P., Shakkotlai, P., Kwack, E.Y., and Bach, L.H., 1989, "Acoustic Temperature Profile Measurement Technique for Large Combustion Chambers," *ASME J. Heat Transfer*, 111, 461–466.

Vere, A.W., 1987, *Crystal Growth, Principles and Progress*, Plenum, New York.

Villers, D., and Platten, J.K., 1992, "Coupled Buoyancy and Marangoni Convection in a Cavity: Experiments and Comparison with Numerical Simulations," *J. Fluid Mech.*, 234, 487–510.

Vortmeyer, D., and Schaefer, R.J., 1974, "Equivalence of One- and Two-Phase Models for Heat Transfer Processes in Packed Beds: One-Dimensional Theory," *Chem. Engrg. Sci.*, 29, 485–491.

Wakao, N., and Kaguei, S., 1982, *Heat and Mass Transfer in Packed Beds*, Gordon and Breach Science Publishers, New York.

Walker, K.L., Homsy, G.M., and Geyling, F.T., 1979, "Thermophoretic Deposition of Small Particles in Laminar Tube Flow," *J. Colloid Interface Sci.*, 69, 138–147.

Walton, A.J., 1989, *Three Phases of Matter*, Second Edition, Oxford University Press, Oxford.

Whalley, P.B., 1987, *Boiling, Condensation, and Gas-Liquid Flow*, Oxford University Press, Oxford.

Whitaker, S., 1991, "Improved Constraints for the Principle of Local Thermal Equilibrium," *Ind. Eng. Chem. Res.*, 30, 983–997.

White, F.M., 1974, *Viscous Fluid Flow*, McGraw-Hill, New York.

Williams, M.M.R., 1988, "The Thermophoretic Force in Knudsen Regime Near a Wall," *Phys. Fluids*, 31, 1051–1057.

Woods, L.C., 1993, *An Introduction to the Kinetic Theory of Gases and Magnetoplasmas*, Oxford University Press, Oxford.

Wu, R.L., Lim, C.J., Grace, J.R., and Brereton C.M.H., 1991, "Instantaneous Local Heat Transfer and Hydrodynamics in a Circulating Fluidized Bed," *Int. J. Heat Mass Transfer*, 34, 2019–2027.

Yamamoto, K., and Ishihara, Y., 1988, "Thermophoresis of a Spherical Particle in a Rarefied Gas of a Transition Regime," *Phys. Fluids*, 31, 3618–3624.

Yao, S.-C., 1993, "Dynamics and Heat Transfer of Impacting Sprays," *Ann. Rev. Heat Transfer*, V, 351–382.

Zappoli, B., 1992, "The Response of a Nearly Supercritical Pure Fluid to a Thermal Disturbance," *Phys Fluids*, A4, 1040–1048.

Zimmerman, G., and Müller, U., 1992, "Bénard Convection in a Two-Component System with Soret Effects," *Int. J. Heat Mass Transfer*, 35, 2245–2256.

Zimmerman, G., Müller, U., and Davis, S., 1992, "Bénard Convection in Binary Mixture with Soret Effects and Solidification," *J. Fluid Mech.*, 230, 657–682.

Part I

Single-Medium Treatments

2

Single-Phase Systems

2.1 Intraphase Heat Transfer

Convective heat transfer within a single-phase flow is considered in this chapter. The *simultaneous* presence of a velocity and temperature gradient field within this single-phase medium and the resulting convective heat transfer can be influenced by the *compressibility*, *chemical reactions*, *turbulence*, *radiation*, *electromagnetic fields*, and *thermobuoyancy*. The effects of these mechanisms on the *intraphasic* heat transfer are discussed by considering the governing equations and by examining some illustrative examples.

This chapter also introduces some of the fundamentals related to these mechanisms which will be used in the following chapters. Evaluation of the effects of these mechanisms on the *bulk* convective heat transfer, i.e., without the inclusion of the complexities associated with the bounding surfaces, allows for an orderly introduction. The effects of these bounding surfaces (solid or another fluid phase) or submerged surfaces are addressed in the later chapters.

2.2 Compressibility

Since the compressibility of the gaseous phase is the largest, in practice only the gas-phase compression and expansion are considered here. The compressibility influences the velocity and temperature fields through the continuity, the momentum, and the energy equations. A gas-phase equation of state relating $\rho - p - T$ is also used. For illustrative purposes of the *intra* gas-phase convective heat transfer, the temperature distribution across *shock waves* and the propagation of *thermoacoustic waves* are considered below.

2.2.1 Shock Waves

Shocks are zones of *steep velocity gradients* and they either propagate into a *low*-pressure region (and are called *compressive waves*) or into a high-pressure region (called *rarefaction waves*). The former occurs, for example, during an explosion where a high pressure wave is created by a thermal energy release and moves into the surrounding undisturbed region. The latter, for example, occurs when due to a sudden leak a depressurized zone is created and the shock propagates into the high pressure surrounding.

A *traveling compressive* shock can *originate* with a rather smooth velocity gradient which would then sharpen. This is because the speed of sound is larger in the high-pressure region [for an isentropic process T is proportional to p to the power of $(c_p/c_v - 1)/(c_p/c_v)$ and the speed of sound is proportional to T to the power of $1/2$], and therefore, as the shocks moves to the low-pressure zone its upstream section moves faster compared to its downstream and the shock steepens (the viscous and conductive effects would tend to bring about an equilibrium). The *traveling rarefaction* waves express the opposite and tend to flatten (Schreier, 1982).

Consider a *viscous, supersonic* flow which becomes *subsonic* as it flows across a *shock* wave. For a *one-dimensional* and *steady* flow of a single-component gas (i.e., stationary *normal* shock), the continuity, momentum and energy equations, i.e., (1.71), (1.76), (1.77), and (1.83), become (Grad, 1952; Gilbarg and Paolucci, 1953; White, 1974), with no radiative heat flux, heat generation or body force,

$$\frac{d}{dx}(\rho u) = 0 \tag{2.1}$$

$$\rho u \frac{du}{dx} = -\frac{dp}{dx} + \frac{d}{dx}(2\mu + \lambda_\mu)\frac{du}{dx} \tag{2.2}$$

$$\rho u \frac{di}{dx} = \frac{d}{dx} k \frac{dT}{dx} + u \frac{dp}{dx} + (2\mu + \lambda_\mu)(\frac{du}{dx})^2 \,. \tag{2.3}$$

By designating the upstream conditions by the subscript u and downstream by d, eliminating the pressure term in the energy equation using the momentum equation, and integrating the three equations we have

$$\rho u = \rho_u u_u = \rho_d u_d = \dot{m} \tag{2.4}$$

$$p + \rho u^2 - (2\mu + \lambda_\mu)\frac{du}{dx} = p_u + \rho_u u_u^2 = p_d + \rho_d u_d^2 \tag{2.5}$$

$$i + \frac{u^2}{2} - \frac{k}{\dot{m}}\frac{dT}{dx} - \frac{2\mu + \lambda_\mu}{2\dot{m}}\frac{du^2}{dx} = i_u + \frac{u_u^2}{2} = i_d + \frac{u_d^2}{2} \,, \tag{2.6}$$

where the condition of zero gradient at the far upstream and downstream locations has been used.

For a perfect gas $di = c_p dT$ and using a *constant* Prandtl number based on $2\mu + \lambda$, i.e., $Pr_\lambda = (2\mu + \lambda_\mu)c_p/k$

$$i + \frac{u^2}{2} - \frac{2\mu + \lambda_\mu}{\dot{m}}\frac{d}{dx}(\frac{i}{Pr_\lambda} + \frac{u^2}{2}) = i_u + \frac{u_u^2}{2} \,. \tag{2.7}$$

For $Pr_\lambda = 1$, this energy equation states that the *stagnation* enthalpy i_o (for $u = 0$) is constant throughout the shock layer. This is a reasonable approximation for monatomic and diatomic gases (White, 1974) and along with a perfect gas behavior and a *constant* c_p gives

$$T + \frac{u^2}{2c_p} = \text{constant}\;, \quad T = T_{st} - \frac{u^2}{2c_p}\;, \quad p = \frac{\rho R_g T}{M} = \frac{\rho_u u_u R_g T}{Mu}\;, \tag{2.8}$$

where T_{st} is the *stagnation* temperature. Now defining the dimensionless variables

$$\begin{aligned} u^* &= \frac{u}{u_u}\;, \quad x^* = \frac{\rho_u u_u x}{\mu_u}\;, \quad 2\mu^* + \lambda_\mu^* = \frac{2\mu + \lambda_\mu}{(2\mu + \lambda\mu)}\;, \\ b &= \frac{(2\mu + \lambda_\mu)_u}{\mu_u} = 2 + (\frac{\lambda_\mu}{\mu})_u\;. \end{aligned} \tag{2.9}$$

Then we have the dimensionless momentum equation as

$$b(2\mu^* + \lambda_\mu^*)u^* \frac{du^*}{dx^*} = \frac{\gamma + 1}{2\gamma}(u^* - 1)(u^* - u_d^*)\;, \tag{2.10}$$

where

$$\gamma = \frac{c_p}{c_v}\;, \quad u_d^* = \frac{u_d}{u_u} = \frac{\gamma - 1 + 2/Ma_u^2}{\gamma + 1}\;, \quad Ma_u = (\frac{u}{a_s})_u\;, \quad a_s = (\frac{\gamma p}{\rho})^{1/2}\;. \tag{2.11}$$

The downstream velocity (i.e., the relationship between the *Mach numbers* Ma_u and Ma_d) is found using the downstream and upstream gradient-free conditions and the energy and momentum equations for an isenthalpic expansion (White, 1974).

Equation (2.10) can be solved for u^* by using a prescribed temperature dependence for μ^* and λ_μ^*. Equation (2.10) can be written as (White 1974)

$$x^* = \frac{2\,\gamma\, b}{\gamma + 1} \int \frac{(2\mu^* + \lambda_\mu^*)u^*}{(u^* - 1)(u^* - u_d^*)}\, du^*\;. \tag{2.12}$$

No experimental results for $\lambda_\mu = \lambda_\mu(T)$ are available and as was mentioned before, the Stokes hypothesis gives $\lambda_\mu = -2\mu/3$ which corresponds to $b = 4/3$. Without using the Stokes constant and instead using a general proportionality between λ_μ and μ, the experimental results for $\mu(T)$ can be used to give a temperature dependence

$$2\mu^* + \lambda_\mu^* = T^{*2/3}\;, \tag{2.13}$$

where from (2.8) the temperature depends on the velocity through (White, 1974)

$$T^* = \frac{T}{T_u} = 1 + \frac{\gamma - 1}{2} Ma_u^2 (1 - u^{*2})\;. \tag{2.14}$$

Equation (2.12) is integrated numerically and the temperature distribution through the shock layer is found. White (1974) gives the details of the numerical integration and the comparison with the available experimental results of Sherman (1955), in which a 6.2-μm-diameter *resistance wire* was used in a *low-density* wind tunnel. An example of the results is reproduced in Figure 2.1 which is for *air* with an upstream Mach number of 3.70,

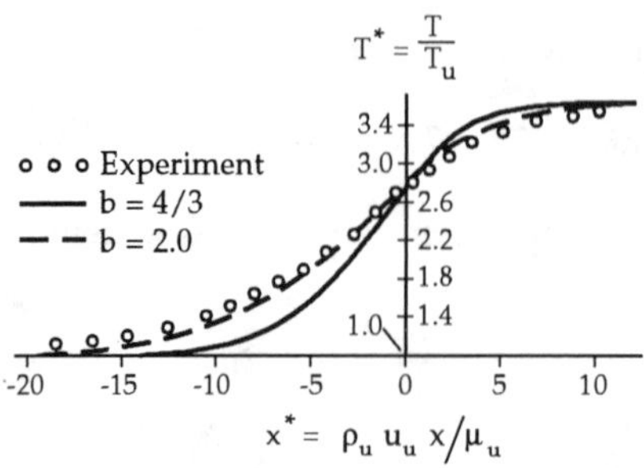

Figure 2.1. Axial distribution of the normalized temperature across a low-density shock wave. (From White, reproduced by permission ©1974 McGraw-Hill.)

and $\gamma = 1.4$. The location of $x^* = 0$ is arbitrarily chosen as where $u^* = (1 + u_d^*)/2$. A good fit to the air data is found when $b = 2.0$ (i.e., $\lambda_\mu = 0$) is used in the momentum equation (2.12).

A *shock-layer thickness* can be defined as a distance $\delta_s^* = \Delta x_s^*$ over which 99 percent of the velocity variations occur. This dimensionless shock-layer thickness is given in White (1974) for several upstream Mach numbers for $\gamma = 1.4$ and is repeated in Table 2.1. Using the mean free path given in terms of the speed of sound and the dynamic viscosity, by combining (1.13), (1.14) and (1.33), the ratio of the dimensional shock-layer thickness to the mean free path is also given in the table. Note that δ_s/λ is the inverse of the *Knudsen number Kn*.

The *continuum* descriptions are valid even for dimensions (over which the significant changes occur) of the order of the mean free path, i.e., $Kn \to 1$. The *first-order kinetic (molecular) theory* applied to this problem has not been as successful (Mott-Smith, 1951; Sherman 1955; White, 1974). The *second-order* kinetic theory has been applied to the transport within the shock front, with more success (Woods, 1993). The effect of the gas radiation has been excluded, but for very strong shocks this can not be justified (Zel'dovich and Raizer, 1969). Recent studies of the one-dimensional shock wave includes the *direct* Monte Carlo simulations for diatomic nitrogen gas which allows for a temperature-dependent *rotational collision* (Boyd, 1993). Ohwada (1993) solves the nonlinear Boltzmann equation for hard spheres molecules and discusses the details of the wave structure.

2.2.2 Attenuation of Thermoconvective Waves

A disturbance in temperature and/or velocity *travels* and *decays* in compressible *viscous* and *heat-conducting* fluids. In relation to the propagation of detonation acoustic waves in the earth's atmosphere, the attenuation of thermoconvective waves in the presence of a vertical (y-direction) temperature gradient dT_o/dy, has been examined by Luikov and Berkovsky (1970)

Table 2.1. Predicted shock-layer thickness.

Ma_u	1.1	1.5	2.0	3.0	4.0	5.0	7.5	10.0
δ_s^*	82.0	22.7	16.1	14.5	15.5	17.4	23.7	31.5
δ_s/λ	49.7	10.1	5.4	3.2	2.6	2.3	2.1	2.1

for *plane* disturbances traveling in the horizontal direction (x-direction). The presence of a vertical temperature gradient may *reduce* the damping of these otherwise *strongly* damped waves. Their analysis is reviewed below in order to examine both the conditions under which the damping is very strong and conditions for less acute damping.

The conservation equations and the equation of state, for a compressible Newtonian fluid with no radiative heat flux or heat generation, are the appropriately simplified forms of (1.71), (1.76), and (1.81), i.e.,

$$\frac{\partial \rho}{\partial t} + u_j \frac{\partial \rho}{\partial x_j} + \rho \frac{\partial u_j}{\partial x_j} = 0 , \tag{2.15}$$

$$\rho \frac{\partial u_i}{\partial t} + \rho u_j \frac{\partial u_i}{\partial x_j} = -\frac{\partial p}{\partial x_i} + \frac{\partial \tau_{ij}}{\partial x_j} + \rho g_i , \tag{2.16}$$

$$\rho(\frac{\partial e}{\partial t} + u_j \frac{\partial e}{\partial x_j}) = k \frac{\partial^2 T}{\partial x_j \partial x_j} - p \frac{\partial u_j}{\partial x_j} + \tau_{ij} \frac{\partial u_i}{\partial x_j} , \tag{2.17}$$

$$f(\rho, p, T) = 0 . \tag{2.18}$$

Consider a fluid at rest, with a *subcritical* (i.e., no motion) constant *vertical* temperature gradient dT_o/dy, being disturbed by perturbations in T, p, ρ and u_i with respect to x and t given by

$$\begin{aligned} &T = T_o(y) + T'(x,t) , \quad \rho = \rho_o + \rho'(x,t) , \\ &p = p_o + p'(x,t) , \quad u_i = u_i'(x,t) . \end{aligned} \tag{2.19}$$

Assuming that the density change with respect to pressure is negligible, we have

$$f_\rho(\rho', T') = 0 , \quad f_p(p', T') = 0 . \tag{2.20}$$

Then the equations governing the disturbances are obtained by substituting these into (2.15) to (2.17). Assuming that the *amplitudes* of the disturbances are small such that their *products* can be neglected and that μ, $\lambda\mu$,

β, k and c_v are constant, these *first-order* governing equations become

$$\frac{\partial T'}{\partial t} + \gamma_o v' = \alpha_o \frac{\partial^2 T'}{\partial x^2} , \tag{2.21}$$

$$\frac{\partial v'}{\partial t} = \nu_o \frac{\partial^2 v'}{\partial x^2} + \beta_o\, g T' , \tag{2.22}$$

$$\rho' = T' \frac{\partial \rho}{\partial T}\bigg|_{T_o} = -\beta_o \rho_o T' , \; p' = T' \frac{\partial p}{\partial T}\bigg|_{T_o} , \; de' = c_{v_o} dT' , \tag{2.23}$$

$$\frac{\partial \rho'}{\partial t} + \rho_o \frac{\partial u'}{\partial x} = 0 , \tag{2.24}$$

$$\frac{\partial u'}{\partial t} = -\frac{1}{\rho_o}\frac{\partial p'}{\partial x} + \frac{2\mu + \lambda_\mu}{\rho_o}\frac{\partial^2 u'}{\partial x^2} , \tag{2.25}$$

where

$$\alpha_o = \frac{k}{\rho_o c_{v_o}(1 + \dfrac{\beta_o p_o}{\rho_o c_{v_o}})} , \quad \gamma_o = \frac{dT_o}{dy}\frac{1}{1 + \dfrac{\beta_o p_o}{\rho_o c_{v_o}}} , \tag{2.26}$$

$\nu_o = \mu/\rho_o$, $\mathbf{g}$ is pointed opposite to the y-axis, u is along x, and v is along y.

Consider sinusoidal disturbances introduced in the *temperature* and in the *vertical* component of the *velocity* at location $x = 0$ in the fluid and given by

$$T' = a_T \cos \omega t , \tag{2.27}$$

$$v' = a_v \cos \omega t , \tag{2.28}$$

i.e., oscillate at an arbitrary *angular frequency* ω with small amplitudes a_T and a_v. Then for the linear equations (2.21) to (2.22) time and space periodic solutions exist of the form

$$T' = a_T\, e^{i(\omega t - \eta x)} , \tag{2.29}$$

$$v' = a_v\, e^{i(\omega t - \eta x)} , \quad u' = a_u\, e^{i(\omega t - \eta x)} . \tag{2.30}$$

These disturbances are substituted into (2.21) to (2.23) and then three linear equations are found for a_T, a_v, and a_u. A nontrivial solution exists only when the determinant of this system of equations is zero. This condition leads to the *characteristic* equation which is the relation between the wave number η and the angular frequency ω, i.e.,

$$\eta^4 \nu_o \alpha_o + \eta^2 i\omega(\nu_o + \alpha_o) - \omega^2 + \gamma_o \beta_o\, g = 0 . \tag{2.31}$$

For complete damping, as $x \to \infty$ the wave number must have a real and an imaginary component $\eta = \eta_r + i\eta_i$ and the imaginary part must be negative, i.e., $\eta_i < 0$. The roots of (2.31) are

$$\text{for} \quad A = \omega^2(\nu_o - \alpha_o)^2 + 4\alpha_o \nu_o \gamma_o \beta_o\, g > 0 , \quad B = \frac{\omega(\nu_o + \alpha_o)}{4\nu_o \alpha_o} , \tag{2.32}$$

$$\eta_r^2 = \eta_i^2 \,, \tag{2.33}$$

$$\eta_i = -[B \pm (4\alpha_o\nu_o)^{-1}A^{1/2}]^{1/2} \quad \text{for} \quad \gamma_o\beta_o\, g < 0 \,, \tag{2.34}$$

$$\eta_i = -[\pm B + (4\alpha_o\nu_o)^{-1}A^{1/2}]^{1/2} \quad \text{for} \quad \gamma_o\beta_o\, g > 0 \,, |\gamma_o\beta_o\, g| \ge \omega^2 \,, \tag{2.35}$$

and for $A < 0$,

$$C = \frac{(-A)^{1/2}}{4\nu_o\alpha_o}, \tag{2.36}$$

$$\eta_r\eta_i = -B, \tag{2.37}$$

$$\eta_i = [\mp C + (B^2 + C^2)^{1/2}]^{1/2} \,. \tag{2.38}$$

These waves travel in the x-direction and perpendicular to the gravity vector and the prescribed vertical temperature gradient, and are called the *transverse waves*.

Strong damping, which also occurs when $dT_o/dy = 0$, occurs for

$$\gamma_o\beta_o\, g > 0 \quad (\text{any } \omega) \tag{2.39}$$

or

$$\gamma_o\beta_o\, g < 0 \qquad \omega \ge \frac{2(-\alpha_o\nu_o\gamma_o\beta_o\, g)^{1/2}}{|\nu_o - \alpha_o|} \,. \tag{2.40}$$

For strong damping, $\eta_r^2 = \eta_i^2$, i.e., the *mean free path* (*attenuation distance*) which is $|\eta_i^{-1}|$ is of the same order of magnitude as the *wavelength* which is $\lambda = |2\pi/\eta_r|$. For gases, $\beta_o > 0$ and for liquids (except for water around 4 °C) $\beta_o < 0$. Then for $-\gamma_o\beta_o\, g > 0$, the vertical temperature gradient for gases should be negative and the reverse for liquids.

Weak damping occurs for $\eta_r/2\pi\eta_i \gg 1$; then using the $A < 0$ conditions given by (2.36) to (2.38) we have

$$\frac{\eta_r}{2\pi\eta_i} \simeq \frac{[-4\alpha_o\nu_o\gamma_o\beta_o\, g - \omega^2(\nu_o - \alpha_o)^2]^{1/2}}{\pi\omega(\nu_o + \alpha_o)} \,. \tag{2.41}$$

Then the larger β and $-\gamma_o$, the lower the frequency, and the closer to unity the Prandtl number ν_o/α_o, the less damped are the waves. The upper limit of the frequency for the weakly damped waves is $\omega \le (-\beta_o\gamma_o\, g)^{1/2}$. Attenuation of the plane acoustic waves has been reviewed by Pierce (1989) and Brown and Churchill (1993), for cases without any buoyancy effects.

2.3 Reacting Flows

The special interest in the *exothermic* chemical reactions, as a volumetric heat source, is in the *temperature dependence* of the rate of reaction. Although in principle both the liquid-phase and the gas-phase reactions can be of interest, most practical cases involve the gaseous phase reactions (two-phase systems will be considered in Sections 3.2, 3.3, and 6.2).

Reactions between molecules of different constituents can occur during their *collisions* and in some cases a third and still different molecule may be needed for a reaction to occur. The *chemical kinetics* or the molecular theory of *homogeneous* chemical reactions (i.e. reactions occurring in the *bulk* of the fluid and not influenced by the interfacial forces and conditions) has been reviewed by Kuo (1986), Williams (1985), Glassman (1987), Kee et al. (1989), and Steinfeld et al. (1989). In the following a summary is given of the dependence of the volumetric reaction rate of species i (volumetric rate of its disappearance) on the local concentration of this species and other species influencing the reaction, and on the local temperature.

2.3.1 Chemical Kinetics

A *stoichiometric reaction* (i.e., all *reactants* turning to *products*) is described by the *chemical equation*

$$\sum_{i=1}^{n} \nu_{ri}\chi_i = \sum_{i=1}^{n} \nu_{pi}\chi_i \,, \tag{2.42}$$

where ν_{ri} is the *stoichiometric coefficient* of species i appearing as a reactant and ν_{pi} is for the product (reactant species have $\nu_{pi} = 0$ and product species have $\nu_{ri} = 0$), n is the the total number of species and χ represent the chemical symbol for the species i. The *order of reaction* is defined as $\sum_i \nu_{ri}$ and for a *simple* reaction of $A + B \rightarrow C + D$ the reaction is second order.

Most reactions are *dominated* by collisions of two different species which have the ability to react, and therefore, are *second order* (Benson, 1976; Glassman, 1987). In equilibrium molecular theory, the rate of collision and reaction of species A and B in a binary mixture has been analyzed and equations such as (1.15) and (1.16) are available for the frequency of collision and rate of reaction. For (2.42) the *rate of disappearance* of species i is given by a modified form of (1.16) which follows the *Arrhenius' postulates*. These are an *exponential* temperature dependence and an *activation energy* for the reaction, and an allowance for the *empirical* adjustment of the collision frequency. For the *production rate* of species i, $\dot{n}_i$, in a *single-step reaction* involving n species, this leads to

$$-\dot{n}_i = M_i(\nu_{pi} - \nu_{ri})a_{1_i}T^{a_{2_i}}e^{-\Delta E_{a_i}/R_gT}\prod_{j=1}^{n}(\frac{\rho_j}{M_j})^{\nu_{rj}} \,, \tag{2.43}$$

where a_{1_i} and a_{2_i} are constants, ΔE_{a_i} is the *molar activation energy* of the reaction involving species, and M_j is the molecular weight of species j. The product $a_1T^{a_2}$ is called the *pre-exponential* factor. For the *simple*, *one-step second-order* reaction between A and B, only one set of a_1, a_2, and ΔE_a are specified and (2.43) can be written as

$$-\dot{n}_A = \frac{\rho_A \rho_B}{M_B} a_1 T^{a_2} e^{-\Delta E_a/R_g T} , \qquad (2.44)$$

where $\dot{n}_A$ is given in kg/m^3-s of species i. Glassman (1987) provides a_1, a_2 and ΔE_a for the hydrogen and some hydrocarbon oxidations. When an species is involved in more than one reaction, then the production rate of this species for all these reactions are *added*.

2.3.2 Premixed Reactants

Because of the low collision frequency and the high activation energy, a mixture of fuel-oxidizer is basically *nonreactive* at room temperature. However, when the temperature at a location in the mixture is raised above a critical value (called the *ignition temperature*) by either *direct heating* (i.e., placement of an *ignitor*), by *external radiation*, or by *compression*, a region of reaction (e.g., *combustion*) moves through the mixture at a propagation (or *flame*) velocity $\mathbf{u}_F$. The magnitude of this velocity can be lower or higher than the speed of sound (Williams, 1985). In most controlled combustion of interest involving the burning of hydrocarbons in air, the propagation is *subsonic* (i.e., the pressure change across the flame is not substantial and this is called a *deflagration wave* as compared to the explosive *detonation waves*). The propagation velocity is then controlled by heat and mass transport, including conduction and mass diffusion, as discussed below for the case of the laminar flow of premixed reactants. In a one-dimensional flow of premixed reactants, when the velocity of the flow (measured with respect to a stationary coordinate) is equal in magnitude but opposite in sign to the reaction-front (or flame) velocity, the flame appears stationary with respect to the stationary frame. If the premixed unreacted gas is stationary, then the flame will be moving into it at a speed u_F.

For a steady, one-dimensional, multicomponent compressible *laminar* flow, with no radiative heat flux and no viscous dissipation and negligible pressure gradient and buoyancy effects, the continuity, species, and energy equations are the simplified forms of (1.71), (1.72), and (1.84) with the momentum equation (1.76) trivial and not usable. These equations for a single-step *zeroth-order* reaction (taken for its simplicity) with a *stoichiometric mixture* of fuel and oxidizer (i.e., no fuel or oxidizer is left in the reacted region) are

$$\frac{d\rho u}{dx} = 0 , \qquad (2.45)$$

$$\rho u \frac{d}{dx}\frac{\rho_i}{\rho} = \frac{d}{dx}\rho D_i \frac{d}{dx}\frac{\rho_i}{\rho} + \dot{n}_i , \quad i = f, o, p , \qquad (2.46)$$

$$\rho_i u_i = \rho_i u - \rho D_i \frac{d}{dx}\frac{\rho_i}{\rho} , \qquad (2.47)$$

$$\rho c_p u \frac{dT}{dx} + \sum_i \rho_i c_{p_i}(u_i - u)\frac{dT}{dx} = \frac{d}{dx}k\frac{dT}{dx} - \Delta i_c \dot{n}_f \ , \tag{2.48}$$

$$\dot{n}_f = -Ae^{-\Delta E_a/R_g T} \ , \tag{2.49}$$

$$p = \frac{\rho R_g T}{M} \ , \tag{2.50}$$

where Δi_c is the heat of combustion per conversion of unit mass of the fuel (fuel is designated with subscript f, oxidizer with o and the products with p), and A is the pre-exponential factor. For zeroth-order reactions, the fuel consumption rate is independent of the fuel concentration. Although for the zeroth-order reactions the local species conservation equation is not used (except for the designation of downstream location where the fuel concentration becomes zero), we will treat the species equation as it also applies to higher-order reactions.

Even for this simple reaction, the simultaneous solution of the above equations, with ρ_i, c_{p_i}, k, and D_i varying with temperature, can only be obtained numerically. Since 1889 *approximations* have been made in order to obtain some closed-form solutions (Kuo, 1986). These solutions are not accurate, but instructive, and one of them is reviewed below.

Assuming that k and c_p (i.e., c_{p_i}) are constant and using a single mass diffusivity D and assuming that $D\rho$ is also constant and equal to k/c_p, we have

$$\frac{k}{c_p} = \rho D \ , \quad Le = \frac{Sc}{Pr} = \frac{k}{\rho c_p D} = 1 \ , \tag{2.51}$$

i.e., the *Lewis number* $Le = Pr/Sc$ is unity. As expected, depending on the species temperature and pressure, Le can be larger or smaller than unity. Using these and neglecting the heat transfer across the plane moving with the mass-averaged velocity, the governing differential equations become

$$\dot{m}\frac{d}{dx}\frac{\rho_f}{\rho} = \rho D\frac{d^2}{dx^2}\frac{\rho_f}{\rho} + \dot{n}_f \ , \tag{2.52}$$

$$\dot{m}c_p\frac{dT}{dx} = k\frac{d^2T}{dx^2} - \dot{n}_f\Delta i_c \ , \tag{2.53}$$

where $\dot{m} = \rho u$.

In 1938 Zel'dovich and Frank-Kamenetsky (Kuo, 1986) introduced the *dimensionless* variables using the conditions in the *nonreacted zone*

$$T^* = \frac{c_p(T - T_n)}{\Delta i_c} \ , \quad \rho_f^* = (\frac{\rho_f}{\rho})_n - \frac{\rho_f}{\rho} \ , \quad \frac{\rho}{\rho_n} = \frac{T}{T_n} \ , \tag{2.54}$$

so the above equations become

$$-\dot{m}\frac{d\rho_f^*}{dx} = -\rho D\frac{d^2\rho_f^*}{dx^2} + \dot{n}_f \ , \tag{2.55}$$

$$\frac{\dot{m}}{c_\rho}\frac{dT^*}{dx} = \frac{k}{c_p}\frac{d^2T^*}{dx^2} - \dot{n}_f \tag{2.56}$$

with the boundary conditions

$$T^* = 0\ , \quad \rho_f^* = 0 \qquad \text{for } x \to -\infty\ , \tag{2.57}$$

$$T^* = \frac{c_p(T_r - T_n)}{\Delta i_c}\ , \quad \rho_f^* = (\frac{\rho_f}{\rho})_n \qquad \text{for } x \to \infty\ . \tag{2.58}$$

The *downstream*, i.e., $x \to \infty$ conditions assume no fuel remains.

Since the differential equations are identical for $Le = 1$, then a *combined* variable is defined as $c_pT + \Delta i_c\rho_f/\rho$. The combined differential equation can be integrated to give

$$c_pT_n + \Delta i_c(\frac{\rho_f}{\rho})_n = c_pT_r = c_pT + \Delta i_c\frac{\rho_f}{\rho}\ , \tag{2.59}$$

i.e., the *sum* of the specific thermal and chemical energies is *uniform* across the reaction zone. This relates T and ρ_f anywhere, and therefore, by solving for T, ρ_f can be found.

Next, the reaction zone is divided into *two* idealized regions, as shown in Figure 2.2. In the upstream region, i.e., $-\infty < x \le 0$, the temperature is below the ignition temperature and *no* reaction occurs (this is called the *preheat* region) and in the downstream region, i.e., $0 \le x < \infty$, the *convection* is negligible and the reaction occurs (this is called the *energy generation* region). These are given by

$$\dot{m}c_p\frac{dT}{dx} = k\frac{d^2T}{dx^2}\ , \qquad -\infty < x \le 0\ , \tag{2.60}$$

$$0 = k\frac{d^2T}{dx^2} - \dot{n}_f\Delta i_c\ , \qquad 0 \le x < \infty\ . \tag{2.61}$$

The temperature at $x = 0$ is taken as the ignition temperature T_i and the conduction heat fluxes are matched at this location, i.e.,

$$T = T_i\ , \quad k\frac{dT}{dx}\bigg|_{0-} = k\frac{dT}{dx}\bigg|_{0+}\ . \tag{2.62}$$

Equation (2.60) and conditions $dT/dx = 0$ and $T = T_n$ for $x \to -\infty$ give

$$\frac{dT}{dx} = \frac{\dot{m}c_p(T - T_n)}{k}\ , \qquad -\infty < x < 0\ . \tag{2.63}$$

Equation (2.61) can be written as

$$\frac{d}{dx}(\frac{dT}{dx})^2 = \frac{2\dot{n}_f\Delta i_c}{k}\frac{dT}{dx}$$

or

$$\left(\frac{dT}{dx}\bigg|_{0+}\right)^2 = -\frac{2\Delta i_c}{k}\int_{T_n}^{T_r}\dot{n}_f\,dT\ , \tag{2.64}$$

where it has been recognized that $\dot{n}_f = 0$ for $x < 0$, i.e., $T < T_i$. Now matching the heat fluxes at $x = 0$, using $\dot{m} = u_F\rho_n$, $T_i - T_n \simeq T_r - T_n$ and

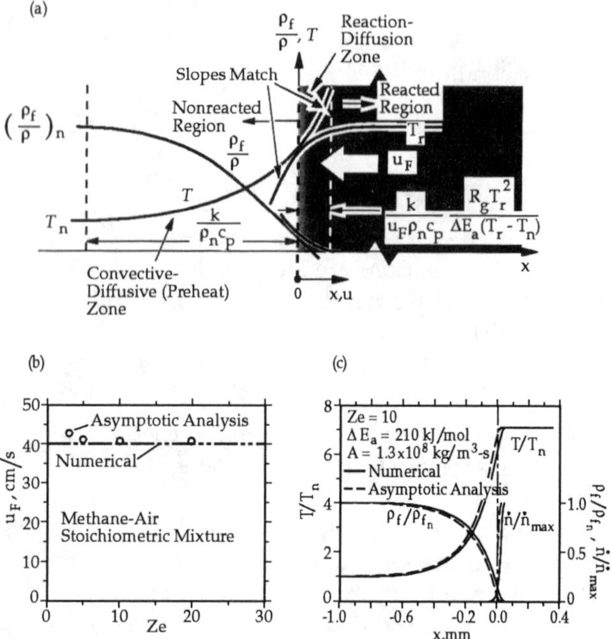

Figure 2.2. (a) Structure of a premixed flame showing the various regions and zones. (b) The predicted flame speed, for stoichiometric methane-air mixture, as a function of the Zel'devich number. (c) Structure for the stoichiometric methane-air premixed flame.

(2.59), we have

$$\frac{\dot{m}c_p(T_r - T_n)}{k} = \frac{u_F\rho_n c_p(T_r - T_n)}{k}$$

$$\frac{u_F\Delta i_c\rho_{fn}}{k} = \left(\frac{2\Delta i_c}{k}\int_{T_n}^{T_r} -\dot{n}_f\, dT\right)^{\frac{1}{2}}$$

or

$$u_F = \left[\frac{2k}{\rho_n c_p(T_r - T_n)}\right]^{1/2}\left(\frac{1}{\rho_{fn}}\int_{T_n}^{T_r} -\dot{n}_f\, dT\right)^{1/2}. \tag{2.65}$$

The integration can be simplified by using the approximations

$$e^{-\Delta E_a/R_gT} \simeq e^{-\Delta E_a/R_gT_r}\, e^{-\Delta E_a(T_r-T)/R_gT_r^2} \tag{2.66}$$

and

$$\int_0^{\frac{\Delta E_a(T_r-T_n)}{R_gT_r^2}} e^{-\frac{\Delta E_a(T_r-T)}{R_gT_r^2}}\, d\frac{\Delta E_a(T_r-T)}{R_gT_r^2} = 1 \tag{2.67}$$

which are valid for small $(T_r - T)/T_r$ and large *Zel'dovich number* $Ze = \Delta E_a(T_r - T_n)/R_gT_r^2$, respectively. These are the *high ignition temperature* and *high activation energy asymptotes* or *approximations* (Williams, 1985).

Then the flame speed is given by

$$u_F = \left(\frac{2k}{\rho_n c_p \rho_{fn}} \frac{Ae^{-\Delta E_a/R_g T_r}}{T_r - T_n} \frac{R_g T_r^2}{\Delta E_a}\right)^{1/2} . \tag{2.68}$$

Since ρ_n and ρ_{fn} are proportional to the pressure, the flame speed is *inversely* proportional to the pressure. The flame speed is also proportional to $k^{1/2}$ and depends on the temperature of the reacted gas *and* its difference with the unreacted temperature. Further discussions, including the extension to the first- and second-order reactions, are given by Glassman (1986). The front speed, which characterizes the premixed gaseous reaction, is *controlled* by the chemical kinetics as well as the mass diffusion and the heat conduction. In *nonpremixed* reactions the mass diffusion *dominates* (this will be discussed in the next section).

Using T_r as the temperature of the reacted and T_n for nonreacted mixture, the flame thickness δ_F can be *estimated* by approximating the balance of heat transport between conduction $k(T_r - T_n)/\delta$ and convection $\rho c_p u_F (T_r - T_n)$ across the reaction zone, i.e., assuming a Péclet number $u_F \delta_F/\alpha$ of unity. This gives

$$\delta_F = \frac{\alpha}{u_F} . \tag{2.69}$$

Under typical atmospheric conditions, for most hydrocarbons u_F=0.40 m/s and for $\alpha = 4 \times 10^{-4}$ m^2/s (approximately for gases around 1300 K), δ_F is about 1 mm. This is much larger than the mean free path which is of the order of 10^{-4} mm. Note that α charges rapidly with temperature and using an average temperature between T_r and T_n is only a rough approximation. More appropriately, by using the proper equations, (2.60) and (2.61), the thickness of the *preheat* zone is nearly $k/u_F \rho_n c_p$ and that for the *heat generation* zone is $(k/u_F \rho_n c_p) R_g T_r^2/\Delta E_a (T_r - T_n)$. The heat generation zone is much smaller than the preheat zone (Williams, 1985).

For the combustion of the premixed methane and air with a zeroth-order reaction, the predicted burning velocity u_F, (2.68), is shown in Figure 2.2(b). The pre-exponential factor and the activation energy are determined by solving (2.53) numerically (Hanamura, 1994) requiring that u_F be equal to 40 cm/s as found experimentally. Note that A and ΔE_a are not unique (i.e., Ze is not unique). The results of the asymptotic analysis are also shown in Figure 2.2(b), where as expected the agreement is very good when $Ze \rightarrow \infty$. The thermophysical properties are evaluated at 1300 K. Then equations (2.67) and (2.68) are rather good approximations for the high activation energies. For $Ze = 10$, a comparison of the flame structures, obtained numerically and from the asymptotical analysis, is given in Figure 2.2(c). The inflection point in the temperature distribution is chosen as the location for $x = 0$, which separates the preheat and reaction zones. In the asymptotic analysis, (2.61) is integrated from an arbitrary location to infinity. By using the assumption given by (2.66), we have

$$\left(\frac{dT}{dx}\right)^2 = \frac{2\Delta i_c}{k} A e^{-\Delta E_a/R_g T_r} \frac{R_g T_r^2}{\Delta E_a}[1 - e^{\frac{-\Delta E_a(T_r - T)}{R_g T_r^2}}] \ . \quad (2.70)$$

This is then solved numerically. In the preheat zone, (2.60) is integrated subject to the proper boundary conditions, and the solution is

$$T = \frac{dT}{dx}|_{x=0} \frac{\lambda}{\dot{m} c_p} e^{\frac{\dot{m} c_p}{k} x} + T_n \ . \quad (2.71)$$

Then by matching the temperatures at $x = 0$, the temperature distribution is obtained over both zones. The distribution of the normalized fuel concentration is made by assuming a unity Lewis number. The distribution of the normalized reaction rate is also shown in Figure 2.2(c). As is evident, the asymptotic analysis predicts the ditstributions rather accurately.

2.3.3 Diffusing Reactants

The fuel and oxidizer can be provided as two *separate* gas *streams*. Then under steady-state conditions, they have to *diffuse* through the products in order to react. The streams can be flowing at various *angles* with each other (e.g. *perpendicular, slanted, parallel,* or *opposing*). In the parallel flow arrangement, the two streams flowing at two different velocities will share a mixing layer. The mixing layer contains the flow instability resulting from the difference in velocity, density and viscosity between the two streams (i.e., *Kelvin-Helmholtz instability*), and the buoyant effect of the chemical reaction. The chemically reacting turbulent mixing layer has been studied numerically or experimentally by Riley et al. (1986), Hemanson and Dimotakis (1989), and Grinstein and Kailasanath (1992).

Assuming that the diffusion time (for the diffusing of the fuel and the oxidizer towards each other) is much *larger* than the reaction time, then an infinitely fast reaction is assumed. This removal of the reaction rate from the analysis simplifies it significantly. This is referred to as the *large Damköhler-number* (ratio of diffusion time to the reaction time) *asymptote* (Williams, 1985).

The case of *coaxial* parallel streams (fuel flowing on the inside and the oxidizer flowing in an annulus surrounding it (i.e., *coaxial jets*), with both streams being *steady* and *laminar*, has been examined and a closed-form solution for the axisymmetric diffusion *flame shape* has been found (Kuo, 1986; Lewis and von Elbe, 1987, Glassman, 1987). For simplicity, the effect of buoyancy is *not* included and the interplay between the *lateral diffusion* of species and their *axial flow* determines the concentration fields (i.e., the flame shape). The inclusion of a finite chemical kinetics, and other complexities, requires numerical integration of the governing equations. An example of such computations is given by Prasad and Price (1992).

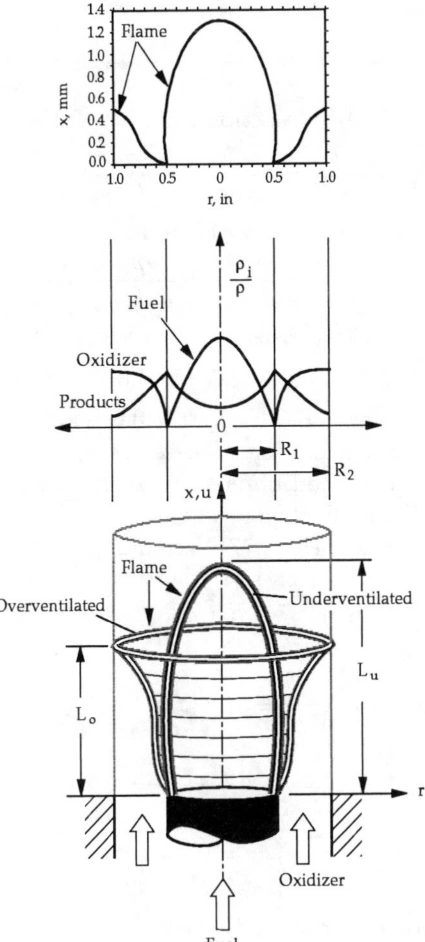

Figure 2.3. An idealized diffusion flame with the fuel flowing inside and the oxidizer in the enclosing annulus. Both underventilated and overventilated flames are shown. (From Lewis and von Elbe, reproduced by permission ©1986 Academic Press.)

Figure 2.3 gives a schematic of the flow arrangement, typical concentration, and temperature distributions at an axial location, and the location of the flames (where the concentrations of the fuel and oxidizer are zero) for a specific set of parameters (Lewis and von Elbe, 1987).

The mass and species conservation equations can be written as

$$\nabla \cdot \rho \mathbf{u} = 0 \,, \tag{2.72}$$

$$\nabla \cdot (\rho \mathbf{u} \frac{\rho_i}{\rho} - \rho D \nabla \frac{\rho_i}{\rho}) = \dot{n}_i \,, \quad i = f, o, p \,. \tag{2.73}$$

By dividing the latter by $M_i(\nu_{pi} - \nu_{ri})$, we will have

$$\nabla \cdot [\rho \mathbf{u} \frac{\rho_i}{\rho M_i(\nu_{pi} - \nu_{ri})} - \rho D \nabla \frac{\rho_i}{\rho M_i(\nu_{pi} - \nu_{ri})}] = \frac{\dot{n}_i}{M_i(\nu_{pi} - \nu_{ri})} \ . \tag{2.74}$$

Note that from the chemical reaction equation (2.15)

$$\frac{\dot{n}_f}{M_f \nu_{rf}} = \frac{\dot{n}_o}{M_o \nu_{ro}} = -\frac{\dot{n}_p}{M_p \nu_p} \ . \tag{2.75}$$

Then by defining (originated by *Shvab* and *Zel'dovich*)

$$X = -\frac{\rho_f}{\rho M_f \nu_{rf}} + \frac{\rho_o}{\rho M_o \nu_{ro}} \tag{2.76}$$

and adding (2.49) for fuel and oxidizer, the result is

$$\nabla \cdot (\rho \mathbf{u} - \rho D \nabla) X = 0 \ . \tag{2.77}$$

The existing closed-form solution for the flame location is for an idealized velocity field and other conditions. These are *equal* and *uniform* velocity u across the inner (fuel) and outer (oxidizer) axial flows (no radial flow $v = 0$), constant ρD, and negligible axial (x-direction) diffusion (compared to the radial direction). Then (2.77) becomes

$$\frac{u}{D} \frac{\partial X}{\partial x} - \frac{1}{r} \frac{\partial}{\partial r} r \frac{\partial X}{\partial r} = 0 \ . \tag{2.78}$$

The boundary conditions are

$$X = -(\frac{\rho_f}{\rho})_{x=0} \frac{1}{M_f \nu_{rf}} \ , \qquad x = 0 \ , \ \ 0 \le r \le R_1, \tag{2.79}$$

$$X = (\frac{\rho_o}{\rho})_{x=0} \frac{1}{M_o \nu_{ro}} \ , \qquad x = 0 \ , \ \ R_1 \le r \le R_2, \tag{2.80}$$

$$\frac{\partial X}{\partial r} = 0 \quad \begin{cases} r = 0, & x > 0 \ , \\ r = R_2, & x > 0 \ . \end{cases} \tag{2.81}$$

Using the separation of variables, the solution is (Kuo, 1986)

$$(\frac{\rho_f}{\rho} - \frac{\rho_o}{\rho} \frac{M_f \nu_{rf}}{M_o \nu_{ro}})(\frac{\rho}{\rho_f})_{x=0} = -(1+a)\frac{R_1^2}{R_2^2} + a -$$

$$2(1+a)\frac{R_1}{R_2} \sum_{n=1}^{\infty} \frac{J_1(\frac{R_1}{R_2}\phi_n)}{\phi_n J_o^2(\phi_n)} J_o(\phi_n \frac{r}{R_2}) \, e^{-(\phi_n^2 D x)/(u R_2^2)} \ , \tag{2.82}$$

where

$$a = (\frac{\rho_o}{\rho_f})_{x=0} \frac{M_f \nu_{rf}}{M_o \nu_{ro}} \ . \tag{2.83}$$

ϕ_n's represent the zeros of the Bessel function of the first kind of the order one $J_1(\phi)$.

Consider a single-step stoichiometric reaction of a fuel and an oxidizer resulting in a product (all gaseous) given by simplified form of (2.42) as

$$\nu_{rf} F + \nu_{ro} O \longrightarrow \nu_p P \ . \tag{2.84}$$

Since under steady flows the fuel and the reactant must diffuse through their common interface, this *diffusion* of the constituents becomes more *limiting* than the temperature dependence of their reaction. Then for the *diffusion-controlled* reactions, the temperature dependence is neglected and if constant properties are also assumed, then the mass and the heat transfers are decoupled. When the velocity field is also simplified and *prescribed*, the species equation can be solved for the concentration distribution. For a stoichiometric mixture, the *flame location* is where the concentrations of the fuel and the oxidizer is zero (and the concentration of the products is the maximum).

For equal fuel and oxidizer velocities, the ratio of the volumetric oxidizer flow to the fuel flow is $(R_2^2 - R_1^2)/R_1^2$. Depending on this ratio and the species molecular weights and densities, there is a *critical* oxidizer flow rate below which the flame is *underventilated* (the flame *height* L_u will be at $r = 0$) and above it the flame is *overventilated* (the flame height L_o will be at $r = R_2$).

At the flame location we have $X = 0$ and the *leading* term of the series given in (2.82) results in a reasonable approximation. For example, the *flame height* L_u for an *underventilated* (i.e., at $r = 0$) flame is found from the first term and is

$$\frac{DL_u}{uR_2^2} = \frac{1}{\phi_1}\ln\frac{2(1+a)\frac{R_1}{R_2}J_1(\frac{R_1}{R_2}\phi_1)}{[a-(1+a)\frac{R_1^2}{R_2^2}]\phi_1 J_o(\phi_1)} . \tag{2.85}$$

Typical predicted flame shapes, i.e., $x = x(r, X = 0)$, are shown in Figure 2.3. These flame shapes have been confirmed experimentally (Lewis and von Elbe, 1986), when the flows are made as close as possible to the idealized conditions assumed in the analysis. A self-similar solution for the *flow field* (away from the entrance) and for the *limits* of $R_2 \to \infty$ and $R_1 \to 0$ and in the case of an overventilated flame, has been obtained by Mahalingam *et al.* (1990a). The effects of *buoyancy, variable properties, turbulence, jets exit conditions*, and the *time dependence* and *three-dimensionality*, must be included for realistic descriptions. Then the governing equations are solved numerically (e.g., Lockwood and Naguib, 1975; Mahalingam *et al.*, 1990b; Grinstein and Kailasanath, 1992; Prasad and Price, 1992; Katta and Roquemore, 1993). The study of Katta and Roquemore (1993) for *vertical* fuel jets, in an otherwise quiescent oxidizer ambient, have unveiled the structure of large-scale buoyancy-driven *vortices outside* the flame surface and smaller Kelvin-Helmholtz vortices in the shear layer of the fuel *inside* the flame structure.

2.4 Turbulence

Turbulence in fluid flow refers to the *time-dependent* velocity fields with the time variations being *nonlinear* superpositions of fluctuations having a very large range of *frequencies* (over several orders of magnitudes). In considering *intramedium* transport, i.e., the turbulent transport *within* a single-phase flow (such as jet plumes and wakes), the attention is on *free-shear* of *boundary-free turbulence*, i.e., the turbulence which is not affected by the bounding or submerged solid surfaces. For these flows, the transition from a laminar flow to a turbulent flow appears at rather *low* Reynolds numbers and is generally caused by an inviscid instability mechanism (in contrast to solid-surface-affected flows). Bejan (1984) discusses the instability of inviscid flows. Reviews of boundary-free turbulent flows are given by Tennekes and Lumley (1972), Schlichting (1979), Batchelor (1982), Cebeci and Bradshaw (1984) and White (1991).

The *study* of *turbulence* can be made from a purely *statistical* view point in which the evolution of the averaged quantities (i.e., characteristics) of the flow are described. This method is originated by the *cascading* phenomena introduced by Taylor and Kolmogorov. Another view point is *deterministic* and seeks coherent structures and behavior (discussions of these views are given by Lesieur, 1986; Landahl and Mollo-Christensen 1987; Sherman, 1990; Sirovich, 1991). The study of turbulence, under the conditions of *local dynamical equilibrium*, where the local input of energy balances the losses and the *initial conditions* are not significant, can be made using the *local scales*. Tennekes and Lumley (1972) discuss these length, velocity, and time scales.

The analysis of turbulent flow and heat transfer can be made by the direct simulation of the velocity, pressure (and if needed density), and temperature fields. This can be done either by using the *highest* time and space resolutions possible, or by making some assumptions about the smaller time- and space-scale phenomena. Alternatively, the time-averaged conservation equations are used. Chart 2.1 gives a classification of methods of turbulent transport simulations. A review of *modeling* of *turbulence* and the computation of the turbulent transport are given by Bradshaw (1978) and by Cebeci and Bradshaw (1984), who take a *practical, approximate,* and *incomplete* approach in order to *predict* the turbulent momentum, heat and mass transfer. Modeling of the turbulent reacting flows (e.g., premixed turbulent combustion) based on the *flamelet structures*, has been discussed by Williams (1985), Kerstein (1992), and Peters (1992). The direct simulation of premixed turbulent flame has been made by Rutland and Trouvé (1993) and the effect of the wrinkles in the flame front on the flame speed for nonunity Lewis number has been analyzed.

In the following, the time averaging of the conservation equations, the modeling of the effect of turbulence on the transport of *mean* momentum

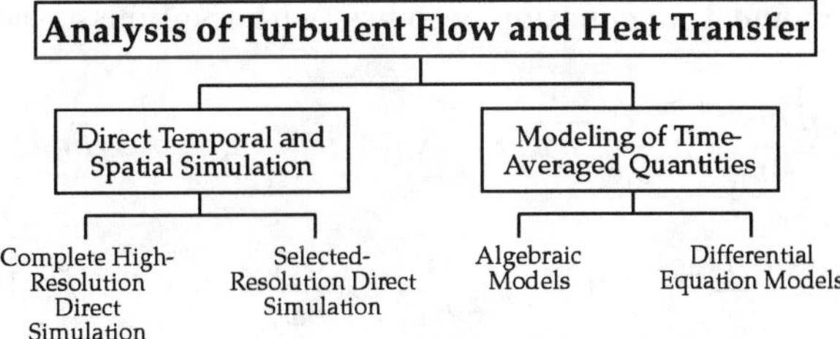

Chart 2.1. A classification of methods of turbulent transport simulations.

and *mean* energy, and some of the boundary-free turbulent flows, will be discussed.

2.4.1 Time-Averaged Conservation Equations

The unsteady continuity, momentum, and energy equations described the instantaneous variations of density, velocity, pressure, and temperature. As we discussed, these time variations in turbulent flows span in time scale over several orders of magnitude. As an alternative to *directly* solving the time-dependent conservation equations, time-averaged conservation equations are used.

The equations of motion (i.e., the mass and momentum conservation) for turbulent flows have been derived by Hinze (1975) and Monin and Yaglom (1979, 1981), for *incompressible Newtonian* fluids. These are obtained by taking the time average (over a time period which in general tends to infinity) of the conservation equations. In order to determine the effects of the velocity and pressure fluctuations on the *mean* (time-averaged) velocity, the magnitude of the time averages of various products of these fluctuations are needed. This leads to a *nonclosure*. In order to overcome this, these time-averaged products are then described using some *closure conditions* (or *statements*) which relate those to the properties of the mean flow.

The *time average* of the quantity ψ over a period τ is defined as

$$\overline{\psi} = \frac{1}{\tau}\int_t^{t+\tau} \psi \, dt \,, \tag{2.86}$$

and this is referred to as *mean* value of quantity ψ. For incompressible flows the variables of interest are decomposed into time-averaged (i.e., mean) and fluctuating components as

$$u_i = \overline{u}_i + u_i' \,, \quad p = \overline{p} + p' \,, \quad T = \overline{T} + T' \,. \tag{2.87}$$

Then the continuity equations become

$$\frac{\partial \overline{u}_i}{\partial x_i} = 0 \,, \quad \frac{\partial u_i'}{\partial x_i} = 0 \,. \tag{2.88}$$

The *mean* momentum equation becomes (with the Boussinesq approximation)

$$\rho_o \frac{D\overline{u}_i}{Dt} + \rho_o \frac{\partial}{\partial x_j}\overline{u_i' u_j'} = -\nabla \overline{p} + \mu_o \nabla^2 \overline{u}_i - \rho_o \beta_o \overline{T} g_i \tag{2.89}$$

or

$$\frac{D\overline{u}_i}{Dt} = -\frac{1}{\rho_o}\nabla \overline{p} - \beta_o \overline{T}\, g_i + \frac{\partial}{\partial x_j}[\nu_o(\frac{\partial \overline{u}_i}{\partial x_j} + \frac{\partial \overline{u}_j}{\partial x_i}) - \overline{u_i' u_j'}]\,, \tag{2.90}$$

where $\overline{u_i' u_j'}$ are called the *Reynolds shear stresses.*

The turbulent Reynolds stresses $-\overline{u_i' u_j'}$, are determined from the equations obtained for the time-averaged products of the velocity fluctuations. Boussinesq introduced the *eddy viscosity* ν_t (a scalar, where in principle it should be a tensor) as a *closure* condition, i.e.,

$$-\overline{u_i' u_j'} \equiv \nu_t(\frac{\partial \overline{u}_i}{\partial x_j} + \frac{\partial \overline{u}_j}{\partial x_i})\,. \tag{2.91}$$

Then we have

$$\frac{D\overline{u}_i}{Dt} = -\frac{1}{\rho_o}\nabla \overline{p} - \beta_o \overline{T} g_i + \frac{\partial}{\partial x_i}(\nu_o + \nu_t)(\frac{\partial \overline{u}_i}{\partial x_j} + \frac{\partial \overline{u}_j}{\partial x_i})\,. \tag{2.92}$$

The equation for the *kinetic energy* of *turbulence* (called the *k-equation* is found by multiplying the momentum equation by u_i' and then taking the time average and by using the continuity equation. The result is (White, 1991)

$$\frac{1}{2}\frac{D\overline{u_i' u_i'}}{Dt} + \frac{\partial}{\partial x_j}\overline{u_j'(\frac{p'}{\rho_o} + \frac{u_i' u_i'}{2})} + \overline{u_i' u_j'}\frac{\partial \overline{u}_j}{\partial x_i} + \beta_o \overline{u_i' T}\, g_i =$$
$$\nu_o \frac{\partial}{\partial x_i}\overline{u_j'(\frac{\partial u_i'}{\partial x_j} + \frac{\partial u_j'}{\partial x_i})} - \nu_o \overline{(\frac{\partial u_i'}{\partial x_j} + \frac{\partial u_j'}{\partial x_i})\frac{\partial u_j'}{\partial x_i}}\,. \tag{2.93}$$

The *negative* of the last two terms on the left-hand side are called the *production rates* of turbulence P_k, and the last term on the right-hand side is called the *viscous dissipation rate* of turbulence ϵ_k. The turbulent kinetic energy along with another quantity, such as the *turbulent mixing length* ℓ, is generally used in the determination of ν_t.

The *mean energy* equation for an *incompressible* fluid ($c_p = c_v$), no viscous dissipation, no heat generation, and no radiation heat flux, is found using the decomposition (2.87) and the energy equation (1.79) and these lead to

$$\rho_o c_{p_o}\frac{D\overline{T}}{Dt} = \frac{\partial}{\partial x_j}(k\frac{\partial \overline{T}}{\partial x_i} - \rho_o c_{p_o}\overline{u_i' T'})\,. \tag{2.94}$$

Introducing the *eddy diffusivity* α_t (a scalar, where again in principle it should be a tensor) as closure condition, we have for the *turbulent heat flux*

$$-\overline{u_i' T'} \equiv \alpha_t \frac{\partial \overline{T}}{\partial x_i}\,. \tag{2.95}$$

The mean energy equation becomes

$$\frac{D\overline{T}}{Dt} = \frac{\partial}{\partial x_i}(\alpha_o + \alpha_t)\frac{\partial \overline{T}}{\partial x_i} \ . \tag{2.96}$$

The eddy diffusivity is generally related to the eddy viscosity by a *turbulent Prandtl number* Pr_t through

$$Pr_t = \frac{\nu_t}{\alpha_t} \ . \tag{2.97}$$

Equations similar to (2.93) can be written for the *transport* of $\overline{u_i'u_j'}$ and $\overline{u_i'T'}$ and they will contain terms with *triple* products of the fluctuation components (e.g., $\overline{u_i'u_j'u_k'}$).

2.4.2 Modeling of Eddy Viscosity and Diffusivity

The momentum equation contains the terms $\overline{u_i'u_i'}$ (i.e., Reynolds shear stresses) and the energy equation contains the terms $\overline{u_i'T'}$ (i.e., turbulent heat flux) which all need to be related to the mean quantities. This relations are central in *turbulence modeling*. Reviews of these models, as related to convection heat transfer, are given by Cebeci and Bradshaw (1984), Launder (1988) and Pletcher (1988), among others, and the effect of buoyancy is discussed by Launder (1975) and Ljuboja and Rodi (1981).

As is pointed out in Chart 2.1, the models can be classified as the *algebraic model* (which uses a prescribed distribution of the mixing length ℓ), the *one-equation model* (which uses a modified turbulent kinetic energy equation along with a prescribed ℓ), the *two-equation model* (which uses the turbulent kinetic energy equation along with a transport equation for ℓ or a transport equation for the turbulent viscous dissipation ϵ_k; this is called the k-ϵ_k model with k standing for the turbulent kinetic energy), and the *Reynolds shear stress model* (which uses elaborate transport equations for the Reynolds shear stresses).

In 1925, Prandtl introduced a mixing length for turbulence ℓ, as a distance in which an eddy (envisioned as a small vortex) can keep its identity (e.g., the heat content or the momentum or kinetic energy) as it moves in a turbulent flow field. The *Prandtl, algebraic* model for eddy viscosity

$$\nu_t = \ell^2[(\frac{\partial \overline{u}_i}{\partial x_j} + \frac{\partial \overline{u}_j}{\partial x_i})\frac{\partial \overline{u}_i}{\partial x_j}]^{1/2} \ . \tag{2.98}$$

This was later modified to the *Prandtl-Kolmogrov model*, using the turbulent kinetic energy, i.e.,

$$\nu_t = c_\nu \ell (\frac{\overline{u_i'u_i'}}{2})^{1/2} \ , \tag{2.99}$$

where c_ν is a constant. As was mentioned, the eddy diffusivity is determined using ν_t and Pr_t where Pr_t is of the nearly unity (Launder, 1978).

The turbulent kinetic energy (2.93) can be *approximated* as

$$\frac{1}{2}\frac{D\overline{u_i'u_i'}}{Dt} = \frac{\partial}{\partial x_j}(\nu_o + \nu_k)\frac{\partial}{\partial x_j}\frac{\overline{u_i'u_i'}}{2} + P_k - \epsilon_k \ , \tag{2.100}$$

where ν_k is the corresponding eddy for the turbulent kinetic energy and is generally related to ν_t through a constant. In the *one-equation* model a relation is given between ϵ_k and the turbulent kinetic energy and ℓ. An example is (Ljuboja and Rodi, 1981) the *algebraic model* for dissipation rate of turbulence

$$\epsilon_k = \frac{c_\epsilon}{\ell}(\frac{\overline{u_i'u_i'}}{2})^{3/2} \ , \tag{2.101}$$

where c_ϵ is a constant.

For *equilibrium* turbulence, the production and the dissipation in (2.100) are balanced, and therefore, further relationships can be developed from this equilibrium condition. Note that the production is defined as

$$P_k = -\overline{u_i'u_j'}\frac{\partial \overline{u}_i}{\partial x_j} - \beta_o\overline{u_i'T'}g_i \ , \tag{2.102}$$

and these are in turn given in terms of ν_t and α_t as in (2.99) and (2.101).

The dissipation rate of the turbulent kinetic energy and the equilibrium condition can be used for local scale studies. When the turbulence is also *isotropic* (i.e., $\overline{u_1'^2} = \overline{u_2'^2} = \overline{u_3'^2}$), then the smallest turbulent *length* ℓ_t, *time* t_t and *velocity* u_t scales (called the *Kolmogrov microscales*) are given by (Tennekes and Lumley, 1972)

$$\ell_t = (\frac{\nu_o^3}{\epsilon_k})^{1/4} \ , \quad t_t = (\frac{\nu_o}{\epsilon_k})^{1/2} \ , \quad u_t = (\nu_o\epsilon_k)^{1/4} \ . \tag{2.103}$$

2.4.3 Boundary-Free Turbulent Flows

As intramedium turbulent transports, the fluid flow and heat transfer in turbulent flows *away* from bounding or submerged surfaces have been studied under the *boundary-free turbulent flows*. For *forced flows*, examples are *wakes* behind objects, *submerged jets* (i.e., a liquid-jet into a liquid or a gas-jet into a gas), and *shear flows* (two streams flowing at an angle to each other). In some cases the *buoyancy also* becomes important, such as in *buoyant jets, thermals,* and *plumes* (plumes are jets driven by buoyancy only) with cross (i.e., at any angle) forced flow. An example of a *purely* buoyant flow is a plume in an otherwise quiescent ambient. The heat transfer aspects of the boundary-free flows (turbulent and laminar) will be discussed in Section 2.8. Here an example of *forced* boundary-free turbulent flows (two-dimensional examples are shown in Figure 2.4), i.e., the *plane* turbulent jet in the *far-field* is considered to illustrate a use of the turbulent mixing length.

The analysis of the spreading of a plane jet (where the far-field entrainment is significant) using the most simple of the turbulence models, i.e.,

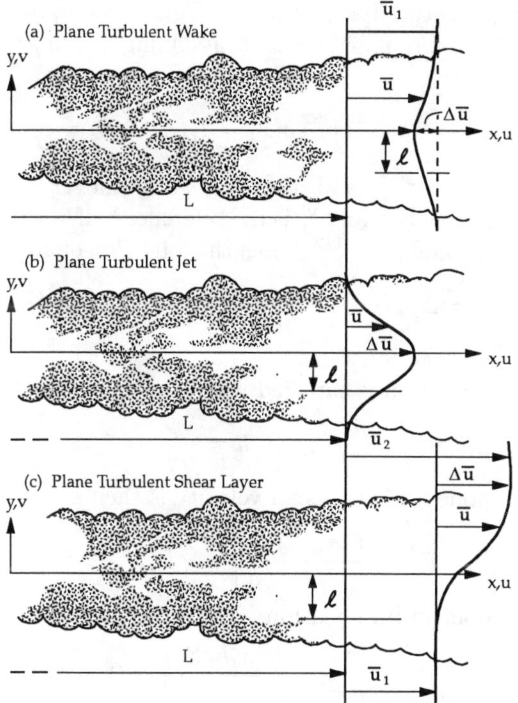

Figure 2.4. Examples of forced turbulent boundary-free flows. (From Tennekes and Lumley, reproduced by permission ©1972 MIT Press.)

the prescribed mixing length, was done initially by Tollmien in 1926 and in 1942 by Reichardt and by Görtler and then reviewed by Schlichting (1979). In the *far-field* region, the *centerline* mean velocity $\Delta\overline{u}$ *decreases* along the jet axis x (with the y-axis perpendicular to it). In the *near-field* region, which will be discussed in Section 2.7.2, the centerline velocity *increases* and the entrainment of the surrounding quiescent fluid is *not* significant. The continuity and mean momentum equations, for this *nearly parallel* flow are

$$\frac{\partial \overline{u}}{\partial x} + \frac{\partial \overline{v}}{\partial y} = 0\,, \tag{2.104}$$

$$\overline{u}\frac{\partial \overline{u}}{\partial x} + \overline{v}\frac{\partial \overline{u}}{\partial y} = \frac{\partial}{\partial y}(\nu_o + \nu_t)\frac{\partial \overline{u}}{\partial y} \simeq \frac{\partial}{\partial y}\nu_t\frac{\partial \overline{u}}{\partial y}\,. \tag{2.105}$$

Using the turbulent mixing length ℓ as a *fraction* of the *half-width* of the jet spread b, which varies along x, and approximating $d\overline{u}/dy$ by $\Delta\overline{u}/b$, the expression for eddy viscosity (2.98) becomes

$$\nu_t = c_t b \Delta\overline{u}\,. \tag{2.106}$$

A *similarity* solution for $\overline{u}$ exists if the half-width (and the mixing length) is assumed to *grow* as the *first* power of x. Furthermore, the *mean momentum flow* rate ρJ must remain constant, assuming a *similarity* in the mean velocity field, leads to

$$J_e = \int_{-\infty}^{\infty} \overline{u}^2 \, dy = a_1 \left[\Delta\overline{u}(x)\right]^2 b(x) = a_1 (\Delta\overline{u}_e)^2 b_e = \text{constant} \,, \quad (2.107)$$

where $\Delta\overline{u}_e$ is evaluated at a distance x_e from the *entrance* and is the reference centerline velocity and b_e is the reference half-width. This requires that $\Delta\overline{u}$ be proportional to $x^{-1/2}$. Then the following quantities are defined

$$\eta = \frac{a_2 y}{x} \,, \quad \overline{u} = \Delta\overline{u}_e (\frac{x_e}{x})^{1/2} \frac{df}{d\eta}(\eta) \,, \quad \nu_t = c_t \Delta\overline{u}_e b_e (\frac{x}{x_e})^{1/2} \,, \quad (2.108)$$

where a_2 is an arbitrary constant.

The continuity equation is satisfied with the *stream function* defined as

$$\psi = a_2^{-1} \Delta\overline{u}_e x_e^{1/2} x^{1/2} f(\eta) \,, \quad \overline{u} = \frac{\partial\psi}{\partial y} \,, \quad \overline{v} = -\frac{\partial\psi}{\partial x} \,. \quad (2.109)$$

The *normal* component of the mean velocity is then given by

$$\overline{v} = \frac{\Delta\overline{u}_e x_e^{1/2}}{a_2 x^{1/2}} (\eta \frac{df}{d\eta} - \frac{1}{2} f) \,. \quad (2.110)$$

Then the *mean* momentum equation (2.105) becomes

$$(\frac{df}{d\eta})^2 + f \frac{d^2 f}{d\eta^2} + \frac{2a_2^2 c_t b_e}{x_e} \frac{d^3 f}{d\eta^3} = 0 \,. \quad (2.111)$$

Then a_2 is chosen such that the coefficient of $d^3 f/d\eta^3$ is 1/2. When (2.111) is integrated twice and $f(0) = 0$, $df/d\eta(0) = 1$ and $df/d\eta(\infty) = 0$ are used we have

$$f^2 + (\frac{df}{d\eta})^2 = 1 \,, \quad a_2 = \frac{1}{2} (\frac{\Delta\overline{u}\, x}{\nu_t})_e^{1/2} = \frac{1}{2} (\frac{x}{c_t b})_e^{1/2} \,, \quad (2.112)$$

which gives

$$f = \tanh \eta \quad (2.113)$$

or

$$\frac{\overline{u}}{\Delta\overline{u}_o} = (1 - \tanh^2 \eta) \,. \quad (2.114)$$

Using x_e, b_e, and by fitting the data to the experimental results, the constant c_t (which is related to a_2 through the above equation) is determined (White, 1991). Figure 2.5 shows the experimental results, which gives $a_2 = 7.67$ and shows that $b \simeq 2y_{\frac{1}{2}}$, where $y_{\frac{1}{2}}$ is the *half-velocity* point. Then from (2.108) and (2.112) we have

$$\eta_b = \frac{a_2 b}{x} \simeq 1.76 \quad (2.115)$$

which gives $b/x = 0.23 = \tan\, 13°$, i.e., a divergence half-angle of 13°. The constant c_t is also found from (2.112), using a_2 and b/x; this gives $c_t = 0.018$. This value for c_t indicates that the turbulent mixing length is a small fraction of the half-width of the jet.

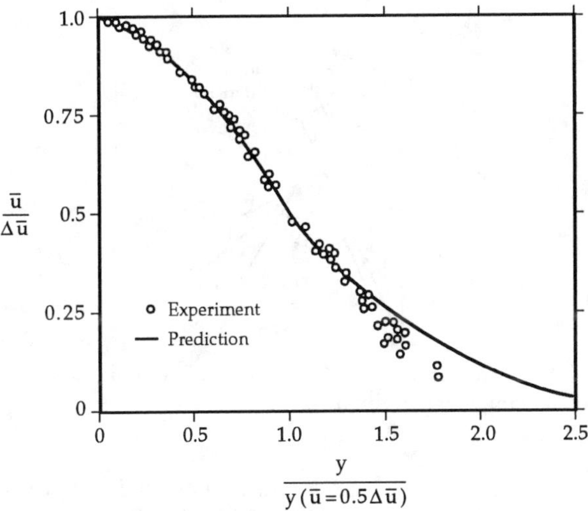

Figure 2.5. The experimental and predicted results for the lateral distribution of the mean velocity of a plane jet. (From White, reproduced by permission ©1974 McGraw-Hill.)

The *actual* (observed) size-range for the eddies is rather *continuous* and *varies* throughout the spread of the jet. Therefore, turbulent mixing lengths smaller than the half-width is expected, and a single mixing length indicating a single eddy size distributed across the jet is not supported by the observations. More *elaborate* turbulent models can be used for this problem and further evaluation of the constants introduced will be required. Recent detail measurements of the axisymmetric jets (Panchapakesan and Lumley, 1993) allows for the evaluation of the models. These include higher-order models which contain averages of the products of the fluctuating components.

2.5 Radiation

Thermal radiation heat transfer in semitransparent fluids occurs due to the local radiation *emission* (which is related to the blackbody emission as an ideal emitter), the local radiation *absorption*, and the local radiation *scattering* (including the *in-scattering* which accounts for radiation arriving as a result of scattering occurring in the surroundings). The fluid has emission, absorption and scattering *properties* which depend on its molecular structure (and the quantum energy states) and the intermolecular interactions. The properties are given in term of the *spectral* (i.e., wavelength dependent) *absorption coefficient* $\sigma_{\lambda a}$ (λ is the *wavelength*) and the *spectral scattering coefficient* $\sigma_{\lambda s}$. Under the condition *molecular local thermal equi-*

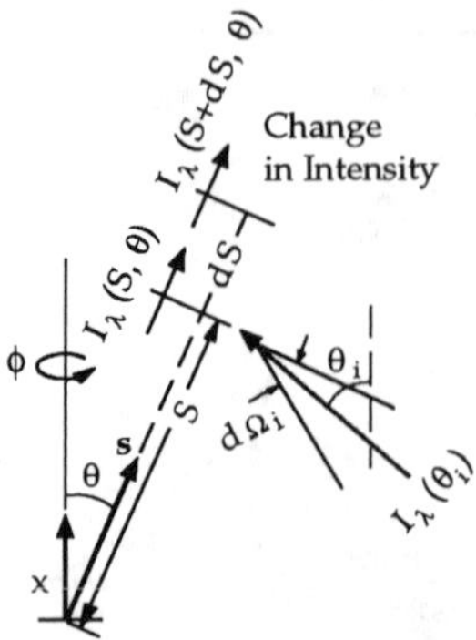

Figure 2.6. A schematic of the change in radiation intensity including in-scattering and showing the coordinates.

librium, the spectral *absorption* and *emission* coefficients are equal (Siegel and Howell, 1992; Brewster, 1992; Modest, 1993). The sum of the spectral absorption and scattering coefficients is called the *spectral extinction* coefficient $\sigma_{\lambda e}$. The scattering by the molecules is nearly *isotropic* (equally distributed in all directions θ around the molecule); however, as the particle size increases the scattering becomes anisotropic and its distribution is shown by the spectral *phase function* $\Phi_\lambda(\theta_i, \theta)$, where θ_i is the incident angle.

The *equation* of *radiative transfer* (a differentio-integral equation) describes the change in the *spectral intensity* I_λ as it travels along a *path* S (shown in Figure 2.6) and is (Siegel and Howell, 1992)

$$\frac{\partial I_\lambda(S)}{\partial S} = -\sigma_{\lambda a} I_\lambda(S) + \sigma_{\lambda a} I_{\lambda b}[T(S)] - \sigma_{\lambda s} I_\lambda(S) + \frac{\sigma_{\lambda s}}{2} \int_{-1}^{1} I_\lambda(S, \theta_i) \Phi_\lambda(\theta_i - \theta)\, d\cos\theta_i \tag{2.116}$$

the *radiative heat flux vector* and its divergence used in the energy equation (1.78) are related to the intensity as

$$\mathbf{q}_r = \int_0^\infty \int_{4\pi} \mathbf{s} I_\lambda \, d\Omega \, d\lambda \,, \tag{2.117}$$

$$\begin{aligned} \nabla \cdot \mathbf{q}_r &= \int_0^\infty \int_{4\pi} \nabla \cdot (\mathbf{s} I_\lambda) \, d\Omega \, d\lambda \\ &= 4\pi \int_0^\infty \sigma_{\lambda a} I_{\lambda b}[T(s)] \, d\lambda \\ &\quad -2\pi \int_0^\infty \sigma_{\lambda a} \int_{-1}^{1} I_\lambda(s, \mu) \, d\mu \, d\lambda \end{aligned} \tag{2.118}$$

where s is the unit vector along S, $d\Omega$ is the incremental *solid angle*, and $\mu = \cos\theta$.

The *diffusion* approximation is made whenever the *mean free path of the radiant energy* $1/(\sigma_a + \sigma_s)$ is small compared to the linear dimension of the system L, i.e., $1/(\sigma_a + \sigma_s)L \ll 1$. Then the *radiant conductivity* k_r is defined such that (Siegel and Howell, 1981)

$$\mathbf{q}_r = -k_r \nabla T = -\frac{16 n^2 \sigma_{SB} T^3}{3\overline{\sigma}_\ell} \nabla T, \tag{2.119}$$

$$\nabla \cdot \mathbf{q}_r = -\nabla \cdot k_r \nabla T = -\frac{16 n^2 \sigma_{SB}}{3\overline{\sigma}_e} \nabla \cdot T^3 \nabla T\,, \tag{2.120}$$

where the radiant conductivity is

$$k_r = \frac{16 n^2 \sigma_{SB} T^3}{3\overline{\sigma}_e} \tag{2.121}$$

and n is the *index of refraction*, σ_{SB} is the Stefan-Boltzmann constant, and $\overline{\sigma}_e$ is the *mean extinction coefficient* and is found by the proper spectral averaging. One example of this averaging will be discussed in Section 4.12.2. This is the *optically* thick limit of the radiative heat flux (Deissler, 1964).

The determination of $\nabla \cdot \mathbf{q}_{r\lambda}$ requires the prior knowledge of $\sigma_{\lambda a}$, $\sigma_{\lambda s}$ and Φ_λ and the solution to the equation of radiative transfer (2.116) which in turn requires the local temperature that is determined from the energy equation. Therefore, the equations for the radiative transfer and the conservation of the total energy are *coupled*. The *properties* and a *solution method* are discussed below.

2.5.1 Radiation Properties of Fluids

The *liquids* absorb *infrared* radiation (the range of radiation most significant in heat transfer) and have spectral absorption coefficients larger than that of the gases (at moderate pressures). Figure 2.7(a) shows the *measured* spectral absorption coefficient for *water* (liquid phase) at standard conditions and in the *near-infrared* wavelength range (Driscoll and Vaughan, 1978). The magnitude of $\sigma_{\lambda a}$ changes with the wavelength (and is not monotonic) from 0.01 to 100 cm^{-1}, i.e., the mean free path of the photons change from 100 at wavelength of 0.70 μm (visible) to 0.01 cm at 2.5 μm.

Absorption by *gases* has been reviewed by Tien (1968), Cess and Tiwari (1972), Edwards (1981), and Brewster, (1992). The *monatomic* gases (e.g., He, Ne, Ar) absorb in the wavelength range $\lambda < 1$ μm, i.e., *visible* and *ultraviolet*. The *symmetric diatomic* gases (e.g., N_2, O_2) act similarly, except at very high pressures. *Asymmetric* diatomic and *polyatomic* gases (e.g., CO, NO, CO_2, H_2O) have active absorption *bands* in the *infrared*. These absorptions are influenced by gas *pressure* and *temperature*. Figure 2.7(b) shows the *measured spectral absorptance* defined as $1-\exp(-L\,\sigma_{\lambda a})$, for CO_2 at 10 atm and 830 K (Siegel and Howell, 1992). The results are for $L = 39$ cm and $\sigma_{\lambda a}$ is in cm^{-1}. Note the strong absorption bands around

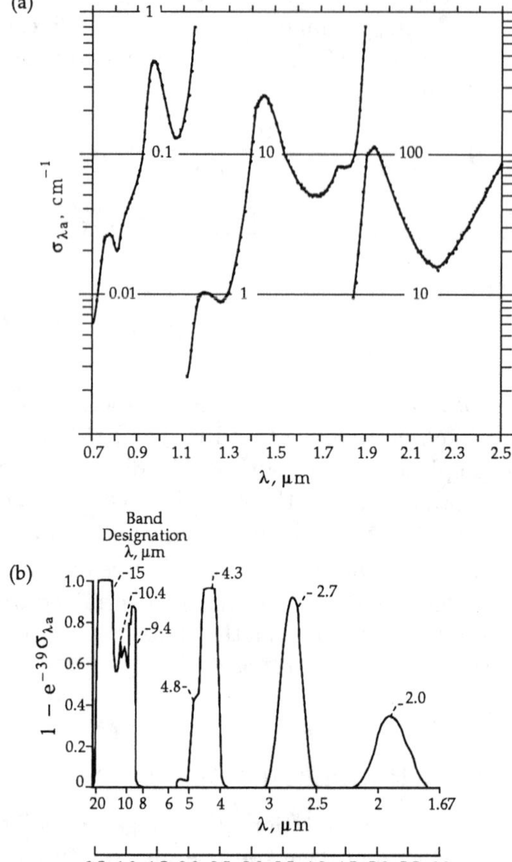

Figure 2.7. (a) Spectral distribution of the absorption coefficient for water (liquid) at standard conditions. (b) Spectral distribution of the absorptance for CO_2 at 830 K and 10 atm (the spectral absorption coefficient is in cm^{-1}). (From Driscoll and Vaughan, reproduced by permission ©1978 McGraw-Hill, and from Siegel and Howell, reproduced by permission ©1992 Taylor and Francis.)

15, 4.3, 2.7, and 2.0 μm. The *wave number* η, which is defined here as $1/\lambda$, is given in cm^{-1} and is related to the wavelength λ in μm through $\eta = 10^4/\lambda$. The infrared absorption of luminous flames has been measured by Hubbard and Tien (1978). The *molecular* descriptions of these bands are discussed by Tien (1968), Cess and Tiwari (1972), Edwards (1981), and Siegel and Howell (1992).

The presence of *particles* in the fluid influences the absorption, emission, and scattering of the thermal radiation. This will be discussed in connection with the solid-fluid systems (Chapters 3, 5, and 7).

2.5.2 Evaluation of Radiative Intensity

Closed-form solutions for I_λ, $\mathbf{q}_r$ and $\nabla \cdot \mathbf{q}_r$ are possible *only* under very simplified conditions. Therefore, these quantities are generally determined by solving the *coupled* equations for the radiative transfer and the energy equation, numerically. One method which maintains the physical features of the radiative transport is the discrete-ordinate method (Carlson and Lathrop, 1968; Fiveland, 1988; Modest, 1993). In this method the radiation field is divided into $2N$ *discrete streams* or *discrete ordinates* and the integral in the equation of radiative transfer is replaced by a summation. In one dimension this reduces the differentio-integral equation into $2N$ coupled ordinary linear differential equation for $I_{\lambda i}$. In three dimensions (Cartesian coordinate) the equation for the ith stream at a location (x, y, z) becomes

$$\mu_i \frac{\partial I_{\lambda i}}{\partial x} + \xi_i \frac{\partial I_{\lambda i}}{\partial y} + \gamma_i \frac{\partial I_{\lambda i}}{\partial z} = -\sigma_{\lambda a} I_{\lambda i} + \sigma_{\lambda a} I_{\lambda b} + \frac{\sigma_{\lambda s}}{4\pi} \sum_j \Delta\Omega_j I_{\lambda j} \Phi_\lambda(\theta_o) , \tag{2.122}$$

where μ_i, ξ_i and γ_i are the directional cosines of the ith coordinate and θ_o is given by

$$\cos \theta_o = \mu_i \mu_j + \xi_i \xi_j + \gamma_i \gamma_j . \tag{2.123}$$

Simplified calculations of gas emission and absorption can be made by using *space*- and *wavelength*-averaged (i.e., *mean* and *total*) emission and absorption coefficients. These are reviewed by Tien (1968).

Gas radiation can be significant in gaseous combustion. The effects of soot and larger particles will be discussed in connection with the single- and two-medium treatments of two-phase systems (Sections 3.3 and 5.3).

2.6 Electro- and Magnetohydrodynamics

Direct-current *electric* fields have been applied across small channels, with forced electrical conducting liquid flowing through them, in order to cause lateral motion and enhance the heat transfer (Jones, 1978; Fujino et. al., 1988). Electric fields have also been used to cause motion (i.e., onset of convection) in liquids under a stabilizing temperature field (Martin and Richardson, 1984). In these examples, the driving force is the *coulombic* (or *electrostatic*) force and the electrical conductivity of the liquid depends strongly on the temperature (as discussed in Section 1.12.4). The convective heat transfer is also influenced, by an applied *magnetic* field to the electrically conducting fluids (liquids and gases), and applications are in pumps, electrical power generators and accelerators, and aerodynamic heating of space- and aircrafts in ionized-gas environments (Romig, 1964). The magnetic field acting on a current carrying fluid results in the *pondermotive* force. The high-temperature ionized gas flows, as in plasmas in arc-welding or in the plasma combustion, are also influenced by the electromagnetic forces (e.g., Tsai and Kou, 1990; Zhao et al., 1990). Other applications are given by Branover et al. (1985).

In the following, the effect of the electrostatic force (in direct-current, dc, electric field applications) and the pondermotive and other forces (in the magnetic field applications) on the fluid flow will be discussed. Then the plasma (i.e., a charged, high-temperature gas) dynamics will be briefly discussed. As part of an *intraphase* transport, the focus is on phenomena not influenced by the bounding surfaces.

2.6.1 Electrohydrodynamics

Electrohydrodynamics deals with the interaction between an applied electrical force (assuming the induced magnetic field effects are negligible) and the motion of a conducting fluid. The *isothermal* aspects of electrohydrodynamics have been reviewed by Melcher and Taylor (1969) and the pertinent *electrodynamic* and *hydrodynamic* laws, the boundary conditions, and the approximations are discussed. Some experimental verifications are also given. For the application of a dc or ac electric field to *nonisothermal* liquids, Turnbull (1981), Jones (1978), and Martin and Richardson (1981) give the following conservation equations (Maxwell's equation for the electromagnetic fields, continuity, momentum, and energy equations) for an incompressible Newtonian-Boussinesq fluid (and by adding the *Joule* heating) as

$$\nabla \cdot \epsilon_e \epsilon_o \mathbf{e} = \rho_e \ , \qquad (2.124)$$

$$\nabla \times \mathbf{e} = 0 \ , \quad \mathbf{e} = -\nabla \varphi \ , \qquad (2.125)$$

$$\frac{\partial \rho_e}{\partial t} + \nabla \cdot \mathbf{j}_e = 0 \ , \qquad (2.126)$$

$$\mathbf{j}_e = \rho_e \mathbf{u} + \sigma_e \mathbf{e} \ , \qquad (2.127)$$

$$\nabla \cdot \mathbf{u} = 0 \ , \qquad (2.128)$$

$$\rho_o \frac{D\mathbf{u}}{Dt} = -\nabla(p - p_\infty) + \mu_o \nabla^2 \mathbf{u} - \rho_o \beta_o (T - T_\infty)\mathbf{g} + \mathbf{f}_e \ ,$$

$$\mathbf{f}_e = \rho_e \mathbf{e} - \frac{\epsilon_o}{2} \mathbf{e} \cdot \mathbf{e} \nabla \epsilon_e - \frac{\epsilon_o}{2} \nabla \left(\rho \left.\frac{\partial \epsilon_e}{\partial \rho}\right|_T \mathbf{e} \cdot \mathbf{e}\right) , \qquad (2.129)$$

$$\rho c_p \frac{DT}{Dt} = k \nabla^2 T + \sigma_e \mathbf{e} \cdot \mathbf{e} \ , \qquad (2.130)$$

where ϵ_e is the fluid *dielectric constant* (or the *relative permittivity*, i.e., the ratio of the *fluid permittivity* to the *free-space permittivity* $\epsilon_o = 8.8542 \times 10^{-12}$ A^2-s^2/N-m^2), ρ_e is the *fluid space-charge density*, σ_e is the *fluid electrical conductivity*, $\mathbf{e}$ is the *electric field vector*, φ is the *electric* (or *voltage*) *potential*, and $\mathbf{j}_e$ is the *current density* vector. The *units* for the symbols are given in the *nomenclature*. Since in the electrohydrodynamics the currents are assumed to be small, the magnetic field is also negligible. Equations (2.124) to (2.127) are the *Gauss law*, the *Faraday law*, the *charge conservation law*, and the *Ohm law*. For *direct current* applications, the last two terms in the volumetric electrical force f_e are negligibly small (because the changes of ϵ_e with temperature are much smaller than those for σ_e). The Joule heating is also very small for *poorly* conducting liquids ($\sigma_e < 10^{-9}$mho/m). Assuming a constant ϵ_e, the *charge relaxation* equation is obtained and is

$$\frac{D\rho_e}{Dt} + \frac{\rho_e \sigma_e}{\epsilon_e \epsilon_o} = 0 \,. \tag{2.131}$$

If the electric field is *alternating* at a frequency f, then for

$$f \gg \frac{\sigma_e}{\epsilon_e \epsilon_o} \tag{2.132}$$

no free electric charge can build up and $\rho_e \mathbf{e}$ in (2.129) will be negligible compared to the other terms in the electrical force.

The distribution of the free space-charge density $\rho_e(\mathbf{x})$ as given by (2.131) is influenced by the variation in σ_e caused by the temperature nonuniformity. In *electrothermal convection*, i.e., when the variation of σ_e with respect to T causes the motion, (this is called the *conductivity mode* or *model*), the empirical σ_e-T relations which were reviewed in Section 1.12.4 are used. The variation in ρ_e can also be caused by the injection of charge through the bounding electrodes and this is called the *mobility mode* or *model*. Since some charge always leaves the electrodes, the electrothermal convection generally is not *purely* in the conductivity mode (Fujino et al., 1989). Because of the uncertainties in the amount of charge injected (from one or both electrodes), at the present the role of the electrical conductivity alone on the fluid motion can not be isolated. The charge injection depends on the distance between the electrodes, among other system parameters (Fujino et al., 1989). These will be further discussed in section 4.13.1.

2.6.2 MAGNETOHYDRODYNAMICS

The distant electrostatic force due to an applied electric field intensity vector $\mathbf{e}$ acting on the free charge-space as given in (2.129) is now replaced by the distant force due to an applied electric and magnetic fields. Shercliff (1965) gives the details of the continuum treatment of magnetohydrodynamics. The *electromagnetic* force in *absence* of any free charge is (Romig, 1964)

$$\mathbf{f}_e = \mathbf{j}_e \times \mathbf{b} = \sigma_e(\mathbf{e} + \mathbf{u} \times \mathbf{b}) \times \mathbf{b} = \sigma_e \mathbf{e} \times \mathbf{b} + \sigma_e(\mathbf{u} \times \mathbf{b}) \times \mathbf{b} \,, \tag{2.133}$$

where $\mathbf{j}_e$ is the current density vector and $\mathbf{b}$ is the magnetic induction vector. The *Hall* current density can be added to $\mathbf{j}_e$ as $-(a_H/b)\mathbf{j}\times\mathbf{b}$, where a_H is the *Hall parameter*.

The Maxwell equations for *time-dependent electric* and *magnetic* fields (in differential forms) which are the appropriate extensions to (2.124) to (2.127) and are given by (Lorrain and Corson, 1970)

$$\nabla\times\mathbf{h}-\frac{\partial\mathbf{d}}{\partial t}=\mathbf{j}_e\ , \tag{2.134}$$

$$\nabla\times\mathbf{e}+\frac{\partial\mathbf{b}}{\partial t}=0\ , \tag{2.135}$$

$$\nabla\cdot\mathbf{b}=\nabla\cdot\mathbf{d}=0\ , \tag{2.136}$$

$$\mathbf{d}=\epsilon_e\epsilon_o\mathbf{e}\ ,\quad \mathbf{b}=\mu_e\mathbf{h}\ ,\quad \mathbf{j}_e=\sigma_e(\mathbf{e}+\mathbf{u}\times\mathbf{b})\ , \tag{2.137}$$

where $\mathbf{d}$ is the *displacement current* vector, $\mathbf{h}$ is the *magnetic field intensity* vector, and μ_e is the *magnetic permeability*. Since in the magnetohydrodynamics the magnetic field is assumed to be strong, it is included in (2.135). The Hall current can be added to $\mathbf{j}_e$ as indicated above. The $\sigma_e\mathbf{e}\times\mathbf{b}$ term can *accelerate* or *decelerate* the flow depending on the directions of $\mathbf{e}$, $\mathbf{b}$, and $\mathbf{u}$. The $\sigma_e\mathbf{u}\times\mathbf{b}$ current, called the *back emf* current, when interacting with $\mathbf{b}$, i.e., $(\sigma_e\mathbf{u}\times\mathbf{b})\times\mathbf{b}$, always *decelerates* the flow. When only a magnetic field b_o is applied perpendicular to the flow with a velocity u_o, then a $\sigma_e u_o b_o^2$ decelerating force hinders its flow. As will be discussed in Section 4.13.2, this volumetric retarding force tends to smoothen the velocity nonuniformities which may be caused, for example, by the no-slip velocity condition on solid surfaces. The Joule heating $\mathbf{j}_e\cdot\mathbf{j}_e/\sigma_e$ can be significant when the velocity or the applied fields are large and σ_e is small.

2.6.3 Plasmas

Plasmas are multicomponent high-temperature gases containing *neutral* particles and atoms and molecules, and *electrons* and *ions* with continuous and extensive interactions among them. The plasma state of matter and plasma physics are described by Shohet (1971) and Chen (1974). Konuma (1992) gives a classification of plasmas based on their electron energy k_BT_e, where T_e is the electron temperature, and the electron density n_e. Figure 2.8(a) shows this classification (1 eV is 1.602×10^{-19} J). The plasmas are further divided into *cold* and *thermal* plasmas. In cold plasmas, the *lighter* electron temperature is much higher than the other *heavier* gas constituents (i.e., ions and neutral atoms or molecules), and therefore, *thermal nonequilibrium* exists between the electrons and the remainder gas molecules (or particles). In thermal plasmas, there exists a near-thermal equilibrium. Konuma (1992) shows that the increase in the pressure (or the molecular density) leads to thermal equilibrium. Figure 2.8(b) shows this trend towards thermal equilibrium with the increase in pressure (one

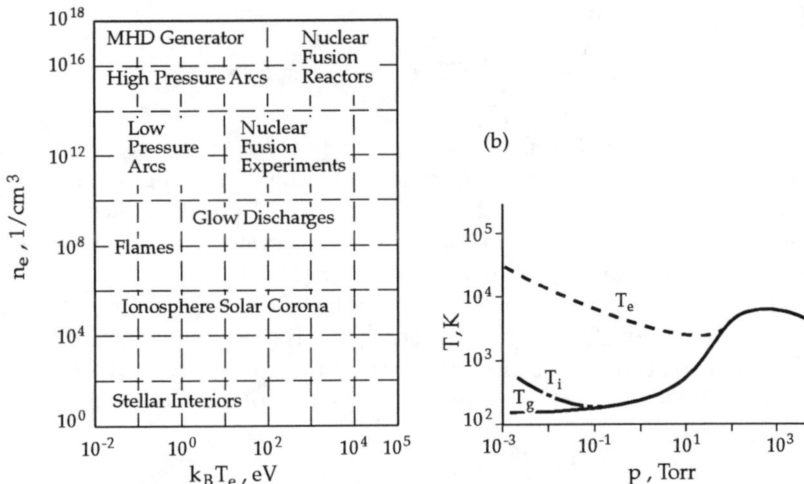

Figure 2.8. (a) A classification of plasmas based on the electron energy and temperature. (b) The thermal equilibrium, or lack of it, between electrons, ions, and other species, as a function of pressure. (From Konuma, reproduced by permission ©1992 Springer-Verlag.)

Torr is equal to 1/760 of one atmosphere pressure and one atmosphere is 1.013×10^5 Pa.), where T_g stands for the gas temperature and T_i is the ion temperature. For $p > 100$ Torr, *thermal equilibrium* is attained. Then the *atmospheric pressure* plasmas are in equilibrium.

Plasmas can be under external applied electric or magnetic fields and they generate (i.e., induce) a magnetic field as they flow. From the heat transfer characteristics, plasmas are a subset of the magnetohydrodynamics, but they are generally treated separately because of the following.

Due to the very high temperatures, the *radiative* heat transfer is generally significant, and the multiple constituents have various constituent size, mass, and initial conditions, and therefore can be in local thermal *nonequilibria* (Eckert and Pfender, 1967). The electrical conductivity of the *dissociated* and *ionized* gases also becomes significantly large with the increase in temperature, and in general, the *thermo-electrophysical property* variations must also be considered. In low-pressure plasmas, the mean free path of the molecules may be of the order of the momentum and thermal boundary-layer thickness, and therefore a *transition flow* regime (*Knudsen regime*, discussed in Section 4.7) or a *molecular flow* regime treatment may be necessary.

The local thermal nonequilibrium of electrons and heavier ions will be discussed in the two-medium treatments in Section 4.13.3. The gas-particle local thermal nonequilibrium, which is of interest in sintering and other aspects of powder and particle processing, will be discussed in Section 5.3.1.

When the *radiation* heat transfer is included and allowance for variation of k is made, (2.130) becomes

$$\rho c_p \frac{DT}{Dt} = \nabla \cdot k \nabla T + \nabla \cdot \mathbf{q}_r + \sigma_e \mathbf{e} \cdot \mathbf{e} \,. \tag{2.138}$$

The *equilibrium population density* of the exited states of the various species are determined from the kinetic theory (Konuma, 1992) and is in general treated similar to the chemical reactions, e.g., (2.44). Treatments of the species equations and other transport and thermodynamic properties of plasmas are reviewed by Eckert and Pfender (1967).

An example of plasma flow under applied field is given by Tsai and Kuo (1990) where allowance for a *velocity slip* is made for electrons flowing at much higher velocities than the rest of the constituents. An example which includes chemical reaction among the constituents is given by Zhao et al. (1990).

2.7 Buoyancy

A stream can flow into an otherwise stagnant ambient with both the stream and the ambient being of the same phase and made of the same species but having two different average temperatures. Then heat exchange will occur between them. Also, due to the temperature difference the motion, and therefore this heat exchange, will be influenced by the thermobuoyancy. Under steady conditions, there are two types of boundary-free thermal mixing flows, one between a finite stream and its infinite surrounding (e.g., plumes, jets) and one between two streams (e.g., two streams flowing at an arbitrary angle to each other). In regards to the driving force, there are two types of forces considered here: one is the purely thermobuoyant flow (e.g., *plumes*) and the other is a combination of thermobuoyant and forced flow (e.g., *thermobuoyant jets*, *nonisothermal shear layers*). A short distance from the source, the flow is almost always *turbulent* (Turner, 1979).

Figure 2.9(a)-(c) shows examples of steady and transient, nonisothermal boundary-free flows. In these examples, the fluid motion is due to *thermobuoyancy only*. A *plume* (i.e., a moving, finite, *negatively* or *positively* buoyant volume of fluid) is formed above (below for negatively buoyant plumes) a heat source (sink) supplying (withdrawing) a heat flow Q for a *finite* time period or *intermittently*. When the heat source is steady (i.e., time-wise continuous), then the early stages of the plume rise (fall) is referred to as the *starting plume*, as shown in Figure 2.9(a). Then as the plume rises far downstream from the source, it will become *developed* (or *self-similar*), i.e., the spatial normalized temperature and velocity distributions will not change in their functional form with respect to a similarity variable which combines the lateral and axial positions. If the source is

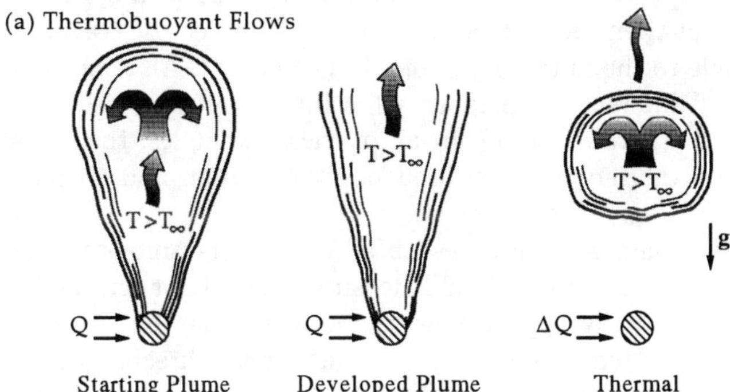

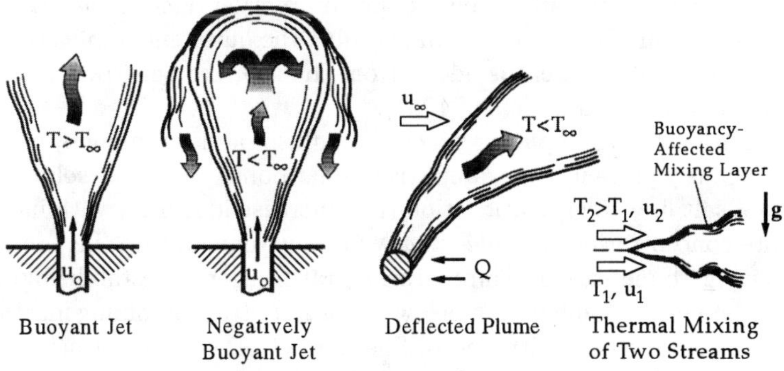

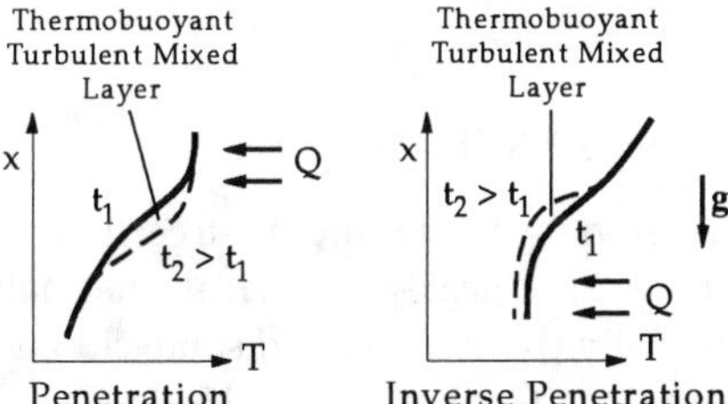

Figure 2.9. Examples of (a) steady and transient, purely thermobuoyant flows, (b) steady, thermobuoyancy-influenced flows, and (c) transient, thermobuoyant flows in heated, stably stratified layers. (Adapted from Jaluria; reproduced by permission ©1980 Pergamon Press.)

intermittent, then as a result of supplying a ΔQ a bulb of buoyant fluid *leaves* the source, and this form of plume is called a *thermal*.

Figure 2.9(b) shows examples of *steady, forced* jets (finite lateral dimensions) which are injected (with an initial momentum) with a temperature which is different than the ambient (which can be otherwise stagnant or be moving at angle to the initial direction of injected jet). Also shown are the two streams of different temperatures flowing at different velocities (the case shown is for parallel flows). In all of these examples, the initial flow inertia and the thermobuoyancy influence the motion and therefore the thermal mixing occurring between the streams.

In both of the examples of Figures 2.9(a) and (b), the ambient is assumed to be at uniform temperature (and density), except for the hydrostatic pressure effect. In many applications, the ambient may be *stratified* (i.e., the density and the temperature may vary along the direction of gravity). Figure 2.9(c) shows examples of *transient*, thermobuoyant flows in *stably*, (i.e., density *increasing* or temperature *decreasing along* the gravity vector) *stratified* ambient. When heat is added from the lower part of a stably stratified layer, the turbulent thermobuoyant motion (i.e., the *mixed layer*) penetrates upward by entraining and eroding the quiescent stably stratified upper ambient. When heat is added from the top, the penetration will be downward (Kaviany and Seban, 1981).

There are many examples in which the diffuso- and thermobuoyancy are simultaneously present. An example is in solar ponds where the solute (e.g., dissolved salt in water) stratification (i.e., increase in density with increase in solute concentration) is *stabilizing*, while the temperature (i.e., decrease in density with increase in temperature) stratification is destabilizing. This occurs when heat in added from below to a water layer containing increasing concentration of solute with depth. Sherman et al. (1978), Turner (1979), Imberger and Hamblin (1982), List (1982), Gebhart et al. (1984), Gebhart et al. (1988), and Mullett(1993). discuss various examples of thermo- and diffusobuoyant flows.

2.7.1 Thermobuoyant Flows

Among the many thermobuoyant boundary-free flows, the axisymmetric turbulent steady plumes and nearly spherical thermals have been analyzed extensively (especially the former) and some closed-form solutions are available for these cases (e.g., Gebhart et al., 1984). These solutions are discussed below for their instructional volumes. Thermobuoyant plumes begin as a laminar flow at the source, but undergo a transition and become turbulent. The transition or the *buckling* of the thermobuoyant plume, which is the magnification of the existing infinitesimal disturbances, has been analyzed by Bejan (1984), Gebhart et al. (1988), and Yang (1992), among others.

(A) Steady Plumes

Assuming a steady axisymmetric turbulent plume discharging upward into an unstratified and otherwise quiescent ambient with x-axis pointing opposite to the gravity vector and the radial and axial components of the velocity designated by u and v, and by neglecting the molecular and axial diffusion and assuming a Boussinesq fluid, (2.88), (2.90), and (2.94) for the *mean quantities* become (Gebhart et al., 1988)

$$\frac{\partial \overline{u}}{\partial x}+\frac{1}{r}\frac{\partial r\overline{v}}{\partial r} = 0\,, \tag{2.139}$$

$$\frac{\partial \overline{u}\,\overline{u}}{\partial x}+\frac{1}{r}\frac{\partial r\overline{u}\,\overline{v}}{\partial r} = -\frac{1}{r}\frac{\partial r\overline{u'v'}}{\partial r}-g\frac{\rho-\rho_\infty}{\rho_o}\,, \tag{2.140}$$

$$\frac{\partial \overline{u}\,\overline{T}}{\partial x}+\frac{1}{r}\frac{\partial r\overline{v}\,\overline{T}}{\partial r} = -\frac{1}{r}\frac{\partial r\overline{v'T'}}{\partial r}\,. \tag{2.141}$$

The boundary conditions are that at $r=0$, $r\overline{u'v'}, r\overline{v'T'}$, and $\overline{v}$ are zero and as $r\to\infty$, $\overline{u'v'}, \overline{v'T'}$, and $\overline{u}$ vanish. The initial conditions are ρ_o, $\overline{u}_o$, and $\overline{T}_o$ at $x=0$. In the approximate *integral method*, the integration is performed in one direction to remove that space variable, as discussed in Section 1.8. Integrating (2.139) to (2.141) perpendicular to x-axis, we have

$$2\frac{d}{dx}\int_o^\infty \overline{u}r dr \equiv \frac{d}{dx}b^2\overline{u}_c \equiv 2\alpha_e b(x)\overline{u}_c(x) = -2\int_o^\infty d(r\overline{v}) = -2b(x)\overline{v}_b(x), \tag{2.142}$$

$$2\frac{d}{dx}\int_o^\infty \overline{u}^2 r\,dr \equiv \frac{d}{dx}b^2\overline{u}_c^2 = -\frac{2g}{\rho_o}\int_o^\infty(\rho-\rho_\infty)rdr \equiv -gb^2(x)\frac{\rho_c(x)-\rho_\infty}{\rho_o}\,, \tag{2.143}$$

$$2\frac{d}{dx}\int_o^\infty \overline{u}\,\overline{T}r\,dr \equiv -\frac{d}{dx}b^2\overline{u}_c{\beta_o}^{-1}{\rho_o}^{-1}[\rho_o-\rho_c(x)] \equiv -2\alpha_e b\overline{u}_c{\beta_o}^{-1}{\rho_o}^{-1}(\rho_o-\rho_\infty)\,, \tag{2.144}$$

where the *entrainment* is defined as

$$e \equiv 2\pi\alpha_e b(x)\overline{u}_c(x) = e(x) \tag{2.145}$$

and $\overline{T}$ and ρ have been related through the thermal expansion coefficient using (1.123). Subscript c indicates the centerline value, α_e is the *entrainment coefficient*, and b is the *effective* radial extent of the plume. If a *uniform* velocity is assumed (called the *top hat* profile), then b is the radius. If any other profile with a maximum at the centerline is assumed, then b will be larger than the radius by a constant that depends on the prescribed velocity distribution.

Then (2.139) to (2.141) can be written as

$$\frac{d}{dx} b^2 \overline{u}_c = 2\alpha_e b \overline{u}_c \ , \tag{2.146}$$

$$\frac{d}{dx} b^2 \overline{u}_c^2 = g b^2 \frac{\rho_\infty - \rho_c}{\rho_o} = g\beta_o b^2 (\overline{T}_c - T_\infty) \ , \tag{2.147}$$

$$\frac{d}{dx} b^2 \overline{u}_c g(\frac{\rho_\infty - \rho_c}{\rho_o}) = \frac{g}{\rho_o} b^2 \overline{u}_c \frac{d\rho_\infty}{dx} = 0 \ , \tag{2.148}$$

where it is assumed that ρ_∞ is constant. Note that (2.1.42) is used in arriving at (2.148). By applying the initial conditions which are $\rho_o, \overline{u}_o$, and T_o, the solutions to (2.146) to (2.148) become

$$\overline{u}_c = \frac{5}{6\alpha_e} (\frac{9}{10} \alpha_e E_o)^{-1/3} x^{-1/3} \ , \tag{2.149}$$

$$g(\rho_c - \rho_o) = -\beta_o \rho_o g(\overline{T}_c - T_o) = \frac{5\rho_o E_o}{6\alpha_e} (\frac{9}{10} \alpha_e E_o)^{-1/3} x^{-5/3} \ , \tag{2.150}$$

$$b = \frac{6}{5} \alpha_e x \ , \tag{2.151}$$

where E_o is defined through

$$\rho_o E_o \equiv 2\pi \int_o^\infty g\overline{u}(\rho - \rho_o) r \, dr \quad \text{at } \ x = 0 \ , \tag{2.152}$$

i.e., $\rho_o E_o$ is the total *weight deficiency* produced per unit time by the source. The experimental results show that α_e=0.116 to 0.153 (Gebhart et al., 1988).

In the integral treatment given above, the specific radial distribution of the velocity has not been sought. The experimental results reviewed by Turner (1979) show that the radial distribution preserves a similarity with respect to a similar variable r^2/x^2. The experimental results are correlated with a Gaussian radial distribution and give

$$\overline{u} = 4.7 {E_o}^{1/3} x^{-1/3} \ e^{(-96\, r^2/x^2)} \ , \tag{2.153}$$

$$-\beta_\infty g(\overline{T}_c - T_\infty) = g \frac{\rho - \rho_\infty}{\rho_\infty} = 11 E_o^{2/3} x^{-5/3} e^{(-71 r^2/x^2)} \ . \tag{2.154}$$

These have the same x-dependence as (2.149) and (2.150). Seban and Behnia (1976) confirm the validity of the integral-Gaussian solution for the variations in r and z. They obtain numerical solutions to the turbulent transport equations (2.139) to (2.141) and by using some turbulence models.

(B) Thermals

A thermal is a finite buoyant volume initiated at an intermittent source. As an example, due to an instability thermals are released at heated horizontal surfaces. Since the index of refraction of fluids changes with temperature, thermal are observed when due to solar irradiation the air adjacent to the ground is heated. They ascend (or descend for negatively buoyant thermals) and grow in size due to their entrainment of the ambient fluid. In general, to their translations, thermals do not remain symmetric and contain internal recirculations of various wavelength (eddy sizes).

Assuming a spherical thermal, we can write the continuity equation by using the entrainment and integrating perpendicular to the translation direction (Gebhart et at., 1988) as

$$\frac{4\pi}{3}\frac{db^3}{dt} = 4\alpha_e b^2 \overline{u} \,, \tag{2.155}$$

where $\overline{u} = dx/dt$ is the translational velocity along the x-axis (opposite to $\mathbf{g}$), b is as before the effective radius, and α_e is the entrainment coefficient. Then the growth in b is found by integrating (2.155), and is (using $b = 0$ at $x = 0$)

$$b = \alpha_e x \,. \tag{2.156}$$

Using the buoyancy velocity and the *conservation* of the total buoyancy input $4\pi(\rho_\infty - \rho_o)b_o^3/3$, the velocity is

$$\overline{u} = a_1 (gb\frac{\rho_\infty - \rho}{\rho_\infty})^{1/2} = a_1 [\frac{g(\rho_\infty - \rho_o)\frac{4}{3}\pi b_o^3}{\rho_\infty \frac{4}{3}\pi b^2}]^{1/2} \,, \tag{2.157}$$

where b_o and ρ_o are for $x = 0$. The experimental results show $a_1 = 0.9$ to 1.2 and $\alpha_e = 0.25$. Then

$$\overline{u} = a_1 [\frac{g(\rho_\infty - \rho_o)b_o^3}{\alpha_e^2 x^2}]^{1/2} = \frac{dx}{dt} \,, \tag{2.158}$$

or

$$x^2 = \frac{2a_1 t}{\alpha_e}(gb_o^3 \frac{\rho_\infty - \rho_o}{\rho_\infty})^{1/2} \,. \tag{2.159}$$

Now using (2.156)

$$b = (2a_1\alpha_e)^{1/2} t^{1/2} (gb_o^3 \frac{\rho_\infty - \rho_o}{\rho_\infty})^{1/4} \,, \tag{2.160}$$

$$\overline{u} = (\frac{a_1}{2\alpha_e})^{1/2} t^{-1/2} (gb_o^3 \frac{\rho_\infty - \rho_o}{\rho_\infty})^{1/4} \,. \tag{2.161}$$

Then the thermal Reynold number $\overline{u}b/\nu$ remains constant during the translation. Thermals and their internal flows are also discussed by Turner (1979).

2.7.2 Thermobuoyancy Influenced Flows

Boundary-free, thermobuoyancy affected flows include *forced flow* of a stream into an ambient (made of the same phase but can be moving relative to a fixed coordinate) which is at a different temperature. The motion in the *mixing layer*, where the stream and the ambient mix and exchange heat, is influenced by the relative velocities, densities, and viscosities. This mixing is central in the *diffusion-controlled* chemical reactions where two streams, one containing the fuel and the other containing the oxidizer, are flown parallel to each other. The thermal mixing of nonreacting injected jets is also of practical significance.

Aspects of thermal mixing in sheared turbulent flows have been studied by Ferchichi and Tavoularis (1992). These include asymmetry of the penetration into the lower and higher velocity sides.

The jets with significant buoyancy effects and their mixing downstream of the injection have been studied extensively (e.g., Madni and Pletcher, 1977; Davis et al., 1978). The jet *initial conditions*, i.e., the conditions within the nozzle and at the exit of the jet, influence the jet spread and mixing significantly.

Strykowski and Russ (1992) examine the mixing of heated axisymmetric jets for initially *laminar* and initially *turbulent* conditions. Their Schlieren photographs for these two initial conditions and for $Re_d = \overline{u}_o d/\nu = 10^4$, where d is the nozzle diameter and $\overline{u}_o$ is the mean area-averaged nozzle exit velocity, are shown in Figure 2.10(a) and (b). The jet exit and ambient temperatures are $\overline{T}_o = 270°\text{C}$, $T_\infty = 25°\text{C}$, and the experiment is for air injected into air. The Richardson number $Ri_d = \Delta\rho g d/\rho\overline{u}_o^2$ is small, indicating that buoyancy is not significant in the large-scale motion. The density ratio ρ_o/ρ_∞ is 0.55. Their results show intermittent spread rates and rapid decay of the mean velocity and temperature downstream for the initially laminar jets [Figure 2.10(a)]. When the boundary layer within the nozzle is disturbed to create an initially turbulent flow, the jet mixing is significantly reduced and no intermittency is observed [Figure 2.10(b)]. The half-angle of the jet spread and the decaying of the mean velocity and temperature are also smaller for the initially turbulent jet. The modeling of intermittency in turbulent shear layers has been examined by Cho and Chung (1992).

In the following, an existing analysis of the vertical, turbulent thermobuoyant jet in the *near-field* region is considered in order to review the various flow regimes existing along the jet and to demonstrate a combination of the turbulence modeling, the entrainment modeling, and the empiricism commonly used to predict the behavior of these various regimes.

A buoyant jet is signified by having equally significant buoyancy $\dot{m}_o(\rho_\infty - \rho_o)$ and momentum $\dot{m}_o\overline{u}_o$ fluxes at the jet exit. When the buoyancy flux dominates, it becomes the already-mentioned plume, and in the absence of any buoyancy flux, it becomes a jet. As with plumes, the buoyant jets are generally turbulent and the effect of the initial turbulence intensity on the downstream turbulence is significant, as was discussed above.

In the *integral method*, the distribution in a given direction is prescribed and it is assumed that this distribution does not change along the flow (i.e., similarity is assumed downstream). However, in the developing regions of the flow, such as the entrance (or near-field) region of a buoyant jet, the ve-

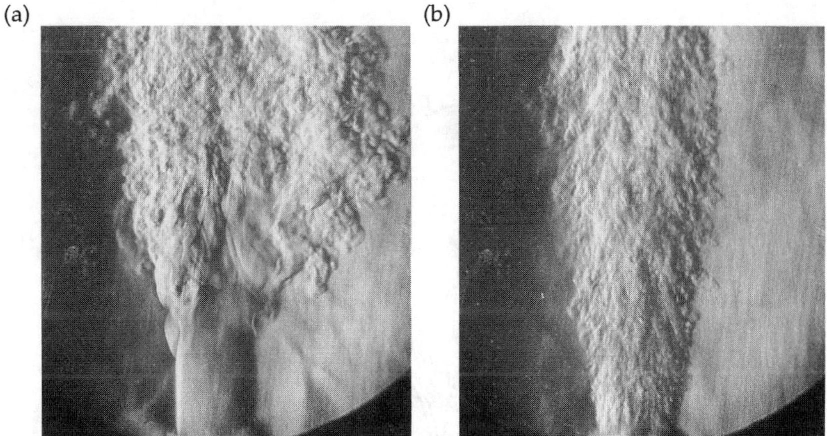

Figure 2.10. Axisymmetric, heated turbulent jets. (a) Initially laminar. (b) Initially turbulent. (From Strykowski and Russ, reproduced by permission ©1992 American Institute of Physics.)

locity and the temperature (or density) distributions change from a nearly *uniform* distribution at the exit to the nearly *Gaussian* distribution downstream. Then a *multidimensional* simulation is required which generally needs to be integrated numerically.

The modeling of the turbulent shear stress $-\overline{u_i'u_j'}$ and the turbulent heat flux $-\overline{u_i'T'}$ in the buoyant jets have been examined by Gibson and Launder (1976), and their models have been applied by Chen and Nikitopoulos (1979), among others. Measurement of the turbulent structure (velocity and temperature) have been made by Sforza and Mons (1978) and Chambers et al. (1985). Recent computations using the elaborate turbulent models are discussed by Martynenko and Korovkin (1992). In the following, the formulation and some of the numerical results of Chen and Nikitopoulos (1979) for the vertical axisymmetric buoyant jet, in the *developing region,* in an otherwise stagnant and uniform temperature ambient, as depicted in Figure 2.11, are reviewed. Then the developed region and its various subregions are examined.

(A) Developing Region

The exiting jet has a uniform mean velocity $\overline{u}_o$ and a uniform mean temperature $\overline{T}_o$ (and density ρ_o). In the developing-flow (or the near-field) region, shown in Figure 2.11, the centerline mean velocity *increases* due to the deceleration of the flow near the edge of the jet. The centerline mean velocity $\overline{u}_c$ no longer increases when the flow development region is reached (the length over which the flow developed is called the *entrance length* x_e), and begins to decrease (i.e., at x_e, $\overline{u}_c$ is maximum) in the development (or

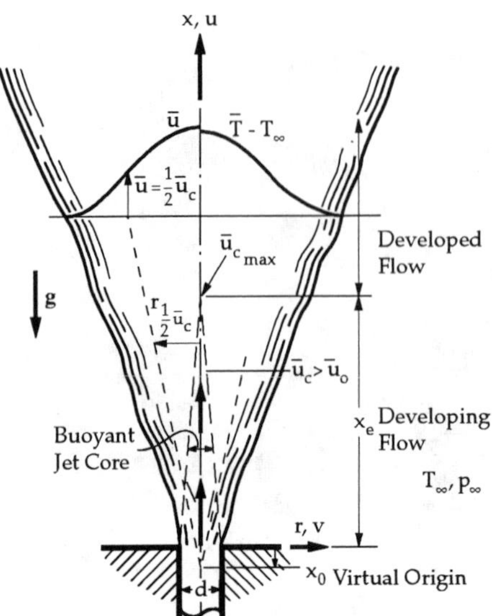

Figure 2.11. The flow and temperature fields downstream of the exit of a heated, vertical turbulent jet. The coordinate system, the developing region, and the profiles in the developed region, are shown.

far-field) region. In the near-field region the entrainment of the surrounding quiescent fluid is not significant compared to the far-field region. For $0 \le x \le x_e$, the effect of the surrounding quiescent flow *penetrates* into the jet and the penetration layer is called the *shear* (or *mixed*) layer. Then at $x = x_e$ the shear layer penetration reaches the centerline.

The simplified forms of the continuity, momentum, and energy equations given by (2.139) to (2.141) are applicable. The radial turbulent shear stresses $-\overline{u'v'}$ and the radial and turbulent axial heat fluxes $-\overline{v'T'}$ and $\overline{u'T'}$ are modeled as (Chen and Nikitopoulos, 1979)

$$-\overline{u'v'} = \frac{1-a_1}{a_2}\frac{\overline{v'}^2}{\frac{1}{2}\overline{u'_iu'_i}}[1+\frac{\frac{1}{2}\overline{u'_iu'_i}g\frac{\partial\overline{T}}{\partial r}}{a_3\epsilon_kT_\infty\frac{\partial\overline{u}}{\partial r}}]\frac{(\frac{1}{2}\overline{u'_iu'_i})^2}{\epsilon_k}\frac{\partial\overline{u}}{\partial r}\,, \tag{2.162}$$

$$\overline{v'}^2 = a_4(\frac{1}{2}\overline{u'_iu'_i})\,, \tag{2.163}$$

$$-\overline{v'T'} = \frac{a_4}{a_3}\frac{(\frac{1}{2}\overline{u'_iu'_i})^2}{\epsilon_k}\frac{\partial\overline{T}}{\partial r}\,, \tag{2.164}$$

$$-\overline{u'T'} = \frac{\frac{1}{2}\overline{u'_iu'_i}}{c_3\epsilon_k}[-\overline{u'v'}\frac{\partial\overline{T}}{\partial r} - \overline{v'T'}(1-a_5)\frac{\partial\overline{u}}{\partial r} + \frac{g(1-a_5)}{T_\infty}\overline{T'}^2]\,. \tag{2.165}$$

The kinetic energy $\overline{u'_iu'_i}/2$, its dissipation rate ϵ_k, and $\overline{T'}^2$ are found from their *transport* equations (this is called the k-ϵ_k-$\overline{T'^2}$ model of turbulence). These are the modified form of (2.100) and two other equations discussed by Gibson and Launder (1976), all given by

$$\frac{\partial}{\partial x}\overline{u}(\frac{1}{2}\overline{u_i'u_i'}) + \frac{1}{r}\frac{\partial}{\partial r}r\overline{v}(\frac{1}{2}\overline{u_i'u_i'}) =$$

$$\frac{1}{r}\frac{\partial}{\partial r}ra_4a_6\frac{(\frac{1}{2}\overline{u_i'u_i'})^2}{\epsilon_k}\frac{\partial}{\partial r}(\frac{1}{2}\overline{u_i'u_i'}) - \overline{u'v'}\frac{\partial\overline{u}}{\partial r} + g\frac{\overline{u'T'}}{T_\infty} - \epsilon_k \,, \tag{2.166}$$

$$\frac{\partial}{\partial x}\overline{u}\epsilon_k + \frac{1}{r}\frac{\partial}{\partial r}r\overline{v}\epsilon_k = \frac{1}{r}\frac{\partial}{\partial r}ra_4a_7\frac{(\frac{1}{2}\overline{u_i'u_i'})^2}{\epsilon_k}\frac{\partial\epsilon_k}{\partial r} +$$

$$a_8\frac{\epsilon_k}{\frac{1}{2}\overline{u_i'u_i'}}(-\overline{u'v'}\frac{\partial\overline{u}}{\partial r} + g\frac{\overline{u'T'}}{T_\infty}) - a_9\frac{\epsilon_k^2}{\frac{1}{2}\overline{u_i'u_i'}} \,, \tag{2.167}$$

$$\frac{\partial}{\partial x}\overline{u}\overline{T'}^2 + \frac{1}{r}\frac{\partial}{\partial r}r\overline{v}\overline{T'}^2 = \frac{1}{r}\frac{\partial}{\partial r}ra_4a_{10}\frac{(\frac{1}{2}\overline{u_i'u_i'})^2}{\epsilon_k}\frac{\partial\overline{T'}^2}{\partial r} -$$

$$2\overline{v'T'}\frac{\partial\overline{T}}{\partial r} - a_{11}\frac{\epsilon_k\overline{T'}^2}{\frac{1}{2}\overline{u_i'u_i'}} \,. \tag{2.168}$$

The eleven constants a_i, $i = 1, 11$, must be prescribed and as was mentioned at the end of Section 2.4.3, the higher-order turbulent models such as the above contain a larger number of constants. However, the magnitude of these constant are obtained from matching the experimental results and in general have relatively tight tolerances and can be used with confidence.

The initial (i.e., at $x = 0$) conditions for the kinetic energy, ϵ_k and $\overline{T'^2}$ and their distributions must also be prescribed. These are given in the normalized forms as

$$\frac{(\frac{1}{2}\overline{u_i'u_i'})_o}{\overline{u}_o^2} = f_k(\frac{r}{d}) \quad \text{at } x = 0 \,, \tag{2.169}$$

$$\frac{\epsilon_{k_o}}{\overline{u}_o^3/d} = f_\epsilon(\frac{r}{d}) \quad \text{at } x = 0 \,, \tag{2.170}$$

$$\frac{\overline{T'}_o^2}{(\overline{T}_o - T_\infty)^2} = f_T(\frac{r}{d}) \quad \text{at } x = 0 \tag{2.171}$$

and specify the initial turbulence strength or *intensity*.

The ratio of the momentum to the buoyant flux is given by the *desimetric Froude number* defined as

$$Fr_d{}^2 = \frac{\overline{u}_o^2 T_\infty}{gd(\overline{T}_o - T_\infty)} \,. \tag{2.172}$$

The numerical result of Chen and Nikitopoulos (1979) for the normalized radial distributions of $\overline{u}$ and $\overline{T}$ in the developing flow region and at various axial locations are given in Figure 2.12. The results are for $Fr^2 = 9.0$ and $(\frac{1}{2}\overline{u_i'u_i'})_o^{1/2}/\overline{u}_o = 0.0125$ with similar intensities for ϵ_k and $(\overline{T'^2})^{1/2}$

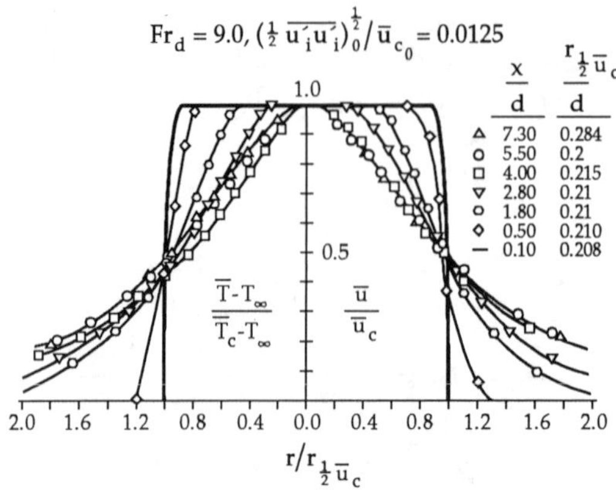

Figure 2.12. Radial distributions of the normalized mean temperature and velocity at various axial locations. (From Chen and Nitikopoulos, reproduced by permission ©1979 Pergamon Press.)

and a Gaussian distribution. The radial location is normalized using the radial location at which $\overline{u}$ drops to half of its value of $r = 0$ (i.e., $\overline{u}_c/2$) and the velocity and the temperature are normalized using the centerline values. The downstream evolution from the flat to a Gaussian distribution is evident. For this Froude number and exit conditions, the normalized entrance length x_e/d is about 4.

(B) Developed Region

The developed or *decaying-entrained region* of the vertical, turbulent buoyant jets follows the developing region. The integral analysis and the entrainment modeling of the *developed* (i.e., decaying or far-field) region of turbulent jets, discussed in Sections 2.4.3 and 2.7.1, has been used for the turbulent buoyent jets with the associated constants determined experimentally. The distance from the entrance to the developed region is given in terms of the dimensionless distance defined below. The developed region consists of the *nonbuoyant, intermediate* (or *transitional*), and plume *subregions*. The results are summarized by Gebhart et al. (1988) and the results for the axisymmetric buoyant jets are given in terms of the densimetric Froude number and the dimensionless distance x^* defined as

$$Fr_d^2 = \frac{\overline{u}_o^2 \rho_o}{gd(\rho_\infty - \rho_o)}, \quad x^* = Fr_d^{-1/2}\left(\frac{\rho_o}{\rho_\infty}\right)^{-1/4}\frac{x}{d}. \tag{2.173}$$

The distributions for various regions (defined below) are given by Gebhart et al. (1988) and summarized below.

Nonbuoyant Subregion:

$x_e^* \leq x^* \leq 0.5$

$$\frac{\overline{u}_c}{\overline{u}_o} = 6.2(\frac{\rho_o}{\rho_\infty})^{-1/2}(\frac{x}{d})^{-1}$$

$$\frac{\overline{T}_c - T_\infty}{\overline{T}_o - T_\infty} = 5(\frac{\rho_o}{\rho_\infty})^{-1/2}(\frac{x}{d})^{-1}$$

Intermediate Subregion:

$0.5 \leq x^* \leq 5$

$$\frac{\overline{u}_c}{\overline{u}_o} = 7.26 Fr_d^{-0.1}(\frac{\rho_o}{\rho_\infty})^{9/20}(\frac{x}{d})^{-1}$$

$$\frac{\overline{T}_c - T_\infty}{\overline{T}_o - T_\infty} = 0.44 Fr_d^{1/8}(\frac{\rho_o}{\rho_\infty})^{-7/16}(\frac{x}{d})^{-5/4}$$

Plume Subregion:

$5 \leq x^*$

$$\frac{\overline{u}_c}{\overline{u}_o} = 3.5 Fr_d^{-1/3}(\frac{\rho_o}{\rho_\infty})^{1/3}(\frac{x}{d})^{-1/5}$$

$$\frac{\overline{T}_c - T_\infty}{\overline{T}_o - T_\infty} = 9.35 Fr_d^{1/3}(\frac{\rho_o}{\rho_\infty})^{-1/3}(\frac{x}{d})^{-5/3} \ . \quad (2.174)$$

The radial distribution is Gaussian (Gebhart et al., 1984) and similar to those given for plumes in the last section. The developed region begins at x_ℓ^*, and as was mentioned in Section (A) above, depending on Fr_d and ρ_o/ρ_∞, this can be many diameters in length. The decay relations for $\overline{u}_c$ and $\overline{T}_c$ has been discussed by So and Aksoy (1993).

2.8 References

Batchelor, G.K., 1982, *The Theory of Homogeneous Turbulence*, Cambridge University Press, New York.

Bejan, A., 1984, *Convection Heat Transfer*, John Wiley and Sons, New York.

Benson, S.W., 1976, *Thermochemical Kinetics*, John Wiley and Sons, New York.

Boyd, I.D., 1993, "Temperature Dependence of Rotational Relaxation in Shock Waves of Nitrogen," *J. Fluid Mech.*, 246, 343–360.

Bradshaw, P., Editor, 1978, *Turbulence*, Springer-Verlag, Berlin.

Branover, H., Lykoudis, P.S., and Mond. M., 1985, *Single- and Multiphase Flows in an Electromagnetic Field: Energy, Metallurgical, and Solar Applications*, Volume 100, Progress in Astronautics and Aeronautics, American Institute of Aeronautic and Astronautics, New York.

Brewster, M.Q., 1992, *Thermal Radiation Transfer and Properties*, John Wiley and Sons, New York.

Brown, M.A., and Churchill, S.W., 1993, "Transient Behavior of an Impulsively Heated Fluid," *Chem. Eng. Technol.*, 16, 82–88.

Carlson, B.G., and Lathrop, K.D., 1968,"Transport Theory: The Method of Discrete Ordinates," in *Computing Methods in Reactor Physics*, Gordon and Breach Science Publishers, New York.

Cebeci, T., and Bradshaw, P., 1984, *Physical and Computational Aspects of Convection Heat Transfer*, Springer-Verlag, New York.

Cess, R.D., and Tiwari, S.N., 1972, "Infrared Radiative Energy Transfer in Gases," *Advan. Heat Transfer*, 8, 229–283.

Chambers, A.J., Antonia, R.A., and Fulachier, L., 1985, "Turbulent Prandtl Number and Spectral Characteristics of a Turbulent Mixing Layer," *Int. J. Heat Transfer*, 28, 1461–1468.

Chen, F.F., 1974, *Introduction to Plasma Physics*, Plenum Press, New York.

Chen, J.C., and Nikitopoulos, P., 1979, "On the Near-Field Characteristics of Axisymmetric Turbulent Buoyant Jets in a Uniform Environment," *Int. J. Heat Mass Transfer*, 22, 245–255.

Cho, J.R., and Chung, M.K., 1992, "A k-ϵ-γ Equation Turbulence Model," *J. Fluid Mech.*, 237, 301–322.

Davis, L.R., Shirazi, M.A., and Slegel, D.L., 1978, "Measurement of Buoyant Jet Entrainment from Single and Multiple Sources," *ASME J. Heat Transfer*, 100, 442-447.

Deissler, R.G., 1964, "Diffusion Approximation for Thermal Radiation in Gases with Jump Boundary Condition," *ASME J. Heat Transfer*, 86, 240–246.

Driscoll, W.G., and Vaughan, W., Editors, 1978, *Handbook of Optics*, McGraw-Hill, New York.

Eckert, E.R.G., and Pfender, E., 1967, "Advances in Plasma Heat Transfer," *Advan. Heat Transfer*, 4, 229–310.

Edwards, D.K., 1981, *Radiation Heat Transfer Notes*, Hemisphere Publishing Company, New York.

Ferchichi, M., and Tavoularis, S., 1992, "Evolution of a Thermal Mixing Layer in Uniformly Sheared Turbulent Flow," *Phys. Fluids*, A4, 997–1006.

Fiveland, W.A., 1988, "Three-Dimensional Radiative Heat Transfer Solutions by the Discrete-Ordinates Method," *J. Thermophys. Heat Transfer*, 2, 309–316.

Fujino, T., Yokoyama, Y., and Mori, Y.H., "Augmentation of Laminar Forced-Convective Heat Transfer by the Application of a Transverse Electric Field," *ASME J. Heat Transfer*, 111, 345–351.

Gebhart, B., Hilder, D.S., and Kelleher, M., 1984, "The Diffusion of Turbulent Buoyant Jets," *Advan. Heat Transfer*, 16, 1-57.

Gebhart, B., Jaluria, Y., Mahajan, R.L., and Sammakia, B., 1988, *Buoyancy-Induced Flows and Transport*, Hemisphere Publishing Corporation, Washington, D.C.

Gibson, M.M., and Launder, B.E., 1976, "On the Calculation of Horizontal, Turbulent, Free Shear Flows Under Gravitational Influence," *ASME J. Heat Transfer*, 98, 81–87.

Gilbarg, D., and Paolucci, D., 1953, "The Structure of Shock Waves in the Continuum Theory of Fluids," *J. Rational Mech. Analys.*, 2, 617–642.

Glassman, I., 1987, *Combustion*, Second Edition, Academic, Orlando.

Grad, H., 1952, "The Profile of a Steady-Plane Shock Wave," *Com. Pure and Appl. Math.*, V, 257–300.

Grinstein, F.F., and Kailasanath, K., 1992, "Chemical Energy Release and Dynamics of Transitional, Reactive Shear Flows," *Phys. Fluids*, A4, 2207–2221.

Hanamura, K., 1994, Gifu University (Private Communication).

Hermanson, J.C., and Dimotakis, P.E., 1989, "Effects of Heat Release in a Turbulent, Reacting Shear Layer," *J. Fluid Mech.*, 199, 333–375.

Hinze, J., 1975, *Turbulence*, Second Edition, McGraw-Hill, New York.

Hubbard, G.L., and Tien, C.-L., 1978, "Infrared Mass Absorption Coefficients of Luminous Flames and Smoke," *ASME J. Heat Transfer*, 100, 235–239.

Imberger, J., and Hamblin, P.F., 1982, "Dynamics of Lakes, Reservoirs and Cooling Ponds," *Ann. Rev. Fluid Mech.*, 14, 153–187.

Jones, T.B., 1978, "Electrohydrodynamically Enhanced Heat Transfer in Liquids: A Review," *Advan. Heat Transfer*, 14, 107–148.

Katta, V.R., and Roquemore, W.M., 1993, "Role of Inner and Outer Structure in Transitional Jet Diffusion Flame," *Combust. Flame*, 92, 274–282.

Kaviany, M., and Seban, R.A., 1981, "Transient Turbulent Thermal Convection in a Pool of Water," *Int. J. Heat Mass Transfer*, 24, 1742–1746.

Kee, R.J., Rupley, F.M., and Miller, J.A., 1989, "CHEMKIN-II: A Fortran Chemical Kinetics Package for the Analysis of Gas-Phase Chemical Kinetics," SAND 89-8009. UC-401, Sandia National Laboratories, Livermore, CA.

Kerstein, A.R., 1992, "Linear-Eddy Modeling of Turbulent Transport. Part 7. Finite-Rate Chemistry and Multi-Stream Mixing," *J. Fluid Mech.*, 240, 289–313.

Konuma, M., 1992, *Film Deposition by Plasma Techniques*, Springer-Verlag, New York.

Kuo, K.K.-Y., 1986, *Principles of Combustion*, John Wiley and Sons, New York.

Landahl, M.T., and Mollo-Christensen, E., 1987, *Turbulence and Random Process in Fluid Mechanics*, Cambridge University Press, New York.

Launder, B.E., 1978a, "On the Effect of a Gravitational Field on the Turbulent Transport of Heat and Momentum," *J. Fluid Mech.*, 67, 569–581.

Launder, B.E., 1978b, "Heat and Mass Transport" in *Turbulence*, Bradshaw, P., Editor, Springer-Verlag, Berlin.

Launder, B.E., 1988, "On the Computation of Convective Heat Transfer in Complex Turbulent Flows," *ASME J. Heat Transfer*, 110, 1112–1128.

Lesieur, M., 1987, *Turbulence in Fluids: Statistical and Numerical Modeling*, Martinus Nijhoff Publishers, Dordrecht.

Lewis, B., and von Elbe, G., 1987, *Combustion, Flames and Explosions of Gases*, Third Edition, Academic, London.

List, E.J., "Turbulent Jets and Plumes," *Ann. Rev. Fluid Mech.*, 14, 189–212.

Ljuboja, M., and Rodi, W., 1981, "Production of Horizontal and Vertical Turbulent Buoyant Wall Jets," *ASME J. Heat Transfer*, 103, 343–349.

Lockwood, F.C., and Naguib, A.S., 1975, "The Prediction of the Fluctuations in the Properties of Free, Round-Jet, Turbulent, Diffusion Flames," *Combust. Flame*, 24, 109–124.

Lorrain, P., and Corson, D.R., 1970, *Electromagnetic Field and Waves*, Second Edition, W.H. Freeman and Company, San Francisco.

Luikov, A.V., and Berkovsky, B.M., 1970, "Thermoconvective Waves, " *Int. J. Heat Mass Transfer*, 13, 741–747.

Madni, I.K., and Pletcher, R.H., 1979, "Buoyant Jets Discharging Nonvertically into a Uniform, Quiescent Ambient: A Finite-Difference Analysis and Turbulence Modeling," *ASME J. Heat Transfer*, 99, 641–646.

Mahalingam, S., Cantwell, B.J., and Ferziger, J.H., 1990b, "Full Numerical Simulation of Coflowing, Asymmetric Jet Diffusion Flame," *Phys. Fluids*, A2, 720–728.

Mahalingam, S., Ferziger, J.H., and Cantwell, B.J., 1990a, "Self-Similar Diffusion Flows," *Combust. Flame*, 82, 231–234.

Martin, P.J., and Richardson, A.T., 1984, "Conductivity Models of Electrothermal Convection in a Plane Layer of Dielectric Liquid," *ASME J. Heat Transfer*, 106, 131–142.

Martynenko, O.G., and Korovkin, V.N., 1992, "Concerning the Calculation of Plane Turbulent Jets of the Basis of $k - \epsilon$ Model of Turbulence," *Int. J. Heat Mass Transfer*, 35, 3389–3395.

Melcher, J.R., and Taylor, G.I., 1969, "Electrohydrodynamics: A Review of the Role of Interfacial Shear Stresses," *Ann. Rev. Fluid Mech*, 1, 111–146.

Modest, M.F., 1993, *Radiative Heat Transfer*, McGraw-Hill, New York.

Monin, A.S., and Yaglom, A.M., 1979, *Statistical Fluid Mechanics: Mechanics of Turbulence, Volume* 1, MIT Press, Cambridge, MA.

Monin, A.S., and Yaglom, A.M., 1981, *Statistical Fluid Mechanics: Mechanics of Turbulence, Volume* 2, MIT Press, Cambridge, MA.

Mott-Smith, H.M., 1951, "The Solution of the Boltzmann Equation for a Shock Wave," *Phys. Rev.* series 2, 82, 885-892.

Mullett, L.B., 1993, "The Role of Buoyant Thermals in Salt Gradient Solar Ponds and Convection More Generally," *Int. J. Heat Mass Transfer*, 36, 1923–1941.

Ohwada, T., 1993, "Structure of Normal Shock Waves: Direct Numerical Analysis of the Boltzmann Equations for Hard-sphere Molecules," *Phys. Fluids*, A5, 217-234.

Ozisik, M.N., 1985, *Radiative Heat Transfer and Interaction with Conduction and Convection*, Werbel & Peck, New York.

Panchapakesan, N.R., and Lumley, J.L., 1993, "Turbulence Measurements in Axisymmetric Jets of Air and Helium. Part 1. Air Jets. Part 2. Helium Jets," *J. Fluid Mech.*, 246, 197–247.

Peters, M., 1992, "A Spectral Closure for Premixed Turbulent Combustion in the Flamelet Regime," *J. Fluid Mech.*, 242, 611–629.

Pierce, A.D., 1989, *Acoustics: An Introduction to Its Physical Principle and Applications*, Acoustical Society of America, Woodbury, CT.

Pletcher, R.H., 1988, "Progress in Turbulent Forced Convection," *ASME J. Heat Transfer*, 110, 1129–1144.

Prasad, K., and Price, E.W., 1992, "A Numerical Study of the Leading Edge of Laminar Diffusion Flames," *Combust. Flame*, 90, 155–173.

Riley, J.J., Metcalfe, R.W., and Orszag, S.A., 1986, "Direct Numerical Simulations of Chemically Reacting Turbulent Mixing Layers," *Phys. Fluids*, 29, 406–422.

Romig, M.F., 1964, "The Influence of Electric and Magnetic Fields on Heat Transfer to Electrically Conducting Fluids, " *Advan. Heat Transfer*, 1, 267–354.

Ruthland, C.J., and Trouvé, A., 1993, "Direct Simulations of Premixed Turbulent Flames with Nonunity Lewis Numbers," *Combust. Flame*, 94, 41–57.

Schlichling, H., 1979, *Boundary-Layer Theory*, Seventh Edition, McGraw-Hill, New York.

Schreier, S., *Compressible Flow*, John Wiley and Sons, New York.

Seban, R.A., and Behnia, M.M., 1976, "Turbulent Buoyant Jets in Unstratified Surroundings," *Int. J. Heat Mass Transfer*, 19, 1197–1204.

Sforza, P.M., and Mons, R.F., 1978, "Mass, Momentum, and Energy Transport in Turbulent-Free Jets," *Int. J. Heat Mass Transfer*, 21, 371–384.

Shercliff, J.A., 1965, *A Textbook of Magnetohydrodynamics*, Pergamon Press, Oxford.

Sherman, F.S., 1955, *A Low-Density Wind-Tunnel Study of Shock-Wave Structure and Relaxation Phenomena in Gases*, NASA Technical Report 3298.

Sherman, F.S., 1990, *Viscous Flow*, McGraw-Hill, New York.

Sherman, F.S., Imberger, J., and Corcos, G.M., 1978, "Turbulence and Mixing in Stably Stratified Waters," *Ann. Rev. Fluid Mech.*, 10, 267–288.

Shohet, J.L., 1971, *The Plasma State*, Academic, New York.

Siegel, R., and Howell, J.R., 1992, *Thermal Radiation Heat Transfer*, Third Edition, Hemisphere Publishing Company, Washington D.C.

Sirovich, L., Editor, 1991, *New Perspectives in Turbulence*, Springer-Verlag, New York.

So, R.M.C., and Aksoy, H., 1993, "On Vertical Turbulent Bouyant Jets," *Int. J. Heat Mass Transfer*, 36, 3187–3200.

Steinfeld, J.I., Francisco, J.S., and Hase, W.L., 1987, *Chemical Kinetics and Dynamics*, Prentice-Hall, Englewood Cliffs, NJ.

Strykowski, P.J., and Russ, S., 1992, "The Effect of Boundary-Layer Turbulence on Mixing in Heated Jets," *Phys. Fluids*, A4, 865–868.

Tennekes, H., and Lumley, J.L., 1972, *A First Course in Turbulence*, MIT Press, Cambridge, MA.

Tien, C.-L., 1968, "Thermal Radiation Properties of Gases," *Advan. Heat Transfer*, 5, 253–324.

Tsai, M.C., and Kou, S., 1990, "Heat Transfer and Fluid Flow in Welding Arcs Produced by Sharpened and Flat Electrodes," *Int. J. Heat Mass Transfer*, 33, 2089–2098.

Turnbull, R.T., 1968, "Electroconvective Instability with a Stabilizing Temperature Gradient. I. Theory, II. Experimental Results," *Phys. Fluids*, 12, 2588–2603.

Turner, J.S., 1979, *Buoyancy Effects in Fluids*, Cambridge University Press, Cambridge.

White, F.M., 1974, *Viscous Fluid Flow*, McGraw-Hill, New York.

White, F.M., 1991, *Viscous Fluid Flow*, Second Edition, McGraw-Hill, New York.

Williams, F.A., 1985, *Combustion Theory*, Addison-Wesley, Redwood City, CA.

Woods, L.C., 1993, *An Introduction to the Kinetic Theory of Gases and Magnetoplamas*, Oxford University Press, Oxford.

Yang, H.Q., 1992, "Buckling of a Thermal Plume," *Int. J. Heat Mass Transfer*, 35, 1527–1532.

Zel'dovich, Ya.B., and Raiser, Yu.P., 1969, "Shock Waves and Radiation," *Ann. Rev. Fluid Mech.*, 1, 385–412.

Zhao, G.Y., Mostaghimi, J., and Boulos, M.I., 1990, "The Induction Plasma Chemical Reactor: Part I. Equilibrium Model," *Plasma Chem. Plasma Process.*, 10, 133–150.

3

Two-Phase Systems

In this chapter we examine the representation (i.e., modeling) of two-phase systems as an effective, single medium. In this effective medium, the two phases are *assumed* to be in *local thermal equilibrium*. The flow and heat transfer in this effective medium is described by models which can be derived from the local volume (and time) averaging. Depending on the complexity of the phase distributions and the velocity fields, various assumptions and simplifications are made to arrive at models which can be used with a reasonable effort.

After reviewing the local thermal equilibrium, the simpler class of two-phase systems in which the solid is stationary is considered first. This will be the porous media. Initially the phase change and reaction will not be addressed, and then both of these are addressed. The case of both phases moving (with the most encountered example being the class of problems where one phase is *dispersed* in another phase which is continuous) is more complex, and the space- and time-averaged equations must be significantly simplified before becoming solvable. The rigorous treatments and approximations and models for dispersed two-phase systems are discussed, and some models for the effective properties and an example with chemical reaction are considered.

3.1 Local Thermal Equilibrium and Intramedium Heat Transfer

For nonisothermal flow in two-phase systems, the condition of local thermal equilibrium among phases i and j has been defined in Section 1.9 as

$$\langle T\rangle^i(\mathbf{x},t) = \langle T\rangle^j(\mathbf{x},t) = \langle T\rangle(\mathbf{x},t) \ , \tag{3.1}$$

where $\langle T\rangle^i$ is the *local phase-volume*-averaged temperature of phase i and $\langle T\rangle$ is the local representative elementary volume- (which contains phases i and j) averaged temperature, as defined in Section 1.8.1. This equilibrium is an idealization, and in practice in nonisothermal two-phase systems the local phase-averaged temperatures are not the same. The extent of this difference depends on the heat generations (in phases i and j), $\langle s\rangle^i$ and $\langle s\rangle^j$, the ratio of the phase conductivities k_i/k_j, the ratio of the heat capacities $(\rho c_p)_i/(\rho c_p)_j$, the phase velocities $\langle\mathbf{u}\rangle^i$, $\langle\mathbf{u}\rangle^j$, the linear dimension of the elements of phases i and j, d_i and d_j, the volume fraction of each phase, the linear dimension of the representative elementary volume ℓ, the linear dimension of the system L, the specific interfacial surface area A_{i_j}/V, the topology of the phases, and the temperature and velocity variation within each phase. In *transient* problems, the penetration distance of the disturbances introduced at the surfaces bounding the two-phase system also influences the extent of the nonequilibrium, and since this penetration distance is time dependent, the extent of the deviation from equilibrium becomes time dependent.

A simple *constraint* for the application of the single-medium treatment to two-phase systems can be a smooth variation of the temperature *across* the *entire system* length, i.e.,

$$\Delta T_d \ll \Delta T_\ell \ll \Delta T_L \ , \tag{3.2}$$

where $d \ll \ell \ll L$. The application can be extended to cases where $d < \ell < L$ and $\Delta T_d < \Delta T_\ell < \Delta T_L$ with reduction in the accuracy of the prediction.

Under the assumption of local thermal equilibrium, the local volume-averaged behavior, using the *effective properties*, of a single medium is analyzed. Also, the behavior far from any bounding surfaces, i.e., the *intramedium* convective heat transfer, is analyzed. Various phase pairs, with any of the phases being continuously distributed, are considered. The case of a stationary and continuous solid phase, with the fluid flow through its interstices, is considered first. The principle of local volume averaging is applied to the relevant conservation equations, and the effective properties are defined after using some closure conditions.

In principle, description of the difference $\langle T\rangle^i - \langle T\rangle^j$ is possible and this is done by obtaining local phase-volume-averaged energy equations (in terms of temperature), taking the difference, applying the necessary closure conditions, and then applying an order-of-magnitude analysis (as done be Whitaker, 1991, for a solid-fluid system with stationary solid phase). However, evaluation of $\langle T\rangle^i - \langle T\rangle^j$ would require an extensive effort due to the large number of parameters and the complex interplay between the transient, the convective, and the heat generation effects. Therefore, a priori determination of some *rigorous constraints* on the validity of the local thermal equilibrium is possible, but very laborious.

In practice, most *transient* two-phase problems with the ratios of thermal conductivities and the heat capacities differing significantly from unity, require nonequilibrium (i.e., two-medium) treatments. Whenever there is a significant *heat generation* within any of the phases, and these ratios are also far from unity, nonequilibrium treatment is also needed. In these cases, most of the temperature drop across the entire system ΔT_L also occurs over a length scale much smaller than L. In the case where the temperature disturbances initiated at the bounding surfaces penetrate a distance of a few representative elementary volumes, then $\Delta T_\ell \approx \Delta T_L$. In addition, when the heat generated by a single element of a dispersed phase is the only source for the change in temperature, then also $\Delta T_d \approx \Delta T_L$.

Both of these cases require nonequilibrium treatment when k_i/k_j and $(\rho c_p)_i/(\rho c_p)_j$ are *far* from unity.

3.2 Continuous Solid–Fluid Systems with Stationary Solid

The case of a *stationary*, structurally *periodic*, and *continuous* solid phase offers a simplicity, i.e., knowing the simple phase distributions *a priori*. This leads to analyses which can be extended rather far by being able to determine the limited effective properties of the medium. Before considering local *volume* averaging, a simpler case of *area* averaging is considered to illustrate how a spatial integration in directions where significant transport occurs results in the appearance of an effective transport mechanism in the nonintegrated direction.

3.2.1 Taylor Dispersion

As an illustration of the consequences of the spatial averaging, consider the *cross-sectional area-averaging* of the energy equation for heat transfer along the flow (longitudinal) in the hydrodynamically fully developed laminar flow through a tube of zero wall thermal conductivity (ideal insulation). This problem was first formulated and solved by Taylor (1953) for species transport, but is equally applicable to heat transfer, and many extensions have since been made to his work (e.g., Yuan et al., 1991). The energy equation integrated laterally includes an added transport term which accounts for the combined effects of the lateral diffusion and the lateral nonuniformity in the velocity distribution. This added transport appears in a diffusion form and is called the *Taylor dispersion*.

Taylor (1953) considers the following problem. For steady, laminar fully developed flow in a tube of radius R, the velocity profile is

$$u(r) = 2\langle u\rangle_A\left(1 - \frac{r^2}{R^2}\right) = \langle u\rangle_A + u'(r) \tag{3.3}$$

where the area-averaged axial (x-direction) velocity $\langle u\rangle_A$ is

$$\langle u\rangle_A = \frac{2}{R^2}\int_0^R ur\,dr \tag{3.4}$$

and $u'(r)$ is the velocity deviation from this average value.

Consider isothermal flow at temperature T_i in the tube for $t < 0$. Then suddenly at $t = 0$ heat is added to a region $0 \le r \le R$, $x_o \le x \le x_o + \delta$ in the fluid to raise its temperature to $T_o > T_i$. This *disklike* temperature *disturbance* travels along the tube and undergoes changes in its boundaries (i.e., its geometry). Figure 3.1(a) shows this initial disturbance and the

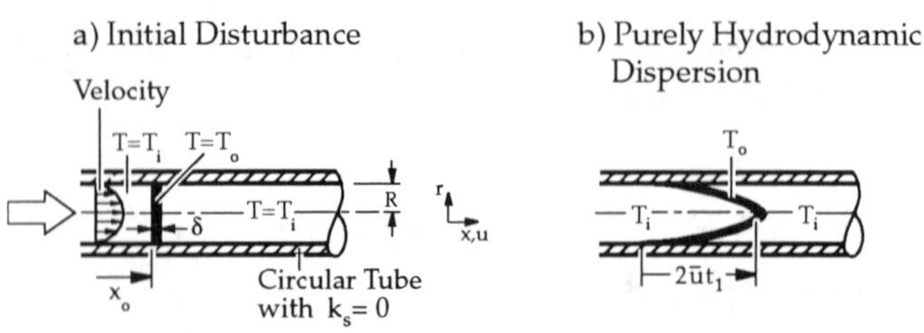

c) Molecular Diffusion - Hydrodynamic Dispersion

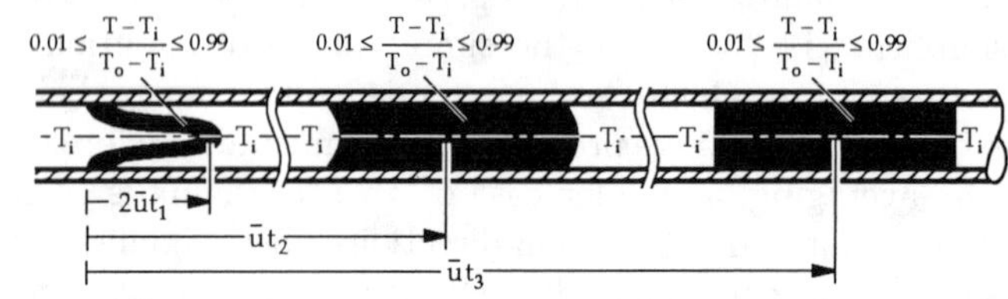

d) Distribution of Asymptotic Temperature and Velocity Deviations

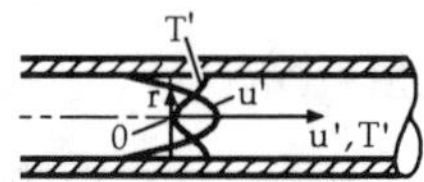

Figure 3.1. Dispersion of a thermal disturbance in a hydrodynamically fully developed tube flow. (a) Initial disk-like disturbance, (b) the disturbance traveling downstream without any molecular diffusion, (c) with molecular diffusion, and (d) distribution of deviations.

anticipated changes for $t > 0$, for when the molecular diffusion is negligible [Figure 3.1(b)], and when it is not [Figure 3.1(c)].

Under negligible fluid molecular conductivity, the nonuniform velocity distribution changes the disk to a *paraboloid* downstream. As shown in Figure 3.1(b) the foremost position of the disturbance at any time t_1 is at $x = 2\langle u\rangle_A t_1 + \delta$. Taylor (1953) shows that the heat content of the initial disk is uniformly distributed in the region $0 \le x \le 2\langle u\rangle_A t_1 + \delta$.

When allowance is made for fluid conductivity, the paraboloid becomes distorted due to the lateral and axial diffusion, shown in Figure 3.1(c). This distortion progresses until the paraboloid shape completely disappears and the disturbed region travels downstream at a velocity $\langle u\rangle_A$, and this disturbed region becomes pluglike again with a symmetry of the temperature distribution around its center. In this large time asymptotic behavior the thickness of the plug is ever increasing with time. This asymptotic behavior is reached when the lateral penetration of the disturbance is nearly completed. From the conduction results this corresponds to a Fourier number Fo_R larger than unity, i.e.,

$$Fo_R = \frac{t\alpha}{R^2} > 1 \quad \text{or} \quad t > \frac{R^2}{\alpha} . \tag{3.5}$$

The asymptotic behavior has many interesting features. The particles that move through the disturbed region, which is moving at $\langle u\rangle_A$, move forward and pass it if they are around the centerline. These particles arrive from $x_1 < 0$ into the disturbed region at T_i and undergo an increase in T and then undergo a decrease in T to leave at T_i again. The fluid particles near the wall arrive into the disturbed region from $x_1 > 0$ and then leave the disturbed region moving to $x_1 < 0$ region, experiencing similar increase and decrease in their temperature. The fluid particles with $u > \langle u \rangle_A$ or $u < \langle u \rangle_A$ exchange heat with each other through lateral diffusion. The extent of the disturbed region increases with time and is *symmetrical* around x_1 (note that the flow is unidirectional, and therefore, *asymmetric*).

Assuming *negligible* axial conduction (i.e., the axial conduction can be added without changing any of the following results, e.g., Aris, 1956) the energy equation (1.84) for this case becomes

$$\rho c_p[\frac{\partial T}{\partial t} + 2\langle u \rangle_A (1 - \frac{r^2}{R^2})\frac{\partial T}{\partial x}] = k\frac{1}{r}\frac{\partial}{\partial r} r\frac{\partial T}{\partial r} \ . \tag{3.6}$$

Using $x_1 = x - \langle u \rangle_A t$, i.e., moving the coordinate at the average fluid velocity, and taking the large time asymptote, $\partial/\partial t \rightarrow 0$, we have

$$u'\frac{\partial T}{\partial x_1} = 2\langle u \rangle_A (\frac{1}{2} - \frac{r^2}{R^2})\frac{\partial T}{\partial x_1} = \frac{\alpha}{r}\frac{\partial}{\partial r} r\frac{\partial T}{\partial r} \ . \tag{3.7}$$

Note that by choosing a moving coordinate that moves with the area-averaged velocity, the portion of fluid near the centerline moves *forward* with respect to this coordinate and the portion near the wall moves *backward*.

The boundary conditions for the r-direction are

$$T(r = 0, x_1) = T(x_1) \ , \tag{3.8}$$

$$\frac{\partial T}{\partial r}(r = R, x_1) = 0 \ . \tag{3.9}$$

Assuming that $\partial T/\partial x_1$ is *constant* (numerical solutions of Nunge and Gill, 1969, has shown that this assumption or lack of it does not alter the final results), the solution to (3.7), subject to the above conditions becomes

$$T = T(x_1) + T'(r) = T(x_1) + \frac{R^2 \langle u \rangle_A}{4\alpha}\frac{\partial T}{\partial x_1}(\frac{r^2}{R^2} - \frac{1}{2}\frac{r^4}{R^4}) \ , \tag{3.10}$$

where $T'(r)$ is local (radial) temperature *deviation* from the centerline value. Equation (3.10) gives the radial variation of the temperature. This variation, which gives rise to the radial (i.e., *lateral*) *molecular* diffusion when *coupled* with the lateral variation of u, leads to the *Taylor dispersion*. The lateral variation of the temperature and velocity deviations are shown in Figure 3.1(d).

The heat transfer rate across x_1 is found by the cross-sectional area averaging of the enthalpy flow and when divided by ρc_p this heat transfer rate Q is

$$\frac{Q}{\rho c_p} = 2\pi \int_0^R u'T'r \, dr = 4\pi \langle u \rangle_A \int_0^R (\frac{1}{2} - \frac{r^2}{R^2})[T - T(x_1)]r \, dr$$

$$= -\frac{\pi R^4 \langle u \rangle_A{}^2}{48\alpha} \frac{\partial T(x_1)}{\partial x_1} = -\frac{\pi}{48} \frac{\langle u \rangle_A^2 R^4}{\alpha} \frac{d\langle T \rangle_A}{dx_1} \tag{3.11}$$

where

$$\langle T \rangle_A = \frac{1}{A} \int_A T \, dA = \frac{2}{R^2} \int_0^R Tr \, dr \ . \tag{3.12}$$

The heat flux across x_1 is

$$\frac{Q}{\rho c_p \pi R^2} = -\frac{\langle u \rangle_A{}^2 R^2}{48\alpha} \frac{d\langle T \rangle_A}{dx_1} \equiv -\alpha \frac{D_{xx}^d}{\alpha} \frac{d\langle T \rangle_A}{dx_1} \ , \tag{3.13}$$

where D_{xx}^d is the axial *Taylor dispersion* coefficient which is similar in form to the diffusion (or conduction) heat flux, and is given for this problem as

$$\frac{D_{xx}^d}{\alpha} = \frac{1}{48} Pe_R^2 \ , \quad Pe_R = \frac{\langle u \rangle_A R}{\alpha} \ . \tag{3.14}$$

Note that for laminar flows $Re_R < 1150$ and that $Pe_R = Re_R Pr$.

Aris (1956) shows that by adding the axial molecular diffusion (3.13) becomes

$$\frac{Q}{\rho c_p \pi R^2} = -\alpha (1 + \frac{D_{xx}^d}{\alpha}) \frac{d\langle T \rangle_A}{dx_1} \ . \tag{3.15}$$

As was mentioned, the constant $\partial T(x_1)/\partial x_1$ assumption can be removed and the Taylor dispersion relation still holds. Then by adding the time-dependent term and moving back to the x coordinate, we have the following area-averaged energy equation for the internal flow considered

$$\frac{\partial \langle T \rangle_A}{\partial t} + \langle u \rangle_A \frac{\partial \langle T \rangle_A}{\partial x} = \frac{\partial}{\partial x} \alpha (1 + \frac{D_{xx}^d}{\alpha}) \frac{\partial \langle T \rangle_A}{\partial x} \ . \tag{3.16}$$

Then by cross-sectional area averaging the energy equation, the interaction of the *lateral* transport and *axial* velocity variation appears as an enhancement in the *axial* diffusion. In the following the spatial averaging is taken over a small representative *volume*, and then the volume-averaged equations will contain the *deviant transport* (due to $\mathbf{u}'$ and T') occurring within this small volume as an *extra* diffusion term.

3.2.2 Local Volume Averaging

As an instructive example of the effective medium description of a two-phase system under the assumption of local thermal equilibrium, consider flow of a fluid through interstices of a stationary solid with periodic- and continuous-phase distributions. The periodicity allows for simple determination of the effective properties using just a unit cell in the structure. There are many applications in which the porous medium has a periodic structure. However, the unit-cell structure may not be simple. Figure 3.2(a) shows the structure of a polyurethane foam, and Figure 3.2(b) shows a *geometric model* of the unit cell. The unit cell is three-dimensional and realistic simulation of fluid flow and heat transfer for these cells are rather involved.

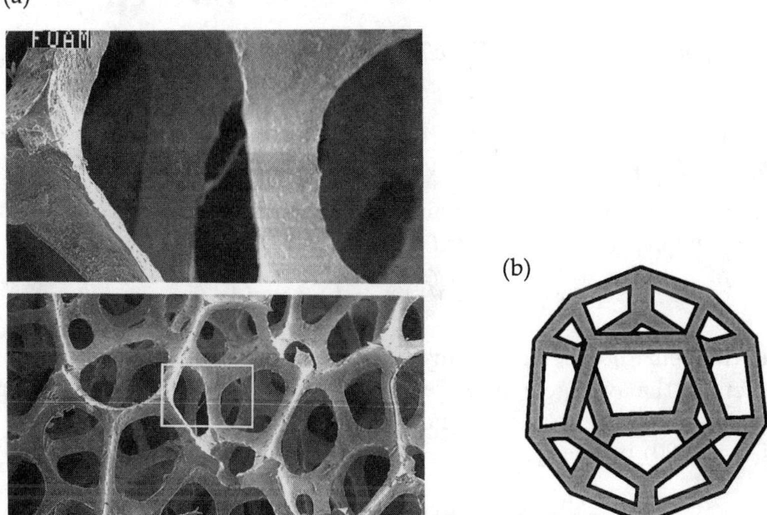

Figure 3.2. (a) Close-up view of a polyurethane foam showing the cellular structure. (b) A geometric model of the unit cell.

There are simpler two- and three-dimensional structures which can be studied for the examination of some of the fundamental features. Consider first the case where the conductivity of the solid is zero, $k_s = 0$. This allows for the simplification of the analysis and for the direct comparison with the Taylor dispersion just discussed. The complete analysis is given by Carbonell and Whitaker (1983). Brenner (1980) also gives a formulation for the local-volume-averaged treatment of flow and thermal energy transport through periodic structures. His treatment of the Taylor dispersion and the use of the closure conditions are similar to those of Carbonell and Whitaker (1983).

The energy equation (1.84) in the fluid phase for the case of no heat generation, radiative heat flux, or viscous dissipation becomes

$$\frac{\partial T}{\partial t} + \nabla \cdot \mathbf{u} T = \alpha_f \nabla^2 T \quad \text{in } V_f \ . \tag{3.17}$$

The boundary conditions on the fluid-solid interface A_{fs} are

$$\mathbf{u} = 0 \ , \quad \mathbf{n} \cdot \nabla T = 0 \quad \text{on } A_{fs} \ , \tag{3.18}$$

where $\mathbf{n}$ is the normal to A_{fs} facing toward the fluid phase.

Following Carbonell and Whitaker (1983), the *fluid-phase* local volume average of a scalar ψ is defined as

$$\langle \psi \rangle^f \equiv \frac{1}{V_f} \int_{V_f} \psi \, dV \ . \tag{3.19}$$

The fluid-phase volume average of the energy equation (3.17) is

$$\frac{\partial \langle T\rangle^f}{\partial t} + \langle \nabla \cdot \mathbf{u}T\rangle^f = \alpha_f \langle \nabla \cdot \nabla T\rangle^f \tag{3.20}$$

where

$$\langle T\rangle^f = \frac{1}{V_f}\int_{V_f} T\,dV \,. \tag{3.21}$$

Slattery (1981) and Gray et al. (1993) give the proof of the *theorem* of the *volume average of a gradient* which states

$$\langle \nabla\psi\rangle^f = \nabla\langle\psi\rangle^f + \frac{1}{V_f}\int_{A_{fs}} \psi\mathbf{n}\,dA \,. \tag{3.22}$$

This theorem shows that the average of a gradient is *different* from the gradient of the average by the area integral of the scalar over the interface. This difference can be substantial.

Then (3.20) becomes

$$\frac{\partial \langle T\rangle^f}{\partial t} + \nabla \cdot \langle \mathbf{u}T\rangle^f = \alpha \nabla \cdot \langle \nabla T\rangle^f \,, \tag{3.23}$$

where the zero gradient condition on A_{fs}, given by (3.18), has been applied to the expansion of the last term in (3.20). Applying the theorem again, the result is

$$\frac{\partial \langle T\rangle^f}{\partial t} + \nabla \cdot \langle \mathbf{u}T\rangle^f = \alpha_f \nabla \cdot \nabla\langle T\rangle^f + \alpha_f \nabla \cdot \frac{1}{V_f}\int_{A_{fs}} T\mathbf{n}\,dA \,. \tag{3.24}$$

To treat $\langle \mathbf{u}T\rangle^f$, define the decompositions

$$T = \langle T\rangle^f + T' \,, \tag{3.25}$$

$$\mathbf{u} = \langle \mathbf{u}\rangle^f + \mathbf{u}' \,, \tag{3.26}$$

where the prime indicates the *pointwise deviation* from the local phase-volume average value.

Then the convective term becomes

$$\frac{1}{V_f}\int_{V_f} \mathbf{u}T\,dV = \langle \mathbf{u}\rangle^f\langle T\rangle^f + \langle \mathbf{u}'T'\rangle^f \,. \tag{3.27}$$

Note that this is similar to the result of the *time* averaging for the case of turbulent flows in Section 2.4.1. Using the energy equation (3.27) and decomposing the area integral term and noting that the spatial variation of T' over ℓ (the linear dimension of the representative elementary volume) are much larger than $\langle T\rangle^f$, we have

$$\frac{\partial \langle T\rangle^f}{\partial t} + \nabla \cdot \langle \mathbf{u}\rangle^f\langle T\rangle^f + \nabla \cdot \langle \mathbf{u}'T'\rangle^f =$$
$$\alpha_f \nabla^2\langle T\rangle^f + \alpha_f \nabla \cdot \frac{1}{V_f}\int_{A_{fs}} T'\mathbf{n}\,dA \,. \tag{3.28}$$

The term $\langle \mathbf{u}'T'\rangle^f$ is due to the deviations of $\mathbf{u}$ and T within the fluid elementary volume. This is similar to the Taylor dispersion given by (3.11). To proceed further, either direct evaluation of $\mathbf{u}'$ and T' is made using a direct simulation or some approximations are introduced. In general, $\mathbf{u}' = \mathbf{u} - \langle \mathbf{u} \rangle^f$ is obtainable by solving the pointwise continuity and momentum equations for periodic structures. The distribution of T' can be found by arriving at a T' equation using the volume-averaged energy equation (3.28) and the pointwise energy equation (3.17) with the substitution of (3.25) and (3.26). The result is

$$\frac{\partial T'}{\partial t} + \mathbf{u} \cdot \nabla T' + \mathbf{u}' \cdot \nabla \langle T \rangle^f = \alpha_f \nabla^2 T' - \alpha_f \nabla \cdot \frac{1}{V_f} \int_{A_{fs}} T' \mathbf{n} \, dA + \nabla \cdot \langle \mathbf{u}' T' \rangle^f . \tag{3.29}$$

Some simplifications are made by dropping terms which are small compared to others. For large time asymptotic behavior, the first term is negligible. The last two terms can also be shown to be negligible using length scales ℓ and L and the order-of-magnitude analysis. It is also convenient to use a *closure condition* to replace T' with a more readily manipulable quantity such as a *transformation vector* $\mathbf{b}$. This is similar to a Taylor series expansion of T' using the first and higher derivatives of $\langle T \rangle^f$. The first-order approximation gives

$$T' = \mathbf{b}(\mathbf{x}) \cdot \nabla \langle T \rangle^f , \tag{3.30}$$

where $\mathbf{b}(\mathbf{x})$ is the *vector transformation function* which has a pointwise variation similar to T' but scales the magnitude of T' to the magnitude of the gradient of the phase-volume-averaged temperature.

Using (3.30) in (3.29), the local volume-averaged equation becomes

$$\frac{\partial \langle T \rangle^f}{\partial t} + \langle \mathbf{u} \rangle^f \cdot \nabla \langle T \rangle^f = \alpha_f \nabla \cdot (\mathbf{I} + \frac{1}{V_f} \int_{A_{fs}} \mathbf{b} \mathbf{n} \, dA) \cdot \nabla \langle T \rangle^f - \nabla \cdot (\langle \mathbf{u}' \mathbf{b} \rangle^f \cdot \nabla \langle T \rangle^f) . \tag{3.31}$$

The $\mathbf{b}$ equation, subject to the simplification made in the T' equation mentioned above, becomes

$$\mathbf{u} \cdot \nabla (\mathbf{b} \cdot \nabla \langle T \rangle^f) + \mathbf{u}' \cdot \nabla \langle T \rangle^f = \alpha_f \nabla^2 (\mathbf{b} \cdot \nabla \langle T \rangle^f) \tag{3.32}$$

subject to

$$-\mathbf{n} \nabla (\mathbf{b} \cdot \nabla \langle T \rangle^f) = \mathbf{n} \cdot \nabla \langle T \rangle^f \quad \text{on } A_{fs} . \tag{3.33}$$

Further order-of-magnitude analysis of the $\mathbf{b}$ equation gives

$$\mathbf{u}' + \mathbf{u} \cdot \nabla \mathbf{b} = \alpha_f \nabla^2 \mathbf{b} \tag{3.34}$$

subject to

$$-\mathbf{n} \cdot \nabla \mathbf{b} = \mathbf{n} \quad \text{on } A_{fs} . \tag{3.35}$$

For periodic structures of period ℓ_i in the direction i, the periodic boundary on $\mathbf{b}$ is represented by

$$\mathbf{b}(\mathbf{x}+\ell_i) = \mathbf{b}(\mathbf{x}) \,. \tag{3.36}$$

For a given periodic structure $\mathbf{u}' = \mathbf{u} - \langle\mathbf{u}\rangle^f$ and then $\mathbf{b}$ are solved for using the continuity, momentum, and the $\mathbf{b}$ equation and the boundary conditions for one spatial period (i.e., on unit cell). This will be discussed in the next section.

The local volume-averaged energy equation can be written as

$$\frac{\partial\langle T\rangle^f}{\partial t} + \langle\mathbf{u}\rangle^f \cdot \nabla\langle T\rangle^f = \nabla \cdot [\frac{\mathbf{K}_e}{(\rho c_p)_f} + \mathbf{D}^d] \cdot \nabla\langle T\rangle^f \,, \tag{3.37}$$

where the fluid-phase *effective thermal conductivity tensor* $\mathbf{K}_e$ and the *dispersion tensor* $\mathbf{D}^d$ are defined as

$$\mathbf{K}_e = k_f(\mathbf{I} + \frac{1}{V_f}\int_{A_{fs}} \mathbf{n}\mathbf{b}\, dA) \qquad \text{for } k_s = 0 \tag{3.38}$$

$$\mathbf{D}^d = -\frac{1}{V_f}\int_{V_f} \mathbf{u}'\mathbf{b}\, dV \,, \tag{3.39}$$

where $\mathbf{I}$ is the identity tensor. Generalization to $k_s \neq 0$ is discussed in the following section.

In this example, the use of the local-volume-averaging of the energy equation, leading to (3.37) and the appearance of the effective medium properties such as $\mathbf{K}_e$ and $\mathbf{D}^d$, is demonstrated. Whenever the assumption of local thermal equilibrium can be justified, the *single-medium treatment* (or the *single-equation model*) of the energy conservation equation in the two-phase systems simplifies the treatment significantly. The *effective* medium *properties* appearing in the local-volume-averaged energy equation are discussed next.

3.2.3 Effective Medium Properties

The effective medium properties $\mathbf{K}_e$ and $\mathbf{D}^d$ appearing in the local-volume-averaged equation need to be determined from the volume average of the $\mathbf{u}'$ and T' distributions. In this section we first show how the formulation of Whitaker-Carbonell reduces to that of Taylor for a circular tube. Then we extend (3.37) to cases where $k_s \neq 0$ and discuss the numerical solutions for $\mathbf{K}_e$ and $\mathbf{D}^d$ for a two-dimensional periodic structure.

(A) Reduction to Taylor Dispersion

For fully developed laminar flow through a tube of zero wall conductivity, the one-dimensional local area-averaged (taking the length of the representative elementary volume along the x-axis to be infinite) energy equation is the simplified form of (3.37), i.e.,

$$\frac{\partial \langle T \rangle^f}{\partial t} + \langle u \rangle^f \frac{\partial \langle T \rangle^f}{\partial x} + [\frac{k_{e_{xx}}}{(\rho c_p)_f} + D^d_{xx}] \frac{\partial^2 \langle T \rangle^f}{\partial x^2} , \tag{3.40}$$

where

$$k_{e_{xx}} = k_f (1 + \frac{1}{V_f} \int_{A_{fs}} n_x b_x \, dA) \tag{3.41}$$

$$D^d_{xx} = -\frac{1}{V_f} \int_{V_f} u' b_x \, dV . \tag{3.42}$$

Since $n_x = 0$ (note that $\mathbf{n} = -\mathbf{s}_r$ in a tube with $r = 0$ at the centerline, i.e., $\mathbf{s}_r$ pointing towards the centerline), then $k_{e_{xx}} = k_f$.

The $\mathbf{b}$ equation (3.34) and the boundary conditions (3.35) become

$$u' + u \frac{\partial b_x}{\partial x} = \alpha_f (\frac{\partial^2 b_x}{\partial x^2} + \frac{1}{r} \frac{\partial}{\partial r} r \frac{\partial b_x}{\partial r}) , \tag{3.43}$$

$$-[(\mathbf{n} \cdot \nabla) b]_x = -\frac{\partial b_x}{\partial r} = n_x = 0 \quad \text{on } A_{fs} , \tag{3.44}$$

or

$$\frac{\partial b_x}{\partial r} = 0 \quad \text{on } r = R . \tag{3.45}$$

Since no axial periodicity exists for this internal flow, then condition (3.36) is interpreted as

$$\frac{\partial b_x}{\partial x} = 0 = \frac{\partial^2 b_x}{\partial x^2} , \tag{3.46}$$

implying an infinite period.

The elementary volume extends laterally from 0 to R and from 0 to 2π and axially has a zero length. Then from the definition of $\mathbf{b}(\mathbf{x})$ given by (3.30), the condition

$$\langle b_x \rangle^f = 0 = \frac{2}{R^2} \int_0^R b_x r \, dr \tag{3.47}$$

is obtained.

Using (3.46), the $\mathbf{b}$ equation (3.43) becomes

$$u' = \alpha_f \frac{1}{r} \frac{\partial}{\partial r} r \frac{\partial b_x}{\partial r} . \tag{3.48}$$

The velocity distribution given in terms of the disturbance (or deviation) is

$$u' = u - \langle u \rangle^f = \langle u \rangle^f (1 - 2\frac{r^2}{R^2}) . \tag{3.49}$$

Substituting this in (3.48) and solving for b_x and using the conditions (3.45) and (3.47) to determine the constants of the integration, the result is

$$b_x(r) = \frac{\langle u \rangle^f R^2}{4\alpha_f} (\frac{r^2}{R^2} - \frac{1}{2}\frac{r^4}{R^4} - \frac{1}{3}) . \tag{3.50}$$

Now the *axial dispersion coefficient* can be evaluated using this in (3.42). The resulting local-volume-averaged energy equation becomes

$$\frac{\partial \langle T\rangle^f}{\partial t} + \langle u\rangle^f \frac{\partial \langle T\rangle^f}{\partial x} = \alpha_f \left(1 + \frac{{\langle u\rangle^f}^2 R^2}{48\alpha_f^2}\right) \frac{\partial^2 \langle T\rangle^f}{\partial x^2} \tag{3.51}$$

which is the same as the Taylor result given by (3.14), where $\langle u\rangle^f = \langle u\rangle_A$ from the averaging volume defined by (3.47).

The dispersion coefficient is a function of the Péclet number based on the linear dimension of the averaging volume. The constant of proportionality and the power of the Péclet number depend on the solid and fluid phase distributions in the elementary volume (including the presence or lack of an order in these phase distributions). Then in general the axial dispersion coefficient is given as

$$\frac{D_{xx}^d}{\alpha_f} = a_1 Pe_R^{a_2} \ . \tag{3.52}$$

(B) Generalization to $k_s \neq 0$

The stationary solid-phase energy equation for the case of negligible radiation, heat generation, and viscous dissipation becomes

$$\frac{\partial T}{\partial t} = \alpha_s \nabla^2 T \quad \text{in } V_s \ . \tag{3.53}$$

When heat transfer through the solid is allowed, then the boundary conditions on A_{fs} are modified to

$$\mathbf{n}_{fs} \cdot k_f \nabla T_f = \mathbf{n}_{fs} \cdot k_s \nabla T_s \quad \text{on } A_{fs} \ , \tag{3.54}$$

$$T_s = T_f \quad \text{on } A_{fs} \ . \tag{3.55}$$

As discussed in Section 1.8.1, the local volume average of a scalar ψ is defined as

$$\langle \psi \rangle = \frac{1}{V} \int_V \psi \, dV \ , \tag{3.56}$$

where $V = V_f + V_s$. Then the local volume average is taken over each phase and the local thermal equilibrium condition is imposed as

$$\langle T\rangle = \frac{1}{V}\int_V T\, dV \equiv \frac{1}{V}\int_{V_f} T\, dV + \frac{1}{V}\int_{V_s} T\, dV =$$
$$\epsilon \langle T\rangle^f + (1-\epsilon)\langle T\rangle^s = \langle T\rangle^f = \langle T\rangle^s \ , \tag{3.57}$$

where the *porosity* ϵ is the fraction of the elementary volume occupied by the fluid, i.e.,

$$\epsilon = \frac{V_f}{V} \ . \tag{3.58}$$

The resulting fluid-solid energy equation is

$$[\epsilon(\rho c_p)_f + (1-\epsilon)(\rho c_p)_s]\frac{\partial \langle T\rangle}{\partial t} + (\rho c_p)_f \langle \mathbf{u}\rangle \cdot \nabla \langle T\rangle = [\epsilon k_f + (1-\epsilon)k_s]\nabla \cdot \nabla \langle T\rangle + \frac{k_f - k_s}{V}\int_{A_{fs}} T_f' \mathbf{n}_{fs}\, dA + (\rho c_p)_f \nabla \cdot \langle \mathbf{u}' T_f'\rangle \,. \tag{3.59}$$

Since the local-volume-averaged velocity $\langle \mathbf{u}\rangle$ is averaged due v_f and v_s and in the solid phase $\mathbf{u} = 0$,

$$\langle \mathbf{u}\rangle = \epsilon \langle \mathbf{u}\rangle^f \equiv \mathbf{u}_D \,, \tag{3.60}$$

i.e., the fictitious *filter* or *Darcean velocity* $\mathbf{u}_D$ flows through the cross section of both phases. This is a convenient velocity to use, along with the actual cross section of the fluid-solid system, for the determination of the volumetric flow rate.

The energy equation can be written as

$$\frac{\epsilon(\rho c_p)_f + (1-\epsilon)(\rho c_p)_s}{(\rho c_p)_f}\frac{\partial \langle T\rangle}{\partial t} + \mathbf{u}_D \cdot \nabla \langle T\rangle = \nabla \cdot [\frac{\mathbf{K}_e}{(\rho c_p)_f} + \epsilon \mathbf{D}^d] \cdot \nabla \langle T\rangle = \nabla \cdot \mathbf{D} \cdot \nabla \langle T\rangle \tag{3.61}$$

where the effective thermal conductivity tensor $\mathbf{K}_e$ and the thermal dispersion tensor $\mathbf{D}^d$ are defined as

$$\mathbf{K}_e = [\epsilon k_f + (1-\epsilon)k_s]\mathbf{I} + \frac{k_f - k_s}{V}\int_{A_{fs}} \mathbf{n}_{fs}\mathbf{b}_f \, dA \tag{3.62}$$

$$\mathbf{D}^d = -\frac{1}{V_f}\int_{V_f} \mathbf{u}'\mathbf{b}_f \, dV \,. \tag{3.63}$$

Then the *total thermal diffusivity tensor* is defined as

$$\mathbf{D} = \frac{\mathbf{K}_e}{(\rho c_p)_f} + \epsilon \mathbf{D}^d \,. \tag{3.64}$$

The closure conditions used are

$$T_f' = T_f - \langle T\rangle = \mathbf{b}_f \cdot \nabla \langle T\rangle \,, \tag{3.65}$$

$$T_s' = T_s - \langle T\rangle = \mathbf{b}_s \cdot \nabla \langle T\rangle \,. \tag{3.66}$$

The $\mathbf{b}$ equations and their boundary conditions become

$$\mathbf{u}' + \mathbf{u}\cdot \nabla \mathbf{b}_f = \alpha_f \nabla^2 \mathbf{b}_f \quad \text{in } V_f \,, \tag{3.67}$$

$$\nabla^2 \mathbf{b}_s = 0 \quad \text{in } V_s \,, \tag{3.68}$$

$$k_f \mathbf{n}_{fs}\cdot \nabla \mathbf{b}_f = k_s \mathbf{n}_{fs}\cdot \nabla \mathbf{b}_s + \mathbf{n}_{fs}(k_f - k_s) \quad \text{on } A_{fs} \,, \tag{3.69}$$

$$\mathbf{b}_f = \mathbf{b}_s \quad \text{on } A_{fs} \,, \tag{3.70}$$

$$\mathbf{b}_f(\mathbf{x} + \boldsymbol{\ell}_i) = \mathbf{b}_f(\mathbf{x}) \,, \quad \mathbf{b}_s(\mathbf{x} + \boldsymbol{\ell}_i) = \mathbf{b}_s(\mathbf{x}) \,. \tag{3.71}$$

Next, as an example, the solution to the $\mathbf{b}$ equations and $\mathbf{K}_e$ and $\mathbf{D}^d$ are reviewed for a simple, two-dimensional periodic structure.

(C) A Two-Dimensional Example

Laminar flow and heat transfer through a two-dimensional, periodic structure made of cylinders, with fluid flowing perpendicular to their axes, has been used as a model. The cylinders can be arranged in-line or be staggered. A *unit-cell*, which contains a single particle with the boundaries chosen such that the symmetries and periodicities are properly prescribed on these boundaries, is used in the analysis. The *unit-cell linear length* ℓ will be larger than the particle diameter d, when the particles are not touching (which is the case for cross flow through these two-dimensional structures). For a two-dimensional problem, the volume-averaged energy equation becomes

$$\frac{\epsilon(\rho c_p)_f + (1-\epsilon)(\rho c_p)_s}{(\rho c_p)_f}\frac{\partial\langle T\rangle}{\partial t} + u_D\frac{\partial\langle T\rangle}{\partial x} + v_D\frac{\partial\langle T\rangle}{\partial y} =$$
$$\frac{\partial}{\partial x}[\frac{k_{e\,xx}}{(\rho c_p)_f} + \epsilon D^d_{xx}]\frac{\partial\langle T\rangle}{\partial x} + \frac{\partial}{\partial y}[\frac{k_{e\,yy}}{(\rho c_p)_f} + \epsilon D^d_{yy}]\frac{\partial\langle T\rangle}{\partial y} \,. \quad (3.72)$$

Because of the flow direction dependence, $\mathbf{D}^d$ is *anisotropic* even for isotropic structures. The properties of $\mathbf{D}^d$ are discussed by Kaviany (1991). The effective directional thermal conductivities $k_{e\,xx}$ and $k_{e\,yy}$ and the directional dispersion coefficients D^d_{xx} and D^d_{yy} would depend on the particle radius R, flow direction θ, Darcean velocities u_D and v_D, the ratio of the thermal conductivities k_f/k_s, the porosity ϵ, etc.

Sahraoui and Kaviany (1993, 1994) have solved the flow and heat transfer in these simple structures, and their results for the distributions of $b_x/|b_{x_{\max}}|$ and for the combined axial conductivity and axial dispersion coefficient are given in Figures 3.3 and 3.4, as a function of the cell Péclet number $Pe_\ell = u_D\ell/\nu$. The results are for unidirectional flow along the principal axis of the structure in the x-direction (i.e., $v_D = 0$). The b fields have more pronounced asymmetry as Pe_ℓ increases, as shown in Figure 3.3. Figure 3.4 shows that the predicted results for the effective properties are in good agreement with the available experimental results for spherical particles. A $Pe_\ell^{1.7}$ high Pe_ℓ asymptotic behavior is found which is similar to the Taylor dispersion behavior of Pe^2. This is because the flow through the in-line arrangement of cylinders is similar to that through a tube, where the fluid-particle path is rather straight and undisturbed. In the staggered arrangement, the power of the Pe_ℓ dependency drops to a value of nearly *near unity* (Sahraoui and Kaviany, 1994). The thermal and hydrodynamic effective properties of porous media (e.g., the permeability tensor) are reviewed by Kaviany (1991).

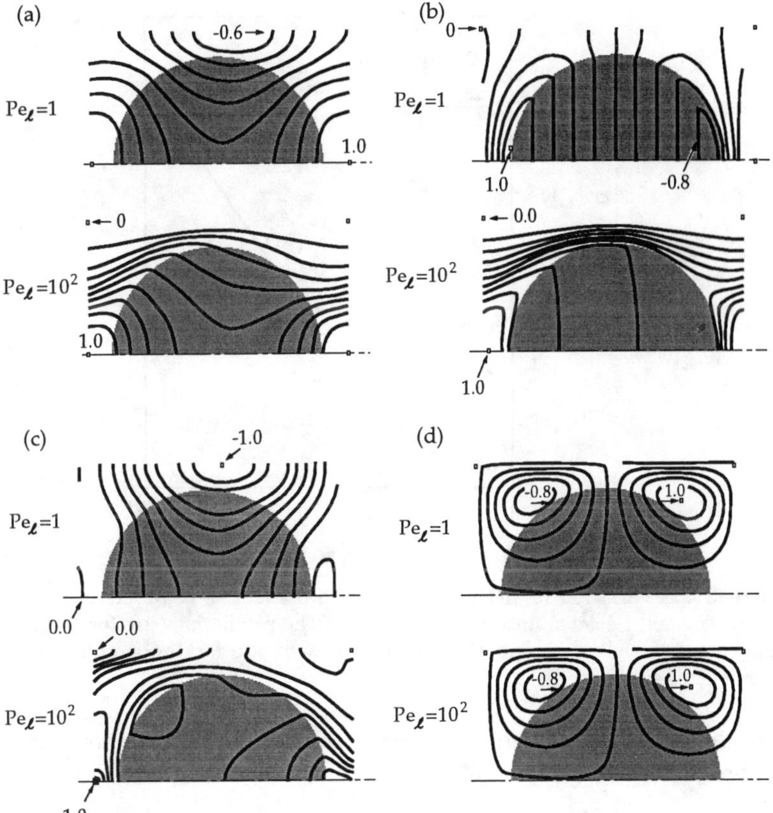

Figure 3.3. (a) Contours of constant $b_x/|b_{x_{max}}|$, in-line arrangement, $\epsilon = 0.5$, $k_s/k_f = 1$, $Re_\ell = 0.01$, (b) for $k_s/k_f = 10^2$, (c) contours of constant $b_y/|b_{y_{max}}|$, $k_s/k_f{=}1$, $\epsilon = 0.5$, $Re_\ell = 0.01$, (d) for staggered arrangement.

3.2.4 Volume-Averaged Treatment of Hydrodynamics

The local volume-averaged momentum equation describing a single-phase flow through a *stationary solid matrix* is given by the *empirical model* of Darcy which relates the external pressure gradient to the local volume averaged *solid-fluid interfacial drag*. The drag is related to the fluid *viscosity*, *specific surface area* (i.e., interfacial area per unit volume), as well as the average *particle size* and the *interstitial clearance distance*. The last three are *geometric parameters* and are combined in a single geometric parameter called the permeability K. When the distribution of the solid in the matrix is not isotropic, the *permeability tensor* $\mathbf{K}$ is used. The Darcy model which is valid when *pore-level* Reynolds number Re_ℓ, where ℓ is the characteristic

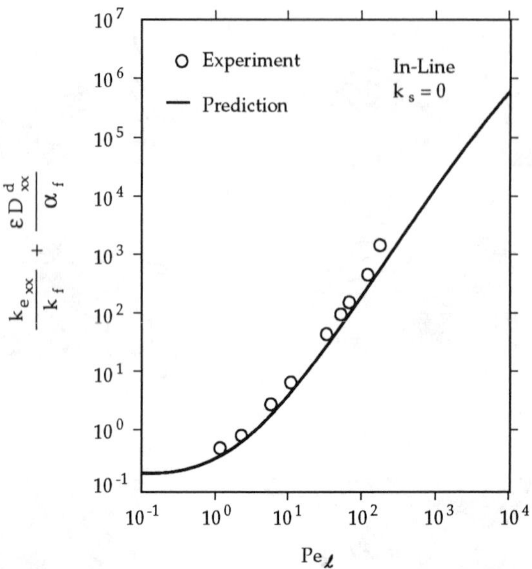

Figure 3.4. Variation of the normalized, axial effective conductivity with respect to the cell Péclet number. The results of the predictions are for the in-line arrangement of cylinders and those of experiment are for the simple-cubic arrangement of spheres.

linear dimension of the pore, approaches zero (i.e., for Stokes flow through the pores), is (Scheidegger, 1974)

$$-\nabla \langle p \rangle^f = -\nabla p = \frac{\mu}{\mathbf{K}} \langle \mathbf{u} \rangle \equiv \frac{\mu}{\mathbf{K}} \mathbf{u}_D \,, \tag{3.73}$$

$$Re_\ell = \frac{\rho |\mathbf{u}_D| \ell}{\epsilon \mu} \,. \tag{3.74}$$

The *Darcean velocity* $\mathbf{u}_D$ is the local volume-averaged velocity and ϵ is the porosity. The permeability tensor $\mathbf{K}$ can be predicted for simple solid phase distributions (i.e., simple structures) such as a single-particle two- or three-dimensional unit cells (Kaviany, 1991).

The Darcy model was originally developed using the results of a *one-dimensional* flow experiment for $Re_\ell \to 0$. Other *one-dimensional* flow experiments at higher values of Re_ℓ show a deviation from the Darcy model and a gradual approach towards an asymptote which relates the pressure gradient to u_D^2.

The experiment of Dybbs and Edwards (1984), in which the index of refraction of the transparent particles is matched with the flowing liquid so that an optical pointwise velocity measurement can be made, show that based on the magnitude of Re_ℓ several flow regimes exists in the pores. For $Re_\ell < 1$, the Darcy or the *creeping flow regime* exists where the viscous force dominate. For 1 to $10 < Re_\ell < 150$ an *inertial flow regime* exists, where the flow is steady but due to the dominance of the inertial force, boundary layers are formed around the solid and an inertial core exists in the midsection of the pores. For $150 < Re_\ell < 300$ an *unsteady laminar flow regime* exists where oscillations of the order of 1 Hz and amplitudes of the order of one-tenth of the particle size are found, but the flow is laminar. For $300 < Re_\ell$ an *unsteady and chaotic flow regime exists* which is not characteristic of a turbulent flow. The flow regimes are further discussed by Kaviany (1991).

Rigorous derivation of the local volume-averaged momentum equation, including the *inertial* and the *boundary* effects, has been discussed by Slattery (1981), Whitaker (1986), and Bear and Bachmat (1991). There are also available some semi-empirical momentum equations. One of these extends the available experimental results for one-dimensional flows and gives the ith component of the momentum equation in the *Cartesian coordinates* as (Vafai and Tien, 1981; Kaviany, 1991)

$$\frac{\rho}{\epsilon}\frac{D}{Dt}u_{D_i} = -\frac{\partial p}{\partial x_i} + \rho g_i + \frac{\mu}{\epsilon}\nabla^2 u_{D_i} - \frac{\mu}{K}u_{D_i} - \frac{C_E}{K^{1/2}}\rho|u_{D_i}|u_{D_i} + \rho g_i\,, \tag{3.75}$$

where C_E is the *Ergun constant.*

3.2.5 Radiation

Radiation heat transfer in porous media becomes important in low-temperature applications (e.g., in refrigeration and in cryogenics) and high-temperature applications (e.g., in combustion). The fluid is generally a gas and the radiative properties of the solid matrix dominates the radiative heat transfer. The optical properties of the solid, i.e., the *index* of *refraction* and *extinction*, n_s and κ_s, vary greatly with the wavelength. The spectral variation of n_s and κ_s for various solids is reviewed by Kaviany (1991). The radiation properties, $\sigma_{\lambda a}$, $\sigma_{\lambda s}$, and Φ, discussed in Section 2.5, need to be evaluated and averaged over the medium to obtain the *effective medium radiation* properties designated by $\langle\sigma_{\lambda a}\rangle$, $\langle\sigma_{\lambda s}\rangle$ and $\langle\Phi_\lambda\rangle$. The procedure for experimental evaluations of these effective properties are in principle similar to those used for gases. The predictions are also similar; however, the *matrix geometric parameters*, in addition to n_s and κ_s, must by used. For spherical particles, the properties have been determined, and a review is given by Kaviany (1991). Considering porous media made of particles, for most porous media the particles are placed close enough such that the scattering from the neighboring particles *interact*. This is called *dependent scattering*, and depending on the porosity can occur even if the ratio of the *average interparticle clearance* C to the wavelength, C/λ, is larger than unity. This can be shown by a Monte Carlo simulation, and since for most porous media the *radiation size parameter* $\alpha_r = \pi d/\lambda$ is much larger than unity, *geometric optics* can be used. The *dependent* scattering, of *high*-porosity *opaque* and *semi-transparent* particles is analyzed by Singh and Kaviany (1991). They also recommend a modification to the numerical discrete-ordinates method of solution of the equation of radiative transfer which allows for the inclusion of this particle-particle interaction (Singh and Kaviany, 1992).

In practice, for most porous media the *optically thick limit* of radiative heat transfer, and therefore, the diffusion approximation discussed in Section 2.5, is applicable. Then the concept of the radiant conductivity k_r, discussed in Section 2.5 and given by (2.119), can be used (Vortmeyer, 1978). For *opaque* spherical particles and by using the *radiation exchange factor* F, this radiant conductivity is approximated as (Brewster and Tien, 1982; Drolen and Tien, 1987; Brewster, 1992)

$$k_r = 4F\, d\, \sigma_{SB}\, T^3$$

or

$$F = \frac{k_r}{4d\, \sigma_{SB}\, T^3}\,, \quad F = F(\epsilon, \epsilon_r, {k_s}^*)\,, \quad {k_s}^* = \frac{k_s}{4d\sigma_{SB}T^3}\,. \tag{3.76}$$

The radiation conductivity has been discussed by Kaviany (1991) and for *opaque* particles depends on the thermal conductivity of the particles, the porosity ϵ, and the particle surface emissivity ϵ_r. It also slightly varies depending on whether the surface is *diffuse* or *specular*. This dependence for *opaque* spherical particles has been examined by Singh and Kaviany (1993). Figure 3.5 shows their results for F (or *dimensionless* k_r) for a packed bed of spherical particles (for porosities between 0.4 and 0.5, the variation of F with respect to the porosity is small). Their results for the Monte Carlo simulation over many particles show that F increases with the particle surface emissivity and that low- and high-k_s (normalized in a manner similar to k_r) asymptotes exist. For $k_s \to 0$, the results are independent of ϵ_r. The result of Figure 3.5 can be correlated with $F = a_1 \epsilon_r \tan^{-1}(a_2 k_s^{*\,a_3}/\epsilon_r) + a_4$. For diffusive surfaces, the constants are (0.5756, 1.5353, 0.8011, 0.1843) and for specular surfaces they are nearly the same.

3.2.6 Exothermic Reaction

Of significance to convective heat transfer is combustion in porous media which is a subset of the exothermic reactions in porous media. A review of the broad field of gas-solid reactions is given by Szekely et al. (1976). Here we consider combustion in porous media and divide this class of problems based on the role of the solid matrix. The solid matrix can be *inert*, *catalytic*, or *combustible*. Chart 3.1 gives a classification of combustion in porous media and the breakdown under these three categories. The catalytic reactions (or *heterogeneous reactions*) in porous media are discussed by Whitaker (1986) and Ochoa et al. (1986), among others. The kinetics of the catalytic reactions are discussed by Gates (1992) and Somorjai (1994).

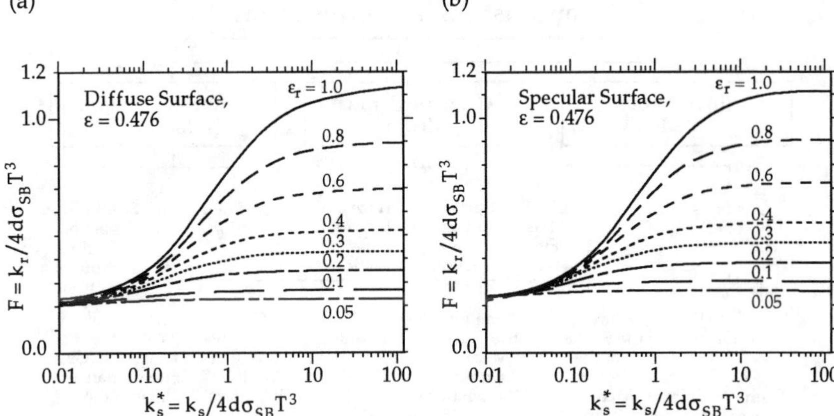

Figure 3.5. The variation of the dimensionless radiant conductivity with respect to the dimensionless solid phase thermal conductivity and for various particle surface emissivities. (a) For particles with a diffuse surface. (b) For specular surface. The porosity is 0.476.

A *catalyst* is a substance that *increases* the rate of approach to equilibrium of a chemical reaction *without* being substantially consumed in the reaction. A catalyst intervenes in forming chemical bonds to one or more reactions, and thereby, facilitates their conversion, but it does not significantly affect the reaction equilibrium. We will refer to the *surface* catalytic reactions in the nonequilibrium treatment of Chapters 4 and 5. In combustion in porous media with surface and gas phase reactions and where the assumption of local thermal equilibrium can be made (i.e., the gaseous surface affected reactions are not analyzed separately), a single chemical kinetic description is used for both phases such as (2.44) is used for the overall reaction. One example is incineration of packed beds of combustible particles (e.g., Fatehi and Kaviany, 1994).

3.2.7 Phase Change in Multicomponent Systems

As an example of phase change in solid-fluid flow systems with the assumption of local thermal equilibrium imposed, consider the formulation of solid-fluid phase change (solidification/melting or sublimation/frosting) of a *binary* mixture. For this problem, the equilibrium condition extends to the local *thermodynamic equilibrium* where the local phasic temperature (*thermal equilibrium*), pressure (*mechanical equilibrium*), and chemical potential (*chemical equilibrium*) are assumed to be equal between the solid and the fluid phases. This is stated as

$$\langle T\rangle^s = \langle T\rangle^f \ , \tag{3.77}$$

$$\langle p\rangle^s = \langle p\rangle^f \ , \tag{3.78}$$

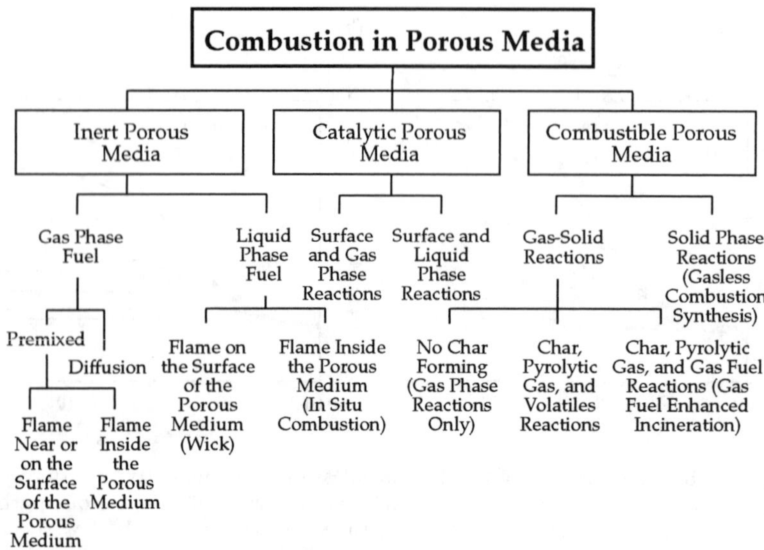

Chart 3.1. A classification of combustion in porous media based on solid matrix and other properties.

$$\langle \mu_A \rangle^s = \langle \mu_A \rangle^f \ , \quad \langle \mu_B \rangle^s = \langle \mu_B \rangle^f \ . \tag{3.79}$$

The local volume averaging of the energy equations, with allowance for a phase change in a binary system, has been discussed by Bennon and Incropera (1987), Rappaz and Voller (1990), Poirier et al. (1991), and Hills et al. (1992). In their *single-medium* treatment, i.e., a local volume-averaged description with the assumption of local thermodynamic equilibrium, a distinct interface between the *region* of *solid phase* and the *region* of *fluid phase* has not been assumed; instead, the fluid-phase volume fraction (i.e., the porosity) is allowed to change *continuously*. This *continuous medium* treatment is in contrast to the *multiple-medium treatment* which allows for separate *solid*, *fluid*, and *mushy* (i.e., solid-fluid) media (e.g., Worster, 1991; Kim and Kaviany, 1992; Vodak et al., 1992). Since the *explicit* tracking of the various distinct interfaces, as defined in the multiple-medium treatment, is *not* needed in the continuous single-medium treatment, it is *easier* to implement. Also, for binary systems the equilibrium temperature depends on the local species concentrations and due to the variation of the concentration within the medium, the phase transition occurs over a range of temperatures, and therefore, the single-medium treatment is even more suitable. In the following, the single-medium treatment of Bennen and Incropera (1987) is reviewed. Many simplifications made in the development of this treatment are discussed by Hills et al. (1992). Alternative derivations and assumptions are discussed by Rappaz and Voller (1990) and Poirier et al. (1991), among others. These will be discussed in Section 4.16.1.

The energy equation for each phase and for the case of negligible viscous dissipation and radiative heat flux is the simplified form of (1.83), i.e.,

$$\frac{\partial \rho_s i_s}{\partial t} + \nabla \cdot \rho_s \mathbf{u}_s i_s = \nabla \cdot k_s \nabla T_f + \dot{s}_s \,, \tag{3.80}$$

$$\frac{\partial \rho_f i_f}{\partial t} + \nabla \cdot \rho_f \mathbf{u}_f i_f = \nabla \cdot k_f \nabla T_f + \dot{s}_f \,. \tag{3.81}$$

The specific enthalpies are defined as

$$i_s = \int_0^T c_{p_s}\, dT + i_{s_o} \,, \tag{3.82}$$

$$i_f = \int_0^T c_{p_f}\, dT + i_{f_o} \,, \tag{3.83}$$

where c_{p_s} and c_{p_f} are the *mixture*-specific heat of the phases (with each phase made of species A and B). The local phase-volume-averaged energy equations are found by the averaging of (3.80) and (3.81) over their perspective phase. Decompositions similar to those defined in (3.25) and (3.26) are used for T_s, i_s, and i_f. When these two equations are added, the result is similar to (3.59) and is

$$\frac{\partial}{\partial t}[(1-\epsilon)\langle\rho\rangle^s\langle i\rangle^s + \epsilon\langle\rho\rangle^f\langle i\rangle^f] + \nabla\cdot[(1-\epsilon)\langle\rho\rangle^s\langle i\rangle^s\langle\mathbf{u}\rangle^s + \epsilon\langle\rho\rangle^f\langle i\rangle^f\langle\mathbf{u}\rangle^f] = \nabla\cdot\{[(1-\epsilon)k_s + \epsilon k_f]\nabla\langle T\rangle + \frac{k_f - k_s}{V}\int_{A_{fs}} T_f' \mathbf{n}_{fs}\, dA\} + \nabla\cdot[(1-\epsilon)\langle\rho\rangle^s\langle\mathbf{u}'i'\rangle^s + \epsilon\langle\rho\rangle^f\langle\mathbf{u}'i'\rangle^f] \,, \tag{3.84}$$

where the porosity ϵ was defined by (3.58) and we have used $\epsilon\dot{s}_f + (1-\epsilon)\dot{s}_s = 0$, for no net energy production. Note that in contrast to the porous media with no change, the local porosity can change significantly due to the phase change. In the treatment of Bennen and Incropera (1987), the last three terms on the right-hand side, which influence the effective thermal conductivity and introduce the dispersion tensor, have been neglected. The dispersion contribution has been assumed *negligible*, due to the assumed low velocities. If the dispersion contributions are included, then (3.84) can be written as

$$\frac{\partial}{\partial t}[((1-\epsilon)\langle\rho\rangle^s\langle i\rangle^s + \epsilon\langle\rho\rangle^f\langle i\rangle^f] + \nabla\cdot[(1-\epsilon)\langle\rho\rangle^s\langle i\rangle^s\langle\mathbf{u}\rangle^s + \epsilon\langle\rho\rangle^f\langle i\rangle^f\langle\mathbf{u}\rangle^f] =$$

$$\nabla\cdot[\mathbf{K}_e + (1-\epsilon)\langle\rho\rangle^f\langle c_p\rangle^f\mathbf{D}_s^d + \epsilon\langle\rho\rangle^f\langle c_p\rangle^f\mathbf{D}_f^d]\cdot\nabla\langle T\rangle \equiv \nabla\cdot\langle\rho\rangle^f\langle c_p\rangle^f\mathbf{D}\cdot\nabla\langle T\rangle \,, \tag{3.85}$$

where $\mathbf{K}_e$, $\mathbf{D}_f^d$, $\mathbf{D}_s^d$ and $\mathbf{D}$ are defined in a manner similar to (3.62) and (3.63).

Then the solid-fluid, i.e., *phase-mass-averaged*, enthalpy is defined using the volume-averaged density $\langle\rho\rangle = (1-\epsilon)\langle\rho\rangle^s + \epsilon\langle\rho\rangle^f$, and we have

$$\langle\rho\rangle\langle i\rangle \equiv (1-\epsilon)\langle\rho\rangle^s\langle i\rangle^s + \epsilon\langle\rho\rangle^f\langle i\rangle^f \,. \tag{3.86}$$

The *mass concentration* of the solid and liquid phases are $\epsilon\langle\rho\rangle^s/\langle\rho\rangle$ and $(1-\epsilon)\langle\rho\rangle^f/\langle\rho\rangle$, respectively.

The phase-mass-averaged velocity is

$$\langle\rho\rangle\langle \mathbf{u}\rangle \equiv (1-\epsilon)\langle\rho\rangle^s\langle \mathbf{u}\rangle^s + \epsilon\langle\rho\rangle^f\langle \mathbf{u}\rangle^f \,. \tag{3.87}$$

The convective term can be written as a phase-mixture component and a phase-relative component as (Bennon and Incropera, 1987)

$$(1-\epsilon)\langle\rho\rangle^s\langle i\rangle^s\langle \mathbf{u}\rangle^s + \epsilon\langle\rho\rangle^f\langle i\rangle^f\langle \mathbf{u}\rangle^f = \langle\rho\rangle\langle \mathbf{u}\rangle\langle i\rangle + \\ (1-\epsilon)\langle\rho\rangle^s(\langle \mathbf{u}\rangle^s - \langle \mathbf{u}\rangle)(\langle i\rangle^s - \langle i\rangle) + \epsilon\langle\rho\rangle^f(\langle \mathbf{u}\rangle^f - \langle \mathbf{u}\rangle)(\langle i\rangle^f - \langle i\rangle) \,. \tag{3.88}$$

Then (3.85) is written as

$$\frac{\partial}{\partial t}\langle\rho\rangle\langle i\rangle + \nabla\cdot\langle\rho\rangle\langle \mathbf{u}\rangle\langle i\rangle = \nabla\cdot\langle\rho\rangle^f\langle c_p\rangle^f\mathbf{D}\cdot\nabla\langle T\rangle - \\ \nabla\cdot[(1-\epsilon)\langle\rho\rangle^s(\langle \mathbf{u}\rangle^s - \langle \mathbf{u}\rangle)(\langle i\rangle^s - \langle i\rangle) + \\ \epsilon\langle\rho\rangle^f(\langle \mathbf{u}\rangle^f - \langle \mathbf{u}\rangle)(\langle i\rangle^f - \langle i\rangle)] \,. \tag{3.89}$$

Now using the definition of the specific enthalpy given by (3.82) and (3.83), the temperature gradient can be replaced with the gradient in the specific enthalpy of the solid as

$$\nabla\langle T\rangle = \frac{1}{\langle c_p\rangle^s}\nabla\langle i\rangle + \frac{1}{\langle c_p\rangle^s}\nabla(\langle i\rangle^s - \langle i\rangle) \,. \tag{3.90}$$

Then the equation of conservation of thermal energy, (3.89), becomes

$$\frac{\partial}{\partial t}\langle\rho\rangle\langle i\rangle + \nabla\cdot\langle\rho\rangle\langle \mathbf{u}\rangle\langle i\rangle = \\ \nabla\cdot\frac{\langle\rho\rangle^f\langle c_p\rangle^f\mathbf{D}}{\langle c_p\rangle^s}\cdot\nabla\langle i\rangle + \nabla\cdot\frac{\langle\rho\rangle^f\langle c_p\rangle^f\mathbf{D}}{\langle c_p\rangle^s}\cdot\nabla(\langle i\rangle^s - \langle i\rangle) - \\ \nabla\cdot[(1-\epsilon)\langle\rho\rangle^s(\langle \mathbf{u}\rangle^s - \langle \mathbf{u}\rangle)(\langle i\rangle^s - \langle i\rangle) + \\ \epsilon\langle\rho\rangle^f(\langle \mathbf{u}\rangle^f - \langle \mathbf{u}\rangle)(\langle i\rangle^f - \langle i\rangle)] \,. \tag{3.91}$$

The last two terms can be written as $-\nabla\cdot(1-\epsilon)\langle\rho\rangle^s(\langle \mathbf{u}\rangle - \langle \mathbf{u}\rangle^s)(\langle i\rangle^f - \langle i\rangle^s)$.

In addition, the equations for the conservation of overall *mass*, *momentum*, and *species* are also derived for this two-phase system by Bennen and Incropera (1987), and are

$$\frac{\partial\langle\rho\rangle}{\partial t} + \nabla\langle\rho\rangle\langle \mathbf{u}\rangle = 0 \,, \tag{3.92}$$

$$\frac{\partial\langle\rho\rangle\langle \mathbf{u}\rangle}{\partial t} + \nabla\cdot\langle\rho\rangle\langle \mathbf{u}\rangle\langle \mathbf{u}\rangle = -\nabla\langle p\rangle + \nabla\cdot\langle\mu\rangle^f\frac{\langle\rho\rangle}{\langle\rho\rangle^f}\nabla\langle \mathbf{u}\rangle - \\ \frac{\langle\mu\rangle^f\langle\rho\rangle}{\mathbf{K}\langle\rho\rangle^f}(\langle \mathbf{u}\rangle - \langle \mathbf{u}\rangle^s) + \langle\rho\rangle\mathbf{f} \,, \tag{3.93}$$

$$\frac{\partial\langle\rho_A\rangle}{\partial t} + \nabla\cdot\langle\rho_A\rangle^f\langle\mathbf{u}\rangle =$$
$$\nabla\cdot\langle\rho\rangle^s\mathbf{D}^s_m\cdot\nabla\frac{\langle\rho_A\rangle^s}{\langle\rho\rangle^s} + \nabla\cdot\langle\rho\rangle^f\mathbf{D}^f_m\cdot\nabla\frac{\langle\rho_A\rangle^f}{\langle\rho\rangle^f} -$$
$$\nabla\cdot[(1-\epsilon)\langle\rho\rangle^s(\langle\mathbf{u}\rangle^s - \langle\mathbf{u}\rangle)(\frac{\langle\rho_A\rangle^s}{\langle\rho\rangle^s} - \frac{\langle\rho_A\rangle}{\langle\rho\rangle}) +$$
$$\epsilon\langle\rho\rangle^f(\langle\mathbf{u}\rangle^f - \langle\mathbf{u}\rangle)(\frac{\langle\rho_A\rangle^f}{\langle\rho\rangle^f} - \frac{\langle\rho_A\rangle}{\langle\rho\rangle})]\,, \tag{3.94}$$

where $\langle\rho\rangle\mathbf{f}$ is the volumetric body force and $\mathbf{K}$ is the permeability tensor which changes with *time* and is *nonuniform* and can be significantly *anisotropic*. The momentum equation (3.93) reduces to the Darcy law (3.73) when $\mathbf{u}_s = 0$, and only the fluid properties, i.e., $\langle p\rangle = \langle p\rangle^f$, $\langle\rho\rangle = \langle\rho\rangle^f$, are used and the inertia, boundary, and body forces are negligible. However, it is slightly different from the semi-empirical, extended momentum equation given by (3.75). A discussion and the derivation of a more general momentum equation, applicable to higher velocities, is given by Ganesan and Poirier (1990). The *total phase, mass diffusivity tensors* $\mathbf{D}^s_m$ and $\mathbf{D}^f_m$ are defined in a manner similar to $\mathbf{D}$ discussed above. For the case of a stationary solid phase, $\mathbf{u}_s = 0$, and we have

$$\mathbf{u} = \frac{\epsilon\langle\rho\rangle^f}{\langle\rho\rangle}\langle\mathbf{u}\rangle^f\,. \tag{3.95}$$

The distribution of *phase mass fractions* $\langle\rho\rangle^s/\langle\rho\rangle$ and $\langle\rho\rangle^f/\langle\rho\rangle$, and the *phasic mass fraction of species* A, $\langle\rho_A\rangle^f/\langle\rho\rangle^f$ and $\langle\rho_A\rangle^s/\langle\rho\rangle^s$, are found using the above governing equations *and* the equilibrium phase diagram. This requires the assumption of local chemical equilibrium, which in turn assumes that the presence of temperature and concentration gradients do *not* alter the phase equilibria.

Consider, as an example, a *solid–liquid* phase change. Then we use the symbol ℓ instead of f. In the phase diagram the liquidus and solidus lines are assumed to be *straight*, as shown in Figure 3.6, compared to the actual diagram which has been depicted in Figure 1.6. Then the *equilibrium partition ratio* k_p is used to relate the solid and liquid phasic mass fraction of species A (on the *liquidus* and *solidus lines*) at a given temperature T and pressure p, i.e.,

$$k_p = \left.\frac{\langle\rho_A\rangle^s/\langle\rho\rangle^s}{\langle\rho_A\rangle^\ell/\langle\rho\rangle^\ell}\right|_{T,p}\,. \tag{3.96}$$

A finite, two-phase region (*mushy region*) can exist for $k_p < 1$ and corresponds to the case where species A has a *limited solubility* in the solid phase. For $k_p = 1$ a *discrete* phase change occurs.

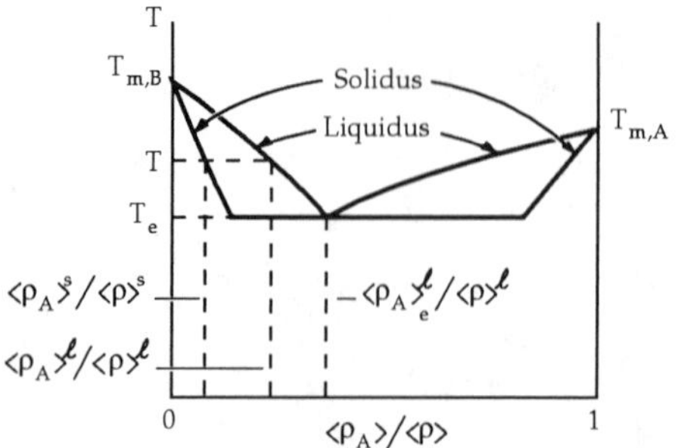

Figure 3.6. The phase diagram for a binary solid–liquid system. The eutectic temperature and species A mass fraction and a dendritic temperature and liquidus and solidus species A mass fractions, are also shown.

Note that

$$\frac{\langle\rho_A\rangle}{\langle\rho\rangle} = \frac{(1-\epsilon)\langle\rho_A\rangle^s}{\langle\rho\rangle} + \frac{\epsilon\langle\rho_A\rangle^\ell}{\langle\rho\rangle} . \tag{3.97}$$

Then the *solid-phase mass fraction*, under *phase equilibrium*, can be expressed in terms of k_p as

$$\frac{(1-\epsilon)\langle\rho\rangle^s}{\langle\rho\rangle} = \frac{\dfrac{\langle\rho_A\rangle^\ell}{\langle\rho\rangle^\ell} - \dfrac{\langle\rho_A\rangle}{\langle\rho\rangle}}{\dfrac{\langle\rho_A\rangle^\ell}{\langle\rho\rangle^\ell} - \dfrac{\langle\rho_A\rangle^s}{\langle\rho\rangle^s}} = \frac{1}{1-k_p}\frac{T-T_\ell}{T-T_{m,B}} , \tag{3.98}$$

where T is the local temperature, T_ℓ is the liquidus temperature (i.e., on the liquidus line) corresponding to $\langle\rho\rangle^s/\langle\rho\rangle$, and $T_{m,B}$ is the melting temperature for $\langle\rho_A\rangle/\langle\rho\rangle \rightarrow 0$ (i.e., pure species B). Then k_p is taken as the *ratio* of the slopes of the liquidus and solidus lines. The phase volume averaged mass fractions are related to the total volume-averaged mass fraction through

$$\frac{\langle\rho_A\rangle^s}{\langle\rho\rangle^s} = \frac{k_p}{1+(1-\epsilon)\dfrac{\langle\rho\rangle^s}{\langle\rho\rangle}(k_p-1)}\frac{\langle\rho_A\rangle}{\langle\rho\rangle} , \tag{3.99}$$

$$\frac{\langle\rho_A\rangle^\ell}{\langle\rho\rangle^\ell} = \frac{1}{1+(1-\epsilon)\dfrac{\langle\rho\rangle^s}{\langle\rho\rangle}(k_p-1)}\frac{\langle\rho_A\rangle}{\langle\rho\rangle} . \tag{3.100}$$

The unknowns ϵ, $\langle i\rangle^f$, $\langle i\rangle^s$, $\langle\mathbf{u}\rangle^f$, $\langle\rho_A\rangle$, $\langle\rho_A\rangle^f$, and $\langle\rho_A\rangle^s$, along with the intermediate variables $\langle i\rangle$, $\langle\rho\rangle$, and $\langle\mathbf{u}\rangle$, are found from (3.92) to (3.94), (3.96) to (3.100), and the definitions of averages. Bennon and Incropera (1987) apply these equations and obtain numerical solutions for a two-dimensional solidification problem. Since these problems consider the confining solid walls which are not in thermal equilibrium with the fluid-solid mixture, this solution is not presented here and instead is discussed in Section 4.16.

3.3 Two-Phase Systems with Both Phases Moving

The hydrodynamics of dispersed two-phase flows addresses the motion of dispersed elements *relative* to the host fluid. This relative velocity is sustained due to the *inertial* and *gravitational* forces, and is opposed by the interfacial drag force between the elements and the fluid. Therefore, the local volume-averaged *hydrodynamic* treatment of the two-phase systems with both phases moving requires a two-medium (or *slip*) treatment. When one of the phases is dispersed in the other, most of the existing hydrodynamic treatments also assume a *continuum* for the dispersed phase and a few maintain a *discrete* phase description with ensemble averaging over many realizations of particle trajectories. Here we begin with the general treatment of two-phase flows with both phases assumed continuous. Then the available descriptions for one of the phases being dispersed is considered. In the *heat transfer* treatment, the local thermal equilibrium is assumed between the phases so that a *single-continuum energy equation* (and a single local temperature) is used. The constraint is as before, i.e., the temperature difference across the system is much larger than that occurring locally between the two phases. Therefore,

$$\Delta T_d \ll \Delta T_\ell \ll \Delta T_L \ , \tag{3.101}$$

where again d is the linear dimension of the phasic elements, ℓ is the linear dimension of the representative elementary volume, and L is the linear dimension of the system. The thermal nonequilibrium of the two-phase systems will be considered in Chapters 5 and 6 for no effects due to the bounding surfaces, and in Chapters 7 and 8 for the lack of thermal equilibrium between the two phases and the bounding or submerged surfaces.

3.3.1 Time and Local Phase-Volume Averaging

The *continuum* treatment of particulate systems has been reviewed by Crowe and Smoot (1979), Meyer (1983), Couderc (1985), Michaelides (1986) and Soo (1989), among others. The treatment is similar to the general treatment of two-phase systems (including the liquid-gas systems) discussed by Lahey and Drew (1989). For solid particles, the a priori knowledge of the particle size and their number density in the host fluid, and the lack of any interfacial deformation and internal particle motion, simplify the treatment compared to the gas-liquid systems (including dispersed flows such as flows of bubbles and droplets).

The phasic energy equations for the continuum treatment of both phases can be obtained by the local *volume averaging* of the energy equation over each phase and by allowing for a local, relative velocity difference between the phases. The local concentration of the second-phase is $1-\epsilon$, where ϵ is the volume fraction of the first phase. In general, ϵ is not known a priori, although the dispersed phase mass flow rate through the system may be prescribed. Therefore, in contrast to the porous media treatment with no phase change, ϵ can *vary significantly* throughout the system (causing *dilute* and *dense* second-phase concentration regions).

Since most two-phase flows of practical importance are turbulent and by using a fixed control (Eulerian) volume the *instantaneous* volume fraction of the phases change rapidly, the *time averaging* of the energy equation (and the associated conservation equations) is needed. This results in a *space-time* averaging and the averaged equations are rather complex. Also, some *closure conditions* for the spatial- and temporal-averaged products of the spatial deviations and temporal fluctuations are needed. The treatment of Lahey and Drew (1989), which is reviewed below, addresses these averages and their complexities.

In Section 1.8 temporal and spatial averaging were discussed. In Section 2.4 temporal averaging was applied to the turbulent flows, and in Section 3.2 the spatial averaging was applied to solid-fluid systems with a continuous and stationary solid phase. In this section, these are all combined and applied to two-phase flows with both phases moving (with respect to a stationary reference frame and with respect to one another).

(A) Space- and Time-Averaging Theorems

Consider an Eulerian (i.e., fixed with respect to a reference) control volume V through which the phases flow. For two-phase flows $V=\sum V_j, j=1,2$, i.e., the control volume is completely filled by the two phases. Lahey and Drew (1989) consider the general case where the volume V is adjacent to a wall. Figure 3.7 shows this Eulerian control volume with no bounding surface adjacent to it. The volume V can *vary* in space (as used in the *nonuniform* finite-volume numerical treatments).

The Leibnitz rule and the Gauss theorem for a control volume enclosing phase j are, respectively,

$$\frac{\partial}{\partial t}\int_{V_j(\mathbf{x},t)}\psi(\mathbf{x},t)=\int_{V_j(\mathbf{x},t)}\frac{\partial\psi}{\partial t}\,dV+\int_{A_i(\mathbf{x},t)}\psi\mathbf{u}_i\cdot\mathbf{n}_j\,dt\,, \tag{3.102}$$

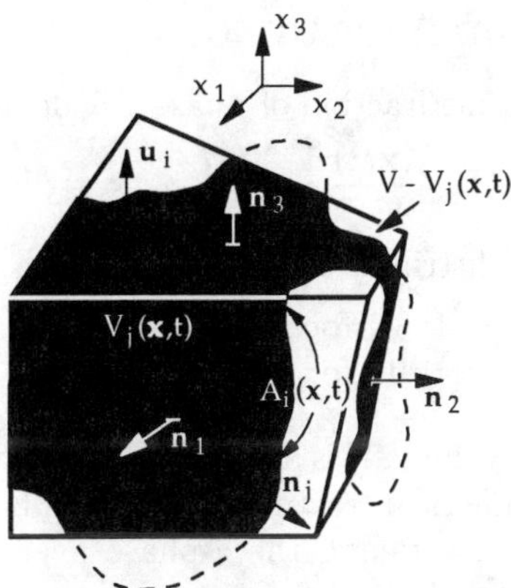

Figure 3.7. The fixed control volume containing, in part, the phase j, the interfacial area i and interfacial velocity $\mathbf{u}_i$. Surface normal unit vectors are also shown.

where A_i and $\mathbf{u}_i$ are the interfacial area and the interfacial velocity, respectively, and $\mathbf{n}_j$ is the interfacial normal unit vector pointing towards phase j, and

$$\int_{V_j(\mathbf{x},t)} \nabla \cdot \psi \, dV = \nabla \cdot \int_{V_j(\mathbf{x},t)} \psi \, dV + \int_{A_i(\mathbf{x},t)} \psi \cdot \mathbf{n}_j \, dA \, . \tag{3.103}$$

The total energy conservation equation is given by (1.78) and for the phase j, using the identity tensor $\mathbf{I}$, is written as

$$\frac{\partial}{\partial t}\rho_j(e_j + \frac{\mathbf{u}_j \cdot \mathbf{u}_j}{2}) + \nabla \cdot \rho_j(e_j + \frac{\mathbf{u}_j \cdot \mathbf{u}_j}{2})\mathbf{u}_j =$$
$$-\nabla \cdot [\mathbf{q}_j - (-p_j\mathbf{I} + \mathbf{S}_j) \cdot \mathbf{u}_j] + \rho_j \mathbf{g} \cdot \mathbf{u}_j + \dot{s}_j \, , \tag{3.104}$$

$$\mathbf{q}_j = -k\nabla T + \mathbf{q}_r \, . \tag{3.105}$$

The frequencies of temporal changes, which are significant to transport, span over many orders of magnitude, and the spatial variations of importance also range from interphasic to that over the system dimensions. As was discussed in Section 1.8, the space- and time-*averaged* energy equation, which is the form containing *some* of the features (i.e., transport mechanisms) associated with the small space- and time-scale variations, is of practical interest. In the following, the spatial averaging is performed first using a treatment similar to that used for the stationary solid phase in Section 3.2. However, here the interfacial area and the phasic volumes change within the fixed control volume, and therefore, these variations must be included in the derivations. Next, temporal averaging is taken, as was done in regards to turbulence in Section 2.4, and this derivation must also allow for the variable interfacial area and the phasic volumes.

(B) Phase-Volume Averaging

The *instantaneous* volume fraction of phase j is defined as

$$\epsilon_j(\mathbf{x},t) = \frac{V_j(\mathbf{x},t)}{V} = \frac{1}{V}\int_V a_j(\mathbf{x},t)\,dV\ , \tag{3.106}$$

where a_j is the phase distribution indicator for phases j, defined as

$$a_j = \begin{cases} 1 & \text{for } \mathbf{x} \text{ in } V_j \\ 0 & \text{for } \mathbf{x} \text{ not in } V_j\ . \end{cases} \tag{3.107}$$

As was mentioned, the volume V is allowed to change in space (for practical use in finite-volume numerical treatments). The j-phase-averaged quantity is defined by (3.19). Now taking the j-phase average of (3.104), Lahey and Drew (1989) use the above definitions and arrive at the *instantaneous* energy conservation equation

$$\begin{aligned}
&\frac{\partial}{\partial t}\epsilon_j\langle\rho_j(e_j+\frac{\mathbf{u}_j\cdot\mathbf{u}_j}{2})\rangle^j + \\
&\frac{1}{V}\nabla\cdot[\epsilon_j V\langle\rho_j(e_j+\frac{\mathbf{u}_j\cdot\mathbf{u}_j}{2})\mathbf{u}_j + \mathbf{q}_j - (-p_j\mathbf{I}+\mathbf{S}_j)\cdot\mathbf{u}_j\rangle^j] - \\
&\epsilon_j\langle\rho_j(\mathbf{u}_j\cdot\mathbf{g}+\frac{\dot{s}}{\rho_j})\rangle^j = -\frac{1}{V}\int_{A_i}\rho_j\mathbf{n}_j\cdot(\mathbf{u}_j-\mathbf{u}_i)(e_j+\frac{\mathbf{u}_j\cdot\mathbf{u}_j}{2})\,dA - \\
&\frac{1}{V}\int_{A_i}\mathbf{n}_j\cdot[\mathbf{q}_j + p_j\mathbf{I}\cdot\mathbf{u}_j - (\mathbf{S}_j-\langle\mathbf{S}_j\rangle_{A_i})\cdot\mathbf{u}_j - \langle\mathbf{S}_j\rangle_{A_i}\cdot\mathbf{u}_j]\,dA\ .
\end{aligned} \tag{3.108}$$

The first area integral is the interfacial internal and kinetic energy flow and the second one is the *net* interfacial flow of the heat flux and external work. The area integral is defined as

$$\langle\psi\rangle_{A_i} = \frac{1}{A_i(\mathbf{x},t)}\int_{A_i(\mathbf{x},t)}\psi\,dA\ . \tag{3.109}$$

(C) Time Averaging

The time average over a period τ is defined by (2.86), and by using this and the theorems of derivatives given by (3.102) and (3.103), Lahey and Drew (1989) arrive at

$$\begin{aligned}
&\frac{\partial}{\partial t}\overline{\epsilon_j\langle\rho_j(e_j+\frac{\mathbf{u}_j\cdot\mathbf{u}_j}{2})\rangle^j} + \frac{1}{V}\nabla\cdot\overline{V\epsilon_j\langle\rho_j(e_j+\frac{\mathbf{u}_j\cdot\mathbf{u}_j}{2})\mathbf{u}_j\rangle^j} = \\
&\quad -\frac{1}{V}\int_{A_i}p_j\mathbf{n}_j\cdot\mathbf{u}_i\,dA + \frac{\partial}{\partial t}\overline{\epsilon_j\langle p_j\rangle^j} - \\
&\quad \frac{1}{V}\nabla\cdot V\overline{\epsilon_j\langle\mathbf{q}_j\rangle^j} + \frac{1}{V}\nabla\cdot V\overline{\epsilon_j\langle\mathbf{S}_j\cdot\mathbf{u}_j\rangle^j} + \\
&\epsilon_j\overline{\langle\rho_j\mathbf{g}\cdot\mathbf{u}_j\rangle^j} + \overline{\epsilon_j\langle s_j\rangle^j} - \frac{1}{V}\int_{A_i}\overline{\mathbf{n}_j\cdot\mathbf{q}_j\,dA} - \overline{\frac{1}{V}\int_{A_i}e_j m_{j_i}\,dA} + \\
&\overline{\frac{1}{V}\int_{A_i}\frac{m_{j_i}}{\rho_j}(p_j-\langle p_j\rangle_{A_i})\,dA} + \overline{\frac{1}{V}\int_{A_i}\mathbf{n}_j\cdot\langle\mathbf{S}_j\rangle_{A_i}\cdot\mathbf{u}_j\,dA} -
\end{aligned}$$

$$\overline{\frac{1}{V}\int_{A_i} \mathbf{n}_j \cdot [(p_j - \langle p_j\rangle_{A_i})^d \mathbf{I} - (\mathbf{S}_j - \langle \mathbf{S}_j\rangle_{A_i})] \cdot \mathbf{u}_j \, dA} +$$

$$\overline{\frac{1}{V}\int_{A_i} (p_j - \langle p_j\rangle_{A_i})^{nd} \mathbf{n}_j \cdot \mathbf{u}_i \, dA} \,, \tag{3.110}$$

where the *interfacial mass flux* for phase j is

$$\dot{m}_{j_i} = \rho_j \mathbf{n}_j \cdot (\mathbf{u}_j - \mathbf{u}_i) \tag{3.111}$$

and A_i/V is the *specific interfacial* area. The drag force on the dispersed phase j has a *pressure* or *form drag* component (shown with superscript d) and an interfacial shear force component (shown with superscript nd). These will be further discussed in Sections 5.1.1. and 5.4.1. The term $p_j - \langle p_j\rangle_{A_i}$, evaluated on the interface, implicitly contains both form drag and nondrag components, but no information on the integrated shear effects. In the above, $p_j - \langle p_j\rangle_{A_i}$ is decomposed into a form drag and a nondrag component.

Introducing time decomposition of quantities, as defined in (2.87), the mean and fluctuation contributions to the mean total energy conservation equation are found. The final form of the *total* energy conservation equation, using the closures described below, is

$$\begin{aligned}
&\frac{\partial}{\partial t}\overline{\epsilon}_j \overline{\langle\rho_j\rangle}^j (\overline{\langle e_j\rangle}^j + \frac{\overline{\langle \mathbf{u}_j\rangle}^j \cdot \overline{\langle \mathbf{u}_j\rangle}^j}{2} + \frac{\overline{\langle \mathbf{u}'_j \cdot \mathbf{u}'_j\rangle}^j}{2}) + \\
&\frac{1}{V}\nabla \cdot [V\mathbf{A} \cdot \overline{\epsilon}_j \overline{\langle\rho_j\rangle}^j (\overline{\langle e_j\rangle}^j + \frac{\overline{\langle \mathbf{u}_j\rangle}^j \cdot \overline{\langle \mathbf{u}_j\rangle}^j}{2} + \frac{\overline{\langle \mathbf{u}'_j \cdot \mathbf{u}'_j\rangle}^j}{2})\overline{\mathbf{u}}_j] = \\
&-\frac{1}{V}\int_{A_i} p_j \mathbf{n}_j \cdot \mathbf{u}_i \, dA + \frac{\partial}{\partial t}\overline{\langle \epsilon_j p_j\rangle^j} - \frac{1}{V}\nabla \cdot [V\overline{\epsilon}_j(\overline{\langle \mathbf{q}\rangle}^j + \langle \mathbf{q}_t\rangle^j)] + \\
&\frac{1}{V}\nabla \cdot [V\overline{\epsilon}_j(\overline{\langle \mathbf{S}_j\rangle}^j + \langle \mathbf{S}_{t_j}\rangle^j) \cdot \overline{\langle \mathbf{u}_j\rangle}^j] + \overline{\epsilon}_j \overline{\langle\rho_j\rangle}^j \overline{\langle \mathbf{u}_j\rangle}^j \cdot \mathbf{g} + \overline{\epsilon}_j s_j + \\
&\overline{\langle \dot{n}_j\rangle}^j (\overline{e}_{j_i} + \frac{\overline{\langle \mathbf{u}_j\rangle_{A_i}} \cdot \overline{\langle \mathbf{u}_j\rangle_{A_i}}}{2} + \frac{\overline{\langle \mathbf{u}'_j \cdot \mathbf{u}'_j\rangle_{A_i}}}{2}) + \\
&\langle \mathbf{f}^d_j\rangle_{A_i} \cdot \langle \mathbf{u}^d_j\rangle_{A_i} + \langle \mathbf{f}^{nd}_j\rangle_{A_i} \cdot \langle \mathbf{u}^{nd}_j\rangle_{A_i} + \langle \frac{A_i}{V}\rangle \langle \mathbf{q}_j\rangle_{A_i} - \\
&\langle \mathbf{u}_j\rangle_{A_i} \cdot \langle \mathbf{S}_j\rangle_{A_i} \cdot \frac{1}{V}\nabla \cdot (V\overline{\epsilon_j}\mathbf{I}) + \langle \dot{m}_{j_i} \frac{p_j - \langle p^d_j\rangle_{A_i}}{\rho_j}\rangle_{A_i} \,,
\end{aligned} \tag{3.112}$$

where

$$\mathbf{q} = -k\nabla T + \mathbf{q}_r \,, \quad \langle \frac{A_i}{V}\rangle \langle \mathbf{q}_j\rangle_{A_i} = -\frac{1}{V}\int_{A_i} \overline{\mathbf{n}_j \cdot \mathbf{q}_j \, dA} \,. \tag{3.113}$$

Note that in (3.112), the *convection* of the total energy (internal and mean and turbulent components of kinetic energy), i.e., the second term on the left-hand side, contains the transformation (or correlation) tensor $\mathbf{A}$ which is generally taken to be equal to the identity tensor $\mathbf{I}$. The first term on the right-hand side is the *pressure work* due to the motion of the *interface*. The second term on the right-hand side is due to the *pressure work* resulting from the change in the *phasic pressure* defined as $\overline{\epsilon}_j\, \overline{p}_j$. The next term on the right-hand side is the *spatial change in the heat flux vector* (mean and turbulent components). The next term is the *viscous stress* (mean and turbulent) *work*. The fifth term is the *body force work* and is followed by the *phasic heat source term*. The seventh term is the *energy* (internal and mean and turbulent kinetic energy) *generation* due to *production* of phase j at the interface. The eighth and ninth terms are the *interfacial drag work* due to form drag and interfacial shear force, respectively. The tenth term is the phasic volumetric *interfacial heat transfer*, and when there is a phase change across the interface the sum of these volumetric interfacial heat transfer terms must *balance* with the heat released in the phase change. The eleventh term is the *interfacial viscous shear work* associated with the change in the phasic volume fraction. The twelfth term is the *interfacial pressure work* associated with the *phase change*.

In order to use (3.112), the quantities (other than the independent variables) introduced above, such as $\mathbf{A}$, $\overline{\langle \mathbf{u}_j' \cdot \mathbf{u}_j' \rangle^j}/2$, $\langle \mathbf{q}_t \rangle^j$, $\langle \mathbf{S}_{t_j} \rangle^j$, $\overline{e}_{j_i}$, $\langle \mathbf{u}_j \rangle_{A_i}$, $\langle \mathbf{S}_j \rangle_{A_i}$, $\langle \mathbf{f}_j^d \rangle_{A_i}$, $\langle \mathbf{f}_j^{nd} \rangle_{A_i}$, $\langle A_i/V \rangle \langle \mathbf{q}_j \rangle_{A_i}$, $\overline{\langle \dot{n}_j \rangle^j}$, $\langle \mathbf{u}_j^d \rangle_{A_i}$, $\langle \mathbf{u}_j^{nd} \rangle_{A_i}$, and $\langle \dot{m}_{j_i} (p - \langle p_j^d \rangle_{A_i})/\rho_j \rangle_{A_i}$, must be known.

Lahey and Drew (1989) show that for *nonisothermal* two-phase flows, in order to determine $\overline{\epsilon}_1$, $\overline{\epsilon}_2 = 1 - \overline{\epsilon}_1$, $\overline{\langle \mathbf{u}_1 \rangle^1}$, $\overline{\langle \mathbf{u}_2 \rangle^2}$, $\overline{\langle p_1 \rangle^1}$, $\overline{\langle p_2 \rangle^2}$, $\overline{\langle e_1 \rangle^1}$, $\overline{\langle e_2 \rangle^2}$, i.e., the 12 *independent* variables, the continuity, momentum, and energy equations involving a total of 337 *closure parameters* must be defined and evaluated. This is rather extensive and as with the turbulent modeling, many *simplifications*, *approximations*, and *mechanistic modelings* are used to reduce the number of these parameters (Lahey and Drew, 1989). Note that all the volume-averaged quantities are influenced by the *phase distributions* in V. Therefore, as with the treatment of the porous media, the phase distributions need to be known a priori before the volume averages can be made. Presently only limited and known phase-distribution regimes are treated with significant details.

Revankar and Ishii (1992) consider the *thermal* energy conservation equation, i.e., (1.83), and suggest the following *enthalpy* formulation *model*, i.e., a time-space-averaged form of (1.83), for a *uniform* control volume distribution. This model is

$$\frac{D}{Dt}\overline{\epsilon}_j \overline{\langle \rho_j \rangle^j}\, \overline{\langle i_j \rangle^j} = -\nabla \cdot \overline{\epsilon}_j (\overline{\langle \mathbf{q}_j \rangle^j} + \langle \mathbf{q}_{t_j} \rangle^j) +$$

$$\overline{\epsilon}_j \frac{D}{Dt}\overline{\langle p_j \rangle^j} + \overline{\langle \dot{n}_j \rangle^j}\, \overline{\langle i_j \rangle}_{A_i} + \overline{\langle \frac{A_i}{V} \rangle} \overline{\langle \mathbf{q}_j \rangle}_{A_i} + \overline{\epsilon}_j \overline{\langle \Phi_{\mu_j} \rangle^j} + \overline{\epsilon}_j \langle \dot{s}_j \rangle^j , \quad (3.114)$$

where $\overline{\langle \dot{n}_j \rangle^j}$ is the volumetric rate of change of phase j to k, and $\overline{\langle i_j \rangle}_{A_i}$ is the enthalpy of phase j at the interface. The turbulent heat flux $\overline{\langle \mathbf{q}_t \rangle}$ and the interfacial heat flux $\overline{\langle \mathbf{q} \rangle}_{A_i}$, which are omitted under the assumption of local thermal equilibrium between phases, are in turn modeled using closure conditions. Adding the two phasic energy equations, we have

$$\frac{\partial}{\partial t}[\overline{\epsilon}\overline{\langle\rho_1\rangle^1}\,\overline{\langle i_1\rangle^1}+(1-\overline{\epsilon})\overline{\langle\rho_2\rangle^2}\,\overline{\langle i_2\rangle^2}]+$$
$$\nabla\cdot[\overline{\epsilon}\overline{\langle\rho_1\rangle^1}\,\overline{\langle i_1\rangle^1}\,\overline{\langle\mathbf{u}_1\rangle^1}+(1-\overline{\epsilon})\overline{\langle\rho_2\rangle^2}\,\overline{\langle i_2\rangle^2}\,\overline{\langle\mathbf{u}_2\rangle^2}]=$$
$$-\nabla\cdot[\overline{\epsilon}(\overline{\langle\mathbf{q}_1\rangle}^1+\langle\mathbf{q}_{t_1}\rangle^1)+(1-\overline{\epsilon})(\overline{\langle\mathbf{q}_2\rangle}^2+\langle\mathbf{q}_{t_2}\rangle^2)]+\overline{\epsilon}\frac{D}{Dt}\overline{\langle p_1\rangle^1}+$$
$$(1-\overline{\epsilon})\frac{D}{Dt}\overline{\langle p_2\rangle^2}+\overline{\langle\dot{n}_1\rangle^1}\,\overline{\langle i_1\rangle_{A_i}}+\overline{\langle\dot{n}_2\rangle^2}\,\overline{\langle i_2\rangle_{A_i}}+\langle\overline{\frac{A_i}{V}}\rangle(\overline{\langle\mathbf{q}_2\rangle_{A_i}}+\overline{\langle\mathbf{q}_1\rangle_{A_i}})+$$
$$\overline{\epsilon}\overline{\langle\Phi_{\mu_1}\rangle}^1+(1-\overline{\epsilon})\overline{\langle\Phi_{\mu_2}\rangle}^2+\overline{\epsilon}\langle\dot{s}_1\rangle^1+(1-\overline{\epsilon})\langle\dot{s}_2\rangle^2\,, \tag{3.115}$$

$$\mathbf{q}=-k\nabla T+\mathbf{q}_r\,, \tag{3.116}$$

where $\overline{\epsilon}$ is the volume fraction of the phase 1. An *interfacial energy balance* gives the relation between interfacial volumetric heat flows by *sensible* heat, i.e., $\overline{\langle A_i/V\rangle}(\overline{\langle\mathbf{q}_1\rangle_{A_i}}+\overline{\langle\mathbf{q}_2\rangle_{A_i}})$, and by *phase change*, i.e., $\langle\dot{n}_1\rangle^1\langle i_1\rangle_{A_i}+\langle\dot{n}_2\rangle^2\langle i_2\rangle_{A_i}$. Note that from an *interfacial mass balance* we have $\langle\dot{n}_1\rangle^1=-\langle\dot{n}_2\rangle^2$.

3.3.2 Two-Medium Treatment of Hydrodynamics

The modeling of the two-medium treatment of the two-phase flow, including the turbulence in both phases, has been reviewed by Elghobashi and Abou-Arab (1983), Lahey and Drew (1989, 1990), Lahey (1991), Revankar and Ishii (1992), and Sangani and Didwania (1993). In Chapter 5, in relation to the two-medium treatment of the heat transfer, the fluid mechanics of the particulate flows (with solid particle) will be discussed in detail.

The phasic continuity and momentum equation derived by Lahey and Drew (1989), using the techniques outlined in the last section, which led to (3.112), are

$$\frac{\partial}{\partial t}\overline{\epsilon}_j\overline{\langle\rho_j\rangle^j}+\frac{1}{V}\nabla\cdot V\overline{\epsilon}_j\overline{\langle\rho_j\rangle^j}\,\overline{\langle\mathbf{u}_j\rangle^j}=\overline{\langle\dot{n}_j\rangle^j}\,, \tag{3.117}$$

$$\frac{\partial}{\partial t}\overline{\epsilon}_j\overline{\langle\rho_j\rangle^j}\,\overline{\langle\mathbf{u}_j\rangle^j}+\frac{1}{V}\nabla\cdot(V\mathbf{B}{:}\overline{\epsilon}_j\overline{\langle\rho_j\rangle^j}\,\overline{\langle\mathbf{u}_j\rangle^j}\,\overline{\langle\mathbf{u}_j\rangle^j})=$$
$$-\overline{\epsilon}_j\nabla\cdot\overline{\langle p_j\rangle^j}\mathbf{I}+(\overline{\langle p_j\rangle^j}-\overline{\langle p_j\rangle_{A_i}})\frac{1}{V}\nabla\cdot V\overline{\epsilon}_j\mathbf{I}-$$
$$\langle\mathbf{S}_j\rangle_{A_i}\frac{1}{V}\nabla\cdot V\overline{\epsilon}_j\mathbf{I}+\frac{1}{V}\nabla\cdot[V\overline{\epsilon}_j(\overline{\langle\mathbf{S}_j\rangle^j}+\langle\mathbf{S}_{t_j}\rangle^j)]+$$
$$\overline{\epsilon}_j\overline{\langle\rho_j\rangle^j}\mathbf{g}+\overline{\langle\dot{n}_j\rangle^j}\langle\mathbf{u}_j\rangle_{A_i}+\langle\mathbf{f}_j^d\rangle_{A_i}+\langle\mathbf{f}^{nd}\rangle_{A_i}\,, \tag{3.118}$$

where the parameters are those defined in association with the energy equation (3.112), except for the *fourth-order correlation tensor* **B** which is the counterpart of **A** in (3.112). In (3.118) the first term on the right-hand side is the phasic pressure gradient; the second and third terms are the effect of change in the volume fraction of the phase; the fourth term is the change in the phasic mean and turbulent shear stresses; the fifth is the phasic body force; the sixth is the momentum exchange due to phase change; and the seventh and eighth terms are the interfacial drag and nondrag forces, respectively. As with (3.118), some of these terms need to be determined using simplified models.

Revankar and Ishii (1992) use simpler forms of the phasic continuity and momentum equations, given by

$$\frac{\partial}{\partial t}\overline{\epsilon}_j\overline{\langle\rho_j\rangle^j} + \nabla\cdot\overline{\epsilon}_j\overline{\langle\rho_j\rangle^j}\,\overline{\langle\mathbf{u}_j\rangle^j} = \overline{\langle\dot{n}_j\rangle^j}\ , \tag{3.119}$$

$$\frac{D}{Dt}\overline{\epsilon}_j\overline{\langle\rho_j\rangle^j}\,\overline{\langle\mathbf{u}_j\rangle^j} = -\overline{\epsilon}_j\nabla\overline{\langle p_j\rangle^j} + \nabla\cdot\overline{\epsilon}_j(\overline{\langle\mathbf{S}_j\rangle^j} + \langle\mathbf{S}_{t_j}\rangle^j) + \overline{\epsilon}_j\overline{\langle\rho_j\rangle^j}\,\mathbf{g} + \overline{\langle\dot{n}_j\rangle^j}\langle\mathbf{u}_j\rangle_{A_i} + \langle\mathbf{f}_j^d\rangle_{A_i} - \nabla\overline{\epsilon}_j\cdot\langle\mathbf{S}_j\rangle_{A_i}\ , \tag{3.120}$$

where $\langle\mathbf{f}^d\rangle_{A_i}$ is the generalized volumetric interfacial drag and $\langle\mathbf{S}\rangle_{A_i}$ is the interfacial shear stress. These are generally modeled using the *velocity slip* between the two phases. We will discuss these models in Chapters 5 and 6.

Elghobashi and Abou-Arab (1987) include the compressibility and give a turbulence model (specifically, the two-equation model for the kinetic energy of turbulence and its dissipation rate) for the determination of the turbulent transport terms. Their treatment assumes small enough particles such the fluid (continuous phase) flow is *not* noticeably influenced by the presence of the particles. This influence, i.e., the *two-way hydrodynamic interaction*, has been addressed by Elghohashi and Truesdell (1993), in the direct numerical simulation for dilute particle suspensions (i.e., $\overline{\epsilon} < 5\times 10^{-4}$) and small particles. They show that the fluid-phase turbulence decays faster when particles are present. This decay is also addressed by Zakharov et al. (1993). Louge et al. (1993) consider relatively large particles and include the effect of the interparticle collisions on the fluid and dispersed phase shear stresses. Anderson and Jackson (1992) also consider these collisions, and their modeling is based on the analogy with the kinetic (molecular) theory of gases. These will be discussed further in Section 5.4.2. Michiyoshi and Serizawa (1986) have measured the turbulent transport in an air-water, two-phase bubbly flow in pipes and have identified macro and micro scales of turbulence for this isothermal flow. For solid-fluid systems, a review of the fluid mechanics is given by Meyer (1983). As is expected, the properties of the phase pair, the concentration of the dispersed phase (i.e., the flow regime), and the direction of gravity with respect to the mean flow direction, all influence the mean and turbulent quantities. In the next section, the available models for the energy equation in solid-fluid two-phase flows, with the solid being the dispersed phase, is considered.

3.3.3 A Simplified Energy Equation for Dispersed Flows

The fluid carrying the particles can be a liquid or a gas, and, as was discussed, in most cases the fluid flow is *turbulent*. The heat transfer aspect of solid-*liquid* two-phase flows has been examined with application to transport of slurries (including coal particle-water slurry), in enhancement of heat transfer by solid particle suspensions, and in liquid fluidized beds. The solid-*gas* systems have been examined more extensively because of the additional relevance to powder and coating technologies, and to the combustion of particulates (Smoot and Pratt, 1979).

For *dilute suspensions*, Soo (1989) suggests the following phasic time-space averaged *thermal* energy equations, which in principle are derived from equation (1.83), but contain many *simplifications* and *approximations* (some of which are heuristic). By dropping the time-averaged symbol, and using c and d for the *continuous* and the *dispersed* phase, respectively, these equations become

$$\frac{D}{Dt}[\epsilon\langle\rho\rangle^c\langle i\rangle^c + \epsilon\langle\rho\rangle^c\frac{\langle\mathbf{u}\rangle^c\cdot\langle\mathbf{u}\rangle^c}{2} + (1-\epsilon)\langle\rho\rangle^d\frac{\langle\mathbf{u}\rangle^d\cdot\langle\mathbf{u}\rangle^d}{2}] = \epsilon\frac{\partial\langle p\rangle^c}{\partial t} + \epsilon\langle\mathbf{u}\rangle^c\cdot\nabla\langle p\rangle^c - \nabla\cdot\epsilon\mathbf{q}_{e_c} + \epsilon\dot{s}_c + \frac{A_i}{V}h_{cd}(\langle T\rangle^d - \langle T\rangle^c) + \langle\dot{n}_c\rangle^c\langle i\rangle^c\,, \tag{3.121}$$

$$\frac{D}{Dt}(1-\epsilon)\langle\rho\rangle^d\langle i\rangle^d = -\nabla\cdot(1-\epsilon)\mathbf{q}_{e_d} + (1-\epsilon)\dot{s}_d + \frac{A_i}{V}h_{cd}(\langle T\rangle^c - \langle T\rangle^d) + \langle\dot{n}_d\rangle^d\langle i\rangle^d \tag{3.122}$$

where the velocities are the phase-averaged velocities. The kinetic energy of the dispersed phase is assumed to be due to the interfacial drag, and therefore, is added to the fluid energy equation.

Now assuming local thermal equilibrium, i.e., $\langle T\rangle^d = \langle T\rangle^c = T$, and by adding these two equations, we have

$$\frac{D}{Dt}[(1-\epsilon)\langle\rho\rangle^d\langle i\rangle^d + \epsilon\langle\rho\rangle^c\langle i\rangle^c + \epsilon\langle\rho\rangle^c\frac{\langle\mathbf{u}\rangle^c\cdot\langle\mathbf{u}\rangle^c}{2} + (1-\epsilon)\langle\rho\rangle^d\frac{\langle\mathbf{u}\rangle^d\cdot\langle\mathbf{u}\rangle^d}{2}] = -\nabla\cdot[(1-\epsilon)\mathbf{q}_{e_d} + \epsilon\mathbf{q}_{e_c}] + (1-\epsilon)\dot{s}_d + \epsilon\dot{s}_c + \epsilon\frac{D\langle p\rangle^c}{Dt} + \langle\dot{n}_d\rangle^d\langle i\rangle^d + \langle\dot{n}_c\rangle^c\langle i\rangle^c\,, \tag{3.123}$$

$$\langle\mathbf{q}_{e_j}\rangle = -\langle k_{e_j}\nabla T\rangle + \langle\mathbf{q}_{t_j}\rangle + \langle\mathbf{q}_{r_j}\rangle\,. \tag{3.124}$$

The conductive, turbulent, and radiative components of the heat flux vector are included. Note that this model does not explicitly allow for the interfacial temperature to be different from the local phasic temperatures. As was discussed in Section 1.10.3, the *simultaneous* inclusion of *phase change* and *nonequilibrium* requires thermodynamic considerations of the interfacial conditions. This will be examined in Sections 6.3 and 6.4 for gas-liquid systems and in Section 7.3.2 for liquid-solid systems. Crowe and Smoot (1979) consider multicomponent gases and recommend energy equations similar to (3.123) and (3.124).

3.3.4 Effective Medium Properties for Dispersed Flows

Although recently rigorous formulations and derivations of the energy equation for two-phase flows have been made, these rigorous forms are not directly applicable. In some simpler models, the heat flux in (3.124) is decomposed to

$$\begin{aligned}(1-\epsilon)\mathbf{q}_{e_d} + \epsilon\mathbf{q}_{e_c} &= -\langle k_e \nabla T\rangle + \langle \mathbf{q}_t\rangle + \langle \mathbf{q}_r\rangle = \\ &\quad [\mathbf{K}_e + \epsilon(\rho c_p)_c \mathbf{D}^d]\cdot\nabla T + \langle \mathbf{q}_r\rangle\,, \end{aligned} \tag{3.125}$$

where $\langle \mathbf{q}_t\rangle$ is the *turbulent heat flux*, $\mathbf{K}_e$ is the *effective medium conductivity* tensor, and $\mathbf{D}^d$ is the *hydrodynamic dispersion* tensor. In the following, the available results and predictive correlations for $\langle \mathbf{q}_t\rangle$, $\mathbf{K}_e$, and $\mathbf{D}^d$ are discussed.

(A) Effective Medium Conductivity

The *isotropic* effective molecular (or stagnant) conductivity k_e of the solid-fluid mixture has been evaluated by using a treatment similar to the molecular theory of gases (a review is given by Soo, 1989), and also by using the continuum heat conduction equation for the solid and the fluid phases with the matching boundary conditions at the solid-fluid interface (a review is given by Kaviany, 1991). For dispersed particles having a conductivity k_d we expect

$$\frac{\langle k\rangle}{k_c} = \frac{k_e}{k_c} = \frac{k_e}{k_c}\left(\frac{k_d}{k_c}, \epsilon\right). \tag{3.126}$$

One of the correlations is (e.g., Kaviany, 1991)

$$\frac{k_e}{k_c} = \left(\frac{k_d}{k_c}\right)^{0.280-0.757\log\epsilon-0.057(k_d/k_c)} \tag{3.127}$$

and is accurate for $k_d \geq k_c$ as well as $k_d < k_c$.

(B) Dispersion Coefficients

As is customary in the closure conditions used in turbulent transport and in dispersion in multiphase systems, the time- and space-averaged effects of the velocity and temperature fluctuations or deviations are expressed in terms of turbulent coefficients and gradients of mean quantities. The dis-

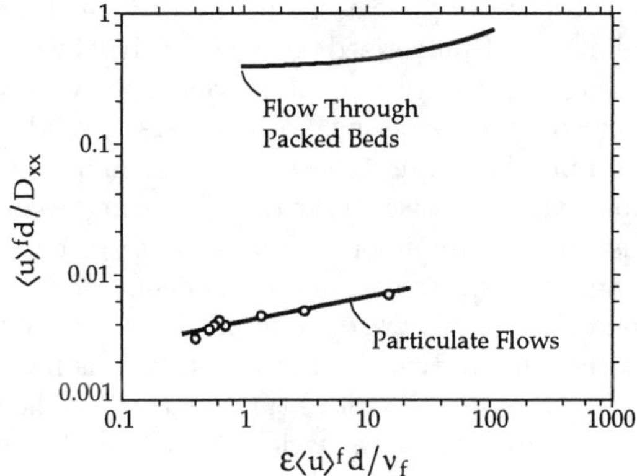

Figure 3.8. The variation of the Péclet number based on the total effect diffusivity D_{xx}, with respect to the particle Reynolds number. (From Davidson et al., reproduced by permission ©1985 Academic Press.)

persion tensor $\mathbf{D}^d$ introduced in (3.63) accounts for this time- and space-averaged fluctuations (as in turbulence) and deviations (as in the effect of the dispersed phase). Presently, no detailed direct theoretical evaluation of $\mathbf{D}^d$ is available. However, experimental determination of $k_e + \epsilon(\rho c_p)_c D^d_{xx}$ for one-dimensional flow along the x-axis is available. In general, the axial dispersion coefficient D^d_{xx} for dispersed solid particles is much larger than its value for packed (stationary) solid particles. Figure 3.8 shows the experimental results reported by Gunn (1968) and also Courdec (1985).

The experimental results for *solid-liquid* systems are compiled by Gunn (1968) and in Figure 3.8. The axial, total effective conductivity $D_{xx} = [k_e/(\rho c_p)_c] + \epsilon D^d_{xx}$ is used in the *effective* Péclet number based on $\langle u \rangle^f$ and is given as a function of the Reynolds number based on $\langle u \rangle$. Only results which are independent of the ratio of the container to particle diameter are reported. As suggested by Gunn (1968), this appears to be the case when the ratio is larger than about 500.

For laminar flow of suspensions, Sohn and Chen (1981) measure the total (molecular and dispersion components) conductivity. They find that for large fluid Péclet numbers based on the local shear rate, the effect of the molecular conductivities is negligible and that the total conductivity depends on the porosity and varies with the Péclet number to the power of 1/2. This increase with $(\langle u \rangle^f)^{1/2}$ is similar to the trend found in Figure 3.8 where the ratio of $\langle u \rangle^f$ to D_{xx} increases as $\langle u \rangle$ increases, with a power less than unity.

(C) Turbulent Heat Flux

The turbulent heat flux for the effective medium has been modeled using the various turbulence models discussed in Section 2.4.2. For example,

Warmeck (1979) discusses applications of the two-equation model of turbulence to combustion of pulverized coal. Similarly, Gosman et al. (1992) extend the derivation of the two-equation model to two-phase flows with one phase dispersed by considering both dispersed solids in liquids and dispersed gas (bubble) in liquids. These involve including or neglecting the effect of the dispersed solid phase on the turbulence in the continuous phase. Here, as an illustrative example of the effect of the dispersed solid on the turbulent heat flux, the *phenomenological* modeling of Michaelides (1986) for dispersed solid phase is reviewed. Solid particles are assumed to be very small such that their hydrodynamic relaxation time is much smaller than the period for the roll-over of a typical turbulent eddy. The turbulent heat flux in the single-medium energy equation (3.124) can be defined as

$$\begin{aligned}\langle \mathbf{q}_t \rangle &= \overline{\rho_e i_e \mathbf{u}_e} - \overline{\rho}_e \overline{i}_e \overline{\mathbf{u}}_e = \overline{\rho_e c_{p_e} T \mathbf{u}_e} - \overline{\rho}_e \overline{i}_e \overline{\mathbf{u}}_e \\ &= \overline{\rho}_e \overline{c_p}_e \overline{T' \mathbf{u}'} + \overline{\rho}_e \overline{T} \overline{c'_{p_e} \mathbf{u}'_e} + \overline{T} \overline{c_p} \overline{\rho' \mathbf{u}'_e} \,. \end{aligned} \tag{3.128}$$

Because a significant turbulent intensity (i.e., turbulent kinetic energy) is assumed, the products of the fluctuations which do not contain $\mathbf{u}'_e$ have been assumed negligible.

Michaelides (1986) uses the concept of the turbulent closure and the mixing length, to relate the fluctuations to the gradients of the mean quantities. Following closures similar to (2.91) and (2.95), and by using a mixing length similar to that given by (2.98), for one-dimensional fields, he recommends the following phenomenological models

$$\overline{\rho'_e u'_e} = -\ell_\rho \frac{d\overline{\rho}_e}{dx} \ell \frac{d\overline{u}_e}{dx} \,, \tag{3.129}$$

$$\overline{c'_{p_e} u'_e} = -\ell_{c_p} \frac{d\overline{c_p}_e}{dx} \ell \frac{d\overline{u}_e}{dx} \,, \tag{3.130}$$

$$\overline{T'_e u'_e} = -\ell_T \frac{d\overline{T}}{dx} \ell \frac{d\overline{u}_e}{dx} \,. \tag{3.131}$$

The last model is not consistent with the definition of the eddy diffusivity given by (2.95) because of the *direct* dependence of the turbulent heat flux on the gradients of both the mean temperature and the mean velocity.

The distribution of the mean *medium* velocity, density, and specific heat is determined from the distribution of the mean void fraction $\overline{\epsilon} = \overline{\epsilon}(\mathbf{x})$, i.e.,

$$\overline{\rho}_e \overline{u}_e = (1 - \overline{\epsilon}) \rho_d \langle \mathbf{u} \rangle^d + \overline{\epsilon} \rho_c \langle \mathbf{u} \rangle^c \,, \tag{3.132}$$

$$\overline{\rho}_e = (1 - \overline{\epsilon}) \rho_d + \overline{\epsilon} \rho_c \,, \tag{3.133}$$

$$(\overline{\rho} \overline{c}_p)_e = (1 - \overline{\epsilon})(\rho c_p)_d + \overline{\epsilon} (\rho c_p)_c \,. \tag{3.134}$$

The mean void fraction distribution $\overline{\epsilon} = \overline{\epsilon}(\mathbf{x})$ is determined from the fluid mechanics (the continuity and momentum equations).

The mixing lengths ℓ_ρ, ℓ_{c_p} and ℓ_T are related to the momentum mixing length by considering the mean path length of the penetration of the particle distribution disturbance and the thermal disturbance. It is a common practice to assume complete similarity between ℓ and ℓ_T, but this simplification is not always justified.

For a one-dimensional flow, Michaelides (1986) suggests an *empirical* relation between $\ell_\rho = \ell_{c_p}$ and ℓ given in terms of the *local solid to fluid mass flux ratio* $(1-\bar{\epsilon})\overline{\langle \rho u\rangle}^d/\bar{\epsilon}\overline{\langle \rho u\rangle}^c = (1-\bar{\epsilon})\rho_d\overline{\langle u\rangle}^d/\bar{\epsilon}\rho_c\overline{\langle u\rangle}^c = \overline{\dot{m}_d}/\overline{\dot{m}_c}$. He applies this model to heat transfer between a solid-gas two-phase flow and a circular tube wall containing this flow and finds satisfactory agreement with the available experimental results. This will be discussed in Section 4.14.

3.3.5 Exothermic Reaction

Under the *homogeneous* flow and heat transfer approximations (or models) *no* velocity and temperature slip is allowed between the two phases. This model continues to be used in the analysis of combustion of solid and liquid sprays (i.e., solid particles or liquid droplets in a continuous gas), e.g., Seshadri et al. (1992) and Silverman et al. (1993), while Mao et al. (1980) and Faeth (1983) show that the velocity and temperature slips (i.e., two-medium treatments of the fluid flow and heat transfer) can become necessary depending on the diameter of the droplets. The two-medium heat transfer treatment is discussed in Chapter 5.

As an example of a single-medium treatment of a two-phase flow, the study by Seshadri et al. (1992) of the flame structure in evaporation-reaction of clouds of *monosize* droplets in a gaseous oxidizer is reviewed below. The study by Silverman et al. (1993) allows for *polysize* droplets and for *fuel-rich* mixtures with a postreaction zone evaporation.

The multicomponent phases require the inclusion of the species equations, and the evaporation rate of the droplets is represented by a kinetic model as a function of the local temperature. The problem is similar to the gaseous premixed combustion examined in Section 2.3.2 and to the combustible porous media of Section 3.2.6. Here the evaporation, which adds to the required preheating, must be included.

Consider a one-step reaction given by (2.84) where fuel f reacts with oxygen O to turn to product p. The reaction rate may be assumed to be *first order* and limited by the availability of the fuel and be given by a simplified form of (2.43), i.e.,

$$-\langle \dot{n}_f\rangle = A\nu_f\rho_f\, e^{-\Delta E_a/R_gT}\,. \tag{3.135}$$

The *Zel'dovich number* Ze defined as

$$Ze = \frac{\Delta E(T_r - T_n)}{R_g T_r^2} \tag{3.136}$$

is assumed to be large to allow for the simplification of the analysis. (As discussed in Section 2.3.2, this is called the high activation energy asymptote.)

Monosized and uniformly distributed droplets of initial radius R_n and number density N_n are assumed. It is further assumed that $N_n/[\epsilon\rho_c+(1-\epsilon)\rho_d]$ and pressure remain constant and that the total effective conductivity $\langle k\rangle = k_e + (\rho c_p)_c D^d_{xx} + k_r$ and the similarly defined *total diffusivity* $\langle D_m\rangle$ also remain constant, where k_r is the radiant conductivity defined by (2.119). The mass continuity, energy, gaseous fuel species, and particle conservation (with no diffusion) equations and the equation of state are

$$\frac{d}{dx}\langle\rho\rangle\langle u\rangle = 0\ , \tag{3.137}$$

$$\langle\rho c_p\rangle\langle u\rangle\frac{dT}{dx} = \langle k\rangle\frac{d^2T}{dx^2} + \langle\dot{n}_f\rangle\Delta i_c - \langle\dot{n}_v\rangle\Delta i_{\ell g}\ , \tag{3.138}$$

$$\langle\rho\rangle\langle u\rangle\frac{d}{dx}\frac{\langle\rho_f\rangle}{\langle\rho\rangle} = \rho\langle D_m\rangle\frac{d^2}{dx^2}\frac{\langle\rho_f\rangle}{\langle\rho\rangle} - \langle\dot{n}_f\rangle + \langle\dot{n}_v\rangle\ , \tag{3.139}$$

$$\langle\rho\rangle\langle u\rangle\frac{d}{dx}\frac{\langle\rho_d\rangle}{\langle\rho\rangle} = -\langle\dot{n}_v\rangle\ , \tag{3.140}$$

$$\langle\dot{n}_v\rangle = a_e N_n 4\pi R^2 T^a\ , \tag{3.141}$$

$$p = \frac{R_g}{M}\rho T = \text{constant}\ , \tag{3.142}$$

where $\langle\dot{n}_o\rangle$ and $\langle\dot{n}_v\rangle$ are the volumetric rates of oxidization of the fuel and vaporization (generation of fuel), respectively, and Δi_c and Δi_{fg} are the *specific heat of combustion* and *evaporation*, respectively. The evaporation rate is modeled similar to the expression given in the molecular treatment by (1.45), except the exponential term containing the activation energy is assumed to be nearly unity. The medium heat capacity and density are

$$\langle\rho c_p\rangle = \epsilon(\rho c_p)_c + (1-\epsilon)(\rho c_p)_d\ , \tag{3.143}$$

$$\langle\rho\rangle = \epsilon\rho_c + (1-\epsilon)\rho_d\ , \tag{3.144}$$

where the solid volume fraction $1-\epsilon$ changes as the radius of the particle decreases, i.e., $1-\epsilon = (1-\epsilon_n)R^3/R_n^3$.

By using the nonreacted and reacted temperatures, T_n and T_r, define the density $\rho_{f_o} \equiv \langle\rho\rangle\langle c_p\rangle(T_r-T_n)/\Delta i_c$, and the nonevaporating mass flux $\rho_n u_{F_o}$, where $u_{F_o} \equiv u_F(\Delta i_{fg}=0)$ is the flame speed for the case of neglected heat absorption due to evaporation, the dimensionless variables are

$$T^* = \frac{T-T_n}{T_r-T_n}\ ,\quad \rho_f^* = \frac{\langle\rho_f\rangle}{\rho_{f_o}}\ ,\quad \rho_d^* = \frac{\langle\rho_d\rangle}{\rho_{f_o}}\ , \tag{3.145}$$

$$\dot{m}^* = \frac{\rho\langle u\rangle}{\rho_n u_F(\Delta i_{fg}=0)}\ ,\quad x^* = \frac{\langle\rho c_p\rangle\rho_n u_{F_o}\int_o^x(\rho/\rho_n)\,dx}{\langle\rho\rangle\langle k\rangle}\ . \tag{3.146}$$

Then the dimensionless forms of equations (3.137) to (3.140) for a *unity*

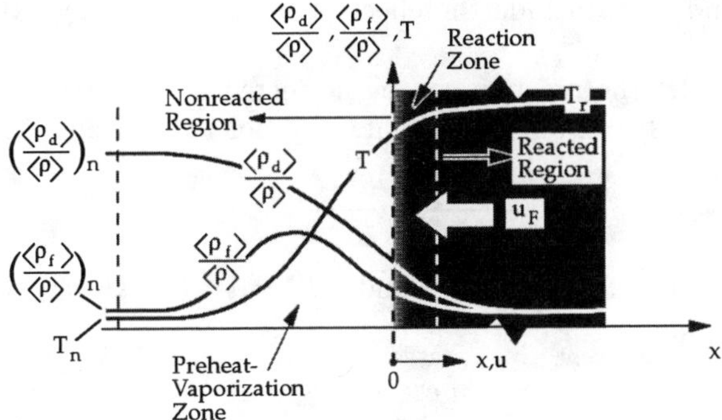

Figure 3.9. Structure of the premixed flame in a two-phase medium. The preheat zone is also shown where the liquid droplets evaporate, supplying the gaseous fuel.

Lewis number $\langle k\rangle/\langle\rho c_p\rangle\langle D_m\rangle = 1$, is

$$\frac{d}{dx^*}\dot{m}^* = 0\,, \tag{3.147}$$

$$\dot{m}^*\frac{dT^*}{dx^*} = \frac{d^2T^*}{dx^{*2}} + \langle\dot{n}_f^*\rangle\frac{\rho_u}{\langle\rho\rangle} - \Delta i_{fg}^*\gamma\rho_s^{*2/3}T^{*a}\,, \tag{3.148}$$

$$\dot{m}^*\frac{d\rho_f^*}{dx^*} = \frac{d^2\rho_f^*}{dx^{*2}} - \langle\dot{n}_f^*\rangle + \gamma\rho_d^{*2/3}T^{*a}\,, \tag{3.149}$$

$$\dot{m}^*\frac{d\rho_d^*}{dx^*} = -\gamma\rho_d^{*2/3}T^{*a}\,, \tag{3.150}$$

where the variable radius R is written in terms of $\langle\rho_d\rangle$ and the evaporation rate is written in terms of $T - T_n$ to allow for simplifications, and this is valid for $T_r - T_n \gg T_n$. The parameters appearing in the above equations are

$$\langle\dot{n}^*\rangle = \frac{\langle k\rangle_n\langle\dot{n}_f\rangle}{u_{F_o}^2\langle\rho c_p\rangle\rho_{f_o}}\,, \qquad \Delta i_{fg}^* = \frac{\Delta i_{fg}}{\Delta i_c}\,, \tag{3.151}$$

$$\gamma = \frac{4.836a_eN_n^{1/3}k_n(T_r - T_n)^a}{u_{F_o}^2\langle\rho c_p\rangle\rho_d^{2/3}\rho_{f_o}^{1/3}}\,. \tag{3.152}$$

By placing the beginning of the reaction zone at $x^* = 0$, as shown in Figure 3.9, the dimensionless boundary conditions are

$$T^* = 1\,, \quad \rho_f^* = 0 \qquad\qquad x^* \to \infty \tag{3.153}$$

$$T^* = 0\,, \quad \rho_f^* = 0\,, \quad \rho_d^* = \frac{\langle\rho_f\rangle_n}{\rho_{f_o}} \qquad\qquad x^* \to -\infty\,. \tag{3.154}$$

Seshadri et al. (1992) initially assume that Δi_{fg}^* is small to construct a solution for temperature T_o^* and the flame speed u_{F_o} corresponding to

$m^* = 1$, and then include the effect of *nonzero* Δi^*_{fg} by examining this solution.

For large Ze, the reaction can be neglected in the *preheat-vaporization zone* and (3.148) can be approximated by (for $\dot{m}^* = 1$ and $\Delta i^*_{\ell g} = 0$)

$$\frac{dT^*_o}{dx^*} = \frac{d^2T^*_o}{dx^{*2}} , \tag{3.155}$$

$$T^*_o = 0 \quad \text{for } x^* \to \infty \quad \text{and} \quad T^*_o = 1 \quad \text{for } x^* = 0 . \tag{3.156}$$

The solution is

$$T^*_o = e^{x^*} \quad x^* \le 0 . \tag{3.157}$$

Using this in the dispersed phase equation (3.150) and the boundary condition given by the last of (3.154), we have

$$\rho^*_d = (\frac{\langle\rho_f\rangle_n}{\rho_{f_o}} + \frac{\gamma}{3a}e^{ax^*})^3 \quad x^* \le 0 . \tag{3.158}$$

The distribution of ρ^*_f and its derivative at 0^- are found by integration of (3.149) using the results for T^*_o and ρ^*_d. This gives

$$-\frac{d\rho^*_f}{dx^*}|_{o^-} = \frac{\gamma}{a}(\frac{\langle\rho_f\rangle_n}{\rho_{f_o}})^{2/3} -$$

$$2(\frac{\gamma}{3a})^2(\frac{\langle\rho_f\rangle_n}{\rho_{f_o}})^{1/3} + \frac{\gamma^3}{27a^3} - \rho^*_f(x^* = 0) . \tag{3.159}$$

In the *reaction zone*, the convection and vaporization terms are negligible, and using the *perturbation parameter* $\delta = Ze^{-1}$ the coordinate x^* is stretched and other variables scaled according to

$$\eta = x^*/\delta , \quad \xi = \frac{\rho^*_f}{\delta} - \frac{\rho^*_f(x^* = 0)}{\delta} , \quad \theta = \frac{1 - T^*_o}{\delta} . \tag{3.160}$$

Then in the limit of $\delta \to 0$, the energy and fuel species equations, (3.148) and (3.149), respectively, become

$$\frac{d^2\theta}{d\eta^2} = \Gamma[\frac{\rho^*_f(x^* = 0)}{\delta} + \xi]e^{-\theta} , \tag{3.161}$$

$$\frac{d^2(\theta + \xi)}{d\eta^2} = 0 , \tag{3.162}$$

where

$$\Gamma = \frac{\nu_f\langle k\rangle_n A\delta^2}{\langle\rho c_p\rangle u^2_{F_o}} e^{\Delta E_a/R_gT_r} . \tag{3.163}$$

The boundary conditions are found by matching with the preheat evaporation zone and the *convective zone* (where derivatives diminish), i.e.,

$$\frac{d\theta}{d\eta} = -1 \quad \eta \to -\infty , \tag{3.164}$$

$$\frac{d\theta}{d\eta} = \frac{d\xi}{d\eta} = 0 \quad \eta \to \infty \,. \tag{3.165}$$

Then the energy equation (3.161) gives

$$2[1 + \frac{\rho_f^*(x^* = 0)}{\delta}]\Gamma = 1 \,. \tag{3.166}$$

Now u_{F_o} can be found from (3.163) as

$$u_{F_o}^2 = \frac{2[1 + \frac{\rho_f^*(x^* = 0)}{\delta}]\nu_f \langle k \rangle_n A \delta^2}{\langle \rho c_p \rangle} e^{-\Delta E_a / R_g T_r} \,. \tag{3.167}$$

Assuming that $\rho_f^*(x^* = 0) = 0$, then T_r and the previously defined $\rho_{f_o}^*$ must be determined. The needed equation is the statement of continuity of fluxes across the preheat-reaction interface, i.e.,

$$\frac{dT_o^*}{d\eta}|_{o^+} + \frac{d\rho_f^*}{d\eta}|_{o^+} = \frac{d\theta}{d\eta}|_{o^-} + \frac{d\rho_f^*}{d\eta}|_{o^-} \,. \tag{3.168}$$

Now using the solutions for the preheat regions (3.157) and (3.159) in (3.168) with $\rho_f^*(x^* = 0) = 0$ and assuming that the gradients at 0^+ are of the order of unity, we have

$$\frac{\gamma}{a}(\frac{\langle \rho_f \rangle_n}{\rho_{f_o}})^{2/3} - \frac{\gamma^2}{3a^2}(\frac{\langle \rho_f \rangle_n}{\rho_{f_o}})^{1/3} + \frac{\gamma^3}{27a^3} - 1 = 0 \,. \tag{3.169}$$

Then ρ_{f_o}, T_r, and u_{F_o} are found from the definition of ρ_{f_o} given just before (3.145), and from (3.167) and (3.169), respectively.

The effect of nonzero Δi_{fg} on u_F is included by adding Δi_{fg} to the activation energy and using the relation for u_{F_o} given by (3.167). This gives

$$u_F = u_{F_o} e^{-\Delta i_{fg} Ze / 2\Delta i_c} \,. \tag{3.170}$$

Seshadri et al. (1992) give numerical results for an evaporation-reaction process.

3.4 References

Anderson, K.G., and Jackson, R., 1992, "A Comparison of the Solutions of Some Proposed Equations of Motion of Granular Materials for Fully Developed Flow Down Inclined Planes," *J. Fluid Mech.*, 241, 145–168.

Aris, R., 1956, "On the Dispersion of a Solute in a Fluid Flowing Through a Tube," *Proc. Roy. Soc. (London)*, A235, 67–77.

Bear, J., and Bachmat, Y., 1991, *Introduction to Modeling of Transport Phenomena in Porous Media*, Kluwer Academic Publishers, Dordrecht.

Bennen, W.D., and Incropera, F.P., 1987, "A Continuum Model for Momentum, Heat and Species Transport in Binary Solid-Liquid Phase Change Systems-I., and -II.," *Int. J. Heat Mass Transfer*, 30, 2161–2187.

Brenner, H., 1980, "Dispersion Resulting from Flow Through Spatially Periodic Porous Media," *Phil. Trans. Roy. Soc. (London)*, 297, 81–133.

Brewster, M.Q., 1992, *Thermal Radiative Transfer and Properties*, John Wiley and Sons, New York.

Brewster, M.Q., and Tien C.-L., 1982, "Examination of the Two-Flux Model for Radiative Transfer in Particulate Systems," *Int. J. Heat Mass Transfer*, 25, 1905–1907.

Carbonell, R.G., and Whitaker, S., 1983, "Dispersion in Pulsed Systems — II. Theoretical Development for Passive Dispersion in Porous Media," *Chem. Engng. Sci.*, 38, 1795–1802.

Couderc, J.-P., 1985, "Incipient Fluidization and Particulate Systems" in *Fluidization*, Second Edition, Davidson, J.F., et al., Editors, Academic, London.

Crowe, C.T., and Smoot, L.D., 1979, "Multicomponent Conservation Equations," in *Pulverized-Coal Combustion and Gasifications*, Smoot, L.D., and Pratt, D.T., Editors, Plenum Press, New York.

Drolen, B.L., and Tien, C.-L., 1987, "Independent and Dependent Scattering in Packed Spheres Systems," *J. Thermophys. Heat Transfer*, 1, 63–68.

Dybbs, A., and Edwards, R.V., 1984, "A New Look at Porous Media Fluid Mechanics: Darcy to Turbulent," in *Fundamentals of Transport Phenomena in Porous Media*, Bear, J., and Corapcigoln, M.Y., Editors, Martinus Nijhoff Publishers, Dordrecht.

Elghobashi, S.E., and Abou-Arab, T.W., 1983, "A Two-Equation Turbulence Model for Two-Phase Flows," *Phys. Fluids*, 26, 931–938.

Elghobashi, S.E., and Truesdell, G.C., 1993, "On the Two-Way Interaction Between Homogeneous Turbulence and Dispersed Solid Particles," *Phys. Fluids*, A5, 1790-1801.

Faeth, G.M., 1983, "Evaporation and Combustion of Sprays," *Prog. Energy Combust. Sci.*, 9, 1-16.

Fatehi, M., and Kaviany, M., 1994, "Adiabatic Reverse Combustion in a Packed Bed," *Combust. Flame* (in press).

Ganesan, S., and Poirier, D.R., 1990, "Conservation of Mass and Momentum for the Flow of Interdendritic Liquid During Solidification," *Metall. Trans.*, 21B, 173–181.

Gates, B.C., 1992, *Catalytic Chemistry*, John Wiley and Sons, New York.

Gosman, A.D., Lekakou, C., Politis, S., Issa, R.I., and Looney, M.K., 1992, "Multidimensional Modeling of Turbulent Two-Phase Flows in Stirred Vessels," *AIChE J.*, 38, 1946-1956.

Gray, W.G., Leijnse, A., Kolar, R.L., and Blain, C.A., 1993, *Mathematical Tools for Changing Spatial Scales in the Analysis of Physical Systems*, CRC Press, Boca Raton, FL.

Gunn, D.J., 1968, "Mixing in Packed and Fluidized Beds," *Chem. Eng. (London)*, CE153–CE172.

Hills, R.N., Loper, D.E., and Roberts, P.H., 1992, "On Continuum Models for Momentum, Heat and Species Transport in Solid-Liquid Phase Change Systems," *Int. Comm. Heat Mass Transfer*, 19, 585–594.

Kaviany, M., 1991, *Principles of Heat Transfer in Porous Media*, Springer-Verlag, New York.

Kim, C.-J., and Kaviany, M., 1992, "A Fully Implicit Method for Diffusion-Controlled Solidification of Binary Alloys," *Int. J. Heat Mass Transfer*, 35, 1143–1154.

Lahey, R.T., Jr., and Drew, D.A., 1989, "The Three-Dimensional Time- and Volume-Averaged Conservation Equations for Two-Phase Flows," *Advan. Nuclear Sci. Tech.*, 20, 1–69.

Lahey, R.T., Jr., and Drew, D.A., 1990, "The Current State-of-the-Art in the Modeling of Vapor/Liquid Two-Phase Flows," ASME paper no. 90-WA/HT-13, American Society of Mechanical Engineers, New York.

Lahey, R.T., Jr., 1991, "Void Wave Propagation Phenomena in Two-Phase Flow (Kern Award Lecture)," *AIChE J.*, 37, 123–135.

Louge, M., Yusof, J.M., and Jenkins, J.T., 1993, "Heat Transfer in the Pneumatic Transport of Massive Particles," *Int. J. Heat Mass Transfer*, 36, 265–275.

Mao, C.-P., Szekely, G.A., Jr., and Faeth, G.M., 1980, "Evaluation of Locally Homogeneous Flow Model of Spray Combustion," *J. Energy*, 4, 78–87.

Martynenko, O.G., and Korovkin, V.N., 1992, "Concerning the Calculations of Plane Turbulent Jets on the Basis of the $k-\epsilon$ Model of Turbulence," *Int. J. Heat Mass Transfer*, 35, 3389-3395.

Meyer, R.E., 1983, *Theory of Dispersed Multiphase Flow*, Academic, New York.

Michaelides, E.E., 1986, "Heat Transfer in Particulate Flows," *Int. J. Heat Mass Transfer*, 29, 265–273.

Michiyoshi, I., and Serizawa, A., 1986, "Turbulence in Two-Phase Bubbly Flow," *Nuc. Eng. Design*, 95, 253–267.

Nunge, R.J., and Gill, W.N., 1969, "Mechanisms Affecting Dispersion and Miscible Displacement," in *Flow Through Porous Media*, American Chemical Society Publication, 180–196.

Ochoa, J.A., Stroeve, P., and Whitaker, S., 1986, "Diffusion and Reaction in Cellular Media," *Chem. Engng. Sci.*, 41, 2999-3013.

Poirier, D.R., Nandapurkar, P.J., and Ganesan, S., 1991, "The Energy and Solute Conservation Equations for Dendritic Solidification," *Metall. Trans.*, 22B, 889–900.

Rappaz, M., and Voller, V.R., 1990, "Modeling of Micro-Macrosegregation in Solidification Processes," *Metall. Trans.*, 21A, 749–753.

Revankar, S.T., and Ishii, M., 1992, "Local Interfacial Area Measurement in Bubbly Flows," *Int. J. Heat Mass Transfer*, 35, 913–925.

Rutland, C.J., and Trouvé, A., 1993, "Direct Simulations of Premixed Turbulent Flames with Nonunity Lewis Number," *Combust. Flame*, 94, 41–47.

Sahraoui, M., and Kaviany, M., 1993, "Slip and No-Slip Temperature Boundary Conditions at Interface of Porous, Plain Media: Conduction," *Int. J. Heat Mass Transfer*, 36, 1019–1033.

Sahraoui, M., and Kaviany, M., 1994, "Slip and No-Slip Temperature Boundary Conditions at Interface of Porous, Plain Media: Convection," *Int. J. Heat Mass Transfer*, 37, 1029–1044.

Sangani, A.S., and Didwania, A.K., 1993, "Dispersed-Phase Stress Tensor in Flows of Bubbly Liquids at Large Reynolds Numbers," *J. Fluid Mech.*, 248, 27-54.

Scheidegger, A.E., 1974, *The Physics of Flow Through Porous Media*, Third Edition, University of Toronto Press, Toronto.

Seshadri, K., Berlad, A.L., and Tangirala, V., 1992, "The Structure of Premixed Particle-Clouds Flames," *Combust. Flame*, 89, 333–342.

Silverman, I., Greenberg, J.B., and Tambour, Y., 1993, "Stoichiometry and Polydisperse Effects in Premixed Spray Flame," *Combust. Flame*, 93, 97-118.

Singh, B.P., and Kaviany, M., 1991, "Independent Theory Versus Direct Simulation of Radiation Heat Transfer in Packed Beds," *Int. J. Heat Mass Transfer*, 34, 2869–2882.

Singh, B.P., and Kaviany, M., 1992, "Modeling Radiative Heat Transfer in Packed Beds," *Int. J. Heat Mass Transfer*, 35, 1397–1405.

Singh, B.P., and Kaviany, M., 1993, "Effect of Particle Conductivity on Radiative Heat Transfer in Packed Beds," *Int. J. Heat Mass Transfer* (in press).

Slattery, J.C., 1981, *Momentum, Energy, and Mass Transfer in Continua*, Robert E. Krieger Publishing Company, Huntington, NY.

Smoot, L.D., and Pratt, D.T., Editors, 1979, *Pulverized-Coal Cumbustion and Gasification*, Plenum Press, New York.

Sohn, C.W., and Chen, M.M., 1981, "Micro-Convective Thermal Conductivity in Disperse Two-Phase Mixtures as Observed in a Low-Velocity Convective Flow Experiment," *ASME J. Heat Transfer*, 103, 47–51.

Somorjai, G.A., 1994, *Introduction to Surface Chemistry and Catalysis*, John Wiley and Sons, New York

Soo, S.-L., 1989, *Particles and Continuum: Multiphase Fluid Mechanics*, Hemisphere Publishing Corporation, New York.

Szekely, J., Evans, J.W., and Sohn, H.Y., 1976, *Gas-Solid Reactions*, Academic, New York.

Taylor, G.I., 1953, "Dispersion of Soluble Matter in Solvent Flowing Slowly Through a Tube," *Proc. Roy. Soc. (London)*, A219, 186–203.

Vafai, K., and Tien, C.-L., 1981, "Boundary and Inertia Effects on Flow and Heat Transfer in Porous Media," *Int. J. Heat Mass Transfer*, 24, 195–203.

Vodak, F., Cerny, R., and Prikryl, P., 1992, "A Model of Binary Alloy Solidification with Convection in the Melt," *Int. J. Heat Mass Transfer*, 35, 1787–1793.

Vortmeyer, D., 1978, "Radiation in Packed Solid," *Proc. 6th Int. Heat Transfer Conf.*, 6, 525-539, Hemisphere Publishing Company, Washington, D.C.

Warmeck, J.J., 1979, "Modeling Multidimensional Systems," in *Pluverized-Coal Combustion and Gasification*, Smoot, L.D., and Pratt, D.T., Editors, Plenum Press, New York.

Whitaker, S., 1986a, "Flow in Porous Media. I: A Theoretical Derivation of Darcy's Law," *Transp. Porous Media*, 1, 3–25.

Whitaker, S., 1986b, "Transient Diffusion, Adsorption and Reaction in Porous Catalysts: The Reaction-Controlled Quasi-Steady Catalytic Surface," *Chem. Engng. Sci.*, 41, 3015–3022.

Whitaker, S., 1991, "Constraints for the Principles of Local Thermal Equilibrium," *Ind. Eng. Chem. Res.*, 30, 983–997.

Worster, M.G., 1991, "Natural Convection in a Mushy Layer," *J. Fluid Mech.*, 224, 335–359.

Yuan, Z.-G., Somerton, W.H., and Udell, K.S., 1991, "Thermal Dispersion in Thick-Walled Tubes as a Model of Porous Media," *Int. J. Heat Mass Transfer*, 34, 2715–2726.

Zakharov, L.V., Ovchinikov, A.A., and Nikolayev, N.A., 1993, "Modelling of the Effect of Turbulent Two-Phase Flow Friction Decrease Under the Influence of Dispersed Phase Elements," *Int. J. Heat Mass Transfer*, 36, 1981–1991.

Part II

Two-Medium Treatments

4

Solid–Fluid Systems with Simple, Continuous Interface

When the assumption of local thermal equilibrium between the *solid* and the *fluid* is not valid, i.e., when the main heat transfer is between the two adjacent phases, the *two-medium* treatment of the heat transfer is needed. In this chapter interfacial geometries that are *simple* and *continuous* (as compared to discrete, which is discussed in Chapter 5), such as plane and curved surfaces, are considered.

4.1 Interfacial Heat Transfer

The primary objective is to identify, and to an extent quantify, the effects of the various flow and surface characteristics on the *interfacial heat flux*

$$q_{sf} = q_s = -k_f \frac{\partial T}{\partial n} \qquad \text{on } A_{fs} \,, \tag{4.1}$$

where n is the coordinate normal to A_{fs} and pointing towards the fluid. The interfacial heat transfer rate is influenced by the *interphasic* heat transfer in each phase. A list of *fluid* and *solid* characteristics which can influence the interfacial heat transfer is given in Chart 4.1. Included for the *fluid* are flow behavior, properties, reaction, and suspensions. For the *solid* they include the geometry, flow arrangement, length, and thickness. Since the examination of all these characteristics is rather too extensive, a selection among them has been made with the aim of including the most significant characteristics. Then only several *fluid*, *surface*, and *flow* characteristics influencing q_{sf} are considered in this chapter. Among the *fluid characteristics* are the fluid *viscosity*, *compressibility*, *rarefaction*, *shear stress-viscosity relation*, *buoyancy*, *reactivity*, *radiation*, *electric charge*, *particulates* (including *solid matrices*), and *phase change*. Among the *surface characteristics* are the *surface roughness*, *permeability*, and *radiation*. Among the *flow characteristics* are *oscillation*, *turbulence*, *direction with respect to the surface*, and *secondary motions*.

4.2 Effect of Viscosity

For relatively high velocities or small kinematic viscosities (i.e., when $Re_x = ux/\nu$ is small), the resistance to flow over surfaces is confined to a small distance δ from the surface. This hydrodynamic resistance is propagated

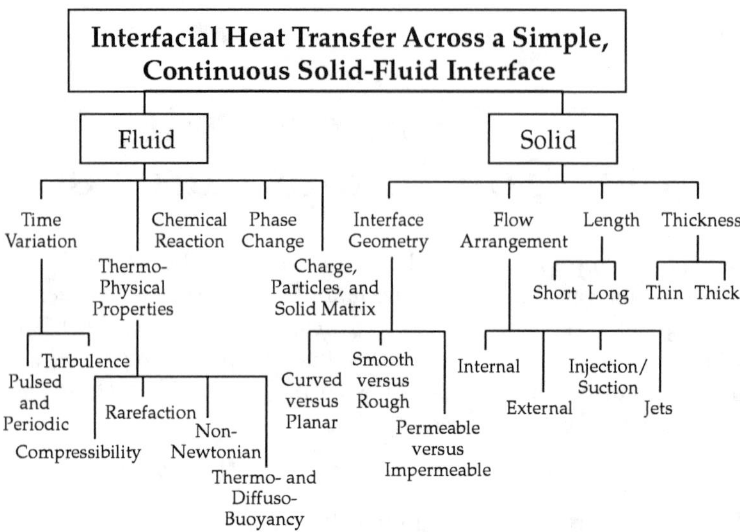

Chart 4.1. A classification of the fluid and solid characteristics influencing the interfacial heat transfer in fluid-solid system with simple and continuous interfacial geometry.

through the fluid by both the no-slip velocity condition on the surface and the fluid viscosity. An analog to this *viscous boundary-layer* resistance (Schlichting, 1979) of *thickness* δ exists when the surface temperature T_s is different from that of the fluid far away from the surface T_∞. This thermal boundary-layer resistance is over a distance δ_t. The ratio of δ to δ_t is proportional to the ratio of the kinematic viscosity to the thermal diffusivity ν/α. This ratio is the *Prandtl number* Pr, and for $Pr \to 0$ the viscous boundary layer resistance is negligible compared to the thermal boundary layer resistance. Then for $Pr \to 0$ the results for a fluid flowing with *zero viscosity*, i.e., the *potential flow*, can be used.

As an example, consider *steady laminar* flow over a wedge with an angle $\beta\pi$, as shown in Figure 4.1, with a free stream temperature T_∞ different from the wedge surface temperature T_s. The *free-stream velocity* (i.e., velocity as $y \to \infty$) for $x \geq 0$ is found from the solution to the potential flow (Eckert and Drake, 1972) and is given by

$$u_\infty = cx^m \,, \quad m = \frac{\beta}{2-\beta} \quad x \geq 0 \,, \tag{4.2}$$

where $\beta = 0$ corresponds to the flow over a *semi-infinite flat plate* and $\beta = 1$ is the *two-dimensional stagnation flow*. In the following, the predicted heat transfer rate from this wedge to an ideal *inviscid* fluid (i.e., $\mu = 0$) and to a viscous fluid are compared for the case of a small Pr (such as liquid

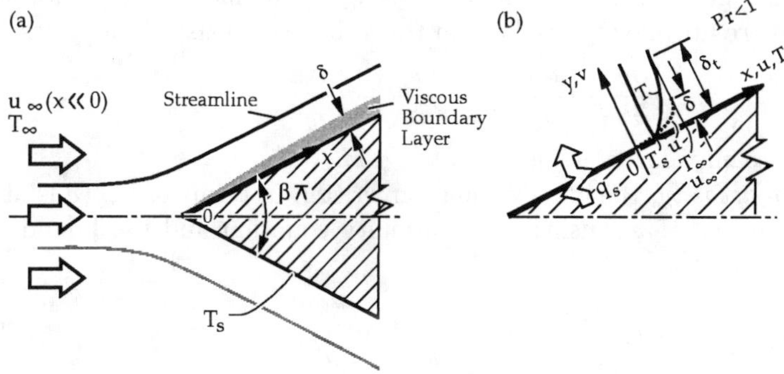

Figure 4.1. A schematic of the laminar, wedge boundary-layer flow and heat transfer. (a) The incoming flow and the wedge free-stream flow. (b) The velocity and temperature profiles at a location x along the wedge.

metals). This demonstrates the effect of the viscosity on the heat transfer rate and shows the *upper bound* of heat transfer corresponding to $\delta \to 0$ (i.e., the fluid velocity slip at the surface).

4.2.1 Zero Viscosity

A review of the solid-fluid interfacial convective heat transfer in *potential flows* is given by Galante and Churchill (1990) and its applicability, which is limited to *laminar, nonseparating* flows with high Reynolds numbers and very small Prandtl numbers (with $Pe_x = Re_x Pr$ being large), is also discussed. For the wedge problem, Chen (1985) considers the case of zero and small viscosities (zero and small Prandtl numbers). Starting from (1.66), (1.74), and (1.84), for a *steady, laminar two-dimensional flow*, the equations for conservation of mass, momentum, and energy, under the assumption of *negligible axial* diffusion and *uniform* pressure across the boundary layer (i.e., the *boundary-layer approximations*) become

$$\frac{\partial u}{\partial x} + \frac{\partial v}{\partial y} = 0 \,, \tag{4.3}$$

$$u\frac{\partial u}{\partial x} + v\frac{\partial u}{\partial y} = -\frac{1}{\rho}\frac{\partial p}{\partial x} + \nu\frac{\partial^2 u}{\partial y^2} \,, \tag{4.4}$$

$$u\frac{\partial T}{\partial x} + v\frac{\partial T}{\partial y} = \alpha\frac{\partial^2 T}{\partial y^2} \,, \tag{4.5}$$

$$T = T_s \,, \quad u = v = 0 \quad \text{at } y = 0 \,, \tag{4.6}$$

$$T = T_\infty \,, \quad u = u_\infty \,, \quad v = 0 \text{ as } y \to \infty \,. \tag{4.7}$$

The pressure gradient is eliminated by using the momentum equation for $y \to \infty$, where no velocity gradient in the y-direction exists. This gives

$$u\frac{\partial u}{\partial x} + v\frac{\partial u}{\partial y} = u_\infty \frac{du_\infty}{dx} + \nu \frac{\partial^2 u}{\partial y^2} \,. \tag{4.8}$$

The free stream velocity u_∞ is given by (4.1).

For *zero viscosity*, the energy equation written in terms of u_∞ (v is also written in terms of u_∞ using the continuity equation and the boundary condition on v at $y = 0$) is

$$u_\infty \frac{\partial T}{\partial x} - y\frac{du_\infty}{dx}\frac{\partial T}{\partial y} = \alpha \frac{\partial^2 T}{\partial y^2} \,. \tag{4.9}$$

Introducing a *dimensionless spatial similarity variable* η as

$$\eta = \frac{y}{2}(m+1)^{1/2}(\frac{u_\infty}{x})^{1/2} \,, \tag{4.10}$$

this energy equation becomes

$$\frac{d^2T}{d\eta^2} + 2\eta\frac{dT}{d\eta} = 0 \,. \tag{4.11}$$

This shows that the temperature profile is *similar* in this spatial variable. The solution subject to the boundary conditions given by (4.6) and (4.7) in the *normalized* form is

$$T^* = \frac{T - T_\infty}{T_s - T_\infty} = 1 - \frac{2}{\pi^{1/2}} \int_o^\eta e^{-\eta^2}\, d\eta \,. \tag{4.12}$$

The *local* heat flux q and its normalized form Nu_x are

$$q = -k\frac{\partial T}{\partial y}\bigg|_{y=0} \,, \tag{4.13}$$

$$Nu_x = \frac{q\, x}{(T_s - T_\infty)k} = [\frac{(m+1)Pe_x}{\pi}]^{1/2} = (m+1)^{1/2}\frac{Pe_x^{1/2}}{\pi^{1/2}} \,, \tag{4.14}$$

where the Peclét number is the product of the Reynolds number Re_x and Prandtl number Pr, i.e., $Pe_x = Re_x Pr = u_\infty x/\alpha$.

This is the local, *normalized* heat transfer rate for the case of flow *slipping* over the solid surface.

4.2.2 Nonzero Viscosity

The solution to the viscous flow, i.e., (4.2), (4.3), and (4.8), has been given by Eckert and Drake (1972). Defining a *new* dimensionless spatial similarity variable (generally referred to as the Falkner-Skan *transformation*) which contains the *viscosity* and defining a *normalized stream function* f as

$$\eta = (\frac{m+1}{2})^{1/2} y (\frac{u_\infty}{\nu x})^{1/2} = (\frac{m+1}{2})^{1/2}(\frac{c}{\nu})^{1/2} x^{(m-1)/2} y \tag{4.15}$$

$$f = (\frac{m+1}{2})^{1/2} \frac{1}{(c\nu)^{1/2}} x^{-(m+1)/2} \psi\,, \quad u = \frac{\partial \psi}{\partial y}\,, \quad v = -\frac{\partial \psi}{\partial x}\,. \tag{4.16}$$

Then the continuity equation (4.3) is satisfied and the transformed momentum equation (4.8) becomes

$$\frac{d^3 f}{d\eta^3} + f \frac{d^2 f}{d\eta^2} - \beta[(\frac{df}{d\eta})^2 - 1] = 0\,. \tag{4.17}$$

The boundary conditions for the velocity become

$$\frac{df}{d\eta} = 0\,, \quad f = 0\,, \qquad \eta = 0\,, \tag{4.18}$$

$$\frac{df}{d\eta} = 0\,, \qquad \eta \to \infty\,, \tag{4.19}$$

where

$$u = u_\infty \frac{df}{d\eta}\,, \quad v = -(\frac{2c\nu}{m+1})^{1/2} x^{(m-1)/2} (\frac{m+1}{2} f + \frac{m-1}{2} \frac{df}{d\eta} \eta)\,,$$

$$-\frac{1}{\rho}\frac{\partial p}{\partial x} = -mc^2 x^{2m-1}\,. \tag{4.20}$$

From (4.15), the *dimensional* viscous boundary thickness δ becomes

$$\delta = (\frac{2}{m+1})^{1/2} (\frac{\nu}{c})^{1/2} x^{(1-m)/2} \eta_\delta\,, \tag{4.21}$$

where η_δ is the value of η at the location of u approaching u_∞ to within *one percent*. The *surface (wall) shear stress* τ_s is given as

$$\tau_s = \mu \frac{\partial u}{\partial y}|_o = \frac{m+1}{2} \rho u_\infty (\frac{\nu u_\infty}{x})^{1/2} \frac{d^2 f}{d\eta^2}|_o \tag{4.22}$$

and in *dimensionless* form it is given as the *friction coefficient* c_f

$$c_f = \frac{\tau_s}{\frac{1}{2}\rho u_\infty^2} = [2(m+1)]^{1/2} \frac{1}{Re_x} \frac{d^2 f}{d\eta^2}|_o\,, \quad Re_x = \frac{u_\infty x}{\nu}\,. \tag{4.23}$$

Solution to the transformed momentum equation is obtained numerically and is given by Eckert and Drake (1972) for various values of m and for the semi-infinite flat plate, i.e., $m = 0$, they give

$$c_f = 0.664 Re_x^{-1/2} \qquad m = 0\,. \tag{4.24}$$

The energy equation (4.5) in terms of a new normalized temperature

$$T^* = \frac{T - T_s}{T_\infty - T_s} \tag{4.25}$$

becomes

$$\frac{d^2 T^*}{d\eta^2} + Pr f \frac{dT^*}{d\eta} = 0\,, \tag{4.26}$$

$$T^* = 1\,, \qquad \eta = 0\,,$$

$$T^* = 0\,, \qquad \eta \rightarrow \infty\,. \tag{4.27}$$

This has a solution

$$1 - T^* = \frac{\int_o^\eta e^{-\int_o^\eta Pr f\, d\eta}\, d\eta}{\int_o^\infty e^{-\int_o^\eta Pr f\, d\eta}\, d\eta}\,. \tag{4.28}$$

The normalized local heat flux Nu_x defined by (4.13) and (4.14) becomes

$$Nu_x = \left(\frac{m+1}{2}\right)^{1/2} \left(\int_o^\infty e^{-Pr \int_o^\eta f\, d\eta}\, d\eta\right)^{-1} Re_x^{1/2}\,. \tag{4.29}$$

Note that compared to (4.14), when the viscosity is included, it appears in the Reynolds number and separately in the Prandtl number. The numerical solution of f is used in the integration appearing above and the solution of f in turn depends on β (or m).

For $Pr \rightarrow 0$, i.e., $\delta \ll \delta_t$, the velocity field can be assumed to be the inviscid limit and (4.14) is recovered, which is independent of ν. For $Pr \rightarrow \infty$, i.e., $\delta \gg \delta_t$ using an approximate linear variation in the velocity, the distributions of the velocity and the stream function can be approximated with

$$\frac{u}{u_\infty} = \left.\frac{d^2 f}{d\eta^2}\right|_o \eta \quad \text{and} \quad f = \left.\frac{d^2 f}{d\eta^2}\right|_o \frac{\eta^2}{2} \tag{4.30}$$

and (4.26) can be integrated. This gives a $Pr^{1/3}$ relation.

For the *semi-infinite flat plate*, $m = 0$ and the inviscid ($Pr = 0$) and the viscous (approximated over a range of Pr) solutions for the normalized heat flux become

$$Nu_x = 0.565 Pe_x^{1/2} \qquad m = 0\,, \quad Pr = 0\,, \tag{4.31}$$

$$Nu_x = 0.332 Pr^{1/3} Re_x^{1/2} \qquad m = 0\,, \quad Pr > 0.6\,. \tag{4.32}$$

Eckert and Drake (1972) and Burmeister (1983) discuss the numerical coefficients for other values of β.

4.3 Oscillation

Heat transfer between a fluid and a solid with the fluid moving with an *oscillatory laminar* motion (with or without a net zero mean velocity) about the stationary solid surface has been studied mostly in applications related to heat transfer enhancement. These applications and those in which the solid surface oscillates have been reviewed by Lemlich (1961). However, in some cases, the oscillations can *reduce* the net heat transfer rate (as will be shown below). The time-averaged (i.e., *net*) heat transfer *across* the solid-fluid interface can be nonzero (as for example in a flow oscillation superimposed on a mean flow with no streamwise temperature gradient) or it can be zero (as in purely oscillatory flows with an axial temperature gradient where the net heat transfer *parallel* to the interfaces is being enhanced). The latter is in some aspects similar to the *Taylor dispersion* discussed in Sections 3.2.1 and 3.2.2, but here the solid and the fluid are not in local thermal equilibrium. Also since both a spatial and a temporal average are needed, there is a similarity with the turbulent heat flow. As examples, in the following one problem dealing with a nonzero net interfacial heat transfer and one dealing with a zero net interfacial heat transfer (but nonzero net longitudinal heat transfer) are discussed. The aim is to show the effect of the amplitude and the frequency of the oscillation on the net heat transfer rate.

4.3.1 Interfacial Heat Transfer in Flow Over a Plate

For the case of the heat transfer between a *flat solid surface* and a laminar fluid flow over it, the effect of the superposition of a sinusoidal motion on the mean flow has been examined by Kestin et al. (1961). The analysis is based on the boundary-layer assumptions, and therefore, no conduction parallel to the surface is allowed. This study leads to the quantitative results on the effect of flow oscillation and also shows that the *analogy* between the momentum and heat transport *breaks down* for oscillatory flows when the temperature and the velocity fields are *out* of phase and can be substantially different. Their analysis is briefly reviewed below.

The two-dimensional conservation equations for an incompressible laminar flow subject to the boundary-layer approximations (i.e., negligible axial diffusion and uniform pressure across the boundary layer) with no buoyancy effects, become

$$\frac{\partial u}{\partial x}+\frac{\partial v}{\partial y}=0\ , \tag{4.33}$$

$$\frac{\partial u}{\partial t}+u\frac{\partial u}{\partial x}+v\frac{\partial u}{\partial y}=-\frac{1}{\rho}\frac{\partial p}{\partial x}+\nu\frac{\partial^2 u}{\partial y^2}\ , \tag{4.34}$$

$$\frac{\partial u_\infty}{\partial t}+u_\infty\frac{\partial u_\infty}{\partial x}=-\frac{1}{\rho}\frac{\partial p}{\partial x}\ , \tag{4.35}$$

$$\frac{\partial T}{\partial t}+u\frac{\partial T}{\partial x}+v\frac{\partial T}{\partial y}=\alpha\frac{\partial^2 T}{\partial y^2}\ . \tag{4.36}$$

The boundary conditions are

$$q_s=-k\frac{\partial T}{\partial y}\ ,\quad u=v=0\ ,\quad T=T_s\quad \text{at}\ y=0 \tag{4.37}$$

$$u=u_\infty\ ,\quad v=0\ ,\quad T=T_\infty\quad \text{for}\ y\to\infty\ . \tag{4.38}$$

Consider a sinusoidal disturbance in the free-stream velocity given by

$$u_\infty=\overline{u}_\infty+u'_\infty=\overline{u}_\infty+u_c\cos\omega\left(\frac{x}{\overline{u}_\infty}-t\right)\ , \tag{4.39}$$

where ω is the angular frequency, and u_c is the amplitude of the sinusoidal motion superimposed on the time-averaged (i.e., mean) velocity $\overline{u}_\infty$. For a small oscillation amplitude, i.e., $u_c/\overline{u}_\infty < 1$, the Taylor series expansion around the time-averaged fields are made as

$$u = \overline{u} + u' = \overline{u} + (\frac{u_c}{\overline{u}_\infty})u_1' + (\frac{u_c}{\overline{u}_\infty})^2 u_2' + \cdots , \tag{4.40}$$

$$v = \overline{v} + v' = \overline{v} + (\frac{u_c}{\overline{u}_\infty})v_1' + (\frac{u_c}{\overline{u}_\infty})^2 v_2' + \cdots , \tag{4.41}$$

$$T = \overline{T} + T' = \overline{T} + (\frac{u_c}{\overline{u}_\infty})T_1' + (\frac{u_c}{\overline{u}_\infty})^2 T_2' + \cdots . \tag{4.42}$$

These are substituted in (4.33) to (4.36), and the equations for the *mean* fields and the *first-* and *second-order* expansions are obtained. The solution to the mean field equations are those discussed in Section 4.2.2 for $m = 0$. The penetration of an oscillatory velocity disturbances into a viscous fluid depends on the frequency of the disturbance ω and the kinematic viscosity ν, i.e., the *penetration depth* is proportional to $(\nu/\omega)^{1/2}$. The penetration of the no-slip boundary condition at the solid surface is the boundary-layer thickness δ and is proportional to $(\nu x/\overline{u}_\infty)^{1/2}$. The *ratio* of these two penetration depths is the *Stokes number Sto*, i.e.,

$$Sto = (\frac{\omega x}{\overline{u}_\infty})^{1/2} . \tag{4.43}$$

For the boundary layer analysis to be valid, the Stokes number must be less than unity, i.e., $Sto < 1$. The details of these solutions are given by Kestin et al. (1961) and their results for the *time-averaged* friction coefficient $\overline{c}_f$ and normalized heat flux $\overline{Nu_x}$, for $Pr = 0.7$, are

$$\overline{c}_f = \frac{2}{Re_x^{1/2}}\frac{d^2 f}{d\eta^2}|_o[1 + \frac{a_u}{\frac{d^2 f}{d\eta^2}|_o}\frac{u_c^2}{\overline{u}_\infty^2} + O(Sto^2\frac{u_c^2}{u_\infty^2}) + \cdots]$$

$$\simeq c_{f_o}(1 + 1.777\frac{u_c^2}{\overline{u}_\infty^2}) , \tag{4.44}$$

$$\overline{Nu} = \frac{dT^*}{d\eta}|_o Re_x^{1/2}[1 + \frac{a_T}{\frac{dT^*}{d\eta}|_o}\frac{u_c^2}{\overline{u}_\infty^2} + O(Sto^2\frac{u_c^2}{\overline{u}_\infty^2}) + \cdots]$$

$$\simeq Nu_{x_o}(1 - 0.0724\frac{u_c^2}{\overline{u}_\infty^2}) \qquad Pr = 0.7 , \tag{4.45}$$

where the constants a_u and a_T (and $d^2 f/d^2\eta|_o$ and $dT^*/d\eta|_o$) are given by Kestin et al. (1961), and c_{f_o} and Nu_{x_o} are the local friction coefficient and the local Nussult number, respectively, for no oscillation (i.e., $u_c = 0$) discussed in Section 4.2.2, and correspond to the case of $m = 0$. These friction coefficients and Nusselt numbers are given by (4.23) and (4.32) and $Re_x = \overline{u}_\infty x/\nu$.

For the case of air (and nearly most of the gases near standard conditions), i.e., $Pr = 0.7$ considered, the oscillations *increase* the friction and *reduce* the heat transfer. Note that this indicates that the analogy between the momentum and heat transport does *not* hold for this case of oscillatory flow. For Prandtl numbers *larger* than unity, where the thermal boundary-layer thickness is *smaller* than the momentum boundary layer thickness, the heat transfer is expected to *increase*. Some results, which include the viscous dissipation, are given by Takhar and Soundalgekar (1977) and point towards this increase.

4.3.2 Axial Dispersion in Flow Through a Tube

As an example of a pure oscillatory flow with a *net* heat transfer only parallel to the solid-fluid interface, consider oscillatory flow in a tube with an *imposed* time- and cross-sectional-averaged axial temperature gradient $d\langle\overline{T}\rangle_A/dx$. The flow oscillation is caused by the imposition of a sinusoidal pressure gradient $(dp/dx)_{max} \cos \omega t$, where $(dp/dx)_{max}$ is the amplitude and ω is the angular frequency. The tube has an inside radius R_i and outside radius R_o and is ideally insulated on the outside (i.e., $r = R_o$, where r is measured from the centerline). A schematic of the problem considered is given in Figure 4.2. Note that the boundary condition at R_o can also represent the thermal condition for a tube in a bundle of capillary tubes. If no flow exists in the interstitial volume between the tubes and when no *net* radial heat flow exists, this zero gradient boundary condition can be applied to an appropriate location in the tube wall. This is discussed by Kaviany (1986) and will be further discussed in the two-medium treatment of porous media in Section 5.3.2. The problem of internal flow oscillation with an axial gradient in the concentration or temperature has been studied by Watson (1983) for enhanced axial mass transfer and by Kurzweg (1985) for heat transfer. This problem has also been studied in connection with the thermoacoustic resonant tube (Rott, 1980).

With *some* similarity to the Taylor dispersion, the presence of a *velocity nonuniformity* and *lateral diffusion* of heat through the fluid and solid, causes a process of heat storage and release in the fluid and the solid. When this is averaged over the entire cycle and across the tube, the result is a significant heat flow along the axis of the tube and towards the lower temperature, without a net fluid motion. This heat transfer is several orders of magnitude larger than that due to molecular diffusion α_f. Kaviany (1990) extends the analysis of Kurzweg (1986) to tubes of finite wall thickness, and the analysis is reviewed below.

For simplicity, assume that the tube is long enough such that both the hydrodynamic and the thermal entrance effects are negligible, and therefore, *axial* diffusions are neglected. Then in the cylindrical coordinate the fluid momentum equation and the fluid and solid phase energy equations

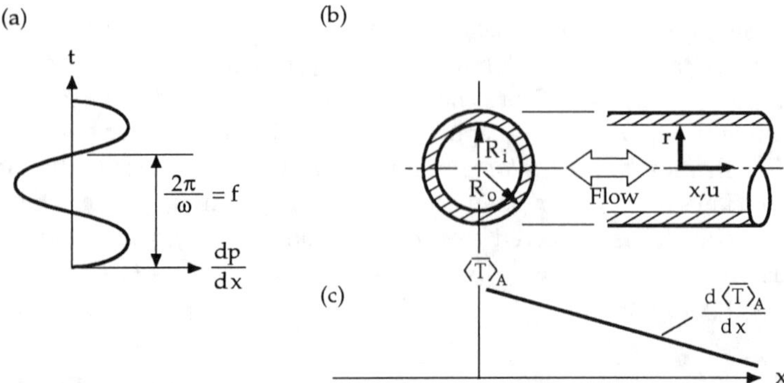

Figure 4.2. A schematic of the oscillatory pipe flow with an area- and time-averaged temperature gradient. (a) Temporal oscillation of the pressure gradient. (b) The tube radii and coordinate. (c) The constant, area- and time-averaged temperature gradient.

become

$$\rho_f \frac{\partial u}{\partial t} = -\frac{\partial p}{\partial x} + \mu_f \frac{1}{r}\frac{\partial}{\partial r} r \frac{\partial u}{\partial r} \qquad 0 \le r \le R_i \ , \tag{4.46}$$

$$(\rho c_p)_f \left(\frac{\partial T}{\partial t} + u \frac{d\langle \overline{T} \rangle_A}{dx}\right) = k_f \frac{1}{r}\frac{\partial}{\partial r} r \frac{\partial T}{\partial r} \qquad 0 \le r \le R_i \ , \tag{4.47}$$

$$(\rho c_p)_s \frac{\partial T}{\partial t} = k_s \frac{1}{r}\frac{\partial}{\partial r} r \frac{\partial T}{\partial r} \qquad R_i \le r \le R_o \ . \tag{4.48}$$

The boundary conditions are

$$\frac{\partial u}{\partial r} = 0 \ , \quad \frac{\partial T_f}{\partial r} = 0 \qquad r = 0 \ , \tag{4.49}$$

$$u = 0 \ , \quad k_f \frac{\partial T_f}{\partial r} = k_s \frac{\partial T_s}{\partial r} \ , \quad T_f = T_s \qquad r = R_i \ , \tag{4.50}$$

$$\frac{\partial T_s}{\partial r} = 0 \qquad r = R_o \ . \tag{4.51}$$

By scaling the length with R_i, time with $1/\omega$, temperature with a ΔT, then the pressure is scaled with $\rho_f R_i^2 \omega^2$ and (4.46) to (4.48) become

$$\frac{\partial u^*}{\partial t^*} = u_c^* \cos t^* + \frac{1}{Sto^2}\frac{1}{r^*}\frac{\partial}{\partial r^*} r^* \frac{\partial u^*}{\partial r^*} \qquad 0 \le r^* \le 1 \ , \tag{4.52}$$

$$\frac{\partial T^*}{\partial t^*} + u^* \frac{d\langle \overline{T^*} \rangle_A}{dx^*} = \frac{1}{Pr Sto^2}\frac{1}{r^*}\frac{\partial}{\partial r^*} r^* \frac{\partial T^*}{\partial r^*} \qquad 0 \le r^* \le 1 \ , \tag{4.53}$$

$$\frac{\partial T^*}{\partial t^*} = \frac{1}{Pr\sigma Sto^2}\frac{1}{r^*}\frac{\partial}{\partial r^*} r^* \frac{\partial T^*}{\partial r^*} \qquad 1 \le r^* \le R^* \ , \tag{4.54}$$

where

$$u_c^* = \left.\frac{dp}{dx}\right|_{\max} \frac{1}{\rho_f R_i \omega^2}\,,\ Sto = (\frac{R_i^2\omega}{\nu_f})^{1/2}\,,$$
$$Pr = \frac{\nu_f}{\alpha_f}\,,\ \sigma = \frac{\alpha_f}{\alpha_s}\,,\ R^* = \frac{R_o}{R_i}\,. \tag{4.55}$$

The *Stokes* (or *Womersley*) *number* Sto is the ratio of the tube radius to the viscous boundary-layer thickness $(\nu_f/\omega)^{1/2}$ and this thickness decreases with an increase in frequency. Then the fraction of tube cross-section filled with flow of nonuniform velocity decreases with an increase in the frequency.

The solution to the momentum equation is discussed by Schlichting (1979). Since the mean area-averaged velocity is zero, the velocity consists of a *deviation component only* and is given by

$$u'^* = iu_c^*[\frac{I_o(i^{1/2}Sto\, r^*)}{I_o(i^{1/2}Sto)} - 1]e^{it^*} = i\,u_c^* F^* e^{it^*}\,, \tag{4.56}$$

where I_o is the Bessel function of the first kind and zero order with complex argument and is given as

$$I_o(i^{1/2}z) = ber(z) + i\,bei(z)\,,\quad i = (-1)^{1/2} \tag{4.57}$$

and ber and bei are the Kelvin functions (Oliver, 1972).

The average displacement of the fluid (along the tube axis) over *half* the period is called the *tidal* displacement. The *dimensionless* tidal displacement is given by

$$\Delta x^* = 2\left|\int_{-\pi/2}^{\pi/2}\int_0^1 u'^* r^*\, dr^*\, dt^*\right|\,. \tag{4.58}$$

This is twice the cross-sectional-averaged oscillation amplitude.

The space and time average of $u'T'$ is taken as

$$\begin{aligned}\frac{D_{xx}^d}{R_i^2\omega} &\equiv -\langle\overline{u'^*T'^*}\rangle_A \\ &= -\frac{\omega}{\pi R_i}\int_0^{2\pi/\omega}\int_0^{R_i} u_{Re}'^* T_{Re}'^* r\, drdt \\ &= -\frac{1}{\pi}\int_0^{2\pi}\int_0^1 u_{Re}'^* T_{Re}'^* r^*\, dr^* dt^*\,,\end{aligned} \tag{4.59}$$

where the *subscript Re* stands for the real part of the complex quantity. The axial dispersion coefficient D_{xx}^d is similar to the Taylor dispersion for continuous flow discussed in Section 3.2.1. Here the dispersion coefficient is scaled using the system parameters R_i and ω instead of the fluid property α_f (thermal diffusivity) used in the Taylor dispersion.

The solution for the temperature distribution in the fluid and the solid are found in terms of both the mean, area-averaged component $\langle\overline{T^*}\rangle_A$ and the *deviation components* T'^*_f and T'^*_s as

$$T^*_f = \langle \overline{T^*} \rangle_A + T'^*_f = \frac{d\langle \overline{T^*} \rangle}{dx^*} x^* + \frac{d\langle \overline{T^*} \rangle}{dx^*} \theta^*_f(r^*, t^*) , \quad (4.60)$$

$$T^*_s = \langle \overline{T^*} \rangle_A + T'^*_s = \frac{d\langle \overline{T^*} \rangle}{dx^*} x^* + \frac{d\langle \overline{T^*} \rangle}{dx^*} \theta^*_s(r^*, t^*) . \quad (4.61)$$

The solutions to θ^*_f and θ^*_s are found by substituting these in (4.53) and (4.54) and applying the boundary conditions (4.49) to (4.51), and are (Kaviany, 1990)

$$\theta^*_f = \{ \frac{u^*_c}{1 - Pr} + \frac{Pr\, u^*_c}{1 - Pr} [\frac{I_o(i^{1/2} Sto\, r^*)}{I_o(i^{1/2} Sto)} - 1] + a_1 I_o(i^{1/2} Sto\, Pr^{1/2} r^*) \} e^{it^*} \equiv g^*_f e^{it^*} , \quad (4.62)$$

$$\theta^*_s = [a_2 I_o(i^{1/2} Sto\, \sigma^{1/2} Pr^{1/2} r^*) + a_3 K_o(i^{1/2} Sto\, \sigma^{1/2} Pr^{1/2} r^*)] e^{it^*} , \quad (4.63)$$

where a_1 and a_2 are constants, and their rather lengthy expressions are given by Kaviany (1990). The case of $Pr = 1$ is treated separately.

In terms of F^* and g^*_f defined in (4.56) and (4.62), respectively, the dimensionless axial dispersion coefficient is

$$\frac{D^d_{xx}}{R_i \omega} = -\frac{1}{2} \int_0^1 (\overline{F^*} g^* - F^* \overline{g^*}) r^*\, dr^* , \quad (4.64)$$

where the bars indicate the complex conjugate of the quantities.

The mean area-averaged heat flux at any location x is

$$\overline{\langle q_x \rangle}_A = -[k_e + (\rho c_p)_f D^d_{xx}] \frac{d\langle \overline{T} \rangle_A}{dx} , \quad (4.65)$$

where k_e is the *effective conductivity*. The Reynolds and Péclet numbers for this flow are defined as

$$Re = \frac{2 u_c R_i}{\nu} = 2 u^*_c Sto^2 , \quad Pe = 2 u^*_c Sto^2 Pr . \quad (4.66)$$

These can be used to estimate the entrance effect and to ensure that flow remains laminar (as assumed). The transition to turbulence for oscillatory and pulsating flows is discussed by Grassman and Tuma (1979).

To illustrate some of the features of this oscillatory flow and heat transfer, consider the flow of water in glass tubes with $R_i = 0.4$ mm, $R_o/R_i = 1.25$, $\Delta x = 20$ cm and with an average temperature of 60°C which results in $Pr = 3$. These conditions are those considered in the experiment of Kaviany and Reckker (1990), where a good agreement was found with the above predictions of D^d_{xx}.

Figure 4.3 shows the distribution of the velocity and the temperature fluctuations for an oscillation at one Hertz ($f = 1$ Hz) and for several elapsed times. The temperature distribution in *both* the solid and the fluid are shown. For this frequency, the velocity distribution is nearly parabolic (characteristic of the fully developed continuous tube flows). Note that over

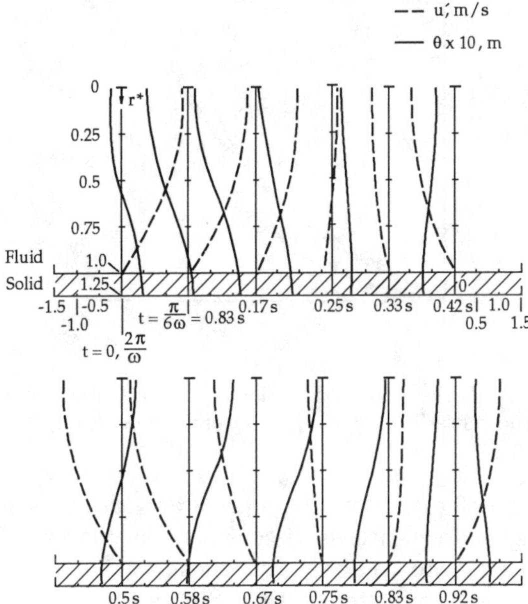

Figure 4.3. Distributions of the instantaneous velocity and deviation in temperature given for every 1/12 increment of the period ($f = 1$ Hz).

one-half of the cycle heat is stored in the solid and in the other half it is removed. Also note that the velocity and the temperature are *not* in phase. Since an axial mean temperature gradient is present, the entire heat stored in the fluid and solid, which is substantial, is released and passed toward the lower temperature side of the tube. This is similar to heat storage and release in regenerators (which have large specific surface area), but here in addition to the solid the fluid also participates in the storage-release of heat content. At higher frequencies, the velocity *distribution* in the tube changes and a *core* (*an inertial core* resulting from the temporal acceleration) of uniform velocity is formed around the centerline.

Figure 4.4 shows the instantaneous velocity distribution for several frequencies and at the elapsed time corresponding to the maximum centerline velocity. The velocity distribution given (4.56) is renormalized using the dimensional tidal displacement and the frequency. The presence of the inertial core reduces the effectiveness of the fluid to store heat, and therefore, causes a degradation in the performance.

Figure 4.5 shows the results for the frequency dependence of the axial dispersion coefficient normalized with respect to the fluid diffusivity. The results are for several tidal displacements. The axial dimensional pressure gradient required for this oscillatory flow is also shown. Note that at low

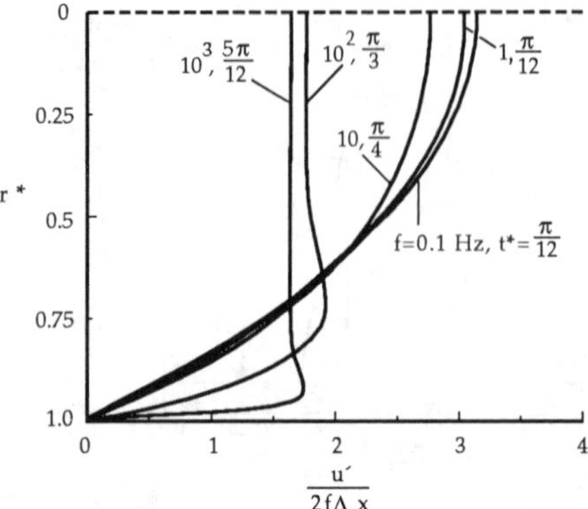

Figure 4.4. Radial distribution of the velocity at the instant of occurrence of the maximum in the centerline velocity, for several frequencies.

frequencies (i.e., $f \to 0$ asymptote) D^d_{xx} increases with f^2, as in the Taylor dispersion. At higher frequencies a viscous boundary layer is formed and this dependence charges to a $f^{1/2}$ (high-frequency asymptote). The transition between the two asymptotes is at the Stokes number of unity, which gives $f_c = \nu/\pi R_i^2$. The results show that axial dispersion coefficient can be as much as 10^5 times larger than molecular diffusion. As the tidal displacement increases, the axial dispersion coefficient and the required pressure gradient also increase.

4.4 Turbulence

In flow over solid surfaces, such as a flat plate, an initially (i.e., in the leading edge region) laminar flow undergoes a *transition* and becomes turbulent (Schlichting, 1979; Sherman, 1990; White, 1991). This *boundary-layer turbulence* can also be caused by the surface roughness (i.e., the *tripped* turbulence). The heat transfer through the solid-fluid interface is affected by the turbulent velocity fluctuations. The heat transfer is also influenced by the *free-stream turbulence* (or fluctuations) when the boundary-layer flow is in the laminar, transitional, or turbulent regime. The *turbulence intensity* (or *level*) of a stream flowing with a mean unidirectional velocity $\overline{u}$ is defined as (Schlichting, 1979)

$$Tu = \frac{(\frac{1}{3}\overline{u_i'^2})^{1/2}}{\overline{u}} \tag{4.67}$$

and for an *isotropic* turbulence $\overline{u_1'^2} = \overline{u_2'^2} = \overline{u_3'^2} = \overline{u'^2}$, and then $Tu = (\overline{u_i'^2})^{1/2}/\overline{u}$. For a *grid* or *screen* generated turbulence, a short distance from the grid the turbulence becomes isotropic.

In channel flows, downstream of an *inertia-influenced entrance region*,

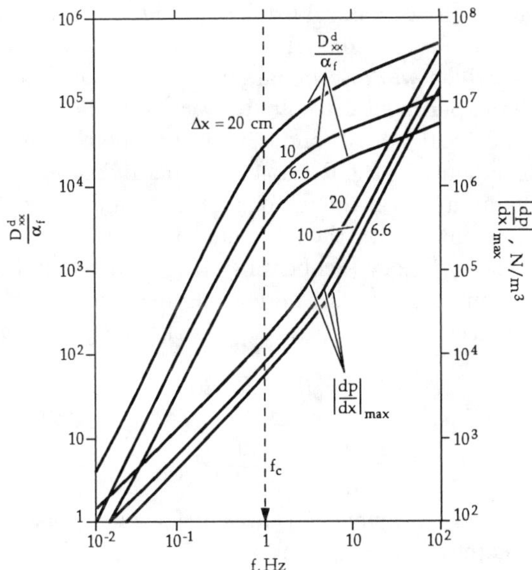

Figure 4.5. Variation of the normalized axial dispersion coefficient and the required axial pressure gradient with respect to the frequency of the oscillation, for several values of the tidal displacement.

there is a fully developed flow region where the lateral transport dominates and this allows for the close examination of the lateral turbulent heat flows.

Reviews of the structure of the small-scale motions of turbulent flow near a surface and its coherent and statistical features are given by Zaric (1982), Pletcher (1988), and Sherman (1990). A review of the prediction of the solid-fluid interfacial heat transfer in turbulent flows is given by Launder (1988) and Pletcher (1988).

In this section some features of the turbulent boundary-layer heat transfer, the effect of the free-stream turbulence on the interfacial heat transfer rate, and some features of the fully developed turbulent channel flows are discussed.

4.4.1 Boundary-Layer Turbulence

Flow disturbances, i.e., velocity fluctuations, are always present in flows and it is not possible to achieve a zero turbulence intensity in the free-stream of flows over surfaces. Experiments show (Schlichting, 1979; Cebeci and Bradshaw, 1984; Sherman, 1990) that even for very low turbulence intensities, the flow over a flat plate will have a laminar behavior (as predicted by the laminar boundary-layer analysis mentioned in Section 4.2.2) only over a finite initial length of the plate, but beyond this length the flow behavior is drastically different. The differences are in the boundary-layer thickness which, when normalized using $\delta/(\nu x/u_\infty)^{1/2}$, will be much larger than the value of about 5 predicted by the laminar flow theory, and in the wall shear stress which will also be larger than that predicted by the laminar theory. The experimental results show a *transition Reynolds number* $Re_x = u_\infty x/\nu$ of 3.5×10^5 to 10^6 and the flow beyond this Reynolds number is considered turbulent. As the Reynolds number indicates, this is when the inertial force dominates over the viscous force to cause flow separation and the dynamics associated with the secondary flows.

The linear stability theory (Schlichting, 1979; Platten and Legros, 1984), applied to the laminar flow with added infinitesimal disturbances, shows that the disturbances begin to grow at (i.e., the *marginal stability state*)

$$\frac{u_\infty \delta_1}{\nu} = Re_{\delta_1} = 520 \ , \tag{4.68}$$

$$\delta_1 = \int_0^\infty (1 - \frac{u}{u_\infty})\, dy \ , \tag{4.69}$$

where the laminar *displacement thickness* δ_1 for a flat plate with zero pressure gradient is found from the solution given in Section 4.2.2 and is

$$\delta_1 = 1.721(\frac{\nu x}{u_\infty})^{1/2} = 1.721 Re_x^{-1/2} \ . \tag{4.70}$$

This value of Re_{δ_1} is smaller than the value measured, but the predicted and measured results are in relatively close agreement.

The transition to turbulence can also be caused by the surface roughness and to some extent by the free-stream turbulence. The solid-fluid interfacial heat transfer (as well as the shear stress) are influenced by the transition.

Turbulence structure near the solid surface shows streaks of low-speed/ high-speed fluid from near the surface, and that these break down periodically in a phenomenon called *bursts*. In a burst the fluid is ejected from the surface layer with a significant normal velocity. These bursts and the resulting *upsweeps* account for 80 percent of the Reynolds stresses in the boundary layers. The *lateral* extent to which these bursts penetrate the fluid, in dimensionless distance, is $y^* = y(\overline{\tau}_s/\rho)^{1/2}/\nu$, where $\overline{\tau}_s$ is the *time-averaged surface shear stress*. This distance is about 100. The *streamwise* (i.e., along the flow) extent of the bursts, in dimensionless distance, is $x^* = x(\overline{\tau}_s/\rho)^{1/2}\nu$. This distance is about 1000. The *period* between the bursts t_b has been related to the *outer* variables such as u_∞ and δ, and to the *inner* variables such as $(\overline{\tau}_s/\rho)^{1/2}$ and ν (Banerjee, 1991).

(A) An Approximate Flow Microstructure Adjacent to the Surface

As an illustrative example, consider the simple *coherent vortical flow* structure of Kasagi et al. (1989) for turbulent flow adjacent to a solid surface. Since the fluctuations in the temperature must depend on the *fluid* conductivity and heat capacity, near the surface the conductivity and heat capacity (including wall thickness) of the *solid* must *also* influence the fluctuations. In principle, it is very difficult to maintain the surface temperature constant along the surface, and therefore the temperature fluctuations right on the surface are *not* zero. Kasagi et al. (1989) include the wall heat transfer into their analysis of turbulent transport through the interface and through the fluid. Their *deterministic*, *periodic* description of the fluid mechanics is rather oversimplified, but it allows for the examination of the turbulent transfer.

Based on their experimental measurements and observations, Kasagi et al. (1989) suggest a model for turbulent flow adjacent to the surface. This microstructure of the flow is called the *streamwise pseudo-vortical motion* and is periodic in one spatial direction and in time. The model is shown schematically in Figure 4.6. Their mathematical model for the instantaneous distributions of the three fluctuating components of the *dimensionless* velocity is

$$u'^* = a_u(y^* + b_u {y^*}^2) \sin \frac{2\pi z^*}{\lambda^*} \sin \frac{2\pi t^*}{t_b^*} , \tag{4.71}$$

$$v'^* = a_v({y^*}^2 + b_v {y^*}^3) \sin \frac{2\pi z^*}{\lambda^*} \sin(\frac{2\pi t^*}{t_b^*} + \phi) , \tag{4.72}$$

$$w'^* = a_w(y^* + b_w {y^*}^2) \cos \frac{2\pi z^*}{\lambda^*} \sin(\frac{2\pi t^*}{t_b^*} + \phi) , \tag{4.73}$$

where the scales used are

$$u'^* = \frac{u'}{(\overline{\tau}_s/\rho)^{1/2}} , \quad y^* = \frac{y(\overline{\tau}_s/\rho)^{1/2}}{\nu} ,$$
$$t^* = \frac{t(\overline{\tau}_s/\rho)}{\nu} , \quad z^* = \frac{z(\overline{\tau}_s/\rho)^{1/2}}{\nu} , \tag{4.74}$$

$\overline{\tau}_s$ is the *time-averaged surface (wall) shear stress*, and $(\overline{\tau}_s/\rho)^{1/2}$ is called the *friction velocity*. The dimensionless period, or time scale t_b^*, is taken to be of the order of the *turbulent burst* period. Note that the time is scaled using the inner variables, and using a constant t_b^* implies that the burst period scales with the inner variables. As was mentioned, this period is not yet well known; for example, in the review of Pletcher (1988) it is suggested that the outer variables be used, i.e.,

$$\frac{\overline{u}_\infty t_b}{\delta} = 6 , \tag{4.75}$$

where δ is the boundary-layer thickness.

The velocity distributions given by (4.71) to (4.73) do *not* include x-direction dependency and satisfy both the boundary conditions at $y = 0$ and the continuity equation. However they do *not* satisfy the momentum equation (but represents the measurements). The boundary conditions at a distance y_∞^* are the *outer* boundary conditions and will be discussed below.

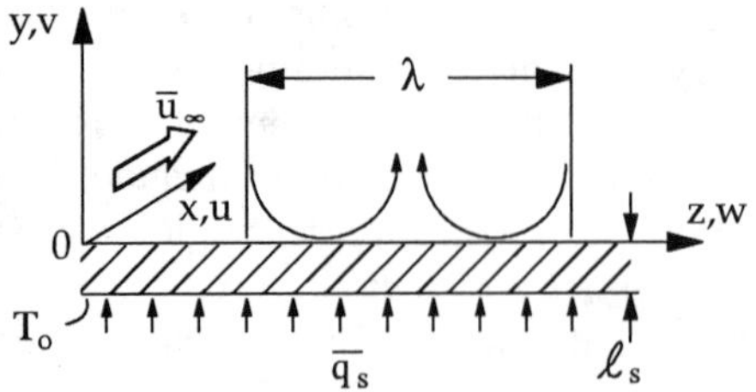

Figure 4.6. The spatially periodic microstructure model for turbulent, fully developed flow adjacent to a solid surface. The structure also undergoes a time-periodic variation.

The instantaneous energy equations for the fluid and the solid phases are written as

$$\frac{\partial T_f^*}{\partial t^*} + v'^* \frac{\partial T_f^*}{\partial y^*} + w'^* \frac{\partial T_f^*}{\partial z^*} = \frac{1}{Pr}(\frac{\partial^2 T_f^*}{\partial y^{*2}} + \frac{\partial^2 T_f^*}{\partial z^{*2}}) \quad \text{for } y^* \geq 0 \tag{4.76}$$

$$\frac{\partial T_s^*}{\partial t^*} = \frac{1}{Pr\sigma^2}(\frac{\partial^2 T_s^*}{\partial y^{*2}} + \frac{\partial^2 T_s^*}{\partial z^{*2}}) \quad 0 \geq y^* \geq -\ell_s^* \, , \tag{4.77}$$

where

$$T_i^* = \frac{T_i(\rho c_p)_i(\overline{\tau}_s/\rho)^{1/2}}{q_s} \, , \qquad i = f \text{ or } s \, , \tag{4.78}$$

$$\sigma = \frac{\alpha_f}{\alpha_s} \, , \quad \ell_s^* = \frac{\ell^*(\tau_s/\rho)^{1/2}}{\nu} \tag{4.79}$$

and $\overline{q}_s = q_s$ is the *prescribed surface heat flux*.

The boundary conditions are

$$T_f^* = T_s^* \, , \quad \kappa \frac{\partial T_f^*}{\partial y^*} = \frac{\partial T_s^*}{\sigma^{1/2} \partial y^*} \quad y = 0 \, , \tag{4.80}$$

where

$$\kappa = [\frac{(k\rho c_p)_f}{(k\rho c_p)_s}]^{1/2}. \tag{4.81}$$

The triple product $k\rho c_p$ is called the *thermal activity* or the *thermal effusivity*, and κ is called the *thermal activity* (or *effusivity*) *ratio*.

The solutions to (4.76) and (4.77) give $T_f^* = T_f^*(y^*, z^*, t^*)$ and $T_s^* = T_s^*(y^*, z^*, t^*)$, respectively, and along with the instantaneous velocity field are used to obtain the various averages. The time- and space-averaged quantities, i.e., the *statistical mean* quantities, are defined $\overline{\langle \ \rangle}_A$ as

$$\langle \overline{\psi} \rangle_A(y^*) = \frac{1}{t_b^* \lambda^*} \int_0^{\lambda^*} \int_0^{t_b^*} \psi(y^*, z^*, t^*) \, dt^* dz^* \, . \tag{4.82}$$

The thermal outer boundary condition at y_∞^* is that $T^* = T_\infty^*$ (i.e., uniform temperature thereafter), where y_∞^* is Prandtl number dependent and is taken to be between 40 and 60 in the computations of Kasagi et al. (1989). The prescribed values for the constants a_i, b_i, ϕ, λ^*, t_b^* are given by Kasagi et al. (1989) and are based on their previous experimental results. As an example, their computed results for the *instantaneous* velocity and temperature fields, for a *prescribed* $\overline{q}_s$ (i.e., constant and uniform heat flux along the lower surface of the solid) and for $Pr = 0.7$, $\lambda^* = 100$, $t_b^* = 200$, $\ell_s^* = 0$, and $y_\infty^* = 39$, are shown in Figure 4.7 for $y^* \le 30$. Note that these values for λ^*, t_b^*, and y^* are characteristics of the turbulent bursts (Banerjee, 1991). The velocity and temperature fields (obtained numerically) are out of phase, as was the case of the time-periodic flow considered in Section 4.3.2. The *velocity vector* is shown on the left-hand side and the *turbulent heat flux vector* on the right-hand side, while the contours of constant temperature are shown on both sides of $z^* = -25$. The turbulent heat flux is largest where the velocity is nearly *normal* to the surface (upward or downward with respect to the surface).

The results of Kasagi et al. (1989) for the variation of the root-mean-square of the *z-averaged surface* temperature normalized with respect to that for $\ell_s = 0$ are shown as a function of the dimensionless wall thickness in Figure 4.8(a). The results are for several values of the *thermal activity ratio* κ and for $Pr = 0.7$. When the *fluid thermal activity* (or *effusivity*), i.e., $(k\rho c_p)_f$, is much smaller than the solid, the surface temperature fluctuations are *damped* by the fluid. The higher fluid thermal activity indicates larger temperature fluctuations in response to the flow fluctuations. Also, as ℓ_s increases the effect of the heat capacitance of the solid wall decreases, and the solution reaches an asymptotic behavior corresponding to that for a *semi-infinite* wall. This coupling between the wall and the fluid, as manifested by the magnitude of $\langle\overline{T'^2(y=0)}\rangle_A$, has also been studied by Sinai (1987).

The distribution of the time- and z-averaged dimensionless y^*-direction heat flux

$$\overline{q}^*_{y^*} = -\langle\overline{v'^*T'^*}\rangle_A \tag{4.83}$$

is given in Figure 4.8 for various Prandtl numbers. As expected, for larger Pr the asymptotic value of unity is approached at smaller y^*. For larger values of Pr the time-space averaged turbulent heat flux shows a negative value and this is due to the larger phase lags between the temperature and velocity fluctuations.

(B) A Mixing-Length Model

A simple semi-empirical—yet relatively accurate in predicting the time- and area-averaged quantities—treatment of heat transfer through the solid-fluid interface is given by Kays and Crawford (1980). This treatment uses the concepts of the *laminar sublayer* adjacent to the wall, the turbulent mixing length ℓ, and the turbulent Prandtl number Pr_t, and predicts the

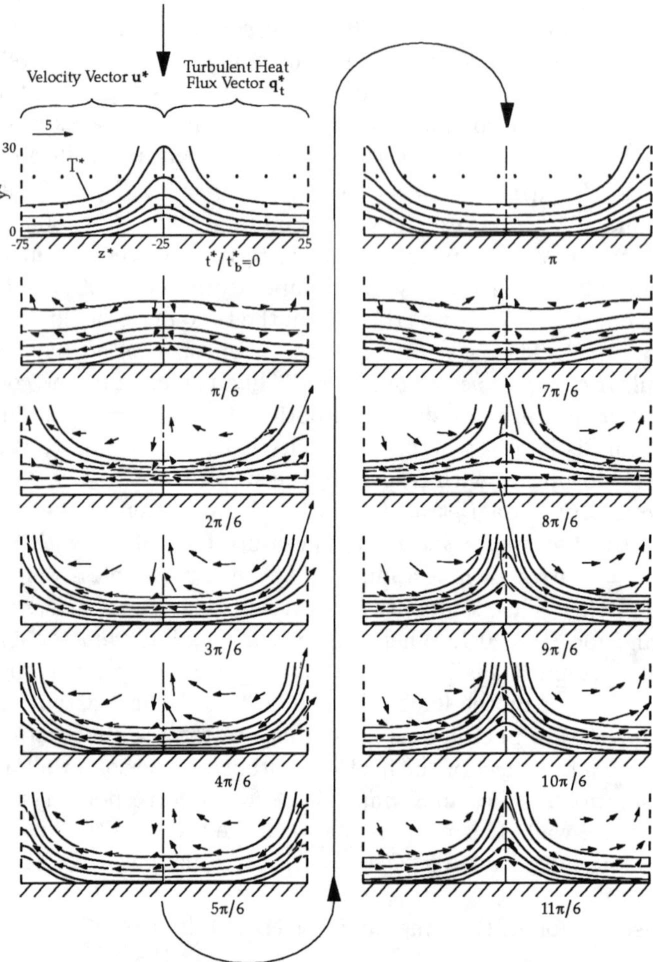

Figure 4.7. Distributions of instantaneous, local temperature (constant T^* contours), velocity vector (left-hand side of each frame), and turbulent heat flux vector (right-hand side of each frame) for the microstructure motion of Figure 4.6. The frames are for various dimensionless elapsed times t^*/t_b^*. (From Kasagi et al., reproduced by permission ©1989 American Society of Mechanical Engineers.)

distribution of the mean temperature $\overline{T}$ and the interfacial mean heat flux $\overline{q}_s$. The results for a constant free-stream velocity u_∞ and temperature T_∞ and a constant and uniform surface temperature $\overline{T}_s$ are reviewed below.

The mean energy equation for a steady two-dimensional boundary-layer flow and heat transfer (i.e., negligible axial conduction) over a flat plate

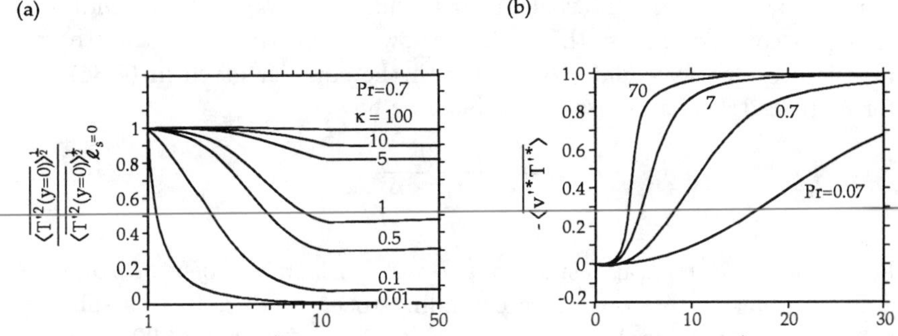

Figure 4.8. (a) Variation of the surface $\overline{\langle T'^2\rangle}^{1/2}$ with respect to the dimensionless solid wall thickness for various thermal conductivity ratios and for $Pr = 0.7$. (b) Distribution of turbulent heat flux for various Prandtl numbers and for $\ell_s = 0$ (From Kasagi et al., reproduced by permission ©1989 American Society of Mechanical Engineers.).

with the x-axis being along the flow, is the simplified form of (2.56), i.e.,

$$\overline{u}\frac{\partial \overline{T}}{\partial x} + \overline{v}\frac{\partial \overline{T}}{\partial y} = \frac{\partial}{\partial y}(\alpha + \alpha_t)\frac{\partial \overline{T}}{\partial y} = -\frac{1}{\rho c_p}\frac{\partial \overline{q}}{\partial y} \,. \tag{4.84}$$

Near the wall, the variation of $\overline{u}$ in the x-direction is much smaller than that in the y-direction, and by *assuming* that $\overline{u} = \overline{u}(y)$ in the continuity the velocity normal to the surface $\overline{v}$ becomes a constant $\overline{v}_o$ (this is called the *Couette flow* assumption). Then from the above equation $\overline{T} = \overline{T}(y)$ and we have

$$\rho c_p \overline{v}_o \frac{d\overline{T}}{dy} + \frac{d\overline{q}}{dy} = 0 \,. \tag{4.85}$$

Integrating once and applying $T = T_s$ and $\overline{q} = \overline{q}_s$ at $y = 0$, we have

$$\rho c_p \overline{v}_o(\overline{T} - T_s) + \overline{q} - \overline{q}_s = 0 \,, \quad \frac{\overline{q}}{\overline{q}_s} = 1 + \frac{\rho c_p \overline{v}_o (T_o - \overline{T})}{\overline{q}_s} \,. \tag{4.86}$$

Defining the dimensionless variables

$$\overline{v}_o^* = \frac{\overline{v}_o}{(\tau_s/\rho)^{1/2}} \,, \quad \overline{u}_o^* = \frac{u^*}{(\tau_s/\rho)^{1/2}} \,, \quad \overline{T}^* = \frac{(T_s - \overline{T})(\overline{\tau}_s/\rho)^{1/2}}{\overline{q}_s/\rho c_p} \tag{4.87}$$

this becomes

$$\frac{\overline{q}}{\overline{q}_s} = 1 + \overline{v}_o^* T^* \,. \tag{4.88}$$

The counterpart of this equation for momentum in the absence of a streamwise pressure gradient is

$$\frac{\overline{\tau}}{\overline{\tau}_w} = 1 + \overline{v}_o^* \overline{u}^* \,. \tag{4.89}$$

For no flow across the interface, $\overline{v}_o^* = 0$ and the energy equation indicates that $\overline{q}$ is constant across this Couette flow region. By integrating the dimensionless energy equation (4.85) with the equality given in (4.84) and for no mass transfer across the surface, we have

$$\overline{T}^* = \int_0^{y^*} \frac{dy^*}{\dfrac{1}{Pr} + \dfrac{\alpha_t}{\nu}} . \tag{4.90}$$

From the experimental results, a *thermal laminar sublayer* or *region* is defined adjacent to the surface extending to $y^* = 13.2$. In this sublayer there is only lateral molecular transport (i.e., $\alpha_t = 0$). Then (4.90) becomes

$$\overline{T}^* = \int_0^{13.2} \frac{dy^*}{\dfrac{1}{Pr}} + \int_{13.2}^{y^*} \frac{dy^*}{\dfrac{1}{Pr} + \dfrac{\alpha_t}{\nu}} . \tag{4.91}$$

The region beyond y^* is called the *turbulent region*, where the molecular transport is negligible.

Now a prescription of α_t is required and this is done using the concept of turbulent mixing length. The turbulent eddy diffusivity α_t is related to the eddy viscosity ν_t and the turbulent Prandtl number Pr_t, as given by (2.98) and (2.97), i.e.,

$$\frac{\overline{\tau}_s}{\rho} = \ell^2 (\frac{d\overline{u}}{dy})^2 = \kappa^2 y^2 (\frac{d\overline{u}}{dy})^2 = \nu_t \frac{d\overline{u}}{dy} , \tag{4.92}$$

$$\frac{\nu_t}{\alpha_t} = Pr_t , \tag{4.93}$$

where κ is the von Kármán constant and is taken as 0.41. The relationship between the turbulent mixing length and the *distance* measured from the *surface* can be intuitively justified. This can be done by considering the no-slip velocity boundary condition at the surface and the growth in the amplitude of the velocity *fluctuations* with respect to the distance from the surface. Then we expect an *increase* in the size of the *eddies* (or fluctuations) as the distance from the surface increases. The *linear* dependence and the constant used in (4.92) are based on empiricism. Taking the square root of the *dimensionless* form of (4.92), we have

$$\frac{d\overline{u}^*}{dy^*} = \frac{1}{\kappa y^*} . \tag{4.94}$$

Then the distribution of the eddy viscosity and eddy diffusivity are given by

$$\frac{\nu_t}{\nu} = \kappa y^* , \tag{4.95}$$

$$\frac{\alpha_t}{\nu} = \frac{\kappa y^*}{Pr_t} . \tag{4.96}$$

Substituting this expression for α_t in (4.91) and assuming that $\alpha_t \gg Pr$

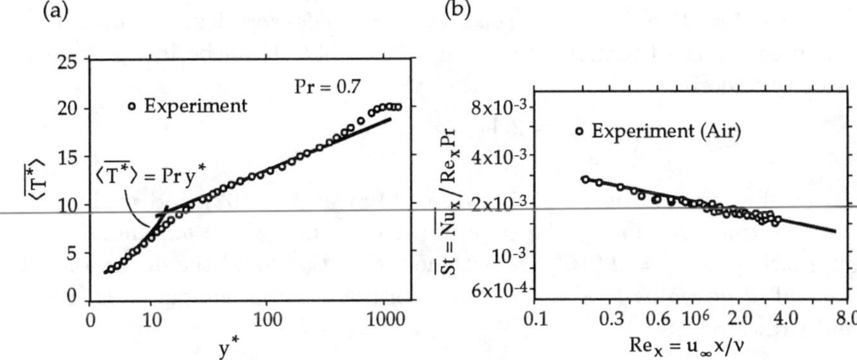

Figure 4.9. (a) The lateral distribution of the time-averaged temperature. The experimental and predicted results are shown. (b) Variation of the Stanton number with respect to the Reynolds number. The experiments for air are shown along with the prediction. (From Kays and Crawford, reproduced by permission ©1993 McGraw-Hill.)

in the turbulent region (i.e., for $y^* \geq 13.2$), we have

$$\overline{T}^* = \int_0^{13.2} Pr\, dy^* + \int_{13.2}^{y^*} \frac{Pr_t\, dy^*}{\kappa y^*}$$

$$= 13.2 Pr + \frac{Pr_t}{\kappa} \ln \frac{y^*}{13.2} . \tag{4.97}$$

The results of this equation for $Pr = 0.7$, $\kappa = 0.41$ and $Pr_t = 0.9$ are compared with the experimental results in Figure 4.9(a). In this semi-logarithm plot, the variation of T^* with y^* in the region outside the laminar sublayer appears linear. The temperature distribution for Pr near unity is given in a simpler approximate form (Kays and Crawford, 1993)

$$\overline{T}^* = 2.195 \ln y^* + 13.2 Pr - 5.66 \qquad Pr \simeq 1 . \tag{4.98}$$

For a free-stream velocity u_∞ and temperature T_∞ and a constant and uniform surface temperature T_s, the boundary conditions at the edge of the boundary layer lead to

$$\overline{u}^*_\infty = \frac{u_\infty}{(\tau_s/\rho)^{1/2}} , \quad T^*_\infty = \frac{(T_s - T_\infty)(\tau_s/\rho)^{1/2}}{\overline{q}_s/\rho c_p} = \frac{(c_f/2)^{1/2}}{\overline{St}} , \tag{4.99}$$

$$\overline{c}_f = \frac{\overline{\tau}_s}{\frac{1}{2}\rho u_\infty} , \quad \overline{St} = \frac{\overline{q}_s}{(T_s - T_\infty)\rho c_p u_\infty} = \frac{\overline{Nu}_x}{Re_x Pr} , \tag{4.100}$$

$$\overline{Nu}_x = \frac{\overline{q}_s x}{(T_s - T_\infty)k} , \quad Re_x = \frac{u_\infty x}{\nu} , \quad Pr = \frac{\nu}{\alpha} , \tag{4.101}$$

where $\overline{c}_f$ is the *friction coefficient*, $\overline{St}$ is the Stanton number, and the ratio $2\,\overline{St}/\overline{c}_f$ is called the *Reynolds analogy factor*.

Using $y^* = 10.8$ for the *viscous, laminar sublayer* thickness (note 13.2 was used for the thermal counterpart), then (4.94) can be integrated for velocity and gives

$$\overline{u}^* = 2.44 \ln y^* + 5.0 \,. \tag{4.102}$$

Now by eliminating ln y^* between (4.98) and (4.102) and evaluating the resulting equation at the edge of the core, i.e., where u_∞ and T_∞ are approached, a $T^*_\infty = T^*_\infty(u^*_\infty)$ is obtained. Using this and the definitions of T^*_∞ and u^*_∞ given by (4.87), the results for the Stanton number St is (Kays and Crawford, 1993)

$$\overline{St} = \frac{\overline{c}_f^{1/2}}{(\overline{c}_f/2)^{1/2}(13.2Pr - 10.16) + 0.9} \,, \tag{4.103}$$

where

$$\frac{\overline{c}_f}{2} = 0.0287 Re_x^{-0.2} \,. \tag{4.104}$$

Equation (4.103) is found by the approximation of the velocity profile and the use of the integral method. The comparison between the prediction of (4.103) and the experimental results is given in Figure 4.9(b) and the agreement is good. For $0.5 < Pr < 1$ and $5 \times 10^5 < Re_x < 5 \times 10^6$, (4.103) can be approximated by

$$\overline{St}\, Pr^{0.4} = 0.0287 Re_x^{-0.2} \,. \tag{4.105}$$

4.4.2 Free-Stream Turbulence

The effect of the free-stream turbulence, as given in terms of Tu defined by (4.67), on the interfacial heat transfer has been recently examined by Kim et al. (1992), and reviews are given by Kestin (1966), Mironov et al. (1982), and Pletcher (1988). The boundary-layer flow can be laminar, transitional, or turbulent. For moderate free-stream turbulence, i.e., $Tu < 0.2$, the effect of the free-stream turbulence is most pronounced in the transitional boundary layers.

Kestin (1966) shows the effect of Tu on the *local* heat transfer in flow of air ($Pr = 0.71$) over a flat isothermal surface. The range of the local Reynolds number is from 4×10^4 to 7×10^5 (i.e., before and beyond the transition). Two ranges of the free-stream turbulence, namely, $Tu = 0.007$ to 0.016 and $Tu = 0.031$ to 0.038 have been used. The results are shown in Figure 4.10. For the limits of the laminar and turbulent flows the experimental results are in agreement with the predictions (also shown) to within ±2.0 percent for the laminar regime and to within ±3.5 percent for the turbulent regime. The effect of the free-stream turbulence on the heat transfer is most pronounced in the range $6 \times 10^4 < Re_x < 4 \times 10^5$ which is the transitional regime. In this regime the effect of a significant (as com-

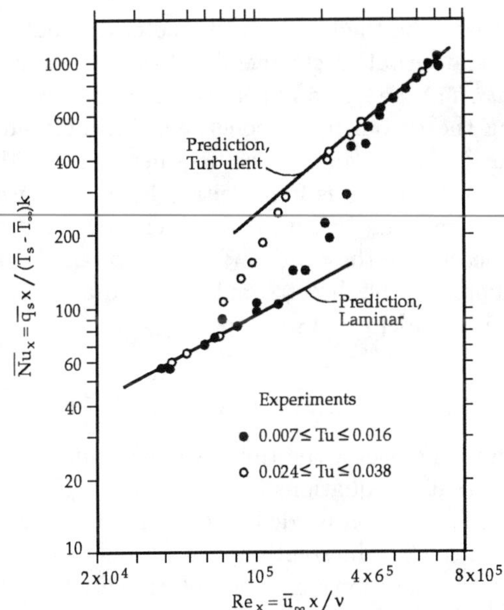

Figure 4.10. Effect of the free-stream turbulence on the heat transfer. The experimental and predicted results are shown for two different ranges of the free-stream turbulence. (From Kestin, reproduced by permission ©1966 Academic Press.)

pared to infinitesimal) free-stream disturbance is to cause the transition to occur at *lower* Reynolds numbers.

Maciejewski and Moffat (1992) show that when the free-stream turbulence is larger than the range considered by Kestin (1966), then the turbulent heat can increase substantially (by as much as 1.8 to 4 times). They define a *modified* Stanton number $\overline{St}_m$ based on the *maximum standard deviation* in the streamwise component of the velocity $u'_{\max}$. Then they correlated this $\overline{St}_m$ with Tu and Pr. This correlation is

$$\overline{St}_m = \frac{\overline{q}}{(T_s - \overline{T}_\infty)\rho c_p u'_{\max}} = 0.024 + 0.012 e^{-(\frac{Tu - 0.11}{0.055})^2} f(Pr) \; , \quad (4.106)$$

$$f(Pr) = \begin{cases} Pr^{-0.6} & Pr > 3 \; , \\ 0.75 & Pr = 0.71 \; , \end{cases} \quad (4.107)$$

which are valid for $Tu \geq 0.2$.

4.4.3 Channel Flow Turbulence

Turbulent channel flows become fully developed hydrodynamically within 10 to 15 hydraulic diameters, and the fully developed turbulent flow is of practical importance. The simpler governing equations also facilitate the solution. The transition Reynolds number (for transition from a laminar to a turbulent flow) based on the hydraulic diameter is about 2300 (Kays and Crawford, 1993). The turbulent channel flow has been studied by the *direct numerical simulation* of the flow and heat transfer (e.g., Kasagi et al., 1992) and also by using the turbulence models (e.g., Antonia and Kim, 1991). Then the distribution of the *lateral* turbulent heat flux (or the distribution of the lateral eddy diffusivity) is found either by a *statistical averaging* of the instantaneous local temperature and velocity distributions or directly from the models describing the time- and space-averaged behaviors. In the following, an example of the direct simulation and an example of the use of the mixing-length model are discussed.

(A) Direct Numerical Simulation

As an alternative to the use of the time-averaged conservation equations, the unsteady conservation equations may be solved directly with the *highest* time- and space-resolution needed to capture the time- and spacewise *fine* and *coarse* structures. The solutions are obtained numerically and, because of the required resolutions, are computationally extensive. Presently, significant progress is being made, and in this section an example is reviewed.

Kasagi et al. (1992) consider flow and heat transfer between parallel plates, as shown in Figure 4.11(a). The heat flux through both sides of the plates is prescribed as q_s, and therefore, for fully developed (thermal and hydrodynamic) fields the axial *gradients* of the surface temperature T_s and the *bulk mixed mean temperature* $\langle\overline{T}\rangle^b_{A_c}$, which is the *velocity-weighted cross-sectional* area averaged temperature, are *constant* along the flow. The plate spacing is 2ℓ, the computational domain in the other direction perpendicular to the flow is taken as $2\pi\ell$, and periodicity is applied on these boundaries. In the direction of the flow a computation domain of $5\pi\ell$, which has proven to make the results independent of this length without any need for further extensions, has been chosen. The dimensionless temperature is defined as

$$T^* = \frac{\langle\overline{T}_s\rangle_A - T(x,y,z,t)}{\overline{q}_s/\rho c_p(\tau_s/\rho)^{1/2}} \; . \tag{4.108}$$

Then the dimensionless surface temperature, i.e., T^* at $y = 0$ and 2ℓ, is taken as

$$T^*(x,0,z,t) = T^*(x,2\ell,z,t) = 0. \tag{4.109}$$

The velocity, time, and the space variables are defined as given by (4.82) and the area averages are over the periodic computational area along the flow (i.e., in the $x - z$ plane).

They solve the continuity, momentum, and energy equations and then take the time and space averages and evaluate the fluctuating components and the turbulent transport quantities. Their results, which are for air

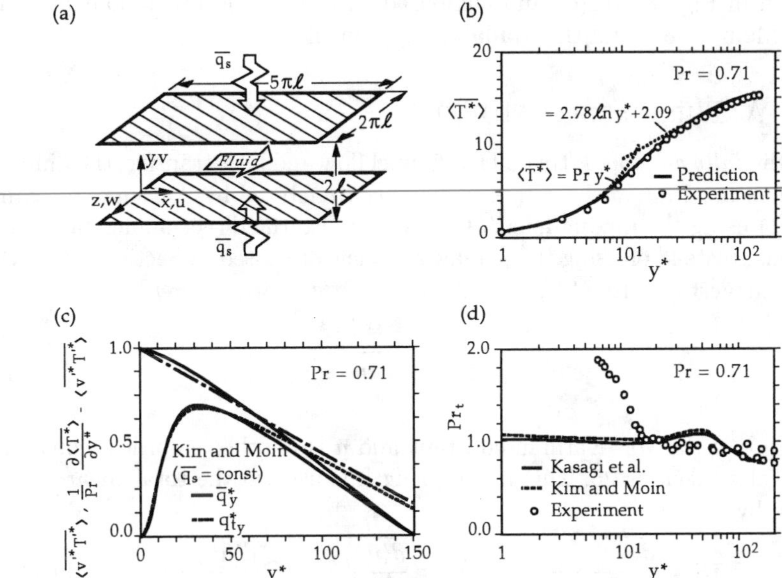

Figure 4.11. (a) The turbulent channel flow and heat transfer modeled with a spatial periodicity. The domain of the computation in the direct simulation are shown as multiples of the channel half-width ℓ. (b) The lateral temperature distribution. (c) Distributions of the turbulent and total lateral heat flux, and (d) the lateral distribution of the turbulent Prandtl number. (From Kasagi et al., reproduced by permission ©1992 American Society of Mechanical Engineers.)

$Pr = 0.71$ for the time- and streamwise-averaged temperature distribution, are given in Figure 4.11(b) and compared with an existing experimental result and also with the prediction of the logarithmic relation discussed in the last section. The agreement with the experimental results is very good. Comparison is also made with the profile $\langle\overline{T^*}\rangle_A = Pr\, y^*$ in the *conductive sublayer region*, $y^* \leq 5$, and the *logarithmic region*, $y^* \geq 20$.

They report their results for the distributions of the dimensionless y-direction turbulent heat flux $q^*_{t_{y^*}} = -\overline{\langle v'^* T'^* \rangle}_A$ and the y-direction total dimensionless heat flux

$$\overline{q}^*_{y^*} = \frac{1}{Pr}\frac{\partial \overline{\langle T^* \rangle}_A}{\partial y^*} - \overline{\langle v'^* T'^* \rangle}_A\,, \tag{4.110}$$

where the time-space average is given by (4.82). These results are given in Figure 4.11(c). They also compare their numerical results with the numerical results for the *isothermal* boundaries obtained by Kim and Moin (1989), for $Pr = 0.71$. Near the surface, the results of the constant heat flux and the constant temperature boundary conditions are nearly identical.

The distribution of the computed turbulent Prandtl number Pr_t is also shown in Figure 4.11(d) and compared with both the predictions of Kim and Moin (1989) and the available experiments.

(B) A Turbulence Model

For the *fully developed*, turbulent channel flow and heat transfer, the dimensionless total *lateral* heat flux $\overline{q}_y^*$ has been analyzed by Antonia and Kim (1991) using the mixing-length theory and also the direct numerical simulations. Instead of using the sublayer models discussed in Section 4.4.1(B), they suggest the use of the modified *van Driest mixing length*

$$\ell^* = \kappa y^*(1 - e^{-\frac{\overline{\tau}^{*1/2} y^*}{a}}) , \qquad (4.111)$$

where κ is the von Kármán constant and a is another empirical constant generally taken as 26. This mixing length is used in the total shear stress given by

$$\overline{\tau}^* = \frac{d\overline{\langle u^*\rangle}_A}{dy^*} - \overline{\langle u'^* v'^*\rangle}_A^* = \frac{d\overline{\langle u^*\rangle}_A}{dy^*} - (\ell^* \frac{d\overline{\langle u^*\rangle}_A}{dy^*})^2 \qquad (4.112)$$

and results in the velocity distribution after the integration. Again, as shown above, the temperature distribution and the turbulent heat flux are obtained similarly, i.e.,

$$\overline{q}_{y^*}^* = \frac{1}{Pr}\frac{d\overline{\langle T^*\rangle}_A}{dy^*} - \overline{\langle v'^* T'^*\rangle}_A , \qquad (4.113)$$

where the turbulent heat flux is given in terms of

$$Pr_t = \frac{\overline{\langle u'^* v'^*\rangle}_A \; d\overline{\langle T^*\rangle}_A/dy^*}{\overline{\langle v'^* T'^*\rangle}_A \; d\overline{\langle u^*\rangle}_A/dy^*} \qquad (4.114)$$

and Pr_t is given by the empirical relation

$$Pr_t = 0.85 + \frac{0.05}{Pr} . \qquad (4.115)$$

Their results for the temperature distribution obtained from the direct numerical simulations and from the use of this modified van Driest mixing length are given in Figure 4.12. The results are for several Prandtl numbers, and good agreements are found between this simple model and the direct simulations. The more elaborate models which are available (e.g., Youssef et al. 1992) cover the span between the *algebraic* models, such as the one discussed here, and the direct numerical simulations.

4.5 Surface Roughness

The surface roughnesses are *anisotropic*, and the average (root-mean-square) roughness δ_r for a *grinded* surface is $1 - 5$ μm; for a *superpol-*

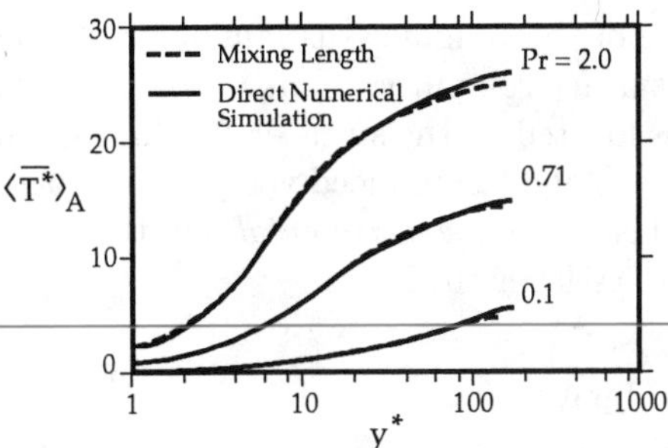

Figure 4.12. The predicted lateral distribution of the normalized temperature by the mixing-length model and by the direct numerical simulation. (From Antonia and Kim, reproduced by permission ©1991 Pergamon Press.)

ished surface is $0.05 - 0.5$ μm; and for an *electrolytic polished* surface is $0.01 - 0.1$ μm. For some heat transfer applications, the surface is intentionally made rough by *surface modification* to enhance the heat transfer rate. The roughness induces secondary flow (i.e., vortices) and can result in the transition to turbulence.

Schlichting (1979) and Kays and Crawford (1993) discuss the regimes of roughness in terms of a *roughness Reynolds number* Re_{δ_r}

$$Re_{\delta_r} = \frac{(\tau_{s_o}/\rho)^{1/2}\delta_r}{\nu} \tag{4.116}$$

where τ_{s_o} is the surface shear stress *without* the roughness and δ_r is the *roughness element size*. For $Re_{\delta_r} < 5$ the surface is considered *smooth* (i.e., τ_s is not affected by roughness). For $5 < Re_{\delta_r} < 70$ there is an increase in τ_s, but some of the smooth-surface characteristic *persists*. This regime is called the *transitional regime*. For $Re_{\delta_r} > 70$ the friction coefficient c_f becomes *independent* of the Reynolds number (i.e., the viscosity does not influence the surface shear stress). This is called the *fully rough* regime.

Klebanoff et al. (1992) postulate a two-region model where in the *inner* region the shed eddies *from* the roughnesses *interact* with the three-dimensional flow around the roughnesses and an *outer* region where the shed eddies *deform* to *generate turbulent* vortex rings.

In order to examine the heat transfer effects of the surface roughness, some of the available results on *isolated* and *distributed* surface roughnesses are reviewed below.

4.5.1 Isolated Roughness and Embedded Vortex

Consider planar surfaces having temperatures different from that of the fluid passing over them. The local flow, without a surface roughness, can be *laminar* or *turbulent*. We consider two examples of surface roughness induced vortical motion adjacent to the surface. In the first one the local flow in the absence of the roughness is laminar and is assumed to remain laminar with the surface roughness. This problem is *analyzed numerically* with the roughness located on the surface of a channel. In the second study the isolated roughness and the induced vortex are embedded in a turbulent boundary layer. This study is *experimental* and the vortex size is the same order as the boundary-layer thickness.

(A) Laminar Flow

Isolated roughnesses, i.e., surface modifications, can be categorized as those where a *second material* is added to the surface material (such as wires adhered to the surface to *trip* the flow and cause secondary motions and even turbulence), and those where *no other* material is added to the surface. The latter includes *surface machining* and, in the case of thin surfaces, the *surface stamping*. Biswas and Chattopadhyay (1992) consider fluid flow and heat transfer adjacent to a channel wall with an isolated *winglet-pair* protuberance on one of the walls. Figure 4.13(a) shows the geometric parameters. The governing equations for the laminar flow and the heat transfer are solved numerically for the steady three-dimensional velocity and temperature fields.

The numerical results of Biswas and Chattopadhyay (1992), for the velocity distribution in the channel cross-section at a location of $x/\ell = 1.69$ downstream of the winglet-pair, are shown in Figure 4.13(b). The results are for $Pr = 0.7$ and $Re_\ell = \langle u \rangle_A \ell/\nu = 500$. The *vortex pair* generated by the winglet pair retains its strength for a distance of the order of the channel height ℓ. This is shown in the distribution of the cross-sectional averaged Nusselt number $\langle Nu_\ell \rangle_A$ (averaged with respect to the top and bottom surfaces) shown in Figure 4.13(c). The vortex-pair produces an increase in this cross-sectional averaged Nussult number, not only in the immediate vicinity of the winglet-pair, but also further downstream.

(B) Turbulent Flow

In the turbulent, two-dimensional boundary-layer study of Eibeck and Eaton (1987), a half-delta wing protuberance is used to generate vortices of the order of the boundary-layer thickness. The distribution of the surface temperature (with a constant surface heat flux) indicates the effect of the induced vortical motion on the heat transfer.

The dimensionless *vortex circulation*

$$\Gamma^* = \frac{\Gamma u_\infty}{\delta} \tag{4.117}$$

used ranged from 0.12 to 0.86 and the turbulent boundary-layer thickness δ ranged from 0.84 to 2.8 cm.

Their measured results for the *spanwise* (*lateral*) distribution of the local Nusselt number, normalized with respect to that for no surface roughness,

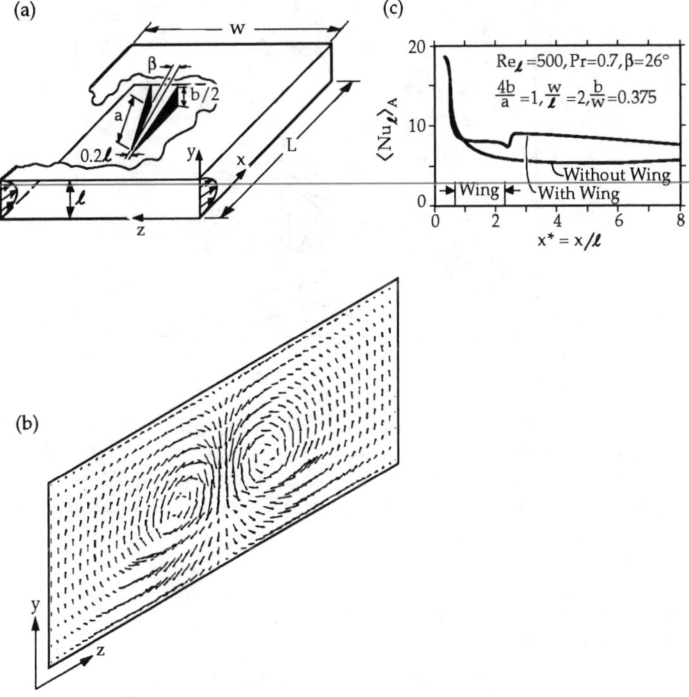

Figure 4.13. (a) An isolated winglet pair placed in the lower surface of a channel. The geometrical parameters and the flow direction are shown. (b) The vortex pair generated behind the wings, at a downstream location. (c) The variation of the area-averaged Nusselt number with respect to the dimensionless downstream axial location. (From Biswas and Chattopadhyay, reproduced by permission ©1992 Pergamon Press.)

is given in Figure 4.14. The results are for an axial location of $x = 1.35$ m from the roughness and for various values of the angle of attack β. Because of the asymmetry of the flow associated with the use of only one wing, the distribution of the normalized heat flux is not symmetric about $y = 0$. As y increases, first the normalized local heat flux *increases* and then *decreases*. The *net* effect is to increase the spanwise-averaged Nusselt number.

4.5.2 Distributed Roughness

The closely placed roughness elements in a distributed surface roughness cause the net fluid mechanics and heat transfer effects to be substantially different from that expected for an isolated roughness. This *interelemental* fluid mechanics and heat transfer *interactions* are generally examined experimentally. The flow over surfaces with a distributed surface roughness

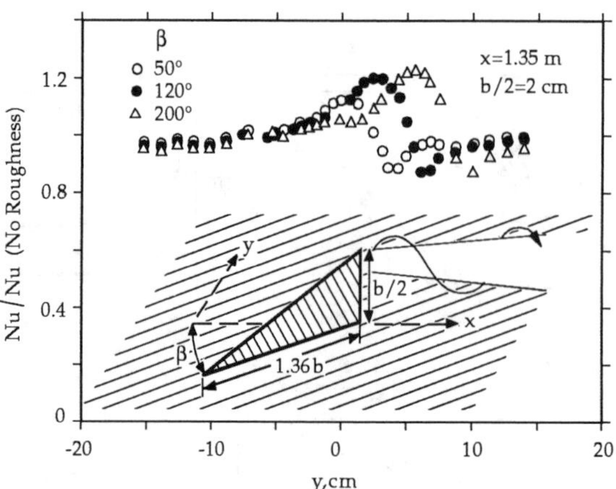

Figure 4.14. The lateral distribution of the normalized Nusselt number at a downstream location from a single, tilted wing. The wing geometrical parameters and the anticipated shedded vortex are also shown. (From Eibeck and Eaton, reproduced by permission ©1987 American Society of Mechanical Engineers.)

generally becomes turbulent, i.e., turbulence is *induced* by surface roughness (Klebanoff et al., 1992).

Analysis of the effect of the surface roughness on the local heat transfer rate in turbulent boundary layer has been made by Kays and Crawford (1980), Coleman et al. (1981), and Taylor et al. (1989). In a *fully rough* turbulence regime, they recommend a *two-layer model*. In the *inner* (adjacent to the surface) regime, a thermal resistance exists, which is empirically found and depends on the *local velocity*, *roughness*, and the *molecular conductivity*. In the *outer* region, fully turbulent transport dominates (no molecular conductivity, i.e., Prandtl number dependence). Using these two layers, the distribution of the dimensionless temperature normal to the surface and the dimensionless local heat flux have been obtained by Kays and Crawford (1993). The temperature distribution is given by

$$\langle \overline{T}^* \rangle = \frac{1}{St_{\delta_r}} + \frac{Pr_t}{\kappa} \ln \frac{32.6 y^*}{Re_{\delta_r}} , \tag{4.118}$$

where

$$St_{\delta_r} = Re_{\delta_r}^{-0.2} Pr^{-0.44} \tag{4.119}$$

and the dimensionless quantities are those defined by (4.74) and (4.87) and κ is the von Kármán constant. The local normalized heat flux, the Stanton number, is defined by (4.100) and is given by

$$St = \frac{c_f/2}{Pr_t + (\frac{c_f}{2})^{1/2} St_{\delta_r}} , \tag{4.120}$$

where

$$\frac{c_f}{2} = \frac{0.168}{\ln \frac{864\,\delta_2}{\delta_r}} , \qquad (4.121)$$

and the *momentum thickness* δ_2 is given by

$$\frac{\delta_2}{x} = 0.036 Re_x^{-0.2} . \qquad (4.122)$$

As expected, the local heat transfer rate is much larger than that for the smooth surfaces. The effect of the *free-stream acceleration* on the heat transfer from rough surfaces has been examined by Coleman et al. (1981) and Taylor et al. (1989). The effect of the roughness *shape* on the heat transfer has been examined by Hosni et al. (1993).

4.6 Compressibility

In many problems, the compressibility of fluids (generally gases), i.e., the change in the density caused by a change in the temperature or pressure, influences the convection heat transfer. One example is the compressibility resulting from *sudden* or *cyclic* changes in the temperature or the pressure which may not cause large velocities but do cause *large velocity gradients.* Another example is in high-speed flows where the *large kinetic energy* also adds to the *compressibility* and the *viscous work effects*. In the high-speed flows, except for very high velocities (i.e., large Mach numbers), the compressibility effect in the momentum equation is not as significant. In this section we first examine the *heat transfer* associated with the density disturbances caused by the change in temperature (or pressure) disturbances, i.e., the *acoustic* aspects of the compressible flow, and then the heat transfer associated with the kinetic energy and the viscous work effects, i.e., the high-speed aspects. The change in density occurring in *exothermic chemical reactions*, *without* considering any acoustic effects, is discussed in Section 4.11. The *thermobuoyancy*, i.e., the fluid motion caused by the change in the fluid density, which in turn is caused by a change in the temperature and results in a spatial variation of the volumetric body force, will be discussed in Section 4.12.

4.6.1 Thermoacoustic Phenomena Due to Standing Waves

Standing acoustic waves generate heat at rates proportional to the square of the amplitude of the pressure disturbance, and this can be substantial in applications related to the thermoacoustic engines used as energy-conversion devices (Wheatley et al. 1984; Swift, 1988). In the following, two of the simple analyses of the heat generation and transport in the presence of standing acoustic waves are reviewed. In the first one the effect of the viscosity is neglected, while in the second one it is included, leading to a Prandtl number effect.

(A) Zero Viscosity

In order to demonstrate the enhanced heat flow due to the presence of a mean temperature gradient in an oscillatory flow field of a standing acoustic wave, Swift (1988) reviews the results for the simple hydrodynamics of zero viscosity. The results are for a *plate* of finite length L (with $L \ll \lambda$ where λ is the wavelength) along the x-axis, *infinite* heat capacity, negligibly *small* thickness, and subject to a mean axial temperature gradient $d\langle\overline{T}\rangle_A/dx$. It is assumed that the *two-dimensional* flow over both sides of the plate is uniform in the y-direction but nonuniform in the x-direction and oscillates with time. The axial heat conduction in the fluid is also neglected compared to the lateral conduction.

The oscillation of the fluid (generally gas) caused by acoustic waves standing or propagating in the x-direction is given by

$$\frac{\partial u}{\partial t} = \frac{1}{\rho}\frac{\partial p}{\partial x} \quad \text{or} \quad \frac{\partial u}{\partial x} = \frac{1}{\rho a_s^2}\frac{\partial p}{\partial t} . \tag{4.123}$$

The pressure, velocity, and temperature fields include time- and area- (perpendicular to the x-axis) averaged components and *deviation* components. For the two-dimensional problem considered these are

$$p = \langle\overline{p}\rangle_A + p' = \langle\overline{p}\rangle_A + p_o\sin(\frac{2\pi x}{\lambda})e^{i\omega t} , \tag{4.124}$$

$$u' = \frac{ip_o}{\langle\overline{\rho}\rangle_A a_s}\cos(\frac{2\pi x}{\lambda})e^{i\omega t} \equiv iu_o\cos(\frac{2\pi x}{\lambda})e^{i\omega t} , \tag{4.125}$$

$$T = \langle\overline{T}\rangle_A + T'(y,t) = \langle\overline{T}\rangle_A + T'(y)e^{i\omega t} , \tag{4.126}$$

where λ is the *wavelength*, ω is the *angular frequency* and p_o and u_o are the pressure and velocity *amplitudes* of the acoustic waves. The *speed* of sound a_s is as defined by (1.13) and is also given in terms of the angular frequency and wavelength as $a_s = \lambda\omega/2\pi$. The energy equation (1.85), after neglecting terms containing the products of the deviation quantities (i.e., first-order approximations) becomes

$$\langle\overline{\rho}\rangle_A c_p(\frac{\partial T'}{\partial t} + u'\frac{\partial\langle\overline{T}\rangle_A}{\partial x}) = k\frac{\partial^2 T'}{\partial y^2} + \beta\langle\overline{T}\rangle_A\frac{\partial p'}{\partial t} , \tag{4.127}$$

$$i\omega\langle\overline{\rho}\rangle_A c_p T'(y) + \langle\overline{\rho}\rangle_A c_p u_o\cos(\frac{2\pi x}{\lambda})\frac{d\langle\overline{T}\rangle_A}{dx}$$

$$= k\frac{d^2T'(y)}{dy^2} + i\omega\langle\overline{T}\rangle_A\beta p_o\sin(\frac{2\pi x}{\lambda}) . \tag{4.128}$$

The solution, subject to $T'(y=0)=0$ and finite T' as $y \to \infty$, is

$$T'(y) = [\frac{\langle\overline{T}\rangle_A\beta}{\langle\overline{\rho}\rangle_A c_p}p_o\sin(\frac{2\pi x}{\lambda}) -$$

$$\frac{u_o}{\omega}\cos(\frac{2\pi x}{\lambda})\frac{d\langle\overline{T}\rangle_A}{dx}][1-e^{-(1+i)y/\delta_t}] , \tag{4.129}$$

$$\delta_t = (\frac{2k}{\langle\overline{\rho}\rangle_A c_p\omega})^{1/2} = (\frac{2\alpha}{\omega})^{1/2} , \tag{4.130}$$

where δ_t is the *diffusion penetration depth* for the period $1/\omega$.

For $y \to \infty$

$$T'(y\to\infty) = \frac{\langle\overline{T}\rangle_A\beta}{\langle\overline{\rho}\rangle_A c_p}p_o\sin(\frac{2\pi x}{\lambda}) - \frac{u_o}{\omega}\cos(\frac{2\pi x}{\lambda})\frac{d\langle\overline{T}\rangle_A}{dx} , \tag{4.131}$$

where u_o/ω is the *oscillation* (or *displacement*) *amplitude*. The deviation in temperature is due to the *adiabatic compression/expansion* (first term) and also due to the convection (second term). At a given location x, these two contributions can exactly nullify each other when a *critical* mean, cross-section-averaged temperature gradient is imposed. This *critical* condition is

$$\begin{aligned} \frac{d\langle\overline{T}\rangle_A}{dy}|_{cr} &= \frac{\langle\overline{T}\rangle_A\beta\omega p_o\sin(\frac{2\pi x}{\lambda})}{\langle\overline{\rho}\rangle_A c_p u_o\cos(\frac{2\pi x}{\lambda})} \\ &= \frac{\langle\overline{T}\rangle_A\beta\omega a_s}{c_p}\tan(\frac{2\pi x}{\lambda}) , \quad y\to\infty . \end{aligned} \tag{4.132}$$

The axial molecular conduction in the fluid has been assumed negligible. The time- and area-averaged convective heat transfer rate along the x-direction only has a *second-order* convection term which is $\langle\overline{\rho}\rangle_A c_p\langle\overline{u'T'}\rangle$. Because the time-averaged product of the velocity deviation and the mean temperature is equal to zero. Then

$$q_x = \frac{1}{\delta_t}\int_0^\infty \langle\overline{\rho}\rangle_A c_p\langle\overline{T'u'}\rangle_A\, dy , \tag{4.133}$$

$$q_x = \frac{1}{2\delta_t}\int_0^\infty \langle\overline{\rho}\rangle_A c_p \mathrm{Re}\langle\overline{T'u'}\rangle_A\, dy = \frac{\langle\overline{\rho}\rangle_A c_p}{\delta_t}\langle\overline{T'u'}\rangle_A \tag{4.134}$$

where the factor 1/2 is from the time averaging and Re indicates real part.

There is some similarity between this heat flux and that used for the Taylor dispersion in Sections 3.2.1 and 4.3.2. Using T' and u' in (4.133) we have

$$q_x = -\frac{1}{2}\langle\overline{\rho}\rangle_A c_p u_o\cos(\frac{2\pi x}{\lambda})[\frac{\langle\overline{T}\rangle_A\beta}{\langle\overline{\rho}\rangle_A c_p}p_o\sin(\frac{2\pi x}{\lambda}) -$$

$$\frac{u_o}{\omega}\cos(\frac{2\pi x}{\lambda})\frac{d\langle\overline{T}\rangle_A}{dx}]\text{Im}(\frac{1}{1+i})$$

$$= -\frac{1}{4}\langle\overline{T}\rangle_A\beta p_o\sin(\frac{2\pi x}{\lambda})u_o\cos(\frac{2\pi x}{\lambda})(\frac{\frac{d\langle\overline{T}\rangle_A}{dx}}{\frac{d\langle\overline{T}\rangle_A}{dx}|_{cr}}-1)$$

$$= -\frac{\langle\overline{T}\rangle_A\beta p_o^2}{\langle\overline{\rho}\rangle_A a_s}\sin(\frac{2\pi x}{\lambda})\cos(\frac{2\pi x}{\lambda})(\frac{\frac{d\langle\overline{T}\rangle_A}{dx}}{\frac{d\langle\overline{T}\rangle_A}{dx}|_{cr}}-1)\,, \tag{4.135}$$

where Im indicates imaginary part.

For an ideal gas $\langle\overline{T}\rangle_A\beta = 1$. For $d\langle\overline{T}\rangle_A/dx < (d\langle\overline{T}\rangle_A/dx)_{cr}$, the heat flux is *from* the *pressure node*, and vice versa. Very high heat flux is possible if p_o^2 is large (Swift, 1988).

(B) Nonzero Viscosity

The inclusion of the viscosity in the momentum equation (the formation of a viscous boundary layer adjacent to the surface) and in the energy equation (viscous dissipation) is addressed by Rott (1980). He considers heat transfer between an acoustically heated fluid and an *isothermal circular tube* of radius R and length L containing this fluid. He examines the distribution of the heat flux in the fluid along the tube axis.

By neglecting the radial pressure gradient and all dissipative terms involving axial differentiation (similar to the boundary-layer approximations), the energy equation written in terms of the total enthalpy, i.e., (1.87), becomes

$$\frac{\partial}{\partial t}\rho(c_pT+\frac{u^2}{2})+\frac{\partial}{\partial x}\rho(c_pT+\frac{u^2}{2})u+\frac{1}{r}\frac{\partial}{\partial r}r\rho(c_pT+\frac{u^2}{2})v$$
$$=\frac{\partial p}{\partial t}+\frac{1}{r}\frac{\partial}{\partial r}rk\frac{\partial T}{\partial r}+\frac{1}{r}\frac{\partial}{\partial r}r\mu u\frac{\partial u}{\partial r}\,, \tag{4.136}$$

where in the kinetic energy the axial component of the velocity u is assumed to be the dominant contributor. Note that the inclusion of the kinetic energy is not important, whenever second- or lower-order approximations are made.

For the case where the viscous boundary-layer thickness δ is much smaller than R, this can be approximated as

$$\frac{\partial}{\partial t}\rho(c_pT+\frac{u^2}{2})+\frac{\partial}{\partial x}\rho(c_pT+\frac{u^2}{2})u+\frac{\partial}{\partial y}\rho v(c_pT+\frac{u^2}{2})$$
$$=\frac{\partial p}{\partial t}+\frac{\partial}{\partial y}k\frac{\partial T}{\partial y}+\frac{\partial}{\partial y}\mu u\frac{\partial u}{\partial y}\,, \tag{4.137}$$

where $y = r - R$. The boundary conditions are $u = v = 0$ at $y = 0$ and the *tube-core* (or free-stream) conditions as $y \to \infty$.

Next, the velocity and temperature fields are decoupled as in (4.40) to (4.42). The time- and area-averaged total enthalpy flow across the boundary layer is

$$-\langle \overline{q_x} \rangle_A = \frac{\partial}{\partial x} \int_0^\infty \overline{\rho u(c_p T + \frac{u^2}{2})}\, dy + [\overline{\rho v(c_p T + \frac{u^2}{2})}]_\infty . \tag{4.138}$$

To the second order, i.e., neglecting terms involving higher products of the deviation components,

$$-\langle \overline{q} \rangle_A = \frac{\partial}{\partial x} \int_0^\infty \langle \overline{\rho} \rangle_A c_p \overline{T'u'}\, dy + \langle \overline{\rho} \rangle_A c_p \overline{T'(\infty) v'(\infty)} . \tag{4.139}$$

The velocity and temperature fields corresponding to this boundary-layer flow and heat transfer are analyzed by Rott (1980). The y-direction velocity variation is also added to a velocity field. This variation in u' is found by solving

$$\frac{\partial u'}{\partial t} = \nu \frac{\partial^2 u'}{\partial y^2} \tag{4.140}$$

subject to $u = 0$ at $y = 0$ and $u' \to u'_\infty e^{i\omega t}$ for $y \to \infty$. This is the Stokes problem and the solution is given by

$$u' = u'_\infty [1 - e^{-(\frac{i\omega}{\nu})^{1/2} y}] e^{i\omega t} . \tag{4.141}$$

The solution to the temperature field and to the lateral velocity for $y \to \infty$, i.e., v'_∞, are also given by Rott (1980). The derivations are rather lengthy and can be found there. The effect of the Prandtl number, in addition to the other parameters, will now appear in the expression for the axial heat flux.

The axial time- and area-averaged heat flux for the case of an acoustic standing wave in a tube of length $L = \lambda/2$ is

$$\langle \overline{q} \rangle_A = 1 + \frac{1-\gamma}{\kappa Pr^{1/2}} (\frac{\langle \overline{\rho} \rangle_A \mu \omega}{8})^{1/2} (1 + a_1 \cos \frac{2\pi x}{L}) u_o^2 , \tag{4.142}$$

where

$$\gamma = \frac{c_p}{c_v} , \quad \kappa = 1 + [\frac{(\rho c_p k)_f}{(\rho c_p k)_s}]^{1/2} ,$$

$$a_1 = \frac{[\gamma - (\kappa - 1) Pr^{1/2}](1 + Pr) + (1 - Pr)(1 + Pr^{1/2})}{[\gamma + (\kappa - 1) Pr^{1/2}](1 + Pr) - (1 + Pr)(1 - Pr^{1/2})} . \tag{4.143}$$

For $Pr \to \infty$ and $\kappa = 1$, then $a_1 = -1$ and $\langle \overline{q} \rangle_A$ vanishes on both ends of the tube. The heat conduction is negligible and the distribution of $\langle \overline{q} \rangle_A$ is *identical* to the distribution of the viscous dissipation.

For $Pr = 1$ and $\kappa = 1$, then $a_1 = 1$, i.e., the maximum in $\langle \overline{q} \rangle_A$ appears on both ends. For $Pr < 1$ and $\kappa = 1$, then $a_1 > 1$ and a *cooling* effect is produced around $x = L/2$.

For $Pr \to 0$ ($\mu \to 0$), $a_1 = (\gamma + 1)/(\gamma - 1)$, and the *maximum* cooling effect is at $x = L/2$ and is $1/\gamma$ *times* the maximum heating at $x = 0$. In this case the viscous dissipation is negligible and the main dissipative effect is the energy exchange between the core and the boundary layer.

The above analysis is applied to the *thermoacoustic resonant tube* by Merkli and Thomann (1975) and Rott (1984). The tube is *closed* in *one end* and is acoustically exited from the other end. In this problem, the Stokes boundary layer thickness is *not* limited to a small fraction of R, i.e., the Stokes number $Sto = (R^2\omega/\nu)^{1/2}$ is not very small.

4.6.2 Thermoacoustical Diffusion

As was discussed in Section 2.2.2, a sudden local change in the temperature of a compressible fluid produces a change in the local pressure and this propagates through the fluid as acoustic waves. In some containers, this sudden local wall disturbance results in waves which are then multiply reflected internally from the container wall. This tends to *enhance* the *thermal mixing* within the container, and the term *thermoacoustic diffusion* is used because there is *no net* motion (although, the term *thermoacoustic convection* has also been used). The case of a fluid contained between two parallel plates with one wall undergoing a change in the temperature has been analyzed by Ozoe et al. (1980) for a *perfect gas* and by Zappoli (1992) for a single-component fluid in a nearly supercritical state. In the latter the analysis is restricted to a slowly varying surface temperature so that some analytical treatments could be made. The former uses a numerical treatment, and therefore, the complete nonlinear transient one-dimensional problem is solved. The *conduction* treatment of traveling waves has been addressed by Churchill and Brown (1987) and Huang and Bau (1993).

Consider a perfect gas contained between two plates separated by a distance L and initially being quiescent and having a uniform temperature gradient dT/dx (which can be zero). We follow the general development of Ozoe et al. (1980); however, the following scalings and numerical results are different from those reported by them. We begin by defining the dimensionless variables using the length L, the temperature at $x = L$, T_L, and the speed $(R_gT_L/M)^{1/2}$. The dimensionless variables for the one-dimensional, transient flow and heat transfer become

$$x^* = \frac{x}{L}, \quad T^* = \frac{T}{T_L}, \quad u^* = \frac{u}{(\frac{R_gT_L}{M})^{1/2}},$$

$$t^* = \frac{t(\frac{R_gT_L}{M})^{1/2}}{L}, \quad \rho^* = \frac{\rho}{\rho_L}, \quad p^* = \frac{p}{p_L} = \frac{p}{\rho_L\frac{R_gT_L}{M}}. \tag{4.144}$$

The continuity, momentum, and energy equation are the simplified forms of the (1.68), (1.76), (1.77), and (1.81). Assuming a perfect gas behavior, we can write $de = c_v\, dT$ and by adding the equation of state, the dimensionless governing equations become

$$\frac{\partial \rho^*}{\partial t^*} + \frac{\partial}{\partial x^*}\rho^* u^* = 0\ , \tag{4.145}$$

$$\frac{\partial}{\partial t^*}\rho^* u^* + \frac{\partial}{\partial x^*}\rho^* u^{*2} = -\frac{\partial p^*}{\partial x^*} + \frac{4}{3}\frac{1}{Re_L}\frac{\partial^2 u^*}{\partial x^{*2}}\ , \tag{4.146}$$

$$\rho^* \frac{\partial T^*}{\partial t^*} + \rho^* u^* \frac{\partial T^*}{\partial x^*} = -(\gamma - 1)p^* \frac{\partial u^*}{\partial x^*} + \frac{\gamma}{Re_L Pr}\frac{\partial^2 T^*}{\partial x^{*2}} + \frac{4}{3}Ec(\frac{\partial u^*}{\partial x^*})^2\ , \tag{4.147}$$

$$p^* = \rho^* T^*\ , \tag{4.148}$$

where the parameter are defined as

$$Re_L = \frac{(R_g T_L/M)^{1/2} L}{\nu_L}\ ,\quad \gamma = \frac{c_p}{c_v}\ ,$$
$$Ec = \frac{\mu(R_g T_L/M)^{1/2}}{c_v \rho_L T_L L}\ ,\quad Pr = \frac{\nu_L}{\alpha_L}\ , \tag{4.149}$$

where Ec stands for the Eckert number generally associated with viscous dissipation. The dimensionless initial and boundary conditions are

$$u^* = 0\ ,\quad T^* = 1 - \left.\frac{\partial T^*}{\partial x^*}\right|_{t^*=0}(1 - x^*) \quad \text{for } t^* = 0 \tag{4.150}$$

$$u^*(x^* = 0) = u^*(x^* = 1) = 0\ ,$$

$$T^*(x^* = 0) = 1 - \left.\frac{\partial T^*}{\partial x^*}\right|_{t^*=0} + \Delta T^*\ ,\quad T^*(x^* = 1) = 1 \quad \text{for } t^* > 0\ , \tag{4.151}$$

where the dimensionless initial temperature gradient $(\partial T^*/\partial x^*)_{t^*=0}$ and the dimensionless temperature jump at $x^* = 0$, ΔT^*, are *prescribed* in addition to Re_L, γ and Ec.

The presence of an initial temperature gradient and the lack of a net motion result in the conditions

$$\int_0^1 \rho^*\, dx^* = 1\ , \tag{4.152}$$

$$p^*(x^*, t^* = 0) = \frac{1}{\int_0^1 dx^*/T^*} = \frac{-(\partial T^*/\partial x^*)_{t^*=0}}{\ln[1 - (\partial T^*/\partial x^*)_{t=0}]}\ , \tag{4.153}$$

$$\rho^*(x^*, t^* = 0) = \frac{-(\partial T^*/\partial x^*)_{t^*=0}}{\ln\left[(1 - (\partial T^*/\partial x^*)_{t^*=0})\right]\left[1 - (\partial T^*/\partial x^*)_{t^*=0}(1 - x^*)\right]}\ . \tag{4.154}$$

For $t^* \to \infty$, the steady-state solution for T^* is

$$T^*(x^*, t^* \to \infty) = 1 + (-\left.\frac{\partial T^*}{\partial x^*}\right|_{t^*=0} + \Delta T^*)(1 - x^*) . \qquad (4.155)$$

In absence of any acoustic waves, the temperature distribution is given by the following nonuniformly converging series

$$T_o^* = 1 + (-\left.\frac{\partial T^*}{\partial x^*}\right|_{t^*=0} + \Delta T^*)(1 - x^*) + \frac{2}{\pi}\sum_{n=1}^{\infty}\frac{\Delta T^*}{n}\sin(n\pi x^*)e^{-n^2\pi^2 t^*/Re_L Pr} , \qquad (4.156)$$

where $Fo = t^*/Re_L Pr = t\alpha_L/L^2$ is the *Fourier number*.

The solution to the finite-difference approximation of the governing equations is found numerically. The distributions of u^* and T^* are obtained for the case of no initial temperature gradient, $\gamma = 1.4$, $Pr = 0.7$, $Ec = 0$, and $\Delta T^* = 0.5$ and for $Re_L = 100$ and 300. The results are shown in Figure 4.15(a) and (b). These results are obtained using a rather coarse grid net, and the results are given only for a *qualitative* illustration. Brown (1992) and Brown and Churchill (1993) show that a *very fine* grid net is required for obtaining accurate quantitative results (as a result he only considers a *semi-infinite* fluid). Figure 4.15(a) shows that the sudden expansion of the gas at $x^* = 0$ initiates the thermoacoustic waves which then propagate towards the other surface and consequently return after reflection. The period for the travel of a distance 2 (twice the length) is $2/\gamma^{1/2}$. Since the Reynolds numbers are low, the viscous force is significant and the amplitude of the thermoacoustic waves decays rapidly. The waves are shown for three different elapsed times corresponding to when the wave has not yet reached the other surface ($t^* = 0.25$), after the first reflection ($t^* = 1.25$), and after the second reflection ($t^* = 2.25$). Note that the period of travel is 1.69. For higher Reynolds numbers, the sudden gas expansion results in *more* than one burst (although the second burst has much smaller amplitude). Note that the velocity and the Reynolds number are both scaled with $(R_g T_L/M)^{1/2}$, and therefore the dimensional velocity amplitude for $Re_L = 300$ is larger than that for $Re_L = 100$.

The temperate distribution is shown in Figure 4.15(b), for $t^* = 5$. The results show that the thermoacoustic waves cause enhanced diffusion (because there is no net motion) and *accelerate* the approach to the steady state.

4.6.3 High-Speed Boundary-Layer Flows

Compressible boundary-layer flows occur in both internal and external high-speed flows, and due to the large kinetic energy of the fluid (generally gases), the variation in the fluid velocity corresponds to a large variation in temperature (this is in addition to the temperature nonuniformity due

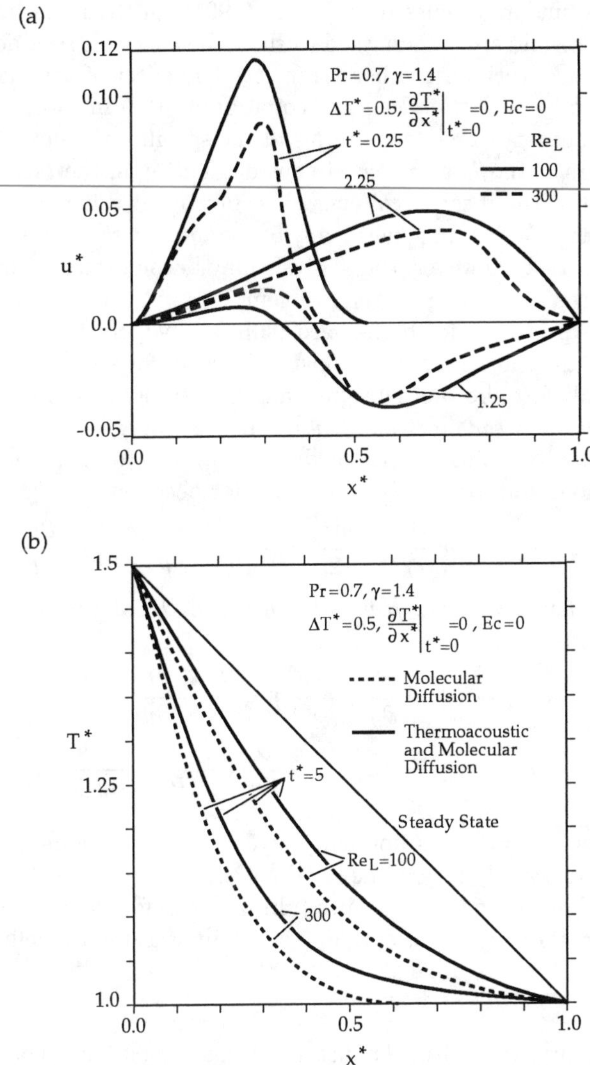

Figure 4.15. (a) Distribution of the velocity for three different elapsed times and two different Reynolds numbers. (b) Distribution of temperature for two elapsed times and two different Reynolds numbers. Also shown is the steady-state distribution.

to any heat transfer across the solid-fluid interface). Due to the presence of large velocity gradients, the viscous work is also significant. The analysis of the compressible boundary-layer flows are given by Eckert and Drake (1972), Kays and Crawford (1993), and White (1991), among others. The formation and the heat transfer aspects of the shock waves are not considered here but are discussed by White (1991). In the following, some of the existing results are summarized to show the effect of the kinetic energy and the viscous work on the heat transfer. The effect of the viscous work appears directly in the *total* energy equation written in terms of the specific internal energy (1.81) or in terms of the specific enthalpy (1.83). This effect appears as the *irreversible* viscous dissipation in the *thermal* energy equation, when the mechanical energy is subtracted using the mechanical energy equation (1.80). In general, any of the *total* or *thermal* energy equations can be used. However, the analysis may become simpler by choosing a specific form, depending on the problem.

Here we expect that for high-speed (sub- and supersonic) flows, the kinetic energy and the work done against the viscous force become significant, while the body force becomes insignificant. In the absence of any volumetric radiation and for *steady laminar flows*, the *total* energy conservation equation (1.87) for two-dimensional fields and subject to the *boundary-layer* approximations and for an *ideal gas* behavior becomes

$$\rho u\frac{\partial}{\partial x}(i+\frac{u^2}{2})+\rho v\frac{\partial}{\partial y}(i+\frac{u^2}{2})=-\frac{\partial}{\partial y}\frac{k}{c_p}\frac{\partial i}{\partial y}+\frac{\partial}{\partial y}\mu u\frac{\partial u}{\partial y}\,. \qquad (4.157)$$

Now noting that $\partial(i+u^2/2)/\partial y=\partial i/\partial y+u\partial u/\partial y$, and using the Prandtl number, this becomes

$$\rho u\frac{\partial}{\partial x}(i+\frac{u^2}{2})+\rho v\frac{\partial}{\partial y}(i+\frac{u^2}{2})=\frac{\partial}{\partial y}\frac{k}{c_p}\frac{\partial}{\partial y}(i+\frac{u^2}{2})$$
$$+\frac{\partial}{\partial y}(1-\frac{1}{Pr})\mu\frac{\partial}{\partial y}\frac{u^2}{2}\,. \qquad (4.158)$$

Note that for $Pr=1$, this boundary-layer equation is similar to the one used in Section 4.2.2, except the *total* specific enthalpy $i+u^2/2$ appears here (instead of the specific enthalpy). In this case for the assumed perfect gas behavior, the *stagnation* enthalpy and temperature are defined as

$$i_{st}=i+\frac{u^2}{2}\,,\qquad T_{st}=T+\frac{u^2}{2c_p}\,. \qquad (4.159)$$

The boundary and initial (x-direction) conditions for a constant free-stream velocity and temperature and for *constant temperature* or for *insulated* ($q_s=0$) surface condition are

$$u=u_\infty\,,\quad T=T_\infty\,,\quad T_{st}=T_{st_\infty}=T_\infty+\frac{u_\infty^2}{2}\qquad y\rightarrow\infty\,, \qquad (4.160)$$

$$u=v=0\,,\quad T=T_s\quad(\text{or }\frac{\partial T}{\partial y}=0)\,,\quad T_{st}=T_s\quad y=0\,, \qquad (4.161)$$

$$u=u_\infty\,,\quad T=T_\infty\,,\quad T_{st}=T_{st_\infty}=T_\infty+\frac{u_\infty^2}{2}\qquad x=0\,. \qquad (4.162)$$

The case of *variable density* is discussed by White (1991) and the similarity solution for the velocity and temperature fields are given. In general, except for very high speeds and when shock waves are formed, the change in the density is not as significant as the effect of the kinetic energy and the viscous work.

Assuming constant properties, the velocity field is found in a manner similar to that discussed in Section 4.2.2. For the case considered here, $m = 0$. The dimensionless temperature is defined using the kinetic energy of the fluid. The space (i.e., the similarity variable) and the stream function are scaled as before, i.e.,

$$T^* = \frac{T_{st} - T_{st_\infty}}{u_\infty^2/2c_p}\ , \quad \eta = (\frac{u_\infty}{2\nu x})^{1/2} y^{1/2}\ , \quad f = \frac{\psi}{(2\nu u_\infty x)^{1/2}}\ . \tag{4.163}$$

The dimensionless form of the energy equation (4.158) for constant properties becomes

$$\frac{d^2T^*}{d\eta^2} + Pr\, f\frac{dT^*}{d\eta} + (Pr-1)\frac{d^2}{d\eta^2}(\frac{df}{d\eta})^2 = 0 \tag{4.164}$$

with the boundary conditions for the dimensionless temperature given by

$$T^* = T^*_s \quad \frac{dT^*}{d\eta} = 0 \quad \eta = 0\ , \tag{4.165}$$

$$T^* = 0 \qquad \eta \to \infty\ . \tag{4.166}$$

For the case of an insulated surface $dT^*/d\eta|_o = 0$, the solution is (Eckert and Drake, 1972)

$$T^* = Pr\int_o^\infty \phi\, d\eta - Pr\int_0^\eta \phi\, d\eta\ , \tag{4.167}$$

$$\phi = e^{-Pr\int_o^\eta f\, d\eta}\int_0^\eta (\frac{d^2 f}{d\eta^2})^2 e^{Pr\int_o^\eta d\eta}\, d\eta\ . \tag{4.168}$$

For this case of an *adiabatic* surface, the surface temperature T_s is called the *recovery temperature* (due to the kinetic energy and the viscous work conversions to thermal energy) and a *recovering factor* r_f is defined as

$$r_f = \frac{T_s - T_\infty}{u_\infty^2/2c_p} \quad \text{for } q_s = 0\ . \tag{4.169}$$

White (1991) shows that the numerical results can be correlated with

$$r_f = Pr^{1/3} \quad 0.1 < Pr < 3\ , \tag{4.170}$$

$$r_f = 1.91Pr^{1/3} - 1.15 \quad 3 < Pr\ . \tag{4.171}$$

The solution for the case of isothermal surface is discussed by both Eckert and Drake (1972) and Kays and Crawford (1993). The lateral temperature distributions for T_{st} and T, at an arbitrary axial location x, are given

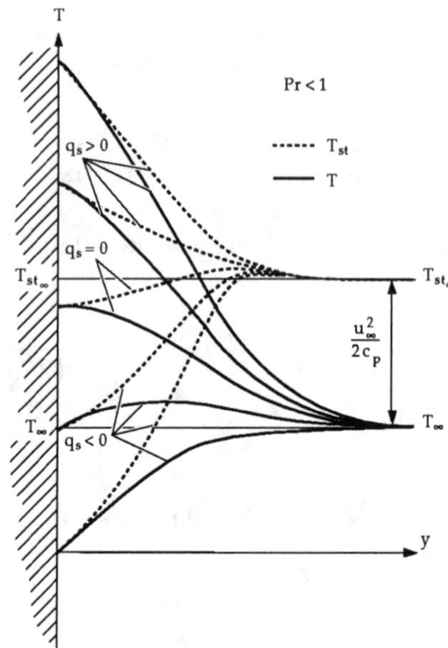

Figure 4.16. Anticipated distributions of the temperature and the stagnation temperature inside a laminar boundary layer of a high-speed flow. The results are for $Pr < 1$.

quantitatively in Figure 4.16 for the case of $Pr < 1$. For the case of $q_s = 0$, i.e., $(\partial T/\partial y)_o = 0$, the resulting T_s is larger than T_∞ (but smaller than $T_{st_\infty} = T_\infty + u_\infty^2/2$, because $Pr < 1$) in accord with the recovery relation given above. For the cases of $q_s > 0$ and $q_s < 0$, typical temperature distributions are also shown in the figure.

4.7 Rarefaction

Rarefied gas dynamics address the dynamics of the flow under the condition where the molecular mean free path λ becomes noticeable compared to ℓ which the local flow clearance distance (for internal flows) or the local characteristic size of the object (for external flows). Then the the discrete molecular aspects of the flow must be recognized. The ratio of the λ/ℓ is called the Knudsen number Kn, and originated with the Knudsen's 1909 experiment on the pressure drop in *isothermal* gas flow through *capillary tubes* (i.e., small ℓ) at *low pressures* (i.e., large λ). His experiment showed that the viscous flow behavior which assumes no velocity slip on solid surfaces *breaks down* when Kn becomes large. Using the molecular mean free path given by (1.14), the speed of sound given by (1.13), the viscosity given by (1.33), and the average clearance distance ℓ, we have

$$Kn = \frac{\lambda}{\ell} = 3(\frac{\pi\gamma}{8})^{1/2}\frac{Ma}{Re_\ell} \,, \tag{4.172}$$

$$Ma = \frac{\overline{u}}{a_s} \,, \quad Re_\ell = \frac{\overline{u}\,\ell}{\nu} \,, \quad \gamma = \frac{c_p}{c_v} \,, \tag{4.173}$$

where Ma is the *Mach number*. Then a large Kn corresponds to either a large Ma or a small Re_ℓ.

In this brief review, some of the gas-dynamic features of the *isothermal* flow is considered first. Then some available results on the *nonisothermal* flows are discussed.

4.7.1 Flow Regimes for Isothermal Flows

To illustrate the role of Kn in marking the transition for a continuum viscous flow with no velocity slip on stationary solid surfaces on the one side and the free molecular flow on the other, consider the internal flow experiment of Knudsen (Cunningham and Williams, 1980). This is flow of an ideal gas through a small-diameter tube of diameter d and length L for which the fully developed viscous flow is given by the Poiseuille-Hagen relation. Consider arranging n tubes per unit *area* with the flow allowed through the tubes only. Then the area porosity ϵ for the fluid flow is $n\pi d^2/4$. The average velocity per unit area $\langle u\rangle_A$ is *smaller* than the average *pore* (or inside-tube) velocity $\langle u\rangle_{A_f} = u_p$ and the relation is $\langle u\rangle_A = \epsilon\langle u\rangle_{A_f}$. The Poiseuille-Hagen relation in terms of $\langle u\rangle_A$ is written as

$$\dot{m} = \rho\langle u\rangle_A = -\frac{d^2}{32\mu}\frac{pM}{R_gT}\frac{dp}{dx} \tag{4.174}$$

or

$$\frac{\dot{m}}{|\frac{dp}{dx}|} = \frac{d^2}{32\mu}\frac{pM}{R_gT} \,. \tag{4.175}$$

This shows that for *isothermal flows* the mass flow rate divided by the pressure gradient increases *linearly* with p.

The experimental results of the Knudsen for $d = 3.3 \times 10^{-2}$ mm and $L = 2$ cm with flow of carbon dioxide at 25°C (he used 24 capillary tubes), are given in Figure 4.17 for various absolute pressures (given in centimeters of mercury). For $Kn > 0.14$, $Kn = \lambda/d$, the linear proportionality is not followed and a deviation from the viscous behavior is observed. The flow rate for a given pressure gradient is larger than that expected for the *viscous regime* (i.e., for *no* velocity slip on the solid surface). Then this deviation is associated with the velocity *slip* at the solid surface. The results show a minimum which occurs around $Kn = 1$, and for yet lower pressures this ratio *increases* with a decrease in pressure. The regime between the

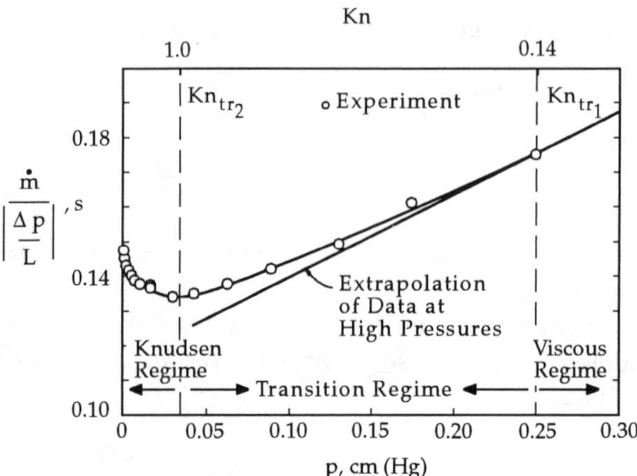

Figure 4.17. The ratio of the mass flux (based on the average velocity per unit area) to the pressure drop across a bundle of capillary tubes (n tubes per unit area) as a function of the absolute pressure. The transitional Knudsen numbers are marked on the top axis. (From Cunnigham and Williams, reproduced by permission ©1980 Plenum Press.)

beginning of the deviation (marked with Kn_{tr_1}) and this minimum (marked with Kn_{tr_2}) is called the *transition regime*. The regime for $Kn > Kn_{tr_2}$ is called the *free-molecular regime or Knudsen regime* (there are other regime classifications and depending on the internal versus external and isothermal versus nonisothermal flows, these transitional Knudsen numbers vary).

The deviation from the viscous regime (i.e., occurrence of a velocity slip on the surface) is modeled using a *velocity slip* at the surface. The velocity slip was discussed in connection with thermophoresis in Section 1.12.4 (C). There are several models for the velocity slip and one of these models is

$$\tau_s = -\mu \frac{\partial u}{\partial n} = \alpha_u u_i \quad \text{on surface}\,, \tag{4.176}$$

where n is the coordinate normal to the surface, α_u is the *dimensional velocity slip coefficient*, and u_i is the *slip velocity* (i.e., the fluid velocity on the surface). The no-slip asymptote is reached for $\alpha_u \to \infty$, which with the constraint on finite τ_s leads to $u_i \to 0$. An alternative velocity slip condition, based on the *Maxwell reflection coefficient*, will be discussed in Section 4.7.2. Using this boundary condition instead of $u = 0$ (on the surface) gives

$$\frac{\dot{m}}{|\frac{dp}{dx}|} = (\frac{d^2}{32\mu} + \frac{d}{4\alpha_u})\frac{pM}{R_g T}\,. \tag{4.177}$$

In general α_u is determined experimentally, but it can be found by the *direct, numerical molecular simulation* (also called *molecular dynamics*) using the kinetic theory of gases. Equation (4.177) does *not* predict the minimum shown in Figure 4.17. Many semi-empirical treatments are available (Cunningham and Williams, 1980) which address this minimum, and one is

$$\dot{m} = -\frac{d^2}{32\mu}\frac{pM}{R_gT}\frac{dp}{dx} - \frac{4}{3}(1 - \frac{Kn}{1+Kn})\frac{d}{4\alpha_u}\frac{pM}{R_gT}\frac{dp}{dx} - \frac{Kn}{1+Kn}D_K\frac{M}{R_gT}\frac{dp}{dx}\,, \qquad (4.178)$$

where the last is called the *Knudsen diffusion flux* and D_K is the *Knudsen diffusivity*.

The gas dynamic aspects of rarefied flows has been reviewed by Sherman (1969), Kogan (1973), Yen (1984), Muntz (1989), and Smith and Jensen (1989). These include the direct molecular flow simulations, which are generally for external flows. Lord (1992) discusses the direct simulation Monte Carlo computation with various assumptions about the reflection/absorption of molecules striking the surface. The fundamental equation of the kinetic theory of gas flow, with two-body collisions included, is the *Boltzmann equation* (Smith and Jensen, 1989)

$$\frac{\partial f}{\partial t} + \mathbf{u}\cdot\nabla f = a - bf\,, \qquad (4.179)$$

where f is the *distribution function*, $a - bf$ is the *collision integral*, a is the *gain term* of the collision integral (i.e., the rate of scattering of molecules into $d\mathbf{u}$), and bf is the *loss term* of the collision integral (i.e., the rate of scattering of molecules out of $d\mathbf{u}$). The Boltzmann equation (4.179) is a conservation (i.e., continuity) equation which states that the rate of increase of f in a given volume element is equal to the difference between the convective transport and collision source/sink. The direct numerical simulation of the Boltzmann equation involves the calculation of f, and $a - bf$, at each velocity point in the discrete velocity space (Yen, 1984). Application of the Monte Carlo method to flow through a slit is discussed by Wadsworth and Erwin (1993). The *transient* motion of low-pressure (i.e., rarefied) gases in a channel, with the channel surface cooled or heated, has been analyzed by Wadsworth et al. (1993). They show that the *transient rarefaction* effects near the surface are very significant and are masked by the continuum treatment.

4.7.2 Nonisothermal Flows

The rate of heat exchange between a surface and a rarefied gas depends on the flux of energy and momentum carried by the *impinging* (on the surface) and the *re-emitted* (from the surface) molecules. A *thermal accommodation coefficient* a_T is defined as (Springer, 1971; Giedt and Willis, 1985)

$$a_T = \frac{q_i - q_r}{q_i - q_{s_o}}\,, \tag{4.180}$$

where q_i and q_r are, respectively, the *incident* and *reflected* energy far from the surface, and q_{s_o} is the ideal energy flux leaving the surface if the molecules were at the surface temperature T_s. For $a_T = 1$, i.e., *complete accommodation* or *diffuse re-emission*, the reflected molecules will have this ideal energy. For $a_T = 0$, i.e., *specular reflection*, the molecules leave with their incident energy. The *surface material*, *ultrafine roughness*, and *contamination* influence a_T. For most surfaces used in practice, $0.8 \le a_T \le 0.98$ (Springer, 1971). Similarly, *tangential* and *normal momentum accommodation coefficient* are defined (Springer, 1971).

For the *transitional regime*, an *alternative* approach uses the temperature slip model similar to the velocity slip given by (4.176). This was discussed in relation to thermophoresis in Section 1.12.4 (C) and the temperature slip boundary condition is

$$T_i - T_s = \alpha_T \lambda \frac{\partial T}{\partial n} \qquad \text{on solid surface}\,, \tag{4.181}$$

where T_i is the interfacial temperature if the gas temperature distribution is extrapolated to the surface, and T_s is the *actual* surface temperature, α_T is the *dimensionless temperature slip coefficient* and $\alpha_T\lambda$ is called the *temperature jump distance*. Note that α_T was also used in connection with the thermophoresis in Section 1.12.4. One of the available correlations for α_T is (Springer, 1971)

$$\alpha_T = \frac{2-\alpha_u}{Pr\alpha_u}\frac{2\gamma}{\gamma+1}\,, \qquad \gamma = \frac{c_p}{c_v}\,. \tag{4.182}$$

Also, alternatively for the velocity the *Maxwell reflection coefficient* F is defined such that the fraction $1 - F$ of impinging molecules are *specularly reflected* (i.e., reversal of the normal velocity component while the tangential velocity component remains unchanged). Then the slip boundary condition (4.176) will be written as $u_i = -(2-F)\lambda(\partial u/\partial n)/F$.

The analysis of solid-rarefied gas heat transfer for internal and external flows in the *transitional regime* is reviewed by Springer (1971) and a recent application to external flows with discrete surface heat source distribution is given by Carey (1992). As an example, consider the internal flow of rarefied gas through a constant surface heat flux q_s tube of diameter d with $Ma < 1$. This problem has been analyzed by Sparrow and Lin (1962). Assuming that the continuum energy equation (1.85) can be used with the slip thermal boundary condition given by (4.181) and the velocity slip given in terms of F, the Nusselt number Nu_d based on the difference between the surface temperature T_s and the *bulk-mean* gas temperature $\langle T\rangle_A^b$, defined below, is given by (Sparrow and Lin, 1962)

$$Nu_d = \frac{q_s d}{(T_s - \langle T \rangle_A^b) k} = \frac{\frac{48}{11}}{1 - \frac{6}{11}\frac{u_i}{\langle u \rangle_A} + \frac{1}{11}(\frac{u_i}{\langle u \rangle_A})^2 + \frac{48}{11}\alpha_T Kn} \tag{4.183}$$

where

$$\langle T \rangle_A^b = \frac{8}{\langle u \rangle_A d^2} \int_0^{d/2} T\, u\, r\, dr \tag{4.184}$$

$$\frac{u_i}{\langle u \rangle_A} = \frac{1}{1 + \frac{F}{16 - 8F}\frac{1}{Kn}} . \tag{4.185}$$

In the limit of $Kn \rightarrow 0$, i.e., the *viscous regime*, the value of $48/11 = 4.36$ for the available solution (Kays and Crawford, 1993) for Nusselt number corresponding to this case is recovered.

4.8 Non-Newtonian Liquids

The interest in the heat transfer aspects of the *non-Newtonian liquids* (i.e., those which do *not* follow the *linear* relation between the viscous shear stress and the deformation rate) has originated from at least three applications. The first is that of the heat transfer (interfacial) characteristics of the *drag reducing* polymer solutions. The addition of small amount of a *high molecular weight* organic compound to water has resulted in a significant reduction in the wall shear stress. However, the interfacial (solid-liquid) heat transfer is *also reduced* (Hartnett, 1992). The second interest is the general need for the knowledge of the interfacial heat transfer rate in cooling/heating of non-Newtonian liquids. The third is in the *heat generation* due to *viscous dissipation*. This can be significant because of the large viscosity of the polymers.

In the following, the rheological aspects of the non-Newtonian liquids is briefly discussed, and an example liquid and a model are reviewed. Then the interfacial heat transfer and the viscous dissipation are discussed.

4.8.1 Rheology

Bird et al. (1987) show that for *macromolecular* liquids (e.g., polymers) the Newton law of viscosity, i.e., (1.32), is *not* adequate. They also show that the alternative viscosity laws used for these liquids would depend on the *manner* the flow is *driven*. According to Böhme (1987), a *simple fluid* is defined as a fluid with the stresses acting on it depending not only on the momentary state of the motion but *possibly* on the past motion (*memory characteristic*), but *not* on the future motion (*deterministic characteristic*)

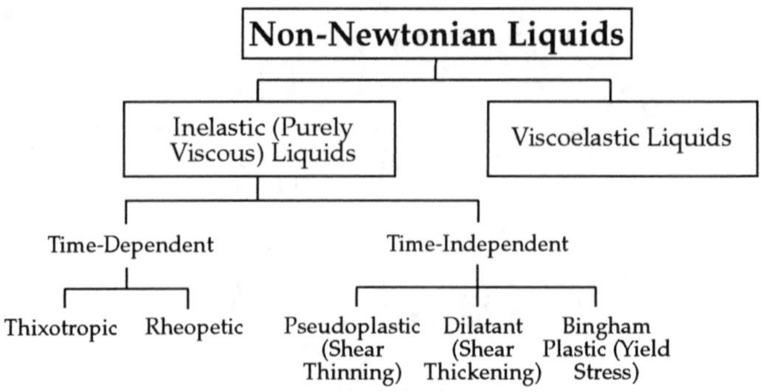

Chart 4.2. A classification of the non-Newtonian liquids.

and *not* on the motion of distant fluid particles (*local action characteristic*). The *Newtonian* fluids have *no memory* and behave according to a *linear constitutive law*, i.e., (1.32), and are a subset of simple fluids. A *rheological classification* of the non-Newtonian liquids is given by Metzner (1965) and also by Shenoy and Mashelkar (1982), and Platten and Legros (1984), and is shown in Chart 4.2.

As an example, consider the *non-Newtonian*, *purely viscous*, *time-independent*, *shear-thinning*, (i.e., *pseudoplastic*) behavior of a macromolecular liquid. Some experimental results given by Winter (1977) for the viscosity of *low-density polyethylene* as a function of the magnitude of the *deformation rate* $\dot{\gamma}$ and for several temperatures, are shown in Figure 4.18(a). For low deformation rates ($\dot{\gamma} \leq 10$ s^{-1}), the measurement is done by a cone-plate viscometer and beyond that by a capillary-tube viscometer. The results show that for $\dot{\gamma} < 0.1$ the viscosity does not significantly vary with $\dot{\gamma}$ and as $\dot{\gamma} \to 0$ an asymptotic *zero-deformation rate behavior* is found. At higher deformation rates, the viscosity (or the *apparent viscosity*) decreases (i.e., *shear* or *deformation thinning*).

One viscosity model for these purely viscous, shear-thinning, non-Newtonian liquids is the *Carreau model* which has been used for heat transfer analysis by Shadid and Eckert (1992). This is

$$\mu^* = \frac{\mu}{\mu_o} = \frac{a_\mu(T)}{[1 + (t_\mu a_\mu \dot{\gamma})^2]^{(1-m)/2}} , \tag{4.186}$$

where μ_o is the *reference* viscosity at *zero* deformation rate and temperature T_o, a_μ describes the *temperature dependence* of the viscosity, m is the *power-law index*, t_μ is the *relaxation time*, and as mentioned before $\dot{\gamma}$ is the *magnitude* of the *deformation rate tensor*, where the deformation tensor is given in terms of the velocity field as $(\nabla \mathbf{u} + \nabla \mathbf{u}^T)/2$. The temperature dependence is based on the semi-empirical Arrhenius description, similar to (1.40), i.e.,

(a)

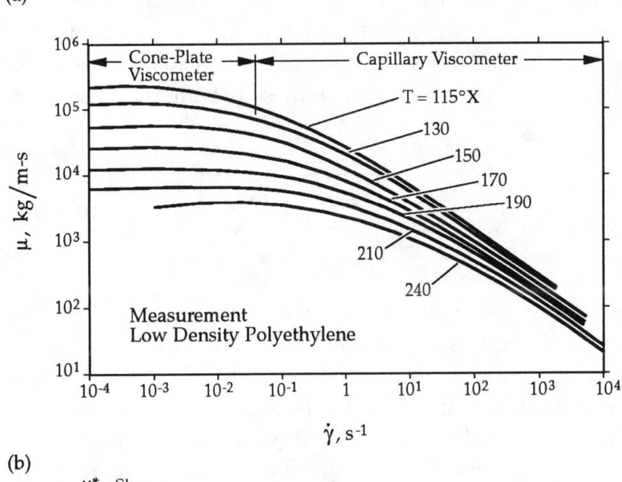

(b)

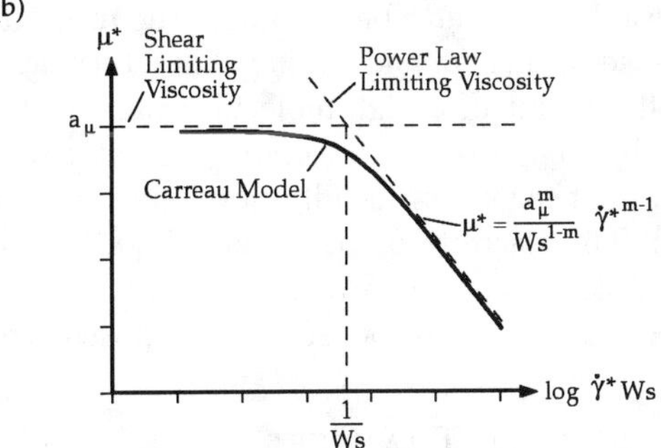

Figure 4.18. (a) The measured viscosity of a low-density polyethylene as a function of the deformation rate and at various temperatures. (b) The Carreau model for a shear-thinning, non-Newtonian liquid. (From Winter, reproduced by permission ©1977 Academic Press.)

$$a_\mu = e^{\frac{\Delta E_\mu}{R_g}\left(\frac{1}{T}-\frac{1}{T_o}\right)} . \tag{4.187}$$

For large deformation rates, the Carreau model leads to the *power-law model* which gives a proportionality between μ and $\dot{\gamma}^{\,m-1}$. The inclusion of the relaxation time t_μ in the model is to further control the transition to the shear thinning regime.

For a flow of characteristic velocity u_o about a surface of characteristic length ℓ, the dimensionless parameters (Shadid and Eckert, 1992)

$$Ws = \frac{t_\mu u_o}{\ell} , \quad Na = \frac{\mu_o}{k}\frac{\Delta E_\mu u_o^2}{R_g T_o^2} , \quad Re = \frac{2\rho u_o \ell}{\mu_o} ,$$
$$Pr = \frac{\mu_o c_p}{k} , \quad \dot{\gamma}^* = \frac{\dot{\gamma}\ell}{u_o} , \quad T^* = \frac{(T-T_o)\rho c_p \ell}{\mu_o u_o} , \tag{4.188}$$

can be defined. The *Weisenberg number* Ws is the ratio of the relaxation time t_μ to fluid residue time (ℓ/u_o). The *Nahme number Na* is the ratio of heat generated by viscous dissipation to that conducted. For *Na* greater than 0.1 to 0.5, the viscous dissipation leads to a *significant* viscosity change (Winter, 1977). Using these dimensionless parameters, the normalized viscosity becomes

$$\mu^* = \frac{a_\mu}{[1 + (Ws\, a_\mu \dot{\gamma}^*)^2]^{(1-m)/2}} , \tag{4.189}$$

$$a_\mu = e^{-\frac{Na Re T^*}{Pr}} \quad \text{for} \quad \frac{T}{T_o} \approx 1 . \tag{4.190}$$

The dimensionless Carreau model is shown in Figure 4.18(b).

For *viscoelastic* liquids rather complex constitutive equations are needed with *many* more constant and variables. The *elastic effect* in liquid flows (i.e., viscoelastic flows) and heat transfer have been discussed by Hartnett (1992). The heat transfer aspects of *slightly elastic* liquids, which are generally *aqueous polymer solutions*, have been studied the most. The interest in these liquids is because of their *drag-reducing* characteristics. For example, addition of a small amount of a high-molecular-weight polymer, such as 10 *parts per million* by weight of *polyacylamide*, in water results in a 40 percent reduction in wall shear stress in fully developed turbulent flow in tubes (Hartnett, 1992). The viscosity of viscoelastic liquids depend on the deformation rate $\dot{\gamma}$. The stresses on the *orthogonal faces* are not equal, and a finite time is required for a strain response to an imposed stress change.

4.8.2 Interfacial Heat Transfer

The *analysis* of the *buoyancy*-induced, or -influenced, flow and heat transfer in external and internal flows has been reviewed by Shenoy and Mashelkar (1982). These include purely viscous and *mildly elastic* drag-reducing solutions. The heat transfer capability of these drag-reducing liquids decrease in nearly the same proportion. The *experimental* results reported by Cho and Hartnett (1982) for *turbulent*, fully developed (thermal and hydrodynamic) flow in a tube of diameter d and for such mildly elastic liquids, show that this drag-reduction characteristic *degrades* as the liquid is *reflown* through the tube. The fluid bulk mean temperature $\langle \overline{T} \rangle_A^b$ is used in the definition of the Nusselt number. The results for 100 parts per million by weight of polyacrylamide in water is shown in Figure 4.19. The results show that after hours of flow, the liquid behaves as a Newtonian fluid. The symbol A stands for the *fresh* sample and D for the most degradation. Hartnett (1992) discusses the chemical reaction between the trace minerals in water and polyacrylamide as the cause of this degradation. The interfacial heat transfer for flow of viscoelastic liquids in tubes of rectangular cross-section has been examined by Hartnett and Kostic (1989).

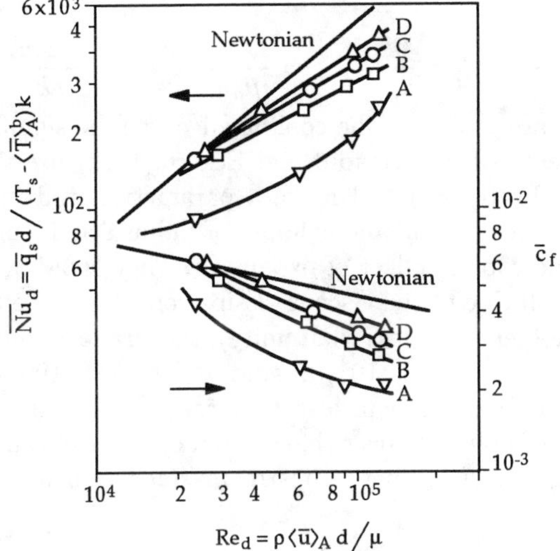

Figure 4.19. Effect of the degradation on the interfacial friction and heat transfer in flow and heat transfer of a mildly elastic drag-reducing solution through tubes. (From Cho and Hartnett, reproduced by permission ©1982 Academic Press.)

4.8.3 Viscous Dissipation

The viscous dissipation term Φ_μ appearing in the energy equation (1.84), becomes significant when the viscosity and the velocity gradients are large. This heat generation in turn influences the temperature distribution and the local viscosity (with the dynamic viscosity decreasing with an increase in temperature). The viscous dissipation can become important in *internal* and *external* flows of polymers. Winter (1977) considers a large class of internal flows of molten polymers, and Shadid and Eckert (1992) consider an external flow over a finite-size tube.

For *axisymmetric* flow along the x-direction, the dimensionless energy equation can be written using the dimensionless quantities defined by (4.188). The result, by maintaining the cooling due to expansion, is (Shadid and Eckert, 1992)

$$\frac{1}{2}RePr(v^*\frac{\partial T^*}{\partial r^*}+u^*\frac{\partial T^*}{\partial x^*}) = \frac{1}{r^*}\frac{\partial}{\partial r^*}r^*\frac{\partial T^*}{\partial r^*} + \frac{\partial^2 T^*}{\partial x^{*2}} + \beta^*(T^*+T_o^*)u^*\frac{\partial p^*}{\partial x^*} + \frac{1}{2}RePr\Phi_\mu^* , \tag{4.191}$$

where the dimensionless viscous dissipation is

$$\Phi_\mu^* = \mu^*\{2[(\frac{\partial v^*}{\partial r^*})^2 + (\frac{v^*}{r^*})^2 + (\frac{\partial u^*}{\partial x^*})^2] + (\frac{\partial u^*}{\partial r^*} + \frac{\partial v^*}{\partial x^*})^2\} \tag{4.192}$$

and

$$u^* = \frac{u}{u_o} \, , \quad v^* = \frac{v}{u_o} \, , \quad T_o^* = \frac{T_o \rho c_p \ell}{\mu_o u_o} \, , \quad \beta^* = \frac{\beta \mu_o u_o^2}{k} \, . \tag{4.193}$$

Other scales are *possible*. The cooling effect can be significant when the pressure gradient is large. Shadid and Eckert (1992) use the constitutive equation given by (4.189) and make a parametric study for an external flow over a finite-length isolated cylinder of radius ℓ and length L. They do not consider the cooling effect of expansion. They show that a significant heating can result due to the viscous dissipation. While the Reynolds number is not large (generally less than unity), the Prandtl number for molten polymers can be as *high* as 10^8 (Shadid and Eckert, 1992). The Weisenberg and Nahme numbers can also by large (of the order of 10^3). Their results for flow of polymers over the isolated cylinder of radius ℓ show that $T^* = (T - T_o)\rho c_p \ell / \mu_o u_o$ in the upstream *stagnation* area can be larger than unity.

4.9 Surface Injection

In internal and external flows, in order to *reduce* the heat transfer to a surface exposed to a high-temperature stream, a lower-temperature fluid is injected through the solid surface. Although in general the injected fluid can be a liquid and a phase change can occur upon interaction with the heat flowing from the stream, the subject of gas (coolant) injection across the interface of a solid exposed to another high-temperature gas has been extensively studied. In this section, we examine the available results for the *local* injection (i.e., *spatially discrete* injection) and for *spatially continuous* injections. Since the injection can be through porous inserts placed in the surface, the thermal fields of the gaseous boundary layer and the porous insert are coupled. As an example, the problem of gas flow through a porous insert subject to a prescribed surface heat flux distribution is also addressed.

4.9.1 Spatially Discrete Injections

Injection of a gas through a permeable surface exposed to a high-temperature gas stream (such as a turbine blade) is referred to as *discrete film cooling*. The injected gas generally has the same constituents as the gas in the external stream. The temperature of the injected gas is nearly the same as the surface temperature (in Section 4.9.3 below, the temperature of the gas while it flows towards the surface is discussed). The objective of the discrete film cooling is to reduce the local heat transfer to the surface. The injections are from discrete sites on the surface called the injection holes, and these are at an *angle* to the surface and have been shown to be most effective when *facing* the downstream (Goldstein, 1971).

The *effectiveness* of the discrete film cooling in reducing the local heat transfer is due to the introduction of the lower-temperature injected gas. However, this is in part *nullified* because the hydrodynamics of this *secondary* flow imposed on the mainstream tend to *enhance* the heat transfer. It is customary (Bergeles et al., 1981) to *separate* these effects by introducing a temperature effect called the *film cooling effectiveness* η

$$\eta = \frac{T_\infty - T_f}{T_\infty - T_s} , \tag{4.194}$$

where T_∞ is the free-stream temperature, T_s is the local surface temperature, and T_f is the local *cross-sectional-averaged* fluid temperature (i.e., the *film temperature*). Right at the injection site η approaches unity and far downstream it approaches zero.

The *hydrodynamic* effect of the film cooling is given through the local dimensionless surface heat flux (i.e., local Nusselt number)

$$(Nu_x)_f = \frac{(q_s)_f x}{(T_f - T_s)k} . \tag{4.195}$$

Now using the case of *no* film cooling

$$Nu_x = \frac{q_s x}{(T_\infty - T_s)k} \tag{4.196}$$

we have

$$\frac{(q_s)_f}{q_s} = \frac{(Nu_x)_f}{Nu_x} \frac{T_f - T_s}{T_\infty - T_s} = \frac{(Nu_x)_f}{Nu_x}(1 - \eta) . \tag{4.197}$$

Then if η is large enough to compensate for the increase in $(Nu_x)_f$, $(q_s)_f / q_s$ will be less than unity, indicating an *effective film cooling*.

Adjacent to the site of injection, η and $(Nu_x)_f$ are the largest and then they decay with $\eta \to 0$ and $(Nu_x)_f \to Nu_x$ far downstream. As an example, consider the experimental results of Mehendale and Han (1992) for a *multisite* injection on a *curved* surface. This example has the complexities of an *interaction* among the discrete injections and also the *flow separation* over a *convex* surface. Figure 4.20 shows the geometrical parameters for injection sites near the stagnation point of a turbulent flow over the convex-planar surface considered by them. They consider two rows of injection holes arranged in a staggered manner with the streamwise separation of 25° and the first row at 15° from the stagnation line and the lateral intersite spacing ℓ of $3d$ and $4d$, where d is the hole diameter of the holes. The radius of curvature is $D/2$ and the results for $D/d = 14$, $\ell/d = 3$, and $Re_D = u_\infty D/\nu = 10^5$ and air as the fluid are shown in Figure 4.21(a) and (b). The Nusselt number based on D is normalized with respect to $Re_D^{-1/2}$ which is suitable for stagnation area heat transfer. The results are

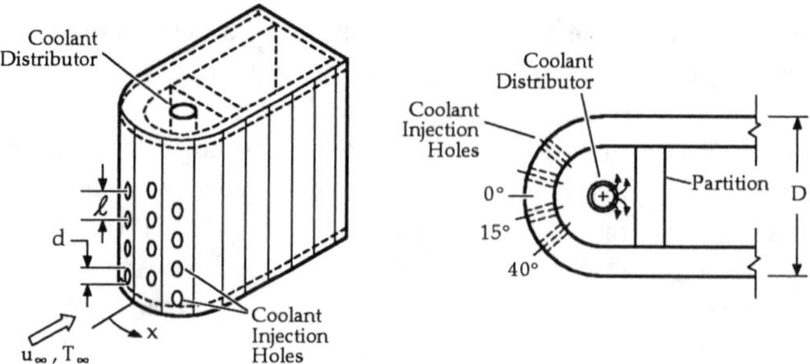

Figure 4.20. The curved surface used and the geometrical parameters of the surface and the injection sites. (From Mehendale and Han, reproduced by permission ©1992 Pergamon Press.)

for various values of the *blowing parameter* B

$$B = \frac{(\rho v)_s}{(\rho u)_\infty} \tag{4.198}$$

which is the ratio of the *injected* mass flux $(\rho v)_s$ to the *free-stream* mass flux $(\rho u)_\infty$. The free-stream turbulence Tu is 0.0075. The results for $B = 0$, i.e., no film cooling, is also shown. Figure 4.21(a) shows the $Re_D^{1/2}$ scaled $(Nu_D)_f$ (based on the film temperature) and Figure 4.21(b) shows the film effectiveness, both as a function of the dimensionless axial location x/d. For $B = 0$, the flow separates downstream of the start of the flat section and this results in a jump in the local Nusselt number. The Nusselt number increases up to the point of *reattachment*, beyond which it begins to decay monotonically. For $B = 0.4$, 0.8, and 1.2, the initial decay of $(Nu_D)_f$ up to the point of separation does not exist and instead $(Nu_D)_f$ increases above the value at $x = 0$. For this experiment the maximum $(Nu_D)_f$ becomes up to three times *larger* due to the local injection. The distribution of η shows peaks near the injection sites and decays thereafter with η being nonzero then up to x/d of 100. As given by (4.192), the product of $(Nu_D)_f$ and η determines the effect of surface cooling.

Analysis of the fluid flow and heat transfer of this turbulent, three-dimensional, multisite, curved surface example is rather involved. The numerical treatment of Bergeles et al. (1981), using the kinetic energy-dissipation rate model of turbulent flows, shows that the main feature in the variation of Nu and η can be predicted using the conservation equations and the turbulent model. However, since the flow of the injected gas is not uniform across the injection area and is influenced by the boundary layer flow, the injected flow must be determined *simultaneously* with the external flow. The ratio of the viscous boundary-layer thickness δ to d also becomes *significant* at high values of B, thus making the boundary-layer and other approximations nonvalid for small δ/d and for large B. For *two-dimensional*, *turbulent* incompressible boundary-layer flows with a *single-slot* injection, one of the

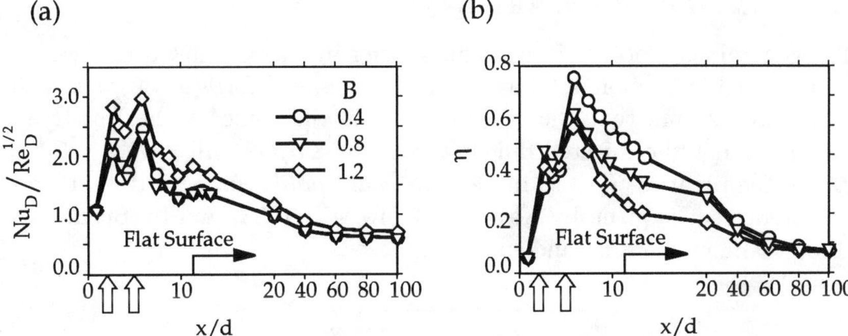

Figure 4.21. The measured axial distribution of (a) the normalized local Nusselt number, and (b) the effectiveness of surface injection for the geometry of Figure 4.20. The results are for three different blowing parameters. (From Mehendale and Han, reproduced by permission ©1992 Pergamon Press.)

earlier models (Goldstein, 1971) gives

$$\eta = 5.76 Pr^{2/3} Re_s^{0.2} (\frac{\mu_s}{\mu_\infty})^{0.2} \frac{c_{p_s}}{c_{p_\infty}} (\frac{d}{x\, B})^{0.8} , \tag{4.199}$$

$$Re_s = \frac{(\rho v)_s d}{\mu_s} , \tag{4.200}$$

where d is the *injection slot width*. Although this predicts the decay in η with $x^{-0.8}$, generally found in the experiments, the predicted magnitude of η is generally higher than the exponential results. Other predictions are discussed by Goldstein (1971) and Kays and Crawford (1993).

4.9.2 Spatially Continuous Injections

As was discussed, the interfacial heat transfer in boundary-layer flows is influenced by flow *through* the interface. Also, in some applications for the *thermal protection* of the surface, injection of fluid through the entire surface (or at least through the entire *critical* surface area) is used. This class of problems with spatially continuous injection is referred to as the *transpiration cooling*. The interfacial flow rate $(\rho v)_s$ may or may not be uniform along the surface and, in principle, can be designed for the local heat flux (or heat load). The control based on the heat load will be discussed in Section 4.9.3. In this section existing treatments of the transpiration cooling of *laminar* and *turbulent* boundary layers are discussed starting with the laminar flows. The problem of *suction* (instead of blowing) is analogous and is used to prevent boundary layer separation and is discussed by Kays and Crawford (1993).

(A) Laminar Boundary Layer

The general class of interfacial heat transfer in laminar flow over a wedge, as discussed in Section 4.2, has also been examined for $v(y=0)=v_s\neq 0$. Again we assume that the injected fluid is the same as the freestream fluid and that the injected fluid *right* at the *surface* will have a velocity v_s and temperature T_s (i.e., in *local thermal equilibrium* with the surface). The treatment is given by Kays and Crawford (1993) and by Burmeister (1983). The momentum and energy equations are

$$u\frac{\partial u}{\partial x}+v\frac{\partial u}{\partial y}=\frac{mu_\infty^2}{x}+\nu\frac{\partial^2 u}{\partial y^2}\,, \tag{4.201}$$

$$u\frac{\partial T}{\partial x}+v\frac{\partial T}{\partial y}=\alpha\frac{\partial^2 T}{\partial y^2}\,. \tag{4.202}$$

As an example, consider the case of constant *surface temperature* T_s and free-stream conditions T_∞ and $u_\infty=cx^m$. Defining the variables

$$v=-\frac{\partial\psi}{\partial x}=-\frac{m+1}{2}(c\nu)^{1/2}x^{(m-1)/2}f-\frac{m-1}{2}cyx^{m-1}\frac{df}{d\eta}\,,\quad u=\frac{\partial\psi}{\partial y}\,,$$

$$\psi=(\nu x u_\infty)^{1/2}f\,,\quad \eta=\frac{y}{(\nu x/u_\infty)^{1/2}}\,,\quad c_f=\frac{d^2f}{d\eta^2}\Big|_o\frac{2}{Re_x^{1/2}}\,,$$

$$Nu_x=\frac{dT^*}{d\eta}\Big|_o Re_x^{1/2}\,,\quad Re_x=\frac{u_\infty x}{\nu}\,,\quad T^*=\frac{T-T_s}{T_\infty-T_s}\,, \tag{4.203}$$

the dimensionless momentum and energy equations become

$$\frac{d^3f}{d\eta^3}+\frac{m+1}{2}f\frac{d^2f}{d\eta^2}+m\Big[1-\Big(\frac{df}{d\eta}\Big)^2\Big]=0\,,\quad \frac{df}{d\eta}\Big|_o=0\,,\quad \frac{df}{d\eta}\Big|_\infty=1\,, \tag{4.204}$$

$$\frac{d^2T^*}{d\eta^2}+\frac{Pr}{2}(m+1)f\frac{dT^*}{d\eta}=0\,,\quad T^*(0)=0\,,\quad T^*(\infty)=1\,. \tag{4.205}$$

The solution for the temperature and Nusselt number are

$$T^*=\frac{\int_0^\eta e^{(-Pr(m+1)/2\int_0^\eta f\,d\eta)}\,d\eta}{\int_0^\infty e^{(-Pr(m+1)/2\int_0^\eta f\,d\eta)}\,d\eta}\,, \tag{4.206}$$

$$Nu_x=\frac{Re_x^{1/2}}{\int_0^\infty e^{(-Pr(m+1)/2\int_0^\eta f\,d\eta)}\,d\eta}\,. \tag{4.207}$$

In the case of flow across the surface, i.e., $v_s\neq 0$, the value of f at the surface, which is related to v_s through

$$f(0)=-\frac{2}{m+1}v_s\Big(\frac{x}{u_\infty\nu}\Big)^{1/2}=-\frac{2}{m+1}v_s\Big(\frac{x^{1-m}}{c\nu}\Big)^{1/2} \tag{4.208}$$

will *not* be a constant, *unless* v_s is proportional to $x^{(m-1)/2}$. For the semi-infinite flat plate $m=0$ and v_s must be proportional to $x^{-1/2}$, i.e., *mono-*

Table 4.1. Effect of blowing on friction and interfacial heat transfer for laminar flow over a semi-infinite flat plate.

B	$\frac{d^2 f}{d\eta^2}\vert_o$	Pr=0.7 $\frac{dT^*}{d\eta}\vert_o$	1.0 $\frac{dT^*}{d\eta}\vert_o$
0	0.332	0.292	0.332
0.25	0.165	0.166	0.165
0.375	0.0937	0.107	0.0937
0.500	0.0356	0.0517	0.0356
0.619	0 (Separation)		0

tonically decreasing with increase in x. We can rewrite (4.208) as

$$f(0) = -\frac{2}{m+1}\frac{v_s}{u_\infty} Re_x^{1/2} \equiv -\frac{2}{m+1} B\,, \quad B = \frac{v_s}{u_\infty} {Re_x}^{1/2}\,. \qquad (4.209)$$

When $f(0)$ is a constant, a similarity solution to the momentum equation exists for various values of m and the dimensionless quantity $v_s Re_x^{1/2}/u_\infty$ which is *also* called the *blowing parameter* B. The magnitudes of $d^2 f/d\eta^2|_o$ and $dT^*/d\eta|_o$ appearing in the expressions for c_f and Nu_x in (4.203) have been determined for various values of B and are given by both Kays and Crawford (1993) and Burmeister (1983). Some of these for the case of $m =$ 0 are repeated in Table 4.1. The hydrodynamic results show that for a *critical* blowing, the lateral gradient of the axial velocity at the surface becomes *zero* (i.e., the flow *separates*). For the flat plate and for $Pr = 1$, the momentum and energy transport become analogous and the derivative of the lateral gradient of the temperature at the surface *also* becomes zero (i.e., adiabatic condition, $q_s = 0$). Table 4.1 shows that for B between zero and this critical value, the friction coefficient and the normalized heat flux *decrease* monotonically with an *increase* in B.

(B) Turbulent Boundary Layers

A two-layer turbulent boundary-layer model of the fluid flow and heat transfer, for the case of a continuous surface injection, has been developed by Kays and Crawford (1993). They also propose a simple *Couette flow* model similar to the one discussed in Section 4.4.1(B). For no pressure gradient, the Couette flow approximation leads to the momentum equation given by

$$v_s \frac{\partial \overline{u}}{\partial y} - \frac{d}{dy}(\nu + \nu_t)\frac{d\overline{u}}{dy} = 0 \; . \qquad (4.210)$$

Subject to the boundary conditions

$$\overline{u} = 0 \, , \quad \frac{\tau_s}{\rho} = \nu \frac{\partial \overline{u}}{\partial y} \quad y = 0 \qquad (4.211)$$

$$\overline{u} = u_\infty \qquad y = \delta \; . \qquad (4.212)$$

Integrating and applying the conditions at $y = 0$, we have

$$v_s \overline{u} - (\nu + \nu_t)\frac{d\overline{u}}{dy} = -\frac{\tau_s}{\rho} \; . \qquad (4.213)$$

Now integrating across the boundary layer, we have

$$\int_0^{u_\infty} \frac{d\overline{u}}{v_s \overline{u} + \tau_s/\rho} = \int_o^{\delta} \frac{dy}{\nu + \nu_t} \qquad (4.214)$$

or

$$\frac{1}{v_s} \ln(1 + \frac{v_s u_\infty \rho}{\tau_s}) = \int_0^{\delta} \frac{dy}{\nu + \nu_t} \; . \qquad (4.215)$$

Using the *friction blowing parameter* B

$$\ln(1 + B) = \frac{c_f}{2} B \, u_\infty \int_0^{\delta} \frac{dy}{\nu + \nu_t} \, , \qquad (4.216)$$

$$B = \frac{v_s u_\infty}{\tau_s/\rho} = \frac{v_s/u_\infty}{c_f/2} \; . \qquad (4.217)$$

Note that the blowing parameter B is now *redefined*. Rearranging this we have

$$\frac{c_f}{2} = \frac{\ln(1 + B)}{B} \frac{1}{u_\infty} (\int_0^{d} \frac{dy}{\nu + \nu_t})^{-1} \; . \qquad (4.218)$$

Since

$$\frac{\ln(1 + B)}{B} = 1.0 \quad \text{as } B \to 0 \, , \qquad (4.219)$$

then based on the available $(c_f)_{v_s=0}$, we have

$$\frac{c_f/2}{(c_f/2)_{v_s=0}} = \frac{\ln(1 + B)}{B} \; . \qquad (4.220)$$

Now using $(c_f/2)_{v_s=0}$ from (4.104), we have

$$\frac{c_f}{2} = 0.0287 \frac{ln(1 + B)}{B} Re_x^{-0.2} \; . \qquad (4.221)$$

The heat transfer treatment is similar. Again assume that v_s is uniform across the thermal boundary layer, and we arrive at the counterpart of

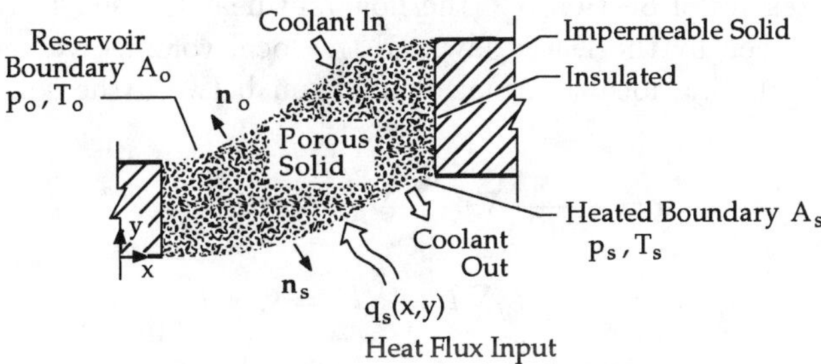

Figure 4.22. The solid porous insert through which the coolant flows to intercept the surface heat flux. The boundary conditions and the coordinate system are also shown. (From Siegel and Snyder, reproduced by permission ©1984 Pergamon Press.)

(4.220), i.e.,

$$\frac{St}{(St)_{v_s=0}} = \frac{\ln(1+B_t)}{B_t}\,, \quad B_t = \frac{v_s/u_\infty}{St}\,, \tag{4.222}$$

where B_t is the *thermal blowing parameter*. Now using the relation for $v_s = 0$, given by (4.105), we have

$$St\, Pr^{0.4} = 0.0257\frac{\ln(1+B_t)}{B_t} Re_x^{-0.2}\,. \tag{4.223}$$

Kays and Crawford (1993) show that both the friction coefficient and the Stanton number predicted by (4.221) and (4.223) agree well with the experimental results for a large range of B and B_t.

Recent studies of Faraco-Medeiros and Silva-Freire (1992) and Puzach (1992) address the effect of the pressure gradient and the surface roughness.

4.9.3 Heat Transfer in Porous Surface Inserts

The local surface heat transfer is generally *not* uniform and in some cases the surface heating is by *irradation*. The problem of designing inserts for *prescribed* $q_s(x,y)$ in two-dimensional curved surfaces has been addressed by Siegel and Snyder (1984). The pressure drop across the porous insert $p_o - p_s$ is generally much higher than the pressure variations along the surfaces o and s. The coolant flows from a reservoir at temperature T_r and uniform pressure p_r towards a surface with surface curvature given by $H_s = H_s(x,y)$, uniform temperature T_s (prescribed for a given heat flux load q_s), and a prescribed surface heat flux distribution $q_s = q_s(x,y)$. Figure 4.22 shows a schematic of the problem. The analysis aims at determining the distribution of the inlet surface curvature $H_o = H_o(x,y)$ to maintain T_s as the prescribed uniform value for the prescribed heat flux load distribution. Using the results of Section 3.2, the flow and heat transfer in the porous medium is given by the Darcy law and the local volume-averaged energy equation (assuming local thermal equilibrium between the solid and gas phases), i.e.,

$$\mathbf{u} = -\frac{k}{\mu(T)}\nabla p\,, \quad \nabla\cdot\rho\mathbf{u} = 0\,, \tag{4.224}$$

$$\nabla\cdot(\rho c_p)_f\mathbf{u}T = -\nabla\cdot\langle k\rangle\nabla T\,, \quad \langle k\rangle = k_e + (\rho c_p)_f\epsilon D^d\,, \tag{4.225}$$

$$p = \rho\frac{R_g}{M}T\,. \tag{4.226}$$

The boundary conditions are

$$p = p_o = p_r\,, \quad \langle k\rangle\mathbf{n}\cdot\nabla T = (\rho c_p)_f(T - T_r)\mathbf{n}\cdot\mathbf{u} \quad \text{on } A_o\,, \tag{4.227}$$

$$p = p_s\,, \quad T = T_s\,, \quad \langle k\rangle\mathbf{n}\cdot\nabla T = q_s(x,y) \quad \text{on } A_s\,, \tag{4.228}$$

$$\frac{\partial u}{\partial x} = \frac{\partial T}{\partial x} = 0 \qquad \text{on side walls}\,. \tag{4.229}$$

Siegel and Snyder (1984) show that the spatial distributions of the temperature, the coolant flow, and the heat flow within the porous medium can be expressed by analytical function terms of a *potential function* ϕ. The potential function is solved for by solving $\nabla^2\phi = 0$ in the two-dimensional domain, where the shape of the coolant inlet boundary H_o is found to satisfy the boundary conditions for the solution of ϕ. They find solutions for a cosine-type surface $H_s(x,y)$ and for several prescribed surface heat flux distributions $q_s(x,y)$. They also discuss the *limitations* of the technique.

4.10 Impinging Jets

In this section we consider the injection of a fluid through a *nozzle* and *impinging* on a solid surface with the ambient and the injected fluids being the *same*. The case of impinging liquid jets in gaseous ambient will be considered in Chapter 8. The effect of impinging jets on the solid-fluid interfacial heat transfer rate is similar to that of flow injection through the surface. The similarity will become more apparent for impinging jets with a *cross flow*, i.e., a mainstream flow parallel to a surface augmented with single or multiple impinging jets. The heat transfer effects are then due to the *hydrodynamic modification* caused by the jet and due to the *thermal modification* caused by the introduction of a fluid of a *different temperature* near the surface (i.e., in the boundary layer in the case of impinging jets with a cross flow). In addition, because of the *high-speed* jets used in practice, the *kinetic energy* of the jet must also be included in the analysis.

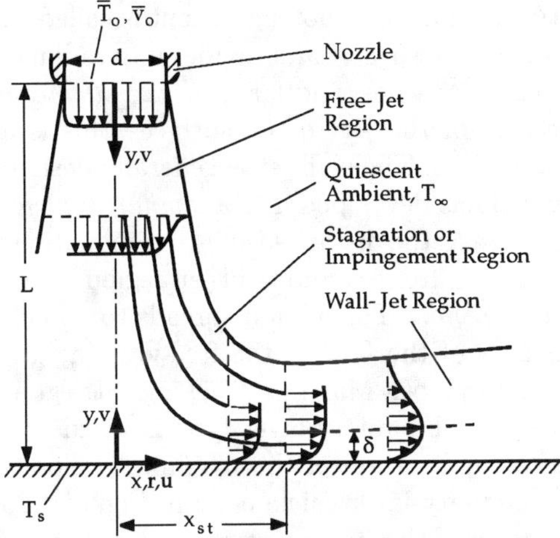

Figure 4.23. Rendering of an isolated impinging jet showing free-jet, stagnation, and wall-jet regions. The geometrical parameters and the coordinate system are also shown. (From Martin, reproduced by permission ©1977 Academic Press.)

In this section, the effects of *isolated jets without cross flow*, isolated jets *with* cross flow, and the effects of an *array* of jets on the interfacial heat transfer rate are examined.

4.10.1 Isolated Jets Without Cross Flow

The *hydrodynamic* aspects of impinging jets have been reviewed by Martin (1977). He also discusses the nozzle shapes, multiple jets, and the applications involving jet impingement. A schematic of an isolated impinging jet is shown in Figure 4.23. The jet can be planar or axisymmetric (round). The nozzle exit dimension is d (*width* of the *slot* or the *diameter*) and for an initially turbulent jet a nearly uniform exit velocity distribution is expected. The surrounding fluid, which is assumed to have the same constituents as the jet, is assumed to be *quiescent*. The behavior of the jet in the far-field region, where the centerline velocity begins to decay and the entrainment of the surrounding quiescent fluid *becomes* significant was discussed in Section 2.4.3, and in the near-field region, where the mass entrainment is *not* significant, was discussed in Section 2.7.2. As indicated, nearly a distance of $4d$ is required for the centerline velocity to begin to decay. Therefore, for impinging jets with the distance between the jet and the surface L smaller than $4d$ (i.e., the developing region shown in Figure 2.11), the jet will have a nearly flat core around the centerline in the velocity distribution (as shown in Figure 2.12). This is depicted in Figure 4.23. As was also shown in Figure 2.23, the jet enlarges (spreads) as it travels towards the surface (due to the reduction of the velocity at and near the edge of the jet). The region where the effect of the presence of the surface on the jet velocity distribution is not significant is called the *free-jet region*. The jet moves towards the surface and divides into radial (for round jets) or axial (for planar jets) branches, and far from the *stagnation point* (a *point* connecting the center of the jet to the surface—this is a *line* for planar jets) it becomes a *wall jet*. The wall jet is generally designated by having a velocity profile which has developed into a *similarity form* (the form of the profile does not change as the distance from the stagnation point increases). The region between the free-jet and wall-jet regions is called the *stagnation* of *impingement region*. This region spreads to about $4d$ (Goldstein et al., 1986). The details of the flow in the free-jet, stagnation, and wall-jet regions are discussed by both Martin (1977) and Striegl and Diller (1984). In the wall-jet region, a boundary-layer type flow exists and the distance from the surface to the location of the maximum velocity are designated with the viscous boundary-layer thickness δ in Figure 4.23.

The *thermal* aspects of the impinging jets can also be divided into free-jet, stagnation, and wall-jet regions. In addition, depending on the presence or lack of a *temperature difference* between the jet and its ambient, *thermal entrainment* must also be addressed. For the case of jet impingement *cooling*, the jet is lower in temperature compared to the surface, but the ambient may be at the jet temperature (no thermal entrainment) or nearly at the surface temperature (called complete thermal entrainment). The reverse occurs in jet impingement *heating*. We begin by considering the case of no thermal entrainment and review the experimental results of Goldstein et al. (1986). Then we review the analysis of Striegl and Diller (1984) involving an *arbitrary* thermal entrainment. Also, we consider *angled* impingement of turbulent jets when azimuthal variation of the local heat transfer rate occurs.

(A) No Thermal Entrainment

In *isolated* high-speed impinging jets, the effect of the kinetic energy of the fluid, in addition to the local hydrodynamic effect, must be addressed. The recovery factor defined by (4.169) needs to be based on the jet exit conditions $\overline{v}_o$ and $\overline{T}_o$ and becomes

$$r_f = \frac{T_s - \overline{T}_o}{\overline{v}_o^2/2c_p} \quad \text{for } q_s = 0 . \tag{4.230}$$

The hydrodynamic effects are shown through the dimensionless local heat flux, i.e., the local Nusselt number Nu_d based on the *adiabatic* surface temperature, i.e., $T_s - T_s(q_s = 0)$. This gives

$$Nu_d = \frac{q_s d}{[T_s - T_s(q_s = 0)]k} , \tag{4.231}$$

where d is the nozzle diameter from the impinging jet originates.

Goldstein et al. (1986) perform an experiment with *round* jets of *air* impinging on heated and adiabatic surfaces. The dimensionless jet to plate separation distance is L/d. They show that their experimental results for r_f and $Nu_d/Re_d^{0.76}$ are independent of the jet Reynolds number $Re_d = \overline{v}_o d/\nu_o$. The jet initial temperature T_o is kept very close to the ambient temperature to *minimize* the *thermal entrainment* of the surrounding *quiescent* air. The effect of the thermal entrainment will be discussed shortly.

Their experimental results for the radial distributions of r_f and $Nu_d/Re_d^{0.76}$ are given in Figure 4.24(a) and (b). The results are for air and $Re_d = 1.24 \times 10^5$ and $L/d = 2$ (i.e., a jet very close to the surface), $L/d = 5$, 7 (or 8), and 10. The range of Re_d in their experiment is from 6.1×10^4 to 1.24×10^5, i.e., *turbulent* jets (the flow is turbulent at least beyond a distance of d from the exit). The results shown in Figure 4.24(a) indicate that outside the stagnation area ($r/d > 4$) the recovery factor is nearly unity. Local minima appear near the stagnation point for $r/d < 2$. These local minima are most noticeable for $L/d = 2$ and the presence of these are associated with the so-called *energy separation* for highly curved flows. In principle, this is the same as the high-speed flow heat transfer discussed in Section 4.6.3. Eckert and Drake (1972) discuss this energy separation (also present in the so-called *vortex* or *Hilsch* tube). They show that when a radial velocity nonuniformity exists in a high-speed flow, the total temperature $T+(u^2/2c_p)$ can be substantially different from the static temperature, and where the velocity is *smaller* the total temperature will be *lower* (here near the stagnation point). Goldstein et al. (1968) indicate that for small L/d, the jet has not yet spread far and the vortices formed in the shear layer (mixing with the quiescent ambient) are not far from the stagnation point and also interact with the surface.

The local Nusselt number Nu_d in this experiment with air is normalized with respect to $Re_d^{0.76}$. Figure 4.24(b) shows that in the stagnation region Nu_d has a maximum, except for $L/d = 2$ where due to the presence of shear layer vortices there are local maxima around r/d of unity. In the wall-jet region a monotonic decrease in the local Nusselt number is expected with the monotonic increase in the viscous and thermal boundary-layer thicknesses.

(B) Thermal Entrainment

In many cases, the ambient fluid is *not* at *uniform* temperature equal to the jet exit temperature $\overline{T}_o$ and in both jet surface cooling or heating applications the ambient is at an intermediate temperature between the jet exit and the surface temperature. Striegl and Diller (1984) define a

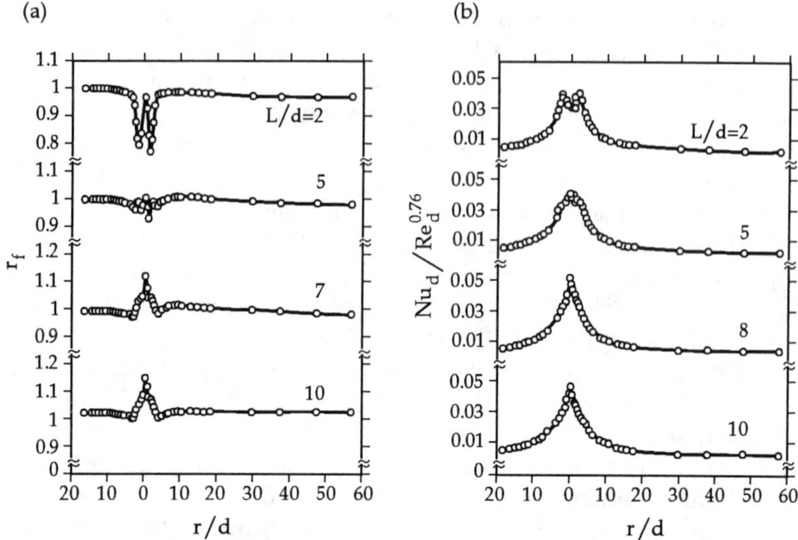

Figure 4.24. (a) Radial distribution of the recovery factor. (b) Radial distribution of the normalized Nusselt number. The results are for a single round jet of air with $Re_d = 1.24 \times 10^5$ and for various L/d. (From Goldstein et al., reproduced by permission ©1986 Pergamon Press.)

thermal *entrainment factor* F as

$$F = \frac{\overline{T}_o - T_\infty}{\overline{T}_o - T_s} \tag{4.232}$$

such that when the *ambient* temperature T_∞ is the same as the jet exit temperature, then $F = 0$, and when it is equal to the surface temperature T_s, then $F = 1$. Experimental results for the distribution of the local Nusselt number (based on $T_s - \overline{T}_o$) have been obtained by Striegl and Diller (1984), Hollworth and Gero (1985), and Goldstein and Seol (1991), for single and multiple as well as planar and axisymmetric jets.

Goldstein and Seol (1991) suggest that the direct use of F can be avoided if the use of the *adiabatic* surface temperature (i.e., the recovery factor) is extended to cases where $\overline{T}_o \neq T_\infty$. They use the experimental results for r_f to show that the local Nusselt number *based* on the local r_f does *not* depend on $\overline{T}_o - T_\infty$ (or F).

(i) Free-Jet Region

For $L \gg d$, it can be assumed that the velocity profile of the jet is fully developed from the nearly uniform exit condition to the bell-type (discussed in Section 2.7.2) and that the centerline velocity $\overline{v}_c$ and temperature $\overline{T}_c$

begin to decay according to the semi-empirical relations

$$\frac{\overline{v}_c}{\overline{v}_o} = a_{v_f} \left(\frac{d}{y_1}\right)^{1/2} , \quad \frac{\overline{T}_c - T_\infty}{\overline{T}_o - T_\infty} = a_{T_f} \left(\frac{d}{y_1}\right)^{1/2} , \tag{4.233}$$

where y_1 is the distance from the jet exit (required that $y_1 \gg d$) and a_{v_f} and a_{T_f} are the velocity and temperature decay coefficients, respectively, for the free-jet region. The jet width is assumed to be linearly increasing with y_1 (as was derived in Section 2.4.3), and the *half-width* b for velocity and temperature are given by semi-empirical relations

$$b_u = 0.114 y_1 , \quad b_T = 0.15 y_1 . \tag{4.234}$$

(ii) Stagnation Region

In the stagnation region a *laminar* boundary-layer flow parallel to the surface is assumed with the velocity distribution approximated by

$$u = a_{u_1} f_1 \left(\frac{y\nu}{a_1}\right)^{1/2} x + 4 a_{u_2} f_2 \left(\frac{y\nu}{a_1}\right)^{1/2} x^3 . \tag{4.235}$$

Along the surface, the *hydrodynamic* stagnation region extents for $x = 0$ to $x - x_{st_u}$, where the spread distance x_{st_u} is found empirically. Similarly an x_{st_T} is defined and then is given as

$$x_{st_u} = 0.34 L , \quad x_{st_T} = 0.45 L . \tag{4.236}$$

The temperature distribution is obtained from the energy equation based on boundary-layer approximations with no velocity normal to the surface, i.e.,

$$\rho c_p u \frac{\partial T}{\partial x} = k \frac{\partial^2 T}{\partial y^2} . \tag{4.237}$$

The approximate solution to this is found for the case of $F = 0$ as

$$\frac{T - T_s}{T_\infty - T_s} = g_o \left(\frac{y\nu}{a_1}\right)^{1/2} + \frac{a_{u_2}}{a_{u_1}} g_2 \left(\frac{y\nu}{a_1}\right)^{1/2} x^2 , \quad F = 0 \tag{4.238}$$

and for the case of $F = 1$ as

$$\frac{T - T_s}{T_\infty - T_s} = a_{T_s} h_o \left(\frac{y\nu}{a_1}\right)^{1/2} + a_{T_s} h_2 \left(\frac{y\nu}{a_1}\right)^{1/2} x^2 , \quad F = 1 . \tag{4.239}$$

The discussion of constants and the functions are given by Striegl and Diller (1984).

(iii) Wall-Jet Region

In the wall-jet region, the free-stream velocity begins to decay and the flow is once again characterized as *turbulent*. The maximum velocity $\overline{u}_{max}$ and half-width of the jet b are given as a function of x by some semi-empirical relations. These relations are

$$\frac{\overline{u}_{\max}}{v_o} = a_{u_w}(\frac{d}{x})^{1/2}\,, \qquad \frac{\overline{T}_{\max} - T_\infty}{\overline{T}_o - T_\infty} = c_{T_w}(\frac{d}{x})^{1/2}\,, \tag{4.240}$$

$$b_u = 0.068x\,, \qquad b_T = 1.01x\,. \tag{4.241}$$

The turbulent velocity and temperature distributions in the wall-jet region are discussed in detail be Striegl and Diller (1984).

(iv) Local Heat Transfer Rate

Based on the linearity of the governing equations and the boundary conditions, Striegl and Diller (1984) construct temperature distributions for the stagnation and wall-jet regions by a linear superposition of the solutions for $F = 0$ and $F = 1$. Then they determine the temperature derivatives at the surface and arrive at the following relations for the local Nusselt number.

$$Nu_d = 1.64a_{L_1}Re_d^{1/2}(\frac{d}{L})^{3/4}[1 - 2.89(\frac{x}{L})^2](1-F) + a_{T_f}(\frac{d}{L})^{1/2}[1 - 3.86(\frac{x}{L})^2]F \quad \frac{x}{L} \le 0.34\,, \tag{4.242}$$

$$Nu_d = 0.00174a_{u_w}Re_d(\frac{d}{x})^{1/2}[(1+F) + 4.85a_{T_w}(\frac{d}{x})^{1/2}F] \quad \frac{x}{L} > 0.34\,, \tag{4.243}$$

where as shown before

$$Re_d = \frac{\overline{v}_o d}{\nu_o}\,. \tag{4.244}$$

The four constants in these equations are found empirically.

Striegl and Diller (1984) compare the predictions of (4.242) and (4.243) to the experimental results for single and multiple jets, and their comparison is shown in Figure 4.25. The axial distance is scaled with the jet separation distance L. Although a large number of constants are present in their analysis, the descriptions of the velocity and temperature distributions in the three regions have been made rather accurately in their analysis. Evaluation of the available turbulence models applied to an isolated impinging jet has been made by Craft et al. (1993). Gross features of the impinging jets can be observed by *laminar* numerical simulations (e.g., Kim and Aihara, 1992). As expected, in *complete* numerical solution using various turbulent models, a similar need for the constants would arise (Craft et al., 1993). A model based on the concept of the *surface renewal* is used by Yuan and Liburdy (1992). This is based on the *intermittent* motion of the eddies from the *turbulent core* into the *wall* (near surface) *region* (while

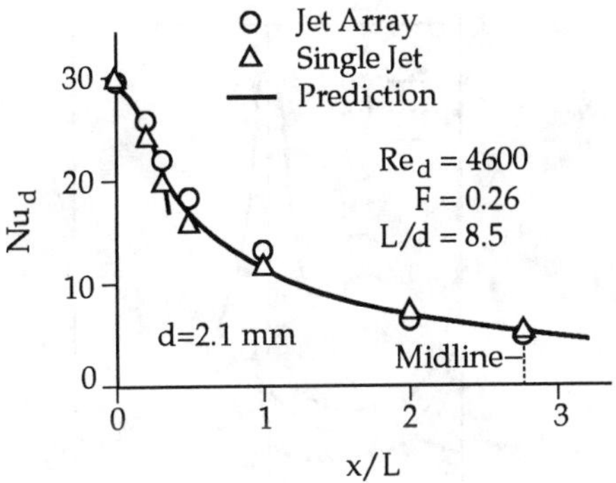

Figure 4.25. Axial distribution of the local Nusselt number for thermally entrained planer jets. The experimental and predicted results are shown. (From Striegel and Diller, reproduced by permission ©1984 American Society of Mechanical Engineers.)

exchanging momentum and energy). The renewal *frequency* for *momentum* is derived from the experimental results, and that for *thermal energy* is modeled by the empirically obtained *displaced temperature spectrum.*

(C) Angled Impingement

The *hydrodynamic* and *thermal* aspects of angled-jet impingement are significantly altered when the angle of impingement (measured from the normal to the surface) is large. Ožd̆emir and Whitelaw (1992) show that for angles larger than 40° the surface pressure contours and the local heat transfer rate change *suddenly* and that the boundary layer approximations made about the flow field are *no longer* valid. They identify *coherent* structures and interactions between *vortices* formed by the *inviscid* instability and the inflowing free jet. The coherent structures influence the heat transfer significantly by enhanced *internal* mixing and by *entrainment.*

4.10.2 Isolated Jets with Cross Flow

The presence of a cross flow (parallel to the surface) can nullify the hydrodynamic effects of the normal impinging jets. A numerical simulation of a two-dimensional *laminar* flow and heat transfer over a finite surface extending to either side of the stagnation point has been performed by Al-Sanea (1992). The ratio of the cross-flow mass flux to the jet mass flux M is

$$M = \frac{\rho_o v_o}{\rho_\infty u_\infty} . \tag{4.245}$$

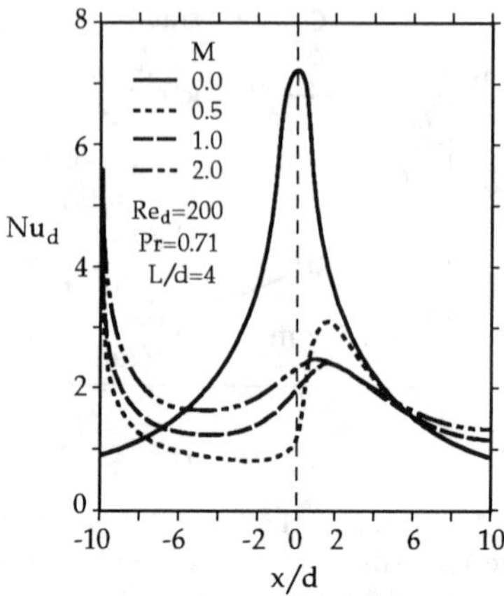

Figure 4.26. Axial distribution of the local Nusselt number around an impinging, planar jet with cross flow. The results are for several values of the ratio of the jet to cross-flow mass fluxes. (From Al-Sanea, reproduced by permission ©1992 Pergamon Press.)

This parameter influences the effectiveness of the normal jet. His numerical results for the distribution of the local Nusselt number around the stagnation point and for a plate of length $20d$ is shown in Figure 4.26. The cross flow is from the left to right, i.e., the leading edge where the boundary-layer thicknesses are zero, is at $x/d = -10$. The fluid has $Pr = 0.71$, and the dimensionless jet parameters are $Re_d = 200$ and $L/d = 4$. The jet and the cross flow have the temperature T_o, the surface is at T_s, and the Nusselt number is given by

$$Nu_d = \frac{q_s d}{(T_s - T_o)k} . \tag{4.246}$$

The computational domain (which is *not* extensive) is given by Al-Sanea (1992). The results shown in Figure 4.26 show that for $M > 2$ the effect of the impinging jet is nearly completely nullified. The cross flow begins to influence the local heat transfer rate for M much less than unity.

4.10.3 Multiple, Interacting Jets

For multiple jets impinging on a surfaces, as the *scaled separation distance* between jets s/d *increases*, the distributions of *local* recovery factor r_f and the normalized heat transfer rate Nu_d become those associated with a *single* jet. Also, when a single row of jets is present, as s/d decreases the distributions of r_f and Nu_d become that for a *planar* (or *slot*) jet. Goldstein and Seol (1991) perform experiments using a *single row* of round jets. The results reported by them for the film cooling effectiveness η defined as

$$\eta = \frac{T_s(q_s = 0) - T_s(q = 0, \overline{T}_o + \dfrac{\overline{v}_o^2}{2c_p} = T_\infty)}{\overline{T}_o - T_\infty} \tag{4.247}$$

is shown in Figures 4.27(a) and (b). The results are for L/d of 2, Figure 4.27(a), and 6, Figure 4.27(b), and for a *single round* jet, a *row* of *round* jets with s/d of 4 and 8, and a *slot jet* (all for $Pr = 0.7$ and Re_d of about 10^4).

Figures 4.27(a) and (b) shows the local distribution of η (in the direction r or x along the flow, i.e., streamwise). The trends towards the planar and single-round asymptotes are apparent. The distribution of the local Nusselt number is expected to follow similar trends.

4.11 Reacting Flows

In nonisothermal flow of a fluid about a continuous solid surface, exothermic chemical reaction can occur within the *fluid* (generally a gas) *phase*, within the *solid phase*, or at the solid-fluid *interface*. As discussed in Section 3.2.6, the *bulk* phase reactions are called *homogeneous* reactions and are generally better understood. While the surface reactions are called *heterogeneous* reactions and generally involve some *catalytic* surface reactions. Boundary layer flows with surface reactions have been reviewed by Chung (1965). Here we consider mainly simple chemical reactions occurring within the fluid phase. In the problem of *combustion synthesis* of *thin solid film* (i.e., a *chemical vapor deposition* method), the surface reaction becomes important and in Section 4.11.4 this will be discussed.

The analysis of the *turbulent* reacting flows addresses the *interaction*, at the *small scales* of turbulence, between the velocity, temperature, density, and species concentration fields. These are discussed by Toong (1983), Williams (1985), Kerstein (1992), and Peters (1992). *Unlike* nonreacting flows, aspects such as the effects of the presence of a solid surface on the interacting have *not* yet been resolved for turbulent reacting flows. Here some simple *laminar* reacting flow examples involving interfacial convective heat transfer will be considered. Initially the effect of *radiation* is not included so that some simple solutions can be found. Then the effect of radiation is included. The effect of *thermobuoyancy* in reacting flows is addressed in Section 4.11.5.

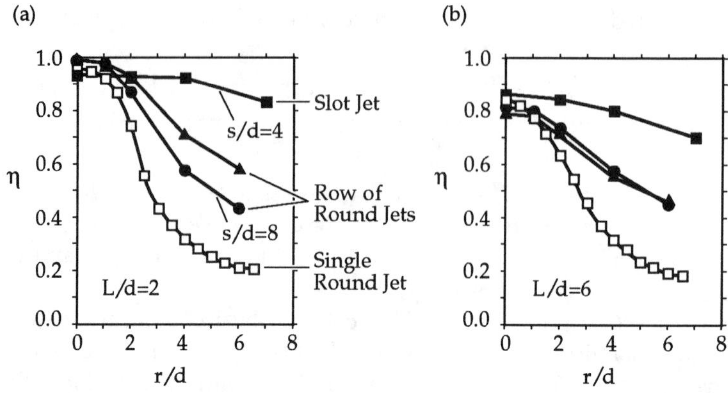

Figure 4.27. The radial distribution of the effectiveness for slot, single-round, and multiple-round jets. The results are for air and a Re_d of about 10^4. (a) $L/d = 2$, and (b) $L/d = 6$. (From Goldstein and Seol, reproduced by permission ©1991 Pergamon Press.)

4.11.1 Spread of a Boundary-Layer Diffusion Flame

Earlier advances in the boundary-layer flows with gas and solid phase reactions have been reviewed by Chung (1965). These include *subsonic* and *supersonic* (including *shocks*) reacting flows, solid surface reactions, dissociation and ionization of gases at high temperatures, diffusion and chemical kinetics-controlled reactions, and ignitions. A *unified* boundary-layer treatment of the momentum, heat, and mass transfer with the mass transfer originating as a uniform *injection* through the surface, has been made by Kays and Crawford (1993) using the formulation of Spalding (1979).

The surface *spread* of the burning front over a combustible solid, with the oxidizer flowing parallel to the surface, has been examined by de Ris (1969). In an idealized boundary-layer flow, as depicted in Figure 4.28, after ignition the heat supplied by the exothermic reaction is partly transferred to the surface, thus sublimating the solid fuel, and this gaseous fuel moves into the gas phase to react with the oxidizer. When this sublimation-reaction is sustained (i.e., stabilized), then the oxidizer flow with velocity u_∞ will result in a constant spread speed (front speed) u_F of the surface sublimation.

The effect of buoyancy (as *part* of the *induced* motion) is neglected and it is assumed that the *diffusion* of species *controls* their reaction. The diffusion *limit* (or *diffusion flame*) assumption is justified when the diffusion time scale is much larger than chemical kinetic time scale. The thermobuoyant diffusion flames will be considered in Section 4.12.2. These assumptions simplify the analysis significantly.

The solid and fluid phases are both assumed to be semi-infinite. The effect

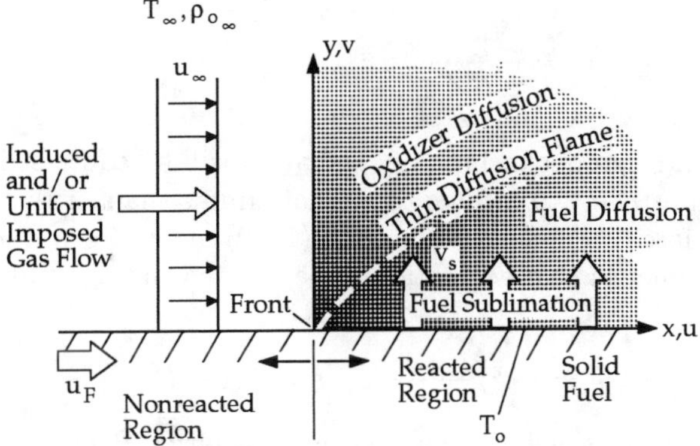

Figure 4.28. A rendering and model of the flame spreading over a surface. The thin diffusion-flame and the coordinate system are also shown.

of the sublimation-reaction, which is manifested through the temperature deviation from the unperturbed ambient temperature T_∞, penetrates as boundary layers into the gas and solid phases. The surface temperature in the *sublimating region* is the phase change temperature T_o.

(A) Formulation

The boundary-layer approximations are applied to the overall mass, species, and energy conservation equations, (1.66), (1.72), and (1.184), respectively. The flow is assumed to be *laminar* and uniform given by u_∞, and the species diffusion velocity is assumed to be negligible compared to the mass-averaged velocity. Then we have (the coordinate system is shown in Figure 4.28)

$$\frac{\partial}{\partial x}\rho_g u_\infty = 0 , \tag{4.248}$$

$$\rho_g u_\infty \frac{\partial}{\partial x}\frac{\rho_i}{\rho} = \frac{\partial}{\partial x}\rho_g D \frac{\partial}{\partial x}\frac{\rho_i}{\rho_g} + \frac{\partial}{\partial y}\rho_g D \frac{\partial}{\partial y}\frac{\rho_i}{\rho_g} + \dot{n}_i$$
$$i = f, o, p \quad y \geq 0 , \tag{4.249}$$

$$(\rho c_p)_g u_\infty \frac{\partial T_g}{\partial x} = k_g\left(\frac{\partial^2 T_g}{\partial x^2} + \frac{\partial^2 T_g}{\partial y^2}\right) + \dot{s} - \left(\frac{\partial}{\partial x} + \frac{\partial}{\partial y}\right) q_r \quad y \geq 0 , \tag{4.250}$$

$$(\rho c_p)_s u_F \frac{\partial T_s}{\partial x} = k_{s_x}\frac{\partial^2 T_s}{\partial x^2} + k_{s_y}\frac{\partial^2 T_s}{\partial y^2} \quad y \leq 0 . \tag{4.251}$$

The boundary conditions are

$$\rho_o = \rho_{o_\infty} , \quad T_g = T_\infty \quad y \to \infty \text{ or } x \to -\infty , \tag{4.252}$$

$$\rho_o = 0 \; , \quad T_g = T_s = T_o \quad y = 0 \; , \tag{4.253}$$

$$T_s = T_\infty \quad y \to -\infty \; . \tag{4.254}$$

The direct and rigorous inclusion of radiation will be addressed in Section 4.11.5. In an attempt to include the radiation in a manner that would yet result in a closed form solution, de Ris (1969) allows for a fraction F_r of the local volumetric heat generation to be emitted through radiation. This gives

$$(\frac{\partial}{\partial x} + \frac{\partial}{\partial y})q_r = -F_r \frac{\dot{n}_o \Delta i_c}{M_o \nu_{ro}} \; , \qquad \dot{s} = -(1 - F_r)\frac{\dot{n}_o \Delta i_c}{M_o \nu_{ro}} \; . \tag{4.255}$$

Note that here Δi_c is in J per $M_o \nu_{ro}$ kg of the oxidizer. Part of this radiation arrives at the solid surface. The amount of radiation absorbed by the surface is modeled assuming a uniform absorption downstream of the front and a decaying absorption upstream. These are given, respectively, by

$$q_{r_s} = q_{r_1} e^{\sigma_e x} \quad x \le 0 \; , \tag{4.256}$$

$$q_{r_s} = q_{r_2} \quad x \ge 0 \; . \tag{4.257}$$

A Lewis number $Le = k_f / \rho c_p D$ of unity and constant properties are assumed. The sublimation occurring on the surface (with a heat of sublimation of Δi_{sg} in J/kg) is governed by the surface mass and heat transfer relations

$$\dot{m}_s = \rho_g v_s = \rho_f v_s - \rho_g D \frac{\partial}{\partial y}\frac{\rho_f}{\rho_g} = \rho_g v_s \frac{\rho_f}{\rho_g} - \rho_g D \frac{\partial}{\partial y}\frac{\rho_f}{\rho_g}$$
$$0 \le x \le \infty \; , \quad y = 0^+ \; , \tag{4.258}$$

$$\Delta i_{sg} \dot{m} = k_f \frac{\partial T}{\partial y} \qquad 0 \le x \le \infty \; , \quad y = 0^+ \; , \tag{4.259}$$

where the mass transfer to the gas phase is by a net interfacial velocity v_s (which is rather small) and by diffusion. de Ris (1969) and Glassman (1987) show that for $Pr = Le = 1$ and for *parallel* flows the interfacial mass flux can be approximated by

$$\dot{m} = -\frac{\Delta B}{\ln(1 + \Delta B)} \rho_g D \frac{\partial}{\partial y}\frac{\rho_f}{\rho_g} \qquad 0 \le x < \infty \; , \quad y = 0^+ \; , \tag{4.260}$$

where

$$\Delta B = -\frac{c_{p_g}(T_o - T_\infty)}{\Delta i_{sg}} + (\frac{\rho_o}{\rho_g})_\infty \frac{\Delta i_c (1 - F_r)}{M_o \nu_{r_o} \Delta i_{sg}} \; . \tag{4.261}$$

The dimensionless *transfer driving force* (or *transfer number*) ΔB (Kays and Crawford, 1993), appears in this *modified* form because of need for the variable transformations discussed below. Note that in Section 4.10 the symbol B was used for the blowing parameter and here we continue its usage.

(B) Transformation

The Schvab-Zel'dovich transformation, discussed in relation to the diffusion flame theory in Section 2.3.3, is used to eliminate the source terms in the conservations equations. The transformed variables, for a reaction of the type given by (2.15), are

$$X_T = \frac{c_{p_g}(T - T_\infty)}{\Delta i_{sg}} + [\frac{\rho_o}{\rho_g} - (\frac{\rho_o}{\rho_g})_\infty]\frac{\Delta i_c(1 - F_r)}{M_o \nu_{r_o} \Delta i_{sg}} , \tag{4.262}$$

$$X_f = \frac{\rho_f \Delta i_c(1 - F_r)}{M_f \nu_{r_f} \Delta i_{sg}} - [\frac{\rho_o}{\rho_g} - (\frac{\rho_o}{\rho_g})_\infty]\frac{\Delta i_c(1 - F_r)}{M_o \nu_{r_o} \Delta i_{sg}} . \tag{4.263}$$

For the solid, we define a dimensionless temperature

$$T_s^* = \frac{c_{p_g}(T - T_\infty)}{\Delta i_{sg}} . \tag{4.264}$$

The transformed form of the conservation equations (4.249) to (4.251) become

$$2\frac{\partial}{\partial \xi}X_i = \frac{\partial^2}{\partial \xi^2}X_i + \frac{\partial^2}{\partial \eta^2}X_i \quad i = T \text{ or } f \quad -\infty < \xi < \infty , \ \eta \ge 0 , \tag{4.265}$$

$$2\gamma\frac{\partial}{\partial \xi}T_s^* - \beta\frac{\partial^2}{\partial \xi^2}T_s^* - \frac{\partial^2}{\partial \zeta^2}T_s^* \quad -\infty < \xi < \infty , \ \zeta \le 0 , \tag{4.266}$$

where the parameters and the dimensionless space variables are

$$\gamma = \frac{(\rho c_p k_y)_s u_F}{(\rho c_p k)_g u_\infty} , \ \beta = \frac{(k_x k_y)_s}{k_g^2} , \ \zeta = \frac{(\rho c_p)_g u_\infty}{2k_{s_y}} y , \tag{4.267}$$

$$\xi = \frac{(\rho c_p)_g u_\infty}{2k_g} x , \ \eta = \frac{(\rho c_p)_g u_\infty}{2k_g} y . \tag{4.268}$$

The boundary conditions for the transformed variables are

$$X_T \to 0 , \ X_f \to 0 \quad \eta \to \infty \text{ or } \xi \to -\infty , \tag{4.269}$$

$$T_s^* \to 0 \quad \zeta \to -\infty \text{ or } \xi \to -\infty , \tag{4.270}$$

$$\frac{\partial}{\partial \eta}X_f = (1 - a_1)\frac{\partial}{\partial \eta}X_T - a_1 a_2 + a_1\frac{\partial}{\partial \xi}T_s^* \quad 0 \le \xi < \infty , \ \eta = 0^+ , \ \zeta = 0^- , \tag{4.271}$$

$$\frac{\partial}{\partial \eta}X_T = \frac{\partial}{\partial \eta}X_f \quad -\infty < \xi \le 0 , \ \eta = 0^+ , \tag{4.272}$$

$$\frac{\partial}{\partial \zeta}T_s^* = \frac{\partial}{\partial \eta}X_T + a_3 e^{a_4 \xi} \quad -\infty < \xi \le 0 , \ \eta = 0^+ , \ \zeta = 0^- , \tag{4.273}$$

$$X_f = T_s^* \quad -\infty < \xi \le 0 , \ \eta = 0^+ , \ \zeta = 0^- , \tag{4.274}$$

$$X_T = -B \quad 0 \le \xi < \infty , \ \eta = 0^+ , \tag{4.275}$$

$$T_s^* = \frac{c_{p_g}(T_o - T_\infty)}{\Delta i_{sg}} \qquad 0 \le \xi < \infty\,, \;\; \zeta = 0^-\,, \tag{4.276}$$

where the constants are

$$a_1 = \frac{\ln(1+\Delta B)}{\Delta B}\frac{\Delta i_c(1-F_r)}{M_f \nu_{rf} \Delta i_{sg}}\,, \quad a_2 = \frac{2q_{r_2}}{\rho_g u_\infty \Delta i_{sg}}\,, \tag{4.277}$$

$$a_3 = \frac{2q_{r_1}}{\rho_g u_\infty \Delta i_{sg}}\,, \quad a_4 = \frac{2k_f \sigma_e}{(\rho c_p)_f u_\infty}\,. \tag{4.278}$$

(C) Example of Results

de Ris (1969) finds the simultaneous solution to the three differential equations by first converting them to three *Wiener-Hopt integral* equations. His exact solution for the front velocity u_F is

$$\gamma = [\frac{(\rho c_p k_y)_s u_F}{(\rho c_p k)_g u_\infty}]^{1/2} = \frac{T_f - T_o}{T_o - T_\infty} + \frac{2q_{r_1} f[\frac{2k_f\sigma_e}{(\rho c_p)_f u_\infty}]}{(\rho c_p)_f u_\infty (T_o - T_\infty)} + \frac{2q_{r_2}}{\pi(\rho c_p)_g (T_o - T_\infty)}\,, \tag{4.279}$$

where T_f and function $f(z)$ are defined as

$$T_f = T_o + \frac{B\Delta i_{sg}}{c_{p_g}}(1 - \frac{1}{a_1}) - (T_o - T_\infty) \tag{4.280}$$

$$f(z) = \begin{cases} \frac{1}{2}\pi - \frac{\sin^{-1}(z-1)}{\pi(2\pi - z^2)^{1/2}} & 0 < z \le 2\,, \\ \frac{1}{2}\pi(z^2 - 2z)^{-1/2}\ln\frac{z - 1 + (z^2-2z)^{1/2}}{z - 1 - (z^2 - 2z)^{1/2}} & 2 \le z\,. \end{cases} \tag{4.281}$$

For the case of *no* surface radiation, (4.279) can be used to determine u_F as

$$u_F = (\frac{T_f - T_o}{T_o - T_\infty})^2 \frac{(\rho c_p k)_g}{(\rho c_p k_y)_s} u_\infty\,, \tag{4.282}$$

i.e., u_F is *directly* proportional to u_∞ and k_{s_x} does *not* influence the spread speed. However, for the case of no radiation the flame is just on the surface and can *extinguish*. The flame stand-off distance depends on the heat loss and gain, and when q_{r_2} is not zero (i.e., when the radiant heat is supplied to the surface), the flame stands away from the surface despite the large heat loss to the semi-infinite solid. Further discussion of the process of flame spread is given by Frey and T'ien (1979) and Williams (1985), and recent treatments of the problem are given by Apte et al. (1991), Wichman and Argawal (1991), Bhattacharjee (1993), and Greenberg and Ronney (1993).

4.11.2 A Steady Boundary-Layer Diffusion Flame

As an example of a forced flow boundary-layer diffusion flame, consider the analysis of a *steady flame* over a solid surface where the fuel is supplied by the *sublimation* of the *solid surface*. Figure 4.29 (a) gives a rendering of this boundary-layer flow, heat transfer, and reaction. In the treatment given below a *thin* reaction zone (i.e., *thin flame*) is assumed. This is typical of diffusion flame analyses (Williams, 1985). It is assumed that the fuel is supplied by sublimation at a steady rate (i.e., the surface ablation effect and other factors tending to violate this assumption are negligible). There are many other restricting assumptions made in the analysis, including the unity Lewis number, constant properties, and lack of buoyancy. However, some main features of the flow, reaction, and phase change are demonstrated. The problem can be considered as an extension of the continuous surface injection problem considered in Section 4.9, and added here are the species diffusion, reaction, and the heat required for the phase change. The original work of Emmons (1956) dealt with a diffusion flame over a *thin liquid film* assuming no significant motion in the *liquid*. Hirano and Kinoshita (1974) have shown the boundary-layer assumptions are not all valid and a velocity maximum is observed at the location of the flame. Tao and Kaviany (1991) have shown that the Emmons (1956) solution, under some conditions, is valid for permeable surfaces and here we consider the fuel being supplied to the gas phase as a result of the *sublimation* of a heated *solid* surface (the heat in turn is supplied by the reaction of this gaseous fuel with a gaseous oxidizer).

For the two-dimensional boundary-layer problem considered [depicted in Figure 4.29 (a)], the variables are scaled using an arbitrary length x along the surface, and the free-stream conditions. This leads to the

$$y^* = \frac{y}{x}\,, \quad u^* = \frac{u}{u_\infty}\,, \quad v^* = \frac{v}{u_\infty}\,, \quad \rho^* = \frac{\rho}{\rho_\infty}\,, \tag{4.283}$$

$$i^* = \frac{i}{c_{p_\infty} T_\infty}\,, \quad \mu^* = \frac{\mu}{\mu_\infty}\,, \quad \Delta i_c^* = \frac{\Delta i_c}{c_{p_\infty} T_\infty}\,, \quad \dot{n}_o^* = \frac{\dot{n}_o x}{\rho_\infty u_\infty}\,, \tag{4.284}$$

$$Re_x = \frac{u_\infty x}{\nu_\infty}\,, \quad Pr = \frac{\mu c_p}{k}\,, \quad Sc = \frac{\nu}{D}\,, \quad Le = \frac{Sc}{Pr} = \frac{k/c_p}{\rho D}\,. \tag{4.285}$$

Here Δi_c is in J per kg of the oxidizer consumed. The *dimensionless boundary-layer* conservation equation for the overall mass, species, momentum, and energy are

$$\frac{\partial \rho^* u^*}{\partial x^*} + \frac{\partial \rho^* v^*}{\partial y^*} = 0\,, \tag{4.286}$$

$$\rho^* u^* \frac{\partial}{\partial x^*}\frac{\rho_o}{\rho} + \rho^* v^* \frac{\partial}{\partial y^*}\frac{\rho_o}{\rho} = \frac{1}{Re_x}\frac{\partial}{\partial y^*}\frac{\mu^*}{Sc}\frac{\partial}{\partial y^*}\frac{\rho_o}{\rho} + \dot{n}_o^*\,, \tag{4.287}$$

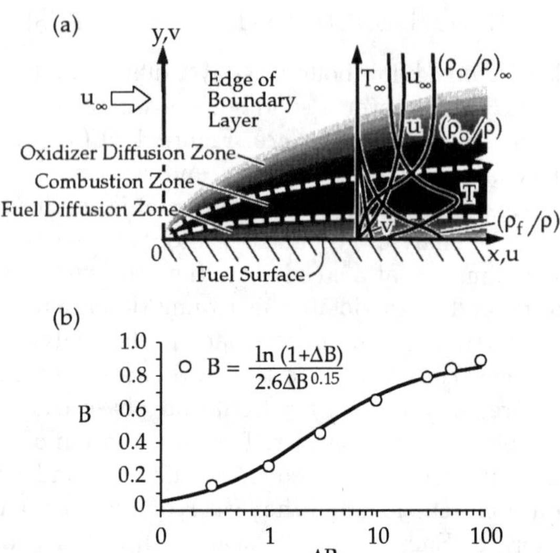

Figure 4.29. (a) A rendering of the two-dimensional steady, boundary-layer diffusion flame with sublimation of the solid fuel. The coordinate system and some of the variables are also shown. (b) The normalized rate of sublimation B as a function of the dimensionless transfer driving force ΔB. (From Glassman, reproduced by permission ©1987 Academic Press.)

$$\rho^* u^* \frac{\partial u^*}{\partial x^*} + \rho^* v^* \frac{\partial u^*}{\partial y^*} = \frac{1}{Re_x} \frac{\partial}{\partial y^*} \mu^* \frac{\partial u^*}{\partial y^*} \, , \tag{4.288}$$

$$\rho^* u^* \frac{\partial i^*}{\partial x^*} + \rho^* v^* \frac{\partial i^*}{\partial y^*} = \frac{1}{Re_x} \frac{\partial}{\partial y^*} \frac{\mu^*}{Pr} \frac{\partial i^*}{\partial y^*} - \dot{n}_o^* \Delta i_c^* \, . \tag{4.289}$$

Now under the conditions assumed for a diffusion flame (i.e., *thin* flame) discussed in Section 2.3.3, including a large Damköhler number, the species and energy equations are assumed. Combining the species and the energy equations by multiplying the former by Δi_c^* and then adding the two, we have

$$\rho^* u^* \frac{\partial}{\partial x^*}(i^* + \Delta i_c^* \frac{\rho_o}{\rho}) + \rho^* v^* \frac{\partial}{\partial y^*}(i^* + \Delta i_c^* \frac{\rho_o}{\rho})$$
$$= \frac{1}{Re_x} \frac{\partial}{\partial y^*} \mu^* \frac{\partial}{\partial y^*} (\frac{i^*}{Pr} + \frac{\Delta i^*}{Sc} \frac{\rho_o}{\rho}) \, . \tag{4.290}$$

The *analogy* between the transport of momentum and the quantity $i^* + \Delta i_c^* \rho_o/\rho$ can become more apparent by noting that

$$\tau_s = \mu \frac{\partial u}{\partial y} = \frac{\mu_\infty u_\infty}{x} \mu^* \frac{\partial u^*}{\partial y^*} \qquad y = 0 \, , \tag{4.291}$$

$$q_s = -k\frac{\partial T}{\partial y} = -\frac{c_{p\infty}T_\infty\mu_\infty}{x}\mu^*\frac{\partial}{\partial y^*}\left(\frac{i^*}{Pr} + \frac{\Delta i_c^*}{Sc}\frac{\rho_o}{\rho}\right)$$
$$= -\dot{m}_s\Delta i_{sg} - q_\ell \qquad y = 0\,. \tag{4.292}$$

This is based on $Pr = 1$ and $\partial\rho_o/\partial y = 0$ on the surface (i.e., surface is impermeable to the oxidizer).

From the comparison of the surface shear stress and heat flux, we have (for $Pr = Sc = 1$)

$$q_s = \frac{c_{p\infty}T_\infty\tau_s}{u_\infty}\left[\left(i^* + \Delta i_c^*\frac{\rho_o}{\rho}\right)_s - \left(i^* + \Delta i_c^*\frac{\rho_o}{\rho}\right)_\infty\right]. \tag{4.293}$$

This can be expressed as

$$\dot{m}_s = \Delta B\frac{\tau_s}{u_\infty}\,, \tag{4.294}$$

where the dimensionless *transfer driving force* (or *transfer number*) ΔB is

$$\Delta B = \frac{i_\infty - i_s + \Delta i_c\left[\left(\frac{\rho_o}{\rho}\right)_\infty - \left(\frac{\rho_o}{\rho}\right)_s\right]}{\Delta i_{sg} + \frac{q_\ell}{\dot{m}_s}}\,, \tag{4.295}$$

where Δi_{sg} is the heat of sublimation and q_ℓ is the *sum* of the *sensible* heat need to raise the solid fuel to T_s (i.e., the sublimation temperature) and the heat *losses* from the solid surface.

Now using the same notation as in Section 4.9 for the boundary-layer surface injection, i.e., using B as the *blowing parameter*, we can write (4.294) as

$$B\,Re_x^{-1/2} = \frac{\dot{m}_s}{\rho_\infty u_\infty} = \frac{(\rho v)_s}{(\rho u)_\infty} = \frac{\Delta B}{2}c_f\,,\quad c_f = \frac{\tau_s}{\rho_\infty u_\infty^2/2}\,. \tag{4.296}$$

The momentum transport is *only* modified by the surface injection, and the solution to the laminar boundary-layer problem discussed in Section 4.9.2 is valid for this diffusion flame problem. Then we have from (4.296), (4.203), and (4.209), and for $m = 0$

$$B = \frac{(\rho v)_s}{(\rho u)_\infty}Re_x^{1/2} = \frac{\Delta B}{2}Re_x^{1/2}c_f = \Delta B\frac{d^2 f}{d\eta^2}\Big|_o = -\frac{f(0)}{2}\,, \tag{4.297}$$

where

$$\frac{d^2 f}{d\eta^2}\Big|_o = \frac{d^2 f}{d\eta^2}\Big|_o(B, Pr) \tag{4.298}$$

and some results for $d^2 f/d\eta^2|_o$ were given in Table 4.1.

Glassman (1987) suggests that a good correlation, which is an *adjusted* correlation of the numerical results by comparing with the available experimental results, is

$$B = \frac{(\rho v)_s}{(\rho_\infty u_\infty)}Re_x^{1/2} = \frac{\ln(1 + \Delta B)}{2.6\Delta B^{0.15}}\,. \tag{4.299}$$

This correlation and the exact results are shown in Figure 4.29(b).

4.11.3 Premixed Flame in a Channel

In connection with the intramedium convective heat transfer, the propagation of the combustion front (i.e., flame propagation) in a premixed fuel-oxidizer medium was considered in Section 2.3.2 (single-phase systems) and Section 3.3.5 (two-phase systems). Premixed combustion can also occur over solid surfaces and inside channels. The latter has proven advantageous for lowering of the pollutant emission. In channel flows, the flame location is determined by the local heat losses (through the wall as well as to the colder portions of the fluid and wall) and under some conditions the flame can move upstream towards the fuel/oxidizer source. This undesirable phenomena is referred to as the *flashback*. The *structure* of the premixed laminar flame and the flashback phenomenon in a circular tube have been examined by Lee and T'ien (1982). Due to the multidimensionality of the flame structure and the heat loss to the wall, the flame speed in a channel will be different than that in the unconfined medium.

An axisymmetric, laminar compressible steady flow with a second-order reaction, i.e., a simple form of (2.44), is assumed. The conservation of overall mass, species, momentum and energy are given by

$$\frac{\partial}{\partial x} r\rho u + \frac{\partial}{\partial r} r\rho v = 0 \,, \tag{4.300}$$

$$\frac{\partial}{\partial x} r\rho v \frac{\rho_f}{\rho} + \frac{\partial}{\partial r} r\rho v \frac{\rho_f}{\rho} = \frac{\partial}{\partial x} r\rho D \frac{\partial}{\partial x}\frac{\rho_f}{\rho} + \frac{\partial}{\partial r} r\rho D \frac{\partial}{\partial r}\frac{\rho_f}{\rho} + r\dot{n}_f \,, \tag{4.301}$$

$$\begin{aligned}\frac{\partial}{\partial x} r\rho u^2 + \frac{\partial}{\partial r} r\rho v u = -\frac{\partial p}{\partial x} + \frac{\partial}{\partial x}\frac{4}{3} r\mu \frac{\partial u}{\partial x} + \frac{\partial}{\partial r} r\mu \frac{\partial u}{\partial r} + \\ \frac{\partial}{\partial r} r\mu \frac{\partial u}{\partial x} - \frac{\partial}{\partial x}\frac{2}{3}\mu \frac{\partial}{\partial r} rv \,,\end{aligned} \tag{4.302}$$

$$\begin{aligned}\frac{\partial}{\partial x} r\rho u v + \frac{\partial}{\partial r} r\rho v^2 = -r\frac{\partial p}{\partial r} + \frac{\partial}{\partial x} r\mu \frac{\partial v}{\partial x} + \frac{\partial}{\partial r} 2r\mu \frac{\partial v}{\partial r} + \\ \frac{\partial}{\partial x} r\mu \frac{\partial u}{\partial r} - 2\frac{\mu v}{r} - r\frac{\partial}{\partial r}\frac{2}{3}\mu\left(\frac{\partial v}{\partial r} + \frac{v}{r} + \frac{\partial u}{\partial x}\right) ,\end{aligned} \tag{4.303}$$

$$\frac{\partial}{\partial x} r\rho u c_p T + \frac{\partial}{\partial r} r\rho v c_p T = \frac{\partial}{\partial x} rk \frac{\partial T}{\partial x} + \frac{\partial}{\partial r} rk \frac{\partial T}{\partial r} - r\dot{n}_f \Delta i_c \,, \tag{4.304}$$

where

$$p = \frac{\rho R_g T}{M} \,, \quad \frac{\rho}{\rho_n} = \frac{T_n}{T} \,, \quad -\dot{n}_f = A\rho_o \rho_f e^{-\Delta E_a / R_g T} \,. \tag{4.305}$$

The *upstream* conditions (*nonreacted*) are designated by subscript n, and M is the *average* molecular weight. Lee and T'ien (1982) consider combustion of the *stoichiometric* mixture of methane (CH_4) and air, i.e., $(\rho_f/\rho)_n = 0.055$, $(\rho_o/\rho)_n = 0.22$, and the remainder being nitrogen and the inlet condition of one atmosphere pressure and $T_n = 298$ K.

Including the nitrogen in the single-step reaction of (2.84), we have

$$\nu_{r_o} \mathrm{O}_2 + \nu_{r_N} \mathrm{N}_2 + \nu_{r_f} \mathrm{CH}_4 \rightarrow \nu_{r_N} \mathrm{N}_2 + \nu_p \ \mathrm{P} \ . \tag{4.306}$$

Allowance is made for the variation of the thermophysical properties with respect to temperature. These are *modeled* for a constant pressure as

$$\frac{\mu}{\mu_n} = (\frac{T}{T_n})^{0.75} \ , \quad \frac{D}{D_n} = (\frac{T}{T_n})^{1.75} \ , \quad \frac{k}{k_n} = (\frac{T}{T_n})^{0.75} \ , \tag{4.307}$$

$$\frac{\rho}{\rho_n} = (\frac{T}{T_n})^{-1} \ , \quad \frac{\alpha}{\alpha_n} = (\frac{T}{T_n})^{1.75} \ , \quad \frac{Le}{Le_n} = 1 \ . \tag{4.308}$$

Note that the thermal diffusivity varies with $T^{1.75}$, and therefore, across the flame, its magnitude can change by an order of magnitude (the adiabatic flame temperature for most of the hydrocarbon burning in air is about 2200 K).

The boundary conditions at the entrance $x = x_1$, at the exit $x = x_2$, at the solid surface $r = R$, and along the centerline $r = 0$, are

$$u = v = 0 \ , \quad T = T_n \ , \quad \frac{\partial}{\partial r}\frac{\rho_i}{\rho} = 0 \quad \text{all } x \text{ and } r = R \ , \ \text{and } x = x_1 \tag{4.309}$$

$$v = 0 \ , \quad \frac{\partial u}{\partial r} = \frac{\partial T}{\partial r} = \frac{\partial}{\partial r}\frac{\rho_i}{\rho} = 0 \quad \text{all } x \text{ and } r = 0 \ , \ \text{and } x = x_2 \ . \tag{4.310}$$

Lee and T'ien (1982) scale the space variables using the flame thickness defined by (2.71) using the upstream conditions, i.e.,

$$\delta_F = \frac{\alpha_n}{u_{F_o}} \ , \tag{4.311}$$

where u_{F_o} is the one-dimensional, adiabatic flame speed taken as 38 cm/s and $\alpha_n = 1.76 \times 10^{-5}$ m^2/s, then $\delta_F = 4.63 \times 10^{-3}$ cm. Note that the actual flame thickness would be larger because α across the flame (and therefore its average value) is much larger than α_n. This scaling of the space dimensions, including the tube radius, has a direct usage in the evaluation of the flashback condition, and the details of this are discussed by Lee and T'ien (1982). The temperature is scaled with T_n, the velocity with u_{F_o}, the mass fraction by $(\rho_i/\rho)_n$, the pressure by $\rho_n u_{F_o}^2$ (and the *difference* in the pressure, $p - p_n$, is used), and the dimensionless quantities are designated by T^*, $\mathbf{u}^*$, ρ_i^*/ρ^*, and p^*.

Their *numerical* results for the distributions of T^*, ρ_f^*/ρ, and p^* for a flow (from left to right) through a tube with a radius $R = 46.5\delta_F$, is given in Figures 4.30(a) and (b). The incoming velocity profile is that of the fully developed isothermal laminar flow with the cross-sectional average velocity $\langle u \rangle_A$ adjusted until the flame is *stable* and *stationary*. (They find that $\langle u \rangle_A$ is larger than u_{F_o}; this will be discussed shortly.) Since the wall temperature is maintained at the upstream temperature T_n (leading to a *nonuniform* heat transfer through the tube wall), and the flame front is *distorted*, as indicated by the constant temperature and fuel concentration contours in Figure 4.30(a). The gas expansion results in a local pressure rise, as shown in Figure 4.30(b). Then the flow approaching the flame is first locally decelerated and then accelerated in some areas. This is shown in Figure 4.30(c), where the velocity field is shown by local vectors. The inlet *parabolic* velocity distribution soon changes downstream, and the local decelerations/accelerations are evident. Far downstream, a nearly fully developed profile is reached while the gas is being cooled (but still at high temperatures and low densities, so that the velocities are large compared to the inlet).

The axial distribution of the cross-sectional-averaged velocity $\langle u \rangle_A$, normalized with respect to u_{F_o}, is shown in Figure 4.30(d) for $R = 23.75\delta_F$. The average flame speed can be *defined* as $\langle u \rangle_A$ evaluated upstream of the flame, and this value is larger than u_{F_o}. Thus the flame speed in the tube is larger for the one-dimensional adiabatic flame speed. Note that u_{F_o} is for an *adiabatic* conditions, but here there is a significant heat loss to the tube wall. The decrease and increase in the area-averaged velocity, along the tube, is also evident in Figure 4.30(d). In the figure, the dimensionless x/δ_F is not extended far enough downstream to show the decay in the average temperature occurring downstream of the reaction zone.

4.11.4 Premixed Flame in Jet Impingement

As an example of an interfacial heat transfer *with* chemical reaction, consider a gas flowing toward the surface in the form of a surface normal jet. The premixed flame over the surface is of interest in the species formed near the surface which may influence surface reactions and depositions (a *chemical vapor surface deposition method*). This method has been used in the synthesis of thin films of diamond (Murayama and Uchida, 1992) and has been analyzed by Meeks et al. (1993). A schematic of the problem is given in Figure 4.31(a), where a jet of premixed fuel and oxidizer impinges on a *substrate* and the flow exits laterally (similar to the impinging jets in Section 4.10). The axisymmetric flow field has been approximated such that the radial velocity v is linearly proportional to the radial location r and v/r changes with the axial location x as determined from a simplified set of continuity and momentum equations with the integration made only in the x direction. These are discussed below.

The simplified equations for overall mass, species, momentum, and energy conservation under steady, laminar compressible axisymmetric flow conditions are (Meeks et al., 1993),

$$\frac{\partial u}{\partial x} + 2(\frac{v}{r}) + \frac{u}{\rho}\frac{\partial \rho}{\partial x} = 0 \,, \tag{4.312}$$

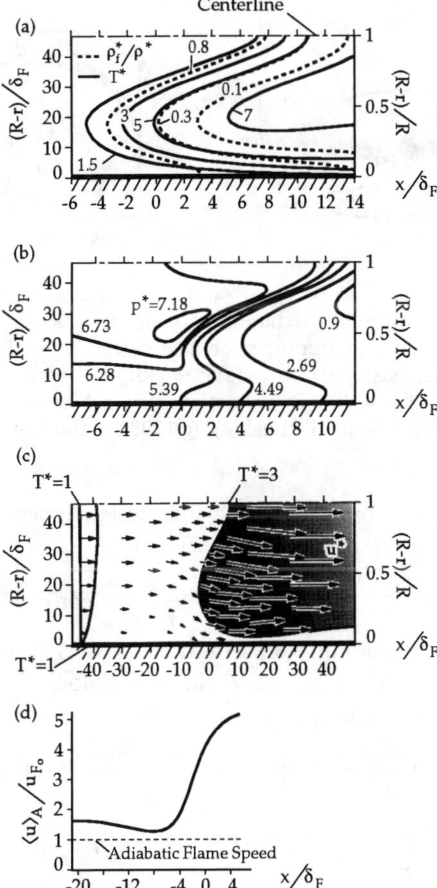

Figure 4.30. (a) Distributions of the temperature and the fuel mass concentration. (b) Distribution of pressure. (c) Distribution of velocity. (d) Axial variation of the normalized area-averaged velocity. All are for a premixed flame in a tube. (From Lee and T'ien, reproduced by permission of Elsevier Science Publishing ©1992 by the Combustion Institute.)

$$\rho u \frac{\partial}{\partial x} \frac{\rho_i}{\rho} + \frac{\partial}{\partial x} \rho \frac{\rho_i}{\rho} (\frac{v_i}{r}) = \dot{n}_i \, , \tag{4.313}$$

$$\rho u \frac{\partial}{\partial x} (\frac{v}{r}) + \rho (\frac{v}{r})^2 = (\frac{1}{r} \frac{\partial \rho}{\partial r}) + \frac{\partial}{\partial x} \mu \frac{\partial}{\partial x} (\frac{v}{r}) \, , \tag{4.314}$$

$$\rho c_p u \frac{\partial T}{\partial x} + \sum_i \rho c_{p_i} \frac{\rho_i}{\rho} (\frac{v_i}{r}) \frac{\partial T}{\partial x} = - \sum_i \dot{n}_i i_i \, . \tag{4.315}$$

The species velocity is given by (1.65), and Meeks et al. (1993) also include the thermodiffusion given by (1.136). The thermodiffusion coefficient is only

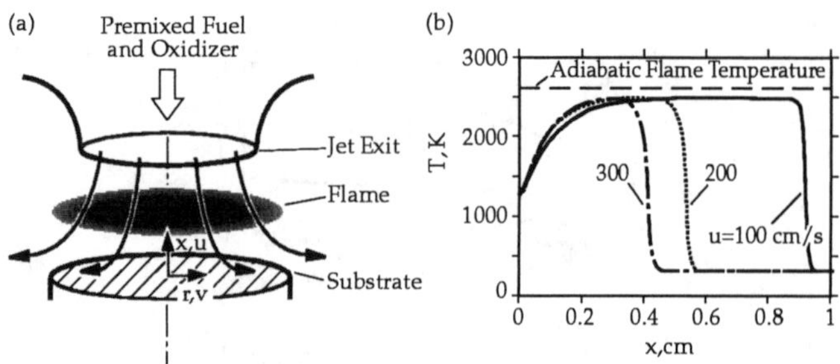

Figure 4.31. (a) A premixed flame adjacent to a surface, the fuel-oxidizer flows through a nozzle and impinges on the surface. (b) The variation of the temperature (one-dimensional) as a function of the distance from the surface and for three different nozzle exit velocities. (From Meeks et al., reproduced by permission of Elsevier Science Publishing ©1993 by the Combustion Institute.)

accurately known for some low molecular weight species. The quantities appearing in the *parentheses* do not vary with r, i.e.,

$$(\frac{v}{r}) = f(x) , \quad (\frac{v_i}{r}) = f_i(x) , \quad \frac{1}{r}\frac{\partial p}{\partial r} = \text{constant} , \tag{4.316}$$

and therefore, the set of equations (4.312) to (4.315) are *ordinary* differential equations in x. The boundary conditions along the x direction, with $x = 0$ located on the surface, are

$$(\frac{v}{r}) = 0 , \quad \rho\frac{\rho_i}{\rho}u + \rho\frac{\rho_i}{\rho}(\frac{v_i}{r}) = \dot{m}_i , \quad u = \sum_i \frac{\dot{m}_i}{\rho} , \quad T = T_s \quad x = 0 , \tag{4.317}$$

$$u = u_n , \quad \frac{v}{r} = 0 , \quad \frac{\rho_i}{\rho} = (\frac{\rho_i}{\rho})_n , \quad T = T_n \quad x = L , \tag{4.318}$$

where $\dot{m}_i$ is the rate of surface production (heterogeneous reaction) of species i and, as with $\dot{n}_i$, is a function of species concentration and temperature, and in general is expressed in a form similar to (2.43).

Their results for a methane-oxygen flame with equal nonreacted volumetric flow rate of each (the *equivalence ratio*, i.e., the fuel to oxidizer mass flow ratio *divided* by the *stoichiometric* fuel oxidizer ratio, of two) is shown in Figure 4.31(b). The gaseous and surface reactions, i.e., the chemical equations for the various reactions and their chemical kinetics are discussed by Meeks et al. (1993). They solve the one-dimensional equation numerically, allowing for variable properties. The axial distribution of the temperature are shown for three different jet exit velocities u_n, a jet-surface separation distance of 1 cm, and exit and surface temperatures of T_n = 298 K and T_s = 1200 K. As expected, the flame thickness *decreases* with an increase in the jet velocity, while the flame stand-off distance does not change. The rate of deposition of carbon on the surface and its nucleation, chemistry, and chemical kinetics are discussed by Meeks et al. (1993). The turbulent, impinging premixed flame has been analyzed using the turbulent models by Bray et al. (1992).

4.11.5 Direct Inclusion of Radiation

In addition to *surface* radiation and the active *absorption bands* in hydrocarbon, carbon dioxide, and water vapor, the formation of *soot* in reacting flows increases the influence of the radiation on the overall heat and mass transfer. The equation of radiative transfer (2.110) describes the transport through an absorbing, emitting, and scattering medium, and the effect of the bounding surfaces can be found by the integration.

Early reviews of the effect of radiation on the boundary-layer flows and heat transfer have been given by Viskanta (1966). Recent discussion are given by Sparrow and Cess (1978), Edwards (1981), Ožisik (1985), and Siegel and Howell (1992).

The radiative heat flux q_r for the case of *one-dimensional* transport within a gas with wavelength-dependent properties and bounded between two parallel surfaces has been formulated by Sparrow and Cess (1978) and Siegel and Howell (1992). Their results have been extended to semi-infinite gaseous (boundary-layer flow and heat transfer) media with *emission* and *reflection* from the solid bounding surface and *emission* and *absorption* within the soot containing gas by Liu and Shih (1980). They consider the case of *negligible* scattering and concentrate on the nine *major* absorption bands in carbon dioxide and water vapor and the spectral absorption of soot particles. For a *gray diffuse* surface of *emissivity* ϵ_{r_s} and temperature T_s, the one-dimensional radiative heat flux at a location y from the surface is given by (Liu and Shih, 1980)

$$\begin{aligned} q_r = 2\epsilon_{r_s} \int_o^\infty E_{b_\lambda}(T_s) E_3(\tau_{r_\lambda})\, d\lambda \, + \\ 4(1-\epsilon_{r_s}) \int_o^\infty E_3(\tau_{r_\lambda}) \int_o^{(\tau_{r_\lambda})_\delta} E_{b_\lambda}(\tau'_{r_\lambda}) E_2(\tau'_{r_\lambda})\, d\tau'_{r_\lambda}\, d\lambda \, + \\ 2\int_o^\infty \int_o^{\tau_{r_\lambda}} E_{b_\lambda}(\tau'_{r_\lambda}) E_2(\tau_{r_\lambda} - \tau'_{r_\lambda})\, d\tau'_{r_\lambda}\, d\lambda \, + \\ 2\int_o^\infty \int_{\tau_{r_\lambda}}^{(\tau_{r_\lambda})_\delta} E_{b_\lambda}(\tau'_{r_\lambda}) E_2(\tau'_{r_\lambda} - \tau_{r_\lambda})\, d\tau'_{r_\lambda}\, d\lambda \,, \end{aligned} \tag{4.319}$$

where E_{b_λ} is the *spectral emissive* power and the *optical thickness*, and the *exponential integrals* are defined as

$$\tau_{r_\lambda} = \int_o^y \sigma_{\lambda a}\, dy \,, \tag{4.320}$$

$$E_n(\tau_{r_\lambda}) = \int_o^1 \mu^{n-2} e^{-\tau_{r_\lambda}/\mu}\, d\mu \,. \tag{4.321}$$

The first term on the right-hand side of (4.319) is the *attenuated surface emission*; the second term is the *attenuated reflected* gas emission; the third term is the *gas emission* in $0 \le y' \le y$; and the fourth term is for $y \le y' \le \delta$, where δ is the boundary-layer thickness.

The numerical results of Liu and Shih (1980) for radiation-affected diffusion flames show that a significant reduction in the computation time can be made by using the *three* most significant bands (with a small correction factor), i.e., 4.3 μm band for carbon dioxide and 6.3 μm and 2.7 μm bands for water vapor. They also introduce and discuss further simplifications in the inclusion of the radiation heat flux.

4.12 Buoyancy

In this section we address the solid-fluid thermal *nonequilibrium* and the interfacial heat transfer with the motion of the fluid being influenced by the heat transfer. As in combustion applications, there can be bulk fluid (homogeneous) and surface (heterogeneous) reactions, and therefore, the source of thermobuoyancy can be *within* the fluid (as compared to the difference between the bulk solid and bulk fluid temperatures). While thermobuoyancy can influence or dominate the fluid motion, the diffusobuoyancy (presence of a species gradient) can hinder or assist this motion.

As will be discussed shortly, the surface geometry and its orientation with respect to the gravity vector greatly influence the interfacial heat transfer and the motion. In order to examine the effects of *surface blowing*, *turbulence*, *reaction* and *forced flow* on the thermobuoyant flow and heat transfer, the most-studied case of a *semi-infinite vertical* flat plate is considered. Some of the results apply to the tilted surfaces for small tilt angles. In this case, the component of the gravity along the surface, i.e., $g \cos \theta$, is used in place of g. Figure 4.32(a) shows a schematic of the thermobuoyant flow adjacent to a tilted surface with $T_s > T_\infty$.

The purely thermobuoyant flows without any reaction are considered first. This is followed by the effect of chemical reactions. The thermo- and diffusobuoyant motions are examined next. Then the thermobuoyancy-influenced motions are considered.

4.12.1 Thermobuoyant Flows Without Reaction

The interfacial heat transfer across the interface of a continuous solid surface and its adjacent moving fluid, with the motion *induced* by thermobuoyancy (i.e., *natural* or *free convection*), has many practical and interesting fundamental features. As is expected, the *solid geometry* and the *direction* of the *gravity* vector with respect to the *interface* both influence the heat transfer and motion. As the ratio of the buoyant inertial force to the viscous forces increases beyond a critical value, flow transitions and then turbulence are observed. Classifications of thermobuoyant flows adjacent to surfaces have been made based on the *time-dependence*, *geometry* (internal

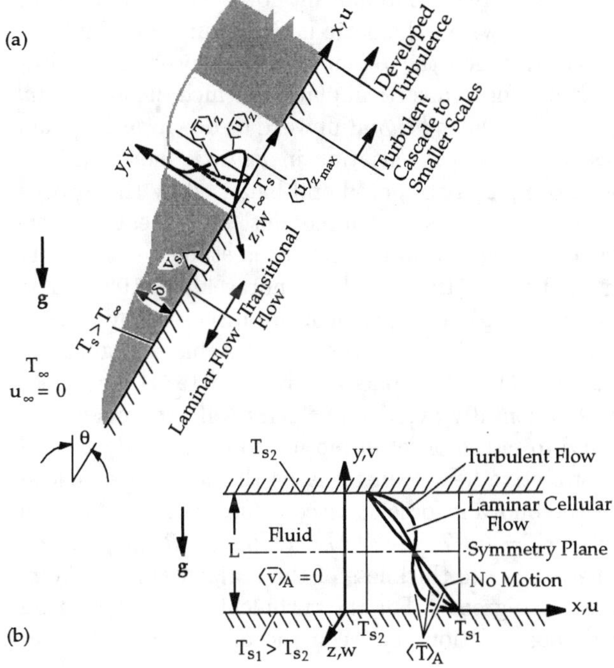

Figure 4.32. (a) Thermobuoyant flow adjacent to a tilted (nearly vertical) isothermal flat plate. The laminar and turbulent boundary layers and the transitions are also shown. (b) Thermobuoyant flow in a fluid layer confined between two plates and heated from below. The vertical temperature distributions for the conduction, laminar, and turbulent regimes are also shown.

and external flows), *surface orientation*, and *flow transitions* (e.g., onset of laminar cellular motion, transition among various cellular motions, transitions from laminar to turbulent flow and to fully turbulent flows). Among the available reviews are Ede (1967), Eckert and Drake (1972), Turner (1979), Jaluria (1980), Bejan (1984), and Gebhart et al. (1988).

As was mentioned, we examine the thermobuoyant flow adjacent to a vertical (or slightly tilted) surface and examine the laminar flow interfacial heat transfer rate. Then we examine the effect of surface blowing, the instability of the laminar flow, and the interfacial heat transfer and the structure of the turbulent thermobuoyant flow. Figure 4.32(a) shows the laminar, transitional, and turbulent flow regimes. Two-dimensional fields are assumed and for the transitional and turbulent regimes the time-averaged (which is equivalent to the line average taken along the z-axis perpendicular to the flow) quantities are shown. The velocity has a maximum near the surface, while the temperature varies monotonically.

We briefly mention another well-studied problem which contains many fundamental features of the purely thermobuoyant flows. That is a horizontal fluid layer heated from below. A schematic is shown in Figure 4.32(b) and this problem is generally called the *Bénard-Rayleigh* problem. Unlike the thermobuoyant flow adjacent to a vertical plate, in which the *horizontal* variation of *density* causes the *unconditional* upward or downward motion of the fluid, in the case of a *vertical* variation in *density* shown in Figure 4.32(b) the motion is completely opposed and balanced by the viscous force, and a *threshold* (or *critical*) buoyant force must be present to overcome the viscous force. This *onset of motion* has been studied extensively and reviews are given by Turner (1979) and Chandrasekhar (1981). The various *modes* of *laminar* flow for this problem are discussed by Turner (1979) and Gebhart et al. (1988). The *turbulent* flow for this configuration has been reviewed by Turner (1979), Arpaci and Larsen (1984), and Gebhart et al. (1988). The horizontally averaged velocity (all components) is zero for this problem, and from the onset of motion to the fully developed turbulent motion the convective heat transfer is by the cellular (or eddy) motion. As the ratio of the buoyant force to viscous force, which is known as the *Grashof number*, $Gr_L = g\beta(T_{s_1} - T_{s_2})L^3/\nu^2$ or the *Rayleigh number*, $Ra_L = Gr_L Pr$ (where L is the thickness of the horizontal fluid layer) increases, these cell sizes vary greatly. For the turbulent flow, a spectrum of sizes is present at various locations between the upper and the lower surface.

(A) Laminar Flow

As an example, consider the case of a vertically oriented *isothermal* semi-infinite plate maintained at T_s in an otherwise quiescent ambient of temperature T_∞. A schematic showing the coordinate system is shown in Figure 4.32(a). The thermobuoyancy-induced flow will be in the form of a boundary-layer flow with a local thickness δ (where the velocity is 0.99 u_{max}). The local surface heat flux q_s has been found for this problem making various approximations. Using the Boussinesq approximations discussed in Section 1.12.1(A) along with the boundary-layer approximations, including that the pressure distribution within the boundary-layer is the *hydrostatic* pressure p_∞, results in

$$\frac{\partial u}{\partial x} + \frac{\partial v}{\partial y} = 0 , \tag{4.322}$$

$$u\frac{\partial u}{\partial x} + v\frac{\partial u}{\partial y} = \nu\frac{\partial^2 u}{\partial y^2} + g\beta(T - T_\infty) , \tag{4.323}$$

$$u\frac{\partial T}{\partial x} + v\frac{\partial T}{\partial y} = \alpha\frac{\partial^2 T}{\partial y^2} . \tag{4.324}$$

The boundary conditions are

$$u = v = 0 \,, \quad T = T_s \quad y = 0 \,, \tag{4.325}$$

$$u = 0 \,, \quad T = T_\infty \quad y \to \infty \,. \tag{4.326}$$

Similarly, solutions for these sets of equations exist (Gebhart et al., 1988). The *normalized* stream function, similarity length scale, temperature, velocity components, and the dimensionless parameters are

$$\psi = 4\nu f(\frac{Gr_x}{4})^{1/4} \,, \quad \eta = \frac{y}{x}(\frac{Gr_x}{4})^{1/4} \,, \tag{4.327}$$

$$T^* = \frac{T - T_\infty}{T_s - T_\infty} \,, \quad u = \frac{4\nu}{x}(\frac{Gr_x}{4})^{1/2}\frac{df}{d\eta} \,, \tag{4.328}$$

$$v = \frac{\nu}{x}(\frac{Gr_x}{4})^{1/4}(\eta\frac{df}{d\eta} - 3f) \,, \quad Gr_x = \frac{g\beta(T_s - T_\infty)x^3}{\nu^2} \,, \quad Pr = \frac{\nu}{\alpha} \,. \tag{4.329}$$

The Grashof number Gr_x is the ratio of the buoyant force to viscous forces and, as was mentioned before, Prandtl number Pr is the ratio of the viscous to thermal boundary-layer thicknesses.

Using these, the *dimensionless* momentum and energy equations become

$$\frac{d^3 f}{d\eta^3} + 3f\frac{d^2 f}{d\eta^2} - 2(\frac{df}{d\eta})^2 + T^* = 0 \,, \tag{4.330}$$

$$\frac{d^2 T^*}{d\eta^2} + 3Pr\, f\frac{dT^*}{d\eta} = 0 \,. \tag{4.331}$$

The boundary conditions become

$$\frac{df}{d\eta} = f = 0 \,, \quad T^* = 1 \quad \eta = 0 \,, \tag{4.332}$$

$$\frac{df}{d\eta} = T^* = 0 \quad \eta \to \infty \,. \tag{4.333}$$

The dimensionless local heat flux, or Nusselt number, Nu_x is given by

$$Nu_x = \frac{q_s x}{(T_s - T_\infty)k} = -\frac{dT^*}{d\eta}|_o(\frac{Gr_x}{4})^{1/4} \,. \tag{4.334}$$

The numerical solutions to (4.330) and (4.331) for $0.01 \leq Pr \leq 10^4$ have been *correlated* (Ede, 1967; Kays and Crawford, 1993) with the function

$$Nu_x = \frac{3}{4}[\frac{2Pr}{5(1 + 2Pr^{1/2} + 2Pr)}]^{1/4} {Ra_x}^{1/4} \,, \tag{4.335}$$

$$Ra_x = Gr_x Pr = \frac{g\beta(T_s - T_\infty)x^3}{\nu\alpha} \,. \tag{4.336}$$

The product of the Grashof and Prandtl numbers is the *Rayleigh* number and includes the effect of the thermal diffusion.

Also, the $Pr \to 0$ and $Pr \to \infty$ *asymptotes* are

$$Nu_x = 0.6(Gr_x Pr^2)^{1/4} \quad Pr \to 0 \,, \tag{4.337}$$

$$Nu_x = 0.503 {Ra_x}^{1/4} \quad Pr \to \infty \,, \tag{4.338}$$

where for $Pr \to 0$ the local dimensionless heat flux is proportional to $Pr^{1/2}$, while for $Pr \to \infty$ the relation is $Pr^{1/4}$.

The local variation of the velocity and temperature are discussed by Gebhart et al. (1988), where the *boundary-layer integral* formulation results in a closed-form solution for the maximum velocity, which occurs at $y = \delta/3$, and the boundary-layer thickness δ. The thickness is given by

$$\frac{\delta}{x} = 3.93(Pr + \frac{20}{21})^{1/4}({Gr_x}^{1/2} Pr)^{-1/2} \,. \tag{4.339}$$

Since *prescribed* profiles of the velocity and temperature are used in the integral method, then δ is (4.339) indicates the edge of the boundary layer where the velocity is zero. A relatively recent accounts of the predictions and the experimental results, including the *leading edge* effects (i.e., deviation from the boundary-layer approximations), are given by Martin (1984).

(B) Surface Injection

Blowing or suction from vertical surfaces into thermobuoyancy-induced flow has an implication in combustion where the *fuel* may be injected into the thermobuoyant boundary layer. Also, if a vertical membrane *separating* fluids with *different* temperatures is permeable and the flow through the membrane is in the opposite direction to the heat flow through the membrane, then through the proper selection of the *seepage* velocity v_s, the net heat transfer between the two fluids can be minimized (Kaviany, 1988). In the following, the existing similarity solution for the case of variable v_s along the surface, as well as the nonsimilar solution for the case of uniform v_s, will be discussed.

As was discussed in connection with the force flows in Section 4.5.2(A), the similarity solution given above can be used for the cases where $v_s = v(y=0)$ is not zero. However, the similarity solution is applicable only if v_s varies with x according to

$$v_s = -\frac{3\nu}{x}(\frac{Gr_x}{4})^{1/4} f(0) \tag{4.340}$$

or

$$f(0) = -\frac{v_s x}{3\nu}(\frac{4}{Gr_x})^{1/2} \equiv -\frac{2}{3}B \,, \qquad B = \frac{v_s x}{2^{3/2}\nu {Gr_x}^{1/4}} \,. \tag{4.341}$$

Then for the injection parameter B to be a constant, the injection velocity should decrease as $x^{-1/4}$.

Assuming that B is a constant, the solution described with no injection

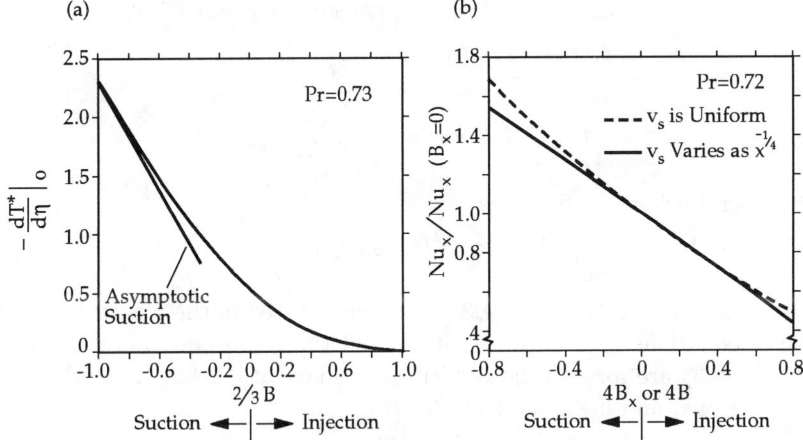

Figure 4.33. (a) Variation of the negative of the dimensionless surface temperature gradient as a function of the blowing (or suction) parameter, for $Pr = 0.73$. (b) Effect of blowing (or suction) on the Nusselt number normalized with respect to the value for no blowing. Both (a) and (b) are for thermobuoyant convection about an isothermal, vertical semi-infinite plate. (From Eichhorn, and Sparrow and Cess, reproduced by permission ©1960, 1961 American Society of Mechanical Engineers.)

can be modified by imposing $f(0) = -2B/3$ as the boundary condition. The resulting equations have been solved numerically by Eichhorn (1960). Also for large $-B$, i.e., *large suction*, $v \gg u$ and the momentum equation (4.330) becomes trivial and the energy equation (4.331) becomes

$$\frac{d^2T^*}{d\eta^2} - 2BPr\frac{dT^*}{d\eta} = 0 \,, \qquad B \ll 0 \,. \tag{4.342}$$

The solution subject to the boundary conditions (4.332) and (4.333) is

$$T^* = e^{2B\,Pr\,\eta} \,, \quad Nu_x = -\frac{dT^*}{d\eta}|_o(\frac{Gr}{4})^{1/4}$$
$$= -2B(\frac{Gr_x}{4})^{1/2}Pr \,, \quad B \ll 0 \,. \tag{4.343}$$

The numerical results of Eichhorn (1960) for v_s varying with $x^{-1/4}$, for both suction $v_s < 0$ and injection $v_s > 0$, along with the above large suction asymptote, are given in Figure 4.33(a). For large blowing, the surface temperature gradient goes to zero and the boundary layer is *blown off*.

Sparrow and Cess (1961) consider *uniform* blowing and solve the boundary-layer problem (4.330) to (4.331) with a *small regular perturbation expansion* in the blowing parameter B_x defined as in (4.341) but with v_s being constant, i.e.,

$$B_x = \frac{v_s}{2^{3/2}\nu}\frac{x}{{Gr_x}^{1/4}} \,. \tag{4.344}$$

Then B_x varies with x as $x^{1/4}$. The expansion in f and T^* are

$$f = f_o(\eta) + B_x f_1(\eta) + \cdots = \sum_i B_x^i f_i(\eta) \,, \tag{4.345}$$

$$T^* = T_o^*(\eta) + B_x T_1^*(\eta) + \cdots = \sum_i B_x^i T_i^*(\eta) \,. \tag{4.346}$$

The lateral velocity v becomes

$$v = \frac{\nu}{x}(\frac{Gr_x}{4})^{1/4}(\eta\frac{\partial f}{\partial \eta} - 3f - B_x\frac{\partial f}{\partial B_x}) \,. \tag{4.347}$$

The expansions (4.345) and (4.346) are substituted in the momentum and energy equations (4.330) and (4.331), and the terms containing identical powers of B_x are sorted. The resulting sorted equations for the zeroth (i.e., no injections) and first power of B_x are

$$\frac{d^3 f_o}{d\eta^3} + 3f_o\frac{d^2 f_o}{d\eta^2} - 2(\frac{df_o}{d\eta})^2 + T_o^* = 0 \,, \tag{4.348}$$

$$\frac{d^2 T_o^*}{d\eta^2} + 3Pr\, f_o\frac{dT_o^*}{d\eta} = 0 \,, \tag{4.349}$$

$$\frac{d^3 f_1}{d\eta^3} + 3f_o\frac{d^2 f_1}{d\eta^2} - 5\frac{df_o}{d\eta}\frac{df_1}{d\eta} + 4\frac{d^2 f_o}{d\eta^2}f_1 + T_1^* = 0 \,, \tag{4.350}$$

$$\frac{d^2 T_1^*}{d\eta^2} + 3Prf_o\frac{dT_1}{d\eta} - Pr\frac{df_1}{d\eta}T_1^* + 4Prf_1\frac{dT_o^*}{d\eta} = 0 \,. \tag{4.351}$$

The boundary conditions for f_o and T_o^* are the same as those given by (4.332) and (4.333). For f_1 and T_1^*, we have

$$f_1 = -1 \,, \quad \frac{df_1}{d\eta} = T_1^* = 0 \quad \eta = 0 \,, \tag{4.352}$$

$$\frac{df_1}{d\eta} = T_1^* = 0 \quad \eta \to \infty \,. \tag{4.353}$$

Then (4.334) is modified to include the first-order effect, and the dimensionless heat flux becomes

$$Nu_x = -(\frac{dT_o^*}{d\eta}|_o + B_x\frac{dT_1^*}{d\eta}|_o)(\frac{Gr_x}{4})^{1/2} \,. \tag{4.354}$$

The ratio of the Nusselt numbers for the case of nonzero and zero blowing, as a function of the blowing parameter, is shown in Figure 4.33(b). For small values of B, the solution for constant v_s (variable B_x) and constant B (variable v_s) are nearly the same. As the magnitude of B increases, the results of Sparrow and Cess (1961) for constant v_s depart from that of Eichhorn (1960) which is for constant B. The results are for $Pr = 0.72$ and 0.73, respectively. The *decrease* in the Nusselt number with *increase* in B (i.e., injection) is evident again.

(C) Instability

The nonlinear dynamics of thermobuoyant has been studied in relation to the *chaotic growth* of disturbances as a desired or undesired feature. Examples of such chaotic thermobuoyant flows are given by Paolucci and Chenoweth (1989), Bau (1992), and Bau and Wang (1992).

As in the forced laminar flow over a semi-infinite surface, the ever-present ambient disturbances will be amplified whenever the interplay between the inertia and viscous forces with the dominance of the inertial force allows for this amplification. The amplified disturbances formed downstream of a stable laminar regime begin with a distinct wavelength and frequency characteristic. Then as they grow downstream they interact with the *base* flow (i.e., the flow resembling the stable laminar flow). These *nonlinear* interactions result in the generation/growth of some new disturbances of *different* wavelengths and frequencies.

The *linear stability theory* studies the initial growth of the ambient disturbances in a base flow (this base condition can also be a quiescent fluid) by introducing periodic disturbances into the conservation equations and determining their growth from the *linearized* equations (these equations are called the characteristic equations). The states of *marginal* or *neutral stability* are those for which the growth rate is zero. These states are indicated by the problem parameters and generally there is a *dominant* wavelength which tends to grow the fastest or with a minimum instability potential. The marginal state is generally given for this wavelength (called the *characteristic wavelength*). The application of the linear stability to thermobouyant flows is discussed by Platten and Legros (1984).

For the thermobuoyant flow adjacent to an isothermal vertical surface, the analysis given by Gebhart et al. (1988) shows that beyond the marginal state a *frequency selection* continues such that a very narrow band of frequencies tend to be significant amplified (i.e., grow to the extent that their amplitude is measurable). In the following, the analysis given by Gebhart et al. (1988) is reviewed and the dependence of this selected frequency on the parameters of the thermobuoyant flow adjacent to a vertical surface (e.g., Gr_x and Pr) is discussed.

The experimental results for the *beginning* (i.e., measurable) of the transitions have been correlated and the results are

$$Gr_x = \begin{cases} 5790(\dfrac{g\,x^3}{\nu^2})^{2/3} & \text{thermal transition ,} \\ 4510(\dfrac{g\,x^3}{\nu^2})^{2/3} & \text{velocity transition ,} \end{cases} \qquad (4.355)$$

where the transition is defined as where the *profile* (temperature for thermal transition and velocity for velocity transition) begins to *deviate* from that associated with the laminar flow.

The end of the transitions is correlated with (Gebhart et al., 1988)

$$Ra_x = 308(\frac{g\,x^3}{\nu^2})^{5/6} \,, \qquad (4.356)$$

where the end of transitions is marked by the position marked where *no* appreciable downstream changes occur in the *intermittency* factor of velocity and temperature distributions.

We now proceed by reviewing the stability of the laminar thermobuoyant base flow following the analysis of Gebhart et al. (1988). The time-dependent, Boussinesq-approximated conservation equations (1.119) to (1.121), for the two-dimensional problem considered, are

$$\frac{\partial u}{\partial x} + \frac{\partial v}{\partial y} = 0 \,, \qquad (4.357)$$

$$\frac{\partial u}{\partial t} + u\frac{\partial u}{\partial x} + v\frac{\partial u}{\partial y} = -\frac{1}{\rho_o}\frac{\partial (p-p_\infty)}{\partial x} + \nu(\frac{\partial^2 u}{\partial x^2} + \frac{\partial^2 u}{\partial y^2}) + g\beta(T-T_\infty) \,, \qquad (4.358)$$

$$\frac{\partial v}{\partial t} + u\frac{\partial v}{\partial x} + v\frac{\partial v}{\partial y} = -\frac{1}{\rho_o}\frac{\partial (p-p_\infty)}{\partial y} + \nu(\frac{\partial^2 v}{\partial x^2} + \frac{\partial^2 v}{\partial y^2}) + g\beta(T-T_\infty) \,, \qquad (4.359)$$

$$\frac{\partial T}{\partial t} + u\frac{\partial T}{\partial x} + v\frac{\partial T}{\partial y} = \alpha(\frac{\partial^2 T}{\partial x^2} + \frac{\partial^2 T}{\partial y^2}) \,. \qquad (4.360)$$

These equations govern the *base* fields (*laminar steady* flow discussed above), as well as the disturbance fields of the velocity (tow components), temperature, pressure, and density. Indicating the base fields with an overbar and the disturbances by a prime, we have the decompositions

$$u = \overline{u} + u' \,, \quad v = \overline{v} + v' \,, \quad T = \overline{T} + T' \,,$$
$$p = \overline{p} - p' \,, \quad \rho = \overline{\rho} - \rho_o \beta T' \,, \qquad (4.361)$$

where the prime quantities are functions of x, y and t while the overbar quantities are function of x and y only (steady base flow). Upon substituting these into (4.357) to (4.360), taking the *time* average and applying the boundary-layer approximations, and neglecting the products of the disturbances, the base-field equations (4.322) and (4.324) are recovered, i.e.,

$$\frac{\partial \overline{u}}{\partial x} + \frac{\partial \overline{v}}{\partial y} = 0 \,, \qquad (4.362)$$

$$\overline{u}\frac{\partial \overline{u}}{\partial x} + \overline{v}\frac{\partial \overline{u}}{\partial y} = \nu\frac{\partial^2 \overline{u}}{\partial y^2} + g\beta(\overline{T} - T_\infty) \,, \qquad (4.363)$$

$$\overline{u}\frac{\partial \overline{T}}{\partial x} + \overline{v}\frac{\partial \overline{T}}{\partial y} = \alpha\frac{\partial^2 \overline{T}}{\partial y^2} \,. \qquad (4.364)$$

The solution of these, subject to the appropriate boundary conditions, have already been given using the dimensionless variables f and T^* and the similarity variable η. The disturbance conservation equations are found by subtracting the base equations from the base plus disturbance equations. The result are

$$\frac{\partial u'}{\partial x} + \frac{\partial v'}{\partial y} = 0 \,, \tag{4.365}$$

$$\frac{\partial u'}{\partial t} + \overline{u}\frac{\partial u'}{\partial x} + v'\frac{\partial \overline{u}}{\partial x} = -\frac{1}{\rho_o}\frac{\partial (p' - p_\infty)}{\partial x} + \nu(\frac{\partial^2 u'}{\partial x^2} + \frac{\partial^2 u'}{\partial y^2}) + g\beta T' \,, \tag{4.366}$$

$$\frac{\partial v'}{\partial t} + \overline{u}\frac{\partial v'}{\partial x} = -\frac{1}{\rho_o}\frac{\partial (p' - p_\infty)}{\partial y} + \nu(\frac{\partial^2 v'}{\partial x^2} + \frac{\partial^2 v'}{\partial y^2}) \,, \tag{4.367}$$

$$\frac{\partial T'}{\partial t} + \overline{u}\frac{\partial T'}{\partial x} + v'\frac{\partial \overline{T}}{\partial y} = \alpha(\frac{\partial^2 T'}{\partial x^2} + \frac{\partial^2 T'}{\partial y^2}) \,. \tag{4.368}$$

Next by assuming *periodic* disturbances in *space* (x, along the flow) and *time* with the amplitudes of the disturbances varying with η, the space similarity variable defined by (4.327), the disturbance stream function and temperature are given by

$$\psi'(x,y,t) = 4\nu(\frac{Gr_x}{4})^{1/4}\phi_\psi(\eta)e^{i(ax+\omega t)} \,, \tag{4.369}$$

$$T'(x,y,t) = (T_s - T_\infty)\phi_T(\eta)e^{i(ax+\omega t)} \,, \tag{4.370}$$

where a is the *wave number* (2π divided by the wavelength), ω is the *angular frequency* (2π times the frequency) of the disturbances, and ϕ_ψ and ϕ_T are the dimensionless *amplitudes*.

The wave number and the frequency can be *scaled* as

$$a^* = \frac{ax}{(Gr_x/4)^{1/2}} \,, \quad \omega^* = \frac{\omega x^2}{4(Gr_x/4)^{3/4}} \,. \tag{4.371}$$

Now substituting (4.369) and (4.370) and ψ and T^* defined in (4.327) and (4.328) into the disturbance equations, the equation for the disturbance amplitudes (called the *Orr-Summerfeld* equation) are found as

$$(\frac{df}{d\eta} - \frac{\omega^*}{a^*})(\frac{d^2\phi_\psi}{d\eta^2} - a^{*2}\phi_\psi) - \frac{d^3 f}{d\eta^3}\phi_\psi = (\frac{d^3\phi_\psi}{d\eta^3} - 2a^{*2}\frac{d^2\phi_\psi}{d\eta^2} + a^{*4}\phi_\psi + \frac{d\phi_T}{d\eta})(4^{3/4}ia^*{Gr_x}^{1/4})^{-1} \,, \tag{4.372}$$

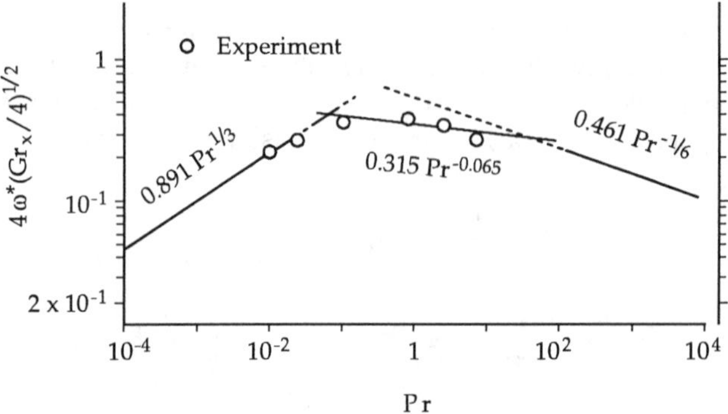

Figure 4.34. The selected frequency of the disturbances which grow in thermobuoyant flow adjacent to vertical, isothermal plates. (From Gebhart et al., reproduced by permission ©1988 Pergamon Press.)

$$(\frac{df}{d\eta} - \frac{\omega^*}{a^*})\phi_T - \frac{dT^*}{d\eta}\phi_\psi = (\frac{d^2\phi_T}{d\eta^2} - a^{*2}\phi_T)(4^{3/4}ia^*PrGr_x{}^{1/4})^{-1} \, . \quad (4.373)$$

The boundary conditions are

$$\phi_\psi = \frac{d\phi_\psi}{d\eta} = \phi_T = 0 \quad \eta \to 0 \quad \text{and} \quad \eta \to \infty \, . \quad (4.374)$$

The solution to the characteristic equations, (4.372) and (4.373), leads to the determination of the amplitudes, when Pr, Gr_x, ω^*, a^* are given. The dimensionless wave number is complex quantity and has a *real* and an *imaginary* component. For *no growth*, or *neutrally* stable state, the imaginary part is zero. Then the neural states are found by selecting $a_i^* = 0$ and determining ω^* as a function of Pr and Gr_x. Also was mentioned, Gebhart et al. (1988) show that the disturbances that grow tend to grow faster through a frequency selection process, i.e., a *filtering* of disturbances are made and only a *small* range of frequencies grow. Their results for this frequency is given in Figure 4.34. Using (4.371), we note that the dimensional frequency is a function of Gr_x, x and also Pr. Close agreement is found between the predicted and measured results.

(D) Turbulence

Boundary-layer turbulence *follows* the instability of the laminar flow caused by the amplification of the ambient disturbances. The interfacial heat transfer (and the mean velocity and temperature distributions) are altered because of the eddy motions (generally characterized as small and large secondary flows). We consider a two-dimensional flow with the mean flow being *steady*, i.e., the thermobuoyant boundary-layer flow. Note that the time-averaged quantities at a location denoted x and y, and the z-direction-averaged quantities are in principle *identical* since the flow is *ideally* two-dimensional (infinite extent in z, $L_z \to \infty$) and the time average is also taken over a very long period, i.e., $\tau \to \infty$. This is stated as

$$\overline{\phi} = \frac{1}{\tau}\int_o^\tau \phi\, dt \ , \ \langle\phi\rangle_z = \frac{1}{L_z}\int_o^{L_z} \phi\, dz \quad \tau \to \infty \ , \ L_z \to \infty \ . \qquad (4.375)$$

In order to examine some features of the turbulent thermobuoyant flows and review some attempts in the prediction of these features, some available experimental results and the existing modeling of the thermobuoyant flow adjacent to a vertical surface are examined below. One feature of the turbulent thermobuoyant convection heat transfer is that the interfacial heat transfer rate q_s across *isothermal* surfaces becomes independent of the system dimensions. This *local* feature results in a dependence of the local Nusselt number on the Grashof number (or Rayleigh number) to the power 1/3. This has been verified experimentally for locally fully developed turbulent thermobuoyant flows.

The *spectral density* of *velocity fluctuation* $E_{u'}$ and *temperature fluctuation* $E_{T'}$ are defined as

$$\overline{u'^2} = \int_{\frac{\omega_1}{2\pi}}^{\frac{\omega_2}{2\pi}} E_{u'}(\omega)\, d\frac{\omega}{2\pi} \ . \qquad (4.376)$$

The *spectral density* of $\overline{u'^2}$ (and $\overline{T'^2}$) over the *frequency interval* $d(\omega/2\pi)$, i.e., $E_{u'}$ (and $E_{T'}$), is expected to *decay* with the increase in the frequency. Furthermore, we expect a *zero* asymptote $E_{u'}$ (and $E_{T'}$) for very *large* frequencies and a frequency-*independent* asymptote for *low* frequencies. In the experiments, the upper bond for the frequency $w_2/2\pi$ is half of the *sampling* rate, and the lower bond for frequency $\omega_1/2\pi$ is the record length of the time series analyzed (Gebhart et al., 1988).

The experimental results for $E_{u'}(\omega)$ and $E_{T'}(\omega)$ for *water* and for the case of *uniform heat flux* q_s and at a location given by the dimensionless parameter $5(g\beta x^4 q_s/5k\nu^2)^{1/5}$ are shown in Figure 4.35(a) and (b). For $\omega/2\pi$ larger than 10 Hz the spectral density approaches the *noise level* of the *hot-wire anemometry* used for measurement. The results are for several values of y (i.e., η) and for low frequencies there is a significant scatter between the results for various values of y. This shows that the large (and slow) eddies are not uniformly distributed across the boundary layer (the largest eddy measurable will be that equal in size to the boundary-layer thickness). The independence from y at higher frequencies indicates independence of the small eddies from the large eddies. The decay of the spectra follows the expected -3 power law (Gebhart et al., 1988).

Analysis of the turbulent thermobuoyant flow adjacent to vertical surfaces has been made using various turbulence models (as compared to the direct simulation). Earlier attempts used the prescribed mixing length with a van Driest modification near the surface, similar to that discussed in

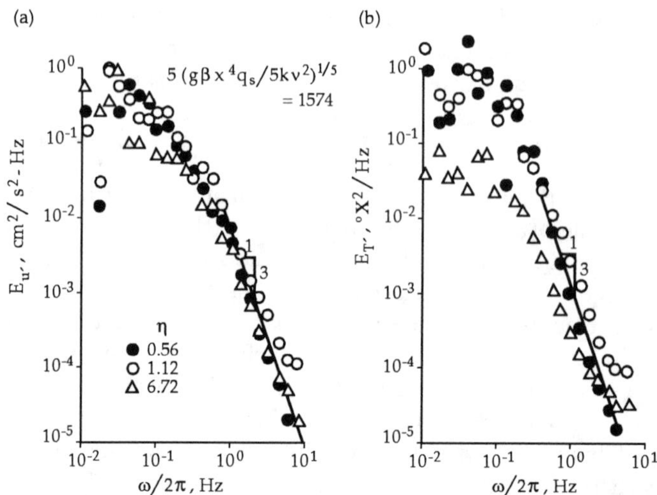

Figure 4.35. The spectral density of (a) velocity and (b) temperature fluctuations for turbulent, thermobuoyant convection adjacent to an isoflux, vertical plate. (From Gebhart et al., reproduced by permission ©1988 Pergamon Press.)

Section 4.4.3(B). Recent turbulent models use the *low Reynolds number*, *buoyancy-adjusted* transport equations for the turbulent kinetic energy and its dissipation rate. These are similar to those discussed in Section 2.7.2, except the presence of the solid surface and the *anisotropy* of the turbulence near the surface are also addressed. Among these are the computations of Lin ad Churchill (1978), To and Humphrey (1986), and Henkes and Hoogendoorn (1989).

In the analysis of the turbulent transport adjacent to surfaces, introductory turbulence models contain a *wall function* which allows for the decay of the turbulent kinetic energy. However, the assumption made about the fluid motion microstructure is that it is *isotropic* (i.e., $\overline{u_i'^2} = \overline{v'^2} = \overline{w'^2}$). To allow for the *anisotropy*, turbulence equations allowing for the evaluation of the fluctuating *components* of velocity (i.e., their root-mean-square) are needed (in addition to transport equation for the turbulent kinetic energy). To and Humphrey (1986) examine the approximate algebraic equations for the components of the *Reynolds stresses* $\overline{u_i'u_j'}$ and turbulent heat flux $\overline{u_i'T'}$. The equations are then added to the equations for the turbulent kinetic energy $\frac{1}{2}\overline{u_i'u_i'}$ and its dissipation rate ϵ_k and are solved simultaneously. Here for simplicity we review *only* the model they use for $\frac{1}{2}\overline{u_i'u_i'}$, ϵ_k, and $\overline{T'^2}$. These contain enough information to illustrate some features of the turbulent thermobuoyant flow modeling and some of its predictions.

The time-averaged conservation equations are found as in Section 2.4.1, but here we allow for the fluctuation of density ρ. These equations, including the assumption of a perfect gas behavior and a uniform pressure distribution across the boundary layer, are

$$\frac{\partial \overline{\rho}}{\partial t} + \frac{\partial}{\partial x_j}(\overline{\rho u_j} + \overline{\rho' u'_j}) = 0 \,, \tag{4.377}$$

$$\frac{\partial \overline{\rho}\,\overline{u}}{\partial t} + \frac{\partial}{\partial x_j}\overline{\rho}\,\overline{u}_j\overline{u}_i + \frac{\partial}{\partial t}\overline{\rho' u'_i} + \frac{\partial}{\partial x_j}(\overline{\rho' u'_j}\overline{u}_i + \overline{\rho' u'_i}\overline{u}_j) =$$
$$-\frac{\partial \overline{p}}{\partial x_i} + (\overline{\rho} - \rho_\infty)g_i + \frac{\partial}{\partial x_j}[\nu(\frac{\partial \overline{u}_i}{\partial x_j} + \frac{\partial \overline{u}_j}{\partial x_i}) - \overline{\rho u'_i u'_j}] \,, \tag{4.378}$$

$$\frac{\partial}{\partial x_j}(\overline{\rho}\,\overline{u}_j\overline{T} + \overline{\rho' u'_j}\,\overline{T} + \overline{\rho' T'}\overline{u}_j) = \frac{\partial}{\partial x_j}(\frac{k}{c_p}\frac{\partial \overline{T}}{\partial x_j} - \overline{\rho u'_j T'}) \,, \tag{4.379}$$

$$\overline{\rho' T'} + \overline{\rho}\overline{T} = \rho_\infty T_\infty \,, \tag{4.380}$$

$$\overline{\rho' u'_i}\,\overline{T} + \overline{\rho u'_i T'} = 0 \,. \tag{4.381}$$

The turbulent fluxes $\overline{\rho}\overline{u'_i u'_j}$ and $\overline{\rho}\overline{u'_i T'}$ are modeled as given by (2.91), (2.95), and (2.99). Since allowance is made for the fluctuation of density, these are written as

$$-\overline{\rho}\overline{u'_i u'_j} = \mu_t(\frac{\partial \overline{u}_i}{\partial x_j} + \frac{\partial \overline{u}_j}{\partial x_i}) - \frac{2}{3}\overline{\rho}(\frac{1}{2}\overline{u'_i u'_i})\delta_{ij} \,, \tag{4.382}$$

$$-\overline{\rho}\overline{u'_i T'} = \frac{\mu_t}{Pr_t}\frac{\partial \overline{T}}{\partial x_i} \,, \tag{4.383}$$

$$\mu_t = a_1\overline{\rho}\frac{(\frac{1}{2}\overline{u'_i u'_i})^2}{\epsilon_k} \,, \tag{4.384}$$

where ϵ_k is dissipation rate of the turbulent kinetic energy and is

$$\epsilon_k = \nu\overline{\frac{\partial u'_i}{\partial x_j}\frac{\partial u'_i}{\partial x_j}} \,. \tag{4.385}$$

The transport equation for the turbulent kinetic energy $\frac{1}{2}\overline{u'_i u'_i}$, i.e., (2.100) (called the k-equation), the transport equation for its dissipation rate, which is similar to the k-equation and is obtained from the Navier-Stokes equations, and an approximate expressions for $\overline{\rho' T'}$ and $\overline{T'^2}$ are (To and Humphrey, 1988)

$$\frac{\partial}{\partial t}\overline{\rho}(\frac{1}{2}\overline{u'_i u'_i}) + \frac{\partial}{\partial x_j}\overline{\rho u_j}(\frac{1}{2}\overline{u'_i u'_i}) =$$
$$\frac{\partial}{\partial x_j}(\mu + \frac{\mu_t}{Pr_k})\frac{\partial}{\partial x_j}\frac{1}{2}\overline{u'_i u'_i} + \overline{\rho}P_k - \overline{\rho}\epsilon_k - \overline{\rho' u'_i}(\frac{\partial \overline{u}_i}{\partial t} + \overline{u}_k\frac{\partial \overline{u}_i}{\partial x_k}) \,, \tag{4.386}$$

$$\frac{\partial}{\partial t}\overline{\rho}\frac{\epsilon_k}{\nu} + \frac{\partial}{\partial x_j}\overline{\rho u_j}\frac{\epsilon_k}{\nu} = \frac{\partial}{\partial x_j}(\mu + \frac{\mu_t}{Pr_\epsilon})\frac{\partial}{\partial x_j}\frac{\epsilon_k}{\nu} +$$
$$\frac{\epsilon_k}{\nu(\frac{1}{2}\overline{u'_i u'_i})}(a_2\overline{\rho}P_k - a_3\overline{\rho}\epsilon_k) - \frac{\epsilon_k}{\nu(\frac{1}{2}\overline{u'_i u'_i})}a_2\overline{\rho' u'_i}(\frac{\partial \overline{u}_i}{\partial t} + \overline{u}_k\frac{\partial \overline{u}_i}{\partial x_k}) \,, \tag{4.387}$$

$$\overline{\rho' T'} = 2a_4 \frac{\frac{1}{2}\overline{u_i' u_i'}}{\epsilon_k} \frac{\overline{\rho}}{\overline{T}} \overline{u_j' T'} \frac{\partial \overline{T}}{\partial x_j} , \tag{4.388}$$

$$2\overline{\rho}\overline{u_j' T'} \frac{\partial \overline{T}}{\partial x_j} = \frac{1}{a_4} \frac{\epsilon_k}{\frac{1}{2}\overline{u_i' u_i'}} \overline{\rho} \overline{T'^2} , \tag{4.389}$$

$$a_1 = a_{1_o} e^{-2.5/(1+Re_t/50)} , \tag{4.390}$$

$$a_3 = a_{3_o}(1 - 0.3e^{-Re_t^2})(1 - e^{-Re_t^2}) , \tag{4.391}$$

where in the expression for a_3 the last term in the parentheses is *only* used near the surface, and a_{1_0}, a_{3_0}, a_2, and a_4 are prescribed constants. The *turbulent Reynolds number* is defined as

$$Re_t = \frac{\overline{\rho}}{\mu} \frac{(\frac{1}{2}\overline{u_i u_i})^2}{\epsilon_k} . \tag{4.392}$$

The *boundary conditions* at the surface are

$$\frac{1}{2}\overline{u_i' u_i'} = 0 \; , \epsilon_k = 2 \quad \nu[\frac{\partial}{\partial y}(\frac{1}{2}\overline{u_i' u_i'})^{1/2}]^2 \qquad y = 0 . \tag{4.393}$$

To and Humphrey (1986) use a *finite* computational domain and at the two other boundaries apply a zero derivative for the mean and turbulent equalities, while for the boundary where the flow enters the domain they use an existing analytical distribution of the *mean* quantities.

Equations (4.386) to (4.391) constitute an *isotropic* eddy viscosity model and are solved numerically by To and Humphrey (1986). As an example, their results for the distribution of $(\overline{T'^2})^{1/2}/(T_s - T_\infty)$ and the variation of Nu_x with respect to Gr_x are shown in Figures 4.36(a) and (b). Figure 4.36(a) shows the prediction of the $k - \epsilon_k - \overline{T'^2}$ model of turbulence, as well as the available experimental results. The model predicts the rise and the decay of the temperature fluctuation fairly well. The computations require the inclusion of the laminar leading edge, and the turbulent models are applied at a rather arbitrary location (or Gr_x) along the plate. To and Humphrey (1986) use various choices of this *transitional* Gr_x, but in the fully developed turbulent region the results are independent of these choices. The existing experimental results are correlated by

$$Nu_x = 0.098{Gr_x}^{1/3}(\frac{T_s}{T_\infty})^{-0.14} \tag{4.394}$$

for air, $1.7\times10^{11} < Gr_x < 1.86\times10^{12}$, and $0.14 < (T_s - T_\infty)/T_\infty < 1.73$. As was mentioned, the 1/3 power is generally expected for the *local* turbulent transport (to make q_s independent of x). The *transition* to turbulence is given by (4.356) and is generally taken as $Gr_x = 1.5 \times 10^8$ for air. The predictions of To and Humphrey (1986) as well as the experimental results for the local Nusselt number are given in Figure 4.36(b). The agreement with the experimental results is improved when they use more elaborate turbulent models.

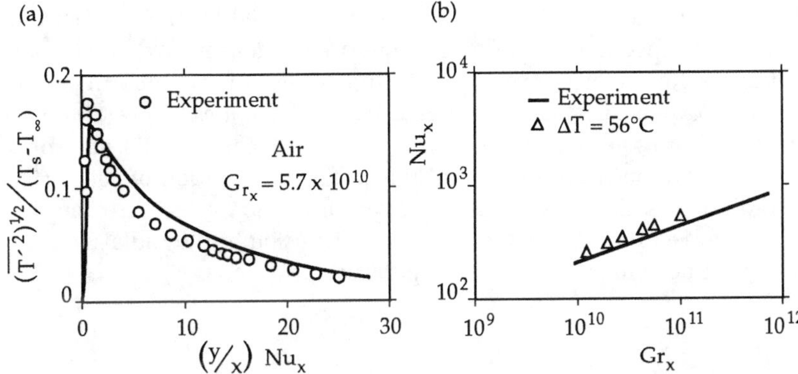

Figure 4.36. (a) Distribution of the root-mean-square of the temperature fluctuation (normalized with respect to the temperature difference between the surface and the free stream) using the similarly variable which contains the Nusselt number. (b) Variation of the local Nusselt number with respect to the Grashof number. The experimental results as well as the predictions based on $\Delta T = 56°\mathrm{C}$ are shown. (From To and Humphrey, reproduced by permission ©1986 Pergamon Press.)

Other turbulence models are also used by To and Humphrey (1986). In their *algebraic stress* model, the transport equations for the Reynolds stresses are approximated by neglecting the convection and diffusion terms. Then they are able to predict the distributions of $(\overline{u'^2})^{1/2}/\overline{u}_{\max}$, $(\overline{v'^2})^{1/2}/\overline{u}_{\max}$, and $(\overline{w'^2})^{1/2}/\overline{u}_{\max}$. They show that the distribution and magnitudes of these turbulent quantities are *vastly different*, indicating *strong anisotropy* of the turbulence adjacent to the surface.

4.12.2 Thermobuoyant Flows with Reaction

Among the examples of thermobuoyant flows adjacent to solid surfaces with reaction at the surface or within the fluid (mostly exothermic chemical gas) are the burning of solids in an otherwise quiescent ambient and the burning of *gaseous fuel* injected through *permeable* solid surfaces also in an otherwise quiescent ambient. Both of these are generally treated as *diffusion* flames because the diffusion of the oxidizer in the ambient and the fuel originating from the surface *controls* the reaction (as compared to the rate of chemical reaction). This *high number Damköhler asymptotic* behavior was discussed in Section 2.3.3 along with the Shvab-Zel'dovich transformation of the variables based on the immediate reaction of the fuel and oxidizer, and, therefore, the *lack* of their *coexistence*.

The problem of the surface injection of a gaseous fuel through a *vertical surface* is considered here. The heat generation occurs *within* the gas, and as is generally the case, the surface injection is *uniform* with velocity v_s. In Section 4.12.1 we examined the problem of the thermobuoyant flows adjacent to a surface, where the motion was caused by the difference between the surface and the otherwise quiescent ambient. We noted that a similarity solution for the velocity field did not exist for a uniform v_s and a first-order expansion was used. Here because of the *high temperatures* and the significant *absorption bands* of some of the gas components, the *surface* and *gas* radiation must also be included. The formulation of the surface-gas radiation is similar to that discussed in Section 4.11.5 in connection with the forced flow diffusion flames. The inclusion of the radiation is also a source of nonsimilarity in the temperature field.

(A) Formulation

Liu et al. (1981,1982) solve the *two-dimensional laminar* thermobuoyant flow diffusion flame with the gaseous fuel injected uniformly through the vertical surface. In their first report they use the local *nonsimilarity* scheme (first- and second-order approximations) to solve the *boundary-layer* governing equations, and in their second report they do not make the boundary-layer approximations and solve the complete two-dimensional equations numerically. They also perform experiments and compare the results with the predictions. In the following, their analysis leading to the local nonsimilarity solution is reviewed.

Consider a *stoichiometric*, *single-step* reaction of methane and oxygen given by

$$\nu_{r_f}\mathrm{CH}_4 + \nu_{r_o}\mathrm{O}_2 \rightarrow \nu_{p_{\mathrm{H_2O}}}\mathrm{H_2O} + \nu_{p_{\mathrm{CO_2}}}\mathrm{CO}_2 \,, \tag{4.395}$$

$$\nu_{r_f} = 1 \,, \ \ \nu_{r_o} = 2 \,, \ \ \nu_{p_{\mathrm{H_2O}}} = 2 \,, \ \ \nu_{p_{\mathrm{CO_2}}} = 1 \,. \tag{4.396}$$

The products can be treated as a mixture with $\nu_p = 3$ and having an average molecular M_p based on the molar fraction of its constituents.

The conservation of the overall mass, species, momentum, and thermal energy, written for a steady, two-dimensional (Cartesian coordinate) *laminar* boundary layer-flow are obtained from (1.66), (1.72), (1.76), (1.77), and (1.83), and are

$$\frac{\partial}{\partial x}\rho u + \frac{\partial}{\partial y}\rho v = 0 \,, \tag{4.397}$$

$$\rho u\frac{\partial}{\partial x}\frac{\rho_i}{\rho} + \rho v\frac{\partial}{\partial y}\frac{\rho_i}{\rho} = \frac{\partial}{\partial y}\rho D\frac{\partial}{\partial y}\frac{\rho_i}{\rho} + \dot{n}_i \,, \quad i = f,o,p \,, \tag{4.398}$$

$$\rho u\frac{\partial u}{\partial x} + \rho v\frac{\partial u}{\partial y} = \frac{\partial}{\partial y}\mu\frac{\partial u}{\partial y} + g(\rho_\infty - \rho) \,, \tag{4.399}$$

$$\rho u\frac{\partial i}{\partial x} + \rho v\frac{\partial i}{\partial y} = \frac{\partial}{\partial y}\frac{k}{c_p}\frac{\partial i}{\partial y} - \frac{d}{dy}q_r + \dot{s} \,. \tag{4.400}$$

With the heat of reaction Δi_c given per *mole produced*, the volumetric heat generation can be written in terms of the production of any of the three *major* species

$$-\frac{\dot{n}_f}{M_f \nu_{r_f}} = -\frac{\dot{n}_o}{M_o \nu_{r_o}} = \frac{\dot{n}_p}{M_p \nu_p} = \frac{\dot{s}}{\Delta i_c} \; . \tag{4.401}$$

The *mass concentrations* ρ_i/ρ are related through

$$\frac{\rho_p}{\rho} = 1 - \frac{\rho_o}{\rho} - \frac{\rho_f}{\rho} \; . \tag{4.402}$$

Assuming a perfect gas behavior and negligible pressure change across and outside the boundary layer, we have

$$i = \int_{T_\infty}^{T} c_p \, dT \; , \quad \rho T = \rho_\infty T_\infty \; . \tag{4.403}$$

It is further assumed that ρu and $\rho k/c_p$ are *constant*.

As discussed in Section 2.3.3, using the Shvab-Zel'dovich transformation the three species conservation equations with their *volumetric mass production rates* can be reduced to two equations with the production rates *not* appearing directly. This variables of the transformations defined by

$$X_i = \{\frac{\rho_i}{\rho}\frac{1}{M_i \nu_i} + [\frac{\rho_o}{\rho} - (\frac{\rho_o}{\rho})_\infty]\frac{1}{M_o \nu_{r_o}}\}\frac{\Delta i_c}{i_\infty} \; , \tag{4.404}$$

$$i = f, p \; , \quad \nu_i = \left\{ \begin{array}{ll} 1 & i = f \; , \\ 3 & i = p \end{array} \right. \tag{4.405}$$

reduces (4.403) to

$$\rho u \frac{\partial}{\partial x} X_i + \rho v \frac{\partial}{\partial y} X_i = \frac{\partial}{\partial y} \rho D \frac{\partial}{\partial y} X_i \; . \tag{4.406}$$

The *heat generation* due to the reaction can also be removed from the energy equation (4.405). The transformation variable and the resulting equation are

$$X_T = \frac{i}{i_\infty} + \frac{\Delta i_c}{M_o \nu_{r_o} i_\infty}[\frac{\rho_o}{\rho} - (\frac{\rho_o}{\rho})_\infty] \; , \tag{4.407}$$

$$\rho u \frac{\partial}{\partial x} X_T + \rho v \frac{\partial}{\partial y} X_T = \frac{\partial}{\partial y}\frac{k}{c_p}\frac{\partial}{\partial y} X_T - \frac{1}{i_\infty}\frac{\partial}{\partial y} q_r \; . \tag{4.408}$$

Allowing for the *variation* of density, the similarity variable η (here a *psuedo-similarity* variable) and an *streamwise* dimensionless variable ξ are defined as

$$\eta = (\frac{g}{4\nu_s^2})^{1/4} x^{-1/4} \int_o^y \frac{\rho}{\rho_s} \, dy \; , \tag{4.409}$$

$$\xi = (\frac{4\nu_s^2}{g})^{1/2} \frac{x^{1/2}}{k_s} \frac{1}{\gamma_r} \; , \quad \gamma_r = \frac{L}{\sigma T_r^3} \; , \tag{4.410}$$

where ν_s, ρ_s, and k_s are evaluated for the mixture at the surface, L is a characteristic length (taken as a location along the x-axis over which the averages are taken), σ is the Stefan-Boltzmann constant, and T_r is the *adiabatic* flame temperature of the *Stoichiometric* premixed flame.

The stream function and its dimensionless form f are related through

$$\psi = 4(\frac{\nu^2 g\, x^3}{4})^{1/2} f(\eta, \xi) \,. \tag{4.411}$$

(B) Thin Diffusion Flame Approximations

The location of the flame η_f is where the concentration of *both* the fuel and oxidizer are *zero* (diffusion flame or thin flame sheet approximation). Then to the either side of the flame only one reactant is present, i.e.,

$$\rho_o = 0 \,, \quad \frac{T}{T_\infty} = 1 + [X_T + (\frac{\rho_o}{\rho})_\infty \frac{\Delta i}{M_o \nu_{r_o} c_{p_\infty} T_\infty}] \frac{c_{p_\infty}}{c_p} \quad 0 \le \eta \le \eta_f \,, \tag{4.412}$$

$$\rho_f = 0 \,, \quad \frac{T}{T_\infty} = 1 + (X_T - X_f) \frac{c_{p_\infty}}{c_p} \,, \tag{4.413}$$

$$X_f = [\frac{\rho_o}{\rho} - (\frac{\rho_o}{\rho})_\infty] \frac{\Delta i}{M_o \nu_{r_o} c_{p_\infty} T_\infty} \quad \eta_f \le \eta < \infty \,. \tag{4.414}$$

The thermobuoyancy term in the momentum equation, using the definition of X_T, will appear as

$$\frac{\rho_\infty}{\rho} - 1 = \frac{T}{T_\infty} - 1 = \{X_T - [\frac{\rho_o}{\rho} - (\frac{\rho_o}{\rho})_\infty] \frac{\Delta i_c}{M_o \nu_{r_o} c_{p_\infty} T_\infty}\} \frac{c_{p_\infty}}{c_p} \,. \tag{4.415}$$

The final form of the species conservation equations is

$$\frac{1}{Sc} \frac{\partial^2}{\partial \eta^2} X_i + 3f \frac{\partial}{\partial \eta} X_i = 2\xi (\frac{\partial f}{\partial \eta} \frac{\partial}{\partial \xi} X_i + \frac{\partial f}{\partial \xi} \frac{\partial}{\partial \eta} X_i) \,. \tag{4.416}$$

The momentum and energy equations for the fuel and the oxidizer regions are:

for $0 \le \eta \le \eta_f$,

$$\frac{\partial^3 f}{\partial \eta^3} - 2(\frac{\partial f}{\partial \eta})^2 + 3f \frac{\partial^2 f}{\partial \eta^2} + (X_T + \frac{\rho_o}{\rho_\infty} \frac{\Delta i_c}{M_o \nu_{r_o} c_{p_\infty} T_\infty}) \frac{c_{p_\infty}}{c_p}$$
$$= 2\xi (\frac{\partial^2 f}{\partial \eta \partial \xi} \frac{\partial f}{\partial \eta} - \frac{\partial^2 f}{\partial \eta^2} \frac{\partial f}{\partial \xi}) \,, \tag{4.417}$$

$$\frac{1}{Pr} \frac{\partial^2}{\partial \eta^2} X_T + 3f \frac{\partial}{\partial \eta} X_T - \frac{1}{Pr} \xi \gamma_r \frac{c_{p_s}}{c_{p_\infty}} \frac{\rho_s}{\rho} [X_T + (\frac{\rho_o}{\rho})_\infty)$$
$$\frac{\Delta i_c}{M_o \nu_{r_o} c_{p_\infty} T_\infty}] \frac{c_{p_\infty}}{c_p T_\infty} \frac{d}{dy} q_r = 2\xi (\frac{\partial f}{\partial \eta} \frac{\partial}{\partial \xi} X_T - \frac{\partial f}{\partial \xi} \frac{\partial}{\partial \eta} X_T) \,; \tag{4.418}$$

for $\eta_f \leq \eta \leq \infty$,

$$\frac{\partial^3 f}{\partial \eta^3} - 2(\frac{\partial f}{\partial \eta})^2 + 3f\frac{\partial^2 f}{\partial \eta^2} - (X_T - X_f)\frac{c_{p\infty}}{c_p} = 2\xi(\frac{\partial^2 f}{\partial \eta \partial \xi}\frac{\partial f}{\partial \eta} - \frac{\partial^2 f}{\partial \eta^2}\frac{\partial f}{\partial \xi}) , \tag{4.419}$$

$$\frac{1}{Pr}\frac{\partial^2}{\partial \eta^2}X_T + 3f\frac{\partial}{\partial \eta}X_T - \frac{1}{Pr}\xi\gamma_r\frac{c_{p_s}}{c_{p\infty}}\frac{\rho_s}{\rho}(X_T - X_f)\frac{c_{p\infty}}{c_p T_\infty}\frac{d}{dy}q_r = 2\xi(\frac{\partial f}{\partial \eta}\frac{\partial}{\partial \xi}X_T - \frac{\partial f}{\partial \xi}\frac{\partial}{\partial \eta}X_T) . \tag{4.420}$$

The boundary conditions are

$$u = \rho_o = 0 , \quad v = v_s , \quad i = i_s , \quad \rho_f = \rho_{f_s} , \quad \rho_p = \rho_{p_s} \quad y = 0 , \tag{4.421}$$

$$u = 0 , \quad i = i_\infty , \quad \rho_o = \rho_{o\infty} , \quad \rho_f = \rho_p = 0 \quad y \to \infty . \tag{4.422}$$

These boundary conditions in transformed variables and coordinates become:

for $\eta = 0$,

$$f_s = -\frac{v_s k_s^{1/2}}{3\nu_s}(3\gamma_r)^{1/2} - \frac{2}{3}\xi\frac{\partial f}{\partial \xi} , \quad \frac{\partial f}{\partial \eta} = 0 , \tag{4.423}$$

$$X_f = -\frac{\frac{\partial X_f}{\partial \eta}}{Sc(3f_s + 2\xi\frac{\partial f}{\partial \xi})} - \frac{\Delta i_c}{M_f \nu_{r_f} c_{p\infty} T_\infty} - (\frac{\rho_o}{\rho})_\infty \frac{\Delta i_c}{M_o \nu_{r_o} c_{p\infty} T_\infty} , \tag{4.424}$$

$$X_p = -\frac{\frac{\partial X_p}{\partial \eta}}{Sc(3f_s + 2\xi\frac{\partial f}{\partial \xi})} - (\frac{\rho_o}{\rho})_\infty \frac{\Delta i_c}{M_o \nu_{r_o} c_{p\infty} T_\infty} , \tag{4.425}$$

$$X_T = \frac{c_{p_s} T_s}{c_{p\infty} T_\infty} - (\frac{\rho_o}{\rho})_\infty \frac{\Delta i_c}{M_o \nu_{r_o} c_{p\infty} T_\infty} ; \tag{4.426}$$

for $\eta \to \infty$,

$$\frac{\partial f}{\partial \eta} = X_f = X_p = X_T = 0 . \tag{4.427}$$

(C) Radiation

The *Planck mean absorption coefficient* $\overline{\sigma}_p$ (Siegel and Howell, 1992) uses the blackbody gas emission spectrum as the weighting function, i.e.,

$$\overline{\sigma}_p(T,p) = \frac{1}{\sigma T^4}\int_o^\infty \sigma_{a\lambda}(\lambda, T, p) E_{b\lambda}\, d\lambda . \tag{4.428}$$

The derivation of the radiative heat flux for an *absorbing-emitting, non-scattering gray* gas confined between two *parallel black* surfaces separated by a distance ℓ at temperatures T_1 and T_2 with the Planck mean absorption coefficient $\overline{\sigma}_p$ is (Siegel and Howell, 1992)

$$\frac{d}{dy}q_r(\tau_r) = 4\overline{\sigma}_p\sigma T^4 - 2\overline{\sigma}_p\sigma_r[T_1^4E_2(\tau_r) + T_2^4E_2(\tau_{r_2} - \tau_r) + \int_o^{\tau_{r_\ell}} T^4(\tau_r')E_1(|\tau_r - \tau_r'|)\,d\tau_r'] \,, \quad (4.429)$$

where τ_r and E_n were defined by (4.320) and (4.321), but here wavelength independence is assumed. For only one bounding black surface with surface temperature T_s and $\ell \to \infty$, we have

$$\frac{d}{dy}q_r(\tau_r) = 4\overline{\sigma}_p\sigma_r T^4 - 2\overline{\sigma}_p\sigma_r[T_s^4E_2(\tau_r) + \int_0^{\infty} T^4(x,\tau_r')E_1(|\tau_r - \tau_r'|)\,d\tau_r'] \,. \quad (4.430)$$

The Planck mean absorption coefficient $\overline{\sigma}_p$ must be allowed to change with the *temperature* variation within the boundary layer. For the H_2O and CO_2 active bands, the *concentration-averaged* absorption coefficient is defined as

$$\overline{\sigma}_p(T,p) = a\left[\frac{\rho_{H_2O}}{\rho}(\overline{\sigma}_p)_{H_2O} + \frac{\rho_{CO_2}}{\rho}(\overline{\sigma}_p)_{CO_2}\right] \,, \quad (4.431)$$

where a is a correction factor.

Liu et al. (1981) treat the radiation using the *eleven* active bands of H_2O, CO_2, and CH_4 and applying the *exponential wide-band* model (Siegel and Howell, 1922). As with the conclusion in 4.11.5, they also show that the *two* H_2O and a *single* CO_2 band absorption account for about 85% of the gas radiation contributions. They also make a gray gas approximation using the *locally scaled* Planck mean absorption coefficient and show that relatively accurate results are found if the coefficient is a function of the local temperature. Recent account of the validation in laminar and turbulent thermobuoyant diffusion flames is given by Orloff et al. (1992).

(D) Example of Results

Liu et al. (1981) also perform experiments and measure the two-dimensional temperature distributions using interferometry. The predicted and measured lateral temperature distributions at an axial location, as well as their predicted distributions of the species (in mole fraction), are shown in Figure 4.37(a). The experimental conditions are not repeated here and since their experiments use air instead of pure oxygen, the concentration distribution of nitrogen is also shown (while the water vapor distribution is not shown, but follows that of CO_2). The *second-order approximation* of

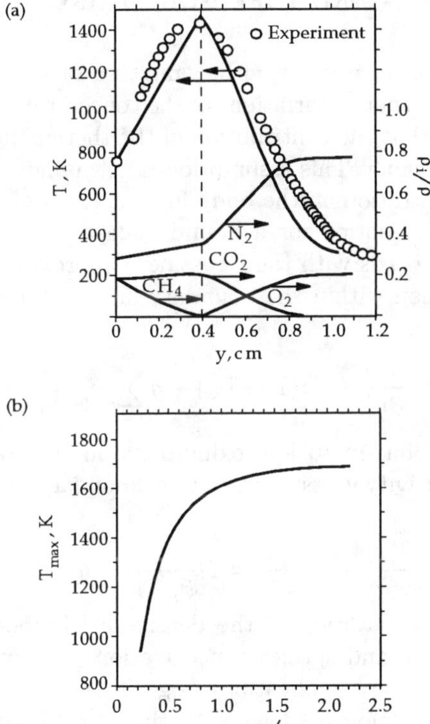

Figure 4.37. (a) Predicted and measured lateral temperature distribution at an axial location. Also shown are the lateral distributions of the species concentrations. (b) Variation of the predicted maximum temperature (i.e., flame temperature) with respect to the blowing velocity. (From Liu et al., reproduced by permission ©1981 Pergamon Press.)

the *local nonsimilarity* (the so-called two-equation model), which results in two ordinary differential equations for each variable and then are all solved simultaneously, results in good agreement with the experimental results. The assumption of a diffusion flame does result in a disagreement with the experimental results only in the regions adjacent to the flame. This is due to the actual, finite reaction rate and its dependence on the temperature. Figure 4.37(b) shows the predicted variation of the maximum temperature T_{max} (occurring at the flame) with respect to the injection velocity v_s at an axial location. An asymptotic maximum temperature is predicted. In practice, at low injection rates the flame quenches, and at higher injection rates the boundary layer becomes unstable.

4.12.3 Thermo- and Diffusobuoyancy

Although so far we have dealt with a mixture of gases (and in general fluids) and with the spatial variations of the concentration of the species, it has been assumed that the contribution of the thermobuoyancy dominates over the diffusobuoyancy. This assumption is now removed by including the effect of the concentration on the body force.

The momentum equation for a boundary-layer, laminar, purely buoyant flow along the x-axis with the Boussinesq approximations and negligible pressure variation within the boundary layer is the simplified form of (1.124), i.e.,

$$u\frac{\partial u}{\partial x}+v\frac{\partial u}{\partial y}=\nu\frac{\partial^2 u}{\partial y^2}+g\beta(T-T_\infty)+g\sum_i \beta_{m_i}[\frac{\rho_i}{\rho}-(\frac{\rho_i}{\rho})_\infty] . \quad (4.432)$$

According to the Boussinesq approximations, in the species equations a constant mass diffusivity is assumed and the boundary-layer form of (1.125) becomes

$$u\frac{\partial}{\partial x}\frac{\rho_i}{\rho}+v\frac{\partial}{\partial y}\frac{\rho_i}{\rho}=D\frac{\partial^2}{\partial y^2}\frac{\rho_i}{\rho}+\dot{n}_i . \quad (4.433)$$

The *simultaneous* variation of the density in the body force with respect to temperature and species concentration (for *binary* systems) has been examined for many applications (e.g., Turner, 1979; Jaluria, 1980). For *laminar*, purely buoyant flows with the density variation caused by the difference in temperature and concentration between a vertical surface and its ambient, similarly solutions to the boundary-layer equations exist (Gebhart et al., 1988). The motion caused by the variation in concentration can *assist* or *oppose* the motion caused by the variation in temperature. In the thermobuoyant flows without reaction examined in Section 4.12.1, the dimensionless parameters combining the imposed overall body force, e.g., $g\beta(T_s-T_\infty)$, the geometry, and the thermophysical properties, were Gr_x and Pr. Here these are extended to include their mass transfer counterparts for a binary system Gr_{x_m} and Sc (*Schmidt number*) defined as

$$Gr_x=\frac{g\beta\Delta T\, x^3}{\nu^2} , \quad Gr_{x_m}=\frac{g\beta_m\Delta\frac{\rho_1}{\rho}\, x^3}{\nu^2} , \quad Pr=\frac{\nu}{\alpha} , \quad Sc=\frac{\nu}{D} , \quad (4.434)$$

$$\Delta=T_s-T_\infty , \quad \Delta\frac{\rho_1}{\rho}=(\frac{\rho_1}{\rho})_s-(\frac{\rho_1}{\rho})_\infty . \quad (4.435)$$

Gebhart et al. (1988) show that the dimensionless thermo- and diffusobuoyancy terms Gr_x and Gr_{x_m}, respectively, can be combined and that similarity solutions exist for *positive* Gr'_x (defined below), with any linear combinations of Gr_x and Gr_{x_m}, i.e.,

$$Gr'_x=a_T Gr_x+a_m Gr_{x_m} \qquad a_T , \; a_m=\begin{cases} 1 & \text{upward flow,} \\ -1 & \text{downward flow.}\end{cases} \quad (4.436)$$

Then the solutions discussed in 4.12.1(A) for laminar thermobuoyant flows can be extended, and the similarity variable (4.327) becomes

$$\eta = \frac{y}{x}(\frac{Gr'_x}{4})^{1/2} . \quad (4.437)$$

The stream function and the velocity components defined in (4.327) to (4.329) are *also* written in terms of Gr'_x in place of Gr_x. Similar to the dimensionless temperature defined in (4.328) a dimensionless species 1 concentration is defined as

$$\rho_1^* = \frac{\frac{\rho_1}{\rho} - (\frac{\rho_1}{\rho})_\infty}{\Delta\frac{\rho_1}{\rho}} . \quad (4.438)$$

Now introducing these into the continuity equation (which is completely satisfied and eliminated by the use of the defined stream function), momentum conservation equation (4.432), energy conservation equation (4.324), and species 1 conservation equation (4.439), we have the transformed equations (which are now ordinary differential equations) as

$$\frac{d^3 f}{d\eta^3} + 3f\frac{d^2 f}{d\eta^2} - 2(\frac{df}{d\eta})^2 + \frac{T^* + \dfrac{\beta_m \Delta\frac{\rho_1}{\rho}}{\beta\Delta T}\rho_1^*}{a_T + a_m\dfrac{\beta_m \Delta\frac{\rho_1}{\rho}}{\beta\Delta T}} = 0 , \quad (4.439)$$

$$\frac{d^3 T^*}{d\eta^3} + 3Pr\, f\frac{dT^*}{d\eta} = 0 , \quad (4.440)$$

$$\frac{d^3 \rho_1^*}{d\eta^3} + 3Sc\, f\frac{d\rho_1^*}{d\eta} = 0 . \quad (4.441)$$

The boundary conditions are

$$\frac{df}{d\eta} = f = 0 , \quad T^* = \rho_1^* = 1 \quad \eta = 0 , \quad (4.442)$$

$$\frac{df}{d\eta} = T^* = \rho_1^* = 0 \quad \eta \to \infty . \quad (4.443)$$

The assumption of $f(0) = 0$ implies that the effect of the interfacial mass transfer (and velocity) on the fluid dynamics is *negligible*. Examining the dependence of $f(0)$ on the mass transfer (Gebhart et al., 1988) gives

$$f(0) = \frac{1}{3Sc}[(\frac{\rho_1}{\rho})_s - (\frac{\rho_1}{\rho})_\infty]\frac{\partial\rho_1^*}{\partial\eta}|_o , \quad (4.444)$$

which shows that for *small* Sc the surface-ambient concentration difference should be *small* for this assumption to be valid. Note that in (4.439) the ratio $(\beta_m \Delta\rho_1/\rho)/\beta\Delta T$ *signifies* the ratio of the concentration and thermal driving buoyant forces. When this ratio is *small*, the thermobuoyancy dominates and (4.439) becomes identical to (4.330) with $a_T = 1$ (note that $Gr_x{}'$ must be positive).

The local Nusselt number Nu_x and its counterpart, the local *Sherwood* number Sh_x, are given by

$$Nu_x = -\frac{dT^*}{d\eta}|_o[\frac{1}{4}(a_T Gr_x + a_m Gr_{x_m})]^{1/4}$$

$$= -\frac{dT^*}{d\eta}|_o(\frac{Gr_x}{4})^{1/4}(a_T + a_m\frac{\beta_m\Delta\frac{\rho_1}{\rho}}{\beta\Delta T})^{1/4} , \qquad (4.445)$$

$$Sh_x = -\frac{\partial\rho_1^*}{\partial\eta}|_o[\frac{1}{4}(a_T Gr_x + a_m Gr_{x_m})]^{1/4} . \qquad (4.446)$$

For $Pr \neq Sc$ the thermal and concentration boundary layers have *different* thicknesses. Gebhart et al. (1981) review the numerical results to (4.439) to (4.441). The results for the heat transfer, given in term of $Nu_x/Gr_x^{1/4}$ and for $Pr = 0.7$ and several values of Sc, are given in Figure 4.38. The results are given with respect to variation in the ratio of the driving buoyant forces. The exact numerical result for $Nu_x/Gr_x^{1/4}$ and $Pr = 0.72$ is 0.357 (Gebhart et al., 1988). For the *assisting* diffusobuoyancy, $a_T = a_m = 1$, the heat transfer rate increases and the increase is *more* noticeable for *smaller* Sc and *larger* concentration to thermal buoyant forces. As the ratio of the Schmidt number to the Prandtl number α/D (i.e., the Lewis number) *increases*, the concentration boundary layer becomes confined to a small region inside the thermal boundary layer and adjacent to the surface, and therefore becomes *less* effective in assisting the buoyant flow.

4.12.4 Thermobuoyancy-Influenced Flows

A fluid flowing over a surface may be driven by an initial or a maintained external pressure gradient (generally called a *forced* flow). This motion can be influenced by a thermobuoyancy caused by a difference in temperature between the surface and the fluid, or by a temperature rise due to exothermic reactions in the fluid. In the purely thermobuoyant flows studied in the previous sections, the *volumetric* body force due to the change in density, which in turn was caused by a change in temperate, was discussed. Now this volumetric force can *oppose* or *assist* a forced flow. Consider the boundary-layer flow and heat transfer problem. Here the available results on the effect of thermobuoyancy on the forced flow will be examined. The *combined* buoyancy and forced flow and heat transfer is also called a *mixed convection*.

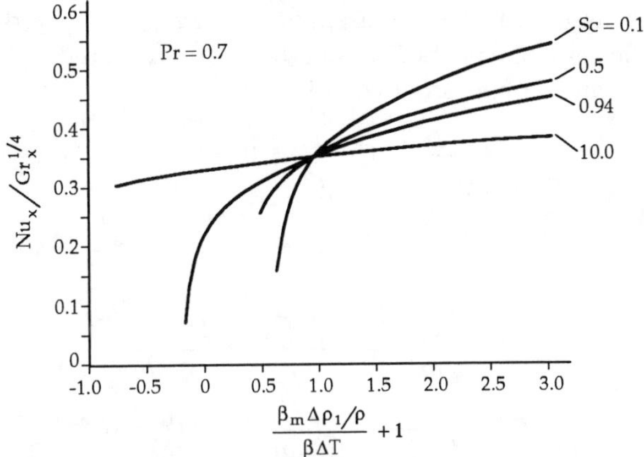

Figure 4.38. Effect of diffusobuoyancy on the normalized interfacial heat transfer rate in laminar, thermobuoyant flow adjacent to isothermal, vertical surfaces. The result are for $Pr = 0.7$ and several Schmidt numbers. (From Gebhart et al., reproduced by permission ©1988 Pergamon Press.)

In order to examine the effect of the thermobuoyancy on forced flow and heat transfer, we consider the *laminar* flow adjacent to an *isothermal* vertical surface. The effect of an *assisting* thermobuoyancy can be examined as a *small* perturbation on the forced flow. This has been done using the standard similarity treatment along with an up to the second order regular perturbation expansion by Merkin (1969) and the results are discussed below. This can demonstrate the extent to which an assisting thermobuoyancy can increase (*enhance*) the interfacial heat transfer. The *complete* governing equations can *also* be solved numerically and the numerical results are also discussed.

As before, in Section 4.2.2, the dimensionless stream function f, the temperature T^*, and the similarity variable η are defined for forced flow as

$$\psi = (2\nu u_\infty x)^{1/2} f\,, \quad T^* = \frac{T - T_\infty}{T_s - T_\infty}\,, \tag{4.447}$$

$$\eta = y\left(\frac{u_\infty}{2\nu\, x}\right)^{1/2} = \frac{y}{x}\left(\frac{Re_x}{2}\right)^{1/2}. \tag{4.448}$$

A regular perturbation in the parameter ϵ which signifies the ratio of the buoyant to inertial forces, i.e.,

$$\epsilon = \frac{g\beta(T_s - T_\infty)x}{u_\infty^2} = \frac{Gr_x}{Re_x^2}\,, \tag{4.449}$$

is made. The expansions, including up to the second order expansions, are

$$f(\eta) = f_o(\eta) + \epsilon f_1(\eta) + \epsilon^2 f_2(\eta) + \cdots\,, \tag{4.450}$$

$$T^*(\eta) = T_o^*(\eta) + \epsilon T_1^*(\eta) + \epsilon^2 T_2^*(\eta) + \cdots\,. \tag{4.451}$$

Then the momentum equation (4.8) for constant u_∞ and the energy equation (4.5) are used with the expansion in f and T^* substituted, and terms having coefficients of common power of ϵ are collected. The resulting zeroth-, first-, and second-order equations are

$$\frac{d^3 f_o}{d\eta^3} + f_o \frac{d^2 f_o}{d\eta^2} = 0 \,, \tag{4.452}$$

$$\frac{d^3 f_1}{d\eta^3} + f_o \frac{d^2 f_1}{d\eta^2} - 2\frac{df_o}{d\eta}\frac{df_1}{d\eta} + 3\frac{d^2 f_o}{d\eta^2} f_1 + 2T_o^* = 0 \,, \tag{4.453}$$

$$\frac{d^3 f_2}{d\eta^3} + f_o \frac{d^2 f_2}{d\eta^2} - 4\frac{df_o}{d\eta}\frac{df_2}{d\eta} + \\ 5\frac{d^2 f_o}{d\eta^2} f_2 + 3f_1 \frac{d^2 f_1}{d\eta^2} - 2\left(\frac{df_1}{d\eta}\right)^2 + 2T_1^* = 0 \,, \tag{4.454}$$

$$\frac{d^2 T_o^*}{d\eta^2} + Pr\, f_o \frac{dT_o^*}{d\eta} = 0 \,, \tag{4.455}$$

$$\frac{d^2 T_1^*}{d\eta^2} + Pr\left(f_o \frac{dT_1^*}{d\eta} - 2\frac{df_o}{d\eta}T_1^* + 3f_1 \frac{dT_o^*}{d\eta}\right) = 0 \,, \tag{4.456}$$

$$\frac{d^2 T_2^*}{d\eta^2} + Pr\left(f_o \frac{dT_2^*}{d\eta} - 4\frac{df_o}{d\eta}T_2^* + \right. \\ \left. 3f_1 \frac{dT_1^*}{d\eta} - 2\frac{df_1}{d\eta}T_1^* + 5f_2 \frac{dT_o^*}{d\eta}\right) = 0 \,. \tag{4.457}$$

The zeroth-order solution is already available and was discussed in Section 4.2.2 (here, $m = 0$).

The boundary conditions are

$$f_o = f_1 = f_2 = \frac{df_o}{d\eta} = \frac{df_1}{d\eta} = \frac{df_2}{d\eta} = T_1^* = T_2^* = 0 \,, \tag{4.458}$$

$$T_0^* = 1 \quad \eta = 0 \,, \tag{4.459}$$

$$\frac{df_1}{d\eta} = \frac{df_2}{d\eta} = T_o^* = T_1^* = T_2^* = 0 \,, \quad \frac{df_o}{d\eta} = 1 \quad \eta \rightarrow \infty \,. \tag{4.460}$$

The dimensionless surface shear stress and heat flux are related to the surface derivatives by

$$c_f = \frac{\tau_s}{\frac{1}{2}\rho u_\infty^2} = \left(\frac{2}{Re_x}\right)^{1/2}\left(\frac{d^2 f_o}{d\eta^2}\Big|_o + \frac{Gr_x}{Re_x^2}\frac{d^2 f_1}{d\eta^2}\Big|_o + \right. \\ \left. \frac{Gr_x^2}{Re_x^4}\frac{d^2 f_2}{d\eta^2}\Big|_o + \cdots\right) , \tag{4.461}$$

$$Nu_x = \frac{q_s x}{(T_s - T_\infty)k} = -\left(\frac{Re_x}{2}\right)^{1/2}\left(\frac{dT_o^*}{d\eta}\Big|_o + \frac{Gr_x}{Re_x^2}\frac{dT^*}{d\eta}\Big|_o + \right. \\ \left. \frac{Gr_x^2}{Re_x^4}\frac{dT_2^*}{d\eta}\Big|_o + \cdots\right) . \tag{4.462}$$

The effect of an assisting thermobuoyancy is to increase the heat transfer rate. The signs of $dT_1^*/d\eta|_o$ and $dT_2^*/d\eta|_o$ may *not* be the same. For example, for $Pr = 1$ the first term results in an *increase* in Nu_x while the second term tends to *reduce* this enhancement (but the sums of the two contributions is a net increase in Nu_x compared to the case of no thermobuoyancy).

The numerical results for the derivatives are available and are discussed by Gebhart et al. (1988). These include the first-order results which are available for a larger range of Pr.

The complete coupled momentum and energy equations have also been solved *numerically*. Lloyed and Sparrow (1970) obtain solutions for the asymptotes $Re_x \rightarrow 0$ and $Gr_x \rightarrow 0$ and also the *mixed* region in between. Their results for the velocity and temperature distributions for $Pr = 0.72$ and various values of Gr_x/Re_x^2 are given in Figure 4.39(a). Note that for $T^* = T^*(\eta)$, the origin of η is on the right. The case of $Re_x \rightarrow 0$ is not shown, i.e., the results are that of the thermobuoyancy-influenced forced flow and heat transfer. Their results for the Nusselt number are given in Figure 4.39(b). Bejan (1984) recommends that the variable $Ra_x^{1/4}/Re_x^{1/2}Pr^{1/3}$ be used as the *regime parameter* and his presentation of the results of Lloyed and Sparrow (1970) is used here. For $Ra_x^{1/4}/Re_x^{1/2}Pr^{1/3}$ much *smaller* than *unity*, forced convection dominates, and for values much *larger* than unity the thermobuoyant convection dominates.

Churchill (1977) correlates the available results for the *local* Nusselt number for *laminar* forced $(Nu_x)_f$ and thermobuoyant, i.e., natural, $(Nu_x)_n$ flows over *isothermal vertical* surfaces and gives

$$(Nu_x)_f = \frac{0.339 Re_x^{1/2} Pr^{1/3}}{[1 + 2(\frac{0.0468}{Pr})^{2/3}]^{1/4}} , \qquad (4.463)$$

$$(Nu_x)_n = \frac{0.503 Ra_x^{1/4}}{[1 + (\frac{0.492}{Pr})^{9/16}]^{4/9}} . \qquad (4.464)$$

Then with these *asymptotes* he recommends the following correlation for the *combined*, *assisting* forced and thermobuoyant flows over vertical isothermal surfaces

$$[\frac{Nu_x}{(Nu_x)_f}]^3 = 1 + [\frac{(Nu_x)_n}{(Nu_x)_f}]^3 . \qquad (4.465)$$

4.13 Electro- and Magnetohydrodynamics

Electro- and magnetohydrodynamics and heat transfer, especially the interfacial convective heat transfer, have been studied in the relation with the *enhancement* of the interfacial heat transfer in the *electrohydrodynamics*, the *hindering* of the interfacial heat transfer in the *magnetohydrodynamics*,

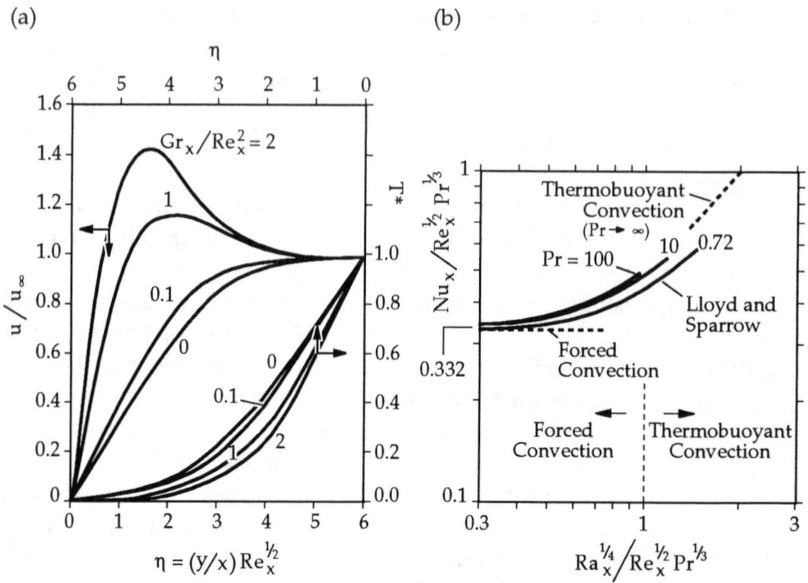

Figure 4.39. (a) Distributions of the normalized velocity and temperatures with respect to the similarity variable for thermobuoyancy-assisted force flow over a vertical, isothermal surface. The results are for several relative thermobuoyancy strengths. (b) Forced and thermobuoyant convection asymptotes and the behavior of the Nusselt number in the transition regime between the asymptotes. (From Gebhart et al., and Bejan, reproduced by permission ©1988 Pergamon Press and ©1984 Wiley and Sons.)

in relation to the interfacial heat transfer resulting *phase change* caused in *welding*, and in interfacial *heat losses* to the bounding surfaces in *plasma processing*. In this section we examine the role of the electromagnetic force on the fluid flow and heat transfer. The electrohydrodynamics is considered first, followed by the magnetohydrodynamics. Then the high-temperature applications in gases, i.e., plasmas, is addressed.

4.13.1 Electrohydrodynamics

As was discussed in Sections 1.12.4(A) and 2.6.1, the fluid electrical conductivity σ_e changes with temperature, and therefore, in nonisothermal electrodynamics the temperature gradient produces a gradient in the electrical conductivity, and in an electrical field this produces a bulk charge density. This charge density in turn interacts with the electric field and as a volumetric force influences (or causes) the fluid motion. As was mentioned, this mode of *electroconvection* is called the conductivity mode. The charge can also flow through the electrodes and the fluid and influences the motion upon its interaction with the field, and this is called the mobility mode. In practice, both of these modes coexist, and while a significant theoretical study of the electrohydrodynamics and heat transfer related to the conductivity mode has been done, the injection mode has not yet been completely studied.

Recent studies have aimed at defining the regimes over which each of these modes dominates. This can be done by examining the generation of the free charge. The conservation of positive and negative charges is described by the *charge transport* equations. The rate of change of the *number density* of *positive* and *negative* charges n_p and n_n are given in terms of the *positive* and *negative* ion *mobility* K_p and K_n, the *dissociation* coefficient κ_d, and the *recombination* coefficient κ_r. This rate of change equations are (Martin and Richardson, 1984)

$$\frac{\partial n_i}{\partial t} + \mathbf{u} \cdot \nabla n_i = -\nabla \cdot \dot{n}_i K_i \mathbf{e} + \kappa_d c - \kappa_r n_i n_j \quad i \neq j \ , \ i = p, n \ . \quad (4.466)$$

In equation (4.466), c is the density of *uncharged dissociable* species. The *space charge density* ρ_e is related to n_p and n_n through

$$\rho_e = e_c(n_p - n_n) \ , \quad (4.467)$$

where e_c is the *charge of a single ion*. The electrical conductivity σ_e is related to these through

$$\sigma_e = e_c(n_p K_p + n_n K_n) \ . \quad (4.468)$$

Using σ_e and ρ_e in (4.71) results in the *charge conservation equation*

$$\frac{D\rho_e}{Dt} = -\nabla \cdot \sigma_e \mathbf{e} \ . \quad (4.469)$$

When the change in ϵ_e is negligible, then (2.124) is used to arrive at (2.131). Using (4.467) and (4.468) we have

$$n_p = \frac{\sigma_e + K_n \rho_e}{e_c(K_p + K_n)} \ . \quad (4.470)$$

Using order-of-magnitude analysis and combining (2.124) and the second of (2.125), we have

$$\rho_e = \frac{\epsilon_e \epsilon_o |\varphi|}{\ell^2} \ , \quad (4.471)$$

where ℓ is the *distance* between the two electrodes, ϵ_o is the *free* space permittivity (discussed in Section 2.6), and $|\varphi|$ is the *voltage* applied across the electrodes. Then the *ratio* of the *conduction current* to the *injection current* appearing in the second term in the numerator of (4.470) is designated as C_o and is

$$C_o = \frac{\sigma_e}{K_n \rho_e} = \frac{\sigma_e \ell^2}{\epsilon_e \epsilon_o K_n |\varphi|} \ . \quad (4.472)$$

As C_o increases, the charge generation due to a thermally induced gradient in the conductivity can become *more* prominent. Fujino et al. (1989)

suggest that for large ℓ (or C_o) the *conductivity mode* (or *conductive current*) dominates and for small ℓ the *mobility mode* (or *injection current*) dominates.

In this section, we review an *analysis* based on the *conduction mode* electrohydrodynamics and heat transfer and then examine an *experiment* designed for operating in the *mobility mode.*

(A) A Conduction-Mode Analysis

The *effectiveness* of the electrohydrodynamics in enhancing the interfacial heat transfer has been in the steady *low* velocity flows (although the problem of the *onset* of motion, i.e., the *onset* of electroconvection has also been studied). One problem that has been analyzed is that of *combined* thermobuoyant-electrohydrodynamics adjacent to a *heated, vertical isothermal* surface. Both dc (Turnbull, 1969, 1970a) and ac (Turnbull, 1970b) electric fields, applied to *liquids*, have been considered. Thermobuoyant electrohydrodynamics is also studied for the horizontal annuli (with a *nonuniform* dc field) filled with a gas (Lykoudis and Yu, 1963) and a review for the various geometries and fluids is given by Lazarenko et al. (1973).

The analysis of Turnbull (1969), for the boundary-layer thermobuoyancy electrohydrodynamics is reviewed below. Laminar, steady, two-dimensional flow with a Buossinesq fluid is assumed. The approximate integral-boundary layer method is used with *prescribed* distributions of the velocity, temperature, and electric potential. In the *boundary layer-integral* analysis of Turnbull (1969), the finite boundary-layer thickness δ is assumed to be the *same* for the three equalities. The coordinate along the vertical surface is x (with velocity component u) and points *opposite* to the gravity. The coordinate perpendicular to the surface is y (with velocity v) and the surface is *maintained* at a temperature T_s, while outside the boundary layer the conditions are uniform T_∞ and e_∞.

Assuming that $\rho = \rho(T)$, i.e., uniform pressure, $\sigma_e = \sigma(T)$, and the remaining thermophysical properties being constant, then (2.129) becomes

$$\rho_o \frac{D\mathbf{u}}{Dt} = -\nabla(p - p_\infty) + \mu\nabla^2\mathbf{u} - \rho_o\beta_o(T - T_\infty)\mathbf{g} + \rho_e\mathbf{e} \,. \qquad (4.473)$$

The variation in the electrical conductivity is assumed to be first order with respect to temperature. Then (1.144) can be written as

$$\sigma_e = \sigma_{e_o}[1 + a_1(T - T_\infty)] \,. \qquad (4.474)$$

The electric field $\mathbf{e}$ is *assumed* to be uniform outside the boundary layer and perpendicular to the surface, while being given by $-\nabla\varphi$ inside the boundary layer ($\varphi = 0$ outside the boundary layer). This can be written as

$$\mathbf{e} = \mathbf{s}_y e_\infty - \nabla\varphi \,, \qquad (4.475)$$

where $\mathbf{s}_y$ is the *unit vector* in the y direction.

The *boundary-layer approximations* of the electrohydrodynamics force $\rho_e \mathbf{e}$ is derived by Turnbull (1969) and the resulting momentum equation is

$$\rho_o(u\frac{\partial u}{\partial x} + v\frac{\partial u}{\partial y}) = \mu\frac{\partial^2 u}{\partial y^2} + \rho_o g\beta(T - T_\infty) +$$
$$\epsilon_e\epsilon_o(e_\infty - \frac{\partial\varphi}{\partial y})\frac{\partial^2\varphi}{\partial x\partial y} + \epsilon_e\epsilon_o\frac{\partial\varphi}{\partial x}\frac{\partial^2\varphi}{\partial y^2} \qquad (4.476)$$

The boundary-layer approximations of the electric potential is also derived by Turnbull (1969), where (2.124) to (2.126) are combined with (4.474) for σ_e and (4.475) for $\mathbf{e}$, and the result is

$$[1 + a_1(T - T_\infty)]\frac{\partial^2\varphi}{\partial y^2} - (e_\infty - \frac{\partial\varphi}{\partial y})a_1\frac{\partial T}{\partial y} +$$
$$\frac{\epsilon_e\epsilon_o}{\sigma_{e_o}}(u\frac{\partial^3\varphi}{\partial x\partial y^2} + v\frac{\partial^3\varphi}{\partial y^3}) = 0\,. \qquad (4.477)$$

The energy equation is accordingly approximated and is the same as (4.5), i.e.,

$$u\frac{\partial T}{\partial x} + v\frac{\partial T}{\partial y} = \alpha\frac{\partial^2 T}{\partial y^2}\,. \qquad (4.478)$$

Assuming a *finite* boundary-layer thickness δ, the boundary conditions are

$$u = v = \varphi = 0\,,\quad T = T_s \quad y = 0\,, \qquad (4.479)$$

$$u = v = \frac{\partial\varphi}{\partial y} = 0\,,\quad T = T_\infty \quad y = \delta\,. \qquad (4.480)$$

The momentum, energy, and electric potential equations (4.476), (4.478), and (4.477), respectively, can be integrated between $y = 0$ and $y = \delta$, and the results are

$$\rho_o\frac{d}{dx}\int_o^\delta u^2\,dy = \mu\frac{\partial u}{\partial y}|_o^\delta + \rho_o\beta_o g\int_o^\delta (T - T_\infty)\,dy +$$
$$\epsilon_e\epsilon_o\int_o^\delta [\frac{\partial\varphi}{\partial x}\frac{\partial^2\varphi}{\partial y^2} + (e_\infty - \frac{\partial\varphi}{\partial y})\frac{\partial^2\varphi}{\partial x\partial y}]\,dy\,, \qquad (4.481)$$

$$\frac{d}{dx}\int_o^\delta u(T - T_\infty)\,dy = \alpha\frac{\partial T}{\partial y}|_o^\delta\,, \qquad (4.482)$$

$$\frac{\epsilon_e\epsilon_o}{\sigma_{e_o}}\frac{d}{dx}\int_o^\delta u\frac{\partial^2\varphi}{\partial y^2}\,dy + \frac{\partial\varphi}{\partial y}|_o^\delta + a_1\rho_\infty(T_s - T_\infty) + a_1(T - T_\infty)\frac{\partial\varphi}{\partial y}|_o^\delta\,. \qquad (4.483)$$

The *assumed* profiles are

$$u = u_o\frac{y}{\delta}(1 - \frac{y}{\delta})^2\,, \qquad (4.484)$$

$$\frac{T - T_\infty}{T_s - T_\infty} = (1 - \frac{y}{\delta})^2\,, \qquad (4.485)$$

$$\phi = a_e y(1 - \frac{y}{\delta} + \frac{y^2}{3\delta^2})e_\infty \,. \tag{4.486}$$

The equations are made dimensionless using the *scales*

$$\delta^* = \frac{\delta}{[\frac{\mu_o^2}{\rho_o^2\beta_o g(T_s - T_\infty)}]^{1/3}} \,, \qquad x^* = \frac{x}{[\frac{\mu_o^2}{\rho_o^2\beta_o g(T_s - T_\infty)}]^{1/3}} \,, \tag{4.487}$$

$$u_o^* = \frac{u_o}{[\frac{\mu_o\beta_o g(T_s - T_\infty)}{\rho_o}]^{1/3}} \,. \tag{4.488}$$

The resulting *dimensionless* parameters are

$$\tau^* = \frac{\epsilon_e\epsilon_o}{\sigma_{e_o}[\frac{\mu_o}{\rho_o\beta_o^2 g^2(T_s - T_\infty)^2}]^{1/3}} \,, \tag{4.489}$$

$$a_{e_o} = \frac{a_1(T_s - T_\infty)}{1 + a_1(T_s - T_\infty)} \,, \tag{4.490}$$

$$El = \frac{\epsilon_e\epsilon_o e_\infty^2}{\rho_o^{1/3}[\mu_o\beta_o g(T_s - T_\infty)]^{2/3}} \,, \tag{4.491}$$

where τ^* is the *normalized electrical relaxation time*, a_{e_o} is the *instantaneous* value for a_e, and El is the *normalized electric field force* (with respect to buoyancy-viscous force). Then El is a measure of the *relative strength* of the electric and the thermobuoyant forces acting on the fluid. The viscous force tends to oppose both of these forces.

After substitutions of (4.484) to (4.491) in (4.481) to (4.483), the dimensionless, y-integrated momentum, energy, and electric potential equations are

$$\frac{1}{105}\frac{d}{dx^*}u_o^{*2}\delta^* = \frac{\delta^*}{3} - \frac{u_o^*}{\delta^*} + El(\frac{1}{3}\frac{d}{dx^*}a_e\delta^* - \frac{1}{5}\frac{d}{dx^*}a_e^2\delta^*) \,, \tag{4.492}$$

$$\frac{1}{30}\frac{d}{dx^*}u_o^*\delta^* = \frac{2}{Pr\delta^*} \,, \tag{4.493}$$

$$\frac{\tau^*}{10}\frac{d}{dx^*}a_e u_e^* + \frac{a_e - a_{e_o}}{1 - a_{e_o}} = 0 \,. \tag{4.494}$$

These equations are solved for u_o^*, δ^*, and a_e, where all of these quantities *vanish* at the surface.

The local heat transfer can be given in terms of the local Nusselt number and is

$$Nu_x = \frac{q_s x}{(T_s - T_\infty)k} = -\frac{\partial T}{\partial y}|_o\frac{x}{T_s - T_\infty} = \frac{2x}{\delta} = \frac{2x^*}{\delta^*} \,. \tag{4.495}$$

Near the *leading edge*, approximations can be made (Turnbull, 1969) and the local Nusselt number is approximated by

$$Nu_x = 0.545 Pr^{1/2} El^{1/6} a_{e_o}{}^{1/6}[(1 - a_{e_o})^{-1/6} \,.$$

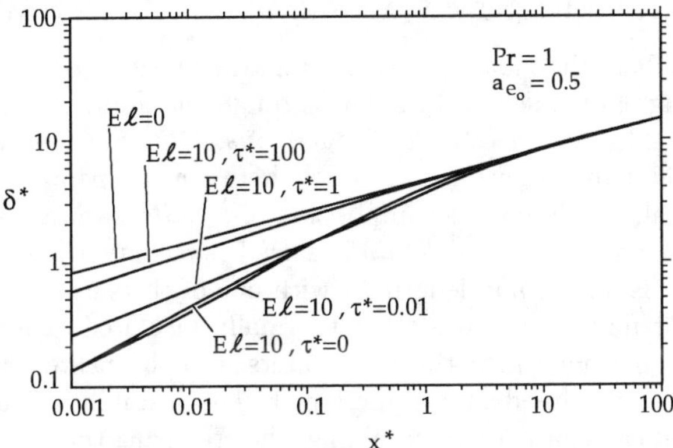

Figure 4.40. Variation of the dimensionless boundary-layer thickness with respect to the dimensionless distance along the surface for the thermobuoyant flow adjacent to an isothermal, vertical plate and effected by an electric field. The results are for various values of the normalized electric force $E\ell$ and normalized electrical relaxation time τ^*. (From Turnbull, reproduced by permission ©1969 American Institute of Physics.)

$$(Pr + \tfrac{6}{7})^{-1/6} \tau^{*\,-1/6} Gr_x^{2/9}] \qquad x \to 0 \, , \tag{4.496}$$

where the Grashof number is as before, i.e.,

$$Gr_x = \frac{g\beta_o (T_s - T_\infty) x^3}{\nu_o^2} \, . \tag{4.497}$$

The numerical results of Turnbull (1969) for $\delta^* = \delta^*(x^*)$ for $Pr = 1$, $a_{e_o} = 0.5$, and various values of $E\ell$ and τ^*, are given in Figure 4.40. The case of $E\ell = 0$ corresponds to the purely thermobuoyant flow and heat transfer. The application of an electric field *reduces* the boundary layer thickness near the leading edge (up to $x^* \simeq 1$), and as $E\ell$ *increases* this become *more* pronounced. The dimensionless electrical relaxation time τ^* *also* influences the heat transfer rate (which is proportional to $1/\delta$). The larger the relaxation time the *less* effective the electrical field becomes. Note that the larger the electrical relaxation time, the longer it takes for the change density to *form*.

The instability of this boundary-layer flow, for strong electric fields, is analyzed by Turnbull (1970a). The instability occurs when the combined thermobuoyancy and electric force can *substantially* overcome the viscous forces such that the disturbances flowing downstream can be amplified. This is similar to the instability studied in Section 4.12.1(C). The threshold voltage for the instability is determined and compared with the experimental results with good agreement. The increase in the heat transfer becomes *substantial* when the boundary layer becomes *unstable*.

(B) An Experiment on Mobility-Mode Instability

As an example of the *mobility mode*, where the injection of charge through the bounding electrodes (and its flow through the fluid) can influence the motion, consider the *electrically induced instability* (i.e., onset of a secondary or cellular motion) in forced flow between *two parallel plates*.

Fujino et al. (1989) consider imposition of a *uniform direct-current* (dc) electric field on an initially *laminar* flow between two parallel surfaces (separation distance ℓ and length L) with one of the surfaces heated with uniform heat flux q_s and the other thermally insulated. There is an unheated section from which the fluid enters into the heated section with the velocity field already fully developed. The instability caused by the imposed field can cause a *lateral* mixing, thus reducing the surface temperature T_s. Fujino et al. (1989) add a small amount of an *electrolyte* solution (mostly xylene) to *increase* the electrical conductivity of the refrigerant fluid (Freon R-113, i.e., trichlorotrifluoroethane) σ_e form 10^{-10} mho/m to 10^{-9} mho/m. Their experiment is rendered in Figure 4.41(a), where the plates are parallel to the gravity vector with the fluid flowing opposite to the direction of gravity and at an area-averaged velocity $\langle u\rangle_A$. The heated plate is electrically heated (with a uniform heat flux q_s entering the fluid). The other plate is thermally insulated, and the two plates make up the positive and the negative electrodes of a uniform electric field $e = \Delta\varphi/\ell$ created by applying a voltage $\Delta\varphi$ across the fluid layer. The *direction* of the electric field can be changed such that the heated surface is either the positive or the negative electrode. The *velocity-weighted area-averaged* fluid temperature (also called the *bulk mixed fluid* temperature) averaged over the flow cross-section A is defined as

$$\langle T_f\rangle_A^b = \frac{1}{A\langle u\rangle_A}\int_A uT_f\, dA\,. \tag{4.498}$$

This temperature increases linearly with x in this constant surface heat flux problem. The flow at $x = 0$ is assumed to be the one-dimensional, *laminar*, Hagen-Poiseuille flow. The start of heating at $x = 0$ results in formation of a thermal boundary layer, and therefore, the *surface* temperature T_s does *not* increase linearly with x in the *thermally developing* region. Shown in Figure 4.41(b) are the predicted $\langle T_f\rangle_A^b = \langle T_f\rangle_A^b(x)$ and $T_s = T_s(x)$, *computed* for a two-dimensional temperature field subject to a constant q_s on one surface, and ideal insulation on the other surface.

Two types of electrodes are used by Fujino et al (1989), one is a Pyrex glass coated by lead oxide, and the other is a Bakelite coated by nickel. The Reynolds number $Re_\ell = \langle u\rangle_A\ell/\nu$ of 191 (laminar) with $\ell = 2$ mm is used. The results shown in Figure 4.41(b) are for the glass plates and show that when a strong *lateral* electric field is applied (positive or negative, although application of a *negative* voltage resulted in a *more* substantial surface temperature lowering), the surface temperature *decreases*, indicating *lateral mixing*. The lateral mixing is uniform along the plate. Their shadow

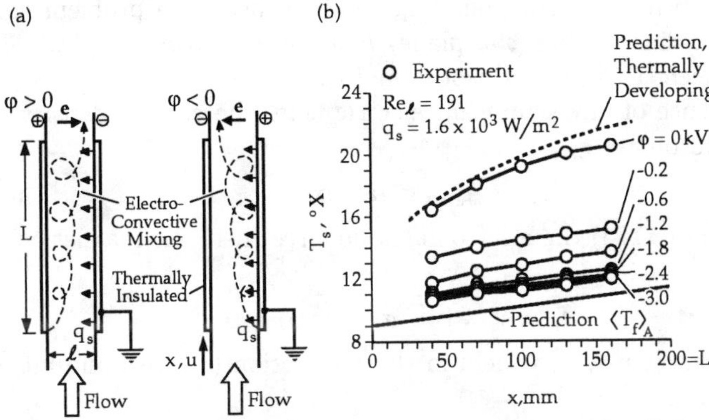

Figure 4.41. (a) A schematic of the upward channel flow of a dielectric liquid with one surface heated and an electric field applied across the channel. (b) The variation of the heated surface temperature with respect to the applied electric field (or voltage). (From Fujino et al., reproduced by permission ©1989 American Society of Mechanical Engineers.)

graph visualization shows that the mixing initiates at the *positively* charged surface, thus when a negative voltage is applied to the insulated surface, the mixing, and therefore, the surface temperature at the heated surface, drops more noticeably. Fujino et al. (1989) conclude that the *mechanism* for mixing is the *injection* of charge from either or both electrodes (as compared to the change in the electric conductivity by temperature variation). The complete theoretical treatment of this mobility-mode instability is *not* yet available.

4.13.2 MAGNETOHYDRODYNAMICS

Magnetohydrodynamics addresses the influence of an imposed *magnetic field* **b** on a *moving*, *electrically conducting* fluid (discussed in Section 2.6.2). The nonisothermal flows are encountered, for example, when high-temperature charged gases or high-temperature liquid metals are used for electrical power generation. The *Joule heating*, $\mathbf{j}_e \cdot \mathbf{j}_e/\sigma_e$, also contributes to the temperature nonuniformity. The flow is generally *turbulent*, and the analysis of the fluid flow and heat transfer requires *modeling* of the electric and magnetic fields and the turbulence (Branover et al., 1985; Bouillard and Berry, 1992). For gases, the *Hall current* and the *ion slip* adjacent to the solid, conducting surface must be addressed (Mittal et al., 1987). As with any high temperature applications, the *variable properties* must also be addressed. As we have discussed in relation with the reactive flows, these (including radiation) can be appropriately modeled. Here we consider a rather *simple* interfacial heat transfer in order to demonstrate the effect of an imposed magnetic field on a moving, conducting fluid not in thermal equilibrium with a bounding solid surface. The problem considered is the two-dimensional (x-y plane) *thermobuoyant-magnetohydrodynamics* flow adjacent to a *vertical* isothermal surface.

In absence of any significant electric field, the current density given by (2.137) becomes

$$\mathbf{j}_e = \sigma_e \mathbf{u} \times \mathbf{b} \,. \tag{4.499}$$

Then from (2.133) the electromagnetic force (here the magnetic force) becomes

$$\mathbf{f}_e = \mathbf{j}_e \times \mathbf{b} = \sigma_e (\mathbf{u} \times \mathbf{b}) \times \mathbf{b} \,. \tag{4.500}$$

For a *uniform* magnetic field in the z direction (z-direction unit vector is $\mathbf{s}_z$), we have

$$\mathbf{b} = \mathbf{s}_z b_o \,. \tag{4.501}$$

Then the *magnetic force* becomes

$$f_{e_x} = -\sigma_e u b_o^2 \,. \tag{4.502}$$

The effect of this *volumetric* force f_{e_x} on the thermobuoyant flow adjacent to a vertical surface (x direction being along the surface and y perpendicular to it and pointing toward the fluid) has been studied (Sparrow and Cess, 1961; Lykoudis, 1962; Cramer, 1963; Romig, 1964) and a general treatment for these boundary-layer flows is given by Rossow (1957). The volumetric force f_{e_x} is a *retarding* (or *decelerating*) force with a hindering strength proportional to the local velocity, i.e., the fluid particles which are flowing at higher velocities experience larger resistance to their flow according to $-\sigma_e u b_o^2$. The analysis of Sparrow and Cess (1961) is reviewed here to demonstrate the extent to which the interfacial heat transfer is *reduced* due to the force f_{e_x}.

The boundary-layer approximations of the thermobuoyant-magnetohydrodynamics flow and heat transfer (with no Joule heating) leads to

$$\frac{\partial u}{\partial x} + \frac{\partial v}{\partial y} = 0 \,, \tag{4.503}$$

$$u\frac{\partial u}{\partial x} + v\frac{\partial u}{\partial y} = \nu\frac{\partial^2 u}{\partial y^2} + g\beta(T - T_\infty) - \sigma_e u b_o^2 \,, \tag{4.504}$$

$$u\frac{\partial T}{\partial x} + v\frac{\partial T}{\partial y} = \alpha\frac{\partial^2 T}{\partial y^2} \,. \tag{4.505}$$

Following the treatment of Section 4.12.1(A), the similarity variable and normalized streamfunction and temperature are introduced as

$$\eta = \frac{y}{x}\left(\frac{Gr_x}{4}\right)^{1/4} , \quad f = \frac{\psi}{4\nu (Gr_x/4)^{1/4}} , \quad T^* = \frac{T - T_\infty}{T_s - T_\infty} \,. \tag{4.506}$$

In addition, the dimensionless parameters are

$$Ly_x = \frac{{Ha_x}^2}{(Gr_x/4)^{1/2}} = \frac{2\sigma_e}{\rho[g\beta(T_s - T_\infty)]^{1/2}} b_o x^{1/2} \,, \tag{4.507}$$

$$Ha_x{}^2 = \frac{\sigma_e b_o^2 x^2}{\mu} \, , \quad Gr_x = \frac{g\beta(T_s - T_\infty)x^3}{\nu^2} \, , \quad Pr = \frac{\nu}{\alpha} \, , \tag{4.508}$$

where Ha_x is the *Hartmann* number, and Ly_x is the *Lykoudis* number (Romig, 1964). The Hartmann number is the ratio of the *magnetic force* to the *viscous force* and the Lykoudis number is the ratio of the *magnetic force* to the *buoyancy force*. Note that for b_o varying with $x^{-1/2}$, the Lykoudis number becomes a *constant*, and therefore, a *similarity solution* can be found. This similarity solution is discussed by Lykoudis (1962). For a *variable* Ly_x, the y-direction velocity would depend on Ly_x. Then in addition to u given by (4.329), we have

$$v = \frac{\nu}{x}(\frac{Gr_x}{4})^{1/2}(\eta\frac{df}{d\eta} - 3f + 2Ly_x\frac{\partial f}{\partial Ly_x}) \, . \tag{4.509}$$

The transformed momentum and energy equations become (Sparrow and Cess, 1961)

$$\frac{\partial^3 f}{\partial \eta^3} + 3f\frac{\partial^2 f}{\partial \eta^2} - 2(\frac{\partial f}{\partial \eta})^2 - 2Ly_x(\frac{\partial^2 f}{\partial Ly_x \partial \eta}\frac{\partial f}{\partial \eta} - \frac{\partial f}{\partial Ly_x}\frac{\partial^2 f}{\partial \eta^2} + \frac{1}{2}\frac{\partial f}{\partial \eta}) + T^* = 0 \, , \tag{4.510}$$

$$\frac{\partial^2 T^*}{\partial \eta^2} + 3Prf\frac{\partial T^*}{\partial \eta} - 2Ly_x(\frac{\partial f}{\partial \eta}\frac{\partial T^*}{\partial Ly_x} - \frac{\partial T^*}{\partial \eta}\frac{\partial f}{\partial Ly_x}) = 0 \, . \tag{4.511}$$

Sparrow and Cess (1961) use *regular* expansions with the *perturbation parameter* Ly_x, i.e.,

$$f = f_o(\eta) + Ly_x f_1(\eta) + \cdots \, , \tag{4.512}$$

$$T^* = T_o^*(\eta) + Ly_x T_1^*(\eta) + \cdots \, . \tag{4.513}$$

The substitutions into (4.510) and (4.511) give the zeroth-order equations which are identical to both (4.330) and (4.331) and also to (4.348) and (4.349). The first-order equations are

$$\frac{d^3 f_1}{d\eta^3} + 3f_o\frac{d^2 f_1}{d\eta^2} - 6\frac{df_o}{d\eta}\frac{df_1}{d\eta} + 5\frac{d^2 f_o}{d\eta^2}f_1 - \frac{df_o}{d\eta} + T_o^* = 0 \, , \tag{4.514}$$

$$\frac{d^2 T_1^*}{d\eta^2} + 3Prf_o\frac{dT_1^*}{d\eta} - 2Pr\frac{df_o}{d\eta}T_1^* + 6Prf_1\frac{dT_o^*}{d\eta} = 0 \, . \tag{4.515}$$

The boundary conditions are

$$f_o = \frac{df_o}{d\eta} = f_1 = \frac{df_1}{d\eta} = T_1^* = 0 \, , \; T_o^* = 1 \quad \eta = 0 \, , \tag{4.516}$$

$$\frac{df_o}{d\eta} = \frac{df_1}{d\eta} = T_o^* = T_1^* = 0 \quad \eta \to \infty \, . \tag{4.517}$$

Numerical solutions to (4.330), (4.331), (4.514), and (4.515) have been found by Sparrow and Cess (1961) for various Prandtl numbers. The local

Nusselt number is

$$Nu_x = -(\frac{Gr_x}{4})^{1/4}\frac{dT^*}{d\eta}|_o \,. \tag{4.518}$$

The ratio of the local Nusselt number with a magnetic field present to when no magnetic is present is given by

$$\frac{Nu_x}{Nu_x(Ly_x=0)} = 1 + \frac{\dfrac{dT_1^*}{d\eta}|_o}{\dfrac{dT_o^*}{d\eta}|_o} Ly_x + \cdots \,. \tag{4.519}$$

The ratio $(dT_1^*/d\eta)_o/(dT_o^*/d\eta)_o$ is *negative* for all Pr, indicating a *decrease* in the interfacial heat transfer rate. Numerical examples are given by Sparrow and Cess (1961).

For *forced flows*, the *retarding* force f_{e_x} also makes the velocity field *uniform* (because f_{e_x} is proportional to u). Note that this volumetric retarding force is *similar* (i.e., *analogous*) to the volume retarding force due to presence of a solid matrix. This bulk viscous force in flow through porous media was discussed in Section 3.2.4, and for a velocity u in the x direction the role of $\sigma_e b_o^2$ is equivalent to μ/K, when K is the permeability. This can be realized by comparing (4.504) with (3.75). Then the solution for forced flow and heat transfer in porous media (and for the case where $C_E = 0$ and $\epsilon \rightarrow 1$) is applicable to the magnetohydrodynamics and heat transfer. An example of a forced boundary-layer flow and heat transfer in the presence of a solid matrix is given by Kaviany (1987) and an example of a channel flow is given by Kaviany (1985). These are discussed in Section 4.15.

4.13.3 Plasmas

Plasmas, and their convective heat transfer aspects, are encountered in welding (Lancaster, 1986), materials removing, vapor-phase synthesis of ultrafine particles (Joshi et al., 1990), and in the deposition of diamond thin films (Stadler and Sharpless, 1990; Ramesham and Ellis, 1992; Tsai et al., 1992). The plasma may be heated by *Joule heating* of an imposed magnetic field or an electric arc, by chemical reaction, or by a combination of these. The interfacial convective heat transfer aspects of plasmas concerns the heat transfer to *surfaces bounding* the plasma. In some applications, due to the high temperatures in the plasma, the surfaces are *externally cooled* to avoid meltdown (while this heat transfer, which is an undesirable loss, is minimized). In other applications, the heat transfer causes sublimation or melting/evaporation of the surface material.

Here we first consider the fluid flow, heat transfer, and the electromagnetic aspects of the *magnetically heated* plasma flowing in a tube (this is a typical induction-heated flowing plasma). The role of the electromagnetic forces acting on the fluid and the Joule heating on the fluid flow and heat transfer will be discussed through the examination of the governing momentum, energy, and electromagnetic field equations. Then some aspects of the *arc plasmas* and the occurrence of chemical reactions in *reactive plasma* are mentioned.

(A) Electromagnetically Heated Plasmas

High-frequency, electromagnetic heated gases and the resulting plasmas have many applications including thin-film deposition and synthesis. The plasmas generally flow, and therefore, in additions to the Joule heating, the force acting on the plasma must also be included and significantly influences the flow field. This *coupled* volumetric heating and magnetohydrodynamics problem, because of the high gas temperatures, is examined here as a convective heat transfer problem involving plasmas. From the convective heat transfer perspective, the objective may be that of *raising* the temperature of a flowing gas to temperatures in access of 10^4 K, while *minimizing* the heat losses to the bounding surfaces and maintaining a steady and stable flow. One example is the *axisymmetric* magnetically heated plasma (called a plasma *torch*), where the magnetic field is due to an imposed current flowing through the *excitation coils*. The *alternating current* passing through these coils generally has high frequencies (radio-frequency range). Then the *time-averaged* effect of the electromagnetic forces (also called the Lorentz forces) and the Joule heating, over a period of excitation, is of interest. Chen (1990) considers this problem and his analysis of the magnetohydrodynamics and heat transfer is reviewed below.

A schematic of the problem considered is given in Figure 4.42. The gas is flown with temperature T_i through an inlet where the *nonuniform* inlet flow is controlled by allowing for a *cascade* of annuli. Here *three* inlet ports are shown and the *innermost* flow is given by an average inlet velocity u_1, the *intermediate* annulus has an average velocity u_2, and the *outermost* annulus has an average velocity u_3. The excitation coils are placed downstream from the inlet starting at a distance $x = \ell_1$, and extending to $x = \ell_2$. The downstream extent used in the computational domain is L and the flowing gas *volume* flowing between $0 \le x \le L$ and $0 \le r \le R_s$ is of interest.

The *time-averaged* governing continuity, momentum, and energy equations are

$$\frac{\partial}{\partial x}\rho u + \frac{1}{r}\frac{\partial}{\partial r} r\rho v = 0 \,, \tag{4.520}$$

$$\begin{aligned} \rho u\frac{\partial u}{\partial x} + \rho v\frac{\partial u}{\partial r} &= -\frac{\partial p}{\partial x} + 2\frac{\partial}{\partial x}\mu\frac{\partial u}{\partial x} + \\ &\frac{1}{r}\frac{\partial}{\partial r} r\mu(\frac{\partial u}{\partial x} + \frac{\partial v}{\partial x}) + \overline{f}_{e_x} - \rho g \,, \end{aligned} \tag{4.521}$$

$$\rho u\frac{\partial v}{\partial x} + \rho v\frac{\partial v}{\partial r} = -\frac{\partial p}{\partial r} + \frac{\partial}{\partial x}\mu(\frac{\partial v}{\partial x} + \frac{\partial u}{\partial r}) +$$

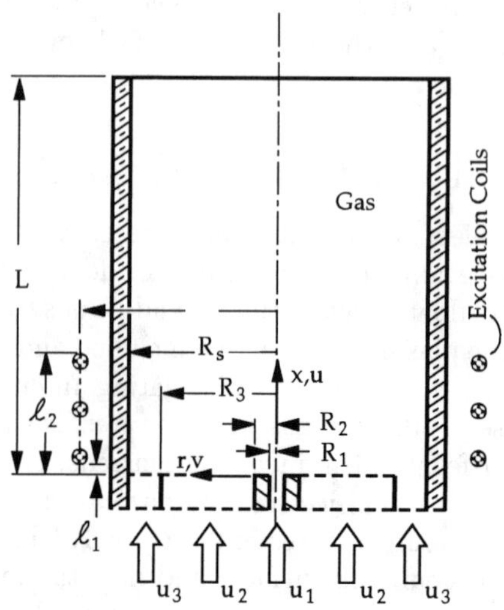

Figure 4.42. A schematic of the electromagnetically heated plasma showing the gas inlet flow distribution through three cascading annuli, the excitation coils, and the cooled, confining cylinder. (From Chen, reproduced by permission ©1990 Pergamon Press.)

$$\frac{2}{r}\frac{\partial}{\partial r}r\mu\frac{\partial v}{\partial r} - \frac{2\mu v}{r^2} + \overline{f}_{e_r} , \tag{4.522}$$

$$\rho u\frac{\partial i}{\partial x} + \rho v\frac{\partial i}{\partial r} = \frac{\partial}{\partial x}\frac{k}{c_p}\frac{\partial i}{\partial x} + \frac{1}{r}\frac{\partial}{\partial r}\frac{rk}{c_p}\frac{\partial i}{\partial r} - \nabla \cdot q_r + \sigma_e\overline{\mathbf{e}\cdot\mathbf{e}} , \tag{4.523}$$

where the volumetric forces $\overline{f}_{e_x}$ and $\overline{f}_{e_r}$ and the volumetric Joule heating $\sigma_e\overline{\mathbf{e}\cdot\mathbf{e}}$ are *time averaged* over one period of the oscillation of the coil excitation.

Assuming no free charge density ρ_e, and that the displacement current $\mathbf{d}$ and the plasma flow induced field $\mathbf{u}\times\mathbf{b}$ are negligible, the Maxwell equations (2.124), (2.136), (2.135), (2.134), and (2.127), become, respectively,

$$\nabla\cdot\mathbf{e} = 0 , \tag{4.524}$$

$$\nabla\cdot\mathbf{b} = 0 , \quad \mathbf{b} = \nabla\times\mathbf{a} , \tag{4.525}$$

$$\nabla\times\mathbf{e} = -\frac{\partial\mathbf{b}}{\partial t} , \tag{4.526}$$

$$\nabla\times\mathbf{b} = \mu_e\mathbf{j}_e , \tag{4.527}$$

$$\mathbf{j}_e = \sigma\mathbf{e} . \tag{4.528}$$

In (4.525) $\mathbf{a}$ is the *vector* (*magnetic*) *potential* and, when working with a magnetic field, has the convenience of satisfying $\nabla\cdot\mathbf{b} = 0$, the same way that the electric potential φ satisfies $\nabla \times \mathbf{e} = 0$ for electric fields. When the electrostatic effects are neglected and $\nabla \cdot \mathbf{a} = 0$ is used, (4.527) and (4.526) lead, respectively, to

$$\nabla^2 \mathbf{a} = -\mu_e \mathbf{j}_e \ , \tag{4.529}$$

$$\mathbf{e} = -\frac{\partial}{\partial t}\mathbf{a} \ . \tag{4.530}$$

For *axisymmetric* coils, only *azimuthal* components of **a**, **e** and $\mathbf{j}_e$ are present. Then (4.529), (4.530), and (4.528) reduce, respectively, to

$$\frac{1}{r}\frac{\partial}{\partial r} r \frac{\partial A_\theta}{\partial r} + \frac{\partial^2 A_\theta}{\partial x^2} - \frac{A_\theta}{r} = -\mu_e j_{e_\theta} \ , \tag{4.531}$$

$$e_\theta = -\frac{\partial A_\theta}{\partial t} \ , \tag{4.532}$$

$$j_{e_\theta} = \sigma_e e_\theta \ . \tag{4.533}$$

Time-periodic variations of the current in the coils I_c and A_θ are assumed with an angular frequency ω and amplitudes I_{c_o}, a_{θ_1} and a_{θ_2}, i.e.,

$$I_c = I_{c_o} \cos \omega t \ , \tag{4.534}$$

$$A_\theta = a_{\theta_1} \cos \omega t + a_{\theta_2} \sin \omega t \ , \tag{4.535}$$

where for A_θ both in- and out-of-phase components with respect to I_c are assumed (due to the presence of the plasma current j_{e_θ}). Then e_θ, b_r and b_θ are determined from (4.532) and (4.525) as

$$e_\theta = \omega(a_{\theta_1} \sin \omega t - a_{\theta_2} \cos \omega t) \ , \tag{4.536}$$

$$b_r = -\frac{\partial}{\partial x} a_{\theta_1} \cos \omega t - \frac{\partial}{\partial x} a_{\theta_2} \sin \omega t \ , \tag{4.537}$$

$$b_x = \frac{1}{r}\frac{\partial}{\partial r} r\, a_{\theta_1} \cos \omega t + \frac{1}{r}\frac{\partial}{\partial r} a_{\theta_2} \sin \omega t \ . \tag{4.538}$$

The components of the *electromagnetic* force $\overline{f}_{e_x}$ and $\overline{f}_{e_r}$ are found from $\overline{f}_e = \overline{\mathbf{j}_e \times \mathbf{b}}$.

The *period-averaged* (i.e., over $2\pi/\omega$) axial and radial volumetric forces $\overline{f}_{e_x}$ and $\overline{f}_{e_r}$, and the volumetric Joule heating are related to a_{θ_1} and a_{θ_2} through

$$\overline{f}_{e_x} = \frac{\sigma_e \omega}{2}\left(a_{\theta_1} \frac{\partial}{\partial x} a_{\theta_2} - a_{\theta_2} \frac{\partial}{\partial x} a_{\theta_1}\right) \ , \tag{4.539}$$

$$\overline{f}_{e_r} = \frac{\sigma_e \omega}{2r}\left(a_{\theta_1} \frac{\partial}{\partial r} r\, a_{\theta_2} - a_{\theta_2} \frac{\partial}{\partial r} r\, a_{\theta_1}\right) \ , \tag{4.540}$$

$$\sigma_e \overline{\mathbf{e} \cdot \mathbf{e}} = \frac{\sigma \omega^2}{2}(a_{\theta_1}^2 + a_{\theta_2}^2) \ . \tag{4.541}$$

The boundary conditions for (4.531) are discussed by Chen (1990), and some useful simplifications, without a significant loss of accuracy, are also

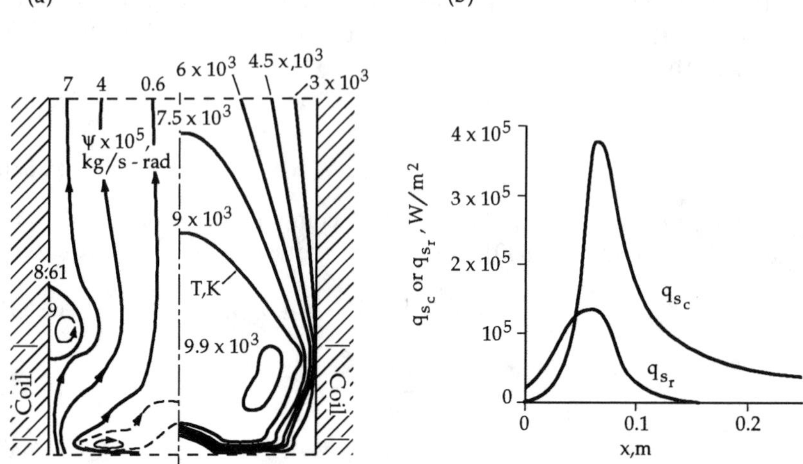

Figure 4.43. (a) The constant stream function and constant temperature contours for the electromagnetically heated plasma shown in Figure 4.42. (b) Distribution of the convective and radiative heat flux on the wall of the confined, heated plasma. (From Chen, reproduced by permission ©1990 Pergamon Press.)

recommended. The *plasma power* (i.e., the volume-averaged Joule heating rate) for the heated gas volume is

$$Q = \int_o^L \int_o^{R_s} \sigma_e \overline{\mathbf{e} \cdot \mathbf{e}}\ 2\pi r\, dr\, dx\ , \tag{4.542}$$

where R_s and L are shown in Figure 4.42.

The numerical results of Chen (1990) for the streamline (integrated using the velocity field) and the temperature distributions are given in Figure 4.43(a). The gas used is *argon* with inlet temperature $T_i = 350$ K, and $\omega/2\pi = 3 \times 10^6$ Hz and the flow and physical dimensions are given by Chen (1990). The plasma power Q is 7×10^3 W. Note that two *secondary flows* (*vortex rings*) are present one near the inlet and radially centered, and one further downstream and next to the bounding tube surface. Temperatures as high at 10^4 K are obtained with this power, geometry, and flow conditions. The surface is generally *actively* cooled by a coolant to prevent meltdown, Chen (1990) applies conductive-convective boundary conditions to the outside of the bounding surface.

The volumetric radiation heat transfer term $\nabla \cdot \mathbf{q}_r$ is *modeled* as *emission* by the *gas* and *absorption* by the *bounding surfaces*. Based on the model used by Chen (1990), the surface *radiation* heat flux q_{r_s} is *less* than the *convective* heat flux q_{r_s} and the computed axial distribution of these fluxes are shown in Figure 4.43(b). For the conditions used in the model computation of Chen (1990), the *fraction* of the total power removed from the surfaces by *convection* is 0.60, that by *radiation* to the surface is 0.22, and the remainder 0.18 *leaves* with the plasma at the exit.

The realistic simulations of plasmas should include *variable* properties and *multiple* species. Mohandi and Gilligan (1990) consider the electrical conductivity and thermodynamic functions for *weakly nonideal* plasmas. The *multicomponent* studies include that of Girshick and Yu (1990), where atmospheric pressure, radio frequency induction plasmas with mixtures of argon, hydrogen, nitrogen, and oxygen are considered. Due to the difference in the thermophysical (including electrical) properties, the *plasma core* (i.e., the *high-temperature core*) size can vary drastically depending on the gas species used.

(B) Arc Plasmas

The fluid flow and heat transfer of arc plasmas between and around an electrode (such as in arc welding) is *affected* by an additional mechanism. This is the *flow* of *electrons* (e.g., welding current) between the anode (positively charged) surface and the cathode (negatively charged) surface. The flow of the *heavier species* (i.e., those other than electrons, which are ions and neutral atoms of the *shielding* gas and other species) is affected by the electromagnetic force acting on the electron-heavier particles. Tsai and Kou (1990) examine the fluid flow and heat transfer in an axisymmetric cathode geometry with a shield gas flowing *parallel* to the cathode and *impinging* on the anode surface. Figure 4.44(a) shows a schematic of the problem. Typical *computed* velocity and temperature distributions are shown in Figures 4.44(b) and (c). In this *simulation*, the anode is water cooled to *prevent* melting and temperatures as high as 2.1×10^4 K are attained by the Joule heating with a current of 200 A. The coupled momentum-energy-electromagnetic equations are solved simultaneously using a *transformed* orthogonal curvilinear coordinate. The *enthalpy* of the electron, generally moving at much *higher* velocities, is accounted for *separately* using the result of the kinetic theory. However, local thermal equilibrium is assumed between the electrons and the heavier particles. The radiation is modeled using the thin-gas approximations. The shield gas is argon and the argon plasma properties, including ρ, c_p, μ, k, and σ_e, are all functions of *temperature*. The arc can be separated into the *cathode region* (which is small), *arc plasma* (which is the mainbody and is about 1-3 mm) and the *anode region* (which is also small). In general, in the welding applications the liquid motion in the *weld pool* (the *melted anode region*) should be simultaneously evaluated.

As evident in Figure 4.44(b), the fluid flow between the cathode and anode is similar to an impinging jet and the velocity at the centerline (the maximum velocity) reaches 206 m/s. The pressure in the stagnation area is 753 N/m^2 above the prescribed far-field pressure of 1 atm (1.103×10^5 N/m^2). The cathode tip can be *optimized* for the desired flow field and the desired anode surface pressure distribution. Figure 4.44(c) shows that the *predicted* and *measured* temperature distributions are in good

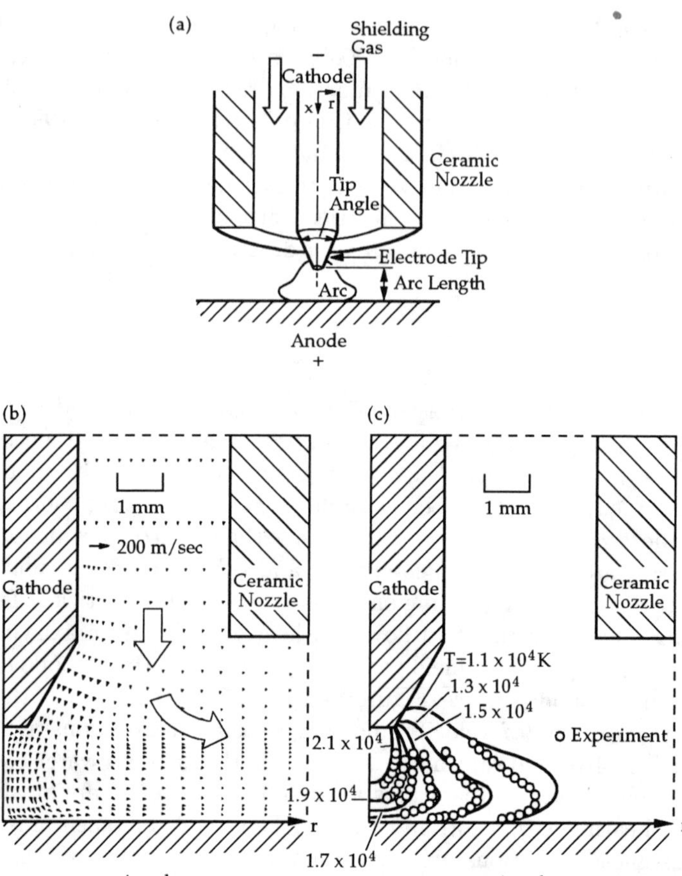

Figure 4.44. (a) A schematic of the arc formed between a cathode and an anode with a shield gas flowing around the arc. (b) Distribution of the gas velocity in and around the arc. (c) Distribution of the temperature. (From Tsai and Kou, reproduced by permission ©1990 Pergamon Press.)

agreement considering the extent of the variation in properties and the idealizations made in the inclusion of the radiation and in modeling the boundary conditions (details are given by Tsai and Kou (1990).

When the phase change (sublimation or melting/evaporation) of the anode is addressed, the flow of the *anode species* into the arc and its influence on the fluid flow and heat transfer must be included. The evaporation of a copper anode and the flow of the copper vapor in the arc as well as its effects on the temperature distribution have been addressed by Zhao et al. (1990). They show that only the anode region is noticeably affected by the evaporation. The effect of the plasma jet on the combined jet shear-thermocapillary-thermobouancy motion of the moving, melted anode in plasma-arc welding has been examined by Keanini and Rubinsky (1993). The plasma jet penetrates the melt pool, and when the anode is thin a hole is created and is called the key hole (Lancaster, 1986).

(C) Chemical Reaction

One of the applications of magnetically heated plasmas is in thin-film deposition, and this may involve chemical reactions in the plasma. Then in addition to fluid flow, heat transfer (including Joule heating and radiation), and electromagnetic fields, the *chemical kinetics*, *nucleation* (both on the surface and in the plasma), and *particle growth* must also be considered. The heat release/absorbed as a result of the reactions must also be included in the energy analysis. An example of *numerical* simulation which includes the *transport of species*, multistep reactions, and zeroth-order chemical kinetics is given by Zhao et al. (1990). In order to reduce the computational effort, fast reactions (chemical equilibrium model) are used and compared with the temperature-controlled reaction (zeroth order kinetic) results, and for the reactions considered, no significant difference was found.

4.14 Particle Suspension

Under *some* conditions, the addition of particles to the fluid flowing over surfaces *increases* the interfacial heat transfer. This depends on the *volume* and *mass fractions* of the particles, the *viscosity* of the fluid, the *charge* on the particles (when an electric field is present), *particle size*, other fluid and solid *thermophysical properties*, and the *solid surface geometry*. Here we treat the fluid-particle system as a *single*, *effective* medium with the fluid and particles in the local thermal equilibrium with each other, but not with the bounding surfaces. In Section 3.3 the local volume-averaged behavior of the fluid-dispersed phase two-phase systems was addressed and the intraphasic heat transfer based on the assumption of the local thermal equilibrium among the phases was examined. In the discussion of the role of the particle suspensions in the effective transport within the medium, in Section 3.3.4 the enhancing role of the particles in the *turbulent* transport in gas-particulate flows was addressed.

In this section, first the suspension of *small* particles in a turbulent *pipe flow* and its role in the interfacial heat transfer is examined using the *phenomenological* model of Section 3.3.4 for the effective medium transport. This will be for a *gas* flow with the suspension of very small particles (such that the particles follow the fluid motion, i.e., their *slip* or *relative* velocity is zero) and without any particle charge effects. Then the effect of the motion of large charged particles, suspended in a insulating liquid, on the heat transfer between two electrodes is examined using the available experimental results. The particle-induced motion results in an increase in the interfacial heat transfer rate.

4.14.1 Internal Flow of Small Neutral Particles

In the two-medium treatment of the turbulent particulate flow and heat transfer in a *tube*, the particle-fluid is treated as a single medium (i.e., local *thermal* and *mechanical* equilibrium are assumed between the particles and the fluid) in thermal *nonequilibrium* with the confining tube surface. Then the interfacial heat transfer *between* the surface and the particulate flow is analyzed. We assume that the particles are small and are not charged such that they will follow the fluid flow even in the small scale of the turbulent eddies. The presence of small suspended particles in a flowing fluid with a velocity gradient results in the *nonuniformity* of the particle distributions. Therefore, the *effective medium properties* discussed in Section 3.3.4, which depend on the particle concentration, vary within the medium. The determination of the particle concentration distribution requires the simultaneous solution to the *particle conservation*, *fluid continuity*, and the *particle-fluid momentum* equations. The *fluid mechanics* of suspensions are discussed by Depew and Kramer (1973) and by Soo (1989). Here we consider a pipe flow for which the *particle distribution* is *known* a priori (through measurement). The particulate flow through tubes is also called a *circulating fluidized bed* and a reference was made to this solid-fluid flow in Section 1.6.1. The interfacial heat transfer in presence of particles has been discussed by Depew and Kramer (1973), Michaelides (1986), Soo (1989), Molodtsof and Muzyka (1992), and Louge et al. (1993). Various analyses have been made depending on the *size* of the particle (*small* versus *massive*, where particle larger than about 200μm are generally considered massive) and particle concentration (*dilute* versus *dense*, porosities larger than about 0.99, are generally considered dilute). The analysis of *massive* particles is based on the local thermal *nonequilibrium* and a separate treatment of the particle trajectory (i.e., a separate momentum equation). The analysis of the *dense* and *massive* particles also include the *interparticle collisions*. These will be further discussed first in Section 5.4 in connection with the *two-medium treatment* of particulate flows without any bounding surfaces (i.e., the *bulk* behavior of particulate flow and heat transfer), and in Section 7.1.1 in connection with the *three-medium treatment* which includes the confining surfaces (i.e., the three-medium treatment of the internal particulate flow and heat transfer). Here we consider the phenomenological model of Michaelides (1986) which was briefly discussed in Section 3.3.4.

In this phenomenological model the concept of an effective medium thermal conductivity is *not* used, and instead the presence of the particles is modeled through the fluctuations in *density* and *heat capacity*. Because *turbulent* flow is assumed with a turbulent Prandtl number of unity, the *molecular* and *dispersive* components of the effective thermal conductivity discussed in Section 3.3.4 are modeled *into* the turbulent transport.

For *large-particle Froude numbers* $Fr_d = \langle u_g \rangle / (gd)^{1/2}$, where d is the particulate diameter, Michaelides (1986) uses an available measured *radial*, *medium* density distribution (i.e., the radial distributions of particles) in particulate flow through tubes of diameter D. For the particular experiment cited by Michaelides (1986), i.e., a particular combination of particle and tube diameters, particle and fluid densities, fluid viscosity, porosity, particle and fluid velocities, and direction of flow with respect to gravity, the results show that *less* particles are present *near* the tube wall compared to the *central core*. For $r = 0$ at the centerline of the tube, this radial distribution is represented by

$$\overline{\rho} = \rho_c (1 + \gamma \frac{D/2 - r}{D/2})^m , \tag{4.543}$$

where $\overline{\rho}$ is the medium density defined by (3.133), but subscript e is dropped for simplicity, ρ_c is the density of the continuous phase (here the gas phase), $0.4 \leq m \leq 0.6$, and γ depends on the *area-averaged* mean mass flux of *particles* $\overline{\langle \dot{m}_d \rangle}_A$. The case of *dilute* suspension flow with very small particles has been reviewed by Soo (1989). The particle concentration measurement of Soo et al. (1964) confirms the distribution suggested by (4.543), except for very small *charged* particles where the electrostatic charge induced by the collision with the tube surface results in a high concentration adjacent to the surface. Particle distribution in tubes will be further discussed in Section 7.1.1, for *massive* particles (i.e., the Froude number is not very large) with high concentrations (dense particulate flows). For large particles the concentrations near the tube wall is larger compared to the central core and *flow reversal* can occur near the tube surface.

The *area-averaged* density for the medium is obtained from (4.543) as is

$$\langle \overline{\rho} \rangle_A = \frac{8}{D^2} \int_0^{D/2} \overline{\rho} r \, dr = \frac{2\rho_c[(1+\gamma)^{m+2} - 1]}{\gamma^2 (m+1)(m+2)} - \frac{2\rho_c}{\gamma(m+1)} . \tag{4.544}$$

The *radial* distribution of the local mean porosity $\overline{\epsilon}$ (local volume fraction of the *continuous* phase) is found from using the local densities, i.e., from (3.137) as

$$1 - \overline{\epsilon} = \frac{\overline{\rho} - \rho_c}{\rho_d - \rho_c} , \tag{4.545}$$

where ρ_d is the density of the dispersed phase (here the solid particles). From the experimental results of Farber and Depew (1963), the range of applicability of (4.543) is for $\epsilon > 0.99$, i.e., very *dilute suspensions*. The area-averaged, mean void fraction $\langle \overline{\epsilon} \rangle_A$ is $\langle \overline{\epsilon} \rangle_A = (\langle \overline{\rho} \rangle_A - \rho_c)/(\rho_d - \rho_c)$. The *local* distribution of the *specific heat capacity* is

$$\overline{c}_p = c_{p_c} [\frac{c_{p_d}}{c_{p_c}} + \frac{\rho_c}{\overline{\rho}} (1 - \frac{c_{p_d}}{c_{p_c}})] . \tag{4.546}$$

The needed derivates of $\overline{\rho}$ and $\overline{c}_p$ are

$$\frac{d\overline{\rho}}{dr} = \frac{\rho_c\gamma_m}{R}(1+\gamma\frac{D/2-r}{D/2})^{m-1} . \tag{4.547}$$

$$\frac{d\overline{c}_p}{dr} = c_{p_c}\frac{\rho_c}{\overline{\rho}^2}\frac{\partial\overline{\rho}}{\partial r}(\frac{c_{p_d}}{c_{p_c}} - 1) . \tag{4.548}$$

In the turbulent phenomenological model of Michaelides (1986), discussed in Section 3.3.4(C), the *mixing lengths* for the turbulent transport of *momentum*, *heat*, *density*, and *specific heat capacity* are defined by (3.132) to (3.134), with the momentum mixing length given by (2.98). The distribution of the *momentum* mixing length is taken from the experimental results which suggest

$$\frac{2\ell}{D} = 0.14 - 0.08\frac{4r^2}{D^2} - 0.06\frac{16r^4}{D^4} , \quad \ell_T = \ell \tag{4.549}$$

and the *heat* and momentum mixing lengths are assumed to be the same. Also, from the experimental results, the *density* mixing length is related to the momentum mixing length using the area-averaged particle to fluid mass flux *ratio* $\langle\dot{m}_d\rangle_A/\langle\dot{m}_c\rangle_A$, i.e.,

$$\frac{\ell_\rho}{\ell} = 0.031 - 0.070\frac{\langle\dot{m}_d\rangle_A}{\langle\dot{m}_c\rangle_A} + 0.045\frac{\langle\dot{m}_d\rangle_A^2}{\langle\dot{m}_c\rangle_A^2} , \quad \ell_{c_p} = \ell_\rho \tag{4.550}$$

and the *specific heat capacity* mixing length is assumed to be equal to the density mixing length.

Assuming a *fully developed* steady turbulent internal flow in a tube, the momentum equations for the x direction (along the tube axis with the velocity component u), with the gravity vector being along the x-axis, becomes

$$\frac{dp}{dx} = \frac{\mu}{r}\frac{d}{dr}r\frac{du}{dr} - \frac{1}{r}\frac{d}{dr}r\tau_t + \rho g . \tag{4.551}$$

The *turbulent shear stress* τ_t is defined and approximated (by neglecting the average of the triple product) by

$$\begin{aligned}\tau_t &= -\overline{u}\overline{\rho' v'} - \overline{\rho}\overline{u' v'} - \overline{\rho' u' v'} \\ &\approx -\overline{u}\overline{\rho' v'} - \overline{\rho}\overline{u' v'} .\end{aligned} \tag{4.552}$$

Then the momentum equation can be integrated and written

$$\frac{r}{2}\frac{dp}{dx} = \mu\frac{du}{dr} - \tau_t + \frac{g}{r}\int_o^r \rho r\, dr . \tag{4.553}$$

Now integrating this momentum equation between the surface $r = D/2$ and any r to eliminated the pressure, we have for the *surface shear stress* τ_s

$$\tau_s\frac{r}{R_o} = \mu\frac{d\overline{u}}{dr} + \overline{u}\overline{\rho' v'} + \overline{\rho}\overline{u' v'} -$$

$$g(\frac{r}{R_o^2}\int_0^{D/2}\overline{\rho}r\,dr-\frac{1}{r}\int_o^r\overline{\rho}r\,dr)\,. \tag{4.554}$$

For a thermally fully developed, steady turbulent pipe flow, the heat flux in the r direction is approximated by

$$q=r\frac{d\overline{T}}{dr}+\overline{\rho}\overline{c}_p\overline{T'v'}+(\overline{T}-T_s)\overline{\rho}\overline{c_p'v'}+(\overline{T}-T_s)\overline{c}_p\overline{\rho'v'}\,, \tag{4.555}$$

where again the average of the triple products of the fluctuations is neglected.

Using the closures discussed in Section 3.3.4 and given by (3.129) to (3.131) and (2.98), the time averages of the fluctuation products are written in terms of the derivatives of the mean quantities. The shear-stress and heat flux equations become

$$\tau_s\frac{r}{R}=\mu\frac{d\overline{u}}{dr}-\overline{u}\ell_\rho\ell\frac{d\overline{\rho}}{dr}\frac{d\overline{u}}{dr}-\overline{\rho}\ell^2\frac{d\overline{u}}{dr}\frac{d\overline{u}}{dr}+$$

$$g(\frac{r}{R^2}\int_o^R\overline{\rho}r\,dr-\frac{1}{r}\int_o^r\overline{\rho}r\,dr)\,, \tag{4.556}$$

$$q=k\frac{\partial\overline{T}}{\partial r}-\ell^2\overline{\rho}\,\overline{c}_p\frac{d\overline{T}}{dr}\frac{d\overline{u}}{dr}-(\overline{T}-T_s)\overline{\rho}\ell_{c_p}\ell\frac{d\overline{c}_p}{dr}\frac{d\overline{u}}{dr}-$$

$$(\overline{T}-T_s)\overline{c}_p\ell_\rho\ell\frac{d\overline{\rho}}{dr}\frac{d\overline{u}}{dr}\,. \tag{4.557}$$

Scaling the variables with the *tube radius* $D/2$, wall friction velocity $(\tau_s/\rho_c)^{1/2}$, surface temperature and heat flux T_s and q_s, and the thermophysical properties of the continuous phase, we have the variables and parameters

$$r^*=\frac{2r}{D}\,,\quad \ell^*=\frac{2\ell}{D}\,,\quad \rho^*=\frac{\overline{\rho}}{\rho_c}\,,\quad c_p^*=\frac{\overline{c}_p}{c_{p_c}}\,,\quad u^*=\frac{\overline{u}}{(\tau_s/\rho_c)^{1/2}}\,, \tag{4.558}$$

$$T^*=(\overline{T}-T_s)\frac{c_{p_c}(\tau_s/\rho_c)^{1/2}\rho_c}{q_s}\,,\quad Re_D=\frac{D(\tau_s/\rho_c)^{1/2}}{\nu_c}\,, \tag{4.559}$$

$$Pr=(\frac{\mu c_p}{k})_c\,,\quad Fr_D^2=\frac{\tau_s/\rho_c}{gD}\,. \tag{4.560}$$

Then the dimensionless shear stress and heat flux equations become

$$-r^*=\frac{2}{Re_D}\frac{du^*}{dr^*}-u^*\ell_\rho^*\ell^*\frac{d\rho^*}{dx}\frac{du^*}{dx}-$$

$$\rho^*\ell^{*2}(\frac{du^*}{dr^*})^2-\frac{4}{Fr_D^2}(r^*\int_o^1\rho^*r^*\,dr^*-\frac{1}{r^*}\int_o^1\rho^*r^*\,dr^*)\,, \tag{4.561}$$

$$-\frac{\int_o^{r^*}c_p^*\rho^*u^*r^*\,dr^*}{\int_o^1 c_p^*\rho^*u^*r^*\,dr^*}=\frac{2}{Re_DPr}\frac{dT^*}{dr^*}-\rho^*c_p^*\ell_{c_p}^*\ell^*\frac{dT^*}{dr^*}-$$

$$-\rho^* T^* \ell_T^* \ell^* \frac{dc_p^*}{dr^*}\frac{du^*}{dr^*} - c_p^* T^* \ell_\rho^* \ell^* \frac{d\rho^*}{dr^*}\frac{du^*}{dr^*} . \tag{4.562}$$

The effect of the suspended particles *manifests* in the *effective* properties for the fluid-particle medium, and this influences the interfacial heat transfer rate by altering the temperature distribution. The normalized heat flux (or the Nusselt number) will include the unknown velocity-weighted, time- and area-averaged medium temperature, also called the bulk-mixed or mean fluid temperature, $\langle \overline{T} \rangle_A$, i.e.,

$$Nu_D = \frac{q_s D}{(T_s - \langle \overline{T} \rangle_A^b) k_c} \tag{4.563}$$

where

$$\langle \overline{T} \rangle_A^b = \frac{8}{\rho_c c_{p_c} (\tau_s/\rho_c)^{1/2} D^2} \int_0^{D/2} \overline{\rho}\, \overline{c}_p \overline{u} (T_s - \overline{T}) r \, dr . \tag{4.564}$$

The parameters of the problem are Fr_D, Re_D, Pr, $\langle \overline{\dot{m}_d} \rangle_A / \langle \overline{\dot{m}_c} \rangle_A$, and the ratio of the heat capacities c_{p_s}/c_{p_c}. Depew and Kramer (1973) and Michaelides (1986) discuss the range of these variables as they are generally encountered in practice. For gases, the *tube Froude number* Fr_D is also assumed to be very large. Here we review the results of Michaelides (1986) which includes a comparison of the computed predictions and the experimental results for the *lateral* distributions of the velocity, temperature, and the eddy diffusivity. His prediction of the variation of the interfacial heat transfer with respect to the solid loading ratio $\langle \overline{\dot{m}_d} \rangle_A / \langle \overline{\dot{m}_c} \rangle_A$ will also be discussed.

The computed radial distributions of the dimensionless velocity and temperature are shown in Figure 4.45(a) and (b), where the dimensionless radial coordinate is

$$y^* = \frac{(D/2 - r)(\tau_s/\rho)^{1/2}}{\nu_c} . \tag{4.565}$$

The results for the velocity are also compared with the available experimental results for *dilute* suspensions. Note that the presence of the particles tends to make *both* the velocity and temperature distributions *more uniform*. The lateral distribution of the dimensionless eddy diffusivity ν_t^* defined as

$$\nu_t = \frac{2\tau}{D(\tau_s/\rho_c)^{1/2} \overline{\rho} \dfrac{d\overline{u}}{dr}} , \tag{4.566}$$

is shown in Figure 4.45(c). Note that the distribution of the eddy diffusivity for heat is *similar* to that for ν_t^*. The results show that initially ν_t^* *increases* linearly with the distance from the surface and then begins to *decrease* before the centerline is reached. Good agreement is found with the experimental results. The variation of the *normalized* Nusselt number (normalized with respect to the results with *no* solid loading) with respect to the solid to fluid loading ratio $\langle \overline{\dot{m}_d} \rangle_A / \langle \overline{\dot{m}_c} \rangle_A$, is shown in Figure 4.45(d)

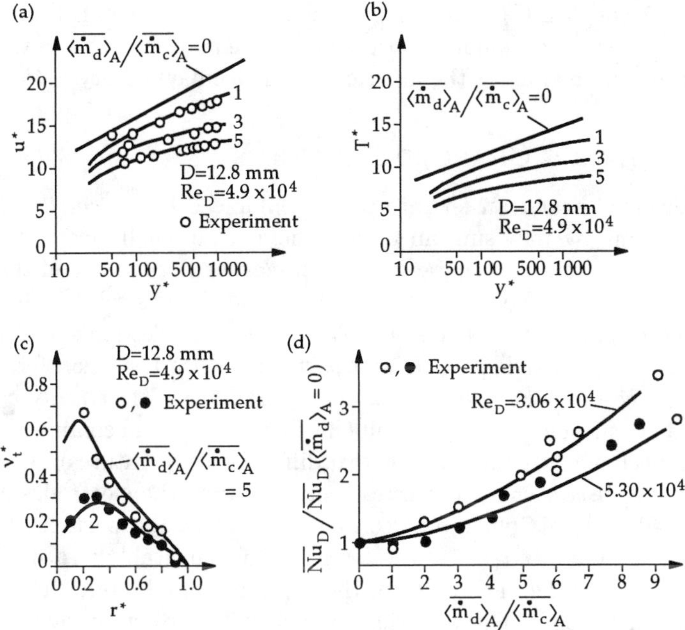

Figure 4.45. (a) Distribution of the local, volume-averaged velocity in a particulate flow through a tube. The predicted and measured results for several values of the particulate loading are shown. (b) The predicted distribution of the local volume-averaged temperature. (c) The radial distribution of the dimensionless turbulent eddy diffusivity. (d) Effect of the suspended particles on the dimensionless interfacial heat transfer rate. (From Michaelides, reproduced by permission ©1986 Pergamon Press.)

for two different values of Re_D. Comparison is also made with the experimental results. Note that as Re_D *increases* the normalized Nusselt number *tends* to unity.

Although very small particles have been assumed and a large temperature gradient exists near the bounding surface, the effect of *thermophoresis* on the particle trajectory has been *neglected*. Thermophoresis was discussed in Section 1.12.4 (C), and, in particulate flows, as related to the physical or chemical vapor deposition, becomes important. Examples are given by Simpkins et al. (1979), Walker et al. (1979), and Kang and Grief (1993).

The problem of internal particulate flow and heat transfer for *massive* particles, where local thermal *nonequilibrium* is assumed among the particles, the fluid, and the bounding surfaces, will be discussed in Section 7.1. There the *two-medium* treatments of the *hydrodynamics* (including *mean* and fluctuating velocity components of both the particle and the fluid phases) and *three-medium* treatment of the heat transfer (including diffusion and convection in fluid and particle phases) will be made. In Sections 5.4.2 and 5.4.3, the *dense* particulate-phase flow and heat transfer are treated using continuous phase transport equation with the transport properties obtained using the kinetic (i.e., molecular) theory.

4.14.2 Large Charged Particles

The interfacial heat transfer rate can be enhanced in *gas-small solid particles* systems by imposing an electric field. The small electrically *non-conducting* particles are charged by *frictional electrication*, and therefore, their motion is influenced by the electric field. In gas-solid systems, because the gas phase heat transfer is *small*, the particles make a *noticeable* contribution by *particle-particle* and *particle-surface* heat transfer.

We now consider *liquid-solid* particle systems with electrically *conducting* particles suspended in electrically insulated liquid and confined between parallel electrodes (maintained at two different temperatures). The particles are *inductively* charged during *collisions* with the electrodes and are accelerated by the Coulomb force. The motion of the particles through the liquid results in *mixing* of the *liquid* and for an stably stratified liquid layer (i.e., *moderately* heated from the top) the liquid motion can be due to the particles only. The particles contacting the lower surface electrode *acquire* the charge of that electrode (in proportion to the electric field) and *experience* a force of acceleration (towards the other electrode) equal to the *square* of the electric field (as discussed in 2.6.1). When this force is large enough to *overcome* the gravitational force, the particles move and reach the other electrode, and once the charge *changes* the direction of the force also changes, and this way the particles can be moving between the electrodes.

Dietz and Melcher (1975) conduct an experiment with 1.6-mm *aluminum spheres* in *transformer oil*. The liquid heat transfer is by *conduction* where no electric field is applied. Then the dimensionless heat transfer rate (i.e., the Nusselt number) is

$$Nu_\ell = \frac{q_s \ell}{(T_{s_1} - T_{s_2})k}, \tag{4.567}$$

where ℓ is the spacing between electrodes (2.7 cm) and T_{s_1} and T_{s_2} are the upper and lower surface temperatures, and for purely conductive heat transfer Nu_ℓ is *unity*. Their experimental results, for no particles, 100, 300, 500, and 700 particles (other spatial dimensions of the fluid layer are given by Dietz and Melcher, 1975), are shown in Figure 4.46. Their results are shown for the variation of Nu_ℓ with respect to the *applied* voltage φ. Note that even without any particles, the measured Nu_ℓ in the *presence* of a large electric field is larger than that with no electric field; however, this increase is *not* as large as that due to the presence of the particles. Then the change in the electrical conductivity caused by the temperature variation, discussed in Section 4.13.1 is *not* very significant in this problem. Also,

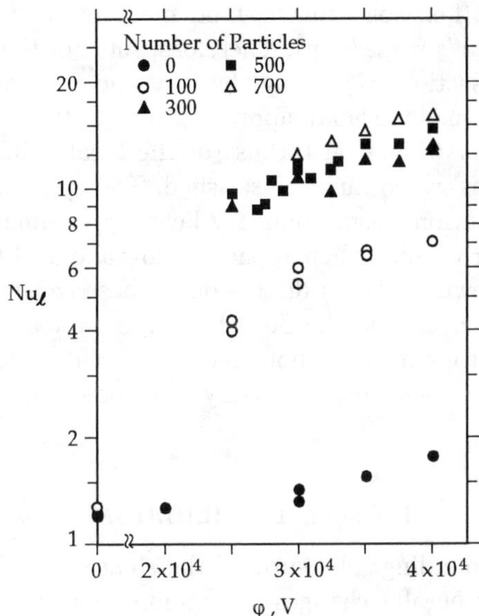

Figure 4.46. Variation of normalized heat transfer rate with respect to the voltage applied and for several number of particles present in a horizontal dielectric liquid layer heated from the top. (From Dietz and Melcher, reproduced by permission ©1975 American Society of Mechanical Engineers.)

for a given applied electric field Nu_ℓ is slightly larger on the φ-increasing *branch* of the *hysteresis* curve. The results also show that as the number of particles increases, an asymptotic behavior is observed and any further increase in the particle numbers does *not* increase the heat transfer rate. This is due to increase in the interparticle collisions.

The heat can also be transferred by the particles which are heated (or cooled) upon contacting the electrodes. Dietz and Melcher (1975) show that the residence time of the particles are too long to allow them to maintain their initial heat content. This is indicative of the *presence* of local thermal equilibrium between the particles and the fluid.

4.15 Porous Media

By assuming that *local thermal equilibrium* exists between a porous medium (rendered as a *rigid solid matrix* through which a fluid can flow) and the fluid flowing through its interstices, the solid matrix-fluid is represented as a *single medium* by using a single temperature the *effective medium properties* discussed in Section 3.2. Here the *interfacial* convective heat transfer between this effective medium and a solid bounding surface is examined. There are some similarities in the conservation equations *representing* this effective medium (i.e., solid matrix-fluid medium) and those used for a *plain fluid* (i.e., a *plain medium*). Therefore, some of the results discussed in the preceding sections regarding the flow and interfacial heat transfer aspects of a plain medium, may apply to this effective medium. However, there are some *constraints* on the use of the local volume-averaged conservation equations which must be satisfied. The approximations used for obtaining field solutions, e.g., boundary-layer approximations, should also be re-examined for validity when applied to flow and heat transfer in porous media. In the following, after a discussion of these *constraints* on the local thermal equilibrium and on the boundary-layer flow and heat transfer in a semi-infinite porous medium bounded by a solid surface, the solutions for *forced* and *thermobuoyant* boundary-layer flows and heat transfer are examined.

4.15.1 Local Thermal Equilibrium

Adjacent to the bounding solid surface, the *solid phase distribution* in the porous medium generally changes to accommodate the geometry of the bounding surface. The thermal *contact* between the solid matrix and the bounding surface varies greatly depending on the solid matrix structure. For example, in the case of a solid matrix made of *packed, nonconsolidated* spherical particles, the *ideal* contact with a nearly planar (large radius of curvature compared to the particle radius) bounding surface, is a *point contact*. For nonconsolidated particles which on their surface have some *planar subsections*, and for *consolidated particles* (e.g., sintered particles) and other consolidated solid matrices, *area contact* but with a noticeable thermal contact resistance is found. These are rendered in Figure 4.47 where the bounding solid surface is assumed to be at a uniform surface temperature T_s and the temperature in the porous medium far from the surface ($y \to \infty$) is T_∞. A third type of contact between a porous medium and the bounding solid surface also shown in the figure is for a solid matrix attached to the surface with no thermal contact resistance. This attached solid matrix is then considered as an *integrated, extended surface*, and because of the possibility of a *large specific interfacial area*, can potentially be used similar to attached *fins*.

When the system temperature difference $|T_s - T_\infty| = \Delta T_\delta$ is across a length δ in the y direction, the constraints for the validity of the *local* thermal equilibrium between the fluid flowing through and the elements of the solid matrix, i.e., for $\langle T\rangle^f = \langle T\rangle^s = \langle T\rangle$, are

$$\Delta T_d \ll \Delta T_\ell \ll \Delta T_\delta \,. \tag{4.568}$$

In these inequality relations, it is assumed that the linear dimension of the representative elementary volume ℓ is much *larger* than d, the particle linear dimension. When the fluid and the solid matrix have nearly equal thermal conductivities, i.e., $k_s/k_f \simeq 1$, these constraints lead to approxi-

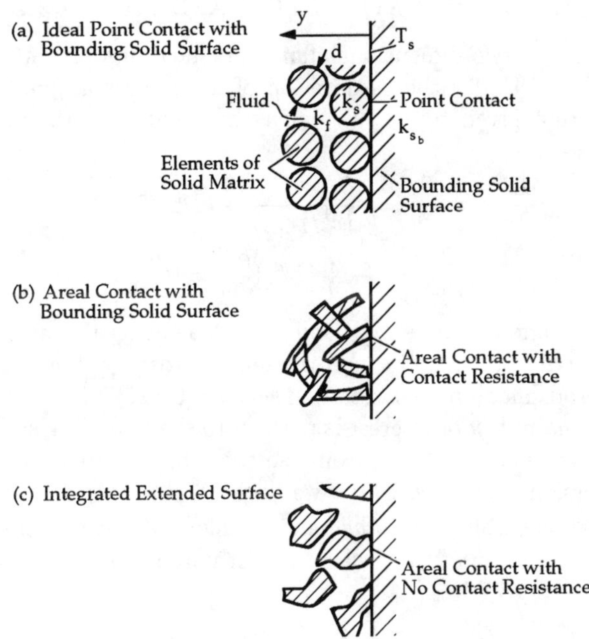

Figure 4.47. Solid phase distributions adjacent to a planar bounding solid surface for (a) a solid matrix made of spherical nonconsolidated particles, (b) a solid matrix made of particles with planar subsections on their surface, and (c) a solid matrix with its particles attached to the bounding solid surface.

mately $\ell/d \geq 5$ and $\delta/\ell > 10$, and are generally satisfied. However, when k_s/k_f is much larger than unity, then unless the contact area is *small* and the contact resistance *large*, the heat flow finds a preferred path through the solid matrix and the difference between the local values of $\langle T\rangle^f$ and $\langle T\rangle^s$ becomes substantial. This *lack* of local thermal equilibrium requires a *two-medium treatment* of the fluid and the solid matrix heat transfer. In Chapter 5 this two-medium treatment is discussed, and in Chapter 7 the effect of the presence of a bounding surface is addressed (but instead of a solid matrix, moving particles are considered and extension to stationary particles can be made readily).

In the following discussions, the local thermal equilibrium is assumed between the *fluid* and the *solid matrix* for $y > 0$ and between the fluid, the solid matrix, and the *bounding surface* at $y = 0$.

4.15.2 Boundary-Layer Flow and Heat Transfer

Consider *steady*, two-dimensional fluid and heat flow in an *anisotropic* porous medium. Under the assumption of local thermal equilibrium, the energy equation given by (3.72), when the volume averaging symbol is dropped, becomes

$$u_D\frac{\partial T}{\partial x}+v_D\frac{\partial T}{\partial y}=\frac{\partial}{\partial x}\left[\frac{k_{e_{xx}}}{(\rho c_p)_f}+\epsilon D^d_{xx}\right]\frac{\partial T}{\partial x}+\frac{\partial}{\partial y}\left[\frac{k_{e_{yy}}}{(\rho c_p)_f}+\epsilon D^d_{yy}\right]\frac{\partial T}{\partial y}=0\,. \quad (4.569)$$

The effective conductivities $k_{e_{xx}}$ and $k_{e_{yy}}$ depend on the anisotropic, local phase distributions, k_s/k_f, and the local porosity ϵ. For the case of an *isotropic* porous medium, a correlation such as (3.127) can be used. Assuming that the main flow of interest is parallel to the bounding surface and in the x direction, i.e., $u_D \gg v_D$, from the available results on the hydrodynamic dispersion (Kaviany, 1991), we have $D^d_{xx} \gg D^d_{yy}$. For example, for spherical particle which are *randomly* arranged, the correlations for D^d_{xx} and D^d_{yy} as a function of the particle Péclet number, $Pe_d = u_D d/2\alpha_f$, and ϵ are

$$\frac{D^d_{xx}}{\alpha_f} = \frac{3}{4}Pe_d+\frac{1}{6}\pi^2(1-\epsilon)Pe_d\ln Pe_d \quad Pe_d \gg 1\,, \quad (4.570)$$

$$\frac{D^d_{yy}}{\alpha_f} = \frac{63(2)^{1/2}}{320}(1-\epsilon)^{1/2}Pe_d \quad Pe_d \gg 1\,. \quad (4.571)$$

In making the boundary-layer approximations, it is assumed that $Pe_x = u_D x/\alpha_f$ is large and the axial diffusion is negligible compared to the lateral diffusion. Here only when D^d_{xx} is small compared to $k_{e_{xx}}/(\rho c_p)_f$, i.e., in the case of negligible hydrodynamic dispersion and large Pe_x, this assumption is valid. This would require that $x/d \gg 1$. Then the *boundary-layer* approximated, thermal energy conservation equation becomes

$$(\rho c_p)(u_D\frac{\partial T}{\partial x}+v_D\frac{\partial T}{\partial y})=\frac{\partial}{\partial y}k_{e_{yy}}\frac{\partial T}{\partial y} \quad Pe_x \gg 1\,,\ Pe_d \to 0\,. \quad (4.572)$$

Because of the difference in the solid and the fluid phase distributions adjacent to the bounding surface, $k_{e_{yy}}$ *varies* with y and this variation must be included for an *accurate* prediction of the heat transfer rate through the bounding surface (Sahraoui and Kaviany, 1993). Using a *uniform* $k_{e_{yy}}$ leads to a *surface temperature slip error*, however, in many analyses uniform $k_{e_{yy}}$ has been used for simplicity.

The description of the fluid flow by the Darcy law (3.73) does not allow for the hydrodynamic (i.e., viscous) boundary-layer formation, i.e., the fluid velocity *slips* over the solid bounding surface. Using other *heuristic* hydrodynamic models, such as (3.75), does allow for a hydrodynamic boundary layer. However, this viscous boundary-layer thickness in the case of forced

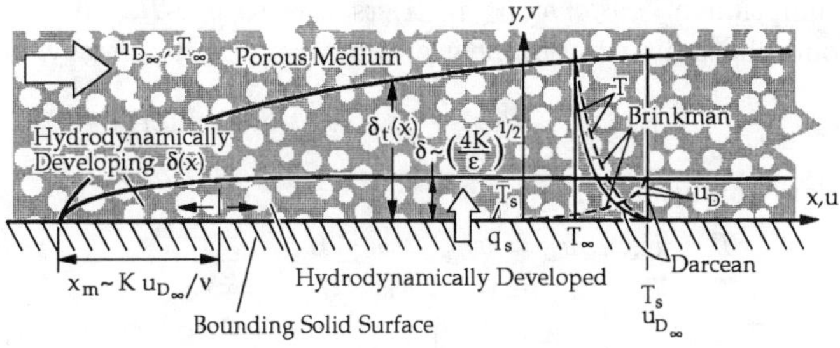

Figure 4.48. A rendering of a semi-infinite porous medium bounded by a solid planar surface. The viscous and the thermal boundary layers adjacent to the surface are shown.

flow is very *small* and can be neglected. This is discussed further in the next section.

4.15.3 Forced Flow

Steady, two-dimensional boundary-layer flow and heat transfer subject to a free-stream Darcean velocity $u_{D\infty}$ and *constant* free-stream and surface temperatures T_∞ and T_s has been examined by Cheng (1978a, b) and Vafai and Tien (1981), among others. A review is given by Nield and Bejan (1992). A schematic of this forced boundary-layer flow and heat transfer is given in Figure 4.48, where the lateral distributions of the temperature and the axial component of Darcean velocity u_D are also shown. The thermal boundary-layer thickness δ_t grows monotonically with the axial distance x. For a Darcean flow, no momentum boundary-layer exists, and for a Brinkman flow, the momentum boundary-layer thickness δ developes to an asymptotic value at a short distance from the the leading edge.

The mass, momentum, and energy equations become, respectively,

$$\frac{\partial u_D}{\partial x} + \frac{\partial v_D}{\partial y} = 0 \,, \tag{4.573}$$

$$\rho_f(u_D \frac{\partial u_D}{\partial x} + v_D \frac{\partial u_D}{\partial y}) = -\epsilon \frac{\partial p}{\partial x} - \frac{\mu_f \epsilon}{K} u_D - \frac{C_E \epsilon \rho_f}{K^{1/2}} u_D^2 + \mu_f \frac{\partial^2 u_D}{\partial y^2} \,, \tag{4.574}$$

$$u_D \frac{\partial T}{\partial x} + v_D \frac{\partial T}{\partial y} = \alpha_e \frac{\partial^2 T}{\partial y^2} \,. \tag{4.575}$$

The assumption of *uniform* α_e restricts this to low-Pe_d flows with a uniform solid phase distribution near and far from the boundary, i.e.,

$$\alpha_e = \frac{k_{e_{yy}}}{(\rho c_p)_f} + \epsilon D^d_{yy} \quad Pe_d \to 0,\ \epsilon,\ k_{e_{yy}} \text{ are uniform.} \tag{4.576}$$

The boundary conditions are

$$T = T_\infty\,,\quad u_D = u_{D_\infty} \qquad x \le 0\,,\ y \ge 0\,, \tag{4.577}$$

$$T = T_\infty\,,\quad u_D = u_{D_\infty} \qquad x > 0\,,\ y \to \infty\,, \tag{4.578}$$

$$T = T_s\,,\quad u_D = v_D = 0 \qquad x > 0\,,\ y = 0\,. \tag{4.579}$$

The *bulk* resistance to the flow is present even away from the boundary, where the momentum equation reduces to

$$0 = -\epsilon\frac{\partial p}{\partial x} - \frac{\mu_f \epsilon}{K} u_{D_\infty} - \frac{C_E \epsilon \rho_f}{K^{1/2}} u^{1/2}_{D_\infty}\,. \tag{4.580}$$

This can be used to replace the pressure gradient term in (4.574) and the result is

$$\rho_f\left(u_D\frac{\partial u_D}{\partial x} + v_D\frac{\partial u_D}{\partial y}\right) = \frac{\mu_f \epsilon}{K}(u_{D_\infty} - u_D) + \frac{C_E \epsilon \rho_f}{K^{1/2}}(u^2_{D_\infty} - u^2_D) + \mu_f \frac{\partial^2 u_D}{\partial y^2}\,. \tag{4.581}$$

Equations (4.573), (4.575), and (4.581) do not admit a similarity solution, but regular perturbation solutions and numerical integrations have been performed. In order to demonstrate some of the flow and heat transfer features, the available results of an approximate integral method (Kaviany, 1987) are reviewed below.

Since for the Darcean flow the axial velocity is *uniform* while for the Brinkman (i.e., non-Darcean) flow the no-slip boundary condition leads to a *nonuniform* velocity distribution, choosing a *simple* velocity profile that satisfies these is not possible. In the interest of satisfying the no-slip condition, the velocity distribution

$$\frac{u_D}{u_{D_\infty}} = \frac{3}{2}\frac{y}{\delta} - \frac{1}{2}\frac{y^3}{\delta^3} \tag{4.582}$$

is chosen, and similarly the temperature distribution is prescribed as

$$T^* = \frac{T - T_s}{T_\infty - T_s} = \frac{3}{2}\frac{y}{\delta_t} - \frac{1}{2}\frac{y^3}{\delta_t^3}\,. \tag{4.583}$$

Using these and the continuity equation, the momentum (non-Darcean) and energy equations (4.575) and (4.581) become, respectively,

$$\frac{39}{280}\frac{d\delta^*}{dx^*} = \frac{3}{2}\frac{1}{\delta^* Re_x} - \frac{3}{8}\xi_x\delta^* - \frac{54}{105}\zeta_x\delta^*$$
$$= \frac{3}{2}\frac{1}{\delta^* Re_x} - \Gamma_x\delta^*\,,\quad \Gamma_x = \frac{3}{8}\xi_x + \frac{54}{105}\zeta_x\,, \tag{4.584}$$

$$\frac{dr}{dx^*} = \frac{10}{Pe_x 2r^2\delta^{*2}(1-\frac{r^2}{7})} - \frac{\frac{r}{2}-\frac{r^3}{28}}{\delta^*(1-\frac{r^2}{7})}\frac{d\delta^*}{dx^*} \,, \quad r = \frac{\delta_t}{\delta} \,. \tag{4.585}$$

The length scale used is the axial location x and (4.584) and (4.585) are solved subject to

$$\delta^*(x^* = 0) = \delta_t^*(x^*) = 0 \,. \tag{4.586}$$

The dimensionless parameters are

$$\xi_x = \frac{\nu_f \epsilon}{K u_{D_\infty}} = \frac{\gamma_x^2}{Re_x} \,, \quad y_x = \frac{C_E \epsilon}{K^{1/2}} x \,, \quad \gamma_x = \frac{x}{(K/\epsilon)^{1/2}} \,,$$
$$Pe_x = Re_x Pr_e \,, \quad Re_x = \frac{u_{D_\infty} x}{\nu_f} \,, \quad Pr_e = \frac{\nu_f}{\alpha_e} \tag{4.587}$$

where γ_x is called the *porous medium shape parameter* and is the ratio of the axial location to the *pore linear dimension* $(K/\epsilon)^{1/2}$, ξ_x is the dimensionless *Darcean resistance*, and ζ_x is the dimensionless *Ergun resistance* to the flow.

The Nusselt number is

$$Nu_x = \frac{q_s x}{(T_s - T_\infty) k_{e_{yy}}} = \frac{3}{2}\frac{1}{\delta_t^*} \,. \tag{4.588}$$

For the *Darcean regime*, the momentum equation becomes trivial and the energy equation can be rewritten using a uniform velocity. Then the results are similar to those for the case of zero viscosity discussed in Section 4.2.1. This integral method results in

$$Nu_x = \frac{3}{2}\frac{1}{8^{1/2}} Pe_x^{1/2} \quad \text{Darcean regime,} \tag{4.589}$$

which is very close to the exact solution given by (4.14).

For $\Gamma_x \to 0$, the solution to the viscous flow discussed in Section 4.2.2 is obtained. The integral method gives

$$Nu_x = 0.33 Re_x^{1/2} Pr_e^{1/3} \quad \Gamma_x \to 0 \quad \text{plain-medium regime,} \tag{4.590}$$

which is very close to the exact solution given by (4.32).

For the *intermediate* or *Darcean regime*, the numerical solutions to (4.584) and (4.585) are obtained. Kaviany (1987) gives the *regime diagram* shown in Figure 4.49. For r *near unity*, (4.584) and (4.585) give

$$\delta^* = (\frac{3}{2}\frac{1}{Re_x \Gamma_x})^{1/2} \,, \qquad r = (\frac{10}{Pr_e}\Gamma_x)^{1/3} \,, \tag{4.591}$$

$$Nu_x = 0.57 \Gamma_x^{1/6} Re_x^{1/2} Pr_e^{1/3} \,. \tag{4.592}$$

Then for $C_E = 0$, we have $\delta = (4K/\epsilon)^{1/2}$, which when noting that $(K/\epsilon)^{1/2}$ can be *smaller* than the pore size, makes the application of the Brinkman equation questionable (Kaviany, 1991). The distance x required for reaching the hydrodynamically, fully developed region, i.e., where $\partial\delta/\partial x = 0$, is nearly equal to $K u_{D_\infty}/\nu_f$.

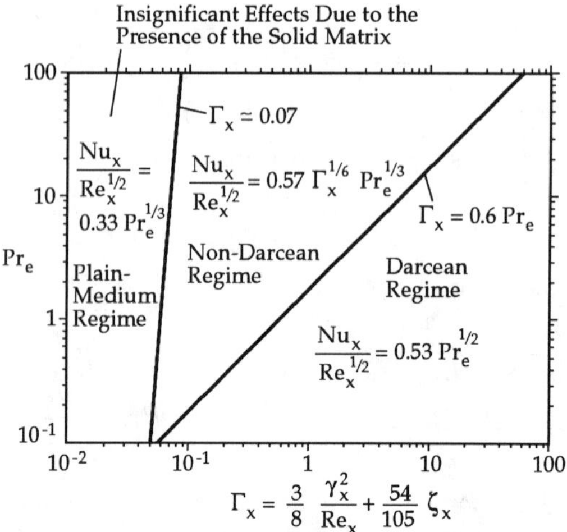

Figure 4.49. Regime diagram for forced boundary-layer flow and heat transfer in a semi-infinite porous medium. The plain-medium, non-Darcean (i.e., Brinkman), and Darcean regimes are shown in a $Pr_e - \Gamma_x$ diagram.

As shown in Figure 4.49, since for most practical cases Γ_x is rather large, the Darcean regime pervails and the slip flow solution is applicable.

4.15.4 Thermobuoyant Flow

The thermobuoyant motion of the fluid in porous medium, *caused* by the heat transfer from the bounding surface to the fluid-solid matrix medium, has been studied with geological and engineering applications. A review of results for various orientations of the bounding surface with respect to the gravity vector is given by Cheng (1978b) and by Nield and Bejan (1992). Here we examine the thermobuoyant flows adjacent to an isothermal, semi-infinite vertical surface bounding a porous medium. The problem has been examined by Cheng and Minkowycz (1977) for the Darcean regime.

The boundary-layer, steady two-dimensional conservation equations for mass, momentum, and energy, are those given by (4.573) to (4.575), except the momentum equation includes the buoyancy term and no external pressure gradient is applied. The momentum equation is then the modified form of (4.323), i.e.,

$$\rho_f \left(u_D \frac{\partial u_D}{\partial x} + v_D \frac{\partial u_D}{\partial y}\right) = -\frac{\mu_f \epsilon}{K} u_D - \frac{C_E \epsilon \rho_f}{K^{1/2}} u_D^2 + \mu_f \frac{\partial^2 u_D}{\partial y^2} + g\beta_f (T - T_\infty) \,. \tag{4.593}$$

The boundary conditions are

$$T = T_\infty \,, \quad u_D = 0 \qquad x \le 0 \,, \quad y \ge 0 \,, \tag{4.594}$$

$$T = T_\infty \,, \quad u_D = 0 \qquad x > 0 \,, \quad y \to \infty \,, \tag{4.595}$$

$$T = T_s \,, \quad u_D = v_D = 0 \quad x > 0 \,, \ y = 0 \,. \tag{4.596}$$

Again, this set of governing equations does not admit a similarity solution, and numerical solutions have been given by Hong et al. (1985), among others. Here the integral method (Kaviany and Mittal, 1987) is applied in order to obtain some approximate closed-form solutions. Since the maximum in the axial velocity changes with the axial location x, the prescribed velocity profile would contain a variable coefficient $U(x)$. Then choosing a profile that satisfies the boundary conditions, the axial velocity and temperature are prescribed with variations

$$u_D = U(x)\frac{y}{\delta}(1 - \frac{y}{\delta})^2 \,, \tag{4.597}$$

$$T^* = \frac{T - T_\infty}{T_s - T_\infty} = (1 - \frac{y}{\delta})^2 \,. \tag{4.598}$$

Upon inserting these into the momentum and energy equations (4.593) and (4.575), respectively, using the mass conservation equation (4.573), and scaling the variables using the length scale x and the velocity scale ν_f/x, we have

$$U^*\delta^*\frac{d}{dx^*}U^*\delta^* = -105U^{*3} - \frac{105}{12}\gamma_x^2 U^{*3}\delta^{*2} +$$

$$35Gr_x U^{*2}\delta^{*2} - \zeta_x U^{*4}\delta^{*2} \,, \tag{4.599}$$

$$U^*\delta^*\frac{d}{dx^*}U^*\delta^* = \frac{60}{Pr_e}U^* \,. \tag{4.600}$$

These are subject to

$$\delta^*(x^* = 0) = U^*(x^* = 0) = 0 \,. \tag{4.601}$$

The dimensionless parameters γ_x, ξ_x and Pr_e are those defined in (4.587) and the Grashof number is

$$Gr_x = \frac{g\beta_f(T_s - T_\infty)x^3}{\nu_f^2} \,. \tag{4.602}$$

In the Darcean regime the velocity slip is not satisfied by (4.597) and instead a profile similar to that for temperature, i.e.,

$$u_D = U(x)(1 - \frac{y}{\delta})^2 \tag{4.603}$$

is used.

The Nusselt number becomes

$$Nu_x = \frac{q_s x}{(T_s - T_\infty)k_{e_{yy}}} = \frac{2}{\delta^*} \,. \tag{4.604}$$

The results of the numerical integration of the above boundary-layer integral formulation (Kaviany and Mittal, 1987), for the condition of $C_E = 0$ (i.e., $\zeta_x = 0$) and $Pr_e = 0.7$, are shown in Figure 4.50. The results are given for $2^{1/2}Nu_x/{Gr_x}^{1/4}$ as a function $\xi_x = 2\gamma_x^2/Gr_x^{1/2}$. The parameter ξ_x has

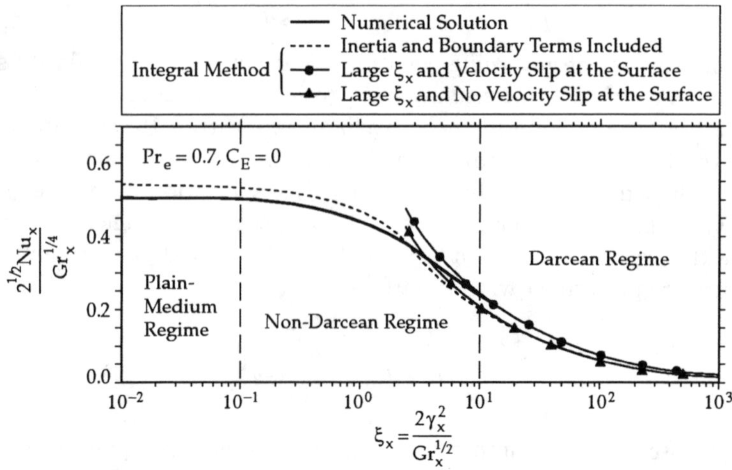

Figure 4.50. Regime diagram for thermobuoyant boundary-layer flow and heat transfer in a porous medium and adjacent to a vertical, isothermal solid surface. The plain-medium, non-Darcean (i.e., Brinkman), and Darcean regimes are shown. The numerical results as well as those from the integral method (both with and without velocity slip) are shown.

the same role as its counterpart in (4.587) for forced flows and increases as the permeability decreases, x increases, or Gr_x decreases. For $\xi_x > 10$, the *Darcean regime* exists. In this regime, the inertia and the boundary terms in the momentum equation are negligible. For this Darcean regime, by using the velocity-slip profile (4.603) in the momentum and energy equations, the solutions are

$$U^* = \frac{Gr_x}{\gamma_x^2} \qquad \xi_x > 10\ , \tag{4.605}$$

$$\frac{\delta^*}{x^*} = 20^{1/2}(\frac{Pr_e Gr_x}{\gamma_x^2})^{-1/2} \qquad \xi_x > 10\ . \tag{4.606}$$

Then the Nusselt number for the *Darcean regime* becomes

$$Nu_x = \frac{1}{5^{1/2}}(\frac{Pr_e Gr_x}{\gamma_x^2})^{1/2} \qquad \xi_x > 10 \tag{4.607}$$

or

$$\frac{2^{1/2}Nu_x}{Gr_x^{1/4}} = (\frac{4}{5})^{1/2} Pr_e^{1/2} \xi_x^{-1/2} \qquad \xi_x > 10\ . \tag{4.608}$$

This is also shown in Figure 4.50.

The *complete* numerical solution of Hong et al. (1985) are also shown in Figure 4.50. For $\xi_x < 0.1$, the *plain-medium regime*, i.e., regime of no significant effects due to the presence of the solid matrix, is found. For $0.1 \leq \xi_x \leq 10$, the *non-Darcean regime* exists where the inertial and boundary effects as well as the bulk Darcean resistance to the flow, influence the heat transfer rate. The results of the application of the integral method with a velocity slip at the surface agree with the numerical results in the Darcean regime, while in the non-Darcean regime the velocity no-slip profile agrees closely with the numerical results.

4.16 Multicomponent Solidification

Here, an example of phase change influenced by the convective heat transfer is considered. The problem is the solidification of a multicomponent melt. Under the condition of local thermal equilibrium, the region of *melt*, the region of *solidified melt*, and the *mushy two-phase* region are all treated as a *single-medium* (as discussed in Section 3.2.7). The bounding solid surface is treated as a seperate medium and is prescribed by a solid surface temperature which is below the melting temperature of the component with the higher melting point. Here, only the solidification of a *binary* melt is considered, and *no* supercooling (i.e., metastability) of the liquid is allowed. As the heat removal continues, the solidification continues as a growing solidified layer over the bounding surface. The bounding surface can be an *enclosure* as in closed models, or an *ideal* planar surface with the melt-mush-solidified melt occupying a semi-infinite domain. The treatment of the melt and the solidified melt as seperate media, including allowance for the liquid supercooling, will be addressed in Sections 7.3.1 and 7.3.2.

Since the single-medium treatment discussed is based on many assumptions, a review of the *local volume averaging* and the *modeling* of the conservation equations over the *mushy region* is made below. This is followed by a discussion of the *instability* of a growing, *horizontal* mushy layer. Then the results of the application of the single-medium formulation to the solidification of a binary liquid in a *closed cavity*, is examined.

4.16.1 Descriptions of Mushy Region

In the *single-medium* treatment, *no* distinct *melt-mush* or *mush-solid interface* is defined, and the liquid volume fraction ϵ changes *continuously* (e.g., Bennon and Incropera, 1987; Hills et al., 1992). In the alternative *discrete* treatment, the liquid, mush, and solid are treated with *distinct interface* (e.g., Kececioglu and Rubinsky, 1989; Worster, 1991, 1992). In any of these models, the general practice is to assume that within the phase or total local volume, the concentration distribution is *uniform* (i.e., a *well-mixed* phase or total local volume). This implies that over the linear dimension of the phasic representative elementary volume, the diffusion time is much shorter than that across the linear dimension of the system. This assumption may *not* be valid whenever the two-phase mushy region, across which most of the change in the concentration occurs, is not much larger than the linear dimension of the phasic representative elementary volume. Then allowance for the variation of the concentration in the phasic representative elementary volumes must be made (Fujii, et al., 1979; Rappaz and Voller, 1990; Poirier, et al., 1991; Sundarraj and Voller, 1993).

Presently the local volume-averaged descriptions of the species, momentum, and energy conservation in multicomponent solidification is *evolving*. The *applicability* of the various proposed *models* to particular *liquids*, *rates of solidification*, geometries of the bounding solid surface, and the melt motions (i.e., thermo- and diffusocapillarity, thermo- and diffusobuoyancy, phase-density buoyancy, forced flow) and *flow structures* (i.e., laminar, transional, turbulent) are *not* yet fully examined and classified. Even when this becomes available, the *model selection* will depend on the particular phenomenon under study. Here we first review the results of a *discrete-media model* on the instability of the liquid and the mushy regions during solidification. Then the results of a *single-medium* model which reveal some, but not all, of the predictions of the discrete-media model are examined. The single-medium model is more convenient to implement for multidimensional systems and predicts some of the relevant *localized* and *gross* features.

4.16.2 Instabilities in Liquid and Mushy Regions

As an example of flow and flow instabilities (i.e., transitions) in solidification of a binary liquid, consider the *layer, unidirectional solidification* occurring given a horizontal solid surface. In the case of *lighter* species being rejected from the solid (i.e., heavier species taken up by the solid), this buoyant residual fluid rises, and a *diffusobuoyant* flow takes place (the density stratification due to the temperature stratification is *stabilizing*). This cooling from below with lighter species ejected is a special case for the possible *directions* of the cooled bounding surface (with respect to gravity) and possible binary systems. These possible combinations have been examined by Huppert (1990).

In the solidification of the *aqueous solutions* of *ammonium chloride* shows some interesting convection patterns, indicative of the *flow transitions*, and has been studied both experimentally and theoretically (e.g., Worster, 1991, 1992). As the solidification proceeds, three stages of the solidification and fluid motion have been identified. The first period is the *dendritic growth* with a nearly *uniform* mushy layer thickness δ and diffusobuoyant motion, above the mushy layer, due to the ejection of the lighter species. Next a few *isolated convective plumes* appear and penetrate into and beyond the buoyant recirculating flow. Eventually, a *chimney* or *vent* is observed beneath each plume and extending through the mushy layer to the solidified surface. In the final stage, the number of chimneys *decreases* after first increasing in both number and strength.

The study of these stages have been attempted by the *single-medium* treatment (Nielson and Incropera, 1993a, b) and by the *discrete-media* treatment (Worster, 1991, 1992). The *discrete-media* model has been used mass transfer in the two-phase mushy region and in the liquid layer above it. His results, which are reviewed below, show that the *disturbances* which grow in the *liquid layer* (called the *boundary-layer mode*), do *not* penetrate into the mushy layer, while those which grow in the mushy layer (called the *mushy-layer mode*) cause perturbation to the solid volume fraction and are interpreted to be the *origin* of the chimneys. But the numerical results of the single-medium treatment, reviewed next in Section 4.16.3, show that the chimney originates from the mushy layer-liquid interface. Since the models and the assumptions and the solution methods are different, no conclusions have yet been made about the most appropriate model for the description of the origin of the chimneys. However, many other predicted features are similar between the models.

Figure 4.51 gives a schematic of the problem considered. The completely solidified layer is shown with $x < 0$. At the interface of this layer and the two-phase mushy layer, the temperature is the eutectic temperature T_e. The liquid temperature far from the mushy layer-liquid located at $x = \delta$ is T_∞. The far-field concentration is $(\rho_A/\rho)_{\ell s_\infty}$. This far-field liquid temperature is above the liquidus temperature $T_{\ell s}(\rho_A/\rho)_{\ell s_\infty}$. The temperature distribution based on the liquidus line is also shown. The temperature stratification results in a *thermal-density* distribution ρ_T which is *stable* (i.e., decreasing with x), while the solute concentration stratification results in an *unstable solutal-density* distribution ρ_S. When the solute contribution to the total density $\rho = \rho_T + \rho_S$ dominates, then an unstable *total* density stratification occurs, as shown in Figure 4.51. The *entire* mushy layer has a unstable stratification, while *only* up to a *short* distance from the interface the liquid has an unstable stratification.

A phase diagram for a binary solid-liquid system in shown in Figure 3.6. Here the concentration on the solidus line is designated by $(\rho_A/\rho)_{s\ell}$ and that on the liquidus line is referred to as $(\rho_A/\rho)_{\ell s}$. The eutectic concentration is $(\rho_A/\rho)_e$.

The equations of conservation of mass (assuming no volume change due to the phase change), species, momentum, and energy, for the *two-phase mushy region*, are (Worster, 1991)

$$\nabla \cdot \epsilon \langle \mathbf{u} \rangle^\ell = 0 \,, \tag{4.609}$$

$$\epsilon \frac{\partial}{\partial t} \frac{\langle \rho_A \rangle^\ell}{\langle \rho \rangle^\ell} + \epsilon \langle \mathbf{u} \rangle^\ell \cdot \nabla \frac{\langle \rho_A \rangle^\ell}{\langle \rho \rangle^\ell} =$$

$$\nabla \cdot \langle D_{me} \rangle \nabla \frac{\langle \rho_A \rangle^\ell}{\langle \rho \rangle^\ell} - \left(\frac{\langle \rho_A \rangle^\ell}{\langle \rho \rangle^\ell} - \frac{\langle \rho_A \rangle^s}{\langle \rho \rangle^s} \right) \frac{\partial \epsilon}{\partial t} \,, \tag{4.610}$$

$$0 = \nabla \langle p \rangle^\ell - \frac{\mu_\ell \epsilon \langle \mathbf{u} \rangle^\ell}{K} + (\rho_\ell - \rho_\infty) \mathbf{g} \,,$$

$$\rho_\ell = \rho_\infty \{ 1 - \beta [\langle T \rangle_e - T_{\ell s} (\frac{\rho_A}{\rho} \Big|_{\ell s_\infty})] + \beta_m (\frac{\langle \rho_A \rangle^\ell}{\langle \rho \rangle^\ell} - \frac{\rho_A}{\rho} \Big|_{\ell s_\infty}) \}$$

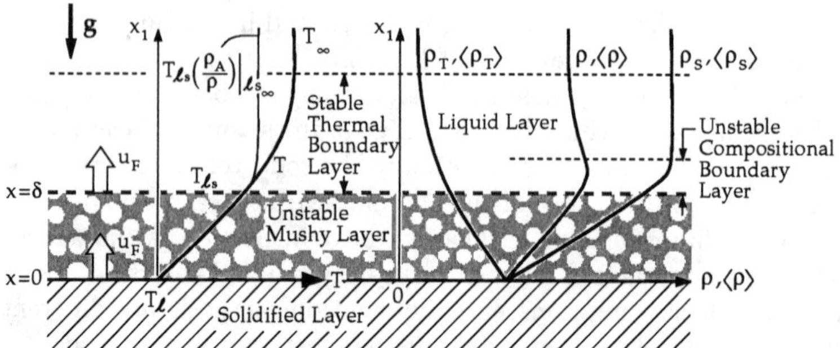

Figure 4.51. Layer solidification of a binary liquid by cooling from below. The solidified layer, mushy layer, and liquid layer are shown with the region of unstable density stratifications in the mushy and liquid layers. The temperature distributions and the thermal- and solutal-density variations are shown. The two interfaces are moving at a velocity u_F. (From Worster, reproduced by permission ©1992 Cambridge University Press.)

$$\equiv \rho_\infty[1 - \beta_1(\frac{\langle\rho_A\rangle^\ell}{\langle\rho\rangle^\ell} - \left.\frac{\rho_A}{\rho}\right|_{\ell s_\infty})] , \tag{4.611}$$

$$[\epsilon(\rho c_p)_\ell + (1-\epsilon)(\rho c_p)_s]\frac{\partial\langle T\rangle}{\partial t} + (\rho c_p)_\ell \epsilon \langle \mathbf{u}\rangle^\ell \cdot \nabla\langle T\rangle = \nabla\cdot\langle k_e\rangle\nabla\langle T\rangle - \rho_s \Delta i_{\ell s}\frac{\partial\epsilon}{\partial t} . \tag{4.612}$$

The local, liquid-phase volume-averaged velocity and concentration are used, and the buoyancy term is due to thermo- and diffusodensity stratifications. The solid phase concentration is different from the liquid, and as the porosity ϵ changes, the species and the energy equations contain source terms describing the corresponding changes in the liquid phase concentration and the temperature (local thermal equilibrium is assumed).

Worster (1992) uses the velocity of the mushy layer-liquid interface u_F, the liquid diffusivity α_ℓ, the concentration $(\rho_A/\rho)_{\ell s_\infty}$, and the phase diagram to define the dimensionless variables

$$\langle\rho_A^*\rangle = \frac{\frac{\langle\rho_A\rangle^\ell}{\langle\rho\rangle^\ell} - \left.\frac{\rho_A}{\rho}\right|_{\ell s_\infty}}{\left.\frac{\rho_A}{\rho}\right|_{\ell s_o} - \left.\frac{\rho_A}{\rho}\right|_e} , \quad \xi = \epsilon\langle\rho_A^*\rangle + (1-\epsilon)\Delta\rho_A^* ,$$

$$\Delta\rho_A^* = \frac{\left.\frac{\rho_A}{\rho}\right|_{s\ell} - \left.\frac{\rho_A}{\rho}\right|_{\ell s_\infty}}{\left.\frac{\rho_A}{\rho}\right|_{\ell s_o} - \left.\frac{\rho_A}{\rho}\right|_e} , \quad x_1^* = x^* - t^* , \quad u^* = \frac{u}{u_F} ,$$

$$\langle T^* \rangle = \frac{\langle T \rangle - T_{\ell s}(\frac{\rho_A}{\rho}\Big|_{\ell s_\infty})}{T_{\ell s}(\frac{\rho_A}{\rho}\Big|_{\ell s_\infty}) - T_e} \, , \quad t^* = \frac{t u_F^2}{\alpha_\ell} \, , \quad x^* = \frac{x u_F}{\alpha_\ell} \, . \tag{4.613}$$

In the moving coordinate x_1^*, the mushy layer-liquid interface is at $x_1^* = \delta^*$. The dimensionless forms of (4.611) to (4.612) become (by *neglecting* the species diffusion compared to the heat diffusion and by using the liquidus relation $\langle T^* \rangle = \langle \rho_A^* \rangle$)

$$(\frac{\partial}{\partial t^*} - \frac{\partial}{\partial x_1^*})\xi + \langle \mathbf{u}_\ell^* \rangle \cdot \nabla \langle T \rangle^* = 0 \, , \tag{4.614}$$

$$(\frac{\partial}{\partial t^*} - \frac{\partial}{\partial x_1^*})(\langle T^* \rangle + Ste_\ell \frac{C^* - \xi}{C^* - \langle T^* \rangle}) + \langle \mathbf{u}_\ell^* \rangle \cdot \nabla \langle T^* \rangle = \nabla^2 \langle T^* \rangle \, , \tag{4.615}$$

$$\langle \mathbf{u}_\ell^* \rangle = Ra_{\ell s} \frac{K_o}{K} (\nabla \langle p^* \rangle^\ell + \langle T^* \rangle x_1^*) \, , \tag{4.616}$$

where the Stefan and Rayleigh numbers are defined

$$Ste_\ell = \frac{\rho_s \Delta i_{\ell s}}{c_{p_\ell} [T_{\ell s}(\frac{\rho_A}{\rho}\Big|_{\ell s_\infty}) - T_e]} \, , \quad Ra_{\ell s} = \frac{\beta_1 g (\frac{\rho_A}{\rho}\Big|_{\ell s_\infty} - \frac{\rho_A}{\rho}\Big|_e) K_o}{\nu_\ell u_F} . \tag{4.617}$$

The reference permability K_o is evaluated at a reference porosity ϵ_o.

For the liquid layer, the dimensionless variables and the dimensionless species, momentum, and energy conservation equations are

$$\rho_{A,\ell}^* = \frac{\frac{\rho_{A,\ell}}{\rho_\ell} - \frac{\rho_A}{\rho}\Big|_{\ell s_\infty}}{\frac{\rho_A}{\rho}\Big|_{\ell s_o} - \frac{\rho_A}{\rho}\Big|_e} \, , \quad T_\ell^* = \frac{T - T_{\ell s}(\frac{\rho_A}{\rho}\Big|_{\ell s_\infty})}{T_{\ell s}(\frac{\rho_A}{\rho}\Big|_{\ell s_\infty}) - T_e} \, , \tag{4.618}$$

$$(\frac{\partial}{\partial t^*} - \frac{\partial}{\partial x_1^*})\rho_{A,\ell}^* + \mathbf{u}_\ell^* \cdot \nabla \rho_{A,\ell}^* = \frac{1}{Le_\ell} \nabla^2 \rho_{A,\ell}^* \, , \tag{4.619}$$

$$\frac{1}{Pr_\ell}(\frac{\partial}{\partial t^*} - \frac{\partial}{\partial x_1^*})\mathbf{u}_\ell^* + \frac{1}{Pr} \mathbf{u}_\ell^* \cdot \nabla \mathbf{u}_\ell^* = \nabla^2 \mathbf{u}_\ell^* +$$

$$Ra_T T_\ell^* \mathbf{x}_1^* + Ra_S(\rho_{A,\ell}^* \mathbf{x}_1^* + \frac{\beta_m}{\beta_1} \nabla p_\ell^*) \, , \tag{4.620}$$

$$(\frac{\partial}{\partial t^*} - \frac{\partial}{\partial x_1^*}) T_\ell^* + \mathbf{u}_\ell^* \cdot \nabla T_\ell^* = \nabla^2 T_\ell^* \, , \tag{4.621}$$

where the Lewis, Prandtl, and Rayleigh (for heat and solute) numbers are defined, respectively, as

$$Le_\ell = \frac{D_\ell}{\alpha_\ell}\,, \quad Pr_\ell = \frac{\nu_\ell}{\alpha_\ell}\,, \quad Ra_T = \frac{\beta g[T_{\ell s}(\frac{\rho_A}{\rho}\Big|_{\ell s_\infty}) - T_\ell]\alpha_\ell^2}{\nu_\ell u_F^3}\,,$$

$$Ra_S = \frac{\beta_m g(\frac{\rho_A}{\rho}\Big|_{\ell s_\infty} - \frac{\rho_A}{\rho}\Big|_e)\alpha_\ell^2}{\nu_\ell u_F^3}\,. \tag{4.622}$$

The boundary conditions are

$$\langle T^* \rangle = -1\,, \quad \mathbf{u} = 0 \qquad x^* = 0\,, \tag{4.623}$$

$$\langle T \rangle^* = T_\ell^*\,, \quad \mathbf{n}\cdot\nabla T_\ell^* = \mathbf{n}\cdot\nabla {\rho^*}_\ell\,, \quad [\mathbf{n}\cdot\mathbf{u}_\ell] = 0\,,$$
$$[T^*] = 0\,, \quad [\mathbf{n}\cdot\nabla T^*] = 0\,, \quad [p^*] = 0\,, \quad \xi = \langle \rho_A \rangle^* \qquad x^* = \delta^* \tag{4.624}$$

$$T_\ell^* = T_\infty^*\,, \quad \rho_{A,\ell}^* = 0 \qquad x^* \to \infty\,, \tag{4.625}$$

where square brackets indicate jump in the quantity across the interface (Worster, 1992).

A set of *steady*-state (i.e., *base* field) solutions for the temperature and species concentration distributions exist in which the interface is at $x^* = \delta_o^*$ and $\mathbf{u}_\ell = \langle \mathbf{u} \rangle^\ell = 0$. These are (Worster, 1991)

$$x^* = \frac{a_1 - Ste_\ell}{a_1 - a_2} \ln \frac{a_1 - 1}{a_1 - \langle T^* \rangle} + \frac{Ste_\ell - a_2}{a_1 - a_2} \ln \frac{a_2 + 1}{a_2 - \langle T^* \rangle}\,, \tag{4.626}$$

$$\langle T^* \rangle = \langle \rho_A^* \rangle \qquad 0 < x^* < \delta_o^*\,, \tag{4.627}$$

$$\langle T^* \rangle = T_\ell^* = -\frac{T_\infty^*}{Le_\ell - 1} = \langle \rho_A^* \rangle \qquad x^* = \delta_o^*\,, \tag{4.628}$$

$$T_\ell^* = T_\infty^* + [T_\ell^*(x^* = \delta_o^*) - T_\infty^*]e^{-(x^* - \delta_o^*)}\,,$$
$$\rho_{A,\ell}^* = T_\ell^*(x^* = \delta_o^*)e^{-Le_\ell(x^* - \delta_o^*)} \qquad x^* > \delta_o^*\,, \tag{4.629}$$

where

$$a_1 = A + B\,, \quad a_2 = A - B\,, \quad A = \tfrac{1}{2}(Ste_\ell + T_\infty^* + \Delta\rho_A^*)\,,$$
$$B^2 = A^2 - \Delta\rho_A^* T_\infty^* - Ste_\ell T_\ell^*(x^* = \delta_o^*)\,. \tag{4.630}$$

The *base* value for the thickness of the mushy layer δ_o^* is found from (4.626) by using $\langle T^* \rangle = 0$.

The base, *total* density distribution are shown in Figure 4.51, where the thermal stratification is destabilizing. Note that in the example shown, the density difference across the mushy layer is larger than that across the semi-infinte liquid layer.

Using these base solutions, Worster (1992) applies the *linear stability theory* to the mushy region and the liquid conservation equations, and obtains the condition for the *marginal stability* of the flow in the mushy region and the liquid. The method applied is similar to that discussed in

Section 4.12.1(C) for the *marginal* stability (i.e., *no* growth or decay in the amplitude of the disturbances introduced into the base fields) of thermobuoyant flows. Note that since some of the *liquid* phase properties are used, *instead* of the *effective* properties, for the mushy region, the Rayleigh numbers are related through

$$Ra_T = a_L a_\Delta Ra_{\ell s}\ ,\quad Ra_S = (1+a_\Delta) a_L Ra_{\ell s}\ , \tag{4.631}$$

where

$$a_L = \frac{(\alpha_\ell/u_F)^2}{K_o}\ ,\quad a_\Delta = \frac{\beta[T_{\ell s}(\left.\frac{\rho_A}{\rho}\right|_{\ell s\infty}) - T_e]}{\beta_1[\left.\frac{\rho_A}{\rho}\right|_{\ell s} - \left.\frac{\rho_A}{\rho}\right|_e]} = \frac{\beta}{\beta_1}\gamma\ . \tag{4.632}$$

In the above, γ is the slope of the liquidus line.

The eigenrelation between the *dimensionless* wave number η^* corresponding to the mushy-region Rayleigh number $Ra_{\ell s}$ is found in a similar manner as that discussed in Sections 2.2.2 and 4.12.1(C), and typical results of Worster (1992) are shown in Figure 4.52(a) to (c). The results are for $Pr_\ell = 10$ (typical of *aqueous solutions*), $Ste_\ell = \Delta\rho_A^* = T_\infty^* = 1$, $a_\Delta = 0$ (i.e., *no* thermobuoyancy), $Le_\ell = 40$, $a_L = 10^5$, and for *uniform* permeability.

As evident in Figure 4.52(a), for $\eta^* = 2.25$ there is an instability which, as shown in Figure 4.52(b), causes motion in the mushy layer and the liquid. This wave number corresponds to a wavelength $\lambda^* = 2\pi/\eta^*$ which is *comparable* to the thickness of the mushy layer. At a higher wave number $\eta^* = 13.3$, there is a flow pattern shown in Figure 4.52(c) which is *in* the liquid boundary layer. This does *not* penetrate far into the mushy layer.

The *diffusobuoyant* flow in the mushy layer causes a change in the porosity (i.e., liquid fraction). These are local *melting* (decrease in porosity) and *densification* of solid matrix (increase in porosity) and Worster (1992) suggests, based on the results of the marginal instability, that these local changes in porosity are the *origin* of the *chimneys*. The later stages of the flow, i.e., beyond the marginal states, can be determined by solving the *nonlinear* conservation equations. These are done numerically and by using a *finite* initial liquid volume (i.e., an enclosure). In the next section, an example of these computaional results is discussed; however, the model used is the single-medium model (which is suitable for numerical solutions).

4.16.3 Thermo- and Diffusobuoyant Motion in Enclosures

As in casting, the solidification can occur in enclosures. The bounding solid surface *completely* encloses the melt and when the solidified melt and the melt have the same density, the solidified melt is also completely enclosed. In order to continue the discussion on chimney formation during the layer

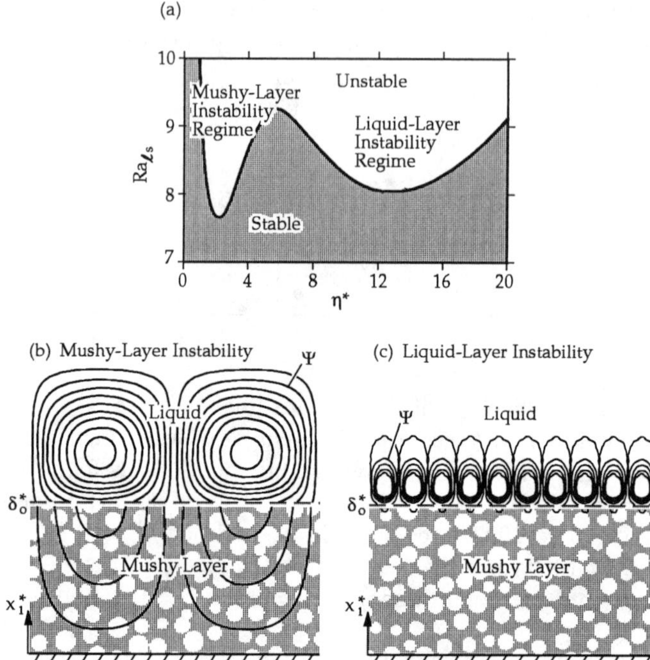

Figure 4.52. (a) Map of marginal instability given in the $Ra_{\ell s} - \eta^*$ diagram showing the mushy-layer and the liquid-layer instability regimes. (b) The flow field in the liquid and mushy layers in the mushy-layer instability regime. (c) The liquid-layer and a smaller mushy-layer flow fields in the liquid-layer instability regime. (From Worster, reproduced by permission ©1992 Cambridge Unversity Press.)

solidification on top of a cooled bounding surface (i.e., *unidirectional solidification*), this problem is now examined with a finite melt volume contained in a *cylinderical* enclosure with its axis parallel to the gravity vector. Using the *single-medium treatment* described in Section 3.2.7, Nielson and Incropera (1993b) solve the *transient, three-dimensional* conservation equations numerically for an aqueous NH_4Cl solution. Their formulation and results are reviewed below and a comparison is made to the results of Worster (1992) discussed above. The melt is assumed to be initially quiescent and the motion induced by the cooling and the solidification is due to the thermo- and diffusobuoyancy.

We begin by rewriting the conservation equations which treat the solidified melt, the two-phase mushy region, and the melt as a single medium with a *variable* liquid volume fraction ϵ (i.e., the porosity). In the analysis of Nielson and Incropera (1993b), the density of the solid and the fluid phase are assumed equal, i.e.,

$$\rho_s = \rho_\ell = \rho \,, \quad \frac{\langle\rho\rangle^\ell}{\langle\rho\rangle} = \epsilon \,, \quad \frac{\langle\rho\rangle^s}{\langle\rho\rangle} = 1 - \epsilon \,. \tag{4.633}$$

The conservation equations are the simplified forms of (3.91) to (3.94), and are written with x, r, and θ as the axes, and u, v, and w as the velocity components. The overall mass, species, momentum (three components), and energy equations are (Nielson and Incropera, 1993b)

$$\frac{\partial\rho}{\partial t} + \nabla\cdot\rho\mathbf{u} = 0 \,, \tag{4.634}$$

$$\frac{\partial}{\partial t}\rho_A + \nabla\cdot\mathbf{u}\rho_A = \nabla\cdot\rho D\nabla\frac{\rho_A}{\rho} + \nabla\cdot\rho D\nabla(\frac{\rho_{A,\ell}}{\rho} - \frac{\rho_A}{\rho}) - \nabla\cdot[\rho\frac{1-\epsilon}{\epsilon}\mathbf{u}(\frac{\rho_A}{\rho} - \frac{\rho_{A,s}}{\rho})] \,, \tag{4.635}$$

$$\frac{\partial}{\partial t}\rho u + \nabla\cdot\rho\mathbf{u}u = -\frac{\partial p}{\partial x} + \nabla\cdot\mu\nabla u - \frac{\mu}{K}u - \rho g \,, \tag{4.636}$$

$$\rho = \rho\{1 - \beta(T - T_o) - \beta_m[\frac{\rho_{A,\ell}}{\rho} - (\frac{\rho_{A,\ell}}{\rho})_o]\} \,, \tag{4.637}$$

$$\frac{\partial}{\partial t}\rho v + \nabla\cdot\rho\mathbf{u}v = -\frac{\partial p}{\partial r} + \nabla\cdot\mu\nabla v - \frac{\mu}{K}v + \frac{\rho w^2}{r} - \frac{\mu u}{r^2} - \frac{2}{r^2}\mu\frac{\partial w}{\partial\theta} \,, \tag{4.638}$$

$$\frac{\partial}{\partial t}\rho w + \nabla\cdot\rho\mathbf{u}w = -\frac{1}{r}\frac{\partial p}{\partial\theta} + \nabla\cdot\mu\nabla w - \frac{\mu}{K}w - \frac{\rho vw}{r} + \frac{\mu w}{r^2} - \frac{2}{r^2}\mu\frac{\partial v}{\partial\theta} \,, \tag{4.639}$$

$$\frac{\partial}{\partial t}\rho i + \nabla\cdot\rho\mathbf{u}i = \nabla\cdot\frac{k}{c_{p_s}}\nabla i + \nabla\cdot\frac{k}{c_{p_s}}\nabla(i_s - i) - \nabla\cdot[\rho\frac{1-\epsilon}{\epsilon}\mathbf{u}(i - i_s)] \,. \tag{4.640}$$

The variation of the permeability with the porosity is modelled using the relation

$$K = K_o\frac{\epsilon^2}{(1-\epsilon)^2} \,. \tag{4.641}$$

Nielson and Incropera (1993b) use a reference permeability $K_o = 5.6 \times 10^{-11}$ m^2, constant thermophysics properties for the aqueous NH_4Cl solution (except for the density), and a cylindrical geometry with a radius of $R = 6.4$ cm and a height $L = 10.2$ cm. For the computational convenience, they use a angular domain of 60° with symmetry applied at the $\theta = 0$ and $\theta = 60°$ boundaries. An initial composition $\rho_{A,\ell}/\rho = 0.32$, which is larger than the eutectic concentration of 0.20, is used in order to achieve a *large* mushy layer. The eutectic temperature is $T_e = -15.4$°C and the

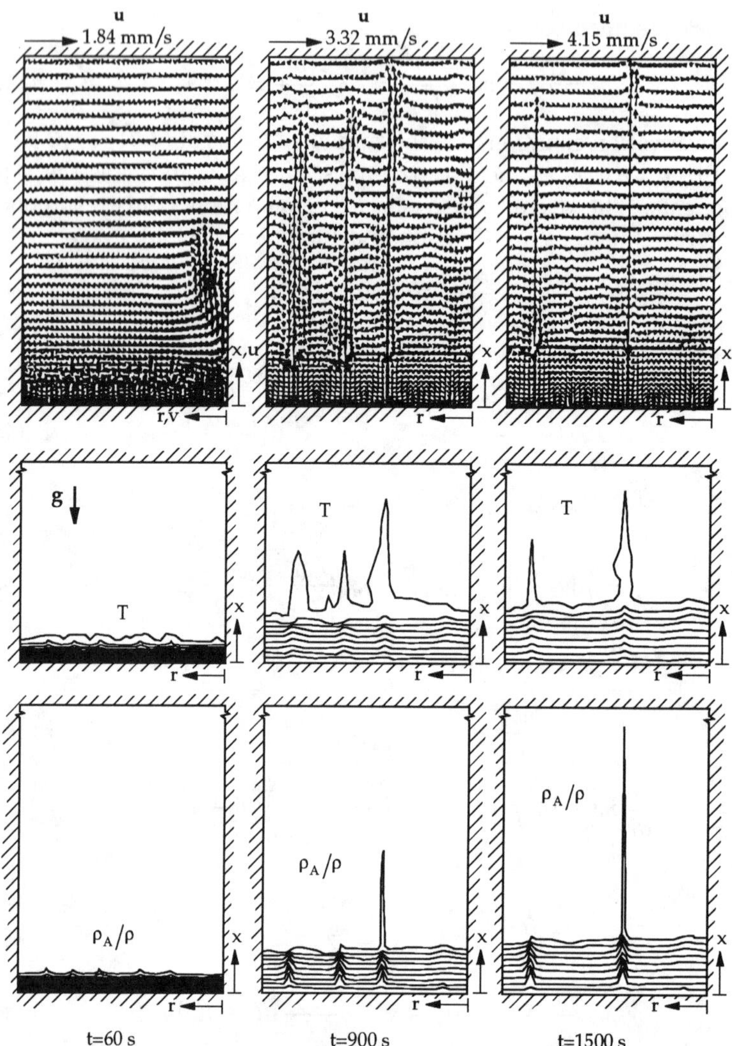

Figure 4.53. Distributions of the velocity (shown with vectors), temperature, and species concentration in a cylindrical mold during solidification of aqueous NH_4Cl. The results are for three different elapsed times and for a x-r plane at $\theta = 37.5°$. (From Nielson and Incorpera, reproduced by permission ©1993 Taylor and Francis.)

initial temperature $T_o = 50°C$. The lower bounding surface is maintained at 14°C while the other bounding surfaces are *adiabatic*. Some of their typical results are shown in Figure 4.53. The results are for a plane crossing at $\theta = 37.5°$ and for three different elapsed times. *Perturbations* in the *porosity* is introduced in order to observe the growth or decay of these disturbances (simulating natural disturbances).

As shown in Figure 4.53, at $t = 60$ s, shortly *after the onset of convection, rigorous* motion is present in the melt near the mushy layer. This motion penetrates into the mushy layer *and* the bulk melt. The velocity vector distribution in the x-r plane at $\theta = 37.5°$ shows a strong upflow adjacent to the cylinder axis. The *colder, water-rich* (i.e., NH_4Cl depleted) melt penetrates into the liquid, as is evident by the distribution of the temperature and the concentration of NH_4Cl (species A). The channels are formed and are large in number. The penetration of the channels into the mushy layer continues.

At $t = 900$ s, the *plume* energing from the channels penetrate *deep* into the melt and *also* into the mushy layer. These are seen in the distributions of $\mathbf{u}, T$ and ρ_A/ρ. The channel in the mushy layer grows by *remelting*. Because of the *low* mass diffusivity, the rising plumes do not undergo noticeable change in their concentration.

At $t = 1500$ s, some of the channels have *disappeared* (i.e., closed) and the growth of the solidification layer has decreased (because of the increase in the *conduction resistance* in the solidified layer). After the onset of convection, there is initially an *increase* in the velocity of the thermo- and diffusobuoyant flows. However, as the melt temperature decreases and the concentration of the lighter species in it increases, the velocities begin to *decrease* (the bulk composition approaches the eutectic composition).

The prediction of this numerical, three-dimensional simulation is in agreement with the three stages mentioned in Section 4.16.2. The origin of the chimney formation in this simulation appears to be the motion *in* the *melt adjacent* to the mushy layer (as compared to the mushy-layer origin predicted by Worster, 1992). The thermo- and diffusobuoyancy in the mushy layer has also been examined by Amberg and Homsy (1993). Both the *stability analysis* based on the *discrete-media* treatment and the *numerical simulation* based on the *single-medium* treatment, show that convection in the melt influences the rate of the solidification and the *quality* of the solid (i.e., crystal) formed. The solidification of single- and multicomponent melts will be further discussed in Sections 7.3.1 and 7.3.2.

4.17 References

Al-Sanea, S., 1992, "A Numerical Study of the Flow and Heat Transfer Characteristics of an Impinging Laminar Slot-Jet Including Crossflow Effects," *Int. J. Heat Mass Transfer*, 35, 2501–2513.

Amberg, G., and Homsy, G.M., 1993, "Nonlinear Analysis of Buoyant Convection in Binary Solidification with Application to Channel Formation," *J. Fluid Mech.*, 252, 79–98.

Antonia, R.A., and Kim, J., 1991, "Reynolds Shear Stress and Heat Flux Calculations in a Fully Developed Turbulent Duct Flow," *Int. J. Heat Mass Transfer*, 34, 2013–2018.

Apte, V.B., Bilger, R.W., Green, A.R., and Quintere, J.G., 1991, "Wind-Aided Turbulent Flame Spread and Burning over Large-Scale Horizontal PMMA Surfaces," *Combust. Flame*, 85, 169–184.

Arpaci, V.S., and Larsen, P.S., 1984, *Convective Heat Transfer*, Prentice-Hall, Englewood Cliffs, NJ.

Banerjee, S., 1991, "Turbulence/Interface Interactions", in *Phase-Interface Phenomena in Multiphase Flow*, Hewitt, G.F., Mayinger, F., and Riznic, J.R., Editors, Hemisphere Publishing Company, New York.

Bau, H.H., 1992, "Controlling Chaotic Convection," in *Theoretical and Applied Mechanics 1992*, Bonder, S., et al., Editors, 187–203, Elsevier, Amsterdam.

Bau, H.H., and Wang, Y.-Z., 1992, "Chaos: A Heat Transfer Perspective," *Ann. Rev. Heat Transfer*, 4, 1–50.

Bejan, A., 1984, *Convection Heat Transfer*, John Wiley and Sons, New York.

Bennen, W.D., and Incropera, F.P., 1987, "A Continuum Model for Momentum, Heat and Species Transport in Binary Solid-Liquid Phase Change Systems. I. and II." *Int. J. Heat Mass Transfer*, 30, 2161–2187.

Bergeles, G., Gosman, A.D., and Launder, B.E., 1981, "The Prediction of Three-Dimensional Discrete-Hole Cooling Processes," *ASME J. Heat Transfer*, 103, 141–145.

Bhattacharjee, S., 1993, "A Comparison of Numerical and Analytical Solutions of the Creeping Flame Spread Over Thermally Thin Material," *Combust. Flame*, 93, 434–444.

Bird, R.B., Armstrong, R.C., and Hassager, O., 1987, *Dynamics of Polymeric Liquids, Volume 1: Fluid Mechanics, Volume 2: Kinetic Theory*, Second Edition, John Wiley and Sons, New York.

Biswas, G., and Chattopadhyay, H., 1992, "Heat Transfer in a Channel with Built-in Wing-type Vortex Generators," *Int. J. Heat Mass Transfer*, 35, 803–814.

Böhme, G., *Non-Newtonian Fluid Mechanics*, 1982, North-Holland, Amsterdam.

Bouillard, J.X., and Berry, G.F., 1992, "Performance of a Multigrid Three-Dimensional Magnetohydrodynamic Generator Calculation Procedure," *Int. J. Heat Mass Transfer*, 35, 2219–2232.

Branover, H., Lykoudis, P.S., and Mond, M., 1985, *Single- and Multiphase Flows in an Electromagnetic Field. Energy, Metallurgical, and Solar Applications*, Volume 100, Progress in Astronautics and Aeronautics, American Institute of Aeronautic and Astronautics, New York.

Bray, K.N.C., Champion, M., and Libby, P.A., 1992, "Premixed Flames in Stagnation Turbulence. Part III— The $\tilde{k} - \tilde{\epsilon}$ Theory for Reactants Impinging on a Wall," *Combust. Flame*, 91, 165–186.

Brown, M.A., 1992, *Thermally Induced Pressure Waves in a Gas: Experimental Observation and Theoretical Prediction of Thermoacoustic Convection*, Ph.D. Thesis, University of Pennsylvania.

Brown, M.A., and Churchill, S.W., "Transient Behavior of an Impulsively Heated Fluid," *Chem. Eng. Technol.*, 16, 82–88.

Burmeister, L.C., 1983, *Convective Heat Transfer*, John Wiley and Sons, New York.

Cancaster, J.F., Editor, 1986, *The Physics of Welding*, Pergamon Press, Oxford.

Carey, V.P., 1992, "Monte Carlo Simulation of the Operating Characteristics of an Ultraminature Hot-Film Sensor in High-Speed Gas Flow," in *Heat Transfer on the Microscale*, HTD-Vol. 200, 45–54, American Society of Mechanical Engineers, New York.

Cebeci, T., and Bradshaw, P., 1984, *Physical and Computational Aspects of Convection Heat Transfer*, Springer-Verlag, New York.

Chandrasekhar, S., 1981, *Hydrodynamic and Hydromagnetic Stability*, Dover, New York.

Chen, X., 1990, "Heat Transfer and Flow in a Radio Frequency Plasma Torch-A New Modeling Approach," *Int. J. Heat Mass Transfer*, 33, 815–826.

Chen, Y.-M., 1985, "Heat Transfer of a Laminar Flow Passing a Wedge at Small Prandtl Number: A New Approach," *Int. J. Heat Mass Transfer*, 28, 1517–1523.

Cheng, P., 1978a, "Convective Heat Transfer in Porous Layers by Integral Methods," *Comm. Heat Mass Transfer*, 5, 243–252.

Cheng, P., 1978b, "Heat Transfer in Geothermal Systems," *Advan. Heat Transfer*, 14, 1–105.

Cheng, P., and Minkowycz, W.J., 1977, "Free Convection About a Vertical Flat Plate Embedded in a Porous Medium with Application to Heat Transfer from a Dike," *J. Geophs. Res.*, 82, 2040–2044.

Cho, Y.I., and Hartnett, J.P., 1982, "Non-Newtonian Fluids in Circular Pipe Flow," *Advan. Heat Transfer*, 15, 59–141.

Chung, P.M., 1965, "Chemically Reacting Nonequilibrium Boundary Layers," *Advan. Heat Transfer*, 2, 109–270.

Churchill, S.W., 1977, "A Comprehensive Correlating Equation for Laminar, Assisting, Forced and Free Convection," *AIChE J.*, 23, 10–16.

Churchill, S.W., and Brown, M.A., 1987, "Thermoacoustic Convection and the Hyperbolic Equation of Convection," *Int. Comm. Heat Mass Transfer*, 14, 647–655.

Coleman, H.W., Moffat, R.J., and Kays, W.M., 1981, "Heat Transfer in the Accelerated Fully Rough Turbulent Boundary Layer," *ASME J. Heat Transfer*, 103, 153–158.

Craft, T.J., Graham, L.J.W., and Launder, B.E., 1993, "Impiniging Jet Studies of Turbulence Model Assessment. II. An Examiniation of the Performance of Four Turbulence Models," *Int. J. Heat Mass Transfer*, 36, 2685–2697.

Cramer, K.R., 1963, "Several Magnetohydrodynamic Free-Convection Solutions," *ASME J. Heat Transfer*, 85, 35–40.

Cunningham, R.E., and Williams, R.J.J., 1980, *Diffusion in Gases and Porous Media*, Plenum Press, New York.

Depew, C.A., and Kramer, T.J., 1973, "Heat Transfer to Flowing Gas-Solid Mixtures," *Advan. Heat Transfer*, 9, 113–180.

de Ris, J., 1969, "Spread of a Laminar Diffusion Flame," *Twelfth Symposium (International) on Combustion*, 241-252, *The Combustion Institute*, Pittsburgh.

Dietz, P.W., and Melcher, J.R., 1975, "Field Controlled Heat Transfer Involving Macroscopic Charged Particles in Liquids," *ASME J. Heat Transfer*, 97, 429–434.

Eckert, E.R.G., and Drake, R.M., Jr., 1972, *Analysis of Heat and Mass Transfer*, McGraw-Hill, New York.

Ede, A.J., 1967, "Advances in Natural Convection," *Advan. Heat Transfer*, 4, 1–64.

Edwards, D.K., 1981, *Radiation Heat Transfer Notes*, Hemisphere Publishing Company, New York.

Eibeck, P.A., and Eaton, J.K., 1987, "Heat Transfer Effects of a Longitudinal Vortex Embedded in a Turbulent Boundary Layer," *ASME J. Heat Transfer*, 109, 16–24.

Eichhorn, R., 1960, "The Effect of Mass Transfer on Free Convection," *ASME J. Heat Transfer*, 82, 260–263.

Emmons, H.W., 1956, "The Film Combustion of Liquid Fuel," *Z. Agnew. Math. Mech.*, 36, 60–71.

Faraco-Medeiros, M.A., and Silva-Freire, A.P., 1992, "The Transfer of Heat in Turbulent Boundary Layers with Injection or Suction: Universal Lawa and Stanton Number Equations," *Int. J. Heat Mass Transfer*, 35, 991–995.

Farber, L., and Depew, C.A., 1963, "Heat Transfer Effects to Gas-Solids Mixtures Using Solid Spherical Particles of Uniform Size," *I & EC Fundamentals*, 2, 130–135.

Fray, A.E., and T'ien, J.S., 1979, "A Theory of Flame Spread over a Solid Fuel Including Finite-Rate Chemical Reactants," *Combust. Flame*, 36, 263–289.

Fujii, T., Poirier, D.R., and Flemings, M.C., 1979, "Macrosegregation in a Multicomponent Low Alloy Steel," *Metall. Trans.*, 10B, 331–339.

Fujino, T., Yokohama, Y., and Mori, Y.H., 1989, "Augmentation of Laminar Forced-Convective Heat Transfer by the Application of a Transverse Electric Field," *ASME J. Heat Transfer*, 111, 345–351.

Galante, S.R., and Churchill, S.W., 1990, "Applicability of Solutions of Convection in Potential Flows," *Advan. Heat Transfer*, 20, 353-388.

Gebhart, B., Jaluria, Y., Mahajan, R.L., and Sammakia, B., 1988, *Buoyancy-Induced Flows and Transport*, Hemisphere Publishing Corporation, Washington, D.C.

Giedt, W.H., and Willis, D.R., 1985, "Rarefied Gases," in *Handbook of Heat Transfer, Fundamentals*, Rohsenow, W.M., et al., Editors, Second Edition, McGraw-Hill, New York.

Girshick, S.L., and Yu, W., 1990, "Radio-Frequency Induction Plasmas at Atmospheric Pressure: Mixtures of Hydrogen, Nitrogen, and Oxygen with Argon," *Plasma Chem. Plasma Process.*, 10, 515–529.

Glassman, I., 1987, *Combustion*, Second Edition, Academic Press, Orlando.

Goldstein, R.J., 1971, "Film Cooling," *Advan. Heat Transfer*, 7, 321–379.

Goldstein, R.J., Behbahani, A.I., and Heppelmann, K.K., 1986, "Streamwise Distribution of the Recovery Factor and the Local Heat Transfer Coefficient for an Impinging Circular Air Jet," *Int. J. Heat Mass Transfer*, 29, 1227–1235.

Goldstein, R.J., and Seol, W.S., 1991, "Heat Transfer to a Row of Impinging Circular Air Jets Including the Effect of Entrainment," *Int. J. Heat Mass Transfer*, 34, 2133–2147.

Grassman, P., and Tuma, M., 1979, "Critical Reynolds Number for Oscillating and Pulsating Tube Flow," (in German), *Thermo-Fluid Dyn.*, 12, 203–209.

Greenberg, J.B., and Ronrey, P.D., 1993, "Analysis of Lewis Number Effects in Flame Spread," *Int. J. Heat Mass Transfer*, 36, 315–323.

Hartnett, J.P., 1992, "Viscoelastic Fluids: A New Challenge in Heat Transfer," *ASME J. Heat Transfer*, 114, 296–303.

Hartnett, J.P., and Kostic, M., 1989, "Heat Transfer to Newtonian and Non-Newtonian Fluids in Rectangular Ducts," *Advan. Heat Transfer*, 19, 247–356.

Henkes, R.A.W.M., and Hoogendoorn, C.J., 1989, "Comparison of Turbulence Models for the Natural Convection Boundary Layer Along a Heated Vertical Plate," *Int. J. Heat Mass Transfer*, 32, 157–169.

Hills, R.N., Loper, D.E., and Roberts, P.H., 1992, "On Continuum Models for Momentum, Heat and Species Transport in Solid-Liquid Phase Change Systems," *Int. Comm. Heat Mass Transfer*, 19, 585–594.

Hirano, T., and Kinoshita, M., 1974, "Gas Velocity and Temperature Profiles of a Diffusion Flame Stabilized in the Stream over Liquid Fuel," *Fifteenth Symposium (International) on Combustion*, 379–387, The Combustion Institute, Pittsburgh.

Hollworth, B.R., and Gero, L.R., 1985, "Entrainment Effect on Impingement Heat Transfer: Part II. Local Heat Transfer Measurements," *ASME J. Heat Transfer*, 107, 910–915.

Hong, J.-T., Tien, C.-L., and Kaviany, M., 1985, "Non-Darcean Effects on Vertical Plate Natural Convection in Porous Media with High Porosity," *Int. J. Heat Mass Transfer*, 28, 2149–2157.

Hosni, M.H., Coleman, H.W., Garner, J.W., and Taylor, R.P., 1993, "Roughness Element Shape Effects on Heat Transfer and Skin Friction in Rough-Wall Turbulent Boundary Layers," *Int. J. Heat Mass Transfer*, 36, 147–153.

Huang, Y., and Bau, H.H., 1993, "Thermoacoustic Convection," in *Heat Transfer in Micro Gravity*, ASME HTD Vol. 269, 1–9, American Society of Mechanical Engineers, New York.

Huppert, H.E., 1990, "The Fluid Mechanics of Solidification," *J. Fluid Mech.*, 212, 209–240.

Jaluria, Y., 1980, *Natural Convection Heat and Mass Transfer*, Pergamon Press, Oxford.

Joshi, S.V., Liang, Q., Park, J.Y., and Batdorf, J.A., 1990, "Effect of Quenching Conditions on Particle Formation and Growth in Thermal Plasma Synthesis of Fine Powders," *Plasma Chem. Plasma Process.*, 10, 339–358.

Kang, S.H., and Grief, R., 1993, "Thermophoretic Transport in the Outside Vapor Deposition Process," *Int. J. Heat Mass Transfer*, 36, 1007–1018.

Kasagi, N., Kuroda, A., and Hirata, M., 1989, "Numerical Investigation of Near-Wall Turbulent Heat Transfer Taking into Account the Unsteady Heat Transfer in the Solid Wall," *ASME J. Heat Transfer*, 111, 385–392.

Kasagi, N., Tomita, Y., and Kuroda, A., 1992, "Direct Numerical Simulation of Passive Scalar Fields in a Turbulent Channel Flow," *ASME J. Heat Transfer*, 114, 598–606.

Kaviany, M., 1985, "Laminar Flow Through a Porous Channel Bounded by Isothermal Parallel Plates," *Int. J. Heat Mass Transfer*, 28, 851–858.

Kaviany, M., 1987, "Boundary Layer Treatment of Forced Convection Heat Transfer from a Semi-Infinite Flat Plate Embedded in Porous Media," *ASME J. Heat Transfer*, 109, 345–349.

Kaviany, M., 1988, "Heat Transfer About a Permeable Membrane," *ASME J. Heat Transfer*, 110, 514–516.

Kaviany, M., 1990, "Performance of a Heat Exchanger Based on Enhanced Heat Diffusion in Fluids by Oscillation: Analysis," *ASME J. Heat Transfer*, 112, 49–55.

Kaviany, M., 1991, *Principles of Heat Transfer in Porous Media*, Springer-Verlag, New York.

Kaviany, M., and Miltal, M., 1987, "Natural Convection Heat Transfer from a Vertical Plate to High Permeability Porous Media: An Experiment and an Approximate Solution," *Int. J. Heat Mass Transfer*, 30, 967–977.

Kaviany, M., and Reckker, M., 1990, "Performance of a Heat Exchanger Based on Enhanced Heat Diffusion in Fluids by Oscillation: Experiment," *ASME J. Heat Transfer*, 112, 56–63.

Kays, W.M., and Crawford, M.E., 1993, *Convective Heat and Mass Transfer*, Third Edition, McGraw-Hill, New York.

Keanini, R., and Rubinsky, B., 1993, "Three-Dimesional Simulation of the Plasma Arc Welding Process," *Int. J. Heat Mass Transfer*, 36, 3283–3298.

Kececioglu, I., and Rubinsky, B., 1989, "A Continuum Model for the Propagation of Discrete Phase-Change Fronts in Porous Media in the Presence of Coupled Heat Flow, Fluid Flow and Species Transport Processes," *Int. J. Heat Mass Transfer*, 32, 1111–1130.

Kerstein, A.R., 1992, "Linear Eddy Modeling of Turbulent Transport. Part 7. Finite-Rate Chemistry and Multi-Stream Mixing," *J. Fluid Mech.*, 240, 289–313.

Kestin, J., 1966, "The Effects of Free-Stream Turbulence on Heat Transfer Rates," *Advan. Heat Transfer*, 3, 1-32.

Kestin, J., Maeder, P.F., and Wang, H.E., 1961, "On Boundary Layers Associated with Oscillating Streams," *Appl. Sci. Res.*, A10, 1-22.

Kim, J.-K., and Aihara, T., 1992, "A Numerical Study of Heat Transfer due to an Axisymmetric Laminar Impinging Jet of Supercritical Carbon Dioxide," *Int. J. Heat Mass Transfer*, 35, 2515–2526.

Kim, J., and Moin, P., 1989, "Transport of Passive Scalars in a Turbulent Channel Flow," in *Turbulent Shear Flows VI*, André, J.-C., et al., Editors, Springer-Verlag, Berlin.

Kim, J., Simon, T.W., and Russ, S.G., 1992, "Free-Stream Turbulence and Concave Curvature Effects on Heated, Transitional Boundary Layers," *ASME J. Heat Transfer*, 114, 338–347.

Klebanoff, P.S., Cleveland, W.G., and Tidstrom, K.D., 1992, "On the Evolution of a Turbulent Boundary Layer Induced by a Three-Dimensional Roughness Element," *J. Fluid Mech.*, 237, 101–182.

Kogan, M.N., 1973, "Molecular Gas Dynamics," *Ann. Rev. Fluid Mech.*, 5, 383–404.

Konuma, M., 1992, *Film Deposition by Plasma Techniques*, Springer-Verlag, New York.

Kurzweg, U.H., 1985, "Enhanced Heat Conduction in Fluids Subjected to Sinusoidal Oscillations," *ASME J. Heat Transfer*, 107, 459–462.

Lancaster, J.F., Editor, 1986, *The Physics of Welding*, Pergamon Press, Oxford.

Launder, B.E., 1988, "On the Computation of Convective Heat Transfer in Complex Turbulent Flows," *ASME J. Heat Transfer*, 110, 1112–1128.

Lazarenko, B.R., Grosu, F.P., and Bologa, M.K., 1975, "Convective Heat Transfer Enhancement by Electric Fields," *Int. J. Heat Mass Transfer*, 18, 1433–1441.

Lee, S.T., and T'ien, J.S., 1982, "A Numerical Analysis of Flame Flashback in a Premixed Laminar System," *Combust. Flame*, 48, 273–285.

Lemlich, R., 1961, "Vibration and Pulsation Boost Heat Transfer," *Chem. Eng.*, 68, 171–176.

Lin, S.-J., and Churchill, S.W., 1978, "Turbulent Free Convection from a Vertical, Isothermal Plate," *Num. Heat Transfer*, 1, 129–145.

Liu, C.N., and Shih, T.M., 1980, "Laminar, Mixed-Convection, Boundary-Layer, Nongray-Radiative Diffusion Flame," *ASME J. Heat Transfer*, 102, 724–730.

Liu, K.V., Lloyed, J.R., and Yang, K.J., 1981, "An Investigation of a Laminar Diffusion Flame Adjacent to a Vertical Flat Plate Burner," *Int. J. Heat Mass Transfer*, 24, 1959–1970.

Liu, K.V., Yang, K.T., and Lloyed, J.R., 1982, "Elliptic Field Calculation of a Laminar Diffusion Flame Adjacent to a Vertical Flat Plate Burner," *Int. J. Heat Mass Transfer*, 25, 863–870.

Lloyed, J.R., and Sparrow, E.M., 1970, "Combined Forced and Free Convection Flow on Vertical Surfaces," *Int. J. Heat Mass Transfer*, 13, 434–438.

Lord, R.G., 1992, "Direct Simulation Monte Carlo Calculations of Rarefied Flows with Incomplete Surface Accommodation," *J. Fluid Mech.*, 239, 449–459.

Louge, M., Yusef, M.J., and Jenkins, J.T., 1993, "Heat Transfer in the Pneumatic Transport of Massive Particles," *Int. J. Heat Mass Transfer*, 36, 265–275.

Lykoudis, P.S., 1962, "Natural Convection of an Electrically Conducting Fluid in the Presence of a Magnetic Field," *Int. J. Heat Mass Transfer*, 5, 23–34.

Lykoudis, P.S., and Yu, C.P., 1963, "The Influence of Electrostrictive Forces in Natural Thermal Convection," *Int. J. Heat Mass Transfer*, 6, 853–862.

Maciejewski, P.K., and Moffat, R.J., 1992, "Heat Transfer with Very High Free-Stream Turbulence. Part I. Experimental Data; Part II. Analysis of Results," *ASME J. Heat Transfer*, 114, 827–839.

Martin, B.W., 1984, "An Appreciation of Advances in Natural Convection Along an Isothermal Vertical Surface," *Int. J. Heat Mass Transfer*, 27, 1583–1586.

Martin, H., 1977, "Heat and Mass Transfer Between Impinging Gas Jets and Solid Surfaces," *Advan. Heat Transfer*, 13, 1–60.

Martin, P.J., and Richardson, A.T., 1984. "Conductivity Models of Electrothermal Convection in a Plane Layer of Dielectric Liquid," *ASME J. Heat Transfer*, 106, 131–142.

Meeks, E., Kee, R.J., Dandy, D.S., and Coltrin, M.E., 1993, "Computational Simulation of Diamond Chemical Vapor Deposition in Premixed $C_2H_2/O_2/H_2$ and CH_4/O_2—Strained Flames," *Combust. Flame*, 92, 144-160.

Mehendale, A.B., and Han, J.C., 1992, "Influence of High Mainstream Turbulence on Leading Edge Film Cooling Heat Transfer: Effect of Film Hole Spacing," *Int. J. Heat Mass Transfer*, 35, 2593–2604.

Merkin, J.H., 1969, "The Effect of Buoyancy Forces on the Boundary-Layer Flow over a Semi-Infinite Vertical Flat Plate in a Uniform Free Stream", *J. Fluid Mech.*, 35, 439–450.

Merkli, P., and Thomann, H., 1975, "Thermoacoustic Effects in a Resonance Tube," *J. Fluid Mech.*, 1970, 161–177.

Metzner, A.B., 1965, "Heat Transfer in Non-Newtonian Fluids," *Advan. Heat Transfer*, 2, 357–397.

Michaelides, E.E., 1986, "Heat Transfer in Particulate Flows," *Int. J. Heat Mass Transfer*, 29, 265–273.

Mironov, B.P., Vasechkin, V.N., Mamonov, V.N., and Yarygina, N.J., 1982, "Transport Processes in Turbulent Boundary Layer Under High-Level Free

Stream Turbulence," in *Structure of Turbulence in Heat and Mass Transfer*, 221–243, Hemisphere Publishing Corporation, Washington, D.C.

Mittal, M.L., Natarja, H.R., and Naidu, V.G., 1987, "Fluid Flow and Heat Transfer in the Duct of an MHD Power Generator," *Int. J. Heat Mass Transfer*, 30, 527–535.

Mohandi, R.B., and Gilligan, J.G., 1990, "Electrical Conductivity and Thermodynamic Functions of Weakly Nonideal Plasma," *J. Appl. Phys.*, 68, 5044–5051.

Molodtsof, Y., and Muzyka, D.W., 1989, "General Probablistic Multiphase Flow Equations for Analyzing Gas-Solid Mixtures," *Int. J. Eng. Fluid Mech.*, 2, 1-24.

Muntz, E.P., 1989, "Rarefied Gas Dynamics," *Ann. Rev. Fluid Mech.*, 121, 387–417.

Murayama, M., and Uchida, K., 1992, "Synthesis of Uniform Diamond Films by Flat Flame Combustion of Acetylene/Hydrogen/Oxygen Mixtures," *Combust. Flame*, 91, 239–245.

Nield, D.A., and Bejan, A., 1992, *Convection in Porous Media*, Springer-Verlag, New York.

Nielson, D.G., and Incorpera, F.P., 1993a, "Three-Dimensional Consideration of Unidirectional Solidification in a Binary Liquid," *Num. Heat Transfer*, 23A, 1–20.

Nielson, D.G., and Incorpera, F.P., 1993b, "Effect of Rotation on Fluid Motion and Channel Formation During Unidirectional Solidification of a Binary Alloy," *Int. J. Heat Mass Transfer*, 36, 489–505.

Oliver, F.W.J., 1972, "Bessel Functions of Integer Order," *Handbook of Mathematical Functions*, Abramowitz, M., and Stegun, X., Editors, U.S. Government Printing Office, Washington, D.C.

Orloff, L., de Ris, J., and Delichatsios, M.A., 1992, "Radiation from Buoyant Turbulent Diffusion Flame," *Combust. Sci. Tech.*, 84, 177–186.

Oždemir, I.B., and Whitelaw, J.H., 1992, "Impingement of an Axisymmetric Jet on Unheated and Heated Flat Plates," *J. Fluid Mech.*, 240, 503–532.

Özisik, M.N., 1985, *Radiative Transport and Interaction with Conduction and Convection*, Werbel and Peck, New York.

Ozoe, H., Sato, N., and Churchill, S.W., 1980, "The Effect of Various Parameters on Thermoacoustic Convection," *Chem. Eng. Commun.*, 5, 203–221.

Paolucci, S., and Chenoweth, D.R., 1987, "Transition to Chaos in a Differentially Heated Vertical Cavity," *J. Fluid Mech.*, 201, 379–410.

Peters, M., 1992, "A Spectral Closure for Premixed Turbulent Combustion in the Flamelet Regime," *J. Fluid Mech.*, 242, 611–629.

Platten, J.K., and Legros, J.C., 1984, *Convection in Liquids*, Springer-Verlag, Berlin.

Pletcher, R.H., 1988, "Progress in Turbulent Forced Convection," *ASME J. Heat Transfer*, 110, 1129–1144.

Poirier, D.R., Mandapurkar, P.J., and Ganesan, S., 1991, "The Energy and Solute Conservation Equations for Dendritic Solidification," *Metall. Trans.*, 22B, 889–900.

Puzach, V.G., 1992, "Heat and Mass Transfer on a Rough Surface with Gas Blowing at the Wall," *Int. J. Heat Mass Transfer*, 35, 981–986.

Ramesham, R., and Ellis, C., 1992, "Selective Growth of Diamond Crystals on the Apex of Silicon Pyramids," *J. Mater. Res.*, 7, 1189–1194.

Rappaz, M., and Voller, V., 1990, "Modeling of Micro-Macrosegregation in Solidification Processes," *Metall. Trans.*, 21A, 749–753.

Romig, M.F., 1964, "The Influence of Electric and Magnetic Fields on Heat Transfer to Electrically Conducting Fluids, " *Advan. Heat Transfer*, 1, 267–354.

Rossow, V.J., 1957, *On Flow of Electrically Conducting Fluids over a Flat Plate in the Presence of a Magnetic Field*, NASA Technical Note 3971.

Rott, N., 1974, "The Heating Effect Connected with Nonlinear Oscillations in a Resonance Tube," *J. Appl. Math. Phys. (ZAMP)*, 25, 619–634.

Rott, N., 1980, "Thermoacoustics," *Advan. Appl. Mech.*, 20, 135–175.

Rott, N., 1984, "Thermoacoustic Heating at the Closed End of an Oscillating Gas Column," *J. Fluid Mech.*, 145, 1–9.

Sahraoui, M., and Kaviany, M., 1993, "Slip and No-Slip Temperature Boundary Conditions at Interface of Porous, Plain Media: Convection," *Int. J. Heat Mass Transfer*, (in Press).

Schlichling, H., 1979, *Boundary-Layer Theory*, Seventh Edition, McGraw-Hill, New York.

Shadid, J.N., and Eckert, E.R.G., 1992, "Viscous Heating of a Cylinder with Finite Length by a High Viscosity Fluid in Steady Longitudinal Flow. II. Non-Newtonian Carreau Model Fluids," *Int. J. Heat Mass Transfer*, 35, 2739–2749.

Shenoy, A.V., and Mashelkar, R.A., 1982, "Thermal Convection in Non-Newtonian Fluids," *Advan. Heat Transfer*, 15, 143–225.

Sherman, F.S., 1955, *A Low-Density Wind-Tunnel Study of Shock-Wave Structure and Relaxation Phenomena in Gases*, NASA Technical Report 3298.

Sherman, F.S., 1969, "The Transition From Continuum to Molecular Flow," *Ann. Rev. Fluid Mech.*, 1, 317–340.

Sherman, F.S., 1990, *Viscous Flow*, McGraw-Hill, New York.

Siegel, R., and Howell, J.R., 1992, *Thermal Radiation Heat Transfer*, Third Edition, Hemisphere Publishing Company, Washington, D.C.

Siegel, R., and Snyder, A., 1984, "Shape of Porous Region to Control Cooling Along Curved Exit Boundary," *Int. J. Heat Mass Transfer*, 27, 243–252.

Simpkins, P.G., Greenberg-Kosinski, S., and MacChesney, J.B., 1979, "Thermophresis: The Mass Transfer Mechanism in Modified Chemical Vapor Deposition," *J. Appl. Phys.*, 50, 5676–5681.

Sinai, Y.L., 1987, "A Wall Function for the Temperate Variance in Turbulent Flow Adjacent to a Diabatic Wall," *ASME J. Heat Transfer*, 109, 861–865.

Smith, H., and Jensen, H.H., 1989, *Transport Phenomena*, Oxford University Press, Oxford.

Soo, S.-L., 1989, *Particulates and Continuum*, Hemisphere Publishing Corporation, New York.

Soo, S.-L., Trezek, G.J., Dimick, R.C., and Hohnstreiter, G.F., 1964, "Concentration and Mass Flow Distribution in a Gas-Solid Suspension," *Ind. Engng. Chem. Fundam.*, 3, 98–104.

Spalding, D.B., 1979, *Combustion and Mass Transfer*, Pergamon Press, Oxford.

Sparrow, E.M., and Cess, R.D., 1961a, "The Effect of a Magnetic Field on Free Convection Heat Transfer," *Int. J. Heat Mass Transfer*, 3, 267–274.

Sparrow, E.M., and Cess, R.D., 1961b, "Free Convection with Blowing or Suction," *ASME J. Heat Transfer*, 83, 387-389.

Sparrow, E.M., and Cess, R.D., 1978, *Radiation Heat Transfer*, Hemisphere Publishing Corporation, Washington, D.C.

Sparrow, E.M., and Lin, S.H., 1962, "Laminar Heat Transfer in Tubes Under Slip-Flow Conditions," *ASME J. Heat Transfer*, 84, 363–369.

Springer, G.S., 1971, "Heat Transfer in Rarefied Gases," *Advan. Heat Transfer*, 7, 163–218.

Stadler, K.R., and Sharpless, R.L., 1990, "Plasma Properties of a Hydrocarbon Arc Jet Used in the Plasma Deposition of Diamond Thin Films," *J. Appl. Phys.*, 68, 6189–6190.

Striegl, S.A., and Diller, T.E., 1984, "An Analysis of the Effect of Entrainment Temperature on Jet Impingement Heat Transfer," *ASME J. Heat Transfer*, 106, 804–810.

Sundarraj, S., and Voller, V.R., 1993, "The Binary Alloy Problem in an Expanding Domain: The Microsegregation Problem," *Int. J. Heat Mass Transfer*, 36, 713–723.

Swift, G.W., 1988, "Thermoacoustic Engines," *J. Acoust. Soc. Am.*, 84, 1145–1180.

Takhar, H.S., and Soundalgekar, V.M., 1977, "Effect of Viscous Dissipation on Heat Transfer in an Oscillating Flow Past a Flat Plate," *Appl. Sci. Res.*, 33, 101–111.

Tao, Y.-X., and Kaviany, M., 1991, "Burning Rate of Liquid Supplied Through a Wick," *Combust. Flame.*, 86, 47–61.

Taylor, R.P., Coleman, H.W., and Hodge, B.K., 1989, "Prediction of Heat Transfer in Turbulent Flow over Rough Surfaces," *ASME J. Heat Transfer*, 111, 569–572.

To, W.M., and Humphrey, J.A.C., 1986, "Numerical Simulation of Buoyant, Turbulent Flow. I. Free Convection Along a Heated, Vertical, Flat Plate," *Int. J. Heat Mass Transfer*, 29, 573–592.

Toong, T.-Y., 1983, *Combustion Dynamics*, McGraw-Hill, New York.

Tsai, C., Nelson, J., and Gerberich, W.W., 1992, "Metal Reinforced Thermal Plasma Diamond Coating," *J. Mater. Res.*, 7, 1967–1972.

Tsai, M.C., and Kou, S., 1990, "Heat Transfer and Fluid Flow in Welding Arcs Produced by Sharpened and Flat Electrodes," *Int. J. Heat Mass Transfer*, 33, 2089–2098.

Turnbull, R.J., 1969, "Free Convection from a Heated Vertical Plate in a Direct-Current Electric Field," *Phys. Fluids*, 12, 2255-2263.

Turnbull, R.J., 1971a, "Instability of a Thermal Boundary Layer in a Constant Electric Field," *J. Fluid Mech.*, 47, 231–239.

Turnbull, R.J., 1971b, "Effect of Non-uniform Alternating Electric Field on the Thermal Boundary Layer Near a Heated Vertical Plate," *J. Fluid Mech.*, 49, 693-703.

Turner, J.S., 1979, *Buoyancy Effects in Fluids*, Cambridge University Press, Cambridge.

Vafai, K., and Tien, C.-L., 1981, "Boundary and Inertia Effects on Flow and Heat Transfer in Porous Media," *Int. J. Heat Mass Transfer*, 24, 195–203.

Viskanta, R., 1966, "Radiation Transfer and Interaction of Convection with Radiation Heat Transfer," *Advan. Heat Transfer*, 3, 175–251.

Wadsworth, D.C., and Erwin, D.A., 1993, "Numerical Simulation of Rarefied Flow Through a Slit. Part I: Direct Simulation Monte Carlo Results," *Phys. Fluids*, 5, 235–242.

Wadsworth, D.C., Erwin, D.A., and Muntz, E.P., 1993, "Transient Motion of a Confined Gas Due to Wall Heating or Cooling," *J. Fluid Mech.*, 248, 219–235.

Walker, K.L., Homsy, G.M., and Geyling, F.T., 1979, "Thermophoretic Deposition of Small Particles in Laminar Tube Flow," *J. Colloid Interface Sci.*, 69, 138–147.

Watson, E.J., 1983, "Diffusion in Oscillatory Pipe Flow," *J. Fluid Mech.*, 133, 233–244.

Wheatley, J., Hofler, T., Swift, G.W., and Migliori, A., 1985, "Understanding Some Simple Phenomena in Thermoacoustics with Applications to Acoustical Heat Engines," *Am. J. Phys.*, 53, 147–162.

White, F.M., 1991, *Viscous Fluid Flow*, Second Edition, McGraw-Hill, New York.

Wichman, I.S., and Argawal, S., 1991, "Wind-Aided Flame Spread over a Thick Solid," *Combust. Flame*, 83, 127–145.

Williams, F.A., 1985, *Combustion Theory*, Addison-Wesley, Redwood City, CA.

Winter, H.H., 1977, "Viscous Dissipation in Shear Flows of Molten Polymers," *Advan. Heat Transfer*, 13, 205–267.

Worster, M.G., 1991, "Natural Convection in a Mushy Layer," *J. Fluid Mech.*, 224, 335–359.

Worster, M.G., 1992, "Instabilities of the Liquid and Mushy Regions During Solidification of Alloys," *J. Fluid Mech.*, 237, 649–669.

Yen, S.M., 1984, "Numerical Solution of the Nonlinear Boltzmann Equation for Nonequilibrium Gas Flow Problems," *Ann. Rev. Fluid Mech.*, 16, 67–97.

Youssef, M.S., Nagano, Y., and Tagawa, M., 1992, "A Two-Equation Heat Transfer Model for Predicting Turbulent Thermal Fields Under Arbitrary Wall Thermal Conditions," *Int. J. Heat Mass Transfer*, 35, 3095–3104.

Yuan, T.D., and Liburdy, J.A., 1992, "Application of a Surface Renewal Model to the Prediction of Heat Transfer in an Impinging Jet," *Int. J. Heat Mass Transfer*, 35, 1905–1912.

Zappoli, B., 1992, "The Response of a Nearly Supercritical Pure Fluid to a Thermal Disturbance," *Phys. Fluids*, A4, 1040–1048.

Zaric, Z.P., Editor, 1982, *Structure of Turbulence in Heat and Mass Transfer*, Hemisphere Publishing Corporation, Washington, D.C.

Zhao, G.Y., Dassanayake, M., and Etemadi, K., 1990, "Numerical Simulation of a Free-Burning Argon Arc with Copper Evaporation from the Anode," *Plasma Chem. Plasma Process.*, 10, 87–98.

Zhao, G.Y., Mostaghimi, J., and Boulos, M.I., 1990, "The Induction Plasma Chemical Reactor: Part I. Equilibrium Model, and Part II. Kinetic Model," *Plasma Chem. Plasma Process.*, 10, 133–167.

5

Solid–Fluid Systems with Large Specific Interfacial Area

The *interfacial* heat transfer rate for a solid-fluid system with an interfacial area A_{sf} and a local interfacial heat flux q_{sf} is

$$Q_{sf} = A_{sf} \langle q_{sf} \rangle_{A_{sf}} . \tag{5.1}$$

Using the *specific interfacial area* A_o (i.e., the solid-fluid interfacial area per unit volume of the combined solid and fluid phases) this can be written as

$$\frac{Q_{sf}}{V} = \frac{A_{sf}}{V} \langle q_{sf} \rangle_{A_{sf}} \equiv A_o \langle q_{sf} \rangle_{A_{sf}} . \tag{5.2}$$

For a solid-fluid system with a *fluid volume fraction* ϵ (i.e., porosity) and for spherical solid particles of diameter d, the specific surface area is

$$A_o = \frac{6(1-\epsilon)}{d} . \tag{5.3}$$

Then as d *decreases* the specific surface area *increases* and, for example, for $\epsilon = 0.5$ the specific surface area A_o for particles of $d = 10^2$ μm is 3×10^4 m^2/m^3, which is rather large. This *large* specific interfacial area of *dispersed systems* has been used for applications requiring large interfacial heat transfer rates Q_{sf}.

The interfacial heat flux q_{sf} depends on the *interphasic* (solid and fluid) heat transfer which is in turn influenced by the fluid- and solid-phase *motions* and their thermophysical properties and on the *interparticle* convective interactions which in turn depends on the average *interparticle clearance distance* C. Figure 5.1 shows two elements of a *dispersed*, solid phase in a flowing, *continuous* fluid phase. The elements are rendered in a simple geometry with a linear dimension d. The *average interelemental distance* C can be very small (e.g., *packed* beds) or very large (e.g., *dilute suspensions*). The solid-fluid interface, where the *main* heat transfer occurs, can change due to *phase change* (*sublimation* of solid or *condensation* (i.e., *crystalization*) or *frosting* of gas or *melting* or *solidification*). There can be *chemical reactions* occurring on the solid surface and/or in the fluid phase. Here the intrasolid reactions are not addressed by assuming impermeable, internally inert solids. The solid elements can be *transparent*, *semitransparent*, or *opaque*, and the radiation heat transfer in *both* the solid and the fluid phases can be significant.

In this chapter we begin by discussing interfacial convective heat transfer in solid-fluid systems with the solid phase *dispersed* as elements in a

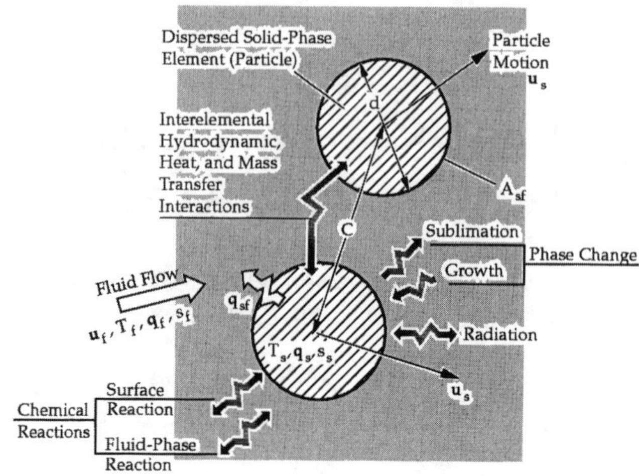

Figure 5.1. A rendering of two neighboring particles moving in a flowing fluid. Each particle can participate in a chemical reaction and undergo growth or decay. The clearance distance C is shown here as the center-to-center distance. Also shown are the symbols used in the analysis.

continuous fluid phase. We begin by considering heat transfer of *isolated* elements and then address the effect of the interelemental interactions on the interfacial convective heat transfer. Then the case of a continuous *stationary* solid phase made of the same elements will be considered. This is the thermal *nonequilibrium* treatment of heat transfer in *porous media*. Next the more complex case of transport in *moving* dispersed elements will be examined. This is the thermal *nonequilibrium* treatment of heat transfer in particulate flows.

5.1 Isolated Dispersed-Phase Elements

The *geometry* of a dispersed element, in practice, can be *complex*, and an element may yet be an *aggregate* of *simple* elements. Here we refer to an element as the *most basic*, *impermeable* solid body which is treatable both using the continuum treatment and applying *well-defined* boundary conditions for its *interface* with the *surrounding* continuous fluid. One of the elements most studied for convective heat transfer is the impermeable *spherical particle* and its two-dimensional counterpart which is the impermeable *cylinder*. For a steady flow and heat transfer, the convective interfacial heat transfer from these elements has been examined subject to a constant interfacial surface temperature (and in some cases a constant interfacial heat flux). The flow inertia, in fluid flow around an isolated element, and its interaction with the viscous forces and the pressure distribution result in various flow *regimes* and *characteristics* around the element. These characteristics change significantly with the magnitude of the fluid velocity.

In this section we concentrate on isolated, impermeable spherical elements and first examine the fluid flow around the element and the variation of this flow structure with respect to the fluid velocity, i.e., description of the flow regimes. Then the effect of the flow structure on the interfacial heat transfer, as well as the variations of its surface-averaged value, are discussed.

5.1.1 Flow Regimes

For the steady flow of a fluid, with a kinematic viscosity ν and a velocity u_∞ over a *stationary solid sphere* of diameter d, the particle Reynolds number defined as

$$Re_d = \frac{u_\infty d}{\nu} \tag{5.4}$$

characterizes the structure of the flow around the sphere. There have been *seven* distinct regimes identified, and the transition between these regimes are marked by the magnitude of Re_d. These regimes are discussed by Clift et al. (1978), and a summary is given below.

(A) Stokes or Creeping Flow Regime, $Re_d \rightarrow 0$

The *low* Reynolds number regime (or the Stokes flow regime) is different from the zero viscosity flow (or potential flow regime). This is because although like the potential flow the *streamlines* have an upstream-downstream *symmetry* (with respect to a plan passing through the center of the particle and perpendicular to the flow), the pressure field is *not* symmetric. Then the *asymmetry* of the pressure field places a net pressure force on the particle which is called the *form drag*. When this drag is added to the *shear stress* (*viscous*) *drag*, the *overall Stokes drag coefficient* (based on the projected area $\pi d^2/4$) becomes

$$c_d = \frac{4F_d/\pi d^2}{\frac{1}{2}\rho u_\infty^2} = \frac{24}{Re_d} \qquad Re_d \rightarrow 0 \, . \tag{5.5}$$

One-third of c_d is due to the form drag.

The improved *Oseen* solution gives

$$c_d = \frac{24}{Re_d}(1 + \frac{3}{16}Re_d) \qquad Re_d \rightarrow 0 \, . \tag{5.6}$$

Higher-order solutions are available and are summarized by Clift et al. (1978).

(B) Unseparated Flow Regime, $1 < Re_d < 20$

The flow asymmetry around the plane perpendicular to the flow and passing through the center of the sphere becomes progressively more pronounced as the Reynolds number increases. The *surface vorticity* $\boldsymbol{\omega} = \nabla \times \mathbf{u}$ has a *maximum* in the upstream half of the sphere surface and as Re_d increases, this moves *further upstream* on the surface. However, the flow remains *attached*, i.e., the *sign* of the vorticity on the surface does *not* change.

(C) Onset of Separation, $Re_d \simeq 20$

The flow *separation* from the surface is indicated by the change in the *sign* of the surface vorticity and begins at a critical Re_d and in the *rear stagnation region* (at the stagnation *point*). The numerical simulations, confined experimentally, give this *critical Reynolds number* for separation to be nearly 20. The *recirculation* region is initially very small, and therefore difficult to detect experimentally (i.e., high-resolution experimental and numerical detections are needed). Then at the onset of separation, the *separation angle* θ_s measured from the upstream stagnation point, as shown in Figure 5.2(a), is 180°.

(D) Steady Wake Regime, $20 < Re_d \le 130$

With increase of Re_d beyond 20, the recirculating *wake* formed in the rear stagnation region becomes larger. Thus the separation point moves upstream and the *length* of the wake *increases*. The wake changes from a *convex* shape to a *concave* shape around $Re_d = 35$. The separation angle θ_s is given by the correlation

$$\theta_s = 180^o - 42.5^o(\ln \frac{Re_d}{20})^{0.483} \qquad 20 < Re_d \le 400 \,. \tag{5.7}$$

The measured, normalized length of the wake ℓ_w/d and its growth with Re_d are shown in Figure 5.2(a). For up to $Re_d = 400$, the length of wake can be defined. Beyond this, steady vortex *shedding* begins and no such length can be used. The wake length grows up to *three* times the diameter.

(E) Onset of Wake Instability, $130 \le Re_d \le 400$

At a Reynolds number of about 130, a weak, long-period oscillation appears in the *tip* of the wake and its amplitude increases with Re_d. At $Re_d \simeq 270$ this amplitude grows to about $0.1d$ and large *vortices* are formed and periodically *released* from the tip. The dimensionless *vortex shedding frequency*, called the *Strouhal number* $Sr_d = fd/u_\infty$, increases with the Reynolds

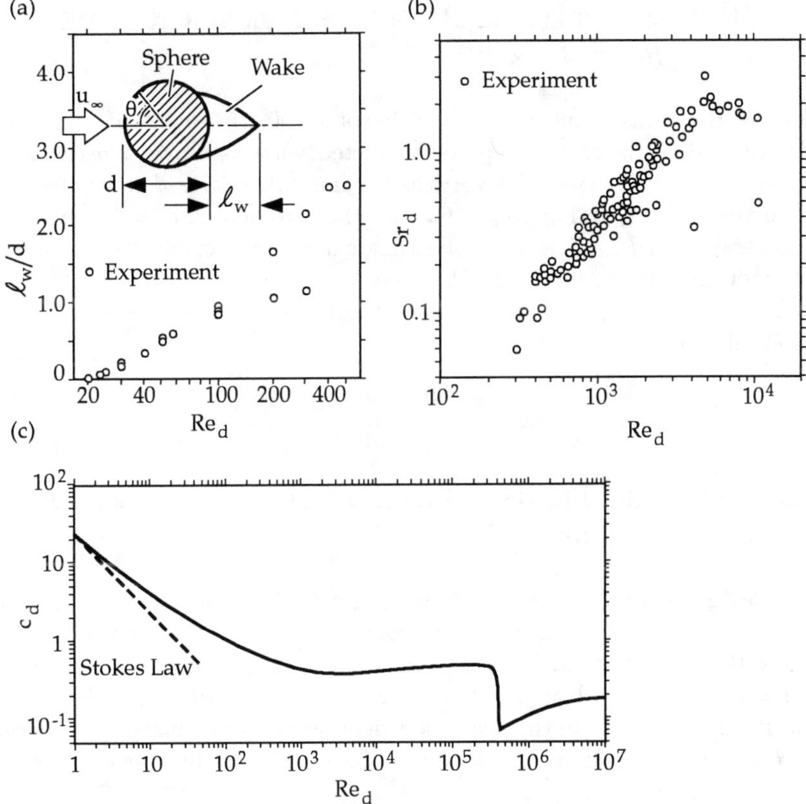

Figure 5.2. (a) The measured variation of the normalized length of the wake behind a solid sphere as a function of the Reynolds number. The polar angle θ is also shown. (b) The measured variation of the normalized oscillation frequency of the wake behind a sphere as a function of the Reynolds number. (c) The measured total drag coefficient as a function of the Reynolds number. Also shown is the Stokes law. (From Clift et al., reproduced by permission ©1978 Academic Press.)

number, and Figure 5.2(b) shows this variation in the vortex shedding frequency.

The relative importance of the *form* drag *increases*, and for $Re_d > 150$ the *overall* drag coefficient begins a move towards *independence* from Re_d. Figure 5.2(c) shows this trend.

For very large Re_d, the separation angle does not change noticeably with Re_d and a correlation, including this asymptote, is given by

$$\theta_s = 78^\circ + 275^\circ Re_d^{-0.37} \quad 400 < Re_d < 3 \times 10^5 \,. \tag{5.8}$$

Asymptotic values of $\theta_s = 81^\circ$ to 83° have been observed experimentally and also from the numerical simulations.

(F) High, Subcritical Reynolds Number Regime, $400 < Re_d < 3.5 \times 10^5$

As the Reynolds number increases beyond 400, the point at which the *detached* shear layer *rolls* up to form shed vortices moves *closer* to the sphere. At $Re_d \simeq 1300$, the vortices become more *three-dimensional* and form vortex *balls*, and at $Re_d \simeq 6 \times 10^3$, the Strouhal number Sr_d becomes independent of Re_d. The wake shedding causes appreciable *fluctuations* in the motion, but the flow is *not* turbulent.

The overall drag coefficient in the *subcritical range* has been correlated as (White, 1991)

$$c_d = \frac{24}{Re_d} + \frac{6}{1 + Re_d^{1/2}} + 0.4 \qquad 0 \le Re_d \le 2 \times 10^5 \,. \tag{5.9}$$

(G) Critical, Transition, and Supercritical Flow, $3.5 \times 10^5 < Re_d$

For $Re_d \simeq 2{\times}10^5$ the position of the separation, fluctuates and the *detached free shear layer* becomes *turbulent* soon after separation, and for $Re_d > 2.8 \times 10^5$ it *reattaches*. This is marked by a large drop in c_d at $Re_d \simeq 2.8 \times 10^5$, as is evident in Figure 5.2(c). The *critical Reynolds number* marks the *transition* in the drag coefficient and depends on the *free-stream turbulence*. A *critical value*, $Re_{d_c} = 3.65 \times 10^5$, has also been suggested (Clift et al., 1978).

5.1.2 Interfacial Heat Transfer

The interfacial heat transfer rate, and its variation over the surface of a dispersed particle, is influenced by the structure of the fluid flow around it. However, as will be shown the effect is *less* drastic than that observed on the drag coefficient c_d. Because of the unsteadiness, i.e., vortex shedding and turbulence, the complete prediction of the interfacial heat transfer for various *regimes* has not yet been attempted. For an azimuthally symmetric flow and heat transfer, the energy equation becomes

$$v\frac{\partial T}{\partial r} + \frac{w}{r}\frac{\partial T}{\partial \theta} = \frac{\alpha}{r^2}\left(\frac{\partial}{\partial r}r^2\frac{\partial T}{\partial r} + \frac{1}{\sin\theta}\frac{\partial}{\partial \theta}\sin\theta\frac{\partial T}{\partial \theta}\right) , \tag{5.10}$$

subject to the boundary conditions (using $R = d/2$)

$$T = T_s \qquad r = R\,, \quad 0 \le \theta \le 180^\circ\,, \tag{5.11}$$

$$T = T_\infty \qquad r \to \infty\,, \quad 0 \le \theta \le 180^\circ\,, \tag{5.12}$$

$$\frac{\partial T}{\partial \theta} = 0 \qquad \theta = 0\,, 180^\circ\,, \quad R \le r < \infty\,. \tag{5.13}$$

The available experimental and theoretical results for the heat transfer, for the various regimes are discussed by Clift et al. (1978) and are reviewed below.

(A) $Re_d \rightarrow 0$ Regime

Since the Péclet number $Pe_d = Re_d Pr$ is indicative of the role of convection compared to conduction, the *conduction-dominated* regime, i.e., $Pe_d \rightarrow 0$, is governed by

$$\frac{\partial}{\partial r} r^2 \frac{\partial T}{\partial r} = 0 \quad Pe_d \rightarrow 0 \, . \tag{5.14}$$

The solution to this equation subject to the above boundary conditions gives the local Nusselt number (which is independence of θ for this case) as

$$Nu_d = \frac{q_{sf} d}{(T_s - T_\infty)k} = -\frac{\partial T}{\partial r}|_R \frac{d}{T_s - T_\infty} = 2 \quad Pe_d \rightarrow 0 \, . \tag{5.15}$$

(B) Attached Wake Regime, $Re_d < 400$

The numerical solutions obtained using the continuity, momentum and energy equations for Reynolds numbers less than those corresponding to the onset of the vortex shedding are discussed by Clift et al. (1978). Figure 5.3(a) shows that results for $Pr = 0.71$ (or $Sc = 0.71$) and several Re_d. The variation of the local Nusselt number (or local Sherwood number) with respect to the *polar* (or *collatitude*) angle ($\theta = 0$ is at the *front* stagnation point) is shown. The $Pe_d \rightarrow 0$ asymptote in which Nu_d does not change with θ is also shown. For $Re_d \geq 20$, where the separation occurs, the separation angle corresponding to the Re_d is also indicated. For $Re_d < 20$, the local Nusselt number decrease *monotocially* with increase in θ. For $30 < Re_d < 57$, a *minimum* in Nu_d appears which *moves* upstream as Re_d *increases*. The *minimum* is behind the separation point, and the *recirculation* causes an increase in the local Nu_d in the *wake region*. The heat transfer from the recirculating region to the free stream is by *diffusion*. The front (or upstream) stagnation region has the *largest* local Nu_d.

(C) Subcritical Regime, $Re_d < 10^5$

For $400 < Re_d$, vortices are *shed* and the flow behind the sphere *oscillates*. For $3000 < Re_d$ the flow in the front stagnation region approaches a *boundary-layer behavior*. This boundary-layer flow and heat transfer has been analyzed using the laminar boundary-layer analysis of the variable free-stream velocity (Section 4.2.2), except that the *axisymmetric* conservation equations need to be first transformed to the *two-dimensional*

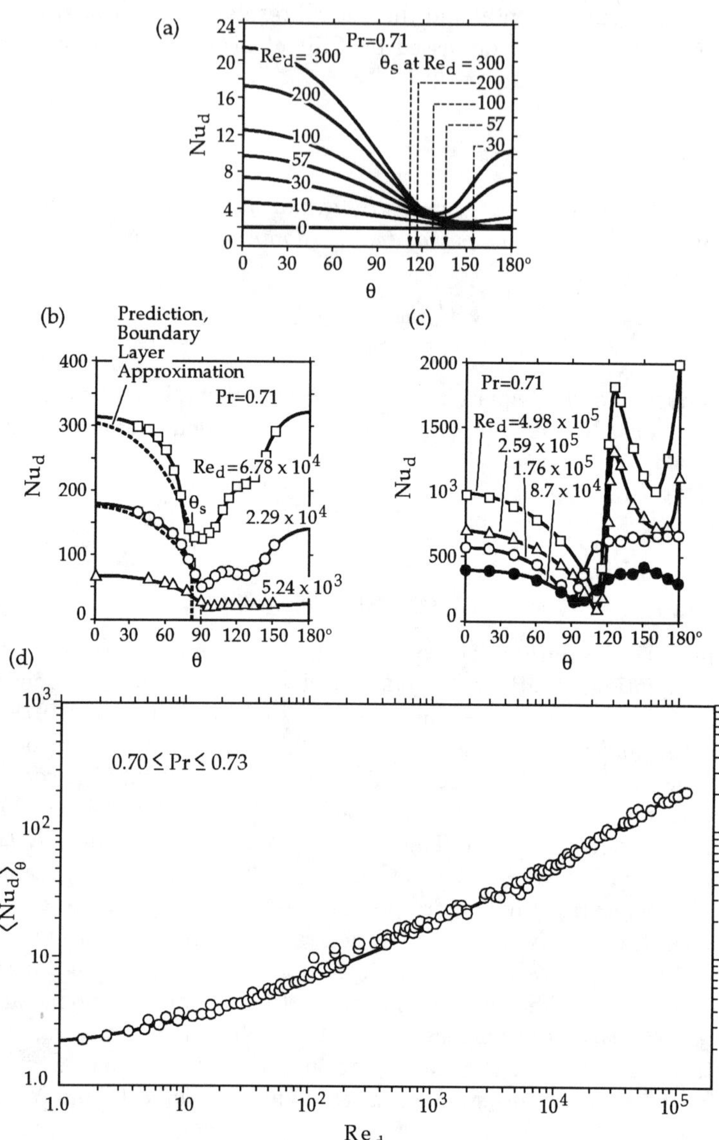

Figure 5.3. (a) The predicted distribution of the local Nusselt number around an isolated isothermal sphere for various (but low) Reynolds numbers and $Pr = 0.71$. The separation angles are also shown. (b) The measured distribution of the local Nusselt number for some subcritical Reynolds numbers. The predictions are based on the boundary-layer approximations, and the separation angles are also shown. (c) The measured local Nusselt number distribution for sub- and supercritical Reynolds numbers. (d) The measured area-averaged Nusselt number as a function of Reynolds number and for $0.70 < Pr < 0.73$. (From Clift et al., reproduced by permission ©1978 Academic Press.)

forms. Cebeci and Bradshaw (1984) discussed this *Mangler* transformation. The front stagnation region will be further studied in Section 6.2.5 (D).

The front stagnation boundary-layer analysis predicts a $Re_d^{1/2}$ dependence for Nu_d and an angular variation which is in good agreement with the experimental results for the stagnation region. Figure 5.3(b) shows the experimental results for $Pr = 0.71$, and $Re_d = 5.24 \times 10^3$, 2.29×10^4, and 6.78×10^4. The predictions, based on the boundary-layer approximation of the front stagnation region are also shown along with the separation angle θ_s corresponding to each Re_d. Note that the boundary-layer theory predicts $Nu_d = 0$ at θ_s, because of the lack of angular diffusion. The local Nu_d begins to increase rapidly with Re_d for $5 \times 10^3 < Re_d$ and for $Re_d = 6.78 \times 10^4$ the front and back maxima are nearly *equal*.

(D) Critical Regime, $10^5 < Re_d$

For yet larger Re_d, approaching and surpassing the critical Reynolds number, the behavior in the back region begins to change even further. Figure 5.3(c) shows the experimental results for $Pr = 0.71$ and $Re_d = 8.7\times10^4$, 1.76×10^5, 2.59×10^5, and 4.89×10^5. For this experiment, at $Re_d = 2.59\times10^5$ *transition* has occurred and a *second* minimum appears. The turbulent flow in the rear follows a $Re_d^{4/5}$ relation while the front stagnation flow gives a $Re_d^{1/2}$ relation, for the local Nusselt number. Therefore, the rear local Nusselt number becomes *larger* than the front as the Re_d increases beyond the transition.

The experimental results for the *polar angle-averaged* Nusselt number, defined as

$$\langle Nu_d \rangle_{A_{sf}} \equiv \langle Nu_d \rangle_\theta = \frac{1}{180^\circ} \int_0^{180^\circ} Nu_d \, d\theta \,, \qquad (5.16)$$

is shown in Figure 5.3(d) for $1 \le Re_d \le 10^5$, and for $0.70 \le Pr \le 0.73$. The results are for several experiments and as noted, the average heat transfer rate is not as sensitive to the flow transitions as the drag coefficient shown in Figure 5.3(b).

The effect of the Prandtl number has been included in some correlations and one of the correlation given by Clift et al. (1978) is

$$\frac{\langle Nu_d \rangle_{A_{sf}} - 1}{(1 + \dfrac{1}{Re_d Pr})^{1/3} Pr^{1/3}} =$$
$$\begin{cases} Re_d^{0.41} & 1 \le Re_d \le 10^2 \,, \\ 0.752 Re_d^{0.472} & 10^2 < Re_d \le 2 \times 10^3 \,, \\ 0.44 Re_d^{1/2} + 0.034 Re_d^{0.71} & 2 \times 10^3 < Re_d \le 10^5 \,. \end{cases} \qquad (5.17)$$

For $2\times10^3 < Re_d$, the correlation can be interpreted as combining the front stagnation flow and wake turbulent region contributions. Other correlations are discussed by Whitaker (1972).

5.2 Convective Interaction Among Dispersed Elements

In examining the interfacial heat transfer between a continuous, flowing fluid and a solid phase distributed in the fluid with a large *specific* interfacial area, we began by discussing the case of *isolated* solid phase elements (i.e., where the *average interparticle clearance distance* C tends towards infinity). We now consider the case of C being *finite* and *small* enough such that the interfacial heat transfer flux q_{sf} for each element is *influenced* by the presence of the *neighboring* elements. As was shown in Figure 5.1, in rendering of the fluid flow and heat transfer around two neighboring particles, the *hydrodynamics* and *heat* and *mass* transfer aspects of each particle can be influenced by the presence of the neighboring particle. For isolated elements, we have been examining

$$\frac{\langle q_{sf}\rangle_{A_{sf}} d}{k(T_s - T_\infty)} = \langle Nu_d\rangle_{A_{sf}} = \langle Nu_d\rangle_{A_{sf}}(Pr, Re_d, \text{particle geometry}) \quad (5.18)$$

and when the interelemental interactions are significant, we expect additional dependence on the *particle concentration* and other parameters, i.e.,

$$\langle Nu_d\rangle_{A_{sf}} = \langle Nu_d\rangle_{A_{sf}}(Pr, Re_d, \text{particle geometry}, \text{particle concentration}, \text{etc.}). \quad (5.19)$$

The particle concentration per unit volume can be given in terms of the porosity, and for *spherical particles* we can write this as

$$\langle Nu_d\rangle_{A_{sf}} = \langle Nu_d\rangle_{A_{sf}}(Pr, Re_d, \epsilon, \text{etc.}) \,. \quad (5.20)$$

This follows when we note that the ratio of the interelemental clearance distance C (which can be defined in various manners including that shown in Figure 5.1) to the particle diameter, C/d, can be given as a function of ϵ (for spherical particles with known, ordered particle arrangements).

In the following, first the case of *dilute* element concentrations, i.e., $\epsilon \to 1$, which is characteristic of solid sprays (as in combustion of coal particles, or drying of particles) will be considered. The existing *numerical* simulations for $2 \le C/d < \infty$ clearly show how q_{sf} and $\langle q_{sf}\rangle_{A_{sf}}$ charges as C/d is decreased. Then we consider lower porosities. For *random* arrangement of particles, the average porosity for the case of a packed bed of spherical particle (stable stationary arrangement of particles) is 0.38 to 0.40. The *smallest* porosity for there packed *ordered* arrangement of spherical particles (without any consolidation) is 0.26 (for the *face-centered cubic* arrangement) and the *simple-cubic* ordered packing arrangement gives $\epsilon = 0.476$.

Note that we so far are discussing the *dispersed* elements and have *not* addressed the modeling of these elements as a continuum. In Section 5.2.3

we will discuss the description of the solid phase as a *collection* of dispersed elements with *no direct* heat transfer among the elements, or alternatively as a continuum with intraphasic heat transfer.

The interelemental interaction has been analyzed using the concept of the unit cell, i.e., a finite-influence volume around an element where the volume of the void space around the particle is chosen to correspond to the average medium porosity. The *geometry* of the unit cell gas been chosen in accord with the type of the treatment. In analytical treatments, *spherical shells* are used around the particles. In *numerical treatments*, the *planar* boundaries of *periodicity* or *symmetry* are used.

In some numerical simulations, a *finite* number of particles are used. Especially where the interest is in isolating and determining the interparticle interactions in the region of *flow entrance* into the region of particle collections. In general, the flow field becomes periodic after a few particles (i.e., the entrance length is only a few particle long).

5.2.1 Dilute Element Concentration

As an illustrative example, consider *tandem, stationary* arrangements of three spheres in a an otherwise uniform, *laminar* steady flow with velocity u_∞. The surface of the spheres is maintained at temperature T_s, while the undisturbed fluid is at temperature T_∞. Ramachandran et al. (1989) solve the steady, *axisymmetric* (i.e., azimuthal symmetry) flow and heat transfer numerically assuming *constant* properties. Figure 5.4(a) shows their results for the distribution of the dimensionless temperature T^* defined as

$$T^* = \frac{T - T_\infty}{T_s - T_\infty} \; . \tag{5.21}$$

The results are for $Pr = 0.2$ and C_{12} (i.e., the *axial* distance *between* the *first* and *second* spheres) and C_{23} (between the *second* and *third*) of $4d$. The top figure is for $Re_d = 20$ and the lower is for $Re_d = 200$. The *lateral* thermal penetration of the effect of the presence of the sphere, i.e., the thermal boundary-layer thickness, *decreases* as Re_d *increases*. The variation of the *local* Nusselt number Nu_d with respect to the polar (or colatitude) angle θ is shown in Figure 5.4(b) for $Re_d = 20$ and $Re_d = 200$. The results are for all the three spheres, i.e., spheres 1, 2, and 3. The *upstream* sphere (i.e., sphere 1) is *unaffected* by the presence of the spheres 2 and 3 downstream. This is evident when comparing the results for sphere 1 to those for isolated spheres given in Figure 5.3(a). The local Nusselt number for spheres 2 and 3 is different than that for sphere 1 by being smaller for all values of θ and having an *extra minimum* at the upstream stagnation point for larger Re_d. This minimum becomes more *pronounced* as Re_d increases. This is due to the *recirulations* between spheres 1 and 2, and 2 and 3. When compared with sphere 1, both spheres 2 and 3 have similar magnitudes and distributions of Nu_d. This indicates that the *thermal entrance length*

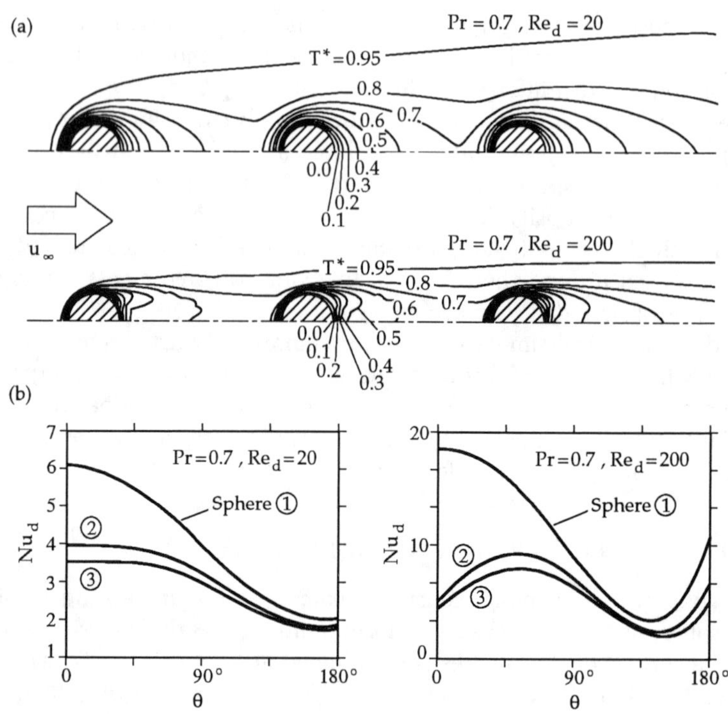

Figure 5.4. (a) Predicted distribution of temperatures around a tandem arrangement of three isothermal spheres for two different Reynolds numbers and for $Pr = 0.7$. (b) Predicted distribution of the local Nusselt number for the three spheres (number 1 is the upstream and number 3 is the downstream sphere). The results are for two different Reynolds numbers and for $Pr = 0.7$. (From Ramachandran et al., reproduced by permission ©1989 Taylor and Francis.)

is nearly *one* interparticle distance C. Note that in determining the local Nusselt number for each sphere, a unit cell (around the particle), velocity-averaged *inlet* fluid temperature is used (Tal et al., 1983; 1984). This is needed because the absolute values of q_{sf} *decreases* substantially for the downstream spheres as the fluid temperature approaches T_s. However, the Nusselt number which is *normalized* with respect to the difference between the surface temperature and this inlet temperature approaches that of the isolated sphere when C/d becomes very large.

As evident in Figure 5.4(b) in the *downstream* stagnation region the local Nu_d does not change substantially whenever there are particles present either *upstream* or *downstream*. The upstream stagnation point, however, is very sensitive to this. The polar-angle-averaged Nusselt number for sphere 2 has been correlated with respect to Pr, Re_d, C_{12}/d, and C_{23}/d by Ramachandran et al. (1989). As expected, for $C_{12} \to \infty$ and $C_{23} \to \infty$, the *isolated* sphere *asymptote* is found. Their correlation for *sphere* 2 is

$$\frac{\langle Nu_d \rangle_\theta}{\langle Nu_d \rangle_\theta |_{C \rightarrow \infty}} = 1 - 4.807(1 + \ln Re_d)^{0.012}(\frac{d}{C_{12}})^{-5.28} -$$

$$0.0697(1 + \ln Re_d)^{0.767}(1 + \ln 10 Pe_d)^{0.095}(\frac{d}{C_{23}})^{-0.13}$$

$$2 \leq C_{12}/d \, , \quad C_{23}/d \leq 10 \, , \quad 0.1 \leq Pr \leq 10 \, , \quad 1 < Re_d \leq 200 \, . \quad (5.22)$$

The Nusselt number for the isolated sphere $\langle Nu_d \rangle_\theta |_{C \rightarrow \infty} = \langle Nu_d \rangle_{A_{sf}}$ is given by (5.17). The presence of spheres 1 and 3 *decreases* the average Nusselt number for sphere 2. The effect of the presence of neighboring particles on the particle drag coefficient has been generally related to the porosity (i.e., the fluid volume fraction). The ratio of these drag coefficients has been correlated as $c_d(\epsilon)/c_d(\epsilon = 1) = f(\epsilon) = \epsilon^{-n}$ and a value of about 4.7 has been suggested for n by Coudrec (1985).

5.2.2 Dense Element Concentration

So far, the free-stream velocity has been used to describe the flow over isolated and dilute-concentration of spherical particles. In dealing with convective interaction for dense particle concentrations, the *cross-sectional* (fluid and solid phases) average velocity, which is also called the *superficial velocity* $\langle u \rangle_A$, is used. For low Reynolds numbers $Re_d = \langle u \rangle_A d/\nu$, using unit-cell models and the direct simulation of laminar, steady flow and heat transfer within them, the interfacial heat transfer rate is determined. For higher Re_d, correlation of the experimental results, guided by the known low Re_d and $\epsilon \rightarrow 1$ behaviors, are used. As the element concentration increases, the *exact* treatments, based on the detail treatment of the fluid dynamics and heat transfer, will be replaced by *approximations* and *models*. These are addressed below.

(A) Unit-Cell Models

A unit cell contains one or more particles with the particle spacing within the cell chosen according to the porosity of the effective medium being modeled. A unit cell containing spherical particles may be a *cubic*, *spherical*, or *cylindrical geometry*. Early models did not include the motion of the fluid (e.g., Tishkoff, 1979; Sangiovanni and Labowsky, 1982), while the recent numerical simulations include the hydrodynamics (e.g., Tal et al., 1983). As was discussed previously, when the wake behind the particle becomes *unsteady*, the numerical simulations become computationally *intensive* because of the required *extra* temporal and spatial *resolutions*.

(B) Semi-Empirical Treatments

As an alternative to the direct numerical simulation, the hydrodynamics of the interelement interaction is treated semi-empirically using a modification of the particle drag coefficient. A semi-empirical treatment of the interelement interactions for *dense* cluster of particles is given by Bellan and Harstad (1987). In this treatment and in the other treatments mentioned above in Section (A) and in Section 5.2.1 for dilute concentrations, no contact has been allowed between the elements, and therefore the solid phase is *discontinuous* or *dispersed*. Yet as the next step in simplifying the treatment, the hydrodynamics and effect on the interfacial heat transfer are *modeled* using a local volume-averaged energy equation for the *fluid phase*. The rigorous derivation of this energy equation under the condition of *thermal nonequilibrium* will be discussed in Section 5.4.1.

In transient experiments designed to determine $\langle q_{sf}\rangle_{A_{sf}}$, the temperature distribution *within* the solid particle also changes with *time*. Therefore, transient heat conduction must be allowed within the solid. In some experiments the solid is heated *inductively* by applying a electromagnetic field, and therefore, the heat generation must be included. For a one-dimensional fluid and heat flow (in terms of the *local volume*-averaged flow), the energy equation for the fluid phase can be written using (3.37). Assuming radial heat flow only in the solid phase, the transient conduction equation and the matching heat flow rate boundary condition can be written. The *distributed* treatment of the conduction within the particle is required whenever the *Biot number* is large. This is (Incropera and DeWitt, 1990)

$$Bi = \frac{h_{sf} d}{6k_s} = \langle Nu_d\rangle_{A_{sf}} \frac{k_f}{6k_s} > 0.1 \, . \tag{5.23}$$

Then the *distributed* treatment of the particle gives the set of energy equations for fluid and solid phases

$$\frac{\partial \langle T\rangle^f}{\partial t} + \langle u\rangle^f \frac{\partial \langle T\rangle^f}{\partial x} = \left[\frac{\langle k\rangle^f}{(\rho c_p)_f} + D^d_{xx}\right] \frac{\partial^2 \langle T\rangle^f}{\partial x^2} + \frac{A_{fs}}{V_f} \frac{\langle q_{sf}\rangle_{A_{sf}}}{(\rho c_p)_f} + \langle \dot{s}\rangle^f \, , \tag{5.24}$$

$$\frac{\partial T_s}{\partial t} = \frac{k_s}{(\rho c_p)_s} \frac{1}{r^2} \frac{\partial}{\partial r} r^2 \frac{\partial T_s}{\partial r} + \dot{s}_s \, , \tag{5.25}$$

$$-k_s \frac{\partial T_s}{\partial r} A_{sf} = \langle q_{sf}\rangle_{A_{sf}} A_{sf} \quad \text{on } A_{sf} \text{ , i.e., } r = R \, . \tag{5.26}$$

The area-averaged interfacial heat flux can be replaced by the Nusselt number using

$$\langle q_{sf}\rangle_{A_{sf}} = \frac{k_f}{d} \langle Nu_d\rangle_{A_{sf}} (T_s - \langle T\rangle^f) \quad \text{on } A_{sf} \, . \tag{5.27}$$

This is a *model* with a *local volume-averaged, one-dimensional continuous* fluid phase and a *pseudo-continuous* (or *pseudo-discrete*) solid phase. Since *no axial* conduction is allowed in the solid phase, the exact location of the dispersed particle for which (5.25) is solved is not specified and it is *assumed* that any location x is the center of a particle. The fluid-phase effective conductivity and axial dispersion coefficient $\langle k \rangle^f$ and D^d_{xx}, depend on porosity ϵ, the ratio of conductivities k_s/k_f, and for D^d_{xx}, the pore-level hydrodynamics. When (5.24) to (5.25) are used to determine $\langle Nu_d \rangle_{A_{sf}}$, this hydrodynamic dependence of D^d_{xx} is *prescribed*. The prescription of D^d_{xx} and $\langle Nu_d \rangle_{A_{sf}}$ will be discussed in Section 5.3. Our knowledge of D^d_{xx} under the local thermal *nonequilibrium* has not advanced far enough to describe D^d_{xx} with the needed confidence. Then the *accuracy* in the prediction of $\langle Nu_d \rangle_{A_{sf}}$ using the measured $\langle T \rangle^f$ and T_s (including at the surface, i.e., $T_s|_R$), would depend on the prescribed D^d_{xx} and $\langle k \rangle^f$.

Wakao and Kaguei (1982) use an *empirical* relation for $\langle k \rangle^f/k_f$ similar to (3.127) and an empirical relation for D^d_{xx}, for high Pe_d, i.e.,

$$\frac{D^d_{xx}}{\alpha_f} = 0.5 Pe_d \quad Pe_d = Re_d Pr = \frac{\epsilon \langle u \rangle^f d}{\alpha_f} = \frac{u_D d}{\alpha_f}, \tag{5.28}$$

which is characteristic of *disordered* arrangement of particles, as discussed in Section 3.2.3(C). They re-evaluated the existing experimental results and obtained transient and steady-state conditions. Their results for $\langle Nu_d \rangle_{A_{sf}}$ are given in Figure 5.5. The results are correlated as

$$\langle Nu_d \rangle_{A_{sf}} = 2 + 1.1 Re_d^{0.6} Pr^{1/3} \quad Re_d = \frac{u_D d}{\nu}, \quad Pr = \frac{\nu_f}{\alpha_f}. \tag{5.29}$$

The heat transfer measurements at low Re_d become very difficult due to required small u_D or d. The $Re_d \rightarrow 0$ *asymptote* is satisfied by this correlation. Note that the power of Re_d suggests a behavior similar to the high Re_d range of (5.17).

Among other correlations are those of Whitaker (1972) and Gunn (1978). The latter includes an explicit porosity dependence. However, in the correlation of Wakao and Kaguei (1982), D^d_{xx}/α_f appears most suitable because it is based on the use of the fluid-phase local volume-averaged energy equation, uses the most rigor in evaluating the measurements, and satisfies the $Re_d \rightarrow 0$ asymptote.

5.2.3 Treatment of Solid Phase as a Continuum

As in Chapter 3 for two-phase systems with local thermal equilibrium, the solid-fluid systems with large specific surface area will be classified depending on whether the solid phase is stationary or flowing. A *stationary* solid phase with a large specific surface area in which the *interstitial volume* between its elements is *well connected*, such that a fluid can *flow* through them, is here referred to as a *porous medium*. For the solid to remain stationary, it is assumed that its *elements* are in some contact. This interelemental *direct* contact influences the interelemental heat transfer which is by

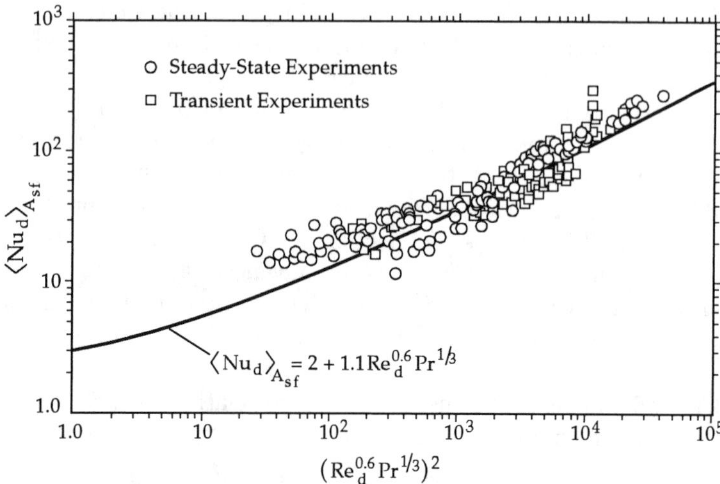

Figure 5.5. Experimental results for the interfacial Nusselt number for packed beds of a sphere. The variation with respect to the product $Re_d^{0.6} Pr^{1/3}$ is also shown with a correlation. (From Wakao and Kaguei, reproduced by permission ©1982 Gordon and Breach Science Publishers.)

conduction (we defer the discussion of *radiation* to Section 5.3). This heat transfer *depends* on the *number* of neighboring elements in contact with a given element (this is called the *coordination number*), the *contact area fraction* (i.e., the ratio of the contact areas to the element surface area, which is zero for point contacts), and the *contact resistance* (due to *coating* of elements, *surface roughness*, etc.). Therefore, in dealing with porous media, we need to address interelemental conduction. In this chapter the solid phase is *not* in local thermal equilibrium with the fluid phase and each phase will have interphasic molecular diffusion. The two-medium treatment of porous media is discussed below in Section 5.3. In *particulate flows*, the particles do collide during their *translation* and *rotation*, and therefore, some *transient interelemental conduction* heat transfer occurs. The particles also *transport* thermal energy by their *travel* between collisions. The particle-particle heat transfer depends on the frequency of collisions, the contact area, duration, etc. For *individual, elastic*, and *plastic* deformation, the heat transfer *during* contact has been determined by Ben-Ammar et al. (1992). Treatments of the thermal energy transport by particles under the condition of local thermal *nonequilibrium* particulate flows (dilute and dense) will be reviewed in Section 5.4.

5.3 Porous Media

In Section 1.10, we briefly discussed the separate continuum energy equations for the solid and fluid phases, in flow and heat transfer through porous media, under the condition of *local thermal nonequilibrium*. In this section these equations are rederived while discussing the intermediate derivations (Section 5.3.1). Then an existing example problem of a traveling thermal pulse in a porous medium is discussed using these energy equations (Section 5.3.2). This is followed by the discussion of the various approximations made to these energy equations and the examination of these *models*.

The two-medium treatment of the heat transfer in porous media is required in the analysis of transient temperature fields, such as in the storage/release of sensible and/or phase-change heat in storage packed beds. It is also required in dealing with some exothermic reactions in porous media (inert, catalytic, or combustible porous media) (Section 5.3.2). As an example, the combustion of a premixed fuel-oxidizer in an inert matrix is considered (Section 5.3.4). The significant role of the solid-fluid interfacial heat transfer (i.e., $\langle Nu_d \rangle_{A_{sf}}$) in the determination of the flame speed will be discussed. When the flame thickness is of the order of the pore linear dimension, the constraints for the validity of the volume-averaged models are not satisfied. This is demonstrated in an example, where the results of the pointwise and volume-averaged treatments are compared.

5.3.1 Local Phase-Volume Averaging

In considering interfacial heat transfer with a large specific interfacial area with the solid phase remaining stationary, while the fluid flows through its interstices, the focus is on a *large* (and nearly the *main*) local temperature difference between the solid and fluid phases. In making a local volume average over a representative elementary volume with a linear dimension ℓ, this would indicate that the main temperature difference in the system ΔT_L is not much larger than that across ℓ, i.e.,

$$\Delta T_\ell \leq \Delta T_L \; . \tag{5.30}$$

In the *absence* of any phasic heat source, since the fluid is flowing and the thermophysical properties of the fluid and solid phases are different, the local phase-volume-averaged temperatures are not equal, i.e.,

$$\langle T \rangle^f \neq \langle T \rangle^s \; . \tag{5.31}$$

When local *heat generation* in *either* of the phases *exists*, and when this is the *main* source for the temperature difference across the system, again the local phase-volume-averaged temperature will be different.

The local phase-volume-averaged treatment of the energy equation in porous media under local thermal nonequilibrium has been addressed by Carbonell and Whitaker (1984). In the following, their formulations and treatments, leading to the two-equation (i.e., two-medium) treatment of the conservation of thermal energy and the heat transfer between the phases, are reviewed. Next their results will be used in a simple, transient problem and later they will be compared with existing semi-heuristic models which lack their physical consistency.

(A) Derivation of the Local Phase-Volume-Averaged Equations

Consider *local, elementary representative volumes* V_f and V_s in continuous fluid and solid phases, respectively. The requirements on the size of these volumes are discussed in Section 1.8.1. Under the condition of no radiation heat flux, no pressure work, no viscous dissipation, no heat generation, and for a single-component fluid, the thermal energy equation (1.84) for any point in the fluid phase, i.e., in V_f, and in the stationary solid phase, i.e., in V_s, is

$$(\rho c_p)_f(\frac{\partial T_f}{\partial t} + \mathbf{u}_f \cdot \nabla T_f) = \nabla \cdot k_f \nabla T_f \quad \text{in } V_f \, , \tag{5.32}$$

$$(\rho c_p)_s \frac{\partial T_s}{\partial t} = \nabla \cdot k_s \nabla T_s \quad \text{in } V_s \, . \tag{5.33}$$

The hydrodynamic and thermal boundary conditions at the fluid-solid interface are given by the no velocity condition and continuity of temperature and the heat flux, i.e.,

$$\mathbf{u} = 0 \, , \quad T_f = T_s \, , \quad \mathbf{n}_{fs} \cdot k_f \nabla T_f = \mathbf{n}_{fs} \cdot k_s \nabla T_s \quad \text{on } A_{fs} \, , \tag{5.34}$$

where surface $\mathbf{n}_{fs}$ is the unit vector pointing from the fluid to the solid and $\mathbf{n}_{fs} = -\mathbf{n}_{sf}$ on A_{fs}.

Now as given by (1.97) to (1.99), designate the local phase-volume-averaged temperatures as $\langle T \rangle^f$ and $\langle T \rangle^s$ and the velocity as $\langle \mathbf{u} \rangle^f$, and *deviations* from these in the phasic elementary representative volumes V_f and V_s as T'_f, T'_s, and $\mathbf{u}'_f$. Now we substitute (1.97) to (1.99) in (5.32) to (5.34), using the definition of phase-volume averaging given by (1.93) and noting that $\langle \langle \phi_i \rangle \rangle^i = \langle \phi_i \rangle^i$, and $\langle \phi'_i \rangle^i = 0$. We also use the theorem given by (3.22). The results are

$$(\rho c_p)_f(\frac{\partial \langle T \rangle^f}{\partial t} + \langle \mathbf{u} \rangle^f \cdot \nabla \langle T \rangle^f) = \nabla \cdot (k_f \nabla \langle T \rangle^f + \frac{k_f}{V_f} \int_{A_{fs}} \mathbf{n}_{fs} T_f \, dA) - (\rho c_p)_f \nabla \cdot \langle \mathbf{u}' T'_f \rangle^f + \frac{1}{V_f} \int_{A_{fs}} \mathbf{n}_{fs} \cdot k_f \nabla T_f \, dA \, , \tag{5.35}$$

$$(\rho c_p)_s \frac{\partial \langle T \rangle^s}{\partial t} = \nabla \cdot (k_s \nabla \langle T \rangle^s + \frac{k_s}{V_s} \int_{A_{fs}} \mathbf{n}_{sf} T_s \, dA) + \frac{1}{V_s} \int_{A_{fs}} \mathbf{n}_{sf} \cdot k_s \nabla T_s \, dA \, , \tag{5.36}$$

$$T'_f = T'_s + \langle T\rangle^s - \langle T\rangle^f \quad \text{on } A_{fs} \,, \tag{5.37}$$

$$\mathbf{n}_{fs}\cdot\nabla T'_f = \mathbf{n}_{fs}\cdot k_s\nabla T'_s + \mathbf{n}_{fs}(k_s\nabla\langle T\rangle^s - k_f\nabla\langle T\rangle^f) \quad \text{on } A_{fs} \,. \tag{5.38}$$

Next using the notation

$$\langle\phi\rangle_{A_{fs}} = \frac{1}{A_{fs}}\int_{A_{fs}}\phi\, dA \tag{5.39}$$

and noting that over A_{fs}, T'_i and $\nabla T'_i$ change *more* substantially than $\langle T\rangle^i$ and $\nabla\langle T\rangle^i$ we have

$$\langle\mathbf{n}_{fs}T_i\rangle_{A_{fs}} \simeq \langle\mathbf{n}_{fs}T'_i\rangle_{A_{fs}} \,, \tag{5.40}$$

$$\langle\mathbf{n}_{fs}\cdot k_i\nabla T_i\rangle_{A_{fs}} \simeq \langle\mathbf{n}_{fs}\cdot k_i\nabla T'_i\rangle_{A_{fs}} \,, \tag{5.41}$$

where i stands for f or s. Then (5.35) and (5.36) become

$$(\rho c_p)_f(\frac{\partial\langle T\rangle^f}{\partial t} + \langle\mathbf{u}\rangle^f\cdot\nabla\langle T\rangle^f) = \nabla\cdot(k_f\nabla\langle T\rangle^f +$$
$$k_f\frac{A_{fs}}{V_f}\langle\mathbf{n}_{fs}T'_f\rangle_{A_{fs}}) - (\rho c_p)_f\nabla\cdot\langle\mathbf{u}'T'_f\rangle^f + \frac{A_{fs}}{V_f}\langle\mathbf{n}_{fs}\cdot k_f\nabla T'_f\rangle_{A_{fs}} \,, \tag{5.42}$$

$$(\rho c_p)_s\frac{\partial\langle T\rangle^s}{\partial t} = \nabla\cdot(k_s\nabla\langle T\rangle^s + k_s\frac{A_{fs}}{V_s}\langle\mathbf{n}_{sf}T'_s\rangle_{A_{fs}}) +$$
$$\frac{A_{fs}}{V_s}\langle\mathbf{n}_{sf}\cdot k_s\nabla T'_s\rangle_{A_{fs}} \,. \tag{5.43}$$

Now by subtracting these local phase-volume-averaged equations from the original equations (5.32) and (5.33), but with the substitution of the decompositions, the equations for the *deviations* T'_f and T'_s are found. These are

$$(\rho c_p)_f(\frac{\partial T'_f}{\partial t} - \nabla\langle\mathbf{u}'T'_f\rangle^f) + \nabla\cdot k_f\frac{A_{sf}}{V_f}\langle\mathbf{n}_{fs}T'_f\rangle_{A_{fs}} = \nabla\cdot k_f\nabla T'_f -$$
$$\frac{A_{sf}}{V_f}\langle\mathbf{n}_{fs}\cdot k_f\nabla T'_f\rangle_{A_{fs}} - (\rho c_p)_f(\mathbf{u}'_f\cdot\nabla\langle T\rangle^f + \mathbf{u}_f\cdot\nabla T'_f) \,, \tag{5.44}$$

$$(\rho c_p)_s\frac{\partial T'_s}{\partial t} + \nabla\cdot k_s\frac{A_{sf}}{V_s}\langle\mathbf{n}_{sf}T'_s\rangle_{A_{fs}} =$$
$$\nabla\cdot k_s\nabla T'_s - \frac{A_{fs}}{V_s}\langle\mathbf{n}_{sf}\cdot k_s\nabla T'_s\rangle_{A_{fs}} \,. \tag{5.45}$$

These two equations for T'_f and T'_f can be simplified by making further examinations and some order-of-magnitude analyses. First, we note that the time response of interest is larger than the time required for the diffusion across the representation elementary (which is required for a local volume average treatment to be valid). This is expressed as

$$\frac{\alpha_f}{\ell^2}t \gg 1 \,, \quad \frac{\alpha_s}{\ell^2}t \gg 1 \,. \tag{5.46}$$

Then the time derivatives are small compared to the remaining terms in (5.44) and (5.45). Next we note that

$$\frac{\Delta T_f'}{\ell} = \frac{\Delta T_\ell}{\ell} = O(\frac{\Delta\langle T\rangle^f}{L}) = O(\frac{\Delta T_L}{L}) \quad \text{or} \quad \frac{\Delta T_f'}{\Delta T_L} \simeq \frac{\ell}{L} , \tag{5.47}$$

and therefore, $\nabla\cdot\langle\mathbf{u}'T'\rangle^f \simeq \langle u\rangle^f\Delta T_f'/L$ and $\mathbf{u}_f'\cdot\nabla\langle T\rangle^f \simeq \langle u\rangle^f\Delta T_L/L$ which is much larger than former. Also, $\nabla^2 T_f' \simeq \Delta T_f'/\ell^2$ while $\nabla\cdot(A_{fs}/V)$ $\langle\mathbf{n}_{fs}T_f'\rangle^{fs} \simeq \Delta T_f'/(\ell L) \ll \Delta T_f'/\ell^2$. Using these, the left-hand side of (5.44) and (5.45) can be neglected leading to

$$(\rho c_p)_f(\mathbf{u}_f'\cdot\nabla\langle T\rangle^f + \mathbf{u}_f\cdot\nabla T_f') =$$
$$\nabla\cdot k_f\nabla T_f' - \frac{A_{fs}}{V_f}\langle\mathbf{n}_{fs}\cdot k_f\nabla T_f'\rangle_{A_{fs}} , \tag{5.48}$$

$$0 = \nabla\cdot k_s\nabla T_s' - \frac{A_{fs}}{V_s}\langle\mathbf{n}_{sf}\cdot k_s\nabla T_s'\rangle_{A_{fs}} . \tag{5.49}$$

These equations describe the variations in T_f' and T_s' within V_f and V_s, respectively, and these deviations are related at A_{fs} through (5.37) and (5.38).

(B) Closure Conditions for Periodic Structures

As with the time-averaged energy equation for turbulent flows, (5.42) and (5.43) contain averages (here local, *phase* volume) of the fluctuations (here referred to as deviations) which in turn can be described by some *predictive relations*. These are *constitutive relations* and can be simple, approximate *algebraic* equation *models* or complex and more complete *differential* equation models. These relations may contain unknowns which are described by yet other predictive relations. As an alternative, the equations for T_f' and T_s' (and $\mathbf{u}'$) can be *directly* solved (generally numerically) using periodic, *unit-cell* models. Except for large unit cell Péclet numbers, the *direct* use of the T_f' and T_s' equations and the use of the local phase-volume averaging gives the *complete* description for periodic structures (this is discussed by Saharoui and Kaviany, 1993, for the case of local thermal equilibrium). However, the use of *closure* and constitutive equations does *reveal* the inherent *anisotropy* of the *dispersion* and in the case of local thermal *nonequilibrium* among the phases allows for the direct *modeling* of the *coupling* between the local phase-volume-averaged energy equations.

Following Carbonell and Whitaker (1984), the coupling of the two energy equations through the boundary conditions (5.37) and (5.38) is used to construct closure constitutive relations. They show that for *periodic structures* a general model which allows coupling through $\langle T\rangle^f - \langle T\rangle^s$ as well as $\nabla\langle T\rangle^f$ and $\nabla\langle T\rangle^s$, and uses a *linear superposition* of these (noting that the T_f' and T_f' equations are *linear*), is

$$T_f' = \mathbf{b}_{ff}\cdot\nabla\langle T\rangle^f + \mathbf{b}_{fs}\cdot\nabla\langle T\rangle^s + f_f(\langle T\rangle^s - \langle T\rangle^f) + g_f , \tag{5.50}$$

$$T_s' = \mathbf{b}_{ss} \cdot \nabla \langle T \rangle^s + \mathbf{b}_{sf} \cdot \nabla \langle T \rangle^s + f_g(\langle T \rangle^s - \langle T \rangle^f) + g_s \,, \tag{5.51}$$

where the *transformation functions* [**b**] and [f] are functions of the position within the unit cell (and are influenced by the *hydrodynamics* and the *thermophysical properties*) and for periodic structures $g_f = g_s = 0$ (Carbonell and Whitaker, 1984). These closures, when written in a matrix form and by using $\Delta\langle T \rangle = \langle T \rangle^s - \langle T \rangle^f$, become

$$[T_f'] = [\mathbf{b}][\nabla \langle T \rangle] + [f]\Delta \langle T \rangle \,. \tag{5.52}$$

The transformation functions relate the distributions of the deviation of temperature in each phase to the *gradients* of the local phase-volume-averaged temperature and the *difference* in the local phase-volume-averaged temperatures. As was mentioned, since T_f' and T_s' equations (5.48) and (5.49) are linear, the *linear superposition* of the coupling *mechanisms* (each phase being influenced by $\nabla\langle T \rangle$ in the other phase), i.e., $\nabla\langle T \rangle^f$, $\nabla\langle T \rangle^s$, and $\Delta\langle T \rangle$, is used. The *dependence* of the distributions of the deviations on the *hydrodynamics* (e.g., *cell Reynolds number*, structure) the *thermophysical properties* (e.g., Pr, k_s/k_f), and *cell geometry* (e.g., structure, ϵ), are also included in the transformation functions. Before insertion of (5.52) into the averaged equations, the equations for [**b**] ad [f], which is similar to the case of local thermal equilibrium in Section 3.2.3, are found from the deviation equations (5.48) and (5.49).

(C) Evaluation of Transformation Functions

The unknowns T_f' and T_s' are now replaced by [**b**] and [f] and the predictive relations for [**b**] and [f] are found by substituting (5.50) and (5.51) into the *linear* T_f' and T_s' equations (5.48) and (5.49). Since the closure conditions are constructed from linear superpositions of the various *interactions*, separate equation for each transformation can be found. This is equivalent to setting the appropriate equations for $[\nabla\langle T \rangle]$ and $\Delta\langle T \rangle$ equal to zero. The resulting equations for each function are:

in V_f,

$$(\rho c_p)_f(\mathbf{u}_f' + \mathbf{u}_f \cdot \nabla \mathbf{b}_{ff}) = \nabla \cdot k_f \nabla \mathbf{b}_{ff} - \frac{A_{fs}}{V_f} \langle \mathbf{n}_{fs} \cdot k_f \nabla \mathbf{b}_{ff} \rangle_{A_{fs}} \,, \tag{5.53}$$

$$(\rho c_p)_f \mathbf{u}_f \cdot \nabla \mathbf{b}_{fs} = \nabla \cdot k_f \nabla \mathbf{b}_{fs} - \frac{A_{fs}}{V_f} \langle \mathbf{n}_{fs} \cdot k_f \nabla \mathbf{b}_{fs} \rangle_{A_{fs}} \,, \tag{5.54}$$

$$(\rho c_p)_f \mathbf{u}_f \cdot \nabla f_f = \nabla \cdot k_f \nabla f_f - \frac{A_{sf}}{V_f} \langle \mathbf{n}_{fs} \cdot k_f \nabla f_f \rangle_{A_{fs}} \,; \tag{5.55}$$

in V_s,

$$0 = \nabla \cdot k_s \nabla \mathbf{b}_{ss} - \frac{A_{fs}}{V_s} \langle \mathbf{n}_{sf} \cdot k_s \nabla \mathbf{b}_{ss} \rangle_{A_{fs}} , \tag{5.56}$$

$$0 = \nabla \cdot k_s \nabla \mathbf{b}_{sf} - \frac{A_{fs}}{V_s} \langle \mathbf{n}_{sf} \cdot k_s \nabla \mathbf{b}_{sf} \rangle_{A_{fs}} , \tag{5.57}$$

$$0 = \nabla \cdot k_s \nabla f_s - \frac{A_{fs}}{V_s} \langle \mathbf{n}_{sf} \cdot k_s \nabla f_s \rangle_{A_{fs}} . \tag{5.58}$$

The boundary conditions (5.37) and (5.38) become, on A_{fs}

$$\mathbf{b}_{ff} = \mathbf{b}_{sf} , \quad \mathbf{b}_{fs} = \mathbf{b}_{ss} , \quad f_f = f_s = 1 , \tag{5.59}$$

$$\mathbf{n}_{fs} \cdot k_f \nabla \mathbf{b}_{ff} = \mathbf{n}_{fs} \cdot k_s \nabla \mathbf{b}_{sf} - k_f \mathbf{n}_{fs} , \tag{5.60}$$

$$\mathbf{n}_{fs} \cdot k_f \nabla \mathbf{b}_{fs} = \mathbf{n}_{fs} \cdot k_s \nabla \mathbf{b}_{ss} + k_s \mathbf{n}_{fs} , \tag{5.61}$$

$$\mathbf{n}_{fs} \cdot k_f \nabla f_f = \mathbf{n}_{fs} \cdot k_s \nabla f_s . \tag{5.62}$$

These linear differential equations for $[\mathbf{b}]$ and $[f]$ are solved along with prescribed $\mathbf{u}_f$ and $\mathbf{u}'_f$ (note that $\mathbf{u}'_f = \mathbf{u}_f - \langle \mathbf{u} \rangle^f$) and the boundary conditions on A_{fs} and on the unit-cell *boundaries*. The unit-cell boundary conditions are found from the *spatial periodicity* (and in some cases where only half or a quarter of the cell is used as the computational domain, the *symmetry* conditions replace some of the periodicity conditions). For periodicity along the unit cell length $\boldsymbol{\ell}_i$, we have

$$[\mathbf{b}(\mathbf{x} + \boldsymbol{\ell}_i)] = [\mathbf{b}(\mathbf{x})] , \tag{5.63}$$

$$[f(\mathbf{x} + \boldsymbol{\ell}_i)] = [f(\mathbf{x})] . \tag{5.64}$$

As was discussed in Section 3.2.3 in dealing with the local thermal equilibrium, $[\mathbf{b}]$ and $[f]$ can be solved for (generally *numerically*) using the above governing and boundary conditions. The *cell geometry*, *hydrodynamics*, and *thermophysical properties* all influence the *magnitude* and the *distributions* of $[\mathbf{b}]$ and $[f]$ within the cell. We will discuss an available example of the evaluation of $[\mathbf{b}]$ and $[f]$ after writing the averaged energy equation in terms of some *spatial integrals* containing $[\mathbf{b}]$ and $[f]$.

(D) Local Phase-Volume-Averaged Energy Equations

The introduction of $[\mathbf{b}]$ and $[f]$ as closure constitutive functions results in the replacement of *spatial integrals* of $[T']$ with the spatial integral integrals containing $[\mathbf{b}]$ and $[f]$. In order to evaluate these integrals, which appear in the phasic thermal energy equations, the distributions of $[\mathbf{b}]$ and $[f]$ *within* periodic unit cells must be known. The equations and boundary conditions for the determination of these distributions was discussed above in Section (C) and here terms containing these spatial integrals of $[\mathbf{b}]$ and $[f]$, in the energy equations, will be examined. Further censorial properties of terms containing these integrals will also be discussed. This is similar to that done for the local thermal equilibrium, where the effective conductivity and dispersion tensors were defined. The phasic thermal energy equations (5.42) and (5.43) contain on their right-hand side terms containing the integrals of $[T']$ which we now replace that with those containing $[\mathbf{b}]$ and $[f]$. As a result every integral containing, for example, T_s' will lead to three integrals containing $\mathbf{b}_{ss}$, $\mathbf{b}_{sf}$ and f_s. Some of these terms can be collected and combined together. The spatial integrals are both *area* (over A_{sf}) and *volume* (over V_f or V_s) integrals. Then by substituting (5.50) and (5.51) in (5.42) and (5.43), respectively, and rearranging, we have the two phasic local-volume-averaged thermal energy equations. Using the order-of-magnitude analysis and the requirement for the local volume averaging, referred to in Section (A), the derivatives of the averaged quantities are neglected compared to the derivatives of the deviations (i.e., $[\mathbf{b}]$ and $[f]$). For example, (Carbonell an Whitaker, 1984)

$$\nabla T_f' \simeq \nabla \mathbf{b}_{ff} \cdot \nabla \langle T \rangle^f + \nabla \mathbf{b}_{fs} \cdot \nabla \langle T \rangle^s + \nabla f_f (\langle T \rangle^s - \langle T \rangle^f) . \qquad (5.65)$$

For the fluid phase, the thermal energy equation (5.42) for constant thermophysical properties and with an added phasic heat generation becomes

$$\begin{aligned} \frac{\partial \langle T \rangle^f}{\partial t} [\langle \mathbf{u} \rangle^f + \alpha_f \frac{A_{fs}}{V_f} (2 \langle \mathbf{n}_{fs} f_f \rangle_{A_{fs}} - \langle \mathbf{n}_{fs} \cdot \nabla \mathbf{b}_{ff} \rangle_{A_{fs}}) - \\ \langle \mathbf{u}' f_f \rangle^f] \cdot \nabla \langle T \rangle^f + [\alpha_f \frac{A_{fs}}{V_f} (-2 \langle \mathbf{n}_{fs} f_f \rangle_{A_{fs}} - \\ \langle \mathbf{n}_{fs} \cdot \nabla \mathbf{b}_{fs} \rangle_{A_{fs}}) + \langle \mathbf{u}' f_f \rangle^f] \cdot \nabla \langle T \rangle^s = \nabla \cdot [\alpha_f (\mathbf{I} + \\ 2 \frac{A_{fs}}{V_f} \langle \mathbf{n}_{fs} \mathbf{b}_{ff} \rangle_{A_{fs}}) - \langle \mathbf{u}' \mathbf{b}_{ff} \rangle^f] \cdot \nabla \langle T \rangle^f + \\ \nabla \cdot (2 \alpha_f \frac{A_{fs}}{V_f} \langle \mathbf{n}_{fs} \mathbf{b}_{fs} \rangle_{A_{fs}} - \langle \mathbf{u}' \mathbf{b}_{fs} \rangle^f) \cdot \nabla \langle T \rangle^s + \\ \alpha_f \frac{A_{fs}}{V_f} \langle \mathbf{n}_{fs} \cdot \nabla f_f \rangle^{fs} (\langle T \rangle^s - \langle T \rangle^f) + \langle \dot{s} \rangle^f . \end{aligned} \qquad (5.66)$$

For the solid phase the thermal energy equation (5.43) for constant thermophysical properties and with an added radiation heat flux vector $\mathbf{q}_r$ and a phasic heat generation becomes

$$\begin{aligned} \frac{\partial \langle T \rangle^s}{\partial t} + [\alpha_s \frac{A_{fs}}{V_s} (2 \langle \mathbf{n}_{sf} f_s \rangle_{A_{fs}} - \langle \mathbf{n}_{sf} \cdot \nabla \mathbf{b}_{sf} \rangle_{A_{fs}})] \cdot \nabla \langle T \rangle^f + \\ [\alpha_s \frac{A_{fs}}{V_s} (-2 \langle \mathbf{n}_{sf} f_s \rangle_{A_{fs}} - \langle \mathbf{n}_{sf} \cdot \nabla \mathbf{b}_{ss} \rangle_{A_{fs}})] \cdot \nabla \langle T \rangle^s = \\ \nabla \cdot [\alpha_s (\mathbf{I} + 2 \frac{A_{fs}}{V_s} \langle \mathbf{n}_{sf} \mathbf{b}_{ss} \rangle_{A_{fs}})] \cdot \nabla \langle T \rangle^s + \\ \nabla \cdot (2 \alpha_s \frac{A_{fs}}{V_s} \langle \mathbf{n}_{sf} \mathbf{b}_{sf} \rangle_{A_{fs}}) \cdot \nabla \langle T \rangle^f + \\ \alpha_s \frac{A_{fs}}{V_s} \langle \mathbf{n}_{sf} \cdot \nabla f_s \rangle_{A_{fs}} (\langle T \rangle^s - \langle T \rangle^f) - \nabla \cdot \langle \mathbf{q}_r \rangle^s + \langle \dot{s} \rangle^s . \end{aligned} \qquad (5.67)$$

The use of the boundary conditions (5.37) and (5.38), containing $[T']$, while $[\nabla T']$ and noting that $\mathbf{u}=0$ on A_{sf}, results in extra conditions containing spatial integrals of $[\mathbf{b}]$ and $[f]$. These are (Carbonell and Whitaker, 1984)

$$k_f\langle \mathbf{n}_{fs}\cdot\nabla f_f\rangle_{A_{fs}} = -k_s\langle \mathbf{n}_{sf}\cdot\nabla f_s\rangle_{A_{fs}}\ , \tag{5.68}$$

$$k_f\langle \mathbf{n}_{fs}\cdot\nabla \mathbf{b}_{fs}\rangle_{A_{fs}} = -k_s\langle \mathbf{n}_{sf}\cdot\nabla \mathbf{b}_{ss}\rangle_{A_{fs}}\ , \tag{5.69}$$

$$k_f\langle \mathbf{n}_{fs}\cdot\nabla \mathbf{b}_{ff}\rangle_{A_{fs}} = -k_s\langle \mathbf{n}_{sf}\cdot\nabla \mathbf{b}_{sf}\rangle_{A_{fs}}\ , \tag{5.70}$$

$$\langle \mathbf{n}_{fs} f_f\rangle_{A_{fs}} = -\langle \mathbf{n}_{sf} f_s\rangle_{A_{fs}}\ , \tag{5.71}$$

$$\langle \mathbf{n}_{fs}\mathbf{b}_{ff}\rangle_{A_{fs}} = -\langle \mathbf{n}_{sf}\mathbf{b}_{sf}\rangle_{A_{fs}}\ , \tag{5.72}$$

$$\langle \mathbf{n}_{fs}\mathbf{b}_{fs}\rangle_{A_{fs}} = -\langle \mathbf{n}_{sf}\mathbf{b}_{ss}\rangle_{A_{fs}}\ . \tag{5.73}$$

The first of the above equations states the equality of the *interfacial conduction heat transfer coefficient* (this coefficient will be discussed shortly).

Note that under the condition of local thermal equilibrium, i.e., $\langle T\rangle^s = \langle T\rangle^f$, the two phasic energy equations can be combined by multiplying the fluid phase equation by ϵ and the solid phase equation by $1-\epsilon$ and then adding the results. Then the results of Section 3.22, i.e., (3.59), are recovered.

The averaged phasic energy equations (5.66) and (5.67) contain *interaction* (coupling) between the phase through the interfacial heat transfer terms containing the difference between the local phasic temperatures, i.e., $\langle T\rangle^s - \langle T\rangle^f$. Further *interactions* occur through the *convection* term which now *also* appears in the *solid* phase equation and through the *diffusion* terms appearing in *both* equations. These interactions between the phases, which appear through spatial integrals of $[\mathbf{b}]$ and $[f]$, can be represented using the concept of *convective* and *diffusive* heat *fluxes* and the concept of the interfacial area-averaged Nusselt number (i.e., dimensionless interfacial area-averaged heat flux). Then following Carbonell and Whitaker (1984), we define *convective velocity vectors* $\mathbf{v}_{ff}$, $\mathbf{v}_{sf}$, $\mathbf{v}_{sf}$, and $\mathbf{v}_{ss}$ as coefficients of terms containing the first derivatives of $[\langle T\rangle]$. Similar to (3.64), we define *total thermal diffusivity tensors* $\mathbf{D}_{ff}$, $\mathbf{D}_{fs}$, $\mathbf{D}_{sf}$, and $\mathbf{D}_{ss}$ as coefficients appearing in terms containing *double* derivatives of $[\langle T\rangle]$. Finally, as in (5.16) and (5.39), the local interfacial area-averaged heat flux and Nusselt number are defined. When these *definitions* are substituted in (5.66) and (5.67), the more *compact* form of the phasic, average thermal energy equations are found.

The local phase-volume-averaged thermal energy equations for the fluid and solid phase, (5.66) and (5.67), respectively, become

$$\frac{\partial\langle T\rangle^f}{\partial t} + \mathbf{v}_{ff}\cdot\nabla\langle T\rangle^f + \mathbf{v}_{fs}\cdot\nabla\langle T\rangle^s = \nabla\cdot\mathbf{D}_{ff}\cdot\nabla\langle T\rangle^f +$$
$$\nabla\cdot\mathbf{D}_{fs}\cdot\nabla\langle T\rangle^s + \frac{A_{fs}}{V_f}\frac{h_c}{(\rho c_p)_f}(\langle T\rangle^s - \langle T\rangle^f) + \langle \dot{s}\rangle^f\ , \tag{5.74}$$

$$\frac{\partial \langle T \rangle^s}{\partial t} + \mathbf{v}_{sf} \cdot \nabla \langle T \rangle^f + \mathbf{v}_{ss} \cdot \nabla \langle T \rangle^s = \nabla \cdot \mathbf{D}_{sf} \cdot \nabla \langle T \rangle^f +$$

$$\nabla \cdot \mathbf{D}_{ss} \cdot \nabla \langle T \rangle^s + \frac{A_{fs}}{V_s} \frac{h_c}{(\rho c_p)_s} (\langle T \rangle^f - \langle T \rangle^s) + \nabla \cdot \langle \mathbf{q}_r \rangle^s + \langle \dot{s} \rangle^s \,. \quad (5.75)$$

From the definition of h_c, which is related to the area-averaged interfacial Nusselt number $\langle Nu_d \rangle_{A_{sf}}$ used in the discussion of the interelemental heat transfer in Section 5.2, we have

$$h_c = \frac{\langle Nu_d \rangle_{A_{sf}} k_f}{d} \equiv k_f \langle \mathbf{n}_{fs} \cdot \nabla f_f \rangle_{A_{fs}} = -k_s \langle \mathbf{n}_{sf} \cdot \nabla f_s \rangle_{A_{fs}} \,. \quad (5.76)$$

We have used the particle size (which is related to the unit cell linear size ℓ through the cell geometry and porosity) and the *fluid* conductivity in the definition of the Nusselt number. The local, area-averaged interfacial *conduction* heat flux is related to the area-averaged Nusselt number and the *conduction heat transfer coefficient* h_c through

$$\langle q_{sf} \rangle_{A_{sf}} = h_c(\langle T \rangle^s - \langle T \rangle^f) = \frac{k_f}{d} \langle Nu_d \rangle_{A_{sf}} (\langle T \rangle^s - \langle T \rangle^f) \,. \quad (5.77)$$

Note that contrary to the conventional usage of the Nusselt number, here $\langle Nu_d \rangle_{A_{sf}}$ accounts for the interfacial *conduction* heat transfer coefficient (this will be further demonstrated in the following example) and is *independent* of the fluid velocity. The effect of the fluid velocity on the *intraphasic* heat transfer is through [**v**] and [**D**]. In the *heuristic* (or semi-heuristic) treatments, simpler *models* (such as the one discussed in Section 5.2.2) are used, and then the hydrodynamic effects are in part *included* in $\langle Nu_d \rangle_{A_{sf}}$. This will be further discussed in Section 5.3.3.

Equations (5.74) and (5.75), along with the required prescriptions of [**v**], [**D**], and $\langle Nu_d \rangle_{A_{sf}}$ constitute a complete description of the *two-medium treatment* of heat transfer in porous media with *periodic structures*. The prescription of [**v**], [**D**] and $\langle Nu_d \rangle_{A_{sf}}$ would require the direct simulation of flow and heat transfer in these periodic structures. For *two-dimensional* and *three-dimensional* structures, no results are yet available. However for a one-dimensional structure, i.e., fully developed flow and heat transfer through a bundle of capillary tubes, a solution is available (Zanotti and Carbonell, 1984), and this is reviewed next.

5.3.2 Example of Capillary Tubes

In order to demonstrate that h_c, appearing in (5.74) and (5.75) as one of the coupling parameters, is independent of the flow (i.e., independent of the hydrodynamics) and depends *only* on the *geometric parameters* and the *thermal conductivities*, we review an example considered by Zanotti and Carbonell (1984). This example also demonstrates the need for a two-medium treatment of heat transfer in porous media when accurate *transient*

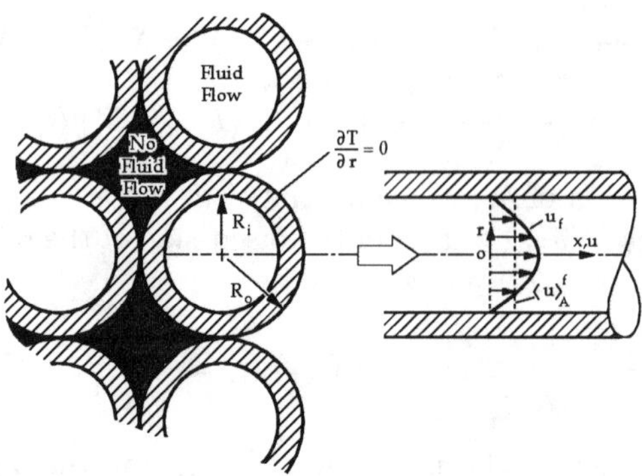

Figure 5.6. A schematic of a fluid flowing through tubes with a square-array arrangement. The one-dimensional velocity profile and the coordinate system used are also shown.

behavior is required, such as in response of the solid-fluid systems to pulses in the fluid inlet temperature.

The problem considered is rendered in Figure 5.6. The bundle of capillary tubes are arranged *orderly* (such as in the simple, in-line array shown). The tube inside and outside the radii are R_i and R_o, respectively, and due to the *lateral* periodicity, the condition of no heat flow applies at $r = R_o$. The axial fluid flow is assumed to be steady, laminar, and fully developed. The heat capacity of the *interstitial volume* between the tubes is neglected (i.e., assuming that a vacuum exists in this volume). Note that the *local* solid-phase-averaged temperature $\langle T\rangle^f_A$ can *only* be of *accurate* axial resolution over distances much larger than $R_o - R_i$. This is because any thermal boundary layer formed in the solid has been *neglected*, thus requiring that the disturbances completely penetrate *radially* such that the temperature across $R_o - R_i$ becomes *nearly uniform*.

(A) Derivation of the Phasic Energy Equations

Under the condition of steady one-dimensional, laminar fully developed flow, transient axisymmetry of the temperature field, and when no volumetric heat generation exists, the *point* thermal energy equations for the fluid and solid phase are

$$\frac{\partial T_f}{\partial t} + u_f \frac{\partial T}{\partial x} = \alpha_f \left(\frac{\partial^2 T_f}{\partial x^2} + \frac{1}{r}\frac{\partial}{\partial r} r \frac{\partial T_f}{\partial r}\right) \quad 0 \le r \le R_i \ , \tag{5.78}$$

$$\frac{\partial T}{\partial t} = \alpha_s \left(\frac{\partial^2 T_s}{\partial x^2} + \frac{1}{r}\frac{\partial}{\partial r} r \frac{\partial T_s}{\partial r}\right) \quad R_i \le r \le R_o \ . \tag{5.79}$$

The *point* boundary conditions are

$$T_f = T_s \,, \quad k_f \frac{\partial T_f}{\partial r} = k_s \frac{\partial T_s}{\partial r} \quad r = R_i \,, \tag{5.80}$$

$$\frac{\partial T_f}{\partial r} = 0 \quad r = 0 \,, \quad \text{and} \quad \frac{\partial T_s}{\partial r} = 0 \quad r = R_o \,. \tag{5.81}$$

For the one-dimensional velocity $u = u(r)$ and time-dependent axisymmetric temperatures $T_f = T_f(t,x,y)$, $T_s = T_s(t,x,y)$, the relevant local *area-averaged* quantities are defined as

$$\langle T\rangle_A^f = \frac{2}{R_i^2}\int_o^{R_i} T_f r\, dr \,, \quad \langle T\rangle_A^s = \frac{2}{R_o^2 - R_i^2}\int_{R_i}^{R_o} T_s r\, dr \,, \tag{5.82}$$

$$T_f = \langle T\rangle_A^f + T_f' \,, \quad T_s = \langle T\rangle_A^s + T_s' \,, \quad u = \langle u\rangle_A^f + u_f' \,. \tag{5.83}$$

Then $\langle T\rangle_A^f = \langle T\rangle_A^f(t,x)$, $\langle T\rangle_A^s = \langle T\rangle_A^s(t,x)$, and $\langle u\rangle_A^f$ is a constant.

Next we substitute (5.83) into equations (5.78) through (5.29). Note that along A_{fs} the derivatives of the temperature fluctuations are uniform, we have for constant properties. For example,

$$\frac{A_{fs}}{(\rho c_p)_f V_f}\langle \mathbf{n}_{fs}\cdot k_f \nabla T_f'\rangle^{fs} = \alpha_f \frac{2}{R_o}\frac{\partial T_f'}{\partial r}|_{r=R_i} \,. \tag{5.84}$$

Then the appropriate forms of (5.42) and (5.43) for the *area-averaged* quantities become

$$\frac{\partial \langle T\rangle_A^f}{\partial t} + \langle u\rangle_A^f \frac{\partial \langle T\rangle_A^f}{\partial x} = \alpha_f \frac{\partial^2 \langle T\rangle_A^f}{\partial x^2} - \frac{\partial}{\partial x}\langle u_f' T_f'\rangle_A^f + \alpha_f \frac{2}{R_o}\frac{\partial T_f'}{\partial r}|_{R_i} \,, \tag{5.85}$$

$$\frac{\partial \langle T\rangle_A^s}{\partial t} = \alpha_s \frac{\partial^2 \langle T\rangle_A^s}{\partial x^2} - \alpha_s \frac{2R_o}{R_i^2 - R_o^2}\frac{\partial T_s'}{\partial r}|_{R_i} \,. \tag{5.86}$$

The T_f' and T_f' equations are also the appropriate forms of (5.44) and (5.45) and are

$$\frac{\partial T_f'}{\partial t} + u_f \frac{\partial T_f'}{\partial x} = \alpha_f \frac{\partial^2 T_f'}{\partial x^2} + \alpha_f \frac{1}{r}\frac{\partial}{\partial r} r \frac{\partial T_f'}{\partial r} - u_f' \frac{\partial \langle T\rangle_A^f}{\partial x} + \frac{\partial}{\partial x}\langle u_f' T_f'\rangle_A^f - \alpha_f \frac{2}{R_o}\frac{\partial T_f'}{\partial r}|_{R_i} \,, \tag{5.87}$$

$$\frac{\partial T_s'}{\partial t} = \alpha_s \frac{\partial^2 T_s'}{\partial x^2} + \alpha_s \frac{1}{r}\frac{\partial}{\partial r} r \frac{\partial T_s'}{\partial r} + \alpha_s \frac{2R_o}{R_o^2 - R_i^2}\frac{\partial T_s'}{\partial r}|_{R_i} \,. \tag{5.88}$$

The boundary conditions (5.80) and (5.81) become

$$T_f' = T_s' + \langle T\rangle_A^s - \langle T\rangle_A^f \,, \quad k_f \frac{\partial T_f'}{\partial r} = k_s \frac{\partial T_s'}{\partial r} \quad r = R_i \,, \tag{5.89}$$

$$\frac{\partial T_f'}{\partial r} = 0 \quad r = 0 \,, \quad \frac{\partial T_s'}{\partial r} = 0 \quad r = R_o \,. \tag{5.90}$$

By assuming that the axial conduction in the $[T']$ equations is negligible compared to that in the radial direction, Zanotti and Carbonell (1984) show that the functions b_{fs} and b_{sf}, defined through the closure conditions, are zero. They also make other order-of-magnitude analyses and their results for the [**b**] and $[f]$ equations are

$$\frac{1}{r}\frac{d}{dr}r\frac{db_{ff}}{dr} = \frac{2}{R_i}\frac{db_{ff}}{dr}|_{R_i} + \frac{u'_f}{\alpha_f} , \tag{5.91}$$

$$\frac{1}{r}\frac{d}{dr}r\frac{db_{ss}}{dr} = -\frac{2R_i}{R_o^2 - R_i^2}\frac{db_{ss}}{dr}|_{R_i} , \tag{5.92}$$

$$\frac{1}{r}\frac{d}{dr}r\frac{df_f}{dr} = \frac{2}{R_i}\frac{df_f}{dr}|_{R_i} , \tag{5.93}$$

$$\frac{1}{r}\frac{d}{dr}r\frac{df_s}{dr} = -\frac{2R_i}{R_o^2 - R_i^2}\frac{df_s}{dr}|_{R_i} . \tag{5.94}$$

The velocity deviation is given, for a Hagen-Poiseuille flow, as

$$u_f = 2\langle u\rangle_A^f(1 - \frac{r^2}{R_i^2}) , \quad u'_f = \langle u\rangle_A^f(1 - \frac{2r^2}{R_i^2}) . \tag{5.95}$$

The boundary conditions for [**b**] and $[f]$ equations are obtained from (5.89) and (5.90), and are

$$f_f = f_s + 1 , \quad b_{ff} = b_{ss} , \quad k_f\frac{db_{ff}}{dr} = k_s\frac{db_{ss}}{dr} ,$$
$$k_f\frac{df_f}{dr} = k_s\frac{df_s}{dr} , \quad r = R_i , \tag{5.96}$$

$$\frac{db_{ff}}{dr} = \frac{df_f}{dr} = 0 \quad r = 0 , \quad \frac{db_{ss}}{dr} = \frac{df_s}{dr} = 0 \quad r = R_o . \tag{5.97}$$

In addition, [**b**] and $[f]$ must satisfy

$$\langle b_{ff}\rangle_A^f = \langle b_{ss}\rangle_A^s = \langle f_f\rangle_A^f = \langle f_s\rangle_A^s = 0 . \tag{5.98}$$

The equations for [**b**] and $[f]$ are solved by Zanotti and Carbonell (1984) and closed-form solutions are found which are then used to determine the terms involving the deviations in (5.85) and (5.86) while noting that

$$T'_f = b_{ff}(r)\frac{\partial\langle T\rangle_A^f}{\partial x} + f_f(r)(\langle T\rangle_A^s - \langle T\rangle_A^f) , \tag{5.99}$$

$$T'_s = b_{ss}(r)\frac{\partial\langle T\rangle_A^s}{\partial x} + f_s(r)(\langle T\rangle_A^s - \langle T\rangle_A^f) . \tag{5.100}$$

The resulting equations for $[\langle T\rangle_A]$ are

$$\frac{\partial\langle T\rangle_A^f}{\partial t} + (1 + \frac{h_cR_i}{6k_f})\langle u\rangle_A^f\frac{\partial\langle T\rangle_A^f}{\partial x} - a_1\frac{h_cR_i}{2k_f}\langle u\rangle_A^f\frac{\partial\langle T\rangle_A^f}{\partial x} =$$
$$\alpha_s[1 + (\frac{1}{12} - \frac{h_cR_i}{36k_f})(\frac{\langle u\rangle_A^f R_i}{2\alpha_f})^2]\frac{\partial^2\langle T\rangle_A^f}{\partial x^2} + \frac{2h_c}{R_i(\rho c_p)_f}(\langle T\rangle_A^s - \langle T\rangle_A^f) , \tag{5.101}$$

$$\frac{\partial \langle T\rangle_A^s}{\partial t} - \frac{(\rho c_p)_f}{(\rho c_p)_s}\frac{R_i^2}{R_o^2 - R_i^2}\frac{h_c R_i}{12k_f}\langle u\rangle_A^f \frac{\partial \langle T\rangle_A^f}{\partial x} =$$
$$\alpha_s \frac{\partial^2 \langle T\rangle_A^s}{\partial x^2} + \frac{2h_c R_i}{(R_o^2 - R_i^2)(\rho c_p)_s}(\langle T\rangle_A^s - \langle T\rangle_A^f) \; . \tag{5.102}$$

Note that the fluid-phase specific surface area is $A_{sf}/V_f = 2/R_i$ and the solid-phase specific surface area is $A_{sf}/V_s = 2R_i/(R_o^2 - R_i^2)$.

The *conduction* heat transfer coefficient h_c is a function of the geometric parameters R_i and R_o and the thermal conductivities, and is given by (Zanotti and Carbonell, 1984)

$$h_c^{-1} = \frac{R_i}{4k_f} + \frac{a(R_o^2 - R_i^2)}{2k_s R_i} \; . \tag{5.103}$$

The interfacial Nusselt number is then

$$\langle Nu_{d_i}\rangle_{A_{sf}} = Nu_{d_i} = \frac{h_c 2R_i}{k_f} = \frac{1}{\dfrac{1}{8} + \dfrac{a(R_o^2 - R_i^2)}{4R_i^2}\dfrac{k_f}{k_s}} \; , \tag{5.104}$$

where

$$a = \frac{4\dfrac{R_o^4}{R_i^4}\ln\dfrac{R_o}{R_i} - 3\dfrac{R_o^4}{R_i^4} + 4\dfrac{R_o^2}{R_i^2} - 1}{2(\dfrac{R_o^2}{R_i^2} - 1)^3} \; . \tag{5.105}$$

Examination of (5.101) and (5.102) shows that hydrodynamics influence the transport through the *diffusion* term in the fluid-phase energy equation and the *convection* term in the solid-phase energy equation. The hydrodynamics do *not* influence h_c as given by (5.103).

Note that the Taylor-Aris dispersion discussed in Section 3.2.1 is recovered by setting $k_s = h_c = 0$. For this case (5.101) reduces to

$$\frac{\partial \langle T\rangle_A^f}{\partial t} + \langle u\rangle_A^f \frac{\partial \langle T\rangle_A^f}{\partial x} = \alpha_f (1 + \frac{\langle u\rangle_A^f \langle u\rangle_A^f R_i^2}{48\alpha_f^2})\frac{\partial^2 \langle T\rangle_A^f}{\partial x^2} \; . \tag{5.106}$$

Then the solid-phase equation becomes trivial.

(B) Phase Lag in a Traveling Pulse

Carbonell and Whitaker (1984) and Zanotti and Carbonell (1984) analyze the travel of a thermal pulse caused by a sudden upstream change in the fluid temperature and use the *method of moments*. The *mean positions* of the *pulses* in the fluid and solid phase are equal to the first absolute moments of the temperature profile in these phases. They determine the pulse velocity as

$$u_p = \frac{\langle u\rangle_A^f}{1 + \dfrac{(\rho c_p)_s}{(\rho c_p)_f}\dfrac{R_o^2 - R_i^2}{R_i^2}} \tag{5.107}$$

with a *constant* phase lag Δx (distance) between the fluid and the solid phase given by

$$\frac{\Delta x}{R_o} = \frac{\langle u \rangle_A^f R_i}{4\alpha_f}\{1 + \frac{1}{\frac{1}{4}\langle Nu_{d_i} \rangle_{A_{sf}}[1 + \frac{(\rho c_p)_f}{(\rho c_p)_s}\frac{R_i^2}{R_o^2 - R_i^2}]}\}. \tag{5.108}$$

Note that the second term in the denominator of (5.108) is the ratio of the phase heat capacities.

Then the requirement for $\Delta x \to 0$ are $Pe_{d_i} = 2\langle u \rangle_A^f R_i/\alpha_f$ tending to zero. Also, when $\langle Nu_{d_i} \rangle_{A_{sf}}$ *increases*, the phase lag *decreases*. Discussion of the interaction between the phases and conditions for the validity of (5.107) and (5.108) is given by Carbonell and Whitaker (1985) and Zanotti and Carbonell (1984).

5.3.3 Approximate Models

As discussed in Section 5.2 and also demonstrated through the more rigorous derivations, in flow and heat transfer in porous media under the condition of local thermal nonequilibrium the effect of the *fluid motion* (i.e., hydrodynamics) must be included in the phasic average energy equations beyond the contribution to the fluid phase convection $(\rho c_p)_f \langle \mathbf{u} \rangle^f \cdot \nabla \langle T \rangle^f$. The first of these contributions has been traditionally included as the fluid phase dispersion (i.e., as enhanced diffusion) and this contribution is important at a high Péclet number $Pe_d = \epsilon \langle u \rangle^f d/\alpha_f$, at least when compared to the fluid phase effective (i.e., the presence of the solid phase accounted for) molecular diffusion which may be small. However, when the solid phase effective molecular diffusion is much larger than that of the fluid, traditionally the effect of the fluid phase dispersion has been neglected. The local interfacial heat transfer between the phases $A_o \langle q_{sf} \rangle_{A_{sf}}$ is central in the nonequilibrium analyses, and this contribution has been traditionally addressed as a hydrodynamically *affected* phenomena. However, as shown by Carbonell and Whitaker (1984), this coupling can be only *conductive* if the hydrodynamic effects are properly included in other mechanisms.

As we discussed in Chapter 4, the Nusselt number, which is a dimensionless interfacial heat flux $\langle q_{sf} \rangle_{A_{sf}}$, is generally used to determine the relative strength of the interfacial *convective* heat transfer compared to conductive heat transfer. There we have examined how the Nusselt number varies with the hydrodynamics (e.g., Re and Pr). Therefore, the use of the Nusselt number in connection with the *conductive* interfacial heat transfer in (5.74) and (5.75) is not in accord with this usage. This is because the Nusselt number in these equations depend on the ratio of the conductivities k_s/k_f as well as the geometrical parameters, but not the hydrodynamics.

In the *simplified* (i.e., approximated or modeled) forms of (5.74) and

(5.75), both the *hydrodynamic* conductivity ratio k_f/k_s and the *geometrical* effects are *lumped*. For example, one of the earliest *models*, i.e., the *Schumann model* (Wakao and Kaguei, 1982) uses only one parameter to include these effects. This *one-dimensional* model, for packed beds of spherical particles of diameter d, is

$$\frac{\partial\langle T\rangle^f}{\partial t}+\langle u\rangle^f\frac{\partial\langle T\rangle^f}{\partial x}=-\frac{A_o\alpha_f}{\epsilon d}\langle Nu_d\rangle_{A_{sf}}(\langle T\rangle^f-\langle T\rangle^s)+\langle\dot{s}\rangle^f\ , \quad (5.109)$$

$$\frac{\partial\langle T\rangle^s}{\partial t}=\frac{A_o\alpha_f}{(1-\epsilon)d}\frac{(\rho c_p)_f}{(\rho c_p)_s}\langle Nu_d\rangle_{A_{sf}}(\langle T\rangle^f-\langle T\rangle^s)+\langle\dot{s}\rangle^s\ . \quad (5.110)$$

This model does *not* include the *axial diffusion* in the fluid and solid phases, among other coupling mechanisms. Then using this model to determine $\langle Nu_d\rangle_{A_{sf}}$, we expect

$$\langle Nu_d\rangle_{A_{sf}}=\frac{h_{sf}d}{k_f}=\langle Nu_d\rangle_{A_{sf}}(Re_d,Pr,\frac{k_s}{k_f},\epsilon,\text{structural parameters})\ , \quad (5.111)$$

where h_{sf} is the convective-conductive interfacial heat transfer coefficient. Later a model which *includes* axial diffusion has been introduced and this is generally called the *continuous-solid model* (Wakao and Kaguei, 1982). This model is

$$\frac{\partial\langle T\rangle^f}{\partial t}+\langle u\rangle^f\frac{\partial\langle T\rangle^f}{\partial x}=\frac{\langle\alpha\rangle^s}{\epsilon}\frac{\partial^2\langle T\rangle^f}{\partial x^2}-$$
$$\frac{A_o\alpha_f}{\epsilon d}\langle Nu_d\rangle_{A_{sf}}(\langle T\rangle^f-\langle T\rangle^s)+\langle\dot{s}\rangle^f\ , \quad (5.112)$$

$$\frac{\partial\langle T\rangle^s}{\partial t}=\frac{\langle\alpha\rangle^s}{(1-\epsilon)}\frac{\partial^2\langle T\rangle^s}{\partial x^2}-\frac{A_o\alpha_s}{(1-\epsilon)d}\langle Nu_d\rangle_{A_{sf}}(\langle T\rangle^f-\langle T\rangle^s)+\langle\dot{s}\rangle^s . \quad (5.113)$$

This model does *not* include the axial dispersion in the fluid and solid phases, among other coupling mechanisms. The *effective conductivities* are expected to be of the form

$$\langle\alpha\rangle^f=\frac{\langle k\rangle^f}{(\rho c_p)_f}=\langle\alpha\rangle^f(k_f,k_s,\epsilon,\text{structure parameters})\ , \quad (5.114)$$

$$\langle\alpha\rangle^s=\frac{\langle k\rangle^s}{(\rho c_p)_s}=\langle\alpha\rangle^s(k_s,k_f,\epsilon,\text{structure parameters})\ . \quad (5.115)$$

Then the effect of dispersion which depends on k_s/k_f and the structural parameters, in addition to Re_d and Pr, must be included in $\langle Nu_d\rangle_{A_{sf}}$. This gives the same expected functional dependence as (5.111).

Yet another model which includes *axial conduction* and *dispersion* in the fluid phase and uses a *distributed* description of the heat transfer in the solid phase was introduced. This is the *dispersion particle-based model* given by (5.24) and (5.25). We can rewrite these equations in terms of

$\langle Nu_d \rangle_{A_{sf}}$ as

$$\frac{\partial \langle T \rangle^f}{\partial t} + \langle u \rangle^f \frac{\partial \langle T \rangle^f}{\partial x} = (\frac{\langle \alpha \rangle}{\epsilon} + D^d_{xx}) \frac{\partial^2 \langle T \rangle^f}{\partial x^2} - \frac{A_o \alpha_f}{\epsilon d} \langle Nu_d \rangle_{A_{sf}} (\langle T \rangle^f - \langle T_s \rangle_{A_{sf}}) + \langle \dot{s} \rangle^f \, , \tag{5.116}$$

$$\frac{\partial T_s}{\partial t} = \alpha_s \frac{1}{r^2} \frac{\partial}{\partial r} r^2 \frac{\partial T_s}{\partial r} + \dot{s}_s \quad \text{in } V_s \, , \tag{5.117}$$

$$-\frac{\partial T_s}{\partial r} = \frac{k_f}{k_s} \frac{1}{d} \langle Nu_d \rangle_{A_{sf}} (T_s - \langle T \rangle^f) \quad \text{on } A_{sf} \, . \tag{5.118}$$

The effective diffusivity $\langle \alpha \rangle = \langle k \rangle / (\rho c_p)_f$ is assumed to be the effective *medium* diffusivity and is obtained from correlations such as (3.127) which are valid for the local *thermal equilibrium* condition. The axial dispersion coefficient is also found from relations obtained under local thermal equilibrium assumptions, such as (5.28). Then it is *expected* that $\langle Nu_d \rangle_{A_{sf}}$ will be a function of Re_d and Pr, as indicated by (5.111). It appears that this last model is able to predict some gross transient (and when a heat source is present, the steady-states) behavior of a packed bed of particles (Wakao and Kaguei, 1982). Any more accurate prediction would require more elaborate models.

5.3.4 Exothermic Reaction

In flow and *exothermic reaction* in porous media the assumption of local thermal equilibrium between the solid and fluid phases is generally *not* valid. Then the local interfacial heat transfer rate $\langle q_{sf} \rangle_{A_{sf}}$ must be addressed. Various applications of such fluid (generally a gas)-solid reactions are discussed by Szekely et al. (1976). Here we consider a recent application related to burning of a premixed gas within (or adjacent to) an inert porous medium. The heat transfer efficiency and the reduction in the pollutant emission are among the attractive features of this combustion technique.

The problem is comparable to the flame propagation in a plain medium discussed in Section 2.3.2. The *differences* are in the two-medium treatment, i.e., treating $\langle T \rangle^g$ and $\langle T \rangle^s$ separately, and in the total thermal conductivity tensor for each phase $\langle \mathbf{K} \rangle^g$ and $\langle \mathbf{K} \rangle^s$. When the solid matrix is *anisotropic*, these tensors will also be anisotropic. The total mass diffusivity tensor $\langle \mathbf{D}_m \rangle^g$, the radiative heat flux vector in each phase, $\langle \mathbf{q}_r \rangle^g$ and $\langle \mathbf{q}_r \rangle^s$, and the volumetric interfacial heat transfer rate $A_o \langle q_{sf} \rangle_{A_{sf}}$ would also need to be addressed. Depending on the choice of the solid material and matrix structure, $\langle \mathbf{K} \rangle^g$, $\langle \mathbf{K} \rangle^s$, $\langle \mathbf{D}_m \rangle^g$, $\langle \mathbf{q}_r \rangle^g$, and $\langle \mathbf{q}_r \rangle^s$ can significantly influence the flame speed u_F within the porous medium. In addition, for *finite* porous media, e.g., porous slabs, the radiation heat loss from the *surface* of the slab can influence the flame speed. In practice the flame can also be placed

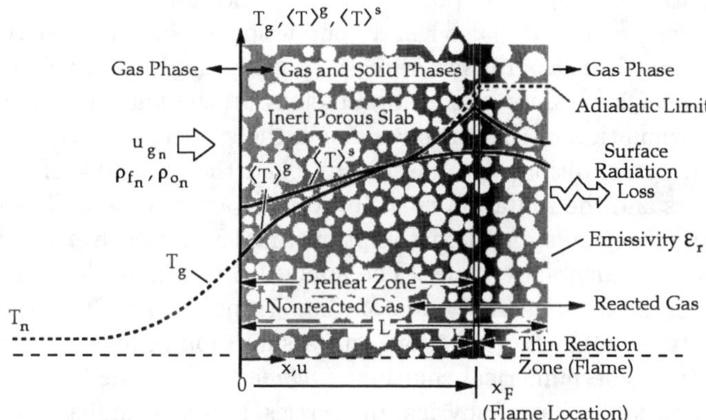

Figure 5.7. A schematic of the premixed gas combustion in a porous medium. The stoichiometric nonreacted gas mixture arrives from the left and after reaction leaves from the right, and the right surface loses heat by radiation. The coordinate system used is also shown.

just outside the porous slab, but here we will consider the structure and the speed of the flame *inside* the solid matrix.

A schematic of the problem considered is shown in Figure 5.7. The premixed gas flowing from the left to the right has the far upstream (gas-phase only) conditions of T_n, u_{g_n} and fuel and oxidant densities ρ_{f_n}, ρ_{o_n} (a stoichiometric or nonstoichiometric ratio). The gas enters into the porous slab at $x = 0$ and inside the slab the local velocity is $\langle u \rangle$. Note that $\langle u \rangle$ is the Darcean velocity, and therefore, across $x = 0$ the velocity is *continuous*, i.e., $u_{g_o} = \langle u \rangle_o$. The velocity changes along x because of the change in temperature which changes the density (the pressure variation across the slab is not noticeable in burner applications). The flame may be *stabilized* at a location x_F and the fluid- and solid-phase heat flow occurs towards the upstream and downstream of the flame. The upstream heating, or *preheating*, is required for *sustaining* the reaction and the downstream heat flow occurs to supply the heat loss (by radiation) from the downstream surface at $x = L$. Since here the reaction occurs in the gas phase, the solid is heated by the gas phase in and around the reaction zone (which is small), and in the upstream of the flame the solid heats the gas. This is an *enhancement* in the preheating. The heat is also transferred by radiation within the porous medium, and this further enhances the preheating. This preheating can result in gas-phase temperatures above the adiabatic flame temperature, and therefore, it is called the *excess enthalpy,* or *excess temperature*, or *excess preheating*, or *superadiabatic combustion* (Takeno and Sato, 1979; Echigo, 1991; Hanamura et al., 1993).

The *magnitude* of the *flame speed*, defined as $u_F = u_{g_n}$, for reaction in the solid matrix (or adjacent to its surface) can be *larger* or *smaller* than the adiabatic flame speed in plain media u_{F_a} (Section 2.3.2). This depends on the extent of the radiation and conduction preheating, on the interfacial volumetric heat transfer rate $A_o\langle q_{sf}\rangle_{A_{sf}}$, and on the downstream radiation heat loss, among other parameters. In the following, the mathematical formulation of this problem, using the conservation equations and the boundary conditions, are discussed. Then the modeling of the transport tensors and the radiation heat flux vectors are discussed. The solution method used depends on the assumptions made, and both *analytical* and *numerical* treatments are made. The analytical treatment is based on the *exclusion* of the *intramedium radiation* heat transfer, use of the Lewis number of unity, a single-step reaction, and assumption of high activation energy kinetics. The numerical solutions *include* the intermedium radiation, the use of variable thermophysical properties, and a large realistic number of reactions. These are discussed below.

(A) Formulation

The two-medium heat transfer treatment of the problem considered requires the *simplification* of the thermal energy equations (5.74) and (5.75) to a set of equations which are closed and a set of coefficients which are prescribable. Present treatments are rather limited, because as was mentioned most tensors in these equations have *not* yet been evaluated. Then in practice the convective velocity vectors $\mathbf{v}_{fs}$, $\mathbf{v}_{sf}$ and $\mathbf{v}_{ss}$ are *excluded* and the hydrodynamic dispersion is included in the *fluid* (here gas) phase. The diffusion term in the gas and solid phases are *both* significant in this problem. The effect of radiation in the gaseous phase can be assumed and justified to be much smaller than that of the solid phase, and therefore, the radiation heat flux is included in the solid-phase energy equation. Note also that in addition to the anisotropy in $\mathbf{D}^d$, the solid matrix structural anisotropy (such as in ordered, anisotropic fiber structures) makes the components of effective conductivity for each phase to be very different. Here the effective species diffusion (including dispersion) is not in principle analogous to the effective heat diffusion because the solid is assumed to be *impermeable* but heat conducting. This influences the Lewis number unity assumption when the transport of heat and species in the gas phase is considered, because the molecular and dispersion coefficients for mass and heat transport will not be the same.

The simplified forms of (5.74) and (5.75) will be used for the thermal energy conservation equations, and the simplified form of (3.93) will be used for the species conservation equation.

For a *one-dimensional*, steady flow of a gas mixture of n species in a *porous medium* and along the x-axis with negligible pressure variation along the flow, the overall mass, species, and energy conservation equations become

$$\epsilon \frac{d}{dx} \langle \rho \rangle^g \langle u \rangle = 0 \,, \tag{5.119}$$

$$\langle \rho \rangle^g \langle u \rangle \frac{d}{dx} \frac{\langle \rho_i \rangle^g}{\langle \rho \rangle^g} = \frac{d}{dx} (D_{m_{xx}} + D^d_{m_{xx}}) \langle \rho \rangle^g \frac{d}{dx} \frac{\langle \rho_i \rangle^g}{\langle \rho \rangle^g} + \epsilon \langle \dot{n}_i \rangle^g \,, \tag{5.120}$$

$$\begin{aligned} \langle \rho \rangle^g c_{p_g} \langle u \rangle \frac{d\langle T \rangle^g}{dx} + \sum_i (\rho c_p)_{g_i} (\langle u_i \rangle - \langle u \rangle) \frac{d\langle T \rangle^g}{dx} = \\ \frac{d}{dx} [(k_{e_{xx}})_g + \langle \rho \rangle^g c_{p_g} D^d_{xx}] \frac{d\langle T \rangle^g}{dx} + \\ A_o h_{sg} (\langle T \rangle^s - \langle T \rangle^g) - \epsilon \sum_i \langle \dot{n}_i \rangle^g \Delta i_c \,, \end{aligned} \tag{5.121}$$

$$0 = \frac{d}{dx} (k_{e_{xx}})_s \frac{d\langle T \rangle^s}{dx} - \frac{d}{dx} \langle q_r \rangle - A_o h_{sg} (\langle T \rangle^s - \langle T \rangle^g) \,. \tag{5.122}$$

The equation of state, stating a perfect gas behavior, for species i becomes

$$p_i = \frac{R_g}{M_i} \langle \rho_i \rangle^g \langle T \rangle^g \,, \quad p = \sum_i p_i \,, \quad \langle \rho \rangle^g = \sum_i \langle \rho_i \rangle^g \,. \tag{5.123}$$

Note that a uniform porosity ϵ is assumed and each phasic conservation equation is multiplied by its volume fraction (ϵ or $1-\epsilon$). Then $(k_{e_{xx}})_g$ and the remainder of the coefficients also include these fraction multiplications. These give

$$(k_{e_{xx}})_g = \epsilon \langle k_{xx} \rangle^g \,, \quad D^d_{xx} = \epsilon \langle D^d_{xx} \rangle^g \,, \tag{5.124}$$

$$(k_{e_{xx}})_s = (1-\epsilon) \langle k_{xx} \rangle^g \,, \quad D_{m_{xx}} = \epsilon \langle D_{xx} \rangle^g \,, \tag{5.125}$$

$$D^d_{m_{xx}} = \epsilon \langle D^d_{m_{xx}} \rangle^g \,. \tag{5.126}$$

The production rate of species i, for *each* reaction involving this species, is given by (2.43), and there may be many coupled reactions involving these species. Then all these production rates of species i must be added. The reactant concentrations for each reaction may not be in the stoichiometric proportions, and therefore, the *prescribed* chemical coefficients will be different from the stoichiometric coefficients assumed in (2.42). The specific nonstoichiometric chemical equations and the appropriated chemical kinetics (i.e., the production rates) must be prescribed for use in (5.120) and (5.121).

Referring to Figure 5.7, the boundary conditions at $x = 0$ (entrance to the porous slab) and $x = L$ (exit from the porous slab) and far upstream $x \to -\infty$ (where no solid matrix is present), are

$$u_{g_o} = \langle u \rangle_o \,, \; \rho_{f_o} = \langle \rho_f \rangle^g_o \,, \; \rho_{o_o} = \langle \rho_o \rangle^g_o \,, \; T_{g_o} = \langle T \rangle^g_o \,, \; \frac{d\langle T \rangle^s}{dx} = 0 \,, \tag{5.127}$$

$$k_g \frac{dT_g}{dx}|_o = [(k_{e_{xx}})_g + \langle\rho\rangle^g c_{p_g} D^d_{xx}]_o \frac{d\langle T\rangle^g}{dx}|_o \ , \tag{5.128}$$

$$(\rho c_p)_{g_o} \langle u\rangle_o \langle T\rangle^g_o - [(k_{e_{xx}})_g + \langle\rho\rangle^g c_{p_g} D^d_{xx}]_o \frac{d\langle T\rangle^g}{dx}|_o = (\rho c_p)_{g_o} u_{g_o} T_{g_o} - k_{g_o} \frac{\partial T_g}{\partial x}|_o \qquad x = 0 \ , \tag{5.129}$$

$$\rho_f = 0 \ , \quad \frac{d\langle T\rangle^g}{dx} = 0 \ ,$$

$$-(k_{e_{xx}})_{s_L} \frac{d\langle T\rangle^s}{dx}|_L = \epsilon_r \sigma_{SB} (\langle T\rangle^{s^4}_L - T^4_\infty) \qquad x = L \ , \tag{5.130}$$

$$\rho_f = \rho_{f_n} \ , \quad \rho_o = \rho_{o_n} \ , \quad u_g = u_{g_n} = u_F \ , \quad T_g \to T_n \quad \text{as} \quad x \to -\infty \ . \tag{5.131}$$

(B) Modeling of Diffusivities and Internal Radiation

In the analysis of the speed of the premixed gas flame in Section 2.3.2, the *adiabatic* flame speed u_{F_a} was found to be proportional to the square root of the gas conductivity. The gas conductivity was important in the reaction zone and in the preheat zone. Here the role of $(D_{xx})_g$ and $(D_{xx})_s$ are expected to be the similar, and therefore, for accurate predictions and when k_s/k_f is much larger than unity accurate prescription of $(D_{xx})_g$ and $(D_{xx})_s$ are needed. The gas-phase *total* effective diffusivities are defined as

$$(D_{xx})_g = \frac{(k_{e_{xx}})_g}{\langle\rho\rangle^g c_{p_g}} + D^d_{xx} \ , \tag{5.132}$$

$$(D_{xx})_s = \frac{(k_{e_{xx}})_s}{\langle\rho\rangle^s c_{p_s}} + (D^d_{xx})_s \simeq \frac{(k_{e_{xx}})_s}{\langle\rho\rangle^s c_{p_s}} \ . \tag{5.133}$$

Note that for large particle-based Péclet numbers, the hydrodynamic dispersion component of (5.132) becomes dominant over the molecular component and this is generally the case in the problem considered. The Péclet number dependence of D^d_{xx} was addressed in Section 3.2.3, and the existing correlations are reviewed by Kaviany (1991). The correlations for $D_{m_{xx}}$ and $D^d_{m_{xx}}$, i.e., the molecular and dispersion components of species diffusion, are available. These are similar to those for $(k_{e_{xx}})_g$ and D^d_{xx}, except they should correspond to $k_s = 0$ (i.e., correspond to impermeable particles), and are also reviewed by Kaviany (1991).

Modeling of $\langle q_r\rangle$ has been done using the emitting-absorbing, *nonscattering* medium treatment (Hsu et al., 1991) or with the *isotropic* scattering included (Sathe et al., 1990). The diffusion approximation, using the radiant conductivity defined by (3.76), is also be applied (Golombok et al.,

1991). This is applicable whenever the optical thickness $\tau_r = L(\langle\sigma_a\rangle+\langle\sigma_s\rangle)$, where $\langle\sigma_a\rangle$ and $\langle\sigma_s\rangle$ are the wavelength and volume-averaged absorption and scattering coefficients, respectively, is near or larger than unity. The inclusion of radiation in the differentio-integral form excludes obtaining closed-form solutions. The magnitude of the radiation heat flux is mostly dependent on the interstitial surface emissivity and the particle conductivity (Section 3.2.5).

(C) An Analytical Treatment for no Internal Radiation

The large activation energy *asymptotics* and the use of the *reaction zone* (with no convection) and the *preheat zone* (with no reaction) have been used by Buckmaster and Takeno (1981), Golombok et al. (1991), and McIntosh and Prothero (1991) for the study of the premixed, gaseous combustion inside and adjacent to a porous slab. In the study of Golombok et al. (1991) the flame location was prescribed to be *at* the downstream end of the slab ($x = L$) and the *internal radiation* was modeled using the diffusion approximation. In the studies of Buckmaster and Takeno (1981) and McIntosh and Prothero (1991), the internal radiation is *not* addressed, but the flame locations *inside* the slab as well as *on* the surface and *downstream* of the slab (called the *flame liftoff*) are considered. In the study of Buckmaster and Takeno (1981) the effective solid-phase conductivity was taken to be infinite (i.e., uniform solid-phase temperature). Here the treatment and the results of the study of McIntosh and Prothero (1991) are briefly reviewed.

We begin by assuming a large porosity, i.e., $\epsilon \to 1$, and a simple, single-step reaction of the type $F+O \to P$, with a single, specific heat of reaction Δi_c and *no* internal radiation. Then (5.121) and (5.122) can be combined to eliminate the interstitial convection and the resultant can be added to the product of (5.120) and $-\Delta i_c$ to eliminate the reaction source term. Then the resulting equation can be integrated to give the relation between the magnitude of the variables and their derivatives at the upstream boundary (inlet) and downstream boundary (outlet) to the porous slab as

$$\langle\rho\rangle^g_o c_{p_g}\langle u\rangle_o(\langle T\rangle^g_L - \langle T\rangle^g_o - (k_{e_{xx}})_{sL}\frac{d\langle T\rangle^s}{dx}|_L +$$
$$\{[(k_{e_{xx}})_g + \langle\rho\rangle^g c_{p_g} D^d_{xx}]\frac{d\langle T\rangle^g}{dx}\}^L_o -$$
$$\epsilon\Delta i_c[\langle\rho\rangle^g_o\langle u\rangle_o\frac{\langle\rho_f\rangle^g}{\langle\rho\rangle^g} - (D_{m_{xx}} + D^d_{m_{xx}})\langle\rho\rangle^g\frac{d}{dx}\frac{\langle\rho_f\rangle^g}{\langle\rho\rangle^g}]^L_o = 0\,. \qquad (5.134)$$

In the treatment of McIntosh and Prothero (1991), $(k_{e_{xx}})_g$, $(k_{e_{xx}})_s$ and $D_{m_{xx}}$ are taken as constant and D^d_{xx} and $D^d_{m_{xx}}$ are assumed to be *zero*. They also assume *stoichiometric* mixtures of fuel (they consider hydrocarbons) and oxidizer (they consider air). Upstream of the slab, there is no reaction, and the *convection* and *diffusion* balance. Then by using T_n for $x \rightarrow -\infty$ and the heat balance $(k_g dT_g/dx)_o = [(k_{e_{xx}})_g d\langle T\rangle^g/dx]_o$, the temperature distribution in the gas phase upstream of the porous slab becomes

$$T_g = T_n + \frac{\alpha_g}{u_F}\frac{(k_{e_{xx}})_g}{k_g}\frac{d\langle T\rangle^g}{dx}\Big|_o e^{u_F x/\alpha_g} \qquad -\infty < x \leq 0 \, . \qquad (5.135)$$

For stoichiometric mixtures, the fuel is completely consumed downstream of the flame location x_F. In the preheat zone, the variation of the fuel density $\langle\rho_f\rangle^g$ is also governed by convection and diffusion and when a *thin reaction zone* is assumed, then at x_F the fuel density will also be *zero*. Then the distribution of $\langle\rho_f\rangle^g$ in the porous slab is found by applying $\langle\rho_f\rangle^g = \rho_{f_o}$ at $x = 0$ and $\langle\rho_f\rangle^g = 0$ at $x = x_F$. The solution to the temperature distributions, $\langle T\rangle^g = \langle T\rangle^g(x)$, $\langle T\rangle^s = \langle T\rangle^s(x)$ are obtained by using various asymptotic approximations and are given by McIntosh and Prothero (1991). We now discuss their results for the temperature distributions and the position of the thin flame.

The results are given in terms of the *dimensionless* variables and parameters defined as

$$T^* = \frac{T}{T_a}\, , \qquad x^* = \frac{x(k_{e_{xx}})_g}{(\rho c_p)_g u_{F_a}}\, , \qquad h^*_{sg} = \frac{A_o h_{sg} k_g}{(\rho c_p)^2_{g_o} u_{F_a} u_F}\, , \qquad (5.136)$$

$$\gamma_r = \frac{\sigma_{SB}\epsilon_r T_a^3}{(\rho c_p)_{g_o} u_{F_a}}\, , \qquad \frac{(D_{xx})_s}{(D_{xx})_g} = \frac{(k_{e_{xx}})_g}{(k_{e_{xx}})_s}\, , \qquad L^* = \frac{L u_{F_a}(\rho c_p)_g}{(k_{e_{xx}})_g}\, . \qquad (5.137)$$

The *adiabatic flame temperature* T_a and *adiabatic flame speed* u_{F_a} are used in the scales given above. Note that the analysis uses the unity Lewis number, i.e.,

$$Le = \frac{(k_{e_{xx}})_g + \langle\rho\rangle^g c_{p_g} D^d_{xx}}{\langle\rho\rangle^g c_{p_g}(D_{m_{xx}} + D^d_{m_{xx}})} = \frac{(k_{e_{xx}})_g}{\langle\rho\rangle^g c_{p_g} D_{m_{xx}}} = 1\, . \qquad (5.138)$$

Figure 5.8 shows the distribution of the dimensionless temperatures $\langle T^*\rangle^g$ and $\langle T^*\rangle^s$ within the porous slab, in terms of the dimensionless distance x^*. The location of the flame x^*_F is also shown.

The results are for a stoichiometric mixture of methane and air, and the magnitudes of the remaining parameters (dimensionless) are given in the figure caption. Note that the solid and fluid phases are *not* in local thermal equilibrium *near* the flame and at the two *ends* of the slab. The effective conductivity of the solid phase is taken to be four times that of the gas phase. This and the heat generation in the gas phase and the radiation loss from the solid phase at $x = L$ cause a substantial difference between the two temperatures throughout the slab. McIntosh and Prothero (1991) use $k_g = (k_{e_{xx}})_g$ in matching the heat flux across the boundary at $x^* = 0$. Note that the radiation heat loss at $x^* = L^*$ causes the maximum temperature to be lower than the adiabatic flame temperature. In their examinations of the flame location, they vary h_{sg} (or h^*_{sg}) and map the location of the flame for several values of h^*_{sg}. Their results are shown in Figure 5.9, where

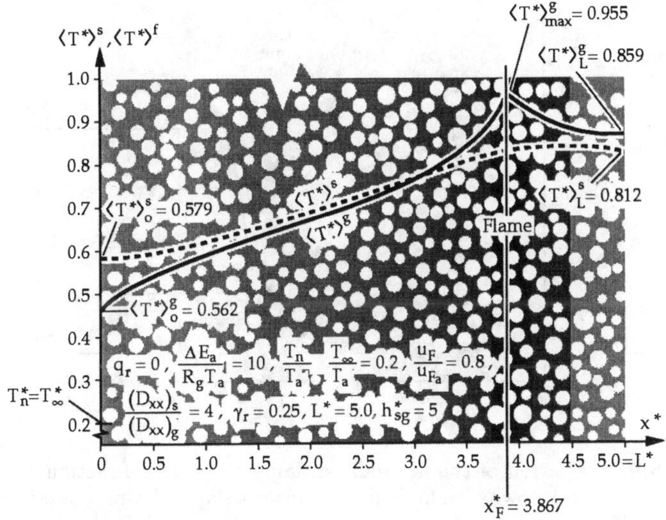

Figure 5.8. The predicted distributions of the gas- and solid-phase temperatures along the porous medium. The location of the flame is also shown for the set of parameters shown in the figure. (From McIntoch and Prothero, reproduced by permission of Elsevier Science Publishing ©1991, The Combustion Institute.)

flame speeds smaller, equal, or larger than the adiabatic flame temperature in plain media are found possible.

They find multiple positions for a given flame speed. Their results give the expected asymptote that when the flame is sufficiently downstream of the slab, the flame speed will correspond to adiabatic flame speed. The flame *liftoff* occurs when $x^* > L^*$ and they also show that for certain flame speeds *flashback* (unstable flame position) occurs. A direct comparison of the flame speed in porous slab can *not* be made with the flame speed in the plain media which is approximated by (2.70). However, (2.70) shows that as T_r (which is equal T_a for the adiabatic flame) decreases, u_F decreases.

In this study the internal radiation is neglected and the ratio of the effective conductivities is not very large (indicating a high porosity, or a low solid conductivity, or a strong anisotropy in the solid matrix structure, resulting in a low axial effective conductivity which in some cases is desirable). Some experimental verifications of these predicted results are given by Williams et al. (1992). They observe that for their 0.4-cm-thick porous, fibrous slab the flame was always located just downstream of the slab. We now consider the inclusion of the internal radiation and a larger value for the ratio of the effective conductivities.

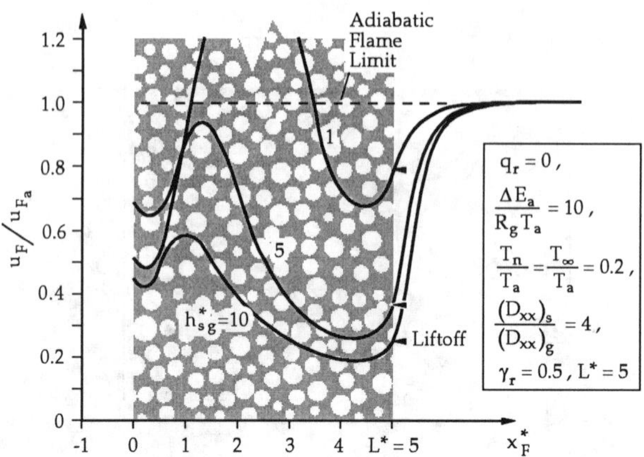

Figure 5.9. Variation of the normalized flame speed as a function of the flame location inside and downstream from the porous slap. The predicted results are for the set of parameters shown in the caption. (From McIntosh and Prothero, reproduced by permission of Elsevier Science Publishing ©1991, The Combustion Institute.)

(D) Numerical Solutions with Internal Radiation

The internal radiation heat transfer can be modeled by considering emission and absorption (Hsu et al., 1991) and by including scattering (Sathe et al., 1990). Here the treatment of Sathe et al. (1990) is reviewed. The equation of radiative transfer (2.110) for *isotropic* scattering, using *wavelength*- and *volume*-averaged properties $\langle \sigma_a \rangle$ and $\langle \sigma_s \rangle$, becomes

$$\mu \frac{dI(x,\mu)}{dx} + (\langle \sigma_a \rangle + \langle \sigma_s \rangle) I(x,\mu) = \sigma_a I_b(\langle T \rangle^s) + \frac{\langle \sigma_s \rangle}{2} \int_{-1}^{1} I(x,\mu)\, d\mu \, , \quad (5.139)$$

$$\langle q_r \rangle = 2\pi \int_{-1}^{1} I(x,\mu) \mu \, d\mu \, . \quad (5.140)$$

The boundary conditions are that the upstream boundary exchanges radiation with a gray diffuse surface with emissivity and temperature ϵ_r and T_n, respectively, and the downstream boundary exchanges radiation with a black surface at T_∞. These are given by

$$I^+(0) = \epsilon_r I_b(T_n) + 2(1-\epsilon_r) \int_o^1 I^-(0,\mu) \mu \, d\mu \quad x = 0 \, , \quad (5.141)$$

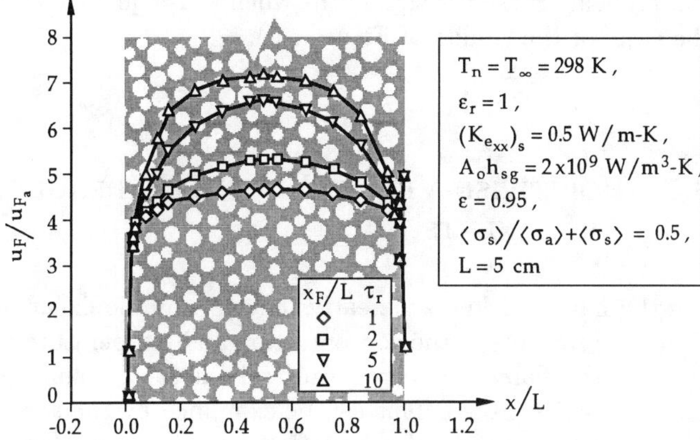

Figure 5.10. Variations of the normalized flame speed as a function of the flame location inside and at the downstream boundary of the porous slap. The predicted results are for the set of parameters shown in the caption. (From Sathe et al., reproduced by permission ©1990 Pergamon Press.)

$$I^-(L) = I_b(T_\infty) \qquad x = L\,. \tag{5.142}$$

The blackbody total intensity $I_b = E_b/\pi = n^2\sigma_{SB}T_\infty^4/\pi$, where n is the fluid index of retraction (taken as unity for gases).

Sathe et al. (1990) solve the equation governing the *single-step*, *zeroth-order* reaction of methane-air, i.e., (5.119) to (5.122), along with the boundary conditions (5.127) to (5.131). They allow for *variable* thermophysical properties and solve the system of equations numerically. In their analysis they also use $D_{xx}^d = D_{m_{xx}}^d = 0$, $D_{m_{xx}} = D$, $(k_{e_{xx}})_g = \epsilon k_g$, and $(k_{e_{xx}})_s = (1-\epsilon)k_s$. Some of their typical results are shown in Figure 5.10 for a stoichiometric mixture and for the set of parameters and properties given in the figure caption. The dimensionless flame location x_f/L and the dimensionless flame velocity u_F/u_{F_a}, are given for various values of the optical thickness $\tau_r = (\langle\sigma_a\rangle + \langle\sigma_s\rangle)L$. The optical thickness limits are *transparent* slab $\tau_r = 0$ and *opaque* slab $\tau_r \to \infty$. Their results are for the *optically thick* slabs $1 \le \tau_r \le 10$. Their numerical results show that u_F/u_{F_a} is *larger* than unity inside the porous slab. For $\tau = 10$ this ratio can be as large as 7 and they obtain *two* stable flame positions for each u_F/u_{F_a}. The ratio $(k_{e_{xx}})_s/(k_{e_{xx}})_g$ in their analysis is about 20. Note that the internal radiation heat transfer is significant when τ_r is equal to unity or larger (as is the case for the results of Figure 5.9).

5.3.5 Pointwise Versus Volume-Averaged Treatments

When the flame thickness is nearly the same as the linear pore dimension ℓ, then the gradients of the concentration and temperature across the pore containing the flame are very large and the local volume-averaging constraints are not satisfied. This can be examined by *comparing* the solution to the *pointwise* conservation equations, pertinent to transport and reaction, to the solution obtained from the local volume-averaged equations. For the case of the *adiabatic*, first-order reaction methane-air flame, in a model two-dimensional porous medium, this is done by Sahraoui and Kaviany (1994). They use the two-medium formulation discussed in Section 5.3.4 (A), where local thermal nonequilibrium is allowed between the solid and gas phases, as well as a single-medium treatment based on the assumption of local thermal equilibrium between the phases (as was discussed in Section 3.2.2). For the single-medium treatment, the two energy equations (5.121) and (5.122) are added using a single temperature. The pointwise solution uses the conservation equations given in Section 3.2.2, but for a two-dimensional geometry with the appropriate boundary and symmetry conditions.

Their results for the temperature distribution for the three models, i.e., a unit-cell model with a periodic in-line arrangement of square solid cylinders, the two-medium model, and the single-medium model, are shown in Figures 5.11 (a) to (c). The results are for $k_s/k_f = 100$, a stoichiometric mixture of methane and air, a unit cell size $\ell =$ 1mm, and a porosity of 0.9. Note that the pointwise solution (i.e., direct simulation) gives a larger flame thickness, compared to the volume-averaged models. The preheating by the solid causes a local gas temperature, at the end of the reaction zone, which is larger than the adiabatic flame temperature (this is called the excess temperature) found for the solid and the gas far downstream from the flame. The direct simulation predicts a higher local excess temperature, compared to the two-medium model. The single-medium is unable to predict an excess temperature because of the local thermal equilibrium assumption employed. In the rendering of the two-medium model, the solid is assumed to be continuous and of uniform distribution along the x-axis. In the single-medium rendering, the solid is randomly distributed and has the same temperature as the gas everywhere.

The Nusselt number in (5.121) and (5.122) is also computed by Sahraoui and Kaviany (1994) and the distribution of the local Nusselt number around

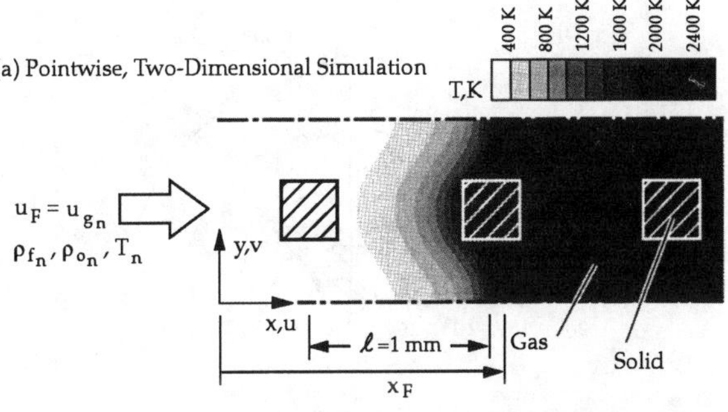

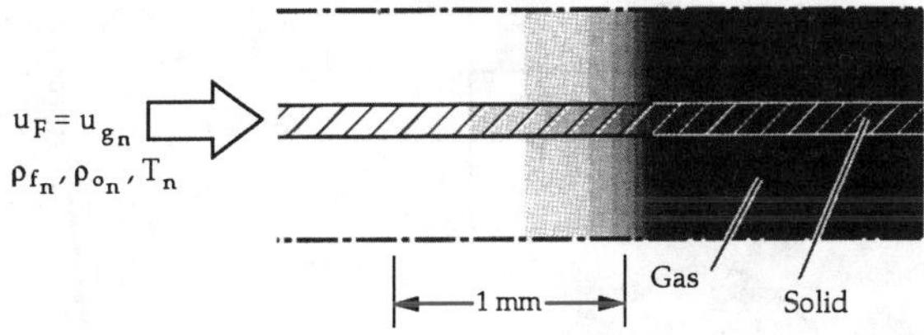

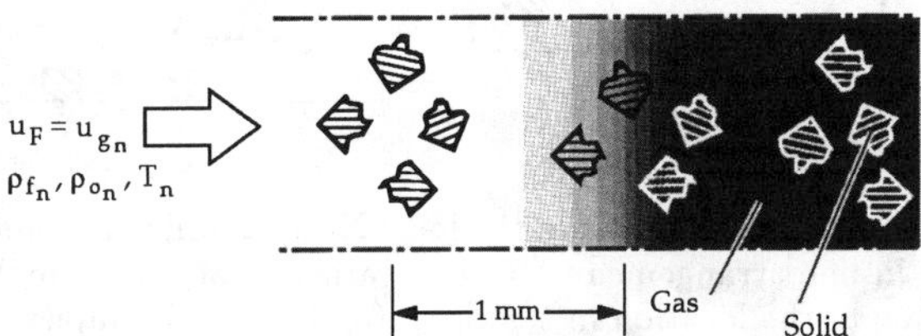

Figure 5.11. Temperature distribution in (a) a two-dimensional direct simulation model, (b) a one-dimensional two-medium model, and (c) a one-dimensional single-medium model, of a porous medium with $\epsilon = 0.9$, $k_s/k_f = 100$. The temperature variation is due to the presence of a premixed methane-air flame.

a solid square cylinder is shown in Figure 5.12 (a). Because of the periodic arrangement of the particles (i.e., cylinders) and in the range of Reynolds number encountered, in the combustion of methane-air in these unit-cell size range, the variation of $\langle Nu_\ell \rangle^{ext}_{A_{sg}}$ does not follow that for spheres given by (5.29). In the front-stagnation region and in the range of Re_ℓ encountered, the local Nusselt number *decreases* with an increase in Re_ℓ, while the change in $\langle Nu_\ell \rangle^{ext}_{A_{sg}}$ is rather small.

For the flame and pore structure (and size) considered, the direct simulation of the flame speed depends on its location within the pore (i.e., the position with respect to solid). This is shown in Figure 5.12(b), where the flame speed is the maximum when it is placed in the center of the pore and

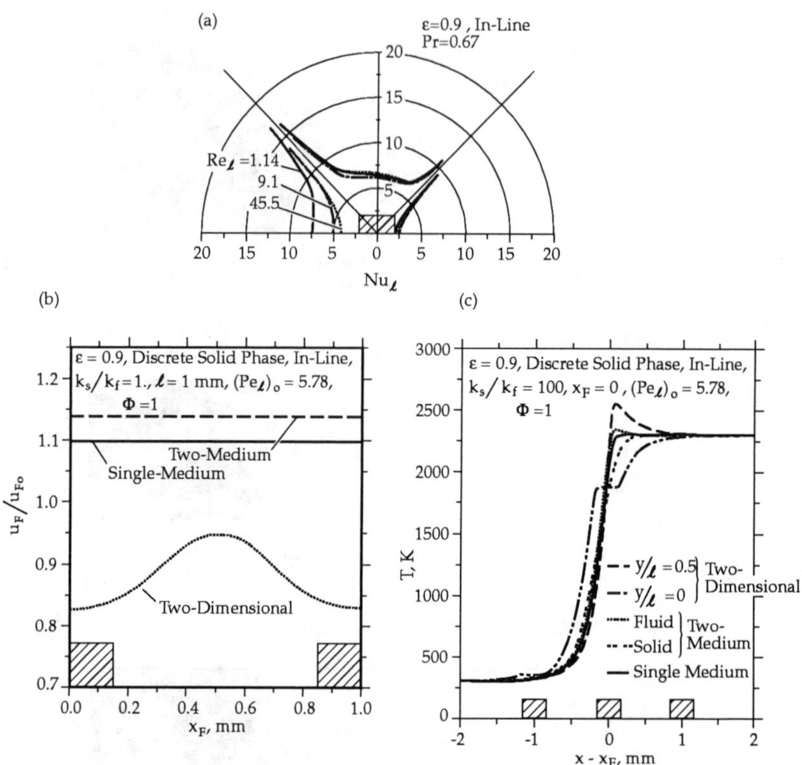

Figure 5.12. (a) Distribution of the local Nusselt number around a square cylinder in the in-line arrangement. (b) The variation of the normalized flame speed as a function of the location in the pore. (c) The axial temperature distributions obtained from various models.

the solid preheating of the gas is maximum (the results are for $k_s = k_f$). Both the one- and two-medium models predict higher velocities which are independent of location.

The axial distribution of the temperature, predicted by the three models, is shown in Figure 5.12(c) for a flame location and for two lateral positions (in the direct simulation). The lateral position $y/\ell = 0$ corresponds to the symmetry plane and the highest, local axial gas velocity. At this location the excess temperature is the largest and is not accurately predicted by the volume-averaged model.

The above example demonstrates the convenience and also the limitation of the volume-averaged treatments and the error expected when the constraints in the applicability of these treatments are not satisfied.

5.4 Particulate Flows

Most solid-particle suspension flows of interest (e.g., pulverized coal combustion, particle heating in plasma and natural gases, powder formation and processing) are *turbulent*. For *very* small particles, such as those assumed in the treatment of Section 4.14.1, the particles are assumed to follow the turbulent fluctuations (velocity and temperature), i.e., *local* momentum and thermal equilibrium. The *larger* (or more *massive*) particles, such as those considered in this section, are influenced by the fluid velocity fluctuations but follow *trajectories* other than the fluid particles. They also are *not* in local thermal equilibrium with the fluid. The *interparticle collisions* also influence this trajectory when the particle concentration is large. The boundary between the *collision-influenced* (i.e., dense suspensions) and *collision-independent* (i.e., dilute suspension) particle *trajectory* is not clearly studied. The experimental results of Farber and Depew (1963), for internal particulate flows, suggest porosities ϵ larger than 0.99 to be in the *collision-independent regime*. In Section 4.14.1, this dilute regime was examined assuming a local thermal equilibrium between the fluid and the particles.

For large particles, *separate* treatments of hydrodynamics of the *particle* and the *fluid* flows are used (i.e., *two-medium treatment* of the *fluid mechanics*). The two-medium treatment (or modeling) of the *collision-independent* (or dilute) *small-particle regime* is done using a *Lagrangian* (i.e., particle follow up) treatment of the motion of the particles. The *collision-dependent regime* for large particles is treated using concepts and models developed in the *kinetic theory* of gases.

A classification of the analysis of the *fluid mechanics* and *heat transfer* aspects of the particulate flows is given in Chart 5.1. In some heat transfer models, for the local thermal nonequilibrium between the particles and the fluid, assume a *continuous* particle phase, and use separate thermal energy equations for the particle and continuous (or fluid) phase. The heat transfer within the continuous is by both turbulent and molecular diffusion and by convection, and there is interfacial heat transfer between the continuous and particle phases. The heat transfer in the particle phase is also modeled as a diffusive-convective transport where the diffusion is due to the interparticle collisions. The temperature distribution within *each* spherical particle may be negligible when the *Biot number* $Bi = \langle Nu_d \rangle_{A_{sf}} k_f / 6k_s$ is small (i.e., $Bi > 0.1$); otherwise, the heat conduction in the particle must be included (as in Section 5.2.2).

The conservation equations (mass, momentum, and thermal energy) for two-phase flows with local thermal and hydrodynamic nonequilibrium (i.e., temperature and velocity slips between the phases) were addressed in Section 3.3. The treatment of Lahey and Drew (1989), which was briefly reviewed, showed that the *rigorous* treatment requires *extensive* closure conditions. Here we consider solid, dispersed particles-fluid flows and re-

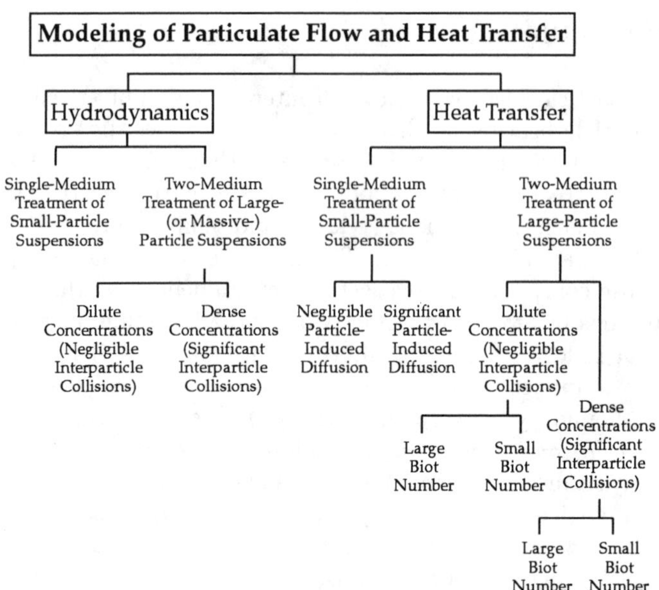

Chart 5.1. A classification of the analysis of the hydrodynamic and heat transfer aspects of particulate flow and heat transfer.

view the recent developments of the phasic conservation equations for these systems. The transport of thermal energy within the *assumed* continuous particle phase is addressed.

In practice, simple models for the transport in each phase are used. After discussing the conservation equations for turbulent flows and heat transfer in particulate flows, *applications* in combustion of small (pulverized) coal particles and in the plasma-particle heat transfer (no reaction) will be examined. The *radiation* heat transfer is important in these applications.

5.4.1 A Generalization of Forces on a Moving Particle

In addressing the fluid flow over *isolated* elements in Section 5.1.1, the element (a spherical particle) was assumed to be *stationary*. When the particle is *moving*, forces other than the viscous and pressure drag force discussed in Section 5.1.1 are exerted upon it and influence its trajectory. For completion we briefly mention these forces. Discussions are given by Michaelides et al. (1992), Elghobashi and Truesdell (1992), and Lu et al. (1993).

Using a force balance on the particle, by equating the product of the particle mass m_s and its acceleration $d\mathbf{u}_s/dt_s$ to the summation of all the *external forces*, we have

$$m_s \frac{d\mathbf{u}_s}{dt_s} = \mathbf{F}_d + m_s \frac{\rho_f}{\rho_s} \frac{D\mathbf{u}_f}{Dt} + \frac{1}{2} m_s \frac{\rho_f}{\rho_s} (\frac{D\mathbf{u}_f}{Dt} - \frac{d\mathbf{u}_s}{dt_s}) +$$

$$m_p(1 - \frac{\rho_f}{\rho_s})\mathbf{g} + 6R^2(\pi \rho_f \mu_f)^{1/2} \int_{t_{s_o}}^{t_s} \frac{\frac{d}{dt}(\mathbf{u}_f - \mathbf{u}_s)}{(t_s - t)^{1/2}} dt \, . \tag{5.143}$$

In (5.143), d/dt_s is the time derivative taken *following* the particle, D/Dt is the *total* derivative evaluated at *particle location*, $m_s\rho_f/\rho_s$ is the fluid mass *displaced* by particle, and t_{s_o} is the time at which the particle motion begins. The first term on the right-hand side of (5.143) is the viscous and pressure *drag force*; the second is the *fluid* pressure gradient and *viscous stress force*; the third is the *inertial force* of *added* mass; the fourth is the *buoyancy force*; and the fifth is the *viscous force* due to *unsteady relative acceleration* (*Basset* force).

The second, third, and fourth terms on the right-hand side of (5.143) are generally negligible when $\rho_f \ll \rho_s$.

The viscous and pressure drag force is given by (5.5) which can be written in term of the slip velocity $\mathbf{u}_f - \mathbf{u}_s$ as

$$\mathbf{F}_d = \frac{1}{2} \pi R^2 c_d \rho_f |\mathbf{u}_f - \mathbf{u}_s| (\mathbf{u}_f - \mathbf{u}_s) \, . \tag{5.144}$$

This is in terms of the drag coefficient c_d which in turn is given by correlation such as (5.9).

The *hydrodynamic response* (or the particle *relaxation*) *time* of a particle t_d is defined through

$$\frac{1}{t_d} = \frac{|\mathbf{F}_d|}{m_s |\mathbf{u}_f - \mathbf{u}_s|} \, , \tag{5.145}$$

and for the case of *Stokes force* only (Section 5.11), i.e., when $c_d = 24/Re_d$, the *Stokes relaxation time* is designated by t_{d_o}.

For *small* particles in a *dilute* concentration and in a turbulent flow, the momentum transport of the particles phase (assuming a continuum phase for the particles) may be assumed to be due to the *turbulent diffusion* of particles. This diffusion is determined by using the particle equation motion (5.143) in a turbulent flow field. Most treatments of small particles assume that the particles do not influence the fluid flow. For these small particles, two-medium treatments are used with the particle diffusion properly modeled. These are addressed by Parthasarathy and Faeth (1990), Elghobashi and Truesdell (1992), Michaelides et al. (1992), Binder and Hanratty (1993), and Lu et al. (1993), among others.

5.4.2 Collision-Included Model of Hydrodynamics

One of the recent efforts in developing equations of motion of particulate flows is in the area of *rapidly deforming* granular materials, and uses the methods of the *kinetic theory of gases*. The method models the particulate phase (i.e., the dispersed phase) as a *continuum* and uses the *averaging* techniques for this phase (Jackson, 1985; Satake and Jenkins, 1988; Louge et al., 1991; Anderson and Jackson, 1992; Roco, 1993). This is similar to the nonequilibrium molecular theory (i.e., kinetic theory) discussed in Section 1.4.3, and its fundamentals are discussed by Woods (1993), and results in the prediction of the *transport* properties for the particulate medium.

For *large* particles the intensity of the particle velocity fluctuations may exceed that of the fluid, and therefore, the treatment of the particles responding to turbulence mentioned in the last section may *not* apply. The *collisions* among the particles, analogous to molecular dynamics in gases, may explain this phenomena. In some treatments, the fluid phase flow is allowed to influence the *mean* velocity of the particles (in a particle continuum) $\overline{\mathbf{u}}_s$ but that it is not allowed to affect their *fluctuating* velocity $\mathbf{u}'_s$. This corresponds to the hydrodynamic relaxation time t_d of the particle fluctuations being *larger* than a typical rollover time of the turbulent eddies. Then the continuum particle phase is treated as a *rapidly flowing granular material* in which momentum and energy are transferred by the velocity fluctuations of the particles (Louge et al., 1992).

In the following, the continuum treatments of the particle phase, the energy of the particle fluctuations, and the modeling of the turbulent, continuous-phase flow, as described by Louge et al. (1991), are reviewed. The mean and the fluctuating components are defined as in (2.86).

(A) Dispersed Phase

It is assumed that the particles are *suspended* due to the drag force $\mathbf{f}_d$ due to the *mean* slip velocity $\overline{\mathbf{u}}_f - \overline{\mathbf{u}}_s$ and their *fluctuation* is the result of their collisions. For the sake of *simplicity*, the volume-averaged symbol is *dropped*; however, $\overline{\mathbf{u}}_f$, $\overline{\mathbf{u}}'_f$, and all other quantities used in the following analysis are *local, phase-volume-averaged quantities*. *Nearly elastic* collisions and the Maxwellian velocity distribution function for the collisions are both assumed. Using t_d defined in (5.145) to represent the drag force, the continuity and *mean* momentum equations for a constant solid density ρ_s become (Louge et al., 1991)

$$\frac{\partial}{\partial t}(1-\overline{\epsilon})\rho_s + \nabla\cdot(1-\overline{\epsilon})\rho_s\overline{\mathbf{u}}_s = 0\ , \tag{5.146}$$

$$\rho_s\frac{D}{Dt}(1-\overline{\epsilon})\overline{\mathbf{u}}_s = \rho_s\frac{\partial}{\partial t}(1-\overline{\epsilon})\overline{\mathbf{u}}_s + \rho_s\nabla\cdot(1-\overline{\epsilon})\overline{\mathbf{u}}_s\overline{\mathbf{u}}_s =$$
$$-\nabla p_s + \nabla\cdot\overline{\mathbf{S}}_s + \frac{\rho_s}{t_d}(1-\overline{\epsilon})(\overline{\mathbf{u}}_f-\overline{\mathbf{u}}_s) - \rho_s(1-\overline{\epsilon})\mathbf{g}\ , \tag{5.147}$$

where the elements of the dispersed-phase *mean* viscous stress tensor are given by

$$\overline{\tau}_{s_{ij}} = \frac{5\pi^{1/2}}{96}\rho_s\, d\,\theta^{1/2}\left(\frac{\partial \overline{u}_{s_i}}{\partial x_j} + \frac{\partial \overline{u}_{s_j}}{\partial x_i} - \frac{2}{3}\frac{\partial \overline{u}_{s_k}}{\partial x_k}\delta_{ij}\right). \tag{5.148}$$

The dispersed phase pressure p_s is given by

$$p_s = \rho_s(1-\overline{\epsilon})\theta\,, \quad \frac{3}{2}\theta = \frac{1}{2}\overline{u_s'^2} \quad \text{or} \quad p_s = \frac{1}{3}\rho_s(1-\overline{\epsilon})\overline{u_s'^2}\,, \tag{5.149}$$

where $(\overline{u_s'^2})^{1/2}$ is the root-mean-square of the *particle fluctuation velocity* u_s' (which is assumed to be *isotropic*) and is used here in a manner analogous to (1.7) in the molecular treatment (i.e., kinetic theory) of gases. The particle-phase viscosity is derived in a manner similar to that for for the gas viscosity as given by (1.33) and (1.34).

In order to *close* the dispersed-phase *mean* momentum equation, the quantity θ (which is called the *granular* or *particulate phase temperature*, because of its *analogy* to the gas temperature) must be specified. This is done by considering a balance of the particle *fluctuation energy*.

(B) Particulate Fluctuation Energy

The equation for the determination of θ, i.e., the *balance* of the local, convective, and diffusive changes of θ with the production and the dissipation rates of θ, is given by

$$\begin{aligned}\frac{3}{2}\rho_s(1-\overline{\epsilon})\frac{D\theta}{Dt} &= \frac{3}{2}\rho_s(1-\overline{\epsilon})\frac{\partial\theta}{\partial t} + \frac{3}{2}\rho_s(1-\overline{\epsilon})\overline{\mathbf{u}}_s\cdot\nabla\theta \\ &= -\nabla\cdot\mathbf{q}_\theta + \overline{\mathbf{S}} : \nabla\overline{\mathbf{u}}_s - \epsilon_{k_s}\,,\end{aligned} \tag{5.150}$$

where ϵ_{k_s} is the *dissipation rate* of the fluctuations in the particle velocity. The diffusive flux of the particle fluctuation energy is given by the diffusion flux vector

$$\mathbf{q}_\theta = -\frac{25\pi^{1/2}}{128}\rho_s d\,\theta^{1/2}\nabla\theta\,. \tag{5.151}$$

The volumetric dissipation rate has a component due to the *inelastic* collisions and a component due to the *interaction* with the continuous phase. These are the first and second terms of the following

$$\epsilon_{k_s} = \frac{24(1-e)}{\pi^{1/2}d}\rho_s\theta^{3/2}(1-\overline{\epsilon})^2 - \frac{\rho_s}{t_d}(1-\overline{\epsilon})(\overline{u_{s_i}'u_{f_i}'} - 3\theta)\,, \tag{5.152}$$

where e is the *coefficient of restitution*. The correlation between the velocity fluctuation of the particles and the fluid is modeled with

$$\overline{u_{s_i}'u_{f_i}'} = \frac{4}{\pi^{1/4}}\frac{d(\overline{u}_{f_i} - \overline{u}_{s_i})^2}{\tau_{d_o}\,\theta^{1/2}}\,, \tag{5.153}$$

where as mentioned before, t_{d_o} is the Stokes relaxation time (i.e., for c_d corresponding to $Re_d \to 0$).

(C) Continuous Phase

The mean continuity and momentum equations for the continuous phase (including the particle-phase drag force contribution) are

$$\frac{\partial}{\partial t}\overline{\epsilon}\rho_f + \nabla \cdot \overline{\epsilon}\rho_f \overline{\mathbf{u}}_f = 0 \,, \tag{5.154}$$

$$\rho_f \frac{D}{Dt}\overline{\epsilon}\,\overline{\mathbf{u}}_f = \rho_f \frac{\partial}{\partial t}\overline{\epsilon}\,\overline{\mathbf{u}}_f + \rho_f \nabla \cdot \overline{\epsilon}\,\overline{\mathbf{u}}_f \overline{\mathbf{u}}_f =$$
$$-\nabla \overline{\epsilon}\,\overline{p}_f + \nabla \cdot \overline{\epsilon}\overline{\mathbf{S}}_f - \frac{\rho_s}{t_d}(1-\overline{\epsilon})(\overline{\mathbf{u}}_f - \overline{\mathbf{u}}_s) \,. \tag{5.155}$$

Note that $\overline{\mathbf{u}}_f$ is the *fluid*-phase *time* and local *volume*-averaged velocity.

The components of the continuous phase *mean*, viscous stress tensor (including the turbulent Reynolds stresses) are

$$\overline{\tau}_{f_{ij}} = \mu_f\left(\frac{\partial \overline{u}_{f_i}}{\partial x_j} + \frac{\partial \overline{u}_{f_j}}{\partial x_i} - \frac{2}{3}\frac{\partial \overline{u}_{f_k}}{\partial x_k}\delta_{ij}\right) - \rho_f \overline{u'_{f_i} u'_{f_j}} \,. \tag{5.156}$$

The Reynolds stresses are given in terms of an *isotropic* eddy viscosity μ_t through

$$\rho\overline{u'_{f_i} u'_{f_j}} = \mu_t\left(\frac{\partial \overline{u}_{f_i}}{\partial x_j} + \frac{\partial \overline{u}_{f_j}}{\partial x_i}\right) + \frac{2}{3}\delta_{ij}\left(\rho\frac{1}{2}\overline{u'_{f_k} u'_{f_k}} + \mu_t \frac{\partial \overline{u}_{f_k}}{\partial x_k}\right) . \tag{5.157}$$

The fluid (or continuous phase) turbulent kinetic energy $\frac{1}{2}(\overline{u'_{f_k} u'_{f_k}})$ is given by an appropriate model below and the eddy viscosity μ_t is modeled using the *fluid turbulent kinetic energy*. The model used by Louge et al. (1991) is

$$\mu_t = c_\mu \rho_f \ell \left(\frac{1}{2}\overline{u'_{f_k} u'_{f_k}}\right)^{1/2} , \tag{5.158}$$

where c_μ is a constant and the mixing length ℓ is in turn *prescribed* as $\ell = \ell(\mathbf{x})$.

(D) Fluid Fluctuation Energy

The equation for the turbulent kinetic energy of the continuous phase is found in a manner similar to (2.93). The result is

$$\frac{\partial}{\partial x_j}\rho_f \overline{\epsilon}\,\overline{u}_{f_j}\left(\frac{1}{2}\overline{u'_{f_k} u'_{f_k}}\right) =$$
$$-\overline{u'_{f_i} u'_{f_j}}\frac{\partial}{\partial x}\rho_f \overline{\epsilon}\,\overline{u}_{f_i} - \frac{\partial}{\partial x_j}\overline{\epsilon p'_f u'_{f_i}} - \overline{p'_f u'_{f_i}}\frac{\partial \overline{\epsilon}}{\partial x_i} -$$
$$\overline{\rho\overline{\epsilon} u'_{f_i} u'_{f_j} \frac{\partial u'_{f_i}}{\partial x_j}} - \rho\overline{\epsilon}\,\overline{u}_{f_i}\overline{u'_{f_i}\frac{\partial u_{f_j}}{\partial x_j}} + \frac{\rho_s}{t_d}(1-\overline{\epsilon})\overline{u'_{f_i}(u'_{s_i} - u'_{f_i})} +$$
$$\overline{\mu_f \overline{\epsilon} u'_{f_i}\frac{\partial}{\partial x_j}\left(\frac{\partial u'_{f_i}}{\partial x_j} + \frac{\partial u_{f_j}}{\partial x_i}\right)} -$$

$$\frac{2}{3}\mu_f\overline{\epsilon u'_{f_i}\frac{\partial^2 u'_{f_j}}{\partial x_i \partial x_j}} - \frac{2}{3}\mu\overline{\frac{\partial \epsilon}{\partial x_i}u'_{f_i}\frac{\partial u'_{f_j}}{\partial x_j}} \,. \tag{5.159}$$

Many closure conditions and models are used for the terms appearing in this equation. One of these closure conditions is

$$\rho_f\overline{\epsilon u'_{f_j}(\frac{1}{2}u'_{f_i}u'_{f_i})} = \overline{\epsilon}\frac{\mu_t}{Pr_k}\frac{\partial}{\partial x_j}\frac{1}{2}\overline{u'_{f_i}u'_{f_i}} \,, \tag{5.160}$$

where Pr_k is the Prandtl number for the turbulent kinetic energy and is generally prescribed as a constant.

5.4.3 Collision-Included Model of Energy Transport

The *two-medium* treatment of the *thermal* energy transport uses the local, mean temperatures $\overline{T}_f$ and $\overline{T}_s$ for the two phases and allows for the interfacial heat transfer. The transport of thermal energy in the particle phase has been modeled by Louge et al. (1993) for particles with *small* Biot number, i.e., $Bi = \langle Nu_d\rangle_{A_{sf}} k_f/6k_s < 0.1$, where the particles are treated as having a *uniform* temperature within them. The turbulent transport in the continuous phase is modeled using the eddy diffusivity for heat which is related to the eddy viscosity through the turbulent Prandtl number. The continuous phase-dispersed phase interfacial heat transfer is modeled similar to that discussed in the treatment of the porous media. This is analogous to the coupling of the momentum transport through the drag force in the momentum equations (5.147) and (5.155).

(A) Dispersed Phase

The thermal energy equation for the particle phase is given in terms of a dispersed phase *effective conductivity* tensor $\mathbf{K}_{e_s}$ which is related to the particle velocity fluctuation θ. This effective conductivity is analogous to the *self-diffusive* transport discussed in relation to (1.37) for gases. The particles-phase energy equation is written as

$$(1-\overline{\epsilon})(\rho c_p)_s\frac{D\overline{T}_s}{Dt} = (1-\overline{\epsilon})(\rho c_p)_s(\frac{\partial \overline{T}_s}{\partial t} + \overline{\mathbf{u}}_s\cdot\nabla\overline{T}_s) =$$

$$\nabla\cdot\mathbf{K}_{e_s}\cdot\nabla\overline{T}_s - \frac{A_o k_f}{d}\langle Nu_d\rangle_{A_{sf}}(\overline{T}_s - \overline{T}_f) \,, \tag{5.161}$$

where the *isotropic* effective conductivity model is suggested (Louge et al., 1993) as

$$\mathbf{K}_{e_s} = \frac{\pi^{1/2}}{16}(\rho c_p)_s d\,\theta^{1/2}\frac{1}{1+\lambda/L}\mathbf{I} \,. \tag{5.162}$$

The effective conductivity has also been examined by Hsiau and Hunt (1993). The *particle mean free path* λ is related to the porosity and the particle diameter through

$$\lambda = \frac{d}{6(1-\overline{\epsilon})2^{1/2}} \tag{5.163}$$

and the characteristic length of the system L is taken as the distance over which the maximum changes in $\overline{T}$ and $\overline{u}$ occurs (such the tube radius).

Louge et al. (1993) discuss the effects of the particle and fluid velocity fluctuations on $\langle Nu_d\rangle_{A_{sf}}$ and state that $\langle Nu_d\rangle_{A_{sf}}$ is dominantly determined by the *local mean slip* velocity $\overline{\mathbf{u}}_f - \overline{\mathbf{u}}_s$. Correlations such as (5.17) with Re_d based on the local mean slip velocity, with some modification due to the interparticle convective interactions, are expected to be applicable.

(B) Continuous Phase

The fluid-phase energy equation can be written with the turbulent heat flux given in terms of the turbulent eddy diffusivity (which is equivalent to the hydrodynamic dispersion used in porous media). Similar to (5.161), the fluid phase energy equation is written as

$$\overline{\epsilon}(\rho c_p)_f\frac{D\overline{T}_f}{Dt} = \overline{\epsilon}(\rho c_p)_f(\frac{\partial \overline{T}_f}{\partial t} + \overline{\mathbf{u}}_f\cdot\nabla\overline{T}_s) =$$
$$\nabla\cdot[\mathbf{K}_{e_f} + (\rho c_p)_f\mathbf{D}_f^d]\cdot\nabla\overline{T}_f + \frac{A_o k_f}{d}\langle Nu_d\rangle_{A_{sf}}(\overline{T}_s - \overline{T}_f)\,. \tag{5.164}$$

The effective conductivity and the dispersion tensors can be modeled as *isotropic* using (Louge et al., 1993)

$$\mathbf{K}_{e_f} = \overline{\epsilon}k_f\mathbf{I}\,, \tag{5.165}$$

$$\mathbf{D}_f^d = \overline{\epsilon}\alpha_t\mathbf{I} = \overline{\epsilon}\frac{\nu_t}{Pr_t}\mathbf{I}\,. \tag{5.166}$$

The mean continuity and momentum equations given by (5.146), (5.147), (5.154), and (5.155); the mean thermal energy equations given by (5.161) and (5.162); the particle and fluid fluctuation velocities given by (5.150) and (5.159), along with the other closure statements—constitute a *complete* problem statement. Using these, the flow and heat transfer within a particulate flow with local thermal nonequilibrium (two-medium treatment) can be described (and when there is a bounding surface with yet another temperature, the three-medium treatment can be made; this is addressed in Section 7.2). In comparison to the phasic time- and local volume-averaged *total* energy conservation equations of Lahey and Drew (1989) given by (3.112), the phasic models of the *thermal* energy conservation equations given by (5.161) and (5.162) are rather readily solvable. The *mechanical energy* components in (3.112) can become significant in some applications, and therefore should be addressed. The *radiation* heat transfer can also become significant and should be included.

5.4.4 Models for Reacting Particles

Combustion of *small coal* particles (pulverized) suspended in a flowing gas, which allows for the supply of oxygen and a steady heat generation, is of industrial importance (e.g., Smoot and Smith, 1985). The coal particles are heated by convection and radiation to the temperature at which the *primary devolatilization* begins (about 1000 K). During this heating period, oxygen is in contact with the particles but because the temperature is below the ignition point, *no heterogeneous* (i.e., surface mediated) reaction occurs. Also, because volatiles are not *yet* formed, *no homogeneous* (i.e., gas phase) reaction also occurs (Lau and Niksa, 1992). The primary devolatilization *expels* hot *gaseous fuels* into the film surrounding the particle. The remaining solid is a *char*. These are in the form of *noncondensible gases* and *tars*. These gases moving away from the particles react with the oxygen from the free stream and because this is a diffusion-limited reaction (the gas temperature is generally much larger than the ignition temperature for homogeneous reactions), a *thin* flame is formed at this reaction *site* (similar to the diffusion flames discussed in Section 2.3.3). Heterogeneous combustion is said to occur when oxygen reaches the particle surface *fast* enough to completely consume all volatiles in an *attached* flame and *simultaneously* result in *char oxidation*. The homogeneous combustion occurs if the surface mass flux of the volatiles (i.e., the *surface blowing*) is *large* enough to *lift* and *sustain* an *envelope flame*. Then, during the combustion of a coal particle and after the preheating period, the gas phase reaction at the *detached flame mode*, with flame stand-off distances of about one particle diameter, occurs. Then as the volatile surface mass flux decreases, the flame approaches the surface and finally the *surface-attached flame mode* (i.e., heterogeneous reactions) and char oxidation occur (Lau and Niksa, 1992).

The behavior of the individual (i.e., isolated) coal particles discussed above is modified when there are neighboring and reacting particles present. The gas flow, which is generally turbulent, is also affected by these particles, and in turn influences the rate of heat and mass transfer from these particles. In this section we briefly discuss the *isolated* reacting particles, and then the existing phasic *transport equations* used for the description of these reacting, particulate flows are discussed.

(A) Isolated Reacting Particles

In order to gain understanding of the devolatilization and the extent of the surface and gas-phase reactions, the behavior of a *single* particle (generally carbon or coal) in a high-temperature gaseous environment has been studied rather *extensively*. Some of these studies and their results are briefly reviewed below, and these results will be later referred to in the examination of the models for reacting, particulate flows.

(i) Chemistry and Devolatilization Kinetics

The *ultimate* (i.e., *most elementary*) chemical constituents of the coal particles are *carbon, hydrogen, nitrogen, sulfur, oxygen*, and *ash* (*minerals*). The *proximate* compositions are grouped into *moisture, volatile water, fixed carbon*, and *ash*. These *chemical* aspects of the coals (significant variations occur depending on the geological origin) are reviewed by both Smoot and Pratt (1979) and Smoot and Smith (1985). The *kinetics* of the devolatilization have been the subject of examination and improvement in modeling, and recent models are given by Niksa and Kerstein (1991). This devolatilization is *thermal conversion* of most of the organic mass, hydrogen, nitrogen sulfur, and oxygen in the coal into gases. The volatiles consist of *noncondensible* gases with high heating value, light oils, and high boiling point *tars*. The major noncondensibles are methane and C_2-C_4 hydrocarbons, hydrogen, water, and oxide of carbon. Tar is the volatile that *condenses* at room temperature. The highly porous solid carbon *residue* remaining after devolatilization is called char and contains most of the original mineral matter. The gas and tar *yield* of the devolatilization depends on the heating rate of the particles (assuming a uniform temperature in the particle, the heating rate is given in K/s).

The modeling of the devolatilization kinetics is discussed by Niksa and Kerstein (1991), and the *rate equations* and model parameters are given. The different coals are described by empirical constants and the effects of temperature, heating rate, pressure, and particle size on the rate equations are included in the models.

(ii) Particle-Size Effect

Assuming a spherical particle and a symmetric distribution of temperature and species concentration around the particles, the combustion of an isolated coal particle can be rendered as shown in Figure 5.13. The *detached flame* mode is shown with a diffusion, *thin* flame sheet approximation. The temperature of the flame is the highest temperature and heat flows from the flame to the ambient and to the particle. The particle surface temperature T_s is above the devolatilization limit, and within the particle a uniform temperature distribution is assumed, i.e., $Bi = \langle Nu_d \rangle_{A_{sf}} k_f / 6k_s < 0.1$. In principle, the char is *porous*, and, depending on the particle diameter and the internal volatile mass flux, nonuniformity in the particle temperature distribution, and therefore in the rate of production of the volatiles, can occur.

The variation in the temperature of the isolated coal particles T_s with respect to time and the rate of production of the products of the combustion have been measured by Beck and Hayhurst (1990), Maloney et al. (1991),

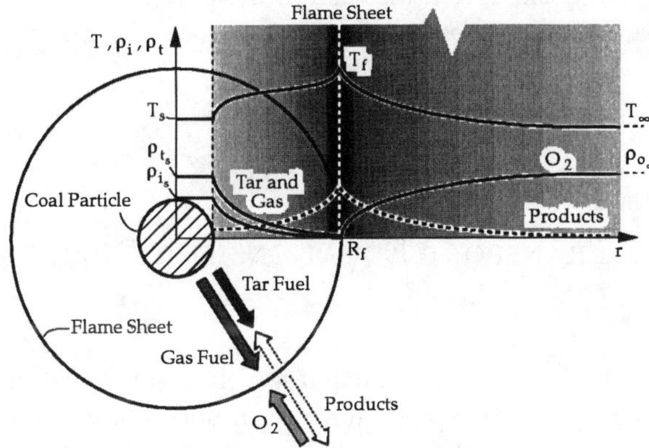

Figure 5.13. The thin detached flame around a devolatilizing coal particle. The species concentrations inside and outside the flame are shown. (From Lau and Niksa, reproduced by permission of Elsevier Science Publishing ©1991, The Combustion Institute.)

Saito et al. (1991), and Zhang (1992), among others. The analysis of an isolated reacting coal particle has been made by Beck and Hayhurst (1990), Blake and Libby (1991), Lau and Niksa (1992), and Makino (1992), among others. Makino (1992) considers a *quiescent* ambient with *carbon* particles and *heterogeneous* reactions

$$2C+O_2 \rightarrow 2CO\ , \tag{5.167}$$

$$C+CO_2 \rightarrow 2CO \tag{5.168}$$

and *homogeneous* reaction

$$CO+\tfrac{1}{2}O_2 \rightarrow CO_2\ . \tag{5.169}$$

His results show that for carbon particles *less* than 100 μm, the particle burns with *no* significant gas-phase reactions, i.e., the flame is *attached* to the surface. For *coal* particles the gas-phase hydrocarbon reactions are included in the model of Beck and Hayhurst (1990) and Lau and Niksa (1991). The latter study shows that particle diameter at which the above-mentioned transition *from* the heterogeneous to homogeneous reaction occurs *depends* on the type of *coal* and the heat of combustion of volatile gases and tar. The particle Reynolds number $Re_d = u_\infty d/\nu$ is generally small, because for high-temperature gases ν is rather *large*, while for the fine particles considered d is very small. The fluid flow adjacent to the particle is significantly altered by the volatile surface mass flux *blowing*, which is the surface condition considered in Sections 4.9.2 and 4.11.2 and is discussed by Blake and Libby (1991). This blowing occurs through open pores on the surface and causes very significant *rotations* of the particles up to

5000 cycle/s during the devolatilization (Saito et al., 1991). The treatment of reacting particles in *turbulent* flows is reviewed by Smoot and Smith (1985).

(B) Phasic Transport Equations for Dilute Suspensions

In Section 3.1 the local, time-, and phase volume-averaged energy equations were discussed for two-phase flows with both phases flowing. There, the coupling between the phases and the *turbulent fluxes* were discussed. In Section 5.4.3, we considered one of these phases being dispersed as solid particles and the *collisions* of these solid particles, and the *turbulence* in the continuous fluid phase (generally gas) was addressed. In considering reacting particulate flows, the practical cases of interest are those where the particles are *small* (such that they are expected *not* to influence the turbulence in the fluid flow), the suspensions are *dilute* (such that the interparticle collisions are neglected), and the particle Biot number is *small* (such that the temperature variation with the particle is *negligible*). Then the fluid-phase transport is generally dominated by turbulent transport (with hydrodynamic, thermal, and chemical *nonequilibrium*). The particle-phase transport is described as a *pseudo-continuum*, where interparticle interactions are neglected, thus resulting in *no diffusion* and *convection* transport within the particle phase (i.e., between the particles). This pseudo-continuum (or *pseudo-discrete*) treatment is similar to that described in Section 5.2.2, where the particles are treated as being present everywhere in space but with a finite size and *uniform* temperature.

The phasic (i.e., nonequilibrium) models of transport related to combustion of pulverized coal are discussed briefly by Smoot and Smith (1985) and are used in the numerical simulations of Ji and Cohen (1992). These models, which allow for the change in the density of the particles (as a result of devolatilization), are also used in the studies of nonreacting, nonequilibrium fluidized-bed flow and heat transfer (e.g., Kuipers et al., 1992).

Because of the chemical reactions, the species conservation equations will be added to the conservation equations discussed in Sections 5.4.2 and 5.4.3.

For monosize, spherical particles of an *instantaneous* diameter d, the porosity and the number concentration of particles n_s are related through the *geometric* relations

$$\epsilon = 1 - \frac{n_s \pi d^3}{6} \quad \text{or} \quad n_s = \frac{6(1-\epsilon)}{\pi d^3} . \tag{5.170}$$

The overall mass conservation for the fluid and solid phases are given by

$$\frac{D}{Dt}\overline{\epsilon}\,\overline{\rho}_f = \frac{\partial}{\partial t}\overline{\epsilon}\,\overline{\rho}_f + \nabla\cdot\overline{\epsilon}\,\overline{\rho}_f\overline{\mathbf{u}}_f = \frac{D}{Dt}n_s\rho_s(\frac{\pi}{6}d^3)\ , \tag{5.171}$$

$$\frac{D}{Dt}(1-\overline{\epsilon})\rho_s = \frac{\partial}{\partial t}(1-\overline{\epsilon})\rho_s + \nabla\cdot(1-\overline{\epsilon})\rho_s\mathbf{u}_s = -\frac{D}{Dt}n_s\rho_s(\frac{\pi}{6}d^3)\ . \tag{5.172}$$

The conservation of *mean* momentum for the fluid phase is written similar to (5.155) and is (Smoot and Smith, 1985)

$$\frac{D}{Dt}\overline{\epsilon}\,\overline{\rho}_f\overline{\mathbf{u}}_f = [\frac{d}{dt}n_s\rho_s(\frac{\pi}{6}d^3)]\mathbf{u}_s - \nabla\overline{\epsilon}\,\overline{p} + \nabla\cdot\epsilon\mathbf{S}_f - \frac{\rho_s}{t_d}(1-\overline{\epsilon})(\overline{\mathbf{u}}_f - \mathbf{u}_s)\ , \tag{5.173}$$

where the viscous shear-stress tensor is decomposed into the mean and the Reynolds shear stresses as

$$\mathbf{S}_f = \overline{\mathbf{S}}_f + \mathbf{S}_{t_f}\ . \tag{5.174}$$

The conservation of momentum for the particle phase is

$$\frac{D}{Dt}(1-\overline{\epsilon})\rho_s\mathbf{u}_s = \frac{\rho_s}{t_d}(1-\overline{\epsilon})(\overline{\mathbf{u}}_f - \mathbf{u}_s)\ . \tag{5.175}$$

The *total* energy conservation for the fluid phase can be, written in terms of the specific internal energy, is similar to (3.112) or (3.121) and becomes (Smoot and Smith, 1985)

$$\frac{D}{Dt}[\overline{\rho}_f\overline{\epsilon}(\overline{e}_f + \frac{1}{2}\overline{\mathbf{u}}_f\cdot\overline{\mathbf{u}}_f) + \rho_s(1-\overline{\epsilon})\frac{1}{2}\mathbf{u}_s\cdot\mathbf{u}_s] = -\nabla\cdot\overline{\epsilon}\mathbf{q}_f - \nabla\overline{\epsilon}\,\overline{p}\,\overline{\mathbf{u}}_f + \nabla\cdot(\overline{\epsilon}\mathbf{S}_f\cdot\overline{\mathbf{u}}_f) - \frac{\rho_s}{t_d}(1-\overline{\epsilon})(\overline{\mathbf{u}}_f - \mathbf{u}_s)\cdot\mathbf{u}_s - \overline{p}\frac{\partial\overline{\epsilon}}{\partial t} + \frac{A_o k_f}{d}\langle Nu_d\rangle_{A_{sf}}(\overline{T}_s - \overline{T}_f) + \sum_i \dot{n}_i i_i\ , \tag{5.176}$$

where $\dot{n}_i$ the volumetric, fluid-phase production rate of species i is given by chemical kinetic models such as (1.16). The heat flux vector is decomposed into the mean and the turbulent components as

$$\mathbf{q}_f = \mathbf{q}_f + \mathbf{q}_{t_f}\ . \tag{5.177}$$

The *thermal* energy equation for the particle phase includes the heat absorbed due to *vaporiztion*, *heterogeneous reaction*, *surface radiation*, and *nonequilibrium* with the fluid phase. Using a simple form for radiation the *single particle* thermal energy equation becomes (Smoot and Smith, 1985)

$$\frac{d}{dt}(\rho c_p)_s(\frac{\pi}{6}d^3)\overline{T}_s = \Delta i_v\frac{d}{dt}\rho_s(\frac{\pi}{6}d^3) + \sum_i \dot{m}_i(\pi d^2)i_i + \epsilon_r\sigma_{SB}F(T_o^4 - T_s^4) - \pi d k_f\langle Nu_d\rangle_{A_{sf}}(\overline{T}_s - \overline{T}_f)\ , \tag{5.178}$$

where $\dot{m}_i$ is the production rate of species i per unit *surface* area of the solid particle and its chemical kinetic model is similar to (1.16). In the radiation term ϵ_r is the *surface* emissivity, and F is the *radiation shape* (or *geometric) factor* between the particle and surrounding *low*-temperature ambient which is assumed to be a blackbody and at T_o.

The fluid-phase species conservation equation becomes (Smoot and Smith, 1985)

$$\frac{D}{Dt}\overline{\epsilon}\,\overline{\rho}_i = \nabla\overline{\epsilon}\,\overline{\rho}_f D \nabla \frac{\overline{\rho}_i}{\overline{\rho}_f} + \dot{n}_i + \frac{6}{d}(1-\overline{\epsilon})\dot{m}_i \; . \tag{5.179}$$

The modeling of the turbulent fluid flow and heat transfer, i.e., $\mathbf{S}_{t_f}$, $\mathbf{q}_{t_f}$, and the interfacial drag, i.e., t_d, are done similar to those discussed in Sections 5.4.2(C) and (D), 5.4.3(B), and 5.4.1. As was discussed in Sections 1.10 and 3.3, the most rigorous descriptions of the *two-phase* flows with one phase *dispersed* and with a simultaneous occurrence of *reaction* and *phase change* leads to rather extensive conservation equations which will not be readily solvable. Note that the above-discussed models are illustrations of the existing extent of rigorous and yet practical simplicity. For accurate predictions, these models are being continuously improved (Smoot and Smith, 1985).

5.4.5 Rarefied Plasma-Particle Flow

Low-pressure *plasma spraying* involves injection of small solid particles into a flowing thermal plasma for the purpose of heating the particle. In coating applications, the *heated* particles, along with the plasma, flow toward the surface which will become the substrate. Then the particle residence time in the plasma and the other process parameters are controlled for the desirable extent of particle melting (or softening, if a distinct melting point does not exist). In this and some other particle processings, some of which also involve reaction, the plasma-particle interfacial heat transfer is of interest.

The particles are generally spherical and may have diameters of the order of 10 μm. Then the Knudsen number defined as $Kn = \lambda/d$ can be large at low pressures.

In Section 4.7, in the discussion of the *low-pressure* gas flow (i.e., large Knudsen number) and heat transfer, the role of the temperature and velocity *slip* in rarefied gas flow and heat transfer was examined. Because of the *high temperatures* existing in *plasmas* (Section 2.6.3) the gas is in part *ionized*. The particles suspended in a plasma can *neutralize* the ions and electrons (by facilitating their recombination on the surface) as these gas species collide with them. The recombination energy is then absorbed by the particles. The extent of this recombination depends on the particle surface charge. This recombination heat transfer, and other mechanisms which will be discussed below, require a *special* treatment of the rarefied plasma-particle interfacial heat transfer transfer. The particle concentration (i.e., *solid loading*) is generally low, and therefore, interparticle collisions and interactions are negligible. Then the interfacial heat transfer rate $\langle q_{sf}\rangle_{A_{sf}}$ is generally the *most* critical variable in the description of the nonequilibrium plasma-particle heat transfer. The drag coefficient and the pressure distribution around the particle also deviate from those discussed in Section 5.1 which are valid for *low* Knudson number flows ($Kn \to 0$). The particle Reynolds number $Re_d = ud/\nu$, where u and ν are evaluated in the free stream, is generally of the order of unity. Then the Nusselt number $\langle Nu_d \rangle_{A_{sf}}$ is not expected to be large. However, the contributions due to high temperatures (i.e., the ionization and recombination along with the lack of local thermal equilibrium between the *atoms, ions*, and *electrons*) and low pressure (i.e., temperature and velocity slip on the particle surface), must be accounted for. The radiation heat transfer is also significant. The plasma gas is generally made of argon, argon-hydrogen, or nitrogen. At elevated temperatures, the atoms, ions, and electron species coexist with *equilibrium* or *nonequilibrium* species *concentrations*. The *prediction* of the equilibrium species concentrations, from the kinetic theory, is discussed by Konuma (1992).

In describing the heat transfer between an ionized, low-pressure gas and a particle capable of carrying or redistributing charge, the *convective heat transfer between* the species (which are not in local thermal equilibrium with each other) and the particle is determined from the gradient of temperature on the particle surface. This requires the simultaneous solution of the species, velocity, and temperature fields. In addition, the *recombination energy released* to the surface and other transfer of energy due to the *electrostatic potential* between the gas and the particle must be included. Theoretical treatments of these heat transfer mechanisms have been given by Godard and Chang (1980), Honda et al. (1981), Chen and Pfender (1983), Chang and Pfender (1990), and Leveroni and Pfender (1990). In the following, the *two-medium treatment* of the electrons and the remaining species (i.e., the heavier species which are the atoms and ions), the hydrodynamics, and the convective heat transfer (with the surface slip boundary conditions) treatments of the interfacial heat transfer, as well as some of the available numerical results, are briefly reviewed.

(A) Thermal Nonequilibrium Among Gas Species

A very small spherical particle of diameter d is assumed and the fluid flows over this particle with some prescribed free-stream conditions. Assuming a *steady* flow, *axisymmetric* (spherical coordinates) *laminar* flow, and by using the coordinate system and parameters shown in Figure 5.14(a), the overall mass conservation is given by

$$\frac{1}{r^2}\frac{\partial}{\partial r}r^2 \rho u + \frac{1}{r\sin\theta}\frac{\partial}{\partial \theta}\rho v \sin\theta = 0 . \qquad (5.180)$$

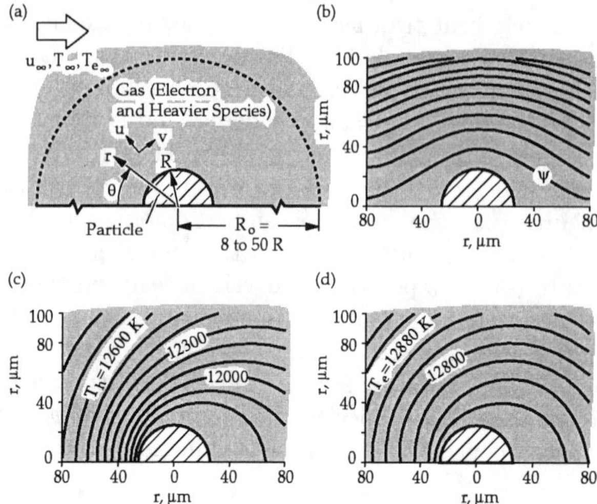

Figure 5.14. (a) A schematic of a particle-gas system showing the coordinate system used and the outer computation radius R_o. (b) Contours of constant values of stream function around the particle. (c) Contours of constant temperature of the heavier species. (d) Contours of constant temperature of the electrons. (From Chang and Pfender, reproduced by permission ©1990 IEEE.)

The species conservation equation for the electrons is

$$\frac{1}{r}\frac{\partial}{\partial r}r^2\rho_e u + \frac{1}{r\sin\theta}\frac{\partial}{\partial\theta}\rho_e v\sin\theta = \frac{1}{r^2}\frac{\partial}{\partial r}\rho D r^2\frac{\partial}{\partial r}\frac{\rho_e}{\rho} + \frac{1}{r^2\sin\theta}\frac{\partial}{\partial\theta}\rho D\sin\theta\frac{\partial}{\partial\theta}\frac{\rho_e}{\rho} + m_e\dot{n}_e\,, \qquad (5.181)$$

where m_e is the mass of an electron and $\dot{n}_e$ is the number of electrons produced per unit time and volume. Then, the *combined* heavy species (i.e., atoms and ions) concentration ρ_h/ρ (i.e., ρ_a/ρ and ρ_i/ρ) can be found using the above two equations.

The momentum conservation equations for a *Newtonian, compressible* fluid and for the r and θ directions are

$$\rho u\frac{\partial u}{\partial r} + \rho\frac{v}{r}\frac{\partial u}{\partial\theta} - \rho\frac{v^2}{r} = -\frac{\partial p}{\partial r} + \frac{\partial}{\partial r}[2\mu\frac{\partial u}{\partial r} + \lambda_\mu(\nabla\cdot\mathbf{u})] + \frac{1}{r}\frac{\partial}{\partial\theta}[\mu(r\frac{\partial}{\partial r}\frac{v}{r} + \frac{1}{r}\frac{\partial v}{\partial\theta}] + \frac{\mu}{r}(4\frac{\partial u}{\partial r} - \frac{2}{r}\frac{\partial v}{\partial\theta} - 4\frac{u}{r} - \frac{2v\cot\theta}{r} + r\cot\theta\frac{\partial}{\partial r}\frac{v}{r} + \frac{\cot\theta}{r}\frac{\partial u}{\partial\theta})\,, \qquad (5.182)$$

$$\rho u\frac{\partial v}{\partial r} + \rho\frac{v}{r}\frac{\partial v}{\partial\theta} + \rho\frac{uv}{r} = -\frac{1}{r}\frac{\partial p}{\partial\theta} + \frac{1}{r}\frac{\partial}{\partial\theta}[\frac{2\mu}{r}(\frac{\partial v}{\partial\theta} + u) + \lambda_\mu(\nabla\cdot\mathbf{u})] + \frac{\partial}{\partial r}[\mu(r\frac{\partial}{\partial r}\frac{v}{r} + \frac{1}{r}\frac{\partial u}{\partial\theta})] +$$

$$\frac{\mu}{r}[2(\frac{1}{r}\frac{\partial v}{\partial \theta} - \frac{v \cot\theta}{r}) \cot\theta + 3(r\frac{\partial}{\partial r}\frac{v}{r} + \frac{1}{r}\frac{\partial u}{\partial \theta})] \, . \tag{5.183}$$

The thermal energy conservation equations for the electrons and for the heavier species are written assuming *monatomic* species. From the kinetic theory of gases the specific heat capacity at constant pressure for ideally behaving monatomic gases is given as

$$c_p = \frac{5}{2}\frac{R_g}{M} \, . \tag{5.184}$$

Then the *enthalpy* of n atoms per unit volume is $5k_B nT/2$. The *electron* energy equation become

$$\frac{1}{r^2}\frac{\partial}{\partial r}r^2\frac{5}{2}k_B n_e T_e u + \frac{1}{r\sin\theta}\frac{\partial}{\partial\theta}\frac{5}{2}k_B n_e T_e v \sin\theta =$$

$$\frac{1}{r}\frac{\partial}{\partial r}k_e r^2 \frac{\partial}{\partial r}T_e + \frac{1}{r^2\sin\theta}\frac{\partial}{\partial\theta}k_e \sin\theta\frac{\partial}{\partial\theta}T_e - \Delta i_i \dot{n}_e + \nabla\cdot\mathbf{q}_r - \dot{s}_{eh} \, , \tag{5.185}$$

where Δi_i is the ionization energy, and $\dot{s}_{eh}$ is the rate of energy transfer *from* the electron to the heavier species. The heaver species energy equation becomes

$$\frac{1}{r^2}\frac{\partial}{\partial r}r^2\frac{5}{2}k_B n_h T_h u + \frac{1}{r\sin\theta}\frac{\partial}{\partial\theta}\frac{5}{2}k_B n_h T_h v \sin\theta =$$

$$\frac{1}{r^2}\frac{\partial}{\partial r}k_h r^2\frac{\partial}{\partial r}T_h + \frac{1}{r^2\sin\theta}k_h \sin\theta\frac{\partial}{\partial\theta}T + \dot{s}_{eh} \, . \tag{5.186}$$

The *volumetric ionization rate* $\dot{n}_e$ is determined from a kinetic model and is a function of the electron *temperature* and the *concentration* of the *electrons* and *atoms*. The interaction between electron and heavier species is divided into the interaction between the electrons and ions and the electrons and atoms and are determined by the proper kinetic models and depends on T_h and T_e as well as on n_a, n_i, and n_e. These kinetic models are discussed by Chang and Pfender (1990).

The two-medium descriptions of the energy transport and the fluid dynamics given above for the electrons and the heavier species, with the viscosity (including the second viscosity), density, and the thermal conductivities being strong functions of the temperature, constitute a continuum description which now requires specification of the boundary conditions. The presumed low pressure of the plasma would require the modeling of these boundary conditions through the slip coefficients. These are discussed next.

(B) Slip Boundary Conditions on Particle Surface

In envisioning the *jump* or slip in the velocity and temperature across the interface of a solid surface and a rarefied gas, a *Knudsen layer* can be defined over which this jump occurs. The thickness of this layer is less than one mean free path λ. For nonequilibrium ionization, Chang and Pfender (1990) consider a jump also in the *degree of ionization*. This is required if the population density (i.e., the concentration) of the electrons in the free stream is not that determined from the equilibrium kinetic theory. Here we will only discuss the jump in the velocity and temperature and assume that the free stream concentrations are in a equilibrium state.

The velocity and temperature slips have been discussed in connection with thermophoresis in Section 1.12.4(C) and in the review of rarefied gas flow and heat transfer in Section 4.7. The velocity at the particle surface, i.e., the slip velocity, is given in terms of the *velocity accommodation factor* for the momentum transfer F discussed in Section 4.7.2, as

$$u_i = -\frac{2-F}{F}\lambda\frac{\partial u}{\partial r} \quad \text{on} \quad A_{sf} \,. \tag{5.187}$$

The jump in the temperatures T_{e_i} and T_{h_i} at the surface are found from the *extrapolation* of temperature across the Knudsen layer (instead of using the thermal accommodation factor) and are given by Giedt and Willis (1985) and Chang and Pfender (1990)

$$T_{e_i} = T_e - \frac{2\gamma}{\gamma+1}\frac{\lambda_e}{Pr_e}\frac{\partial T_e}{\partial r} \quad \text{on} \quad A_{sf} \,, \tag{5.188}$$

$$T_{h_i} = T_h - \frac{2\gamma}{\gamma+1}\frac{\lambda_h}{Pr_h}\frac{\partial T_h}{\partial r} \quad \text{on} \quad A_{sf} \,, \tag{5.189}$$

where λ_e and λ_h are the mean free path of the electrons and the heavier species, $\gamma = c_p/c_v$ for the mixture, and

$$Pr_e = \frac{(\mu c_p)_e}{k_e} \,, \qquad Pr_e = \frac{(\mu c_p)_h}{k_h} \,. \tag{5.190}$$

The heating of the particles due to the *ion recombination* at the surface is given by

$$q_{rec} = a_T n_i (\frac{k_B T_h}{2\pi m_i})^{1/2} \frac{\Delta i_i - \Delta w_b}{a_T} \tag{5.191}$$

where n_i is the concentration of the ions, a_T is the *thermal accommodation coefficient* (discussed in Section 4.7.2) of ions, and Δw_s is energy due to the *work function* (Chang and Pfender, 1990).

(C) Results for Interfacial Heat Transfer

The overall mass, species, momentum, and energy conservation equations and boundary conditions described above are solved numerically using variable properties by Chang and Pfender (1990) with $F = 0.9$ and $a_T = 0.8$. Some of their typical results for *argon* plasma with equilibrium concentration of ions, electrons, and atoms are given in Figures 5.14(b) to (d) and 5.15(a) and (b). The distributions of the stream function, heavier species temperature, and electron species concentrations are shown in Figure 5.14(b) to (d) for a total pressure of 80 *mbar* in the free stream, $d = 50\ \mu$m, $T_{h_\infty} = T_{e_\infty} = T_\infty = 1.3 \times 10^4$ K, $T_s = 10^3$ K and $Re_d = 2.5$.

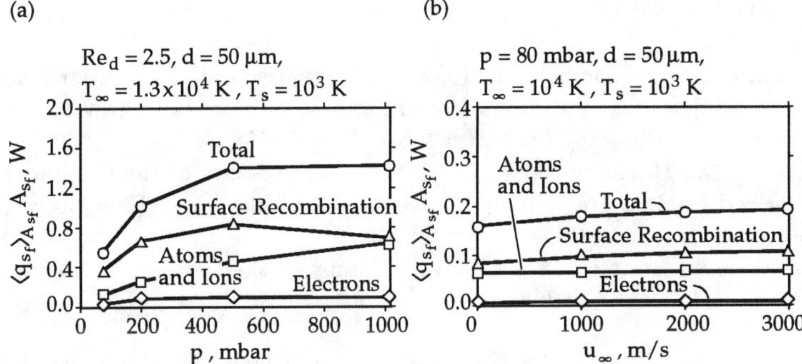

Figure 5.15. (a) Variation of the interfacial heat transfer rate with respect to the total gas pressure. (b) Variation of the interfacial heat transfer rate with respect to the gas free-stream velocity. (From Chang and Pfender, reproduced by permission ©1990 IEEE.)

Note that for this low Reynolds number flow, the velocity field has a *nearly* upstream downstream *symmetry* (due to the large variation in temperature, the thermophysical properties change significantly). The temperature field does not have this symmetry and behind the sphere the plasma is cooler.

The interfacial heat transfer which has contributions from the electron and heavier species convection and from the recombination is shown in Figure 5.15(a) and (b) for various pressures and free-stream velocities. Figure 5.15(a) shows that the electron convection makes the least contribution to the interfacial heat transfer. The atoms and ions, which are larger in number, convect more heat to the particle and this contribution increases with pressure. The recombination contribution is highest at *low* pressures and after peaking, begins to decrease when the atmospheric pressure is reached. At these high pressures, the heavier species convection and recombination make equal contributions. Note that the area-averaged interfacial heat flux $\langle q_{sf} \rangle_{A_{sf}}$ is very large (because the surface area πd^2 is very small). The variation of the total and individual interfacial heat transfer rates, with respect to the free stream velocity, are shown in Figure 5.15(b). The results show a *weak* velocity dependence over the range of the velocities considered. Note that typical plasma spray velocities are rather large. The surface recombination contribution increases *most* noticeably with velocity, as compared to the other two contributions.

5.5 References

Anderson, K.G., and Jackson, R., 1992, "A Comparison of the Solutions of Some Proposed Equations of Motion of Granular Materials for Fully Developed Flow Down Inclined Planes," *J. Fluid Mech.*, 241, 145–168.

Beck, N.C., and Hayhurst, A.N., 1990, "The Early Stages of the Combustion of Pulverized Coal at High Temperatures. I: The Kinetics of Devolatilization," *Combust. Flame*, 79, 47–74.

Bellan, J., and Harstad, K., 1987, "The Details of the Convective Evaporation of Dense and Dilute Clusters of Drops," *Int. J. Heat Mass Transfer*, 30, 1083–1093.

Ben-Ammar, F., Kaviany, M., and Barber, J.R., 1992, "Heat Transfer During Impact," *Int. J. Heat Mass Transfer*, 35, 1495–1506.

Binder, J.L., and Hanratty, J.J., 1993, "Use of Lagrangian Statistics to Describe Slurry Transport," AIChE J., 39, 1581–1591.

Blake, T.R., and Libby, P.A., 1991, "Combustion of a Spherical Carbon Particle in Slow Viscous Flow," *Combust. Flame*, 86, 147–161.

Buckmaster, J., and Takeno, T., 1981, "Blow-off and Flashback of an Excess Enthalpy of an Excess Enthalpy Flame," *Combust. Sci. Technol.*, 25, 153–158.

Carbonell, R.G., and Whitaker, S., 1984, "Heat and Mass Transfer in Porous Media," in *Fundamentals of Transport Phenomena in Porous Media*, Bear, J., and Corapcioglu, M.Y., Editors, Martinus Nijhoff Publishers, Dordrecht.

Cebeci, T., and Bradshaw, P., 1984, *Physical and Computational Aspects of Convection Heat Transfer*, Springer-Verlag, New York.

Chang, C.H., and Pfender, E., 1990, "Heat and Momentum Transport to Particulates Injected into Low-Pressure ($\sim$ 80 mbar) Nonequilibrium Plasmas," *IEEE Transactions on Plasma Science*, 18, 958–967.

Chen, X., and Pfender, E., 1983, "Behavior of Small Particles in a Thermal Plasma Flow," *Plasma Chem. Plasma Process.*, 3, 351–366.

Clift, R., Grace, J.R., and Weber, M.E., 1978, *Bubbles, Drops and Particles*, Academic Press, New York.

Coudrec, J.-P., 1985, "Incipient Fluidization and Particulate Systems," in *Fluidization*, Davidson, J.F., et al., Editors, Second Edition, Academic Press, London.

Echigo, R., 1991, "Radiation-Enhanced/Controlled Phenomena of Heat and Mass Transfer in Porous Media," *Proceedings ASME-JSME Thermal Engineering Conference*, Volume 4, xxi–xxxii.

Elghobashi, S., and Truesdell, G.C., 1992, "Direct Simulation of Particle Dispersion in a Decaying Isotropic Turbulence," *J. Fluid Mech.*, 242, 655-700.

Farber, L., and Depew, C.A., 1963, "Heat Transfer Effects to Gas-Solids Mixtures Using Solid Spherical Particles of Uniform Size," *I & EC Fundamentals*, 2, 130–135.

Giedt, W.H., and Willis, D.R., 1985, "Rarefied Gases," in *Handbook of Heat Transfer: Fundamentals*, Rohsenow, W.M., Editor, Second Edition, McGraw-Hill, New York.

Goddard, R., and Chang, J.-S., 1980, "Local and Total Heat Transfer on a Sphere in a Free Molecular Ionized Gas Flow," *J. Phys. D: Appl. Phys.*, 13, 2005–2012.

Golombok, M., Prothero, A., Shirvill, L.C., and Small, L.M., "Surface Combustion in Metal Fibre Burner," *Combust. Sci. Technol.*, 77, 203–223.

Gunn, D.J., 1978, "Transfer of Heat or Mass to Particles in Fixed and Fluidized Beds," *Int. J. Heat Mass Transfer*, 21, 467–476.

Hanamura, K., Echigo, R., and Zhdank, S.A., 1993, "Superadiabatiac Combustion a Porous Medium," *Int. J. Heat Mass Transfer*, 36, 3201–3209.

Honda, T., Hayashi, T., and Kanzawa, A., 1981, "Heat Transfer from Rarefied Ionized Argon Gas to a Biased Tungsten Fine Wire," *Int. J. Heat Mass Transfer*, 24, 1247–1255.

Hsiau, S.S., and Hunt, M.L., 1993, "Kinetic Theory Analysis of Flow-Induced Particle Diffusion and Thermal Conduction in Granular Material Flows," *ASME J. Heat Transfer*, 115, 541–548.

Hsu, P.-F., Howell, J.R., and Mathews, R.D., 1991, "A Numerical Investigation of Premixed Combustion within Porous Inert Media," *Proceedings ASME-JSME Thermal Engineering Conference*, Volume 4, 225–231.

Incropera, F.P., and DeWitt, D.P., 1990, *Introduction to Heat Transfer*, Second Edition, John Wiley and Sons, New York.

Jackson, R., 1985, "Hydrodynamic Stability of Fluid-Particle Systems," in *Fluidization*, Second Edition, Davidson, J.F., et al., Editors, Academic Press, London.

Ji, C.-C., and Cohen R.D., 1992, "An Investigation of the Combustion of Pulverized Coal-Air Mixture in Different Combustor Geometries," *Combust. Flame*, 90, 307–343.

Kaviany, M., 1991, *Principles of Heat Transfer in Porous Media*, Springer-Verlag, New York.

Konuma, M., 1992, *Film Deposition by Plasma Techniques*, Springer-Verlag, New York.

Kuipers, J.A.M., Prins, W., and van Swaaij, W.P.M., 1992, "Numerical Calculations of Wall-to-Bed Heat Transfer Coefficients in Gas-Fluidized Beds," *AIChE J.*, 38, 1079–1091.

Lahey, R.T., Jr., and Drew, D.A., 1989, "The Three-Dimensional Time- and Volume-Averaged Conservation Equations for Two-Phase Flows," *Advan. Nuclear Sci. Tech.*, 20, 1–69.

Lau, C.W., and Niksa, S., 1991, "The Combustion of Individual Particles of Various Coal Types," *Combust. Flame*, 90, 45–70.

Leveroni, E., and Pfender, E., 1990, "A Unified Plasma-Particle Heat Transfer Under Noncontinuum and Nonequilibrium Conditions," *Int. J. Heat Mass Transfer*, 33, 1497–1509.

Louge, M., Yusef, M.J., and Jenkins, J.T., 1993, "Heat Transfer in the Pneumatic Transport of Massive Particles," *Int. J. Heat Mass Transfer*, 36, 265–275.

Louge, M.Y., Mastorakos, E., and Jenkins, J.T., 1991, "The Role of Particle Collisions in Pneumatic Transport," *J. Fluid Mech.*, 231, 345–359.

Lu, Q.Q., Fontaine, J.R., and Aubertin, G., 1993, "Numerical Study of the Solid Particle Motion in Grid-Generated Turbulent Flows," *Int. J. Heat Mass Transfer*, 36, 79–87.

Makino, A., 1992, "An Approximate Explicit Expression for the Combustion Rate of a Small Carbon Particle," *Combust. Flame*, 90, 143–154.

Maloney, D.J., Monazam, E.R., Woodruff, S.D., and Lawson, L.O., 1991, "Measurement and Analysis of Temperature Histories and Size Changes for Single Carbon and Coal Particles During the Early Stages of Heating and Devolatilization," *Combust. Flame*, 84, 210–220.

McIntosh, A.C., and Prothero, A., 1991, "A Model of Large Heat Transfer Surface Combustion with Radiant Heat Emission," *Combust. Flame*, 83, 111–126.

Michaelides, E.E., and Lasek, A., 1987, "Fluid-Solids Flow with Thermal and Hydrodynamic Nonequilibrium," *Int. J. Heat Mass Transfer*, 30, 2663–2669.

Michaelides, E.E., Liang, L., and Lasek, A., 1992, "The Effect of Turbulence on the Phase Change of Droplets and Particles under Nonequilibrium Conditions," *Int. J. Heat Mass Transfer*, 35, 2069–2076.

Niksa, S., and Kerstein, A., 1991, "FLASHCHAIN Theory for Rapid Coal Volatilization Kinetics. 1. Formulation" *Energy Fuels*, 5, 647–665.

Parthasarathy, R.N., and Faeth, G.M., 1990, "Turbulent Modulation in Homogeneous Dilute Particle-Laden Flows, and Turbulent Dispersion of Particles in Self-Generated Homogeneous Turbulence," *J. Fluid Mech.*, 220, 485–537.

Ramachandran, R.S., Kleinstreuer, C., and Wang, T.-Y., 1989, "Forced Convection Heat Transfer of Interacting Spheres," *Num. Heat Transfer*, A15, 471–487.

Roco, M.C., Editor, 1993, *Particulate Two-Phase Flows*, Butterworth-Heinemann, Boston.

Sahraoui, M., and Kaviany, M., 1993, "Slip and No-Slip Temperature Boundary Conditions at Interface of Porous, Plain Media: Convection," *Int. J. Heat Mass Transfer* (in press).

Sahraoui, M., and Kaviany, M., 1994, "Direct Simulation Versus Volume-Averaged Treatment of Adiabatic, Premixed Flame in a Porus Medium," *Int. J. Heat Mass Transfer* (submitted).

Saito, M., Sadakata, M., Sato, M., Soutome, T., Murata, H., and Ohno, Y., 1992, "Combustion Rates of Pulverized Coal Particles in High-Temperature/High-Oxygen Concentration Atmosphere," *Combust. Flame*, 87, 1–12.

Sangiovanni, J.J., and Labowsky, M., 1982, "Burning Times of Linear Fuel Droplet Array: A Comparison of Experiment and Theory," *Combust. Flame*, 47, 15–30.

Satake, M., and Jenkins, J.T., Editors, 1988, *Micromechanics of Granular Materials*, Elsevier, Amsterdam.

Sathe, S.B., Peck, R.E., and Tong, T.W., 1990, "A Numerical Analysis of Heat Transfer and Combustion in Porous Radiant Burners," *Int. J. Heat Mass Transfer*, 33, 1331–1338.

Smoot, L.D., and Pratt, D.T., 1979, *Pulverized-Coal Combustion and Gasification*, Plenum Press, New York.

Smoot, L.D., and Smith, P.J., 1985, *Coal Combustion and Gasification*, Plenum Press, New York.

Soo, S.-L., 1989, *Particulates and Continuum*, Hemisphere Publishing Corporation, New York.

Szekely, J., Evans, J.W., and Sohn, H.Y., 1976, *Gas-Solid Reactions*, Academic Press, New York.

Takeno, T., and Sato, K., 1979, "An Excess Enthalpy Flame Theory," *Combust. Sci. Technol.*, 20, 73–84.

Tal (Thau), R., Lee, D.N., and Sirignano, W.A., 1983, "Hydrodynamics and Heat Transfer in Sphere Assemblages-Cylindrical Cell Modes," *Int. J. Heat Mass Transfer*, 26, 1265–1273.

Tal (Thau), R., Lee, D.N., and Sirignano, W.A., 1984, "Heat and Momentum Transfer Around a Pair of Spheres in Viscous Flow," *Int. J. Heat Mass Transfer*, 27, 1953–1962.

Tishkoff, J.M., 1979, "A Model for the Effect of Droplet Interactions on Vaporization," *Int. J. Heat Mass Transfer*, 22, 1407–1415.

Wakao, N., and Kaguei, S., 1982, *Heat and Mass Transfer in Packed Beds*, Gordon and Breach Science Publishers, New York.

Whitaker, S., 1972, "Forced Convection Heat Transfer Correlations for Flow in Pipes, Passed Flat Plates, Single Cylinders, Single Spheres, and for Flow in Packed Beds and Tube Bundles," *AIChE J.*, 18, 361–371.

White, F.M., 1991, *Viscous Fluid Flow*, Second Edition, McGraw-Hill, New York.

Williams, A., Wooley, R., and Lawes, M., 1992, "The Formation of NO_x in Surface Burners," *Combust. Flame*, 89, 157–166.

Woods, L.C., 1993, *An Introduction to the Kinetic Theory of Gases and Magnetoplasmas*, Oxford University Press, Oxford.

Xavier, A.M., and Davidson, J.F., 1985, "Heat Transfer in Fluidized Beds: Convective Heat Transfer in Fluid Beds," in *Fluidization*, Davidson, J.F., et al., Editors, Second Edition, Academic Press, London.

Zanotti, F., and Carbonell, R.G., 1984, "Development of Transport Equations for Multiphase System-III," *Chem. Engng. Sci.*, 39, 299–311.

Zhang, D.-K., 1992, "Laser-Induced Ignition of Pulverized Fuel Particles," *Combust. Flame*, 90, 134–142.

6

Fluid–Fluid Systems

In this chapter, we consider fluid-fluid two-phase flows under the condition that the *interfacial heat transfer* between the two fluids is the principal convective heat transfer. We also examine the models describing *phasic energy conservation* for cases where one of the fluids is *dispersed* within the other. The heat transfer between a pair of *nonvolatile immiscible liquids* is of interest in the direct contact heat exchange, and generally one of the liquids is dispersed. We first consider these liquid-liquid systems and the condition for a liquid-liquid interfacial *vapor nucleation*. The *gas-liquid systems* are considered next. These are the most widely encountered fluid-fluid system and occur with or without a *continuous interface*, with or without a *phase change*, and with or without a *chemical* reaction (for multicomponent systems). Since the gases from two different streams brought together *mix*, the two-medium treatment of the *gas-gas systems* is limited to the application in the modeling of the *low-pressure plasmas* heat transfer. There the heat exchange between the electrons and the heavier species must be addressed. Although this heat exchange is through *molecular collisions* and is dealt with using the result of the kinetic theory of gases, the two-medium treatments of the electrons and the heavier species discussed briefly in Section 5.4.5 has similarities with the phasic energy conservation equations used in this chapter and in Chapter 5. This special gas-gas system is also considered in this chapter.

6.1 Liquid–Liquid Systems

In most liquid-liquid direct-contact heat exchanges, one of the *immiscible* liquid phases is dispersed in the other and the temperature of one of the liquids can be *higher* than the *saturation temperature* of the other. Examples are the injection of liquid *drops* into an otherwise quiescent *host* liquid (i.e., a *batch* heat-exchange process) or into a host liquid stream (i.e., a *continuous* heat-exchange process). Another example is the combustion of *composite liquid drops* (i.e., *emulsions*) such as in liquid water-hydrocarbon *spray combustion*. As we discussed in Section 1.5.1(C), in the absence of any impurities and gas-filled cavities the homogeneous nucleation may be delayed far beyond the saturation temperature. The *thermodynamic* upper limit of this *superheat* was discussed there and here in order to proceed with the interfacial heat transfer between two immiscible liquids without any vapor nucleation and growth. We first re-examine this superheat limit for the homogeneous vapor nucleation.

6.1.1 Vapor Nucleation at Liquid–Liquid Interface

The superheat limit of a liquid in a system of two *single-component*, immiscible liquids and its vapor nucleation and growth are addressed below. This is followed by the effect of a *dissolved gas* (i.e., a multicomponent liquid) on this superheat limit. In addition to a superheat resulting from the *operating temperature* above the saturation temperature of one of the liquids, the superheat can be a result of a *depressurization* (i.e., *decompression*) of the system or decompression of a liquid injected into a host liquid.

(A) Single-Component Immiscible Liquids

The vapor-bubble nucleation, and the superheat limit at the interface between two immiscible liquids in intimate contact, have been examined experimently and also by the use of the molecular theory of liquids. A recent review is given by Avedisian (1992) and treatments of the subject are given by Blander et al. (1971), Jorgenelen et al. (1978), and Avedisian and Glassman (1981). The experimental observations of the homogeneous nucleation are generally made by a rapid *isothermal decompression* of the system. In the experiments of Avedisian and Glassman (1981), water drops of about 1 mm in diameter were released in n-alkane liquids, and the pressure was lowered at a rate of 1.4×10^6 Pa/s. The decompression rates are high enough to allow for small residence time of droplets before nucleation, and therefore negligible dissolution. This is because at higher temperatures strict immiscibility does *not* exist. Their experimental results for the *nucleation temperature* (i.e, *maximum superheat* T_{sp}) as a function of pressure agree with the prediction of the kinetic theory of homogeneous nucleation.

The kinetic theory prediction of the relationship between the *nucleation temperature* (i.e., superheat limit) T_{sp}, around which the volumetric nucleation rate changes significantly and becomes very large, and the *liquid pressure* p_ℓ is given in an approximate *explicit* form by (Avedisian, 1992)

$$T_{sp} = \frac{16\pi\sigma^3}{3(p_v - p_\ell)^2}(k_B \ln \frac{n_\ell A K_f}{\dot{n}_{\rm hom}})^{-1} , \qquad (6.1)$$

where n_ℓ is the *number of molecules per unit interfacial area* when the nucleation occurs, A is the *pre-exponential* factor and accounts for the formation of a critical nucleus for cavitation, k_f is the *rate constant*, and $\dot{n}_{\rm hom}$ is the homogeneous *nucleation rate* which will be discussed in Section 6.2.2(C) and in turn depends on the liquid temperature (here T_{sp}). In practice a value of $\dot{n}_{\rm hom}$ is *prescribed*. The *surface tension* σ of the vapor nucleus and the *vapor pressure* p_v within the nucleus are functions of the temperature and also depend on the *fluid pairs*. The methods of computation of the variables in (6.1) are discussed by Avedisian and Glassman (1981). The results obtained from (6.1) are expected to be similar to those obtained from (1.57) using the thermodynamic considerations (Lienhard and

Karimi, 1981). Other approximate explicit and implicit expressions for T_{sp} are given by Cole (1974) and Lienhard and Karimi (1981).

(B) Liquids Containing Dissolved Gas

When a dissolved gas is present in the droplet, then during the isothermal depressurization the nucleation occurs at *higher* pressures (Forest and Ward, 1977). This *decrease* in the superheat caused by the dissolved gas is the result of the reduction in the critical nucleus radius. The modification of (6.1) to allow for the effect of the dissolved gas has been discussed by Forest and Ward (1977). In practical liquid-liquid heat exchange, both dissolved gas and the dissolving of the liquids into each other (i.e., miscibility) occur. Therefore, the superheats occurring in these systems are much lower than the limits given by the thermodynamic, Section 1.5.1(A), or the kinetic theory predictions. In the following, we consider the heat exchange between a liquid dispersed in an immiscible host liquid and by assuming that no vaporization will occur.

6.1.2 Isolated Dispersed-Liquid Elements

In order to analyze the interfacial heat transfer between a liquid phase dispersed in a host liquid, where these liquids are assumed immiscible and nonvolatile, some *fluid mechanic* and *heat transfer* aspects of the idealized, isolated moving liquid drops are examined. In Section 1.6.1(A), the fluid-pair properties determining the *shape* of the drops (mostly under the free fall or rise) were contained, and in Section 1.12.3(B), the effect of heat transfer on the motion of such drops through the *thermocapillary* effect was indicated. In the following, we will review the existing studies of the *shape* and the *internal circulation* of drops, the *interfacial* drag force and heat transfer, and the modeling of the *intradroplet* heat transfer for large Biot number drop-flow conditions. These are for isolated drops where this idealization allows for detailed experimental and theoretical studies. The interactions among drops and their modeling will be addressed in Section 6.1.3.

(A) Shape

The steady-state or transient behavior of a *finite-volume* liquid moving into an *unbounded* immiscible, *isothermal* host liquid is analyzed by solving the continuity and momentum equations, (1.67) and (1.74), along with the interfacial force balance given by (1.131), the continuity of velocities across this interface, and the absence of mass transfer across the interface (called the kinematic condition). The further assumption of an *incompressible* and *Newtonian* fluid pair, simplifies the analysis.

As an example, consider the numerical results of Koh and Leal (1989) and Hervieu et al. (1992) for the evolution (i.e., the deformation) of a drop with a dynamic viscosity μ_d and density ρ_d moving at the *terminal velocity* under the action of *gravity*, i.e., *phase-density buoyant motion*, in a host (i.e., a continuous) liquid of viscosity μ_c and density ρ_c. Very slow flow is assumed (i.e., $Re_d = u_d d/\nu_c \rightarrow 0$). The terminal velocity for a spherical drop with radius R is (Clift et al., 1978)

$$u_d = \frac{1}{3}\frac{R^2 g}{\nu_c}\left|\frac{\rho_d - \rho_c}{\rho_c}\right|\frac{1 + \dfrac{\mu_d}{\mu_c}}{1 + \dfrac{3}{2}\dfrac{\mu_d}{\mu_c}} . \tag{6.2}$$

This velocity and the drag on the fluid particle will be discussed in the following section. The *capillary number* is given in terms of u_d as

$$Ca = \frac{\mu_c u_d}{\sigma} , \tag{6.3}$$

where σ is the liquid-liquid interfacial tension.

The *initial* shape of the drop is taken as *oblate* or *prolate ellipsoids* with the ratio of the two *principal radii* given by R_1/R_2, with R_1 being the radius along the gravity vector. When the surface tension dominates, i.e., the capillary number is small, the droplet shape will become spherical. The ratio of the viscosities does not influence the final shape. However, the time of the evolution is influenced by μ_d/μ_c. Using the dimensionless time, scaled with μ_d and R, the numerical results of Hervieu et al. (1992) for the evolution of initially *prolate* drops ($R_1/R_2 = 2$), Figure 6.1(a), and initially oblate drops ($R_1/R_2 = 0.5$), Figure 6.1(b) are shown for $\mu_d/\mu_c = 0.5$ and various capillary numbers. The results shown in Figure 6.1(a) indicates that for $Ca \leq 1.25$ the surface tension dominates and the final form is *spherical*, while for $Ca \geq 1.25$ the falling drop forms a *tail* (axisymmetric) and gradually a *pinching* of the tail occurs and the drop will *break* into two. For the viscous dominated drops dynamics, i.e., $Ca \rightarrow \infty$, the elongation at the tail is accompanied by the penetration of the surrounding host liquid, causing an *indentation*. For the oblate drops, Figure 6.1(b), for $Ca \leq 3.5$ after an initial indentation in the rear of the drop the final shape is spherical. For $Ca > 3.5$ this indentation grows into a *dimple* (axisymmetric) while the front remains hemispherical. This is magnified with increase in Ca and the phenomenon of the *host-fluid entrapment* occurs. For $Ca \geq 10$ elongated toroidal lips are formed which can gradually pinch and break off from the initial drop.

Recent experimental observations of drops support the results of the numerical simulations (Koh and Leal, 1990; Baumann et al., 1992). The effect of the thermocapillarity, i.e., the variation of the surface tension with temperature (in a nonisothermal host liquid) on the drop motion and shape has been addressed in the numerical simulations of Ascoli and Lagnado (1992), and the effect of the presence of a surfactant concentration gra-

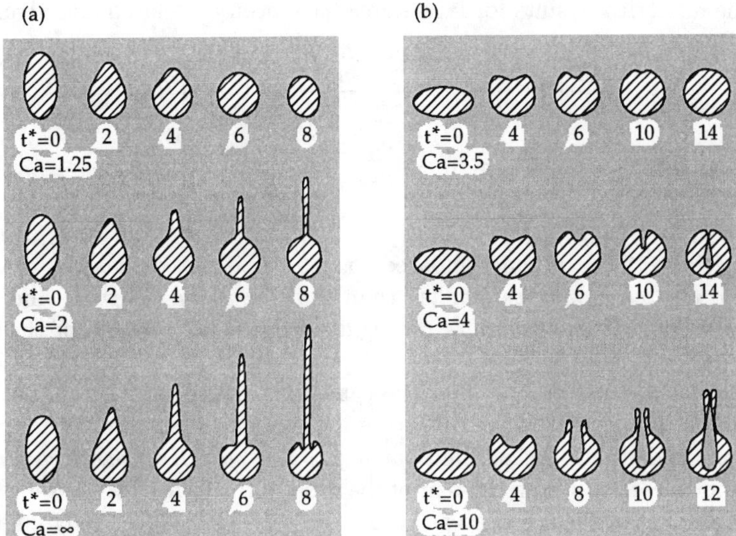

Figure 6.1. (a) Evolution of an initially prolate drop with respect to the dimensionless time and for three different capillary numbers. (b) An initially oblate drop. (From Hervieu et al., reproduced by permission ©1992 American Nuclear Society.)

dient (i.e.,diffusocapillarity) is examined by Bohran and Mao (1992). By examining the results of Figures 6.1(a) and (b) for $Re_d \rightarrow 0$, the classification given in Figure 1.7 for $Re_d \neq 0$, and considering the influence of the variation in the surface temperature, we note that except for small *Bond numbers*, i.e., $Bo < 0.4$, the *shape* of the drop and its evolution is determined by a *large* number of parameters.

(B) Internal Circulation and Interfacial Drag

For *spherical* drops and $Re_d = u_d d/\nu_c \rightarrow 0$, the closed-form solution to the *isothermal* flow of a host fluid over a fluid particle is available. The solution due to Hadamard and Rybczynski is reviewed by Clift et al. (1978), and the drag coefficient defined by (5.5) is given by

$$c_d = \frac{8}{Re_d} \frac{2 + 3\dfrac{\mu_d}{\mu_c}}{1 + \dfrac{\mu_d}{\mu_c}} \qquad Re_d \rightarrow 0 \,. \tag{6.4}$$

By equating this drag to the buoyancy force, the terminal velocity given by (6.2) is found.

The numerical results for larger Re_d have been correlated and given by

$$c_d = \frac{48}{Re_d}[1 + \frac{3}{2}\frac{\mu_d}{\mu_c} + \frac{(2+3\frac{\mu_d}{\mu_c})^2}{Re_d^{1/2}}(a_1 + a_2 \ln Re_d)]$$

$$Re_d \geq 50 \ , \quad \frac{\mu_d}{\mu_c} \leq 2 \ , \tag{6.5}$$

where $a_1 = a_1(\mu_d\rho_d/\mu_c\rho_c)$ and $a_2 = a_2(\mu_d\rho_d/\mu_c\rho_c)$ are functions given by Clift et al. (1978). When compared to (5.9), it is noted that the drag on a *fluid* particle is *always smaller* than that for a *rigid* particle and that the high-Re_d plateau found for the rigid spheres is not present for the fluid spheres.

The shear stress on the interface causes an interfacial motion, i.e., the tangential component of the velocity is not zero on the surface. The vortical motion inside the spherical fluid particles (called the *Hill spherical vortex*) transports vorticity from the rear of the drop to the front. The presence of a small amount of surface of an active species tends to reduce the circulation significantly. For *nonisothermal* flows and for relatively large drops, the thermocapillary and thermobuoyancy also influence the internal motion of the drops (i.e., Zhang et al., 1991). These are expected to also influence the interfacial drag.

As an example, the flow fields inside and outside a fluid particle are shown in Figure 6.2. The numerical results are for $Re_d = 100$, $\mu_d/\mu_c = 55$, and $\rho_d/\rho_c = 790$, where the viscosity and the density ratios are not representative of liquid-liquid systems, and the figure is used here to be just suggestive. The details are given by Clift et al. (1978). Note the *vortical structure* of the motion inside the fluid sphere. The center of the vortex is about *one-third* of a radius distance from the surface. The point of separation and the vortex in the wake of the fluid particle are shown with the host fluid vorticity and stream function distributions.

(C) Interfacial Heat Transfer

As with the drag coefficient, because of the surface motion of the fluid particles the interfacial convective heat transfer rate is also different from that for the rigid particles. The interfacial, area-averaged Nusselt number $\langle Nu_d\rangle_{A_{sf}}^{\text{ext}}$ for rigid spheres was discussed in Section 5.1.2. As was discussed there, various external flow regimes exist and these in turn influence the *local* and the area-averaged Nusselt number. For $Re_d \to 0$ the Stokes flow solution for *large* Pr_c gives (Clift et al., 1978) the relation for $\langle Nu_d\rangle_{A_{\ell\ell}}^{\text{ext}}$ for the case of *spherical fluid particles* as

$$\frac{\langle Nu_d\rangle_{A_{\ell\ell}}^{\text{ext}}}{Pe_d^{1/2}} = \frac{q_{\ell\ell}d}{(T_s - T_\infty)k_c Pe_d^{1/2}} = 0.651(\frac{1}{1+\kappa})^{1/2} \quad Re_d \to 0, \ \frac{\mu_d}{\mu_c} < 2. \tag{6.6}$$

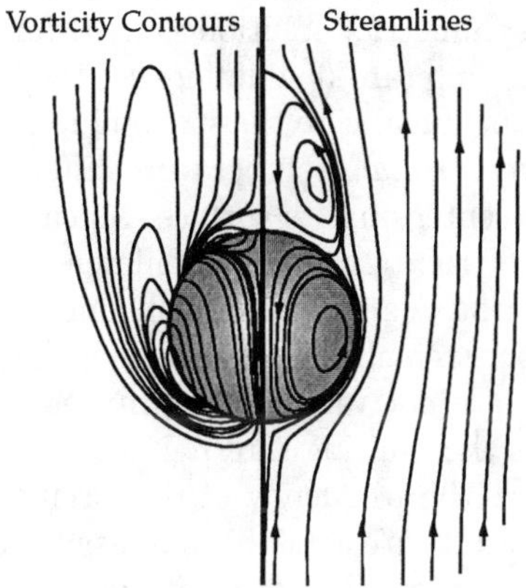

Figure 6.2. Flow structure inside a drop and in its host fluid. The results are for $Re_d = 100$, $\mu_d/\mu_c = 55$, and $\rho_d/\rho_c = 790$. The drop Hill's vortex and the host fluid wake vortex are shown. (From Clift et al., reproduced by permission ©1978 Academic Press.)

For larger Re_d, a correlation is given by Clift et al. (1978) as

$$\frac{\langle Nu_d\rangle^{\rm ext}_{A_{\ell\ell}}}{Pe_d^{1/2}} = \frac{2}{\pi^{1/2}}\{1 - \frac{\dfrac{2+3\kappa}{3(1+\kappa)}}{1+[\dfrac{(2+3\kappa)Re_d^{1/2}}{(1+\kappa)(8.67+6.45\kappa^{0.64})}]^n}\}^{\frac{1}{2}} \quad \frac{\mu_d}{\mu_c} < 2\ , \tag{6.7}$$

where

$$\kappa = \frac{\mu_d}{\mu_c}\ ,\ n = \frac{4}{3} + 3\kappa\ . \tag{6.8}$$

(D) Intradroplet Heat Transfer for Large Biot Numbers

In dealing with the heat transfer from finite-volume droplets, the temperature distribution within the droplets (and therefore their heat content) changes with time. This transient heat transfer is addressed using the *lumped* or *distributed* treatments, depending on the ratio of the external to the internal resistance to the heat flow. When the *droplet Biot number* $Bi = \langle Nu_d\rangle^{\rm ext}_{A_{\ell\ell}} k_c/(6k_d)$ is *small*, the temperature distribution in the droplet can be assumed to be *uniform* (*thermally lumped droplet*). However, when Bi is large (the *critical* value for droplets is not accurately known, but for rigid particles is taken as 0.1), the internal heat flow (or the resistance to the heat flow) must be addressed.

In some analyses, the fluid motion is ignored, and therefore, a transient conduction (radially symmetric) solution is used for the temperature distribution. However, the interanl flow influences the heat flow even though it primarily occurs perpendicular to the streamlines. For no vortical motion inside the droplet, the *maximum* temperature difference across the droplet is between its droplet center and its surface. As can be deduced from Figure 6.2(a), at large Re_d the temperature difference between the *center* of the vortex and the surface is the maximum and is the driving force for the diffusion across the streamlines. Therefore, the purely conductive temperature distributions are not valid. This diffusion from the vortex center to the surface remains the limiting internal heat transfer mechanism even when $Re_d \rightarrow \infty$, where the frequency of the circulation tends to be very large. The numerical results are available and asymptotic, $Re_d \rightarrow 0$, for the internal flow field. They show that the diffusion solution can be used if a *correction* is made. This is by using an *effective thermal diffusivity* which is a_α times that of the dispersed-phase thermal diffusivity α_d (Sideman, 1966; Clift et al., 1978). Then the *transient internal* conductance can be written as

$$\frac{\langle Nu_d \rangle^{\rm int}_{A_{\ell\ell}}}{Pe_d^{1/2}} = \frac{q_{\ell\ell} d}{(T_s - \langle T \rangle^d) k_d} = \frac{2\pi^2}{3} \frac{\sum_{i=1}^{\infty} e^{-n\pi^2 \frac{t a_\alpha \alpha_d}{d^2}}}{\sum_{i=1}^{\infty} \frac{1}{n^2} e^{-n\pi^2 \frac{t a_\alpha \alpha_d}{d^2}}} . \tag{6.9}$$

Values of a_α between 2.5 and 3.2 have been recommended.

The *internal droplet* motion and its effect on the transient intradroplet heat transfer will be further discussed in Section 6.4.1(E) and (F), in connection with the droplet evaporation.

6.1.3 Continuous- and Dispersed-Phase Energy Equations

In the last section some of the transport characterization of isolated drops in a host liquid were reviewed. The two-phase flow and heat transfer involving many drops is addressed below.

Modeling of the liquid-liquid heat exchange, using a continuous description for the host liquid and a discrete description for the dispersed phase, has been discussed by Crowe (1988) and Jacobs and Golafshani (1989). It is assumed that the drops do *not* collide and coalesce. The presence of the drops influences the thermal energy transport through the continuous phase and this can be modeled as an enhanced diffusion. When the flow of the continuous phase is turbulent, the interaction of the droplets and this turbulence has not yet been studied, and therefore, models of this enhanced diffusion in the continuous phase are rather incomplete (Crowe, 1988). Depending on the magnitude of Bi, the dispersed-phase energy equation is treated as lumped (small Bi) or distributed (large Bi). Jacobs and Golafshani (1989) consider both of these, along with the effect of the variation in the properties. For the systems considered, the *distributed* approach appears as being more appropriate.

Following a treatment similar to that given for *low Bi* for the dispersed solid particles in Section 5.2.1, the thermal energy equation (with *no* phase change) for the continuous and the dispered phase can be modeled with

$$\epsilon(\rho c_p)_c \frac{\partial \langle T \rangle^c}{\partial t} + (\rho c_p)_c \langle \mathbf{u} \rangle \cdot \nabla \langle T \rangle^c = \nabla \cdot [\langle \mathbf{K}_e \rangle + \epsilon(\rho c_p)_c \mathbf{D}_c^d] \cdot \nabla \langle T \rangle^c +$$

$$\frac{6(1-\epsilon)}{d} \frac{\langle T \rangle^d - \langle T \rangle^c}{\dfrac{1}{\langle Nu_d \rangle_{A_{\ell\ell}}^{\text{ext}} \dfrac{k_c}{d}} + \dfrac{1}{\langle Nu_d \rangle_{A_{\ell\ell}}^{\text{int}} \dfrac{k_d}{d}}} + \langle \dot{s}_c \rangle , \tag{6.10}$$

$$\frac{\pi d^3}{6} \frac{\partial}{\partial t} (\rho c_p)_d \langle T \rangle^d = \pi d^2 \frac{\langle T \rangle^c - \langle T \rangle^d}{\dfrac{1}{\langle Nu_d \rangle_{A_{\ell\ell}}^{\text{ext}} \dfrac{k_c}{k_d}} + \dfrac{1}{\langle Nu_d \rangle_{A_{\ell\ell}}^{\text{int}} \dfrac{k_d}{d}}} , \tag{6.11}$$

$$\langle T \rangle^d = \frac{1}{V_d} \int_{V_d} T \, dV , \quad \langle T \rangle^c = \frac{1}{V_c} \int_{V_c} T \, dV , \tag{6.12}$$

where V_d is $\pi d^2/3$, while V_c is a *representative* elementary volume in the continuous phase. The velocity used in the determination of the Nusselt number is the local, *relative* velocity. The hydrodynamics of this two-phase system are treated similar to that discussed in Section 5.4.4(B) and the drag coefficients given by (6.4) and (6.5) are used. The *effective conductivity* and the *dispersion* tensors for the continuous phase must be prescribed, and the effect of turbulence on the heat transfer is *included* in $\mathbf{D}_c^d$. When a variation exists in the element diameter, the *ensemble averages* over the element diameters can be taken, if a diameter distribution is known a priori.

6.2 Gas–Liquid Systems

The gas-liquid interfacial local *convective heat flux* $q_{g\ell}$ is of interest in a very large range of applications. In general, the interfacial local *mass flux* $\dot{\mathbf{m}}_{g\ell}$, due to phase change and/or species transport, is also of interest. The interface may be *continuous* or one of the phases can be *dispersed* in the other. At this interface, the liquid-vapor *phase change* (*condensation/evaporation*) may occur. The interface is generally *wavy* and the phasic flows are *turbulent*, and therefore, even the simplest phase distributions (i.e., the continuous interface) without any heat transfer (i.e., isothermal) has a *complex* hydrodynamics. Figure 6.3 gives a rendering of the gas-

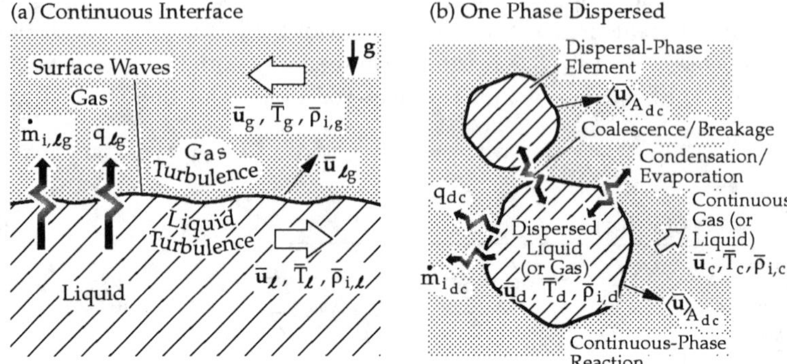

Figure 6.3. A rendering of the hydrodynamic and heat and mass transfer aspects of the liquid-gas interface. (a) Continuous interface, and (b) discontinuous interface (one phase dispersed).

liquid flow and heat transfer with the distiction made between systems with a continuous interface and systems with one of the phases dispersed.

The hydrodynamics of the gas–liquid interface addresses the recent advances in the turbulence-interface interactions for continuous interfaces and the elemental breakup and coalescence, among the examples. Some of these recent advances are discussed by Hewitt et al. (1991). A classification of some aspects of the gas–liquid *interfacial hydrodynamics* is given in Chart 6.1. For *isothermal* flows, these include the effect of the gas and liquid flow direction with respect to each other and also with respect to the gravity vector, the surface waves and the interfacial instabilities, the *decay* of *bulk* turbulence and the *generation* of *surface* turbulence near the interface, and the effect of the presence of a *solid surface* near the interface. The *nonisothermal* aspects include the effect of the phase change and the variation in the thermophysical properties. For dispersed systems, the *shape* of the dispersed element and the *element-element* interactions are also addressed. These aspects of the motion influence the heat transfer, and throughout this section some of the findings with their relation to the interfacial heat transfer are reviewed.

Some aspects of the gas–liquid *interfacial heat transfer* are mentioned in the classification given in Chart 6.2. These include the *multicomponent*

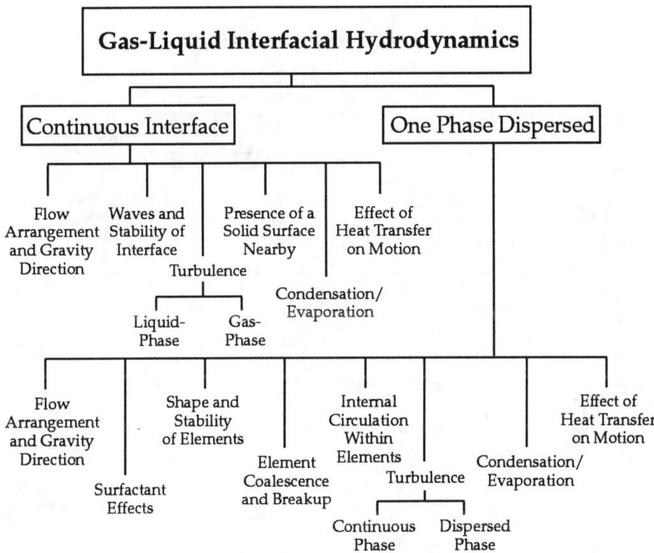

Chart 6.1. Some hydrodynamic aspects of the gas-liquid interface.

systems, the gas phase formation by *bulk nucleation* caused by a sudden depressurization, and the chemical reactions. The heat transfer of gas-liquid systems generally involves phase change and the *spray combustion* applications have resulted in many recent advances.

In this section, some heat transfer aspects of the *continuous* interface are discussed first. Then before examining the elements of a dispersed phase, the formation of *vapor bubbles* in the bulk liquid is discussed in order to note that *superheated* liquids can continue to exist and flow as a single phase (without phase change) up to a superheat limit. Then some interfacial heat transfer aspects of *isolated* elements (liquid or drops) are discussed, followed by the examination of *multielement* systems (including interaction among elements). Finally, the existing models for the transport of thermal energy in the continuous and dispersed phases are reviewed.

6.2.1 Continuous Interface

As indicated in Chart 6.2, even with no phase change the presence of interfacial *waves* and/or turbulence and the various *geometries* of the interface encountered in applications, results in an extensive combination of mechanisms influencing the interfacial heat transfer. Also, as will be demonstrated, despite its many process applications the heat transfer across a gas-liquid *turbulent* interface has not been as extensively studied as that of the fluid-solid interface. To demonstrate some of the features of the tur-

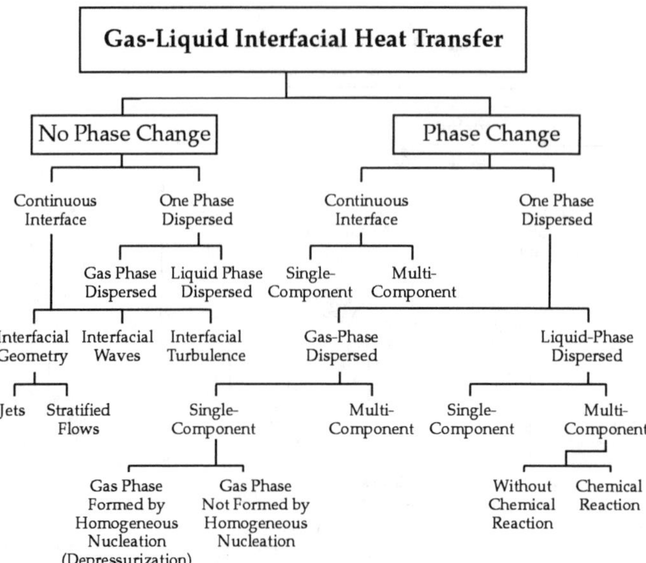

Chart 6.2. Some heat transfer aspects of the gas-liquid interface.

bulent gas-liquid interfacial transport, the following two examples are reviewed. The first deals with the interfacial heat transfer with phase change (here, condensation) between a *horizontal*, semi-infinite turbulent liquid layer and a semi-infinite vapor layer above it. The second example examines heating of a turbulent, axisymmetric liquid *jet* with its surrounding vapor (including condensation).

(A) Condensation on a Horizontal Turbulent Interface

The condensation on thin liquid film flowing over a solid surface will be discussed in Section 8.2.1 as part of the three-medium (solid-liquid-gas) treatment. Here we consider a problem in which a solid surface does *not* influence the heat transfer (and the condensation rate). This class of problems has been reviewed in part by Sideman and Moalem-Moran (1982), Whalley (1987), and Gerner and Tien (1989). Here we discuss the effect of the interfacial *turbulence* on the condensation rate for the case of *no mean* fluid motion. This is the case of a horizontal subcooled liquid surface with *no mean* horizontal motion which is *agitated* by a submerged liquid jet from beneath, while an otherwise saturated vapor on top of this liquid condenses at the interface. This is a limiting case of the general problem of condensation on a moving interface (i.e., a stratified flow) that has been studied by Lim et al. (1984), among others; a review is given by Chelata et al. (1989).

In considering the heat transfer across a *nonwavy*, turbulent gas-liquid interface, we note that the *source* of the turbulence can significantly affect the interfacial transfer rate. The turbulence can be generated by an imposed longitudinal shear rate on the interface and this leads to bursts near the interface. The period of these bursts would then depend on the imposed shear rate (Hewitt, et al., 1991; Banerjee, 1991). When a solid surface is nearby and a shear rate is imposed, the bursts may originate from the solid surface and travel to the gas-liquid interface to be *reflected*. Here we consider the experiments of Brown et al. (1990), where the interfacial turbulence is a result of a *submerged* turbulent *liquid* jet placed in the liquid phase and away from the interface. Other mechanisms of interfacial turbulence generation are discussed by Komori (1991) and Perkins et al. (1991).

Consider *direct-contact* vapor condensation on a shear-free and relatively wave-free horizontal interface with a liquid-side turbulence which is *isotropic* in the *horizontal* plane and with *no mean* motion (except for that due to the condensation). Sonin et al. (1986) examine this problem experimentally in order to determine the dependence of the condensation rate on the turbulence intensity (i.e., functuations in the velocity components). Isolation of the effect of a net flow (although the turbulent intensity is affected by the mean flow) is desirable, because earlier attempts to examining the combined effects have not been successful (i.e., Lim et al., 1984).

The liquid-side resistance to the heat flow dominates, and therefore, the one-dimensional heat flow through the subcooled liquid phase is examined. Using the turbulent eddy diffusivity α_t, we have the mean thermal energy equation (2.96) reduced to

$$\overline{u}\frac{d\overline{T}}{dy} = \frac{d}{dy}(\alpha + \alpha_t)\frac{d\overline{T}}{dy} , \tag{6.13}$$

where y is measured from the interface. At the interface, where condensation occurs with a mass flux $\dot{m}_{g\ell} = \rho u_c$ and at a known saturation temperature T_s for a *single-component* fluid, the interfacial boundary condition become

$$\overline{T} = T_s \ , \ q_s = -\rho c_p(\alpha + \alpha_t)\frac{d\overline{T}}{dy} = \dot{m}_{g\ell}\Delta i_{\ell g} = \rho u_c \Delta i_{\ell g} \qquad y = 0. \tag{6.14}$$

The turbulent eddy diffusivity α_t is given in terms of the *vertical* component of the turbulent heat flux using (2.95) and we have

$$\alpha_t = -\frac{\overline{v'T'}}{\dfrac{d\overline{T}}{dy}} . \tag{6.15}$$

The anticipated vertical distributions of the mean temperature $\overline{T}$, the root-mean-square of the velocity fluctuations $(\overline{u'^2})^{1/2}$, $(\overline{v'^2})^{1/2}$, and $(\overline{w'^2})^{1/2}$, for u and w in the horizontal plane and v in the y direction, and the turbulent eddy diffusivity, are shown in Figure 6.4. At the interface, the tur-

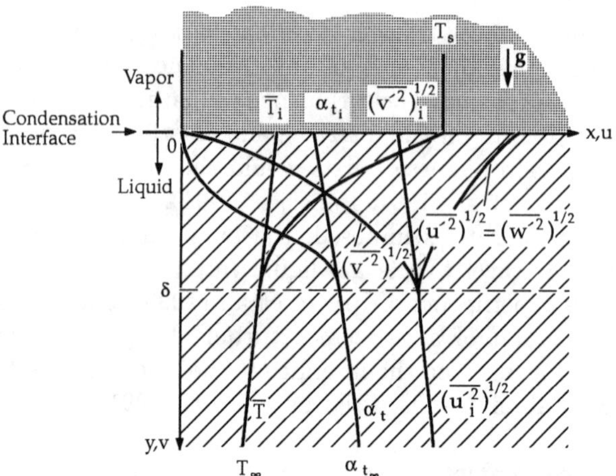

Figure 6.4. Distributions of the mean temperature, eddy diffusivity, and the root-mean-square of the velocity fluctuations, near a nonwavy liquid-vapor interface with condensation. Also shown are extrapolation of the bulk liquid-phase distributions.

bulence is expected to be *anisotropic*, and $(\overline{v'^2})^{1/2}$ is expected to be zero. Then at the interface $\alpha_t = 0$. The depth (or thickness) of the boundary layer on the liquid side, where because of the decrease in α_t the temperature gradient is large, is shown with δ. The vapor is at the saturated state and no noticeable temperature variation is expected near the interface in the vapor side. Also shown in Figure 6.4 are the *jump interfacial conditions* which result if the reduction in $(\overline{v'^2})^{1/2}$ and α_t are ignored and their values are taken by the *extrapolations* of their bulk (far from the interface) distributions. Using the bulk-extrapolated value of α_t, i.e., α_{t_i}, results in a lower surface temperature (for a given heat flux), and this temperature jump interfacial condition is also shown as $\overline{T}_i$.

In order to relate the interfacial heat flux (using its normalized form in terms of the Nusselt or the Stanton number) to the velocity functuations and other properties of the bulk and interfacial turbulence, the distribution of α_t is needed in (6.13). Sonin et al. (1986) and Brown et al. (1990) examine the existing *large*- and *small*-eddy models for the magnitude and the variation of α_t near the interface. Here a model for the variation of α_t with respect to y is examined, followed by the correlation of the experimental results for the Stanton number.

(i) Surface Renewal Model

Expansion of the functuation quantities in (6.15) with respect to y leads to (Brown et al., 1990)

$$\alpha_t = \frac{y^2}{t_e} + O(y^3) + \cdots , \tag{6.16}$$

where t_e is interpreted as the time a typical eddy spends in contact with the interface during which the transport is by the molecular diffusion. This is the *surface renewal model* of eddies periodically *sweeping* the interface, and as t_e *decreases* the intensity of the eddy transport *increases*. This time period is defined as

$$\frac{1}{t_e} = (\overline{\frac{dT'}{dy}\frac{dv'}{dy}})_{y=0}\ (-\frac{d\overline{T}}{dy})^{-1}\ . \tag{6.17}$$

In order to satisfy the $y \to \infty$ asymptote, (6.16) is modified to give

$$\frac{1}{\alpha_t} = \frac{t_e}{y^2} + \frac{1}{\alpha_{t_\infty}}\ . \tag{6.18}$$

If a homogeneous (i.e., uniform) turbulence is assumed far from the interface, α_{t_∞} will be a known constant. When the turbulence away from the interface is not uniform, then the value at a *finite* location y outside the boundary layer, i.e., $y > \delta$, is used.

Integrating (6.13) twice and using the boundary conditions, gives

$$\ln[1 + \frac{c_p(T_s - \overline{T})}{\Delta i_{\ell g}}] = \overline{u}_c \int_0^y \frac{dy}{\alpha + \alpha_t}\ . \tag{6.19}$$

Substitution of (6.18) into the above and the integration to $y \to \infty$, where the bulk conditions are valid, gives the expression for the Stanton number in terms of $(\overline{v'^2})_i^{1/2}$ (Brown et al., 1990)

$$\begin{aligned} St &= \frac{\dot{m}_{g\ell}\Delta i_{\ell g}}{\rho c_p (T_s - T_\infty)(\overline{v'^2})_i^{1/2}} \\ &= \frac{\ln(1+Ja)}{Ja}\frac{1}{(\overline{v'^2})_i^{1/2}}\frac{1}{\int_0^\infty \frac{\alpha_{t_\infty} - \alpha_t}{(\alpha+\alpha_t)(\alpha+\alpha_{t_\infty})}\,dy}\ , \end{aligned} \tag{6.20}$$

where the *Jakob number* is defined as $Ja = c_{p_\ell}(T_s - T_\infty)/\Delta i_{\ell g}$. Now using (6.18) in this equation and assuming that $\alpha_{t_\infty} \gg \alpha$, the result is (Brown et al., 1990)

$$\begin{aligned} St &= \frac{\ln(1+Ja)}{Ja}\frac{1}{(\overline{v'^2})_i^{1/2}}\frac{1}{\int_0^\infty (\alpha + \frac{y^2}{t_e})^{-1}\,dy} \\ &= \frac{\ln(1+Ja)}{Ja}\frac{2}{\pi(\overline{v'^2})_i^{1/2}}(\frac{\alpha}{t_e})^{1/2}\ . \end{aligned} \tag{6.21}$$

For small Ja, which is typical for practical applications, an expansion in Ja leads to

$$St = \frac{2}{\pi(\overline{v'^2})_i^{1/2}}(\frac{\alpha}{t_e})^{1/2}(1 - \frac{Ja}{2})\ . \tag{6.22}$$

Then if $(\overline{v'^2})_i^{1/2}$ and t_e are known functions of the interfacial turbulence, the dimensionless condensation rate St can be determined. Presently, $(\overline{v'^2})_i^{1/2}$ is *measured* and t_e is *correlated* using the experimented results for St.

(ii) Correlation for Stanton Number

Using their experimental results for *water vapor* condensing on a turbulent horizontal *water* interface, Brown et al. (1990) developed a correlation, with guidance from (6.22), and gave

$$St = \begin{cases} 0.0198 Pr_\infty^{-1/3}(1 - Ja/2) & Ri_\ell \rightarrow 0 \, , \\ 0.0136 - 8.1 \times 10^{-4} Ri_\ell (1 - Ja/2) & 3.5 < Ri_\ell < 15 \, , \end{cases} \tag{6.23}$$

where the *Richardson number* Ri_ℓ includes the turbulence characteristic *integral length scale* ℓ_t, and is defined as

$$Ri_\ell = \frac{\beta_\infty g (T_s - T_\infty)\ell_t}{(\overline{v'^2})_i} \, . \tag{6.24}$$

The Richardson number is a measure of the ratio of the buoyant to the inertia forces. Since *no* mean characteristic velocity and time scales are available in this problem the integral length scale is taken as

$$\ell_t = [(\overline{v'^2})^{1/2} \overline{t}_t \,]_{y=0} \, , \tag{6.25}$$

where turbulence *integral time scale* $\overline{t}_t$ is given by

$$\overline{t}_t = \int_0^\infty R(t) dt = \int_0^\infty \frac{v'(t'+t)v'(t)}{\overline{v'^2}} dt \tag{6.26}$$

and $R(t)$ is the Eulerian time *correlation* of the velocity. Brown et al. (1990) find that ℓ_t remain independent of $(\overline{v'^2})^{1/2}$ in their experiment.

The above illustration of the condensation on a turbulent interface indicates the role of the interfacial turbulence on the condensation rate. The condensation rate, when it is large, can also influence the interfacial turbulence and the flow structure. This has been addressed by Chun et al. (1986). Their experimental results show that when the liquid-side turbulence intensity *exceeds* a *threshold value*, the interface becomes *unstable* with very short, high-intensity *bursts* of condensation occurring *intermittently*.

(B) Heating of a Turbulent Liquid Jet by Its Vapor

Another practical example of interfacial convective heat transfer across a continuous gas-liquid interface is that of heating a liquid jet. The jets are generally *turbulent*, and at a critical distance from the jet exit they become *unstable* and *wavy*. This leads to the jet *breakup* and *droplet formation*. For a *subcooled* liquid jet *released* in a gasous ambient made of only the vapor of the liquid constituent (i.e., single-component liquid and gas), the *condensation* of the vapor on the liquid surface is *controlled* by the liquid side heat transfer.

The analysis of the interfacial heat transfer between a continuous, nonwavy turbulent liquid jet discharged *vertically downward* into its otherwise *quiescent* vapor was initiated by Kutateladze (1963) and later was expanded by Hoang and Seban (1988), among others. Here some aspects of the analysis dealing with the liquid jet breakup, the eddy diffusivity model for the liquid phase turbulence, and the local interfacial heat transfer are discussed.

(i) Liquid Jet Breakup

For a liquid jet of density ρ exiting *downward* from a nozzle of diameter d_o (radius R_o) with an area-averaged velocity $\langle u\rangle_{R_o}$, in principle the downstream behavior depends on the level of turbulence at the exit $\langle\overline{u'^2}\rangle_{R_o}^{1/2}$, the liquid viscosity μ_ℓ, the interfacial tension σ, the magnitude of the gravitational acceleration g, and the downstream location x. The effect of the gas viscosity is assumed negligible. The length of the nozzle (or its ratio to d_o) also influences the downstream behavior. One of the most significant dimensionless parameters influencing the liquid-jet hydrodynamics is the *Weber number* which is the ratio of inertial to surface tension forces, i.e.,

$$We = \frac{\rho_\ell d_o \langle u\rangle_{R_o}^2}{\sigma} \ . \tag{6.27}$$

Empirical results (Hoang and Seban, 1988) show that for $We^{1/2} < 3$ drops are *formed* at the nozzle. For $3 < We^{1/2} < We_{cr}^{1/2}$, drops are formed from axisymmetric waves at a *transition location* x_{tr} given by

$$(\frac{x}{d_o})_{tr} = a_1 We^{1/2} \ , \tag{6.28}$$

where a_1 is about 12 and

$$We_{cr}^{1/2} = 3.1 + 790\frac{\mu_\ell}{(\rho d_o \sigma)^{1/2}} \ . \tag{6.29}$$

This ratio of the viscous to the surface tension forces is named the *Ohnesorge number*. Hoang and Seban (1988) discuss the *limitations* on these correlations and *jet regimes*. Here we note that the following analysis applies to the *nonwavy, turbulent* liquid jets, and therefore is valid only over the subcritical (or subtransitional) regime of the liquid jets.

(ii) Eddy Diffusivity Model

The description of the heat and momentum transport within the turbulent liquid jet requires the specification of the *turbulent heat flow* and the *turbulent momentum flux*. For the lateral transport, the concept of the eddy diffusivity and viscosity are used. The steady, *axisymmetric mean* mass, momentum, and thermal energy conservation equations are the simplified forms of (2.88), (2.92), and (2.94). These are written for $\overline{u}$ and $\overline{v}$ along x

and r, with g also being along x, and by assuming that the lateral diffusion dominates over the axial diffusion. These become

$$\frac{\partial \overline{u}}{\partial x} + \frac{1}{r}\frac{\partial}{\partial r}\overline{v}r = 0 , \tag{6.30}$$

$$\overline{u}\frac{\partial \overline{u}}{\partial x} + \overline{v}\frac{\partial \overline{u}}{\partial r} = \frac{1}{r}\frac{\partial}{\partial r}r(\nu_\ell + \nu_t)\frac{\partial \overline{u}}{\partial r} + g , \tag{6.31}$$

$$\overline{u}\frac{\partial \overline{T}}{\partial x} + \overline{v}\frac{\partial \overline{T}}{\partial r} = \frac{1}{r}\frac{\partial}{\partial r}r(\alpha_\ell + \alpha_t)\frac{\partial \overline{T}}{\partial r} . \tag{6.32}$$

The eddy viscosity and diffusivity are assumed to be equal (i.e.,$Pr_t = 1$) and a model similar to that used for the jets being of the phase as their surrounding, i.e., (2.106), is used. These give

$$\nu_t = c_t R\overline{u} = \alpha_t , \tag{6.33}$$

where R is the *local* radius of the jet and in general $\overline{u}$ varies with x and r. This eddy diffusivity model for liquid jets was first used by Kutateladze and the analysis of the jet heating rate is named after him (Hoang and Seban, 1988).

As is expected, this *simple*, *algebraic* model of turbulence is not capable of accurately describing the turbulence both in the *core* and *near* the *interface* of the liquid jet. Therefore, one expects that c_t will depend on the *flow characteristics* and also on the rate of *condensation*. This shortcoming of the model has been demonstrated by Hoang and Seban (1988).

(iii) Local Interfacial Heat Transfer

By allowing for condensation and shear stress at the liquid-gas interfaces, the boundary conditions for (6.30) to (6.32) become

$$T = T_s \ , \ -\rho_\ell\overline{v}\Delta i_{\ell g} = -\dot{m}_{g\ell}\Delta i_{\ell g} = [k_\ell + (\rho c_p)_\ell \alpha_t]\frac{\partial \overline{T}}{\partial r} , \tag{6.34}$$

$$-\rho_\ell\overline{uv} = \tau_{g\ell} = \rho(\nu_\ell + \nu_t)\frac{\partial \overline{u}}{\partial r} \quad r = R , \tag{6.35}$$

$$\frac{\partial \overline{u}}{\partial r} = \frac{\partial \overline{T}}{\partial r} = \overline{v} = 0 \quad r = 0 , \tag{6.36}$$

$$R = R_o \ , \ \overline{u} = \langle u \rangle_{R_o} \ , \ \overline{v} = 0 \ , \ \overline{T} = T_o \quad x = 0 . \tag{6.37}$$

The conservation equations (6.30) to (6.32), subject to the boundary conditions (6.34) to (6.37) with the eddy diffusivity and viscosity model (6.33) and a precribed c_t, can be solved for $\overline{u}$, $\overline{v}$, R, and $\overline{T}$. When $\dot{m}_{g\ell} = 0$, a closed-form solution is possible; otherwise, a numerical integration is required. Hoang and Seban (1988) show that the *average* (over $0 \le x \le x$ and $0 \le r \le R$) Stanton number can be expressed as

$$\langle St \rangle_{A_{\ell g}} = \frac{\langle q_s \rangle_{A_{\ell g}}}{(T_s - \langle \overline{T} \rangle_R^b)(\rho c_p)_\ell \langle u \rangle_{R_o}} = \langle St \rangle_{A_{\ell g}}(\frac{x}{R_o}, Pr_\ell, Re_{d_o}, Ar_{d_o}) , \tag{6.38}$$

where the bulk-mixed temperature $\langle \overline{T} \rangle_R^b$ and the parameters are defined as

$$\langle \overline{T} \rangle_R^b = \frac{2}{R^2 \langle \overline{u} \rangle_R} \int_0^R \overline{T}\overline{u} r dr \ , \tag{6.39}$$

$$Pr_\ell = \frac{\nu_\ell}{\alpha_\ell} \ , \quad Re_{d_o} = \frac{\langle u \rangle_{R_o} d_o}{\nu_\ell} \ , \quad Ar_{d_o} = \frac{g d_o^3}{\nu_\ell^2} \ . \tag{6.40}$$

The *Archimedes number* Ar_{d_o} is the ratio of the buoyant to viscous forces (similar to the Grashof number for the thermobuoyant flows) and here the vapor density is assumed to be negligible compared to the liquid density.

It is evident that a large number of parameters, both directly and indirectly (through jet hydrodynamics), influence the interfacial heat transfer rate. Hoang and Seban (1988) show that a more accurate description of the turbulence and the wave-free jet length (over which the waves do not influence the transport) are needed in order to accurately predict the large body of the experimental results available.

6.2.2 Bulk Bubble and Droplet Nucleation

We begin discussion of the *single-component*, gas-liquid two-phase systems, with one phase dispersed, by considering the *formation* of bubbles in *bulk superheated liquids* and droplets in *bulk supersaturated vapors*. These are the processes of vapor-bubble and liquid-drop formations in the *bulk* of a liquid or a vapor, and constitute two of the *simplest* and yet practical problems of liquid-gas phase change and heat transfer (without considering the bounding solid surfaces). The bulk nucleation of vapor bubbles is generally caused by a sudden *decrease* in the initially single-phase liquid. The bulk nucleation of liquid drops is generally caused by a sudden *increase* in the initially single-phase vapor. These *thermodynamic* phenomena of *second-phase formation*, along with the growth *dynamics* determined by the *transport* described by the principles of the conservation of mass, momentum, and heat transport, leads to the *phase change dynamic* predictions.

In this section we re-examine the *metastable* thermodynamic states of *bulk* fluids (liquids and vapors) for single-component systems discussed in Section 1.5.1(C). There we mentioned upper *limits* to the *bulk* superheat and supersaturation, beyond which the nucleation of the second phase begins at a very *fast* rate. In order to be able to use the results of the *kinetic theory* for the *nucleation rates* in a phase dynamic and heat transfer analysis, the *thermodynamic equilibrium* of *curved surfaces* (such as bubbles and drops) is first discussed. Then a general account is made of the bubble (or droplet) *number density* in which the nucleation rates are introduced. The results of the kinetic theory for these rates are considered next, followed by a reference to the phase dynamic and interfacial heat transfer analysis under *nonequilibrium* conditions.

(A) Thermodynamic Equilibrium Radius of Bubbles and Droplets

In relation to the bulk nucleation, the initial bubble or droplet size is assumed to be small such that the *shape* of these dispersed elements is taken as spherical. Also, it is assumed that there is *no motion* inside and outside of these elements. This allows for the thermodynamic analysis of the elements using *mechanical*, *thermal*, and *chemical equilibrium*. In terms of the dispersed-phase properties designated by a subscript d and the continuous properties designated by c, these *thermodynamic equilibrium conditions* are

$$\begin{aligned} &\text{mechanical equilibrium} \qquad p_d = p_c + \Delta p_\sigma \ , \\ &\text{thermal equilibrium} \qquad T_d = T_c \ , \\ &\text{chemical equilibrium} \qquad \mu_d = \mu_c \ , \end{aligned} \tag{6.41}$$

where Δp_σ is the pressure difference caused by the surface tension and μ is the chemical potential. The equilibrium relations between the dispersed phase radius, the *saturation* condition within the dispersed phase, and the *metastable conditions* in the continuous phase are discussed below.

(i) Bubbles

For bubbles, the static mechanical (i.e., force balance) equilibrium between the vapor (dispersed phase) and the liquid (continuous phase) is given by the simplified form of (1.131). In (1.131) the gas and liquid motions are also included. When there is no motion and for a spherical interface, the two principal radii are equal, i.e., $2H = 2/R$. This equilibrium radius is designated as R_e, and then (1.131) gives

$$p_{v_e} = p_\ell + \frac{2\sigma}{R_e} \ , \tag{6.42}$$

which is called the *Laplace equation*. This shows that the equilibrium vapor pressure p_{v_e} is *larger* than the liquid pressure p_ℓ by $2\sigma/R_e$, where σ is the gas-liquid interfacial tension.

The thermal and chemical equilibrium also require that

$$T_{ve} = T_\ell \ , \tag{6.43}$$

$$\mu_{ve} = \mu_\ell \ . \tag{6.44}$$

Now using the Gibbs-Duhem relation (Carey, 1992) which is

$$d\mu = -s dT + v dp \tag{6.45}$$

and integrating it for a *constant temperature* process, the *saturation pressure* $p_s(T_\ell)$ to p gives

$$\mu - \mu_s = \int_{p_s(T_\ell)}^{p} v \, dp \ . \tag{6.46}$$

For the vapor phase and with an ideal-gas behavior, this gives

$$\mu_{ve} = \mu_{vs} + R_g T_\ell \ln \frac{p_{ve}}{p_s(T_\ell)} . \tag{6.47}$$

For the liquid phase with a constant specific volume, integration of (6.45) gives

$$\mu_\ell = \mu_{\ell_s} + v_\ell [p_\ell - p_s(T_\ell)] . \tag{6.48}$$

By using (6.42) and (6.47) in (6.48), we have

$$p_{v_e} = p_s(T_\ell) \exp\{\frac{v_\ell [p_\ell - p_s(T_\ell)]}{R_g T_\ell}\} . \tag{6.49}$$

Now using the mechanical equilibrium condition (6.42) in this and solving for R_e, we have

$$R_e = \frac{2\sigma}{p_s(T_\ell) \exp\{\frac{v_\ell [p_\ell - p_s(T_\ell)]}{R_g T_\ell}\} - p_\ell} . \tag{6.50}$$

This expression gives the *equilibrium* (*thermodynamic*) *radius* R_e of a bubble. The bubble is at a state given by p_{v_e} and T_ℓ, while the liquid must be in a *superheated* (*metastable*) state given by p_ℓ and T_ℓ.

(ii) Droplets

For droplets, the counterparts of (6.42), (6.49), and (6.50) are found using the *saturation pressure* $p_s(T_v)$ and these are

$$p_{\ell e} = p_v + \frac{2\sigma}{R_e} , \tag{6.51}$$

$$p_v = p_s(T_v) \exp\{\frac{v_\ell [p_{\ell_e} - p_s(T_\ell)]}{R_g T_v}\} , \tag{6.52}$$

$$R_e = \frac{2\sigma}{\frac{R_g T_v}{v_\ell} \ln[\frac{p_v}{p_s(T_v)}] - p_v - p_s(T_v)} . \tag{6.53}$$

The liquid is in a state given by p_{ℓ_e} and T_v with an equilibrium radius R_e, while the vapor is *supersaturated* with the state given by p_v and T_v.

(B) Number Density of Bubbles and Droplets

The dispersed-phase elements are created by *nucleation* or by *other* phenomena. Here we consider the presence of a continuous phase (liquid or vapor) with the second phase (vapor or liquid) dispersed as elements *created* by nucleation (homogeneous and heterogeneous), which is a thermodynamically controlled process, or by *dissociation*, which is a thermally and hydrodynamically controlled process, and *destroyed* by *coalescence*, a hydrodynamically controlled process, or by complete *condensation* (or *evaporation*).

The *number density* of the dispersed element n_d, can change within a control volume and these *source* and *sink* terms are balanced by the *local* and *convective* (with velocity $\mathbf{u}_d$) rate of change. Kocamustafaogullari and Ishii (1983) discuss this formulation for bubbles, and here we make a general treatment for any dispersed phase.

The *source* of the *volumetric generation rate* of the dispersed-phase elements can be the *homogeneous nucleation* $\dot{n}_{hom}$, heterogeneous nucleation $\dot{n}_{het}$, and the *dissociation* (i.e., *breakup*) of elements $\dot{n}_{dis}$. The *sink* in the number density of the elements $\dot{n}_{sin}$ can occur due to the *coalescence* or complete *condensation* of bubbles or *evaporation* of droplets. Then a dispersed-phase element number density *conservation equation* can be written as (Kocamustafaogullari and Ishii, 1983)

$$\frac{\partial n_d}{\partial t} + \nabla \cdot \mathbf{u}_d n_d = \dot{n}_{hom} + \dot{n}_{het} + \dot{n}_{dis} - \dot{n}_{sin} \equiv \dot{n}_d \ . \tag{6.54}$$

For two-phase liquid-gas flows, without considering the presence of any bounding surfaces, the rates on the right-hand side of (6.54) can be determined from the thermodynamics (including kinetic theory), hydrodynamics, and heat transfer consideration. In the following, the predictions of $\dot{n}_{hom}$, based on the kinetic theory, for bubbles and droplets, are discussed. The treatment of $\dot{n}_{het}$ requires the specification of the impurities (heterogeneous nucleation sites) and along with $\dot{n}_{dis}$ and $\dot{n}_{sin}$ depends on the particular application and is briefly mentioned next in connection with the interfacial heat transfer.

(C) Homogeneous Nucleation Rates

The *thermodynamic limit* of superheat in liquids, based on the use of the van der Waals equation of state, was discussed in Section 1.5.1(C) and more details are given by Lienhard and Karimi (1981), Skripov (1974), and Avedisian (1992). The limit of supersaturation of the vapor can be obtained in a similar manner (Springer, 1978; Carey, 1992). From the *kinetic theory*, at these limits the predicted nucleation rate of the bubbles or droplets begins to grow very fast (super-exponentially) making the phase change highly probable. These predictions of the volumetric nucleation rate, as a function of the *continuous* phase temperature and pressure, the surface tension, and some molecular properties are reviewed below.

(i) Bubbles

The *kinetics* of the *vapor embryo* formation and growth (or decay) begins with the equilibrium distribution of the embryo size (Carey, 1992). The rate of the growth of the embryo (from n molecules to $n+1$) is obtained and for a *steady-state* process is taken as the rate at which bubbles of a critical radius are generated. This rate of homogeneous nucleation leading to a *spontaneous* growth is given by (Carey, 1992)

$$\dot{n}_{\text{hom}} = n_\ell(\frac{3\sigma}{\pi m})^{1/2} \exp\{-\frac{16\pi\sigma^3}{3k_B T_\ell[\eta p_s(T_\ell) - p_\ell]^2}\} , \tag{6.55}$$

$$\eta = \exp\{\frac{v_\ell[p_\ell - p_s(T_\ell)]}{R_g T_\ell}\} , \tag{6.56}$$

where n_ℓ is the number of liquid molecules per unit volume, σ is the liquid-vapor interfacial tension, m is the mass of a single molecule, k_B is the Boltzmann constant, v_ℓ is the specific volume of the liquid, p_ℓ is the superheated liquid pressure, T_ℓ is the liquid (and vapor) temperature, and $p_s(T_\ell)$ is the saturation pressure at temperature T_ℓ.

Since T_ℓ appears in the exponential term, the homogeneous nucleation rate increases very rapidly as the *kinetic limit* of the *superheat* T_{sp}, given by (1.57) or (6.1), is reached, i.e., $\dot{n}_{\text{hom}}$ given in numbers per m^3 per s becomes very large. Note that (6.1) is a simplified form of (6.55) given for determination of T_{sp}, and a large value of $\dot{n}_{\text{hom}}$ is then prescribed.

Because the liquid is contained or at least in part is in contact with a solid surface, this homogeneous nucleation, which requires a large superheat, is generally *not* observed, and instead the heterogeneous nucleation which requires much *smaller* superheat (for wetting liquids) occurs. The heterogeneous nucleation will be discussed in the three-medium treatment in Section 8.5.1.

(ii) Droplets

The kinetic theory of formulation of *molecular aggregates* leading to liquid embryos and then droplets in supersaturated vapor has been reviewed by Springer (1978) and Carey (1992). In dealing with supersaturated vapors, the *extent* of the supersaturation is generally given by the *supersaturation ratio* S defined as

$$S = \frac{p(T_v)}{p_s(T_v)} . \tag{6.57}$$

This ratio may become as large as 8 for water vapor and is generally smaller for the hydrocarbon (Springer, 1978). As we mentioned earlier, supersaturated (metastable) states can be reached by *isothermal compression*, *isobaric cooling*, *isometric cooling*, or *isentropic expansion.*

The detail derivation and the advances made in the kinetic theory of droplet nucleation and the prediction of the volumetric droplet nucleation rate are given by Springer (1978). The volumetric droplet nucleation rate is given by (Carey, 1992)

$$\dot{n}_{\text{hom}} = (\frac{2\sigma M}{\pi N_A})^{1/2}(\frac{p_v}{k_B T_v})^2 v_\ell \exp\{-\frac{16\pi(\frac{\sigma}{k_B T_v})^3(\frac{M v_\ell}{N_A})^2}{3[\ln\frac{p_v}{p_s(T_v)}]^2}\} , \tag{6.58}$$

where M is the molecular weight, N_A is the Avogadro's number, p_v is the vapor pressure, and T_v is the vapor temperature.

The supersaturation limits obtained from (6.58), by noting the temperature at which $\dot{n}_{\text{hom}}$ begins to grow super-exponentially, is in agreement with both the thermodynamic supersaturation limit (using the *vapor* spinodal limits) and with the experimental results. These as well as a simplified form of (6.58) are discussed by Carey (1992).

The above discussions of the nucleation rates have been limited to single-component fluids. The kinetic theory of bubble formation in the presence of a dissolved gas has been examined by Kwak and Panton (1983). Any impurities, including the dissolved gas in the case of bubble nucleation, tend to *decrease* the superheat and supersaturation limits.

(D) Nonequilibria and Interfacial Heat Transfer

So far under the assumption of no fluid and heat flow, *thermodynamic equilibrium* has been assumed between the dispersed and the continuous phases. In the presence of a fluid flow and a gradient in the temperature, once a dispersed-phase element is formed and moves within the continous phase it exchanges mass (i.e., condensation and evaporation and change in R), momentum (i.e., interfacial drag), and heat with its surroundings (continuous phase and other dispersion-phase elements). The *nonequilibrium* treatment of dispersed gas-liquid systems will be made in detail in Sections 6.2.3 to 6.2.5. Here we deal only with some aspects of the nonequilibrium as a follow-up on the formulation of the dispersed elements through bulk nucleation. The nonequilibrium treatment is made using transport models which allow for *interfacial transport* of mass, momentum, and heat. Among possible models are those allowing for the local vapor-liquid interface to be at the *saturation temperature* (which is a function of the local pressure). Then for the interface being at $T_{\ell g}$, heat exchange can occur between the interface and the dispersed phase as well as between the interface and the continuous phase. The difference between these two heat transfer rates results in a phase change. This modeling of the thermal nonequilibrium was discussed in Section 1.10.3. For example, the volumetric heat from the dispersed phase and the continuous phase to the interface can be modeled using known interfacial Nusselt numbers. If a local dispersed-phase volume-averaged temperature $\langle T\rangle^d$ and a local (but away from the interface) continuous phase temperature T_{c_∞} are used with a spherical dispersed-phase element of radius $R = d/2$ and a continuous-phase volume fraction ϵ, these volumetric interfacial heat transfer rates become

$$\frac{A_{\ell g}}{V} q_{dc} = n_d \frac{6\epsilon}{d} \langle Nu_d\rangle_{A_{\ell g}}^{\text{int}} \frac{k_d}{d} \left(\langle T\rangle^d - T_{\ell g}\right) , \tag{6.59}$$

$$\frac{A_{\ell g}}{V} q_{cd} = n_d \frac{6\epsilon}{d} \langle Nu_d\rangle_{A_{\ell g}}^{\text{ext}} \frac{k_c}{d} \left(T_{c_\infty} - T_{\ell g}\right) . \tag{6.60}$$

The average interfacial Nusselt numbers on the dispersed-phase side

$\langle Nu_d \rangle^{\rm int}_{A_{\ell g}}$ and on the continuous-phase side $\langle Nu_d \rangle^{\rm ext}_{A_{\ell g}}$ are found from the single- or multiple-element analyses (depending on the concentration of the elements). Since *simultaneous* with the interfacial heat transfer, the interfacial mass transfer (i.e., *condensation* or *evaporation*) occurs, then allowance must be made for any change in R. This aspect of the dispersed-phase dynamics requires the consideration of the mass and momentum conservation within each phase as well as on the interface. As expected, a rigorous treatment of the element dynamics is rather involved and complex. Simpler treatments have been made, for example, under assumptions of a *quiescent* continuous phase and an *isolated* element, the elemental growth (or decay) has been studied (e.g., Cha and Henry, 1981). However, for multi-elemental, flowing systems with a temperature gradient within the system, no rigorous treatment is available.

A treatment of the nonequilibrium vapor-liquid flow and heat transfer for *rapid depressurization* of a subcooled liquid, using a constant R (equal to its inception equilibrium value R_e and $\langle T \rangle^d = T_{\ell g}$ (i.e., assuming that $Bi = \langle Nu_d \rangle^{\rm ext}_{A_{\ell g}} k_c/(6k_d)$ is smaller than 0.1) is given by Deligiannis and Cleaver (1990). The interfacial temperature $T_{\ell g}$ is generally taken as the saturation temperature at the local p_c. They assume that the dispersed-phase elements move with the *same* velocity as the continuous phase (i.e.,$\mathbf{u}_c = \mathbf{u}_d$); therefore, the *interfacial drag* is taken as *zero*. Then the mass conservation for both phases, the continuous phase momentum conservation, and the energy conservation equations, i.e., (3.119), (3.120), (3.121), and (3.122), for a dilute concentration of the dispersed-phase elements can be modified as

$$\frac{\partial}{\partial t}\epsilon\rho_c + \nabla\cdot\epsilon\rho_c\mathbf{u} = \frac{A_{\ell g}}{V}\dot{m}_{cd} - \dot{n}_d(\rho V)_d \, , \tag{6.61}$$

$$\frac{\partial}{\partial t}(1-\epsilon)\rho_d + \nabla\cdot(1-\epsilon)\rho_d\mathbf{u} = \dot{n}_d(\rho V)_d - \frac{A_{\ell g}}{V}\dot{m}_{cd} \, , \tag{6.62}$$

$$\frac{\partial}{\partial t}\epsilon\rho_c\mathbf{u} + \nabla\cdot\epsilon\rho_c\mathbf{u}\cdot\mathbf{u} = -\epsilon\nabla p + \nabla\cdot\epsilon\mathbf{S}_c + \epsilon\rho_c\mathbf{g} + \frac{A_{\ell g}}{V}\dot{m}_{cd}\mathbf{u} - \dot{n}_d(\rho V)_d\mathbf{u} \, , \tag{6.63}$$

$$\frac{\partial}{\partial t}\epsilon\rho_c(i_c + \frac{\mathbf{u}\cdot\mathbf{u}}{2}) + \nabla\cdot\epsilon\rho_c(i_c + \frac{\mathbf{u}\cdot\mathbf{u}}{2})\mathbf{u} = \epsilon\frac{\partial p}{\partial t} + \nabla\cdot\epsilon\mathbf{q}_c + \frac{A_{\ell g}}{V}\langle Nu_d\rangle^{\rm ext}_{A_{\ell g}}\frac{k_c}{d}(T_{c_\infty} - T_{\ell g}) + \frac{A_{\ell g}}{V}\dot{m}_{cd}i_c \, , \tag{6.64}$$

$$\frac{\partial}{\partial t}(1-\epsilon)\rho_d i_d + \nabla\cdot\epsilon\rho_d i_d\mathbf{u} = \frac{A_{\ell g}}{V}\langle Nu_d\rangle^{\rm ext}_{A_{\ell g}}\frac{k_d}{d}(T_{\ell g} - T_d) - \frac{A_{\ell g}}{V}\dot{m}_{cd}i_d \, . \tag{6.65}$$

The volume and density of elements, V_d and ρ_d, can in general change with time and space. The modeling of $\dot{m}_{cd}$, $\mathbf{S}_c$, E_d and $\mathbf{q}_c$ and other assumptions (including the treatment of $\dot{n}_d$ discussed above) are needed in order to make these equations solvable. Efforts are continuing in the understanding and the modeling of the equilibrium and nonequilibrium aspects of the bulk nucleation and transport (e.g., Elias and Chambré, 1993).

6.2.3 Isolated, Dispersed-Phase Elements

The heat transfer between the continuous phase (liquid or gas) and a dispersed-phase element (i.e., a *bubble* or a *droplet*) results in the *growth* or *decay* of that element. This phase growth or decay occurs as a result of the interfacial phase change which in turn is a result of the heat and species flows on either side of the interface. For *single-component* systems, this interfacial evaporation or condensation at the interface is governed by the heat transfer *only* with the interfacial temperature generally taken as the *saturation* temperature at the *local continuous* phase pressure $T_{dc} = T_s(p_c)$. For multicomponent systems, the *interfacial concentration* of *species* must be also addressed using the *chemical equilibrium conditions*.

A rendering of the growth or decay of a dispersed-phase element in a *stream* of the continuous phase is given in Figure 6.5. The *change* in the size of the dispersed-phase element, here shown as $R(t)$ for a *spherical* element, is of practical interest and also central to the analysis. There can exist an *internal, cellular motion* within the dispersed-phase element. Then in general the velocity, temperature, and species concentration fields within the dispersed-phase element, $\mathbf{u}_d(\mathbf{x}, t)$, $T_d(\mathbf{x}, t)$, and $\rho_{i_d}(\mathbf{x}, t)$, are time dependent and *nonuniform*. The *local* interfacial heat flux $\mathbf{q}_{dc}$ and $\dot{\mathbf{m}}_{i_{dc}}$ can also vary along the interface, due to the *asymmetry* of the flow, temperature, and concentraion fields *within* and *outside* the element. The interfacial surface-averaged quantities are $\langle q\rangle_{A_{dc}}$, $\langle \dot{m}_{i_{dc}}\rangle_{A_{dc}}$, $\langle T_{dc}\rangle_{A_{dc}}$, and $\langle \mathbf{u}_d\rangle_{A_{dc}}$, and the dispersed-phase volume-averaged quantities are $\langle T\rangle^d$, $\langle \rho_i\rangle^d$, and $\langle \dot{s}\rangle^d$. Around the dispersed phase element, the velocity, temperature, and concentration fields, $\mathbf{u}_c(\mathbf{x}, t)$, $T_c(\mathbf{x}, t)$, and $\rho_{i_c}(\mathbf{x}, t)$, respectively, are different from the free-stream conditions $\mathbf{u}_{c_\infty}$, T_{c_∞}, ρ_{i_∞}, because of the *interaction* of the continuous and dispersed phases. There can also be heat generation sources, which can be nonuniform, within the element and the continuous phase, $\dot{s}_d(\mathbf{x})$ and $\dot{s}_c(\mathbf{x})$. These can be due, for example, to a chemical reaction or an electromagnetic heating.

The treatment of this two-phase system begins with the *pointwise* descriptions (overall mass, species, momentum, and energy conservation equations) in *each* phase and the description of the *interfacial conditions* (overall mass, species, momentum, and energy balances, and generally the thermodynamic equilibrium conditions). However, because of the required *simultaneous hydrodynamic heat transfer* and *thermodynamic considerations*, as well as the *movement* of the *interface*, a *general* treatment of the muticomponent systems with phase growth or decay, even for an isolated element, is rather involved. Here we consider some available and rather specific

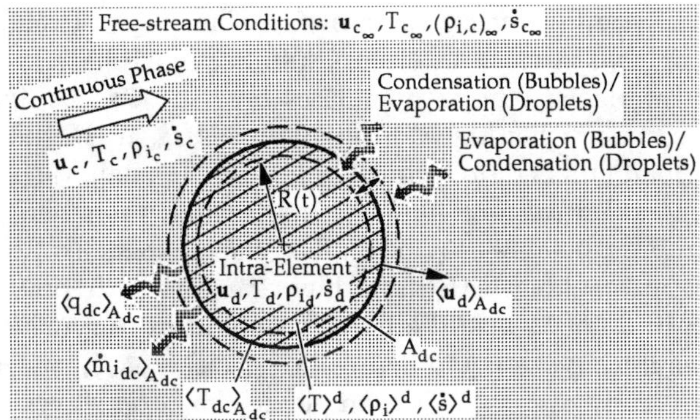

Figure 6.5. A rendering of the growth or decay of a dispersed element (bubble or droplet) in the stream of a continuous phase (liquid or gas).

treatments (and results) of the *bubble* and *droplet phase dynamics*, to illustrate the role of the interfacial heat transfer on the phase dynamics, i.e., $R = R(t)$. In Section 6.3 the phase dynamics of the *bubbles* are discussed, and Section 6.4 is on the *droplets*. The *interactions* between the dispersed elements (bubbles or droplets) are also discussed in these sections, and the bubble- and droplet-phase energy equations are also examined.

6.3 Bubbles

In this treatment of the convective heat transfer aspects of the bubble flow in a host continuous liquid, the behavior of a *single* bubble is discussed first, and this isolation allows for a systemetic inclusion of the various effects such as the relative motion and turbulence. Then the *interbubble* interactions are considered and *local volume-averaged* models for the energy conservation in each phase are given and discussed.

6.3.1 Growth or Decay of Isolated Bubbles

In this section we consider the phase dynamic of an *isolated* bubble element, and in Section 6.2.6 the interactions among the elements will be addressed. Among the *hydrodynamic* aspects of the *continuous phase* (i.e., liquid) is the presence of *turbulence*. The *internal, recirculating motion* of the bubbles are generally *negligible* because of the low Reynolds number (based on the bubble diameter and liquid kinematic viscosity). The bubble can be moving within the continuous phase and this can be a *combined translational* (including *oscillations*) and *rotaional* motion. The bubble can be *growing* due to the *vaporization* in a *superheated* liquid (including *breakup*), or *decaying* due to the *condensation* (including *complete collapse*) in a *subcooled* liquid. The bubble-liquid can be *single-component* or *multicomponent* (and therefore, mass transfer *affects* the growth). The *shape* of the bubble can be spherical or *nonspherical*, as discussed in Section 1.6.1(A).

When the interfacial tension dominates over the viscous, inertial, and buoyancy forces, i.e., when the Morton number Mo, the Reynolds number Re, and the Bond number Bo, defined in (1.61), are small, the shape of the droplet will be *spherical*. For bubbles, the Reynolds number is generally small and for $Bo \leq 0.4$, a *spherical* bubble shape is assumed.

For a spherical bubble with its radius changing with time $R = R(t)$ while flowing in a liquid with the liquid temperature varying in *time* and *space* (such as during the liquid depressurization), i.e., $T_\ell = T_\ell(\mathbf{x}, t)$, only a few detailed analyses exist. For example, in the *early* stages of a bubble growth in an otherwise *quiescent* and *superheated* liquid, the *expansion* of the bubble causes a significant liquid motion which in principle controls the bubble growth during the early stages (Mikic et al., 1970). This *liquid-inertial regime* is later followed by a *heat-transfer-controlled regime* and the *transition* between the regimes is marked by a *critical* bubble radius. Here we briefly consider some features and mechanisms of the bubble growth or decay with the emphasis mostly on those aspects of the phase dynamic controlled by the interfacial heat transfer. These are the heat *diffusion-convection* in the *liquid* phase, the effect of *bubble translation*, and the effects of the bulk liquid-phase *motion* and *turbulence*. Finally, the role of the mass transfer in multicomponent systems is briefly mentioned.

(A) Spherically Symmetric Bubble

For a bubble *growing* (without a translational motion) in an otherwise *quiescent* superheated liquid with a uniform liquid pressure (i.e., uniform liquid superheat), analytical solutions exist for the case of either simple hydrodynamic effects or no hydrodynamic effects. For single-component systems, the growth is due to the convective interfacial heat transfer rate $A_{\ell q} q_{\ell q}$ with the liquid flow pointing away from the interface (due to the vapor expansion) and the heat flow pointing toward the interface (due to the liquid superheat). The interfacial heat flux is given by

$$q_{\ell g} = -k_\ell \frac{\partial T_\ell}{\partial r} = \dot{m}_{\ell g} \Delta i_{\ell g} = \rho_v \frac{dR}{dt} \Delta i_{\ell g} \quad \text{on } A_{\ell g} \, . \tag{6.66}$$

A Nusselt number can be defined as

$$\langle Nu_d \rangle^{\text{ext}}_{A_{\ell g}} = \frac{q_{\ell g} d}{(T_{\ell_\infty} - T_s) k_\ell} = \frac{2\rho_v \Delta i_{\ell g} R}{k_\ell} \frac{1}{T_{\ell_\infty} - T_s} \frac{dR}{dt} \, , \tag{6.67}$$

which as will be shown, will be *constant* during the *heat-transfer-controlled regime* of the bubble growth.

The spherical, *radially symmetric* (i.e., r-direction variations only) forms of the overall mass, species, momentum, and energy conservation equations applicable to a control volume around the moving interface have been discussed by Burmeister (1983) and Arpaci and Larsen (1984). The general treatment of the liquid momentum equation, including the liquid viscous force, the surface tension, the gas-phase density change, and the liquid inertial force is discussed by Scriven (1959), van Stralen (1968), Cho and Seban (1969), Prosperetti and Plesset (1978), Payvar (1987), and Arefmanesh et al. (1992).

Here we first consider the bubble growth regime *controlled by* the *heat transfer* and only use the continuity equation for the *liquid phase* given by

$$\frac{1}{r^2}\frac{\partial}{\partial r}r^2 v_\ell = 0 \quad r \geq R\,. \tag{6.68}$$

The mass balance at the bubble surface gives

$$\frac{d}{dt}\rho_v\frac{4}{3}\pi R^3 = \rho_\ell 4\pi R^2[\frac{dR}{dt} - v_{\ell_i}(R)]\,. \tag{6.69}$$

As is generally the case, the rate of change in ρ_v is much *smaller* than the rate of change in R, and therefore, this equation can be used to give the *liquid* radial interfacial velocity v_{ℓ_i} as

$$v_{\ell_i}(R) = \frac{dR}{dt}\frac{\rho_\ell - \rho_v}{\rho_\ell}\,, \tag{6.70}$$

where in general $\rho_v \gg \rho_g$

Now taking the liquid interfacial radial velocity v_{ℓ_i} as dR/dt and when this is used as the liquid velocity at R, the continuity equation (6.68) gives

$$v_\ell = (\frac{R}{r})^2\frac{dR}{dt}\,. \tag{6.71}$$

The thermal energy equation for the liquid phase can be written as

$$\frac{\partial T_\ell}{\partial t} + v_\ell\frac{\partial T_\ell}{\partial r} = \frac{\alpha_\ell}{r^2}\frac{\partial}{\partial r}r^2\frac{\partial T_\ell}{\partial r} \quad r \geq R\,. \tag{6.72}$$

The boundary and initial conditions are

$$T_\ell = T_s(p_\ell) \quad r = R\,,\ t > 0\,, \tag{6.73}$$

$$T_\ell = T_{\ell_\infty} \quad r \to \infty\,,\ t > 0\,, \tag{6.74}$$

$$T_\ell = T_{\ell_\infty}\,,\ R = 0 \quad r \geq 0\,,\ t = 0\,. \tag{6.75}$$

A *large* Jakob number Ja (defined below) asymptotic solution to (6.71) and (6.72), subject to (6.73) to (6.75) and (6.66), exists and is (Prospretti and Plesset, 1978)

$$R(t) = (\frac{12}{\pi})^{1/2}Ja(\alpha_\ell t)^{1/2}\,,\quad Ja = \frac{(T_{\ell_\infty} - T_s)(\rho c_p)_\ell}{\rho_v\Delta i_{\ell g}}\,. \tag{6.76}$$

Note that since the radius grows as $t^{1/2}$, the Nusselt number defined by (6.67) remains constant in the heat-transfer-controlled regimes, i.e.,

$$\langle Nu_d\rangle^{\rm ext}_{A_{\ell g}} = \frac{12}{\pi}Ja \quad \text{heat-transfer-controlled regime.} \tag{6.77}$$

A more general solution that satisfies *both* the *liquid-inertial* regime and the *heat-transfer* regime *asymptotes* is derived by Mikic et al. (1970) and is given in *dimensionless* form as

$$R^*(t^*) = \tfrac{2}{3}[(t^*+1)^{3/2} - t^{*3/2} - 1] , \tag{6.78}$$

where

$$R^* = \frac{a_1}{{a_2}^2} R \ , \ \ t^* = \frac{{a_1}^2}{{a_2}^2} t \ ,$$
$$a_1 = [\frac{2(T_{\ell_\infty} - T_s)\Delta i_{\ell g}\rho_v}{3T_s\rho_\ell}]^{1/2} \ , \ \ a_2 = (\frac{12\alpha_\ell}{\pi})^{1/2} Ja \ . \tag{6.79}$$

From (6.78) the liquid-inertial-controlled regime (i.e., $t^* \ll 1$) gives (Mikic et al., 1970)

$$R^* = t^* \tag{6.80}$$

and for the heat-transfer-controlled regime (6.78) reduces to (6.76). The above general solution includes the *variation* in the *vapor density* and this effect has also been included in the analysis of Theofanous and Patel (1976).

The above discussions, formulations, and solutions were restricted to the case of constant T_{ℓ_∞} with no translational motion of the bubble. These are rather restrictive assumptions since in general the liquid temperature is not uniform, nor does the bubble remain stationary.

The decay (and the ultimate collapse) of the bubble in single-component systems, with the liquid *subcooled* by *pressurization*, has been analyzed by Cho and Seban (1969). In their analysis, the *motion* of the *vapor* inside the bubble is allowed and the inclusion of the vapor motion shows that the temporal decrease of the radius is *not* monotonic, but oscillatory. With this allowance in the motion of the vapor, the conservation equations for the vapor phase (again assuming radial symmetry) become (Cho and Seban, 1969)

$$\frac{\partial \rho_v}{\partial t} + \frac{1}{r^2}\frac{\partial}{\partial r} r^2 \rho_v v_v = 0 \ , \tag{6.81}$$

$$\rho_v(\frac{\partial v_v}{\partial t} + v_v \frac{\partial v_v}{\partial r}) = -\frac{\partial p_v}{\partial r} + \frac{4}{3}\mu_v(\frac{\partial^2 v_v}{\partial r^2} + \frac{2}{r}\frac{\partial v_v}{\partial r} - 2\frac{v_v}{r^2}) \ , \tag{6.82}$$

$$(\rho c_p)_v \frac{\partial T_v}{\partial t} + (\rho c_p)_v v_v \frac{\partial T_v}{\partial r} = \frac{k_v}{r^2}\frac{\partial}{\partial r} r^2 \frac{\partial T_v}{\partial r} + \frac{\partial p_v}{\partial t} + v_v \frac{\partial p_v}{\partial r} \ . \tag{6.83}$$

The liquid momentum equation, in addition to the continuity and energy equations given by (6.68) and (6.72), is

$$\frac{\partial v_\ell}{\partial t} + v_\ell \frac{\partial v_\ell}{\partial r} = -\frac{1}{\rho_\ell}\frac{\partial p_\ell}{\partial r} \ , \tag{6.84}$$

where the viscosity terms are eliminated using the liquid phase continuity equation.

The interfacial conditions are (Cho and Seban, 1969)

$$\rho_v(v_{v_i} - \frac{dR}{dt}) = \rho_\ell(u_{\ell_i} - \frac{dR}{dt}) \ , \tag{6.85}$$

$$p_v - p_\ell = \frac{2\sigma}{R} + \rho_v(v_v - \frac{dR}{dt})(v_\ell - v_v) + \frac{4}{3}\mu_v(\frac{\partial v_v}{\partial r} - \frac{v_r}{r}) - \frac{4}{3}\mu_\ell(\frac{\partial v_\ell}{\partial r} - \frac{v_\ell}{r}) , \tag{6.86}$$

$$k_\ell \frac{\partial T_\ell}{\partial r} - k_v \frac{\partial T_v}{\partial r} = \rho_v(\frac{dR}{dt} - v_v)[\Delta i_{\ell g} + \frac{(v_v - \frac{dR}{dt})^2}{2} - \frac{(u_\ell - \frac{dR}{dt})^2}{2} + \frac{4}{3}\frac{\mu_\ell}{\rho_\ell}(\frac{\partial v_\ell}{\partial r} - \frac{v_\ell}{r}) - \frac{4}{3}\frac{\mu_v}{\rho_v}(\frac{\partial v_v}{\partial r} - \frac{v_v}{r})] . \tag{6.87}$$

In the above, the force balance is the simplified and appropriate form of (1.131). A *perfect-gas* behavior is assumed, and the Clapeyron equation (1.51), when applied to the *interface* and by assuming a saturated state condition, gives

$$p_{v_i} = p_{v_o} \exp\{\frac{\Delta i_{\ell g}}{R_g T_o}(1 - \frac{T_o}{T_i})\} . \tag{6.88}$$

where p_{v_o} and T_o are at a *reference* state.

Using the liquid momentum equation along with the liquid continuity equation and the interfacial force balance and integrating the resultant from $r = R$ to $r \to \infty$, Cho and Seban (1969) obtain the *bubble dynamic equation*

$$(1 - \frac{\rho_v}{\rho_\ell})R\frac{d^2R}{dt^2} + [2 - \frac{1}{2}(1 - \frac{\rho_{v_i}}{\rho_\ell})](1 - \frac{\rho_{v_i}}{\rho_\ell})(\frac{dR}{dt})^2 + \frac{d}{dt}(1 - \frac{\rho_{v_i}}{\rho_\ell})R(\frac{dR}{dt} - v_{v_i}) + \frac{\rho_{v_i}}{\rho_\ell}R\frac{\partial}{\partial t}v_{v_i} + \frac{\rho_{v_i}}{\rho_\ell}(1 + \frac{\rho_{v_i}}{\rho_\ell})\frac{dR}{dt}v_{v_i} - \frac{1}{2}(\frac{\rho_{v_i}}{\rho_\ell})^2 {v_{v_i}}^2 = \frac{p_{\ell_i} - p_{\ell_\infty}}{\rho_\ell} . \tag{6.89}$$

Note that for $\rho_v/\rho_\ell \to 0$ this equation becomes

$$R\frac{d^2R}{dt^2} + \frac{3}{2}(\frac{\partial R}{\partial t})^2 = \frac{p_{\ell_i} - p_{\ell_\infty}}{\rho_\ell} , \tag{6.90}$$

which is known as the *Rayleigh equation*

The complete set of equations (6.68), (6.72), (6.81) to (6.88) , and (6.89), subject to several approximations and simplifications, have been solved numerically for a spherical bubble collapsing in a subcooled liquid, by Cho and Seban (1969), where the results and discussions, including the *oscillatory* behavior of the decay, can be found. Note that unlike the recirculating motion discussed in Sections 6.1.2(B), only *radial motion* is allowed in the bubble. Based on this radial diffusion-convection, an *internal* Nusselt number $\langle Nu_d \rangle^{\text{int}}_{A_{\ell g}}$ can be defined. However, since the interfacial temperature is taken as $T_s = T_s(T_{\ell_\infty})$, this internal conductance of the heat flow does *not* directly influence the heat flow *into* the liquid, but influences the external conductance through the change in R which depends on the change with sensible heat content of the bubble.

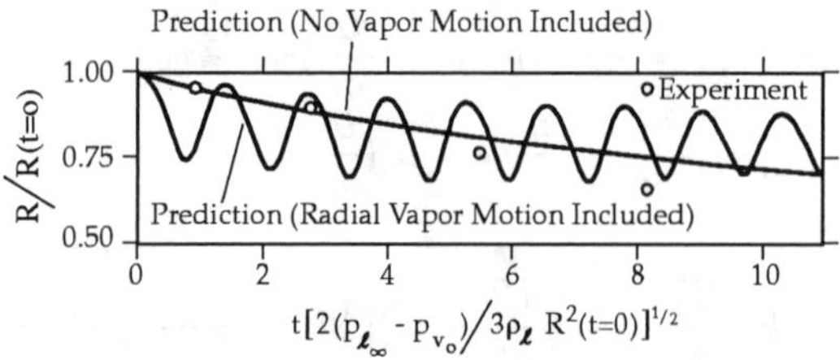

Figure 6.6. Predicted oscillatory and nonoscillatory decay of a vapor bubble in a subcooled liquid. The available experimental results are also shown. The variation of the normalized bubble radius is given with respect to the normalized time. (From Cho and Seban, reproduced by permission ©1969 American Society of Mechanical Engineers.)

As an example, some typical results for the *oscillatory* bubble decay obtained by Cho and Seban (1969) and their comparison with an available experiment are shown in Figure 6.6. The *normalized* bubble radius (with respect to the initial radius) is shown as a function of the normalized time (using, among other parameters, the initial vapor pressure p_{v_o}). The results are for a steam-water system and the conditons are given by Cho and Seban (1969). The experimental results agree more closely with the analysis based on *no* vapor motion (also shown). This can be due to the *departure* of the bubble shape from spherical which can in turn change the interfacial heat transfer rate and *prevent* oscillation.

(B) Effect of Relative Motion

The growing or decaying, single-component bubble discussed above was assumed to be stationary, while in practice (when the gravitational acceleration is significant) the bubble *ascends* due to phase-density buoyancy. As a result the interfacial heat transfer rate will be altered and the above assumed radial symmetry will *not* be valid. The effect of the bubble translation on the interfacial heat transfer rate (i.e., the *increase* in the interfacial heat transfer rate, and therefore, the growth or decay rate) is expected to be significant for high *translation* velocities u_d, i.e., high bubble Reynolds numbers $Re_d = u_d d/\nu_\ell$. The *terminal velocity* for spherical bubbles of constant radius is the same as that given for droplets by (6.2). The *dynamic* bubble velocity can be found using the general particle force balance given by (5.143), and a sample computation is made by Gopalakrishna and Lior (1992). Because of the interfacial *tangential* and *radial* velocities associated with the fluid particle phase change, modifications of (5.143) are needed. However, presently these are *not* available. The liquid flow around a rising (and growing or decaying) spherical bubble has been approximated by a *potential-flow* solution (Tokuda et al., 1968), where the error in the interfacial velocity associated with the zero liquid viscosity is expected to be of the order of $Re_R^{-1/2}$.

The potential-flow solution for the distributions of the *tangential* and *radial* components of the liquid velocity, w_d and v_d, are given as functions of u_d, r, and θ (assuming azimuthal symmetry) by Tokuda et al. (1968) as

$$w_\ell = u_d(1+\frac{R^3}{2r^3})\sin\theta \ , \tag{6.91}$$

$$v_\ell = u_d(1-\frac{R^3}{r^3})\cos\theta + (\frac{R}{r})^2\frac{dR}{dt} \ . \tag{6.92}$$

The liquid thermal energy equation, with assumed azimuthal symmetry, becomes

$$\frac{\partial T_\ell}{\partial t} + \frac{w_\ell}{r}\frac{\partial T_\ell}{\partial \theta} + v_\ell\frac{\partial T_\ell}{\partial r} = \frac{\alpha_\ell}{r^2}\frac{\partial}{\partial r}r^2\frac{\partial T_\ell}{\partial r} + \frac{\alpha_\ell}{r^2\sin\theta}\frac{\partial}{\partial\theta}\sin\theta\frac{\partial T_\ell}{\partial\theta} \ . \tag{6.93}$$

The initial and boundary conditions are

$$T(r,\theta) = T_{\ell_\infty} \qquad t=0 \ , \tag{6.94}$$

$$T(\theta,t) = T_{\ell_\infty} \qquad r\rightarrow\infty \ , \tag{6.95}$$

$$\begin{aligned} T(\theta,t) = T_s \ , \quad & \frac{\partial R}{\partial t}\rho_v = (\frac{dR}{dt} - v_{\ell_i})\rho_\ell \ , \\ & \frac{dR}{dt}\rho_v\Delta i_{\ell g} = \frac{1}{2}\int_0^\pi k_\ell\frac{\partial T_\ell}{\partial r}\sin\theta d\theta = \\ & = \langle Nu_d\rangle_{A_{\ell g}}^{\text{ext}}\frac{(T_s - T_{\ell_\infty})k_\ell}{d} \qquad r = R \ . \end{aligned} \tag{6.96}$$

The solutions to above hydrodynamic-heat transfer formulation, along with the thermodynamic condition (6.88), have been obtained using the boundary-layer *approximations* (i.e., neglecting the diffusion along the θ-direction). Also assumed is a *thermal boundary-layer thickness* $\delta = \delta(\theta,t)$ around the bubble in the liquid phase, and a similarity space variable $\eta = r/\delta(\theta,t)$. These solutions have been discussed by Tokuda et al. (1968), Ruckenstein and Davis (1971), Moalem and Sideman (1973), and Dimić (1977). *Long* and *short* time asymptotic solutions, *quasi-steady* (neglecting the effect of temporal change in R) solutions, as well as solutions for *large* values of the Péclet number $Pe_d = Re_d Pr_\ell = u_d d/\alpha_\ell$, have been found. In the quasi-steady solution and for $Pe_d \gg Ja$, the solution becomes nearly that for the *solid* spheres. For inviscid flows, as was discussed in Section 4.2, the liquid Prandtl number (i.e., the ratio of the thermal to viscous boundary layer thicknesses) does *not* appear as a dimensionless parameter and the normalized, average interfacial heat transfer rate $\langle Nu_d\rangle_{A_{\ell g}}^{\text{ext}}$ depends on Pe_d.

Moalem and Sideman (1973) assume $Pe_d \gg 1$ and $dR/dt \ll u_d$, and give

$$\langle Nu_d\rangle_{A_{\ell g}}^{\text{ext}} = \frac{2}{\pi}Pe_d^{1/2} \qquad Pe_d \gg Ja \ . \tag{6.97}$$

Then once u_d is known from the bubble force balance, this normalized interfacial heat transfer rate can be used in (6.96) to determine dR/dt. This is done by numerical integration starting from a known R at $t=0$. Further discussions on the effect of Ja and Pe on the interfacial heat transfer rate are given by Mayinger et al. (1991).

(C) Turbulent Liquid Motion

In the above treatments, first no translation motion of the bubble was allowed and then this translational motion (which is generally due to phase-density buoyancy) was included. Laminar (and generally inviscid) *liquid motion* has been assumed. Here, first a *combined* bubble buoyancy-bulk liquid motion is addressed as a simple extension of the laminar translational motion and then the effect of *turbulence* in the liquid motion (which is characteristic of multibubble systems) is addressed. The goal is to point out the rather significant effect that the bulk liquid-phase flow characteristics can have on the interfacial heat transfer rate, i.e., on $\langle Nu_d \rangle^{\text{ext}}_{A_{\ell g}}$.

For a *bubble* rising with the velocity $\mathbf{u}_b$ in a liquid with a bulk liquid motion $\mathbf{u}_{\ell_\infty}$, the *relative* velocity is $\mathbf{u}_d = \mathbf{u}_b - \mathbf{u}_{\ell_\infty}$. Then the Péclet number in (6.97) can be used with $\mathbf{u}_d = |\mathbf{u}_b - \mathbf{u}_{\ell_\infty}|$. This extension has been discussed by Moalem and Sideman (1973) for *cross flows* (i.e., $\mathbf{u}_d$ being upward against the gravity vector and $\mathbf{u}_{\ell_\infty}$ being in the horizontal direction).

For turbulent liquid flows, the heat transfer in the liquid phase adjacent to the bubble is enhanced by the turbulent fluctuations. This is similar to the turbulent liquid-gas interfacial heat transfer (and phase change) discussed in Section 6.2.1 for the continuous interface. In the experiments of Bankoff and Mason (1962), a liquid (subcooled water) jet of *mean* velocity $\overline{\mathbf{u}}_{\ell_\infty}$ impinges on a orifice through which bubbles (steam) are released and then bubbles move upward with a *mean* velocity $\overline{\mathbf{u}}_b$. After a growth due to the depressurization, the bubbles begin to decay and gradually collapse. The interfacial heat transfer rate for this turbulent liquid jet flow and the ascending bubbles have been correlated by Bankoff and Mason (1962) using a general relation

$$\langle \overline{Nu_d} \rangle^{\text{ext}}_{A_{\ell g}} = \langle \overline{Nu_d} \rangle^{\text{ext}}_{A_{\ell g}}(Pe_d, Sr) , \tag{6.98}$$

$$Sr = \frac{|\overline{\mathbf{u}}_b|}{|\overline{\mathbf{u}}_{\ell_\infty}|} , \quad Pe_d = \frac{|\overline{\mathbf{u}}_b| d}{\nu_\ell} . \tag{6.99}$$

For *ellipsoidal* bubbles with *smooth surfaces* and for this counterflow of the liquid jet and bubbles, Bankoff and Mason (1962) suggest

$$\langle \overline{Nu_d} \rangle^{\text{ext}}_{A_{\ell g}} = 3.74 \times 10^{-4} (Pe_d)^{1.74} Sr . \tag{6.100}$$

This shows a strong dependence on Pe_d and, when compared to (6.97), results in a much *larger* interfacial heat transfer rate. *Oscillations* of the bubbles with frequencies up to 2500 Hz have been observed by Bankoff and Mason (1962). This can be due to the vapor hydrodynamics, as discussed in Section (A) above in relation to the bubbles with no translational motion. No analysis of the effect of the liquid turbulence on the bubble turbulent liquid interfacial heat transfer is available, although in principle treatments such as those in Section 6.2.1 can be extended to single bubbles (with the vapor-side hydrodynamics and heat transfer included).

(D) Multicomponent Systems

The bubble and the liquid may contain more than one constituent, and the average species concentration in the bubble and liquid may be different than the *equilibrium* concentration. This *lack* of *chemical equilibrium* results in the flow of species within the gas and the liquid as well as across the interface. In general, the presence of a second and less volatile species tends to *decrease* the rate of growth or decay of the bubble. In Section 1.5.2 equilibrium aspects of the multicomponent, multiphase systems were discussed. Here, we first discuss the relation between the *equilibrium* concentrations of a species (in a multicomponent system) in the gas and liquid phases. As in the above discussions of the single-component systems in which the saturation temperature $T_s = T_s(p_{\ell_\infty})$ was assumed at the interface, for the multicomponent systems an equilibrium concentration relation is assumed at the interface between the gas phase (i.e., a *partial or vapor pressure*) and the liquid phase (i.e., a *solute concentration*). After examining the interfacial equilibrium condition, the deriving force for the mass transfer from the bulk of each phase towards the interface will be formulated using *nonuniform* distributions of the species concentration in each phase.

The thermodynamic equilibrium partial pressure of the solute species A in the gas phase for a *dilute solution* is considered. For a dilute solution the concentration of solute species A in the liquid phase is *small* such that force experienced by the molecules of the solute species A is *different* from if there were only molecules of the species A. This was mentioned in Section 1.5.2. Then this partial pressure at a temperature T and a pressure p (total pressure) is p_i and is determined from the equilibrium conditions and the assumed ideal gas behavior. We consider a *binary* system (a more general treatment is given by Levine, 1988) with an *ideally dilute solution.* The chemical equilibrium conditions are given by (1.58). For an ideally *dilute solution*, the *solute* A chemical potential is given by (Levine, 1988)

$$\mu_{A,\ell} = (\mu_{A,\ell})_o + R_g T \ln y_{A,\ell} \ , \tag{6.101}$$

where $y_{A,\ell}$ is the *mole* fraction of the species A in the liquid phase and the subscript $_o$ refers to a *reference* state. The chemical potential for the gas phase with an ideal behavior is given by

$$\mu_{A,g} = (\mu_{A,g})_o + R_g T \ln \frac{p_A}{p_o} \ . \tag{6.102}$$

Using the chemical equilibrium among the phases, as given by (1.58) and by inserting the above two equations, we have the *phase equilibrium relation*

$$(\mu_{A,\ell})_o + R_g T \ln y_A = (\mu_{A,g})_o + R_g T \ln \frac{p_A}{p_o} \tag{6.103}$$

or

$$\frac{p_A}{p_o y_A} = e^{\frac{\mu_{A,\ell}-\mu_{A,g}}{R_g T}} \tag{6.104}$$

or

$$p_A \equiv K_A(T) y_{A,\ell} \, , \tag{6.105}$$

where the last expression is called the *Henry law* and the coefficient K_A is called the *Henry law constant* (Levine, 1988). The Henry law (6.105) is similar to the Raoult law (1.60), and both state that the partial pressure of the species A is *proportional* to the mole fraction in the solution. However, in the Henry law the proportionality constant is *not* the vapor pressure of the pure species A. The Henry law can be written in terms of the *mass fraction* for a binary system by using a *modified* proportionality constant $K_A^{'}$ in place of K_A. Then the phase equilibrium relation can be written as

$$p_A = K_A^{'} (\frac{\rho_A}{\rho})_\ell \, . \tag{6.106}$$

A further discussion, including the nonideal gas behavior, will be given in Section 6.2.5(A).

The *isothermal*, *mass-transfer-controlled* bubble growth has been analyzed by Payvar (1987) and Arefmanesh et al. (1992), among others. For the isothermal liquid flow and mass transfer the liquid velocity v_ℓ is obtained (for the case of no translational motion) in the same manner as discussed in Section (A). Using this velocity, the mass transfer is analyzed starting from the species conservation equation in the liquid phase given by

$$\frac{\partial}{\partial t} \frac{\rho_{A,\ell}}{\rho_\ell} + v_\ell \frac{\partial}{\partial r} \frac{\rho_{A,\ell}}{\rho_\ell} = \frac{D_\ell}{r^2} \frac{\partial}{\partial r} r^2 \frac{\rho_{A,\ell}}{\rho_\ell} \qquad r \geq R \tag{6.107}$$

with the boundary and initial conditions

$$\rho_{A,\ell} = (\rho_{A,\ell})_\infty \qquad r \to \infty, \ t > 0 \, , \tag{6.108}$$

$$\rho_{A,\ell} = p_A \frac{\rho_\ell}{K_A^{'}} \, , \quad \frac{d}{dt} \frac{4}{3} \pi \rho_g R^3 = \rho_\ell D (\frac{\partial}{\partial r} \frac{\rho_{A,\ell}}{\rho_\ell}) 4\pi R^2 \quad r = R, \ t > 0 \, , \tag{6.109}$$

$$p = p_o \, , \quad p_A = p_{A,o} \, , \quad \rho_{A,\ell}(r) = (\rho_{A,\ell})_\infty \, , \quad R = R(t=0) \quad t = 0. \tag{6.110}$$

A *sudden* change in the pressure $p = p(t)$ causes $p_A = p_A(t)$ and this in turn leads to $\rho_{A,\ell} = \rho_{A,\ell}(R,t)$. Then through the Henry law, the mass transfer between the interface and the bulk liquid begins. The species concentration boundary layer is found from (6.107) and is similar in behavior to the thermal boundary layer discussed in Section (A) above. In the problem formulated above the mass transfer in the *liquid side* has been analyzed for the bubble growth. The *combined* heat and mass transfer during the growth requires the simultaneous solution of the energy and species conservation equations with the inclusion of the requirements for the thermal, mechanical, and chemical equilibria.

Moalem and Sideman (1993) consider the *nonisothermal* mass transfer in the decay of bubbles in subcooled liquids, but assume that one of the gas components (in a binary gas mixture) is *noncondensible*. This simplifies the analysis, and the *gas side* resistance to the condensation rate of the volatile component becomes the mass-transfer *control* mechanism.

6.3.2 Interbubble Interactions

The presence of other moving bubbles influences the liquid flow field around a moving bubble, and in a nonisothermal field, where the bubbles undergo growth or decay, the temperature field around a bubble is also influenced by the neighboring bubbles. For the case of phase-density buoyant motion only, where the bubbles travel vertically upward against the gravitational vector, the *condensation* of *bubble trains* (i.e., an in-line arrangement of bubbles) has been studied by Moalem et al. (1973). For single-component systems, the bubbles traveling in a *subcooled liquid* (of the same species) will eventually *collapse*. The distance from the point of release of a bubble of radius $R = R(t = 0)$ to the location when the bubble reaches a *prescribed* fraction of this initial radius is designated with L. Then if the bubbles are released at a *frequency* f and if the *modified* (i.e., due to the presence of the other bubbles) *ascending velocity* of the bubbles is designated with u_{b_m}, and also assuming that they *all* travel with the *same* modified velocity, then the *distance* between the center of two neighboring bubbles is u_{b_m}/f and *remains* constant. Note that from (6.2) the *terminal* velocity depends on the radius, and the assumption of constant u_{b_m} implies that the *inertial* and *viscous* forces will *prevent* the *deceleration* of the bubbles to lower velocities as they travel upward. These and other assumptions described by Moalem et al. (1973) lead to the modeling of the ascending bubble trains as n *stationary* bubbles with the subcooled *liquid flowing* over them along the gravity vector with a velocity $|u_{\ell_\infty}| = |u_{b_m}|$. This model is shown in Figure 6.7(a). The modified bubble velocity is determined from a *geometric* model which allows for a velocity *increase* since each bubble is partly in the wake of the bubble leading it. The modified phase-density buoyant bubble velocity u_{b_m} is given by

$$\frac{u_{b_m}}{u_b} = \frac{R^2}{R^2 - 0.99r_w^2} , \tag{6.111}$$

where r_w is the radius of the projected contact area between the wake of the *leading* bubble and the *rear* bubble. The rear bubble is taken as the bubble with $R = R(t = 0)$, and r_w is determined from f, $R(t = 0)$, and R_{N-1}. This is done by iteration (along with the entire hydrodynamic and heat transfer solutions), because R_{N-1} is not known a priori. Note that

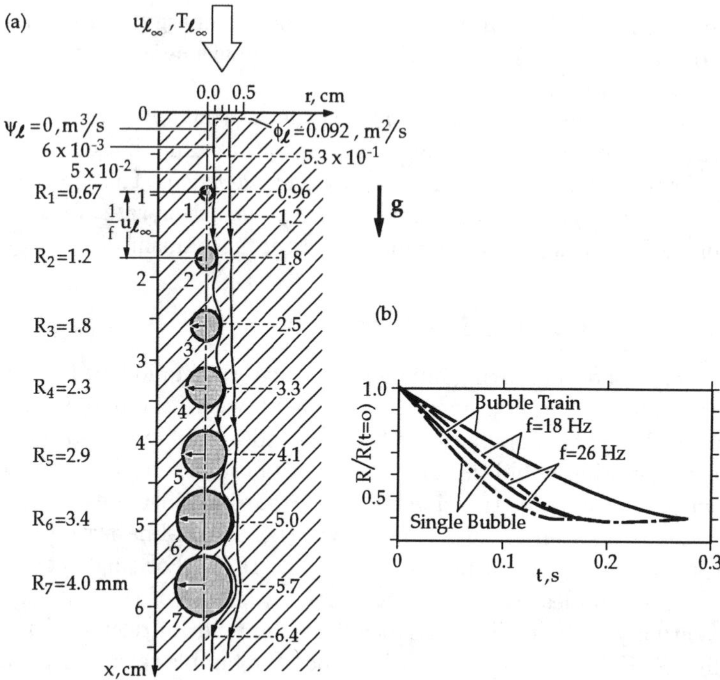

Figure 6.7. (a) A bubble train condensing in a subcooled liquid. The bubbles are treated as stationary and the liquid is moving. The distance between the bubbles is uniform and constant. The computed distributions of the liquid stream function and the velocity potential and radii of the bubbles are also shown. (b) Decay of the bubble radius for a single bubble and a bubble in a train of six bubbles at two different bubble release frequencies. (From Moalem et al., reproduced by permission ©1973 Pergamon Press.)

$u_{b_m} > u_b$, i.e., in this model the wake allows the bubbles to have higher terminal velocities.

For this quasi-steady treatment, the flow field around *stationary* bubbles is analyzed by assuming an *inviscid* liquid flow. The methods of *doublets* and *dipoles* employed in the potential flow theory have been used by Moalem et al. (1973) to determine the distributions of the *liquid stream function* ψ_ℓ and *liquid potential function* ϕ_ℓ around the bubble assemblage (i.e., the bubble train). Figure 6.7(a) shows their computed potential flow field for an assemblage of seven bubbles. Since the bubble radius depends on the interfacial heat transfer, which in turn depends on the hydrodynamics, the determination of the radius requires the *simultaneous* solutions to the *momentum* and *energy equations* in the liquid phase. The *iteration* procedure is described by Moalem et al. (1973). For the results shown in Figure 6.7(a), the bubble radius has reached about 1/6 of its initial value in less than 5 cm. The heat transfer is determined by using the quasi-steady and axisymmetric assumptions and the boundary layer approximations (expected to be valid for Péclet numbers, $Pe_d = u_{b_m} d/\alpha_\ell$, larger than 1000). The details of the *numerical* solution given by Moalem et al. (1973).

When compared to the results for *single bubbles*, i.e., the results discussed in Section 6.3.1(B), the effect of the presence of the neighboring bubbles can be examined through the reduction in the radius with respect to time (or the distance in the model used here). Figure 6.7(b) shows typical results of Moalem et al. (1973) for a single-component system with the *final* value of $R/R(t=0) = 0.4$ and $N = 6$. The results are for pentane and for two different frequencies, $f = 18$ and 26 Hz. At low frequencies (up to 12 to 14 Hz), the bubble interaction is through heat transfer only as the trailing bubbles travel through the region heated by the leading bubbles. This results in a *reduction* in the condensation rate. As the frequency increases, the bubbles enter the wake regions of the leading bubbles, and as given by (6.111), the modified velocity is *higher* than that for a single bubble. This hydrodynamic *enhancement* tends to *nullify* the local increase in the temperature. Then as shown in Figure 6.7(b), for $f = 16$ Hz the condensation rate is *lower*, but for $f = 26$ Hz the condensation rate is nearly *equal* to that for a single bubble.

6.3.3 Bubble and Liquid-Phase Conservation Equations

A generalization of the above interacting, single column of bubbles can be made to the volumetric distribution of the bubbles, and then a local volume-averaged treatment such as that given in Section 6.1.3 can be made. For the *gravity-assisted bubble-liquid heat exchangers*, the bubbles can be moving upward due to buoyancy while the liquid can flow in the *same* direction (i.e., *co-current* flow) or opposite to them (i.e., *counter* flow). Then the *relative* motion of the bubbles can be used for the determination of the interfacial heat transfer rate (or the Nusselt number). Sideman and Moalem (1974) consider multibubble systems and for a swarm of bubbles they first *modify* the bubble velocity to obtain $\mathbf{u}_{b_m}$ and then alter for the *relative* motion of the liquid using the volume-averaged liquid velocity $\langle \mathbf{u}_\ell \rangle$. Their result for the *relative* bubble velocity is

$$|\mathbf{u}_{b_m} - \mathbf{u}_\ell| = \frac{u_b[1-(1-\epsilon)^{5/3}][a \pm (1-\epsilon)u_\ell]}{\epsilon^2(1-\epsilon)}$$
$$+ \text{ counter } - \text{ co-current flow}, \qquad (6.112)$$

where u_b is the terminal bubble velocity given by (6.2), ϵ is the liquid (continuous phase) volume fraction, and a is a *geometric* constant.

Then they recommend using this relative velocity along with the results for the single bubble given in Section 6.3.1(B) to arrive at the surface area-averaged Nusselt number

$$\langle Nu_d \rangle_{A_{\ell g}}^{\text{ext}} = \frac{2}{\pi} Pe_d^{1/2} \qquad Pe_d \gg Ja\,, \qquad (6.113)$$

where

$$Pe_d = \frac{|\mathbf{u}_{b_m} - \mathbf{u}_\ell| d}{\alpha_\ell} \,. \tag{6.114}$$

As was discussed in Section 6.3.1(B), this result for the Nusselt number is also based on the inviscid liquid-phase flow and the axisymmetric, boundary-layer treatment of the heat transfer.

Note that in the presence of *turbulence* in the liquid phase, as discussed in Section 6.3.1(C), the Nusselt number given by (6.113) is not valid and use must be made of the applicable correlations, for example, (6.100).

Now similar to the treatment of the dispersed *monosize* liquid drops in an immiscible liquid, given in Section 6.3.1, energy equations for the liquid and the bubble (i.e., dispersed) phase can be written by noting that here the radius changes with time (or position). Also assume that there is *no* internal (i.e., *intrabubble*) resistance to the heat flow and that all bubbles are at the *saturation* temperature T_s such that the entire interfacial heat transfer results in phase change. Based on these, the liquid phase energy equation and the interfacial energy balance which are the appropriate forms of (3.121), (6.66), and (6.69) become

$$\epsilon(\rho c_p)_\ell \frac{\partial \langle T\rangle^\ell}{\partial t} + (\rho c_p)_\ell \langle \mathbf{u}_\ell \rangle \cdot \nabla \langle T\rangle^\ell = \nabla \cdot [\langle \mathbf{K}_e \rangle + \epsilon(\rho c_p)_\ell \mathbf{D}_\ell^d] \cdot \nabla \langle T\rangle^\ell + \frac{6(1-\epsilon)}{d} \langle Nu_d \rangle_{A_{\ell g}}^{\text{ext}} \frac{k_\ell}{d} (T_s - \langle T\rangle^\ell) - \frac{6(1-\epsilon)}{\pi d^3} \frac{d}{dt} \frac{\pi d^3}{6} \rho_v i_\ell \,, \tag{6.115}$$

$$\frac{d}{dt} \frac{\pi d^3}{6} \rho_v \Delta i_{\ell g} = \pi d^2 \langle Nu_d \rangle_{A_{\ell g}}^{\text{ext}} \frac{k_\ell}{d} (\langle T\rangle^\ell - T_s) \,. \tag{6.116}$$

If the spatial radiation of the saturation temperature is allowed, then the bubble phase energy equation should also be included.

The liquid mass and momentum conservation equations are used for the determination of ϵ and $\langle \mathbf{u}_\ell \rangle$. These can be the simplifid forms of the equations discussed in Section 3.3.2 with the phase change properly included. For multicomponent systems, the volume-averaged species equations and the extra thermodynamic relations must be added. Sideman and Moalem (1974) solve the above equations for a quasi-steady bubble flow and neglect the diffusion term in the liquid phase. They also examine the role of the liquid flow *direction* on the heat exchanger performance and determine the role of the *noncondensibles* in the *multicomponent systems*.

The general *hydrodynamic* treatment of the bubble-liquid flows can be made using the space-time averaged momentum equations discussed in Section 3.3.2, and a model for the bubble-phase stress tensor, for *laminar* high Reynolds number flows, is given by Sangani and Didwania (1993). For the case of *turbulent* liquid flow, additional equations describing the Reynolds shear stress tensor and other averaged turbulent quantities are required. An example of the two-equation model of turbulence applied to the bubble-liquid flow is given by Gosman et al. (1992).

6.4 Droplets

As was done in the treatment of the bubbles, here the discussion of the convective heat transfer aspects of the droplets moving in a continuous host gas begins with the examination of a *single* droplet. Then after discussing the related thermodynamic aspects, the growth or decay of a stationary (spherically symmetric) droplet is considered. This is followed by the examination of the effects of the *droplet rotation*, *relative motion*, *internal droplet circulation*, *transient behavior*, *simultaneous evaporation/condensation*, and *layered droplet structure*, on the growth or decay of an *isolated* droplet. Then the *interdroplet* interactions are discussed, and the *local volume-averaged* models for the conservation equations, for each phase, are discussed.

6.4.1 Growth or Decay of Isolated Droplets

Although *both* the droplet growth (caused by *condensation*) and decay (caused by *evaporation*) are encountered in practice, *droplet evaporation* has been studied most extensively. This is in connection with the liquid-fuel sprays (droplet size between 10 and 1000 μm) in combustion systems (Sirignano, 1983). Here, we examine *isolated* droplets (elements), and in Section 6.2.6 the interaction among the elements will be addressed. In general, the *gas* phase is *multicomponent*, while the *liquid* droplets can be initially *single* component. When the evaporation rate is *very high*, the penetration (this is not only by diffusion, because of the internal droplet motion) of the other species from the gas phase into the liquid phase may be *negligible*. At high temperatures and pressures, the *nonideal*, *thermodynamic* behavior (i.e., nonideal behavior models) for multicomponent gas-liquid systems, must be addressed. We have briefly mentioned this in Section 6.2.4(D) and will further address this below. Then the analysis of the droplet evaporation begins by assuming quasi-steady, *stationary* droplets and examining the normalized interfacial heat transfer rate $\langle Nu_d \rangle^{\text{ext}}_{A_{\ell g}}$ for *spherically symmetric* droplets. The effect of these droplet *rotation* on $\langle Nu_d \rangle^{\text{ext}}_{A_{\ell g}}$ is then added (rotation with *frequencies* of order of 10^3 Hz occurs in spray combustion applications) to this spherically symmetric limit.

Next the relative motion and its effect on *convection* in the *gas* phase is addressed ($Re_d = u_d d/\mu_g$ is between zero and a few hundreds in spray combustion applications) beginning with no internal droplet circulation and the analysis of the *front stagnation region*. Then the effects of the *internal droplet* circulation on $\langle Nu_d \rangle^{\text{ext}}_{A_{\ell g}}$ and the *internal* resistance to heat flow $\langle Nu_d \rangle^{\text{int}}_{A_{\ell g}}$ are addressed. Since the droplet size changes during the evaporation (droplet life time of the order of 10^{-3} s in spray combustion), and because of the initially subcooled state of the liquid, the *transient effects* can become important, and therefore the *transient behavior* of evaporating droplets is examined next. Some special aspects of multicomponent systems, i.e., *combined* (or *simultaneous*) *evaporation* and *condensation* and *layered* droplets, are discussed at the end of this section.

(A) THERMODYNAMIC MODEL OF MULTICOMPONENT GAS–LIQUID SYSTEMS

The perfect gas behavior assumed in the discussion of Section 6.2.4(D) is not realized in dealing with the high-temperature and high-pressure multicomponent droplet evaporation, and therefore in *some* analyses a *nonideal* gas behavior is introduced (Curtis and Farrell, 1992; Delplanque and Sirignano, 1993). The *fugacity* of species A in the gas phase $f_{A,g}$ is related to the chemical potential through (Levine, 1988)

$$\mu_{A,g} = (\mu_{A,g})_o + R_g T \ln \frac{f_{A,g}}{p_o} . \tag{6.117}$$

When compared to (6.102) which is for ideal gas behavior and valid for $p \to 0$, we have the *asymptotic* condition which is the equality of the fugacity and the partial pressure for $p \to 0$, i.e.,

$$f_{A,g} \to p_A \quad \text{or} \quad \frac{f_{A,g}}{p_A} \to 1 \quad p \to 0 . \tag{6.118}$$

Deviation from this perfect gas behavior is shown through the departure of $f_{A,g}/p_A$ from unity, and this ratio can be *smaller* than unity at lower pressures and much *larger* than unity at high pressures (Levine, 1988). For convenience, the chemical equilibrium condition (1.58) for gas–liquid systems can be replaced with the equality of fugacities, i.e., for a *binary system*

$$f_{A,\ell} = f_{A,g} \; , \quad f_{B,\ell} = f_{B,g} \; . \tag{6.119}$$

The relation between f and the temperature and the volume is (Levine, 1988; Delplanque and Sirignano, 1993)

$$R_g T \ln \frac{f_{A,g}}{p_A} = \int_V^\infty \left(\frac{\partial p}{\partial N_A} - \frac{R_g T}{V}\right) dV - R_g T \ln Z \; , \tag{6.120}$$

where Z is the *compressibility factor* defined by (1.19), and N_A are the number of moles of species A in the volume V. At *high pressures* (especially near the *critical* point) the heat of vaporization for the vapor formation from a *liquid mixture* becomes *larger* than that for a *pure* liquid (Delplanque and Sirignano, 1993) and the specific *heat* of *vaporization* of species A found from

$$\Delta i_{\ell g} = \frac{R_g T^2}{M_A} \frac{\partial}{\partial T} \ln \frac{f_{A,g}/Y_{A,g}}{f_{A,\ell}/Y_{A,\ell}} . \tag{6.121}$$

Instead of the van der Waals equation of state, (1.17) or (1.52), other equations of state are used to relate p, T, and V (or v). One of these is the

Peng-Robinson equation of state (Curtis and Farrell, 1992), i.e.,

$$p = \frac{R_g T/M}{v - b_1} - \frac{a_1(T)}{v(v+b_1) + b_1(v - b_1)} . \tag{6.122}$$

The other is the modified *Redlich-Kwong* equation of state (Delplanque and Sirignano, 1993)

$$p = \frac{R_g T/M}{v - b_1} - \frac{a_1}{T^{1/2} v(v+b_1)} , \tag{6.123}$$

where $a_1(T)$, a_1, and b_1, are related to fundamental molecular properties. The thermodynamic relation (6.120), the evaluation of $\Delta i_{\ell g}$ using (6.121), and the equation of state (6.122) are used mostly in the *numerical simulations* of the droplet evaporation. The results of these simulations will be discussed in Section (F) below and in Section 6.4.2. In *simplified* analyses leading to *closed-form* solutions, the ideal gas behavior is used. Examples are the treatments in Sections (B) to (E) below.

(B) Spherically Symmetric Droplets

For the case of a droplet evaporating in a *quiescent gas* with the center of the droplet remaining *stationary* and the phase-density buoyant and thermobuoyant motions neglected, a *quasi-steady* (i.e., assuming that the change in the radius of the droplet does *not* affect the transport processes) treatment is made leading to the determination of the rate of evaporation. In droplet combustion, the reaction and the heat generation which supplies the heat for the evaporation must be included. Here we assume that the ambient is at a higher temperature than the droplet. For a *radially* symmetric and steady droplet evaporation of species A, the simplified forms of the conservation equations for the overall mass (1.66), species (1.72), and energy (1.83) for a perfect gas (assuming that the gas flow is small enough such that an isobaric condition can be imposed) becomes

$$\frac{1}{r^2}\frac{d}{dr} r^2 \rho_g v_g = 0 \ , \ \ r^2 \rho_g v_g = r^2 \dot{m}_g = \text{constant} \ , \tag{6.124}$$

$$\frac{1}{r^2}\frac{d}{dr} r^2 v_g \rho_{A,g} = \frac{1}{r^2}\frac{d}{dr} r^2 \rho_g D_g \frac{d}{dr}\frac{\rho_{A,g}}{\rho_g} , \tag{6.125}$$

$$\frac{1}{r^2}\frac{d}{dr} r^2 v_g (\rho c_p)_g T_g = \frac{1}{r^2}\frac{d}{dr} r^2 k_g \frac{d}{dr} T_g , \tag{6.126}$$

$$\rho_g T_g = \text{constant} \ , \tag{6.127}$$

where the enthapy transport due to the diffusion of species, as discussed in connection with (1.85), has been assumed to be *negligible*. As will be discussed in Section (D) below, this contribution can become *significant* in high-temperature applications. The boundary conditions at the droplet surface and far into the quiescent ambient are

$$T_g = T_s \ , \ \ \rho_{A,g} = \rho_{A,g}[T_s, (Y_{A,\ell})_s] \equiv (\rho_{A,g})_s$$
$$q_{g\ell} = -k_g \frac{dT_g}{dr} = \Delta i_{\ell g} \rho_g v_g = \Delta i_{\ell g} \dot{m}_{\ell g} \ ,$$
$$\dot{m}_{\ell g} = \dot{m}_g = \rho_{A,g} v_g - \rho_g D \frac{d}{dr} \frac{\rho_{A,g}}{\rho_g} \qquad r = R \ , \tag{6.128}$$

$$T_g = T_\infty \ , \ \ \rho_{A,g} = (\rho_{A,g})_\infty \ , \ \ \rho_g = \rho_{g\infty} \qquad r \rightarrow \infty \ . \tag{6.129}$$

For single-component liquids, at the liquid surface the vapor density $\rho_{A,g}$ will be a function of T_s only. The gas density and the temperature at any location r are related the rough (6.127). Also assumed in the boundary conditions at $r = R$ is that the liquid droplet is at a *uniform* temperature such that the entire heat flowing from the gas results in the evaporation.

The general solution uses the Shvab-Zel'dovich transformation discussed in Section 2.3.3, and is discussed by Williams (1985). For the case of only evaporation (i.e., no reaction) and constant properties, the quasi-steady evaporation rate $4\pi R^2 \dot{m}_{\ell g}$ for this case of a stationary droplet, $u_d = 0$, is given by

$$4\pi R^2 \dot{m}_{\ell g} = 4\pi \rho_g D_g R \ln(1 + \Delta B) \qquad u_d = 0 \ , \tag{6.130}$$

where the *transfer number* is similar to that used in Section 4.11.1(A), and here it is defined as

$$\Delta B = \frac{c_{p_g}(T_\infty - T_s)}{\Delta i_{\ell g}} \ . \tag{6.131}$$

The normalized heat flux given as the Nusselt number, which is uniform around the drop, is

$$\langle Nu_d \rangle_{A_{\ell g}}^{\mathrm{ext}} = \frac{q_{g\ell} d}{(T_\infty - T_s) k_g} = \frac{2 c_{p_g} \rho_g D_g}{k_g} \frac{\Delta i_{\ell g}}{c_{p_g}(T_\infty - T_s)} \ln(1 + \Delta B) =$$
$$\frac{2}{\Delta B Le_g} \ln(1 + \Delta B) = \frac{2}{Ja \, Le_g} \ln(1 + Ja) \ , \tag{6.132}$$

where the *Jakob number* Ja becomes the same as the transfer number and the gas *Lewis number* Le_g is as before,

$$Le_g = \frac{\alpha_g}{D_g} = \frac{Sc_g}{Pr_g} \ . \tag{6.133}$$

When the liquid is not at a uniform temperature (i.e., subcooled droplets), then the transfer number ΔB can be *modified* to include the heat flowing in the liquid (if this quantity is known). This is done similarly to that in Section 4.1.2 and is given (Sirignano, 1983)

$$\Delta B = \frac{c_{p_g}(T_\infty - T_s)}{\Delta i_{\ell g} + \dfrac{k_\ell}{\dot{m}_{\ell g}} \left.\dfrac{dT_\ell}{dr}\right|_R} \; . \tag{6.134}$$

A more detailed treatment of the heat and mass transfer aspects of the spherically symmetric droplet, which includes the nonideal gas behavior and require a numerical integration, is given by Curtis and Farrell (1992) and Delplanque and Sirignano (1993).

(C) Droplet Rotation

Although the inevitable *rotation* of droplets moving *relative* to the ambient gas may influence the heat transfer and the evaporation rate, the effect of rotation of a droplet with its center *remaining* stationary is considered here. This problem has been analyzed by Lozinski and Matalon (1992) as a *perturbation* around the spherically symmetric solution just discussed and provides for addressing the effect of the droplet spin. However, when the effects of the droplet translation, Section (D), the internal motion of droplet, Section (E), the internal and external transient behaviors, Section (F), and the gas-phase turbulence are added to the effect of the rotation, the effect of rotation on the evaporation rate is not expected to be very noticeable.

Consider a single evaporating spherical liquid droplet rotating at an angular frequency ω about its axis with its surface at temperature T_s and placed in otherwise quiescent ambient gas at T_∞. The flow within the droplet interior corresponds to a *rigid body* rotation and is given by (Lozinski and Matalon, 1992)

$$u_\ell = \omega r \sin\theta \, , \tag{6.135}$$

$$p_\ell = \rho_\ell \omega^2 r^2 \frac{\sin^2\theta}{2} + p_o \, , \tag{6.136}$$

where u_ℓ is the tangential velocity and p_o is determined by balancing the normal stress components at the interface. The details of the gas-phase motion induced by the droplet rotation is given by Lozinski and Matalon (1992). The coordinate system and a typical result for the streamlines corresponding to the motion caused by the rotation *only* (i.e., the radial motion caused by evaporation should be superimposed on this motion), are shown in Figure 6.8. The velocity far from the droplet shows how *heat* is *convected towards* the droplet at the *poles* and *convected away* from it at the *equator*. This *enhances* the evaporation rate at the poles while hindering it at the equators. Lozinski and Matalon (1992) show that the spatially *averaged* (integrated over the surface area) rate of vaporization is *increased* with a ω^4 dependency. The results for the average enhancement of the vaporization

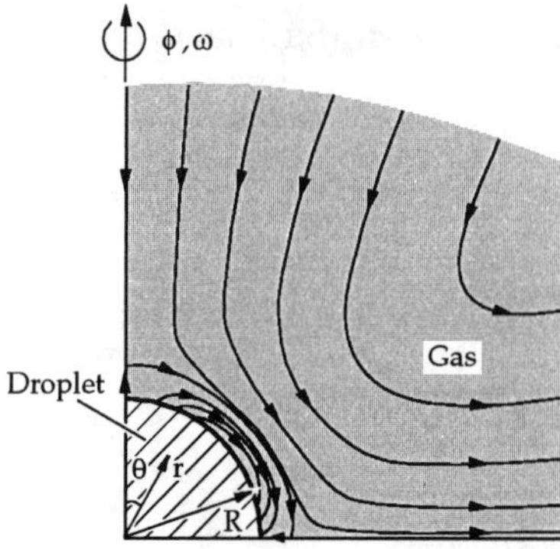

Figure 6.8. The streamlines around a droplet rotating at an angular frequency ω with its center remaining stationary. The coordinate system used is also shown. (From Lozinski and Matalon, reproduced by permission ©1992 The Combustion Institute.)

rate is given by

$$\frac{\langle \dot{m}_{\ell g} \rangle_{A_{\ell g}}}{\langle \dot{m}_{\ell g} \rangle_{A_{\ell g}}(\omega = 0)} = 1 + (\frac{R^2 \omega}{\nu_g})^4 {Pr_g}^{7/5} f(\Delta B, \frac{T_s}{T_\infty}) . \quad (6.137)$$

The magnitude of $f = f(\Delta B, T_s/T_\infty)$ is not large, and therefore rather large angular frequencies are needed for a noticeable enhancement.

(D) Effect of Relative Motion on Case-Phase Convection

The interfacial heat transfer between an isolated solid sphere and a fluid, with a *relative velocity* $\mathbf{u}_\infty$, was discussed in Section 5.1.2. The *front stagnation region* was analyzed assuming a *steady, laminar boundary-layer* flow and heat and mass transfer (i.e., high Reynolds numbers). Here that discussion is *extended* to droplet evaporation by noting that the interfacial mass transfer is *coupled* to the heat transfer through the *normal interfacial gas-phase velocity* which is similar to *surface injection* velocity (discussed in Section 4.9). The variations in the thermophysical properties and the heat transfer by radiation to the droplet can also be included. Harpole (1981) has analyzed the *front stagnation region* by assuming a *zero velocity* at the drop surface [the internal flow will be discussed in Section (E) below], and by using empiricism has arrived at the *surface-area*-averaged interfacial heat transfer rate. His analysis is reviewed below.

Neglecting thermo- and diffusobuoyancy and using the *relative* velocity

$$\mathbf{u}_\infty = \mathbf{u}_{g_\infty} - \mathbf{u}_d \ , \quad (6.138)$$

the azimuthally symmetric temperature and velocity fields are analyzed subject to the boundary-layer approximations. Figure 6.9 gives a schematic of the problem considered and shows a gas of *relative velocity* $\mathbf{u}_\infty$, temper-

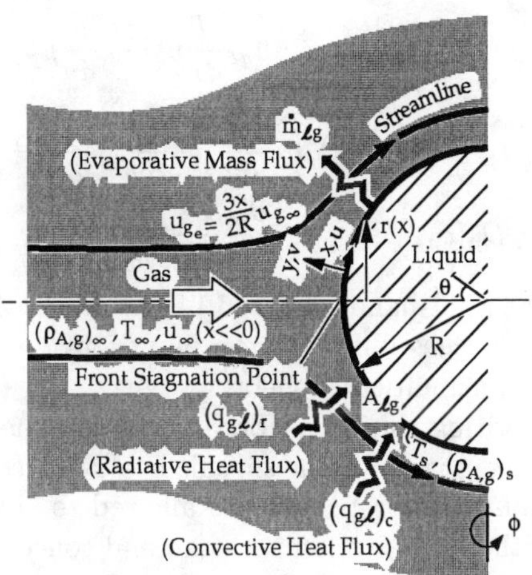

Figure 6.9. A rendering of the front stagnation region of a droplet in a gas stream. The potential flow solution for velocity around the droplet, along with the coordinate system, is also shown.

ature T_∞, and species concentration (assuming a *binary* system) $(\rho_{A,g})_\infty$, approaching a sphere with a radius R, a surface temperature T_S, and a gas-phase surface concentration $(\rho_{A,g})_s$. By assuming a zero velocity at the surface, a uniform surface temperature, and that *all* the heat transferred to the surface results in evaporation, the liquid phase transport is *decoupled* from the gas phase. The effect of the change in R on the gas-phase transport is also neglected so that a *quasi-steady* treatment is possible. The transient behavior is discussed in Section (F) below. For the front stagnation region, a free-stream (at the edge of the viscous boundary layer) velocity is given by the potential flow solution as

$$u_{g\ell} = \frac{3x}{2R} u_{g\infty} \, . \tag{6.139}$$

As also shown in Figure 6.9, the coordinate axis x runs along the surface, y is perpendicular to the surface, and r is measured from the line passing through the center of the sphere and parallel to $\mathbf{u}_\infty$. The convective heat transfer $q_{\ell g}$ varies along x and so does the evaporation mass flux $\dot{m}_{\ell g}$. Symmetry with respect to the azimuthal axis is assumed and the variations with respect to polar (or collatitude) angle θ are now shown with respect to x. The overall mass, species, momentum, and energy conservation equations in this coordinate system and subject to the laminar boundary-layer approximations become (Harpole, 1981)

$$\frac{\partial}{\partial x} r \rho_g u_g + \frac{\partial}{\partial y} r \rho_g v_g = 0 \, , \tag{6.140}$$

$$\rho_g u_g \frac{\partial}{\partial x} \frac{\rho_{A,g}}{\rho_g} + \rho_g v_g \frac{\partial}{\partial y} \frac{\rho_{A,g}}{\rho_g} = \frac{\partial}{\partial y} \rho_g D_g \frac{\partial}{\partial y} \frac{\rho_{A,g}}{\rho_g} \, , \tag{6.141}$$

$$\rho_g u_g \frac{\partial u_g}{\partial x} + \rho_g v_g \frac{\partial u_g}{\partial y} = \rho_{g\infty} u_{ge} \frac{d}{dx} u_{ge} + \frac{d}{dy} \mu_g \frac{\partial u_g}{\partial y} \,, \tag{6.142}$$

$$(\rho c_p)_g u_g \frac{\partial T_g}{\partial x} + (\rho c_p)_g v_g \frac{\partial T_g}{\partial y} = \frac{\partial}{\partial y} k_g \frac{\partial T_g}{\partial y} + \rho_g D_g (c_{p_{A,g}} - c_{p_{B,g}}) \frac{\partial T_g}{\partial y} \frac{\partial}{\partial y} \frac{\rho_{A,g}}{\rho_g} \,. \tag{6.143}$$

The last term on the right-hand side of (6.143) is due to the enthalpy transport caused by the species diffusion in a binary gas, as discussed in connection with (1.85) for multicomponent systems. Also, (1.64) for a *binary* system is used in writing this transport in terms of the species diffusion. Note that for equal constant-volume specific heats this contribution diminishes. The thermophysical properties are allowed to change with respect to position (i.e., with respect to temperature and concentration, because a *uniform pressure* is assumed since the pressure variation determined from u_{g_e} is not significant).

Assuming a single-component liquid, then $(\rho_{A,g})_s = (\rho_{A,g})_s(T_s)$ and no mass flux of species B occurs across the interface. The local interfacial mass flux due to the evaporation of species A is $\dot{m}_{\ell g}$ and is given by the boundary condition (6.128) which can be written as

$$v_{g_s} = \frac{q_{\ell g}}{\rho_g \Delta i_{\ell g}} = -\frac{D_g}{1 - \dfrac{\rho_{A,g}}{\rho_g}} \frac{\partial}{\partial y} \frac{\rho_{A,g}}{\rho_g} \qquad y = 0 \,. \tag{6.144}$$

The *local* interacial heat flux has a convective component $(q_{g\ell})_c$ and a radiative component $(q_{g\ell})_r$, i.e.,

$$q_{\ell g} = (q_{\ell g})_c + (q_{\ell g})_r = k_g \frac{\partial T_g}{\partial y} + (q_{\ell g})_c \quad y = 0 \,, \tag{6.145}$$

$$\frac{q_{\ell g}}{(q_{\ell g})_c} = 1 + \frac{(q_{\ell g})_r}{(q_{\ell g})_c} \equiv 1 + \beta \,. \tag{6.146}$$

The ratio β defined above is *prescribed*. The remainder of the boundary condition given in (6.128) apply here and the velocity boundary conditions

$$\mathbf{u} = 0 \text{ at } y = 0 \,, \quad \mathbf{u} \to \mathbf{u}_\infty \text{ for } y \to \infty \,, \tag{6.147}$$

are added.

Similarity solutions can be obtained when the spatial transformation (Mangler transformation) is made to reduce the above to a set of two-dimensional boundary-layer equations with r, x, and y transformed to two variables. Harpole (1981) discusses this and the introduction of the dimensionless stream function which allows for the density variation. This solution allows for the variation of $\rho_g k_g$ through a variable γ defined as

$$\gamma = \frac{\ln \dfrac{(\rho_g k_g)_\infty}{(\rho_g k_g)_s}}{\ln \dfrac{T_s}{T_\infty}} . \tag{6.148}$$

Heat transfer results for the front stagnation region have been obtained by Harpole (1981) and by comparison with the existing experimental results, and assuming the *dominance* of the front portion heat transfer, an extension has been made to the *entire* surface area. The results for the *interfacial area-averaged Nusselt number* (for convective heat transfer only), i.e,

$$\langle Nu_d \rangle_{A_{\ell g}}^{\rm ext} = \frac{(q_{\ell g})_c d}{(T_\infty - T_s) k_{g_\infty}} \tag{6.149}$$

is given as a function of various variables. One of these variables is the *modified transfer number* defined as

$$\Delta B = \frac{c_{p_\infty}(T_\infty - T_s)}{\Delta i_{\ell g}}(1+\beta) \quad 0 \le \beta \le 1 , \tag{6.150}$$

where, as was mentioned, β is the ratio of the radiative to the convective heat flux, (6.146), and is prescribed. The others are the Reynolds, and Prandtl numbers, i.e.,

$$Re_d = \frac{u_\infty d}{\nu_{g_\infty}} , \quad Pr = \frac{\nu_{g_\infty}}{\alpha_{g_\infty}} . \tag{6.151}$$

As was discussed in Section 5.1.2, in the front stagnation region the local Nu_d varies with $Re_d{}^{1/2}$ and Harpole suggests a $Pr^{0.38}$ relation (compared to the $Pr^{1/3}$ characteristic of the flat plate boundary layer flow and heat transfer). He correlates the results for $\langle Nu_d \rangle_{A_{\ell g}}^{\rm ext}$ using Pr, Re_d, ΔB, γ, and T_s/T_∞. His correlation is

$$\langle Nu_d \rangle_{A_{\ell g}}^{\rm ext} = (1.56 + 0.626 Pr^{0.38} Re_d^{1/2})(1 +$$

$$a\Delta B)^{-0.7}[1 + (0.327\gamma - 0.0844)(1 - \frac{T_s}{T_\infty})] \quad 100 \le Re_d \le 1000 , \tag{6.152}$$

where a is a constant (which is 1.60 for *water* evaporating in the air). This relationship shows the effects of the surface *mass transfer* (i.e., $v_{g_s} > 0$, which is similar to the *blowing effect* discussed in Section 4.5.2) as signified by ΔB, and the *property variations* as signified through γ and T_s/T_∞. The radiative heat transfer changes ΔB through β in (6.150).

Note that (6.152) is valid for Re_d of the order of 100 and that the $Re_d \to 0$ and $\Delta B \to 0$ asymptote (i.e., *heat diffusion* only) for $\langle Nu_d \rangle_{A_{\ell g}}^{\rm ext}$, which is 2, is *not* obtained from (6.152). The $Re_d = 0$ asymptote with $\Delta B \ne 0$ is given by (6.132).

(E) Internal Droplet Circulation

Since the interfacial temperature (and concentration) is determined from the thermodynamics conditions, the effect of the internal droplet motion on the gas-droplet heat transfer is through the *fraction* of heat transferred from the gas to the droplet used for raising the temperature of the droplet (compared to the fraction resulting in the surface evaporation). To determine this *sensible heat* fraction, the *simultaneous* treatment of the liquid and the gas phase transport is required. In Section 6.1.2(B), the internal motion of an isolated fluid particle in relative motion with a host, immiscible fluid was briefly mentioned. In droplet evaporation, the *transient* distribution of the droplet temperature depends on the strength of this internal circulation. As was mentioned in Section 6.1.2(C), the internal vortical motion causes rapid circulation (along the streamlines), but because the streamlines are closed the core of the vortex can only be heated by *diffusion*. This internal flow and its effect on the evaporation rate is further discussed below.

The transient temperature distribution within the droplet, during evaporation, is influenced by the liquid motion. A *complete* transient solution to the coupled gas and liquid phase hydrodynamics and heat and mass transfer (for laminar flows) has been made, and the numerical results will be discussed in Section (F). Here are some *approximate* descriptions of the internal droplet flow.

The gas flow around the droplet is axisymmetric, and the internal flow of the liquid is also axisymmetric. For $Re_d = u_\infty d/\nu_g$ of order of 100, which is characteristic of the spray combustion applications (based on the initial droplet diameter), there will *not* be a *before-after* symmetry (with respect to a plane perpendicular to the flow and passing through the center of the droplet). This is true for both the gas- and the liquid-phase hydrodynamics as shown in Figure 6.2, which is for a set of parameters, $Re_d = 100$, $\mu_\ell/\mu_g = 55$, and $\rho_\ell/\rho_g = 790$, and representative of a droplet in relative motion with a gas. However, in the existing analyses the liquid flow has been assumed to *have* a before-after symmetry (e.g., Prakash and Sirignano, 1978). The *Hill spherical vortex* has this symmetry, and a *model* of the droplet hydrodynamics which has this symmetry is shown in Figure 6.10.

The stream function ψ_ℓ and the *vorticity* ω_ℓ in the coordinate shown in Figure 6.10 are given by (Clift et al.,1978; Prakash and Sirignano, 1978)

$$\psi_\ell = -\tfrac{1}{2}\Gamma_\ell r^2(R^2 - r^2)\sin\theta\,, \qquad (6.153)$$

$$\omega_\ell = 5\Gamma_\ell r\sin\theta\,, \qquad (6.154)$$

where Γ_ℓ represents the *strength* of the vortex. Note that because a spherical coordinate system is used, ψ_ℓ is in m^3/s and Γ_ℓ is in 1/m-s. The magnitude of Γ_ℓ has been determined for liquid-gas systems and is (Prakash and Sirignano, 1978) *approximated* by

$$\frac{u_\infty}{\Gamma_\ell R^2} = 4 \text{ to } 10\,. \qquad (6.155)$$

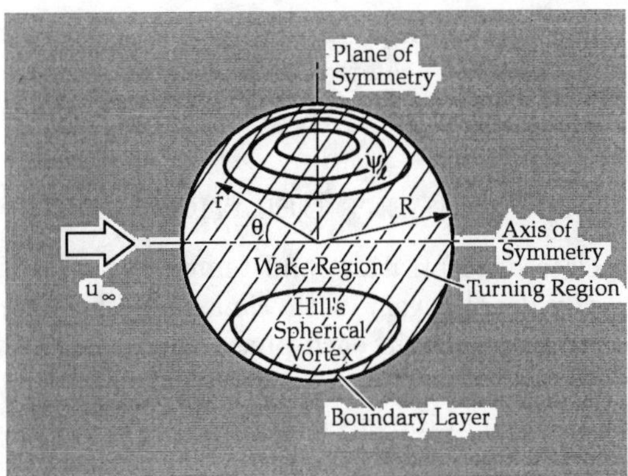

Figure 6.10. The internal motion of a droplet in relative motion with the surrounding gas. The Hill's symmetrical vortex model and the various hydrodynamic regions in the droplet are also shown. (From Prakash and Sirignano, reproduced by permission ©1978 Pergamon Press.)

The characteristic liquid velocity (in the droplet) is *approximated* by

$$u_\ell = \Gamma_\ell R^2 = u_\infty/(4 \text{ to } 10)\ . \tag{6.156}$$

As is shown in Figure 6.10, a viscous (and thermal) boundary layer is formed adjacent to the surface of the droplet, and after a *turning region* there is a *wake region* in the middle of the *vortex ring*. Further discussions, including the before-after *asymmetry*, are given by Prakash and Sirignano (1978).

For liquids (including light fuels) the Prandtl number is of the order of 10, and therefore, $Pe_d = Re_d Pr$ of the order of 10^3 are encountered. From the analysis of Section 4.2 for solid surfaces, the thermal boundary-layer thickness is proportional to $Pe_d^{-1/2} Pr^{1/6}$, and therefore, it can be fairly *small*. Across this thin thermal boundary layer Prakash and Sirignano assume a *quasi-steady* heat flow, and in the *thermal core* (in the vortex). They allow for the *temporal* and *spatial* changes of the temperature. Then using the liquid flow model mentioned above, and by assuming that the iso-streamlines and isothermal lines coincide, they examine the solution to the energy equations

$$\mathbf{u}_\ell \cdot \nabla \Gamma_\ell = \alpha_\ell \nabla^2 \Gamma_\ell \quad \text{thermal boundary layer}\ , \tag{6.157}$$

$$\frac{\partial T}{\partial t} - \mathbf{u}_\ell \cdot \nabla \Gamma_\ell = \alpha_\ell \nabla^3 \Gamma_\ell \quad \text{thermal core.} \tag{6.158}$$

The boundary conditions at the surface and at the interface between the two regions are written accordingly.

As was mentioned in Section 6.1.2(D), the presence of the liquid motion *reduces* the droplet heat-up time, which is estimated by R^2/α_ℓ for purely diffusive heating, by a factor of about 1/10. Prakash and Sirignano (1978) show that after $0.05R^2/\alpha_\ell$ the temperature distribution within the droplet becomes nearly linear. Typical predictions of the temperature distribution based on this flow and the heat transfer model will be examined in Section (F) below.

(F) Predicted Transient Behavior

The velocity, species concentration, and temperature distributions in the gas and within the droplet vary with *time*, and the *time scales* associated with *significant* changes in each phase and in any of these variables also vary greatly. Scaling analysis based on u_∞, R, and the thermophysical properties of the liquid and the gas have been made by Prakash and Sirignano (1978, 1980), Sirignano (1983), and Haywood et al. (1989). Based on the results of these analyses, various approximations have been made to the governing conservation equations. Some of these approximations have been reviewed in Sections (B) to (E) above. For *single-component liquids* evaporating into a *binary gas*, the overall mass, species, momentum (r and θ directions, with the assumption of axisymmetry), and thermal energy conservation equations which can be solved numerically in the primitive variables (i.e., ρ_g, $\rho_{A,g}$, u_g, v_g, p_g, u_ℓ, v_ℓ, p_ℓ, and T_g). For the *liquid* phase, the use of the stream function and the vorticity reduces the conservation equations to those for the vorticity-stream function and the thermal energy which are solved numerically for ψ_ℓ, ω_ℓ and T_ℓ. These gas and liquid conservation equations have been solved *without* the use of the boundary-layer approximations and in the *transient* form by Haywood et al. (1989), Chiang et al. (1992), and Chiang and Kleinstreuer (1993). They also allow for the variations in the thermophysical properties. Full sets of equations, including the *droplet force balance* (i.e., the Lagrangian treatment of the droplet momentum conservation) giving the transient variation of the relative velocity of the droplet (because R decreases and also because the drag force on the droplet retards the gas flow) are given by Haywood et al. (1989). These include the initial and boundary conditions for the *computational domain*. Here we rewrite the short and vectorial form of these equations (Chiang and Kleinstreuer, 1993). The *dimensionless* conservation equations for the *gas* phase, in vectorial form, are

$$\frac{D\rho_g^*}{Dt^*} + \rho_g^* \nabla \cdot \mathbf{u}_g^* = 0 , \tag{6.159}$$

$$\frac{D\rho_{A,g}^*}{Dt^*} = \frac{1}{Pe_g Le_g} \nabla \cdot \rho_g^* D^* \nabla \frac{\rho_{A,g}^*}{\rho_g} , \tag{6.160}$$

$$\frac{D}{Dt^*} \rho_g^* \mathbf{u}_g^* = -\nabla p_g^* + \frac{1}{Re_g} \nabla \cdot \mu_g^* \nabla \mathbf{u}_g^* , \tag{6.161}$$

$$\frac{D}{Dt^*} \rho_g^* c_{p_g}^* T_g^* = \frac{1}{Pe_g} \nabla \cdot kg^* \nabla T_g^* +$$

$$\rho_g^*(c_{p_{A,g}}^* - c_{p_{B,g}}^*)\frac{1}{Pe_g Le_g} D^*\nabla T_g^* \cdot \nabla \frac{\rho_{A,g}^*}{\rho_g} \, . \tag{6.162}$$

The dimensionless conservation equations for the *liquid* phase are

$$\frac{D\rho_\ell^*}{Dt^*} + \rho_\ell^* \nabla \cdot \mathbf{u}_\ell^* = 0 \, , \tag{6.163}$$

$$\frac{D}{Dt^*}\rho_\ell^* \mathbf{u}_\ell^* = -\nabla p_\ell^* + \frac{1}{Re_\ell}\nabla \cdot \mu_\ell^* \nabla \mathbf{u}_\ell^* \, , \tag{6.164}$$

$$\frac{D}{Dt^*}\rho_\ell^* c_{p_\ell}^* T_\ell^* = \frac{1}{Pe_\ell}\nabla \cdot k_\ell^* \nabla T_\ell^* \, , \tag{6.165}$$

where

$$\mathbf{u}^* = \frac{\mathbf{u}}{u_\infty(t=0)} \, , \quad p_i^* = \frac{p_i}{(\rho_i)_r u_\infty^2(t=0)} \, , \quad T^* = \frac{T}{T_\infty} \, , \tag{6.166}$$

$$\mathbf{x}^* = \frac{\mathbf{x}}{d(t=0)} \, , \quad t^* = \frac{td(t=0)}{u_\infty(t=0)} \, , \quad Re_i = \frac{u_\infty d(t=0)}{(\nu_i)_r} \, , \tag{6.167}$$

$$Pe_i = Re_i Pr_i \, , \quad Le_g = \frac{Sc_g}{Pr_g} \, , \quad i = \ell \text{ or } g \, . \tag{6.168}$$

These and other properties are scaled using the *reference* state r which is the *upstream* condition for the *gas* and the *initial* (i.e., $t = 0$) condition of the droplet for the *liquid*.

Some of the typical transient numerical results obtained by Chiang et al. (1992) are shown in Figures 6.11(a) to (c) and Figures 6.12(a) to (c). Since they allow for the transient *reduction* in d and u_∞ (i.e., transient reduction in relative velocity), the magnitude of Re_d decreases with time. Their results are for pure n-octane (species A) evaporation in air with $p_\infty =$ 10 atm, $T_\infty = 1250$ K, $Pr_g = 0.74$, $Pr_\ell = 8.6$, $Sc_g = 2.36$, $T_d(t = 0) = 300$ K, and $Re_d(t = 0) = 100$ (corresponding to $u_\infty = 25$ m/s).

Figures 6.11(a) and (b) show the constant temperature and constant concentration contours in the *gas* phase at $t^* = 25$ ($Re_d = 23.9$). As is evident, due to the *large* initial Reynolds number (as also shown in Figure 5.4), the gradients of the temperature T_g and the mass fraction $\rho_{A,g}/\rho_g$ are large near the droplet, and the effect of the temperature and concentration disturbances do not penetrate very far into the gas phase. Figure 6.11(c) shows the contours of constant stream functions in the liquid phase. The structure of the flow resembles the Hill vortex, and a *slight* asymmetry exists with respect to a plane passing through the center of sphere and perpendicular to the flow. Note that at this elapsed time the diameter is *reduced* to 60 percent of its initial value. The gas-phase flow has already separated (separation angle of about 130°) and as evident from Figures 6.11(a) and (b), the front stagnation region has a boundary-layer behavior, while a wake is formed behind the droplet. Because of the droplet deceleration and other *transient* effects, and also because of the surface evaporation, some hydrodynamic aspects of the gas and the liquid phase are *different* compared to

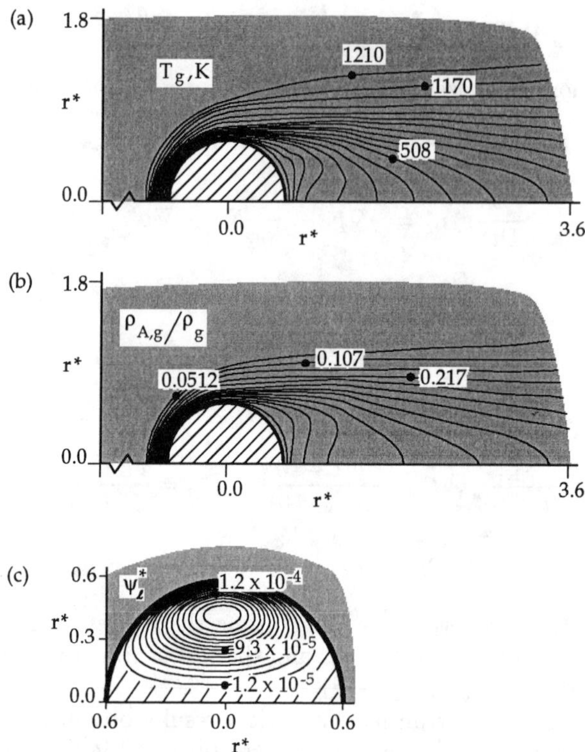

Figure 6.11. Distributions of (a) gas-phase temperature, (b) gas-phase concentration, and (c) liquid-phase dimensionless stream function, for $t^* = 25$, $Re_d = 23.9$. (From Chiang et al., reproduced by permission ©1992 Pergamon Press.)

the results obtained under a quasi-steady treatment. These are discussed by Chiang et al. (1992).

The transient temperature distribution in the *droplet* is shown in Figures 6.12(a) to (c). The results are $t^* = 0.50$ (corresponding to the instantaneous Reynolds number of $Re_d = 96.5$), $t^* = 5.0$ ($Re_d = 76.0$), and $t^* = 25$ ($Re_d = 23.9$). Since the residence time $d(t = 0)/u_\infty$ is used as the time scale, for $t^* = 0.5$ the *thermal* and *viscous* penetrations into the droplet are not significant and the dominant mode of heat transfer over most of the droplets is *diffusion*. For $t^* = 5.0$, the *internal circulation* is significant and *convection* dominates, except for the heat flow *across* the vortex towards the core, where the *diffusion* dominates and a large temperature gradient is present. For $t^* = 25$, the gas-phase shear stress decreases, the circulation *decays* in strength, and a convection-diffusion mode prevails over most of the drops. For these later elapsed times, the droplet is nearly at a uniform temperature.

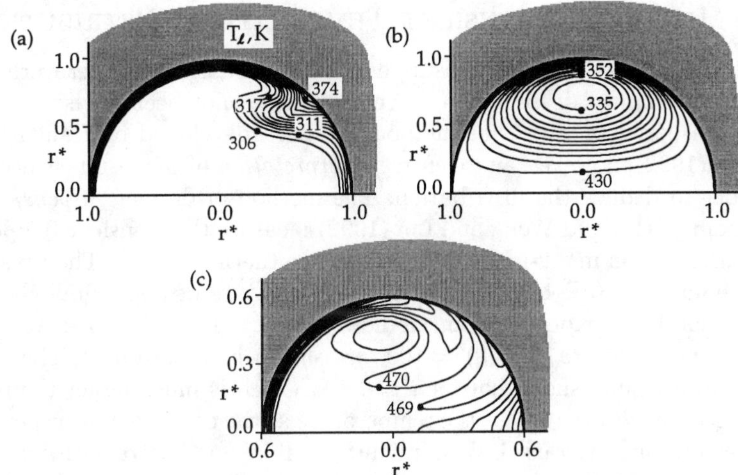

Figure 6.12. Distributions of the liquid-phase temperature for (a) $t^* = 0.50$, $Re_d = 96.5$, (b) $t^* = 5.0$, $Re_d = 76.0$, and (c) $t^* = 25$, $Re_d = 23.9$. The results are for $T_\infty = 1250$ K and $T_d(t = 0) = 300$ K. (From Chiang et al., reproduced by permission ©1992 Pergamon Press.)

The numerical results of Chiang et al. (1992) have been correlated to include the effects of evaporation, transient behavior, and variable properties on the drag coefficient, and the interfacial area-averaged Nusselt and Sherwood numbers. These are similar to (6.152) and are given as

$$c_d = \frac{2\psi}{Re_d}(1 + 0.325 Re_d^{0.474})(1 + \Delta B)^{-0.32} , \tag{6.169}$$

$$\langle Nu_d \rangle_{A_{\ell g}}^{\text{ext}} = 2 + 0.454 Re_d^{0.615} Pr_g^{0.98} (1 + \Delta B)^{-0.7} , \tag{6.170}$$

$$\langle Sh_d \rangle_{A_{\ell g}}^{\text{ext}} = 2 + 0.39 Re_d^{0.54} Sc_g^{0.76} (1 + \Delta B_m)^{0.557} , \tag{6.171}$$

$$0.4 \le \Delta B \le 13 \ , \ \ 0.2 \le \Delta B_m \le 6.4 \ , \ \ 30 \le Re_d \le 200 \ , \tag{6.172}$$

where ΔB is defined by (6.131), and

$$\Delta B_m = \frac{(\frac{\rho_{A,g}}{\rho_g})_s - (\frac{\rho_{A,g}}{\rho_g})_\infty}{1 - (\frac{\rho_{A,g}}{\rho_g})_s} . \tag{6.173}$$

The specific heat capacity in (6.162) and the other properties in $\langle Nu_d \rangle_{A_{\ell g}}^{\text{ext}}$, $\langle Sh_d \rangle_{A_{\ell g}}^{\text{ext}}$, Pr_g, and Sc_g are evaluated at the *film* temperature (i.e., the arithmetic average between T_s and T_∞), except for $\rho_{g\infty}$ used in the *instantaneous* Re_d. These correlations reduce to the proper asymptotes for $Re_d \to 0$. Similar correlations are given by Chiang and Kleinstreuer (1993).

(G) Measured Transient Temperature Distribution

A reliable method for the measurement of the transient temperature distribution *within* a droplet in relative motion has not been devised yet. A radiation-based, *noninvasive* method has been developed by Hanlon and Melton (1992); however, an extensive *interpretation* of the results is needed in order to deduce the distribution. The method of droplet *suspension* is commonly used, and Wong and Lin (1992) measure the transient temperature distribution in a 2-mm droplet using fine thermocouples. Their results for a liquid-fuel (JP-10), initially at a temperature of approximately 310 K and suddenly exposed to an air flow of $T_\infty = 1000$ K with a velocity corresponding to $Re_d(t=0) = 100$, are shown in Figure 6.13. Their experimental results show that for $t = 0.4$ s, which is much larger than the residence time $d(t=0)/u_\infty$, the temperature at the center of the suspended droplet has already raised *above* its initial value, and a *core* with *uniform* temperature exists around the center. The gradient of the temperature at the surface is very *large*, indicating a significant heat inflow for droplet heating. The experimental results follow a similar trend for $t = 0.8$, 1.2, and 2.4 s. While the surface temperature rises, the uniform-temperature core becomes *larger* and the gradient at the surface *decreases* with an increase in time. Also shown in Figure 6.13 are *predictions* based on purely transient *diffusion* heat flow within the droplet and the predictions of Prakash and Sirignano (1980) based on the Hill vortex core with convection and diffusion and a quasi-steady boundary-layer transport adjacent to the surface, discussed in Section (E) above.

The purely *molecular* diffusion does follow the trend in the measured temperature distributions, but the molecular diffusion is too *slow* to result in the magnitudes corresponding to the measurement. As was mentioned in Section 6.1.2(D), Wong and Lin (1992) also find that if allowance is made for an *enhanced* diffusion, the experimental results can be predicted with a transient diffusion model when $a_\alpha = 3.2$ is used in (6.9). The vortex model predicts vortex-core temperatures which are too low and droplet-center temperatures which are too high compared to the measurements. Wong and Lin (1992) measure the internal droplet velocity and find it to be lower than that predicted by (6.156). They argue that the combination of this lower velocity and the *actual mismatch* between the streamlines and the isotherms are the cause of the difference between the prediction of the vortex model and the experimental results. Further comments about the difference between the measured and the predicted internal temperature distribution and the effect of the suspension device on the internal liquid motion are given by Megaridis (1993).

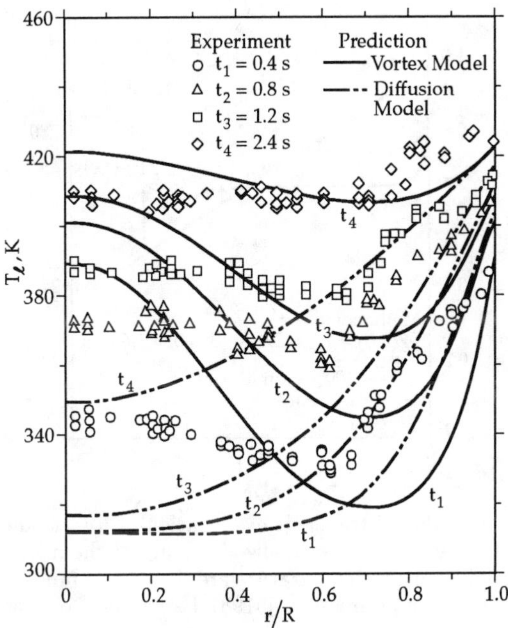

Figure 6.13. The measured and predicted (diffusion and convection-diffusion models) transient temperature distributions in a droplet for four different elapsed times. (From Wong and Lin, reproduced by permission ©1992 Cambridge University Press.)

(H) Simultaneous Evaporation and Condensation

In evaporation of an initially single-component droplet of species A (in general, initially a *subcoupled liquid*) in a high-temperature multicomponent gas containing a condensable (at the droplet temperature) species B (generally a *superheated gas*), with liquid A being *more* volatile than species B, *condensation* of species B can occur simultaneously with the evaporation of species A. This would require that the droplet surface temperature be at or below the saturation temperature of species B (at the ambient pressure). For the case when the liquids of A and B are *miscible*, then a binary diffusion (and when the droplet is in relative motion, a combined convection-diffusion) occurs within the droplet. Also, the heat released during the condensation of species B is *added* to the interfacial heat flow and can *raise* the surface temperature and/or cause *extra* evaporation of species A. Since the condensed species B *dissolves* in the liquid, the evaporation of species A is *hindered* by the mass diffusion *resistance* to the flow of species A in the liquid. The *net* effect of the condensation on the evaporation depends on the system variables and, as will be shown, changes with time. For the case of miscible liquids and a stationary droplet, an extension to the results of the spherically symmetric droplet evaporation has been made by including the condensation of species A present in the multicomponent

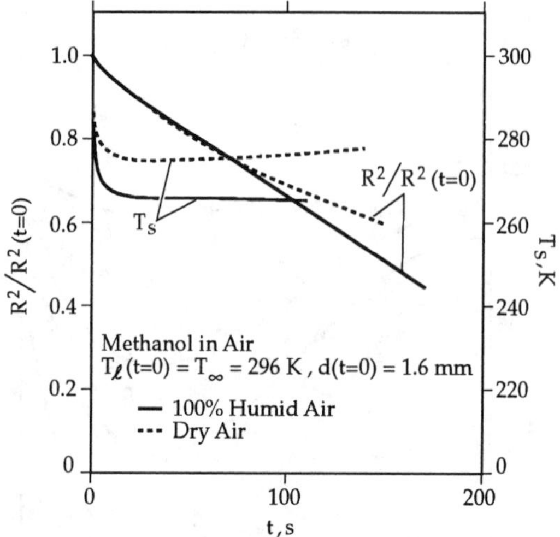

Figure 6.14. The predicted transient evaporation of a methanol droplet in dry and 100 percent humid air. The normalized, square of the instantaneous droplet radius and the droplet surface temperature are shown as a function of time. (From Law et al., reproduced by permission ©1987 Pergamon Press.)

ambient gas. Then in addition to the evaporation rate of species B, i.e., $\dot{m}_{A,\ell g}$, the condensation rate of species B, i.e., $\dot{m}_{B,\ell g}$, and the associated *thermodynamics* (i.e., interfacial conditions), *heat* and *mass transfer* (in the gas and liquid phase), are addressed. Law and Binak (1979), Law et al. (1987), Vesala (1993), and Mani et al. (1993) analyze this problem, and some typical results of Law et al. (1987) for evaporation of methanol droplets in dry and 100% humid air at 296 K are shown in Figure 6.14.

Their numerical results show that after an initial period, the 1.6-mm methanol droplet continues to evaporate at a constant rate in dry air, while in the 100% humid air after an initial period of *enhanced* evaporation, the evaporation rate becomes less than that for the case of dry air. This *decrease* in the evaporation is accompanied by an *increase* in the surface temperature T_s. The experimental results of Law et al. (1987) show that for 100% humid air the surface temperature *does* increase substantially (larger than that predicted), while the evaporation rate does *not* differ substantially from that for the dry air. They suggest that the variable thermophysical properties should be included in the numerical integrations and that the heat loss through the droplet suspension needle can also influence the experimental results.

(I) Layered Droplet

Considered a *two-component* liquid, in a *layered* droplet the *core* (i.e., the central region) of the droplet may be *rich* in species A while the shell (the region around this central region) may be rich in species B. This is an *emulsion droplet* which is *not* completely mixed and can be formed by growing the shell of species B around the core. However, due to diffusion some traces of one species will be present in the region dominated by the other species. Then during the evaporation *both* the *core-shell interface* and the *droplet-gas interface* are *mobile*. An analysis of this evaporation for a spherically symmetric droplet has been made by Bharat and Ray (1992). They show that when *both* species evaporate, the core can *either grow* or *decay* depending on the relative *volatility* of the species, the *phase equilibria* (thermodynamics) and the transport properties. Also when the core evaporates, *either* the *shell* or *core disappear*, again depending on the system parameters. The species and thermal energy conservation equations in the *liquid* and the *gas* phase and the core-shell and droplet-gas interfacial boundary conditions (including the thermodynamics) are given by Bharat and Ray (1992). The two-moving interface problem is solved numerically. Their results show that the *relative volatility* of the species plays the *most* significant role.

6.4.2 Interdroplet Interactions

In process applications and in spray combustion applications, the droplet *number density* is generally large and the isolated behavior so far considered is *altered* by the presence of the neighboring droplets. This modification in behavior, compared to the isolated droplet, is *time dependent*, and this makes the detailed inclusion of the *interparticle interactions* rather difficult. However, using simple models and direct simulations, some of these modifications have been identified and correlated with the particle number density or spacing. Here a simple, spherically symmetric *unit-cell model* directly applicable to the multidroplet systems (when the assumptions can be justified) is discussed. This is followed by the results for a *direct simulation* of the hydrodynamics, and heat and mass transfer, for a *tandem* arrangement of *two* droplets.

(A) Droplet-in-Bubble Model for Evaporation

One of the earliest and simpliest models that allows for the effect of the presence of the neighboring droplets on the droplet evaporation rate is the *quasi-steady, droplet-in-bubble* model. This is an extension of the *spherically symmetric* droplet evaporation discussed in Section 6.4.1(B). In this model a bubble *shell* is placed around the droplet with the relation between the droplet to the bubble radii given in terms of the porosity (gas phase is the continuous phase and its volume fraction is ϵ) as

$$\frac{V_d}{V_d + V_b} = 1 - \epsilon \quad \text{or} \quad \frac{R_d}{R_b} = (1 - \epsilon)^{1/3} , \tag{6.174}$$

where R_d is the radius of the droplet, R_b is the radius of the *bubble*, and $V_b = 4\pi(R_b^3 - R_d^3)/3$. The *quasi-steady* treatment begins with the conservation equations (6.124) to (6.127) and the boundary conditions (6.128) and (6.129).

For $\epsilon < 1$, i.e., when a *finite* V_b is used, the heating of the droplet *changes* the heat content of V_b and the evaporation of the droplet *changes* the species concentration in V_b. Then *simultaneously* T_∞ and $(\rho_{A,g})_\infty$, which are now evaluated at $r = R_b$, and T_s and $(\rho_{A,g})_s$ undergo change. The surface temperature and concentration T_s and $(\rho_{A,g})_s$ change because of *thermodynamic* relation with T_∞ and $(\rho_{A,g})_s$ while the *instantaneous* T_∞ and $(\rho_{A,g})_\infty$ are evaluated from the approximate heat and mass balances

$$(\rho_g c_p)_\infty \frac{4}{3}(R_b^3 - R_d^3)\frac{dT_\infty}{dt} = -4\pi R_d^2 k_g \left.\frac{dT}{dr}\right|_{R_d} , \tag{6.175}$$

$$(\frac{\rho_{A,g}}{\rho_g})_\infty = \frac{(m_{A,g})_b}{(m_{A,g} + m_{B,g})_b} ,$$

$$(m_{A,g})_b = (m_{A,g})_{t=0} + \int_0^t \dot{m}_{\ell g} A_{\ell g} dt , \tag{6.176}$$

where m_g is the mass content of V_b.

The quasi-steady solutions to (6.124) to (6.127), with the modified boundary conditions for a *finite* bubble, along with the transient evaluation of the surface and ambient temperatures and concentrations, have been obtained by Tishkoff (1979). For *small* porosities, i.e., when R_b/R_d is not very large, the droplet may *not* be completely evaporated because $(\rho_{A,g})_s$ becomes equal to $(\rho_{A,g})_\infty$ and this is called the *saturation limit* of the bubble. This *critical* R_b/R_d depends on the initial conditions and the thermophysical properties. Some typical results of Tishkoff (1979) are shown in Figure 6.15 for heptane with $p_\infty = 572$ Pa, $T_\infty(t = 0) = 625$ K, $(\rho_{A,g})_\infty(t = 0) = 0$. The left-hand vertical axis shows the variation of the normalized *saturation radius*, i.e., the radius at which $(\rho_{A,g})_s$ becomes equal $(\rho_{A,g})_\infty$, with respect to the ratio R_b/R_d. The *critical* R_b/R_d for this example is 4. The right-hand vertical axis shows the variation of the normalized elapsed time required for the *complete* evaporation (normalized with respect to that for $R_b \to \infty$), with respect to the ratio R_b/R_d. For small R_b/R_d (but larger than the critical value), because in the *initial* period the temperature gradient applied across $R_b - R_d$ is large compared to the quasi-steady solution, and when R_b/R_d is sufficiently large such that the saturation effects are small, the evaporation rate is *larger* than that for $R_b \to \infty$. Then $t_e(R_b \to \infty)$ is larger than that for a finite R_b and this is evident in Figure 6.15, where this ratio reaches a maximum of about 1.02 around $R_b/R_d = 1.5$.

Tishkoff (1979) shows that a general correlation relating the interacting and the isolated droplet evaporation rates can be given as (for R_b/R_d larger

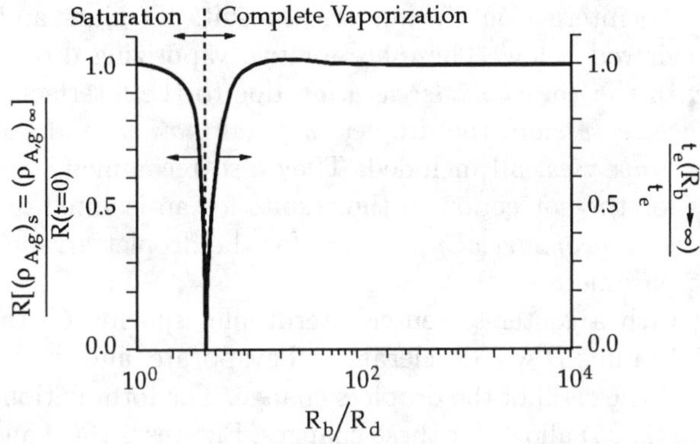

Figure 6.15. The prediction of the droplet-in-bubble model for the variations of the normalized saturation radius and the normalized elapsed time for complete evaporation, with respect to the ratio of the bubble to droplet radii. (From Tishkoff, reproduced by permission ©1979 Pergamon Press.)

than the critical value)

$$\frac{\langle Nu_d \rangle^{\text{ext}}_{A_{\ell g}}}{\langle Nu_d \rangle^{\text{ext}}_{A_{\ell g}}(\epsilon \to 1)} = e^{(1-\epsilon)^{0.28}} , \tag{6.177}$$

where $\langle Nu_d \rangle^{\text{ext}}_{A_{\ell g}}(\epsilon \to 1)$ is the isolated-droplet result given by (6.132). He also suggests that this correlation can be used for *moving* droplets and can *account* for the interaction due to the presence of the neighboring droplets. Then other isolated-droplet results given in Section 6.4.1 can also be used for $\epsilon \to 1$ in the above correlation. Wong and Chang (1992) experimentally examine the predictions of the droplet-in-bubble model for R_b/R_d *less* than the critical value and find good agreements with the predictions.

An extension to the droplet-in-bubble model has been in the *droplet-cluster* model for evaporation, as discussed by Bellan and Cuffel (1983) and Bellan and Harstad (1987a,b). The bubble-cluster moves with a *velocity* relative to the ambient, and the ambient gas can *penetrate* into the cluster by *convection* (for a *dilute* cluster), by diffusion (for a *dense* cluster), or by a *combination* of these two mechanisms.

(B) Direct Simulation of Interacting Multidroplet Evaporation

The hydrodynamic and heat and mass transfer aspects of a *moving, tandem* arrangement of a pair of droplets have been examined experimentally (e.g., Choi and Lee, 1992) and by solving the conservation equations numerically (e.g., Raju and Sirignano, 1990; Chiang and Kleinstreuer, 1992; Chiang and Sirignano, 1993). The results show that the drag coefficient and the rate of heat and mass transfer *decrease* when the droplets *interact*. To demonstrate this interaction, the numerical results of Chiang and Sirignano (1993) are reviewed below. These are for two vaporizing droplets moving in tandem with the *transient deceleration* due to the interfacial drag, the droplet *surface regression*, the droplet *internal motion*, and the *transient* and *variable properties*, all included. They also recommend a correlation which allows for the correction to the results for an isolated droplet. The correlations use a *geometrical parameter* for the droplet arrangement and other system parameters.

Beginning with a center-to-center interdroplet spacing C, the droplets with an initial radius R will decelerate and evaporate, and the interdroplet distance C and the radii of the droplets change. The formulation of Chiang and Sirignano (1993) allows for these changes. Figures 6.16(a) and (b) show their computed distributions of the velocity, vorticity, and temperature in the liquid and the gas phases and the distribution of the n-decane in the gas phase (air). The results are for the initial conditions of $Re_d = 100$, ambient temperature 1000 K, and droplet initial temperature of 300 K. The *initial* spacing C/R is 4. The results of Figures 6.16(a) and (b) are for $t\nu_\infty/R(t = 0)$ equal 3. These correspond to instantaneous Reynolds numbers of 80.31 and 85.84 for the *lead* and *downstream* droplets, respectively. The instantaneous spacing $C/R(t = 0)$ is 3.71 (indicating that the droplets may eventually *collide* and *coalesce*). The downstream droplet, being in the wake of the leading droplet, experiences a *lower* drag force and *lower* evaporation rate.

The variation of the surface area-averaged Nusselt number with respect to the dimensionless time, scaled with $R^2(t = 0)/\nu_\infty$, is shown in Figure 6.17. The results are for the initial Re_d of 100 and for three different values of the initial interdroplet spacing $C/R(t = 0)$. The results for an isolated droplet are also shown. For the *lead* droplet, the Nusselt number is similar in behavior to the isolated droplet (but has a lower magnitude), and for $C/R(t = 0)$ larger than 4, *no* significant difference is found. Note that because of the limitation on the selection of extent of the computational domain, the isolated droplet behavior is *not* completely reached even for $C/R(t = 0) = 16$. The *downstream* droplet experiences a *significant* reduction in the average Nusselt number and the results do not reach that of the lead droplet even for $C/R(t = 0) = 16$. The results of Figure 6.17 again show the *time dependence* of $\langle Nu_d \rangle^{\text{ext}}_{A_{\ell g}}$ and the need for the inclusion of the interdroplet interactions. In arriving at correlations for the effect of $C/R(t = 0)$ on the derivation from the isolated-droplet behavior, this *transient* behavior should be incorporated. Chiang and Sirignano (1993) suggest some correlations for the drag coefficient, and Nusselt and Sherwood numbers for the *early stages* of the evaporation.

The effect of *turbulence* in the gas phase on the evaporation rate has been examined experimentally by Gökalp et al. (1992). Their results show that the *vaporization kinetics* should be considered in addition to the turbulent temporal and spatial changes in the velocity around the droplet.

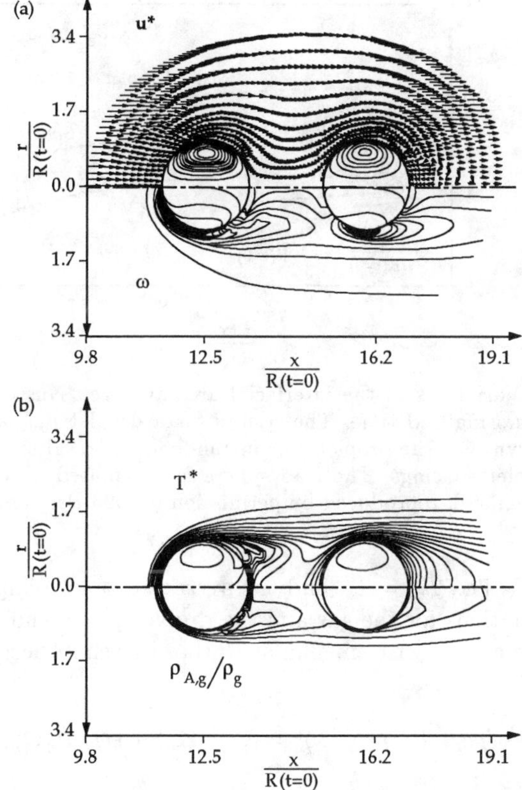

Figure 6.16. Instantaneous distributions of the dimensionless (a) gas-phase velocity vector, liquid-phase stream function, and gas- and liquid-phase vorticity, and (b) gas- and liquid-phase temperature and gas-phase species concentration. The results are for $t^* = 3$. (From Chiang and Sirignano, reproduced by permission ©1993 Pergamon Press.)

6.4.3 Droplet and Gas-Phase Conservation Equations

The description of species, overall mass, momentum, and energy conservation in the dispersed liquid phase and the continuous gas phase are made through *time-space* averaged treatments outlined in Section 3.3. The gas-phase motion is generally *turbulent*, and the *phase change* makes these time- and volume-averaged equations very complex. The energy equation for a *monosize* droplet evaporation can be written using *effective* transport properties $\langle \mathbf{K}_e \rangle$ and $\mathbf{D}_g^d$ for the molecular and dispersion (including turbulent) components of diffusion and the interfacial Nusselt numbers. For a monosize droplet system with diameter d, the gas-volume (or void) fraction is related to the droplet number density n and d through (5.170). Assuming that the *instantaneous* droplet surface temperature is the *saturation*

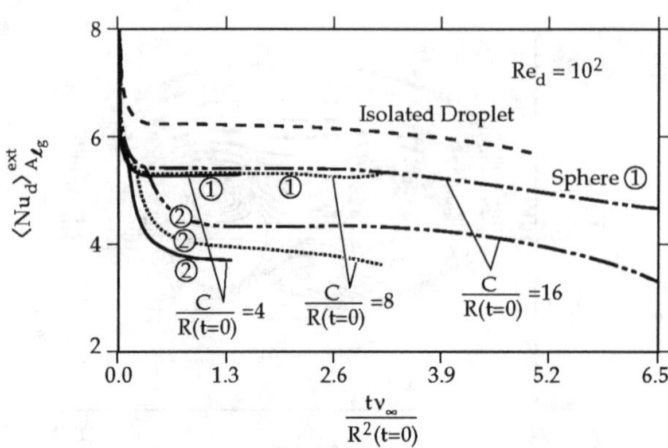

Figure 6.17. Variations of the interfacial area-averaged Nusselt number with respect to the normalized time. The results for a droplet pair and an isolated droplet are shown. For the droplet pair in tandem, the results are for three different interdroplet spacings. The lead sphere is designated as sphere 1. (From Chiang and Sirignano, reproduced by permission ©1993 Pergamon Press.)

temperature T_s, i.e., $T_{\ell g} = T_s$, and the average droplet temperature is T_ℓ, the energy equations for the gas and the droplet phase and the interfacial energy balance can be written, similar to those given in Sections 6.1.3 and 6.3.3, as

$$\epsilon(\rho c_p)_g \frac{\partial \langle T\rangle^g}{\partial t} + (\rho c_p)_g \langle \mathbf{u}_g\rangle \cdot \nabla \langle T\rangle^g = \nabla \cdot [\langle \mathbf{K}_e\rangle + \epsilon(\rho c_p)_g \mathbf{D}_g^d] \cdot \nabla \langle T\rangle^g +$$

$$\frac{6(1-\epsilon)}{d} \langle Nu_d\rangle^{\text{ext}}_{A_{\ell g}} \frac{k_g}{d}(T_s - \langle T\rangle^g) - \frac{6(1-\epsilon)}{\pi d^3} \frac{d}{dt}\frac{\pi d^3}{6}\rho_\ell i_g \, , \tag{6.178}$$

$$\frac{\pi d^3}{6}(\rho c_p)_\ell \frac{dT_\ell}{dt} = \pi d^2 \langle Nu_d\rangle^{\text{int}}_{A_{\ell g}} \frac{k_\ell}{d}(T_s - T_\ell) + \frac{d}{dt}\frac{\pi d^3}{6}\rho_\ell i_\ell \, , \tag{6.179}$$

$$\frac{d}{dt}\frac{\pi d^3}{6}\rho_\ell \Delta i_{\ell g} = \pi d^2 [\langle Nu_d\rangle^{\text{ext}}_{A_{\ell g}} \frac{k_g}{d}(T_s - \langle T\rangle^g) - \langle Nu_d\rangle^{\text{int}}_{A_{\ell g}} \frac{k_\ell}{d}(T_s - T_\ell)] \, . \tag{6.180}$$

In writing (6.178) to (6.180), the interfacial mass balance $\langle \dot{n}_g\rangle^g = -\langle \dot{n}_\ell\rangle^\ell$ has been used.

The external Nusselt number includes the interdroplet interaction, which can be significant, and the internal Nusselt number discussed in Section 6.1.2 (D) can be used. The gas-volume fraction ϵ and the superficial gas-phase velocity $\langle \mathbf{u}_g\rangle$ are obtained through appropriate gas-phase mass and momentum equations, such as those described in Section 3.3.2. A more general treatment, including the droplet motion, will be addressed in Section 8.1.3 in connection with internal, evaporating dispersed droplet flow.

The species concentration distributions which influence the external Nusselt number, is also described by a time-space average equation, similar to that discussed in Section 3.2.7. Using the external Sherwood number, this can be written as

$$\epsilon \frac{d}{dt}\langle \rho_A \rangle^g + \langle \rho \rangle^g \langle \mathbf{u}_g \rangle \cdot \nabla \frac{\langle \rho_A \rangle^g}{\langle \rho \rangle^g} = \nabla \cdot (\langle \mathbf{D}_{me} \rangle + \epsilon \mathbf{D}^d_m) \langle \rho \rangle^g \cdot \nabla \frac{\langle \rho_A \rangle^g}{\langle \rho \rangle^g} +$$

$$\frac{6(1-\epsilon)}{d} \langle Sh_d \rangle^{\text{ext}}_{A_{\ell g}} \frac{D_g}{d} [(\frac{\rho_{A,g}}{\rho_g})_s - \frac{\langle \rho_A \rangle^g}{\langle \rho \rangle^g}] , \tag{6.181}$$

$$\frac{d}{dt} \frac{\pi d^2}{6} \rho_\ell = -\pi d^2 \langle Sh_d \rangle^{\text{ext}}_{A_{\ell g}} \frac{D_g}{d} [(\frac{\rho_{A,g}}{\rho_g})_s - \frac{\langle \rho_A \rangle^g}{\langle \rho \rangle^g}] . \tag{6.182}$$

Again the effective molecular and dispersion (including turbulent) components of diffusion are prescribed through simple or complex models.

The force balance on the droplet (including the effect of the interdroplet interaction on the drag coefficient) is used to determine the slip velocity $|\mathbf{u}_d - \langle \mathbf{u}_g \rangle|$ and for the determination of the Nusselt and Sherwood numbers. Descriptions of evaporating multidrop systems, including the modeling of the gas-phase turbulence, are discussed by Gyarnuthy (1982) and Michaelides et al. (1992). Droplet evaporation in superheated vapor and in the presence of a solid surface will be discussed in Section 8.1.3.

6.5 Gas–Gas Systems

In Sections 6.1 and 6.2, the fluid-fluid *interfacial* as well as *intraphasic* convective heat transfer were examined. Here we briefly consider the gas-gas heat exchange under the condition of *thermal nonequilibrium* between the *electrons* and the *heavier species*. The role of convective heat transfer is limited to the intracontinuum (the electrons and the heavier species are treated as seperate continua). The heat exchange between the electrons and the species is the result of *recombination*, and then the energy of the recombination (i.e., the negative of the *ionization* energy) is *released*. The physics of *time-dependent* (i.e., *ultrafast*) collisional-radiative *atomic processes* such as ionization and spontaneous emission and absorption of radiation has been discussed by Foord et al. (1990). Under steady-state flow, the treatment is simplified and has been formulated by Bose (1990), Chang and Pfender (1990), and Mostaghimi and Boulos (1990). The energy equations for the electrons and the heavier species were discussed in Section 5.4.5, where the *number density* of the electrons n_e, *neutral atoms* n_a, and *ions* n_i (with the number of *heavier species* being $n_h = n_a + n_i$) are used along with a *single* velocity $\mathbf{u}$ for both electrons and the heavier species. The local volumetric Joule heating $\sigma_e \mathbf{e} \cdot \mathbf{e}$, the volumetric radiation energy $-\nabla \cdot q_r$, and the volumetric ionization energy $-n_e \Delta i_i$, are added to the electron energy conservation equation. In addition, the energy *exchange* (similar to heat transfer) *between* the *electrons* and the *ions*, and the *electrons* and the *neutral atoms*, are included. The energy equation for the heavier species includes the convective and conductive transport, the pressure work, and the energy exchange with the electrons. These two energy equations are (Chang and Pfender, 1990)

$$\nabla \cdot \frac{5}{2} k_B n_e T_e \mathbf{u} = \nabla k_e \cdot \nabla T_e + \mathbf{u} \cdot \nabla p_e - \nabla \cdot \mathbf{q}_r + \sigma_e \mathbf{e} \cdot \mathbf{e} - \dot{n}_e \Delta i_i - k_B (T_e - T_h) \dot{n}_{ei} - k_B (T_e - T_h) \dot{n}_{ea} \,, \tag{6.183}$$

$$\nabla \cdot \frac{5}{2} k_B n_h T_h \mathbf{u} = \nabla k_h \cdot \nabla T_h + \mathbf{u} \cdot \nabla p_h + k_B (T_e - T_h) \dot{n}_{ei} - k_B (T_e - T_h) \dot{n}_{ea} \,. \tag{6.184}$$

The *electron-ion* $\dot{n}_{ei}$ and the *electron-neutral atom* $\dot{n}_{ea}$ *collision rates* are modeled from the kinetic theory (Chang and Pfender, 1990) and are

$$\dot{n}_{ei} = \frac{2 n_e n_i e_c^4}{3 m_h k_B T_e (4\pi\epsilon_o)^2} \left(\frac{8\pi m_e}{k_B T_e}\right)^{1/2} \ln \frac{9 (4\pi \epsilon_o k_B T_e)^3}{4\pi n_e e_c^6} \,, \tag{6.185}$$

$$\dot{n}_{ea} = -\frac{3 n_e m_e (1 - a_i) \rho}{m_h^2} \left(\frac{8\pi m_e}{k_B T_e}\right)^{1/2} A_{ea} \,, \tag{6.186}$$

where m_h is the mass of the heavier species, m_e is the mass of an electron (9.109×10^{-31}kg), e_c is the electron charge (1.602×10^{-19}C), ϵ_o is the free space permittivity, a_i is the *degree* of *ionization*, and A_{ea} is the electron-neutral atom collision cross section.

Examples of computations (one- and two-dimensional) are given by Bose (1990), Chang and Pfender (1990), and Mostaghimi and Boulos (1990). Chang and Pfender (1990) also consider *turbulent* gaseous flows and use an algebraic turbulence model. In applications in high-frequency, inductively heated plasmas, even at high pressure (i.e., atmospheric) the deviation from the local thermal equlibrium becomes very *important* as the frequency increases. For example, this *nonequilibrium* results in a *lower* temperature in the plasma torch discussed in Section 4.13.3(A).

6.6 References

Arefmanesh, A., Advani, S., and Michaelides, E.E., 1992, "An Accurate Numerical Solution for Mass Diffusion-Induced Bubble Growth in Viscous Liquids Containing Limited Disolved Gas," *Int. J. Heat Mass Transfer*, 35, 1711–1722.

Arpaci, V.S., and Larsen, P.S., 1984, *Convective Heat Transfer*, Prentice-Hall, Englewood Cliffs, NJ.

Ascoli, E.P., and Lagnado, R.R., 1992, "The Linear Stability of a Spherical Drop Migrating in a Vertical Temperature Gradient," *Phys. Fluids*, A4, 225–233.

Avedisian, C.T., 1992, "Homogeneous Bubble Nucleation Within Liquids: A Review," *Two-Phase Flow and Heat Transfer*, Kim, J.-H., et al., Editors, *ASME HTD-Vol.*197, American Society of Mechanical Engineers, New York.

Avedisian, C.T., and Glassman, I., 1981, "Superheating and Boiling of Water in Hydrocarbons at High Pressure," *Int. J. Heat Mass Transfer*, 24, 695–706.

Banerjee, S., 1991, "Turbulence/Interface Interactions," in *Phase-Interface Phenomena in Multiphase Flow*, Hewitt, G.F., et al., Editors, Hemisphere Publishing Corporation, Washington, D.C.

Bankoff, S.G., and Mason, J.P., 1962, "Heat Transfer from the Surface of a Steam Bubble in Turbulent Subcooled Liquid Stream," *AIChE J.*, 8, 30-33.

Baumann, N., Joseph, D.D., Mohr, P., and Renardy, Y., 1992, "Vortex Rings of One Fluid in Another in Free Fall," *Phys. Fluids*, A4, 567–580.

Bellan, J., and Cuffel, R., 1983, "A Theory of Nondilute Spray Evaporation Based upon Multiple Drop Interactions," *Combust. Flame*, 51, 55-67.

Bellan, J., and Harstad, K., 1987a, "Analysis of the Convective Evaporation of Nondilute Clusters of Drops," *Int. J. Heat Mass Transfer*, 30, 125–136.

Bellan, J., and Harstad, K., 1987b, "The Details of the Convective Evaporation of Dense and Dilute Clusters of Drops," *Int. J. Heat Mass Transfer*, 30, 1083–1093.

Bharat, V., and Ray, A.K., 1992, "Evaporation and Growth Dynamics of a Layered Droplet," *Int. J. Heat Mass Transfer*, 35, 2389-2401.

Blander, M., Hengstenberg, D., and Katz, J.L., 1971, "Bubble Nucleation in n-Pentane, n-Hexane, n-Pentane+Hexadecane Mixtures, and Water," *J. Phys. Chem.*, 75, 3613–3619.

Borhan, A., and Mao, C.-F., 1992, "Effect of Surfactants on the Motion of Drops Through Circular Tubes," *Phys. Fluids*, A4, 2628–2640.

Bose, T.K., 1990, "One-Dimensional Analysis of the Wall Region for a Multiple-Temperature Argon Plasma," *Plasma Chem. Plasma Process.*, 10, 189–206.

Brown, J.S., Knoo, B.C., and Sonin, A.A., 1990, "Rate Correlation for Condensation of Pure Vapor on Turbulent, Subcooled Liquid," *Int. J. Heat Mass Transfer*, 33, 2001–2018.

Burmeister, L.C., 1983, *Convective Heat Transfer*, John Wiley and Sons, New York.

Carey, V.P., 1992, *Liquid-Vapor Phase-Change Phenomena*, Hemisphere Publishing Corporation, Washington, D.C.

Cha, Y.S., and Henry, R.E., 1981, "Bubble Growth During Decompression of a Liquid," *ASME J. Heat Transfer*, 103, 56–60.

Chang, C.H., and Pfender, E., 1990, "Nonequilibrium Modeling of Low-Pressure Argon Plasmas—Jets. Part I: Laminar Flow; Part II: Turbulent Flow," *Plasma Chem. Plasma Process.*, 10, 473–500.

Chelata, G.P., Cumo, M., Farello, G.E., and Focardi, G., 1989, "A Compression Analysis of Direct Contact Condensation of Saturated Steam on Subcooled Liquid Jets," *Int. J. Heat Mass Transfer*, 32, 639–654.

Chiang, C.H., Raju, M.S., and Sirignano, W.A., 1992, "Numerical Analysis of Convecting, Vaporizing Fuel Droplets with Variable Properties," *Int. J. Heat Mass Transfer*, 35, 1307–1324.

Chiang, H., and Kleinstreuer, C., 1993, "Numerical Analysis of Variable-Fluid-Property Effects in the Convective Heat and Mass Transfer of Fuel Droplets," *Combust. Flame*, 92, 459–464.

Chiang, C.H., and Sirignano, W.A., 1993, "Interacting, Convecting, Vaporizing Fuel Droplets with Variable Properties," *Int. J. Heat Mass Transfer*, 36, 875–886.

Cho, S.M., and Seban, R.A., 1969, "On Some Aspects of Steam Bubble Collapse," *ASME J. Heat Transfer*, 91, 537–542.

Choi, K.J., and Lee, H.J., 1992, "Experimental Studies on the Dynamics and Evaporation of Tandem Liquid Droplets in a Hot Gas Flow," *Int. J. Heat Mass Transfer*, 35, 2921–2929.

Chun, J.-H., Shimko, M.A., and Sonin, A.A., 1986, "Vapor Condensation onto a Turbulent Liquid. II. Condensation Burst Instability at High Turbulence Intensities," *Int. J. Heat Mass Transfer*, 29, 1333–1338.

Clift, R., Grace, J.R., and Weber, M.E., 1978, *Bubbles, Drops, and Particles*, Academic Press, New York.

Cole, R., 1974, "Boiling Nucleation," *Advan. Heat Transfer*, 10, 85–166.

Crowe, C.T., 1988, "Computational Techniques for Two-Phase Flow and Heat Transfer," *Direct Contact Heat Transfer*, Kreith, F., and Boehm, R.F., Editors, 41–59, Hemisphere Publishing Corporation, Washington, D.C.

Curtis, E.W., and Farrell, P.V., 1992, "A Numerical Study of High-Pressure Droplet Vaporization," *Combust. Flame*, 90, 85–102.

Deligiannis, P., and Cleaver, J.W., 1990, "The Role of Nucleation in the Initial Phase of a Rapid Depressurization of a Subcooled Liquid," *Int. J. Heat Mass Transfer*, 16, 975–984.

Delplanque, J.-P., and Sirignano, W.A., 1993, "Numerical Study of the Transient Vaporization of an Oxygen Droplet at Sub- and Super-Critical Conditions," *Int. J. Heat Mass Transfer*, 36, 303–314.

Dimić, M., 1977, "Collapse of One-Component Vapor Bubble with Translatory Motion," *Int. J. Heat Mass Transfer*, 20, 1325–1332.

Elias, E., and Chambré, P.L., 1993, "Flashing Inception in Water During Rapid Decompression," *ASME J. Heat Transfer*, 115, 231–235.

Foord, M.E., Maron, Y., and Sarid, E., 1990, "Time-Dependent Collisional-Radiative Model for Quantitative Study of Nonequilibrium Plasmas," *J. Appl. Phys.*, 68, 5016–5027.

Forest, T.W., and Ward, C.A., 1977, "Effect of a Dissolved Gas on the Homogeneous Nucleation Pressure of a Liquid," *J. Chem. Phys.*, 66, 2322–2330.

Gerner, F.M., and Tien, C.-L., 1989, "Axisymmetric Interfacial Condensation Model," *ASME J. Heat Transfer*, 111, 503–510.

Gökalp, I., Chauveau, C., Simon, O., and Chesneau, X., 1992, "Mass Transfer from Liquid Fuel Droplets in Turbulent Flow," *Combust. Flame*, 89, 286–298.

Gopalakrishna, S., and Lior, M., 1992, "Analysis of Bubble Translation During Transient Flashing Evaporation," *Int. J. Heat Mass Transfer*, 35, 1753–1761.

Gosman, A.D., Lekaku, C., Politis, S., Issa, R.I., and Looney, M.K., 1992, "Multidimensional Modeling of Turbulent Two-Phase Flows in Stirred Vessels," *AIChE J.*, 38, 1946–1956.

Gyarmuthy, G., 1982, "The Spherical Droplet in Gaseous Carrier Streams: Review and Synthesis," in *Multiphase Science and Technology*, Hewitt, G.F., et al., Editors, volume 1, 99–279, Hemisphere Publishing Corporation, Washington, D.C.

Hanlon, T.R., and Melton, L.A., 1992, "Eciplex Fluorescence Thermometry of Falling Hexadecane Droplet," *ASME J. Heat Transfer*, 114, 450–457.

Harpole, G.M., 1981, "Droplet Evaporation in High-Temperature Environments," *ASME J. Heat Transfer*, 103, 86–91.

Haywood, R.J., Nafziger, R., and Renksizbulut, M., 1989, "A Detailed Examination of Gas and Liquid Phase Transient Processes in Convective Droplet Evaporation," *ASME J. Heat Transfer*, 111, 495–502.

Hervieu, E., Courtis, N., and Tavares, M., 1992, "Deformation of a Fluid Particle Falling Freely in an Infinite Medium," in *ANS Proceedings*, HTC-Vol.6, 201–207, American Nuclear Society, LaGrange Park, IL.

Hewitt, G.F., Mayinger, F., and Riznic, J.R., 1991, *Phase-Interface Phenomena in Multiphase Flow*, Hemisphere Publishing Corporation, Washington, D.C.

Hoang, T.C., and Seban, R.A., 1988, "The Heating of a Turbulent Water Jet Discharged Vertically into a Steam Environment," *Int. J. Heat Mass Transfer*, 31, 1199–1209.

Jacobs, H.R., and Golafshani, M., 1989, "A Heuristic Evaluation of the Governing Mode of Heat Transfer in a Liquid-Liquid Spray Column," *J. Heat Transfer*, 111, 773–779.

Jorgenelen, F.C.H., Groenewg, F., and Gouda, J.H., 1978, "Effects of Interfacial Forces on the Evaporation of a Superheated Water Droplet in Hot Immiscible Oil," *Chem. Engng. Sci.*, 33, 777–781.

Kocamustafaogullari, G., and Ishii, M., 1983, "Interfacial Area and Nucleation Site Density in Boiling Systems," *Int. J. Heat Mass Transfer*, 26, 1377–1387.

Koh, C.J., and Leal, L.G., 1989, "The Stability of Drop Shape for Translation at Zero Reynolds Number Through a Quiescent Fluid," *Phys. Fluids*, A1, 1309–1313.

Koh, C.J., and Leal, L.G., 1990, "An Experimental Investigation on the Stability of Viscous Drops Translating Through a Quiescent Fluid," *Phys. Fluids*, A2, 2103–2109.

Komori, S., 1991, "Surface-Renewal Motions and Mass Transfer Across Gas-Liquid Interfaces in Open-Channel Flows," *Phase-Interface Phenomena in Multiphase Flow*, 31–40, Hewitt, G.F., et al., Editors, Hemisphere Publishing Corporation, Washington, D.C.

Kutateladze, S.S., 1963, *Fundamentals of Heat Transfer*, English Edition, Academic Press, New York.

Kwak, H.-Y., and Panton, R.L., 1983, "Gas Bubble Formation in Nonequilibrium Water-Gas Solutions," *J. Chem. Phys.*, 78, 5795–5799.

Law, C.K., and Binak, M., 1979, "Fuel Spray Vaporization in a Humid Environment," *Int. J. Heat Mass Transfer*, 22, 1009–1020.

Law, C.K., Xiong, T.Y., and Wang, C.H., 1987, "Alcohol Droplet Vaporization in Humid Air," *Int. J. Heat Mass Transfer*, 30, 1435–1443.

Levine, I.N., 1988, *Physical Chemistry*, Third Edition, McGraw-Hill, New York.

Lienhard, J.H., and Karimi, A., 1981, "Homogeneous Nucleation and the Spinodal Line," *ASME J. Heat Transfer*, 103, 61–64.

Lim, I.S., Tankin, R.S., and Yuen, M.C., 1984, "Condensation Measurement of Horizontal Concurrent Steam/Water Flow," *ASME J. Heat Transfer*, 106, 425–432.

Lozinski, D., and Matalon, M., 1992, "Vaporization of a Spining Fuel Droplet," *Twenty-Fourth Symposium (International) on Combustion*, 1483–1491, Combustion Institute, Pittsburgh.

Mani, S.V., Kulmala, M., and Vesala, T., 1993, "Evaporation of Polydisperse Ethanol Aerosols in a Humid Environment," *Int. J. Heat Mass Transfer*, 36, 705–711.

Mayinger, F., Chen, Y.M., and Nordmann, D., 1991, "Heat Transfer at the Phase Interface of Condensing Bubbles," in *Phase Interface Phenomena in Multiphase Flow*, Hewitt, G.F., et al., Editors, Hemisphere Publishing Corporation, New York.

Megaridis, C.M., 1993, "Comparison Between Experimental Measurements and Numerical Predictions of Internal Temperature Distributions of a Droplet Vaporizing Under High-Temperature Convective Conditions," *Combust. Flame*, 93, 287–302.

Michaelides, E.E., Liang, L., and Lasek, A., 1992, "The Effect of Turbulence on the Phase Change of Droplets and Particles Under Nonequilibrium Conditions," *Int. J. Heat Mass Transfer*, 35, 2069–2076.

Mikic, B.B., Rohsenow, W.M., and Griffith, P., 1970, "On Bubble Growth Rates," *Int. J. Heat Mass Transfer*, 13, 657–666.

Moalem, D., and Sideman, S., 1973, "The Effect of Motion on Bubble Collapse," *Int. J. Heat Mass Transfer*, 16, 2321–2329.

Moalem, D., Sideman, S., Orell, A., and Hestroni, G., 1973, "Direct Contact Heat Transfer with Change of Phase: Condensation of a Bubble Train," *Int. J. Heat Mass Transfer*, 16, 2305–2319.

Mostaghimi, J., and Boulos, M.I., "Effect of Frequency on Local Thermodynamic Equilibrium Conditions in an Inductively Coupled Argon Plasmas at Atmospheric Pressure," *J. Appl. Phys.*, 68, 2643–2648.

Payvar, P., 1987, "Mass Transfer-Controlled Bubble Growth During Rapid Decompression of a Liquid," *Int. J. Heat Mass Transfer*, 30, 699–706.

Perkins, R.J., Carruthers, D.J., Drayton, M.J., and Hunt, J.C.R., 1991, "Turbulence and Diffusion at Density Interfaces," *Phase Interface Phenomena in Multiphase Flow*, 21–43, Hewitt, G.F., et al., Editors, Hemisphere Publishing Corporation, Washington, D.C.

Prakash, S., and Sirignano, W.A., 1978, "Liquid Fuel Droplet Heating with Internal Circulation," *Int. J. Heat Mass Transfer*, 21, 885–895.

Prakash, S., and Sirignano, W.A., 1980, "Theory of Convective Droplet Vaporization with Steady Heat Transfer in the Circulating Liquid Phase," *Int. J. Heat Mass Transfer*, 23, 253–268.

Prosperetti, A., and Plesset, M.S., 1978, "Vapor-Bubble Growth in a Superheated Liquid," *J. Fluid Mech.*, 85, 349–368.

Raju, M.S., and Sirignano, W.A., 1990, "Interaction Between Two Vaporizing Droplets in an Intermediate Reynolds Number Flow," *Phys. Fluids*, A2, 1780–1796.

Ruckenstein, E., and Davis, J., 1971, "The Effect of Bubble Translation on Vapor Bubble Growth in a Superheated Liquid," *Int. J. Heat Mass Transfer*, 14, 939–952.

Sangani, A.S., and Didwania, A.K., 1993, "Dispersed-Phase Stress Tensor in Flows of Bubbly Liquids at Large Reynolds Numbers," *J. Fluid Mech.*, 248, 27–54.

Scriven, L.E., 1959, "On the Dynamics of Phase Growth," *Chem. Engng. Sci.*, 10, 1–13.

Sideman, S., 1966, "Direct Contact Heat Transfer Between Immiscible Liquids," *Advan. Chem. Engng.*, 6, 207–280.

Sideman, S., and Moalem, D., 1974, "Direct Contact Heat Exchangers: Comparison of Counter- and Co-Current Condensers," *Int. J. Multiphase Flow*, 1, 555–572.

Sideman, S., and Moalem-Maron, D., 1982, "Direct Contact Condensation," *Advan. Heat Transfer*, 15, 227–281.

Sirignano, W.A., 1983, "Fuel Droplet Vaporization and Spray Combustion Theory," *Prog. Energy Combust. Sci.*, 9, 291–322.

Skripov, V.P., 1974, *Metastable Liquids*, John Wiley and Sons, New York.

Sonin, A.A., Shimko, M.A., and Chun, J.-H., 1986, "Vapor Condensation onto a Turbulent Liquid-I. The Steady Condensation Rate as a Function of Liquid-side Turbulence," *Int. J. Heat Mass Transfer*, 29, 1319–1332.

Spinger, G.S., 1978, "Homogeneous Nucleation," *Advan. Heat Transfer*, 14, 281–346.

Theofanous, T.G., and Patel, P.D., 1976, "Universal Relation for Bubble Growth," *Int. J. Heat Mass Transfer*, 19, 425–429.

Tishkoff, J.M., 1979, "A Model for the Effect of Droplet Interactions on Vaporization," *Int. J. Heat Mass Transfer*, 22, 1407–1415.

Tokuda, N., Yang, W.J., and Clark, J.A., 1968, "Dynamics of Moving Gas Bubbles in Injection Cooling," *ASME J. Heat Transfer*, 88, 371–378.

van Stralen, S.J.D., 1968, "The Growth Rate of Vapor Bubbles in Superheated Pure Liquids and Binary Mixtures, Part I: Theory," *Int. J. Heat Mass Transfer*, 11, 1467–1489.

Vesala, T., 1993, "On Droplet Evaporation in the Presence of a Condensing Substance: The Effect of Internal Diffusion," *Int. J. Heat Mass Transfer*, 36, 695–703.

Whalley, P.B., 1987, *Boiling, Condensation, and Gas-Liquid Flow*, Oxford University Press, Oxford.

Williams, F.A., 1985, *Combustion Theory*, Addison-Wesley, Redwood City, CA.

Wong, S.-C., and Chang, J.-C., 1992, "Evaporation of Nondilute and Dilute Nondisperse Droplet Clouds," *Int. J. Heat Mass Transfer*, 35, 2403–2411.

Wong, S.-C., and Lin, A.-C., 1992, "Internal Temperature Distributions of Droplets Vaporizing in High-Temperature Convective Flows," *J. Fluid Mech.*, 237, 671–687.

Zhang, N., Xu, Y., and Yang, W.-J., 1991, "Natural Convection and Drop Morphology in Liquid-Liquid Direct Contact," *Exp. Heat Transfer*, 4, 309–317.

Part III

Three-Medium Treatments

7

Solid–Solid–Fluid Systems

In this chapter the heat transfer in fluid-solid-solid systems with thermal *nonequilibrium* between the phase pairs will be examined. In the simplest of these systems, *one* of the solid phases has a *simple* geometry and *bounds* (in part) the other solid phase and the fluid phase. The second solid phase can be *dispersed* (as elements) into the fluid, and when there is *no phase change* this will be the *particles-fluid flow around solid surfaces*. These particles-fluid flows are divided into two classes. One is the *particulate flows*, where the fluid velocity is *high* enough such that there is a *net* (in the cross-section of the tube or channel) flow of particles. The second class is the *fluidized beds*, where the fluid velocity is *low* enough such that *ideally* there is *no net* (in the cross-section of the bed) flow of particles (i.e., particles *recirculate* in the bed). Chart 7.1 is based on this classification and gives further divisions in each class of particles-fluid flow systems. Figures 7.1(a) and (b) render these two systems and shows that in practice there are also surfaces *submerged* in these flows. Figure 7.1(a) is for an internal particulate flow and heat transfer, where the net flow of particles and the fluid is *upward* against the gravity vector. The solid surface bounding the particulate system is designated with s_b and the submerged solid with s_s, while the solid particles phase is designated with s and the fluid phase with f. Figure 7.1(b) shows a *short* fluidized bed with a *bubble* (i.e., an isolated volume of *low* particle concentration), which is a result of the *hydrodynamic instability*, moving upward in the direction of the fluid flow. The particle concentration is *high* (called the *dense phase*) up to a height referred to as the *bed height* across which there is an *abrupt* change in the particle concentration leading to a *dilute phase* above the bed (called the *freeboard-phase*). In this chapter, Section 7.1 deals with the particulate flows and Section 7.2 with the fluidized beds.

When there is a *phase change*, as in the *solidification/melting* or in *frosting/sublimation*, the three phase (fluid-solid-solid) thermal nonequilibrium treatment may also become necessary. For example, in the multicomponent solidification, solid particles may nucleate and grow within a *supercooled* liquid confined in a cavity with the heat removal from the surfaces of this cavity. The liquid motion may be caused by the thermo- and diffusobuoyancy, and the cavity surface, the liquid, and the solid particles are in thermal nonequilibrium. For frosting over *surfaces*, the temperature across the frost layer is *not* uniform, again requiring a nonequilibrium treatment. These phase change problems are discussed in Section 7.3.

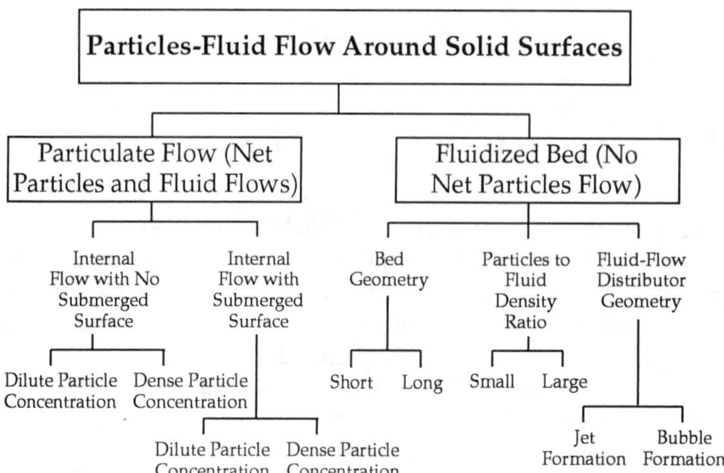

Chart 7.1. A classification of the particles-fluid flow around solid surfaces.

7.1 Particulate Flow Around Solid Surfaces

In this section the *simultaneous* flow of a fluid and particles *suspended* within it, over solid surfaces, is addressed by allowing for *thermal nonequilibrium* among the three phases, namely, the *fluid*, the *particles*, and the bounding *solid surface*. The case of *small* particles in thermal equilibrium with the fluid was reviewed in Section 4.14.1. The allowance for the difference between the local temperature of the particles $\langle T\rangle^s$ and the fluid $\langle T\rangle^f$ is required when relatively *large* (or *massive*) particles are considered such that the *thermal relaxation* time of the particles is much larger than their *translation time* (i.e., based on the relative particle velocity) in the fluid. This was discussed in Section 5.4.3, where particles-fluid heat transfer was considered in the absence of any bounding solid surface. In using $\langle T\rangle^s$, it is generally assumed that the temperature within the particle is *uniform*, i.e., $Bi = \langle Nu_d\rangle^{\rm ext}_{A_{sf}}(k_f/6k_s) < 0.1$.

The three-medium treatment of the heat transfer in particulate flow around solid surfaces requires examination of the transport *within* these three phases, as well as the transport *across* the three interfaces. Chart 7.2 gives a short list of the *interphasic* and *intraphasic* (i.e., *interfacial*) transport features of this three-phase system. The *fluid* turbulence is affected by the presence of the particles and the solid surface. The *concentration, size*, and *shape* of the *particles*, and the *length* and the *geometry* of the *solid surface*, will influence the intra- and interphasic hydrodynamics and heat transfer.

The simultaneous flow of the fluid and the suspended particles over a solid surface and the various mechanisms for heat transfer are rendered in Figure 7.2. The *fluid* is treated as a continuum, and then local, fluid-phase

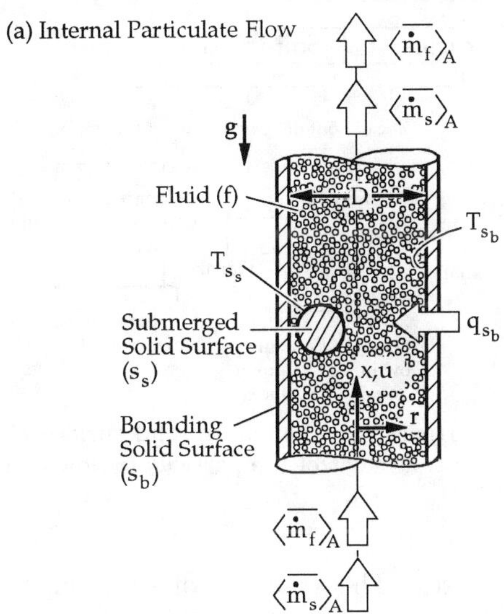

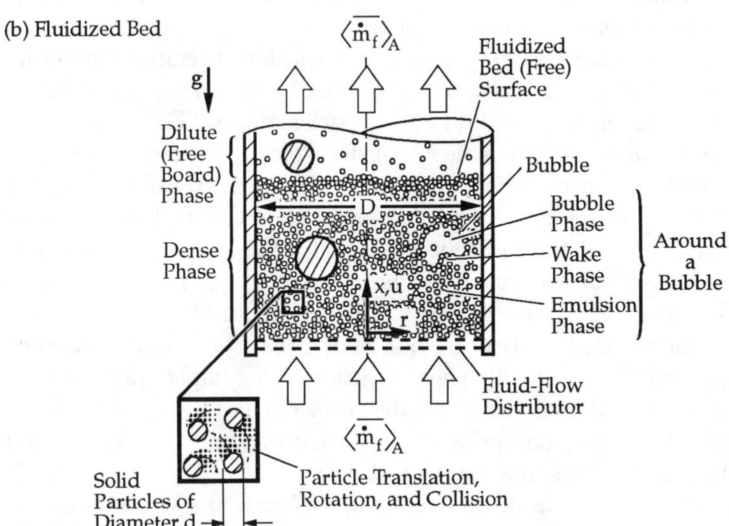

Figure 7.1. Rendering of the particulate flow and heat transfer. (a) An internal particulate flow with net flows of solid particles and fluid. (b) A fluidized bed with no net flow of solid particles but with a net flow of fluid.

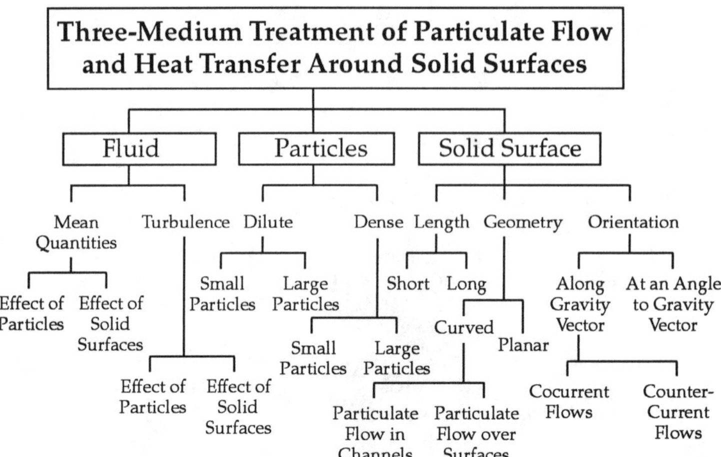

Chart 7.2. Description of fluid, particles, and bounding solid surface characteristics required for a three-medium treatment of the particulate flow and heat transfer.

volume-averaged quantities are used. The variation of the temperature *within* each *particle* is *neglected*, and local, particles-volume or *particles-phase* averaged quantities are also used. The *bounding solid surface* is treated either as a *prescribed* boundary condition or as a nonuniform temperature solid phase, where a *pointwise* variation in temperature is allowed. Then the *intraphasic* transport is allowed by the local, phase-volume averaged treatment for the fluid and particles phases. The heat flux vectors include the *particle-particle* heat transfer which is represented by $\mathbf{q}_{ss}$ and is by *convection* and *radiation*. The *interphasic* heat transfer occurs between *any* of these *phase pairs*. The *fluid-particles* interfacial heat transfer $\mathbf{q}_{fs}$ was discussed in detail in Sections 5.2 and 5.4.3, and is by *convection*. The *fluid-bounding surface* interfacial heat transfer $\mathbf{q}_{fs_b}$ is also by convection and is *influenced* by the *presence* of the particles. The *particles-bounding surface* interfacial heat transfer $\mathbf{q}_{ss_b}$ is by *conduction* (during contact) and by *radiation*. The conductive heat transfer during impact (among the particles or between the particles and the surface) can be an *elastic* or a *plastic* collision depending on the material properties and the relative velocity (Kaviany, 1988; Ben-Ammar et al., 1992).

The heat transfer from the bounding surface is the sum of $\mathbf{q}_{fs_b}$ (*convection*) and $\mathbf{q}_{ss_b}$ (*conduction*, including particles slip, and *radiation*), and is designated as $\mathbf{q}_{s_b}$, i.e.,

$$\mathbf{q}_{s_b} = \mathbf{q}_{fs_b} + \mathbf{q}_{ss_b} \quad \text{on the bounding solid surface.} \tag{7.1}$$

The combination of heat transfer due to particles motion and collision with the surface is treated as the *convective* heat transfer with the particles phase.

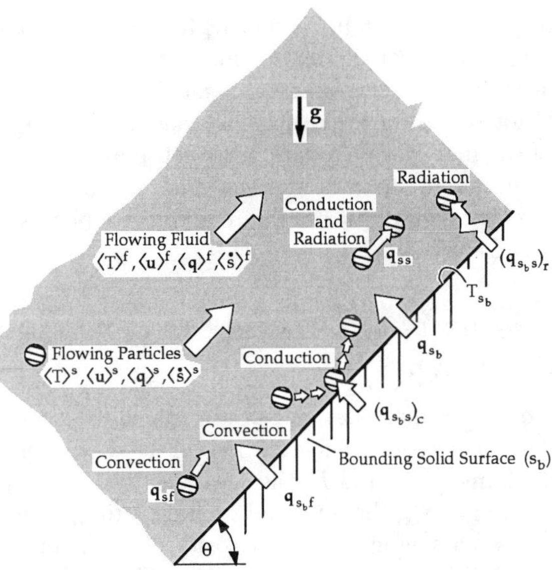

Figure 7.2. Various modes of heat transfer among particles, fluid, and bounding surfaces. For clarity, only a few particles are shown.

7.1.1 Hydrodynamics of Internal Particulate Flow

The internal particulate flow, i.e., the fluid-particles flow in a *channel*, is also called the *circulating fluidized bed* and in a steady operation it requires introduction of the particles before the *entrance* to the channel and their removal (for reintroduction at the entrance) at the *exit*. This is done using particle separators called *cyclones* (they use the centrifugal force for the separation). The particulate flow is generally *upward* against the gravity (Grace, 1986), although *downflow* applications are also found (Kim and Seader, 1983) and the case of inclined channels is considered by Ocone et al. (1993). A typical particle diameter d is between 50 and 500 μm, a mean superficial gas velocity $\langle \overline{u}_f \rangle$ is between 3 and 12 m/s, and the porosity (i.e., void or fluid volume fraction) ϵ ranges between 0.85 and 0.99. The range of the circulating beds, in terms of the *dimensionless* superficial *gas* velocity $\langle \overline{u}_f^* \rangle = \overline{u}^*$ and the *dimensionless* particle diameter d^* was shown in Figure 1.9 in connection with the discussion of the two-phase flows in Section 1.6.1. Further discussions on the gross *hydrodynamic* features of the solid-gas flows are given by Grace (1986). Here in order to evaluate the heat transfer between the solid-fluid two-phase flow and the confining solid surface, the variation of the *mean* porosity $\overline{\epsilon}$ and the *mean* particle velocity $\langle \overline{u_s} \rangle^s$ across the channel (e.g., a tube) is examined. These distributions are different for the *dilute* (e.g., $\overline{\epsilon} > 0.99$) and the *dense* (e.g., $\overline{\epsilon} < 0.99$)

concentration of the particles. In Section 4.14.1, the case of a dilute particle concentration (and in local thermal equilibrium with the host fluid) was considered. Here, first the experimental results on the *radial* distributions of $\overline{\epsilon}$ and $\langle \overline{u_s} \rangle^s$ for dense concentrations are examined. This is followed by an overview of the momentum equations presently used, for the dilute and dense concentrations, in the heat transfer analyses. In Sections 7.1.2 and 7.1.3, some specific analyses for the dilute and dense phases, with thermal nonequilibrium, are reviewed.

(A) Measurements in a Dense Concentration of Small Particles

In the upflow of a fluid-particles mixture in tubes, the radial distribution of the *instantaneous* particle concentration (or $1 - \epsilon$) depends on the particles and tube diameters d and D, the area-averaged mean fluid and solid velocities $\langle \overline{u_f} \rangle_A$ and $\langle \overline{u_s} \rangle_A$, the fluid and solid densities ρ_f and ρ_s, the fluid viscosity μ_f, the area-averaged mean porosity $\langle \overline{\epsilon} \rangle_A$, the distance from the tube entrance x, and the fluid-phase turbulence. The local particle concentration may change with time, with periods of these fluctuations as small as those associated with the fluid turbulence and as large as those associated with porosity waves traveling the entire tube. A complete *diagram* of the *radial* particle distribution regimes is not presently available. For example, for *very small* particles and *high velocities*, the mean *medium* density, which is related to the mean porosity through (4.545), was described by (4.543) which predicts a *larger* particle concentration around the tube *centerline*. In the experimental results of Miller and Gidaspow (1992), which are for *lower* velocities, the particle concentration was found to be *larger* near the *tube surface*. This recent experiment is reviewed below.

In the experiment of Miller and Gidaspow (1992), a vertical clear tube 6.6 m long with an inside diameter $D = 7.5$ cm is used for the *upflow* of $d = 75$ μm spherical particles in *air*, under *isothermal* (room temperature) conditions. Using the particles *hydrodynamic* considerations (i.e., the particles relaxation time compared with the fluid flow fluctuations period), particles with diameters *smaller* than 200 μm are generally categorized as *small* and moving with fluid with *no* velocity slip. Using similar *heat transfer* considerations, particles less than 100 μm are generally categorized as *small* and are in local thermal equilibrium with the fluid. The particles flow rate is measured by an extraction probe, and the local porosity is measured by a radiation attenuation technique. Their results are shown in Figures 7.3(a) to (c). Figure 7.3(a) shows the measured *radial* distribution of the *mean solid-phase volume fraction*, i.e., $1-\overline{\epsilon}$, for a *downstream location* x of 5.5 m (measured from the entrance), an *air superficial mean velocity*, $\langle \overline{u_f} \rangle_A = 2.9$ m/s, and for three different values of the cross-sectional averaged, mean *solid-phase mass flux* $\langle \overline{\dot{m}_s} \rangle_A$, where A is the tube cross-section area. The results show a nearly complete symmetry with respect to the tube

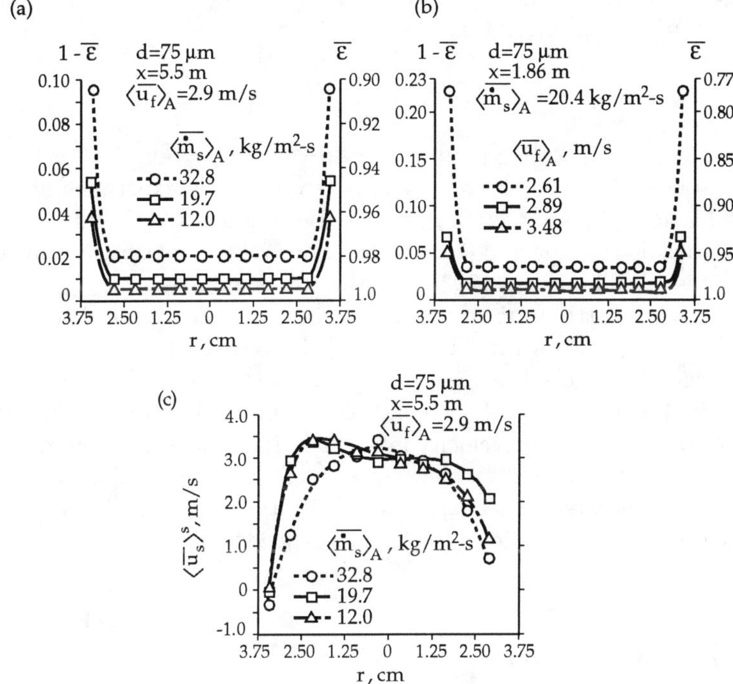

Figure 7.3. The measured radial distributions of (a) solid-phase volume fraction for a given area-averaged, mean gas-phase velocity, (b) solid-phase volume fraction for a given area-averaged, solid-phase mean mass flux, and (c) solid-phase, mean phase velocity. (From Miller and Gidaspow, reproduced by permission ©1992 American Institute of Chemical Engineers.)

centerline and that the particle concentration is *higher* adjacent to the tube surface compared to the *central core*. This difference in concentration increases with an increase in the solid-phase mass flux. For the range of mass flux shown in the figure, the particle concentration distribution does *not* change. In the regime of very *dilute* particulate flows, discussed in Section 4.14.1 and reviewed by Michaelides (1986) and Soo (1989), this larger particle concentration near the tube surface was *not* observed. Due to this difference, we *refer* to the results shown in Figures 7.3(a) and (b) as belonging to the the regime of *dense* particulate flows. The maximum solid-phase volume fraction in Figure 7.3(a) is about 0.10. For a given solid-phase mass flux, as the air superficial velocity is *reduced*, the local solid-phase volume fraction *increases*. This is shown in Figure 7.3(b), where solid-phase volume fractions up to 0.22 are found near the surface as the fluid-phase velocity is *lowered*. This distribution of the particles is in agreement with the existing data, except others have found a smaller gradient of the porosity near the surface. For example, the results compiled by Rhodes et al. (1992) follow a correlation

$$\frac{1-\overline{\epsilon}}{1-\langle\overline{\epsilon}\rangle_A} = 2(\frac{2r}{D})^2 \,. \qquad (7.2)$$

Note that this correlation predicts a solid volume fraction $1-\overline{\epsilon}$ at the tube surface which is *twice* the area-averaged value. The experimental results of Miller and Gidaspow (1992) presented in Figures 7.2(a) and (b) do not follow this distribution and (7.2), which is derivable from simple momentum equations (Rhodes et al., 1992), is rather too simplistic.

The radial distribution of the mean, local volume-averaged solid-phase velocity $\langle\overline{u_s}\rangle^s$, is shown in Figure 7.3(c), for a given $\langle\overline{u_f}\rangle_A$ and three different values of the solid-phase mass flux. This velocity distribution does *not* have the symmetry of the solid-phase volume fraction shown in Figures 7.3(a) and (b). The velocity is larger in the central core and decreases towards the surface, and for the *largest* $\langle\overline{\dot{m}_s}\rangle_A$ a mean solid-phase *velocity reversal* is measured. The solid-phase velocity in the core is rather independent of the magnitude of $\langle\overline{\dot{m}_s}\rangle_A$ and depends on $\langle\overline{u_f}\rangle_A$.

(B) Prediction of Velocity and Concentration Distributions

The hydrodynamics of particulate flows have been the subject of intense theoretical treatments. These were discussed in Sections 3.3 and 5.4, and reviews are given by Anderson and Jackson (1992), Gidaspow (1993), and Roco (1993). Even in the case of *internal*, *upward* flows, because of the *large* number of parameters influencing the velocity and porosity distributions, presently the *transition* from the *dilute*- to *dense*-concentration behavior has *not* yet been completely resolved. Then *separate* descriptions are used for the dilute- and dense-concentration particulate flows, and these are discussed below.

7.1.2 Dilute, Internal Particulate Flow

For *massive* particles (diameters larger than about 200 μm) with a Biot number $Bi = \langle Nu_d\rangle^{\rm ext}_{A_{sf}} k_f/6k_s$ smaller than 0.1, a uniform temperature can be assumed in the particles, but the difference in the local temperatures between the solid and the fluid phases, $\langle T\rangle^s - \langle T\rangle^f$, may *not* be negligible. For a *dilute* particle concentration (here we *arbitrarily* assume $\epsilon > 0.99$), the momentum and energy equations can be *simplified*. Then the *dilute*, *internal* particulate flow and heat transfer with *massive* particles becomes a thermal *nonequilibrium* (between the particles and the fluid) *extension* of the treatment given in Section 4.14.1 for *small* particles. The fluid flow is *turbulent*, and even at high porosities (or void fractions) associated with the dilute suspensions the fluid turbulence *can* be affected. This is because the particles are large enough such that the particles relaxation time given by (5.145) can be larger than the typical eddy turnover time. One of the earliest analyses of the thermal nonequilibrium, turbulent particulate flows in tubes was done by Tien (1961), and fully developed fields, an eddy diffusivity model for the fluid phase, and no heat diffusion in the particles phase are assumed. In the following, a *semi-empirical* treatment of the internal, turbulent particulate flow and heat transfer is reviewed where the effect of the particles flow rate on the heat transfer through the tube surface is examined. This does *not* include the effect of *radiation* which is briefly mentioned at the end.

In a turbulent particulate flow through a tube of diameter D, as rendered in Figure 7.1(a), assuming that a *fully developed* flow field (for both the fluid and the particles) exists, the *one-dimensional* mean *fluid* and *particles* velocities and mean porosity distributions can be written as

$$\epsilon(r) \equiv \langle\overline{\epsilon}\rangle_A f_\epsilon(r) , \tag{7.3}$$

$$u_f(r) \equiv \langle\langle\overline{u}\rangle^f\rangle_A f_f(r) , \tag{7.4}$$

$$u_s(r) \equiv \langle\langle\overline{u}\rangle^s\rangle_A f_s(r) , \tag{7.5}$$

where the functions f_i, $i = \epsilon, f, s$, are unknown and describe the radial distributions and the *tube cross-sectional area-averaged quantities* are defined as

$$\langle\psi\rangle_A = \frac{4}{\pi D^2}\int_0^{D/2} 2\pi\psi r dr . \tag{7.6}$$

These functions, and their counterparts for the fluctuation components, can be determined for the fluid and particle mass and momentum energy equations. We consider in this section a *simplified description* of the particulate flow for a *dilute* concentration of particles and follow the analysis of Molodtsof and Muzyka (1992) for the turbulent internal particulate flow and heat transfer. As will be shown, they *assume* that the self-similar (Molodtsof and Muzyka, 1991) distributions of the variables exists. A more rigorous treatment, using the the dense concentration conservation equations given in Section 5.4.2, will be discussed in Section 7.1.2. The following semi-empirical treatment shows the overall effect of the presence of particles on the heat transfer from the tube surface.

For a *prescribed heat flux* at the tube surface q_{s_b}, and under a thermally fully developed field, the surface temperature T_{s_b} *increases* linearly with the axial location x, i.e., dT_{s_b}/dx is constant and through an energy balance is given by

$$[(\rho c_p)_f\langle\langle\overline{u}\rangle^f\rangle_A + (\rho c_p)_s\langle\langle\overline{u}\rangle^s\rangle_A]\frac{\pi D^2}{4}\frac{dT_{s_b}}{dx} \equiv$$

$$(\langle\overline{\dot{m}_f}\rangle_A c_{p_f} + \langle\overline{\dot{m}_s}\rangle_A c_{p_s})\frac{\pi D^2}{4}\frac{dT_{s_b}}{dx} = \pi D q_{s_b} . \tag{7.7}$$

Molodtsof and Muzyka (1992) assume that there are *no* particles at the surface, i.e., $\overline{\epsilon}(r = D/2) = 1$. Then the heat transfer at the surface is between the fluid and the surface *only*, i.e.,

$$q_{s_b} = -\langle k\rangle \frac{\partial \langle \overline{T}\rangle^f}{\partial r} \qquad r = D/2 \, . \tag{7.8}$$

They also assume that $d\overline{\epsilon}/dr(r = D/2) = 0$.

The time- and phase volume-averaged energy equation for two-phase flows were discussed in Section 3.3.1, and (3.112) is a rather *rigorous* phasic energy equation which contains many unknowns, and (3.114) is a rather *simplified* form. For no phase change and radiation, and by neglecting the pressure work and the viscous dissipation, Molodtsof and Muzyka (1989) suggest that the variation in the porosity should be included and give the time- and phase volume-averaged energy equation as

$$\frac{\partial}{\partial t}(\rho c_p)_j \overline{\epsilon}_j \langle \overline{T}\rangle^j + \nabla \cdot (\rho c_p)_j \overline{\epsilon}_j (\langle \overline{\mathbf{u}}\rangle^j \langle \overline{T}\rangle^j + \langle \overline{\mathbf{u}'T'}\rangle^j) =$$

$$\nabla \cdot (\langle k\rangle^j \nabla \overline{\epsilon}_j \langle \overline{T}\rangle^j - \langle k\rangle^j \langle \overline{T}\rangle_{A_{sf}} \nabla \overline{\epsilon}_j) + \langle \frac{\overline{A_{sf}}}{V}\rangle \langle q_j\rangle_{A_{sf}} \qquad j = s, f \, , \tag{7.9}$$

where the second term in the parentheses on the right-hand side describes the change in the porosity as a diffusive heat transfer based on the *particles-fluid interfacial temperature* $\langle \overline{T}\rangle_{A_{sf}}$. Alternative, *simplified* forms for the local phase-volume-averaged energy equation are given by Michaelides and Lasek (1982) and Soo (1989). The latter includes the radiative heat transfer exchange between the particles and the tube surface.

For the thermally fully developed fields, the *reduced temperatures* are defined as

$$T_f(r) \equiv T_{s_b}(x) - \langle \overline{T}\rangle^f(x,r) \, , \tag{7.10}$$

$$T_s(r) \equiv T_{s_b}(x) - \langle \overline{T}\rangle^s(x,r) \, , \tag{7.11}$$

$$T_{sf}(r) = T_{s_b}(x) - \langle \overline{T}\rangle_{A_{sf}}(x,r) \, . \tag{7.12}$$

Then the phasic energy equations are obtained from (7.9) and are

$$(\rho c_p)_f (\epsilon u_f \frac{dT_{s_b}}{dr} + \frac{1}{r}\frac{d}{dr} r\epsilon \langle \overline{u'T'}\rangle^f) =$$

$$-\langle k\rangle^f (\frac{1}{r}\frac{d}{dr} r\epsilon T_f - T_{sf}\frac{d\epsilon}{dr}) - \langle \frac{\overline{A_{sf}}}{V}\rangle \langle \overline{q}\rangle_{A_{sf}} \, , \tag{7.13}$$

$$(\rho c_p)_s [(1-\epsilon)u_s \frac{dT_{s_b}}{dx} + \frac{1}{r}\frac{d}{dr} r(1-\epsilon)\langle \overline{u'T'}\rangle^s] =$$

$$-\langle k\rangle^s [\frac{1}{r}\frac{d}{dr} r(1-\epsilon)T_s - T_{sf}\frac{d(1-\epsilon)}{dr}] + \langle \frac{\overline{A_{sf}}}{V}\rangle \langle \overline{q}\rangle_{A_{sf}} \, . \tag{7.14}$$

Now by noting that *at* the surface the fluid turbulent heat flux in (7.13) is zero, and by integrating (7.14) across the tube cross-section, we have

$$\langle \langle \frac{\overline{A_{sf}}}{V}\rangle \langle \overline{q}\rangle_{A_{sf}}\rangle_A = (\rho c_p)_s \langle 1-\epsilon\rangle_A \langle \langle \overline{u}\rangle^s\rangle_A \frac{dT_{s_b}}{dx}$$

$$= \langle \overline{\dot{m}_s}\rangle_A c_{p_s} \frac{dT_{s_b}}{dx} \, . \tag{7.15}$$

Then noting that the particles-fluid interfacial heat transfer rate can *only* change radially, a function $g(r)$ is introduced to rewrite the *local*, volumetric interfacial heat transfer rate as

$$\langle\frac{\overline{A_{sf}}}{V}\rangle\langle\overline{q}\rangle_{A_{sf}} = [(\rho c_p)_s(1-\epsilon)\frac{dT_{s_b}}{dx}]g(r) \ . \tag{7.16}$$

Now substituting for the volumetric interfacial heat transfer in (7.13) and (7.14) and combining the radial variations using

$$\xi_f = -\langle k\rangle^f(\frac{d}{dr}\epsilon T_f - T_{sf}\frac{d\epsilon}{dr}) - (\rho c_p)_f\epsilon\overline{\langle u'T'\rangle}^f \ , \tag{7.17}$$

$$\xi_s = -\langle k\rangle^s[\frac{d}{dr}(1-\epsilon)T_s - T_{sf}\frac{d(1-\epsilon)}{dr}] - (\rho c_p)_s(1-\epsilon)\overline{\langle u'T'\rangle}^s \ , \tag{7.18}$$

we have

$$(\rho c_p)_f\epsilon u_f = \frac{1}{r}\frac{d}{dr}r\frac{\xi_f}{dT_{s_b}/dx} - (\rho c_p)_s(1-\epsilon)g(r) \ , \tag{7.19}$$

$$(\rho c_p)_s(1-\epsilon)u_s = \frac{1}{r}\frac{d}{dr}r\frac{\xi_s}{dT_{s_b}/dx} + (\rho c_p)_s(1-\epsilon)g(r) \ . \tag{7.20}$$

For the case of no suspensions, the fluid energy equation (7.19) becomes

$$(\rho c_p)_f u_f = \frac{1}{r}\frac{d}{dr}r\frac{\xi_f(\epsilon=1)}{\frac{dT_{s_b}}{dx}(\epsilon=1)} \ . \tag{7.21}$$

For the *constant heat flux* case, the presence of the solid particles (i.e., $\epsilon < 1$) *changes* dT_{s_b}/dx and $dT_f/dr(r = D/2)$. Since generally $(\rho c_p)_s > (\rho c_p)_f$, then from (7.7) dT_{s_b}/dx is smaller when $\epsilon < 1$. Also dT_f/dr changes *only* due to the change in $\langle k\rangle^f$, because as given by (7.8) the porosity is assumed to be *unity* at $r = D/2$. The surface Nusselt number $\overline{Nu}_D$ is defined as

$$\overline{Nu}_D = \frac{q_{s_b}D}{(T_{s_b} - \langle\overline{T}\rangle_A^b)k_f} \ , \tag{7.22}$$

where in the porosity- and velocity-weighted, *area-averaged* temperature (i.e., the *bulk mixture* mean temperature), $\langle\overline{T}\rangle_A^b$, is defined through

$$(\langle\overline{\dot{m}_f}\rangle_A c_{p_f} + \langle\overline{\dot{m}_s}\rangle_A c_{p_s})(T_{s_b} - \langle\overline{T}\rangle_A^b) =$$

$$\frac{4}{\pi D^2}\int_0^{D/2} 2\pi r[(\rho c_p)_f\epsilon u_f T_f + (\rho c_p)_s(1-\epsilon)u_s T_s]dr \ . \tag{7.23}$$

Then for a given q_{s_b}, $\overline{Nu}_D$ changes as $T_{s_b} - \langle\overline{T}\rangle_A^b$ changes.

The distributions of T_f and T_s are determined from the solution to (7.19) and (7.20). Since ϵ, u_f, and u_s are only functions of r, the ratio of the Nusselt numbers with and without the particles is

$$\frac{\overline{Nu}_D}{\overline{Nu}_D(\epsilon=1)} = \frac{T_{s_b} - \langle\overline{T}\rangle_A^b(\epsilon=1)}{T_{s_b} - \langle\overline{T}\rangle_A^b} = f(\frac{\langle\overline{\dot{m}_s}\rangle_A}{\langle\overline{\dot{m}_f}\rangle_A}) \ . \tag{7.24}$$

Molodtsof and Muzyka (1992) use a first-order perturbation around the solution for $\epsilon = 1$ and re-average to arrive at

$$\frac{\overline{Nu}_D}{\overline{Nu}_D(\epsilon = 1)} = \frac{(1 + \langle\overline{\dot{m}_s}\rangle_A/\langle\overline{\dot{m}_f}\rangle_A)^2}{1 + a_1(\langle\overline{\dot{m}_s}\rangle_A/\langle\overline{\dot{m}_f}\rangle_A) + a_2(\langle\overline{\dot{m}_s}\rangle_A/\langle\overline{\dot{m}_f}\rangle_A)^2}, \qquad (7.25)$$

where a_1 and a_2 depend on the radial profiles of f_ϵ, f_f, and f_s, where these functions are in (7.3) to (7.5), and the radial profiles of T_f and T_s are as determined from (7.19) and (7.20). Then a_1 and a_2 are *geometric constants* which can be determined from some assumed radial profiles. The analysis of Molodtsof and Muzyka does not allow for the determination of the radial distributions of ϵ, u_f, u_s, T_f, T_s, and T_{sf}, but they show that depending on the magnitudes of a_1 and a_2, an *increase* or a *decrease* in the ratio of Nusselt numbers in (7.25) can be found. They examine the available experimental results for *dilute* suspension of 200-μm-diameter glass beads in air and determine the constants a_1 and a_2 through the *best curve fit* for several values of $\langle\overline{\dot{m}_s}\rangle_A/\langle\overline{\dot{m}_f}\rangle_A$. Their results are shown in Figure 7.4(a), where the available experimental results of two different investigations for four values of the Reynolds number $Re_D = \langle\overline{u_f}\rangle_A D/\nu_f$ are used. The plot of $a_1 + a_2\langle\overline{\dot{m}_s}\rangle_A/\langle\overline{\dot{m}_f}\rangle_A$ versus $\langle\overline{\dot{m}_s}\rangle_A/\langle\overline{\dot{m}_f}\rangle_A$ is nearly linear as predicted based on *constant* a_1 and a_2. Figure 7.4(b) shows that depending on a_1 and a_2 a *monotonic increase*, a *monotonic decrease*, a *minimum*, or a *maximum* in the ratio of the Nusselt numbers can be found. The range of a_1 and a_2 corresponding to these *regimes* is given in Figure 7.4(c). Note that the results of the local thermal equilibrium treatment shown in Figure 4.45(d) correspond to the monotonically *increasing* Nusselt number ratio, and therefore, are for a specific set of a_1 and a_2.

The above analysis does *not* address the *details* of the hydrodynamics and heat transfer which lead to the determination of a_1 and a_2. These details, i.e., the radial distributions, will be considered for the case of *dense* particles concentrations in Section 7.1.3.

The effect of *radiation*, neglected above, is examined experimentally by Hasegawa et al. (1983). Their results show that the *high surface emissivity* $(\epsilon_r)_s$ particles (e.g., graphite) *further increase* the ratio of the Nusselt numbers (with and without suspension) at *elevated* temperatures. The theoretical treatment of Tamehiro et al. (1973) uses a *uniform* porosity and the available results for the *turbulent* flow and heat transfer without any suspensions for the velocity, the eddy viscosity, and the diffusivity distributions. They also use a prescribed Nusselt number for the fluid-particles interfacial heat transfer $\langle Nu_d\rangle^{\rm ext}_{A_{sf}}$ and assume a *nonscattering* medium. They consider the case of a *constant surface temperature*. They show that an asymptotic value for $\overline{Nu}_D$ is *not* reached (unlike the fully developed flow and heat transfer *without* radiation). They also show that when $\langle Nu_d\rangle^{\rm ext}_{A_{sf}}$ is high (i.e., a highly turbulent flow), then by assuming a local thermal equilibrium between the particles and the fluid the analysis can be signifi-

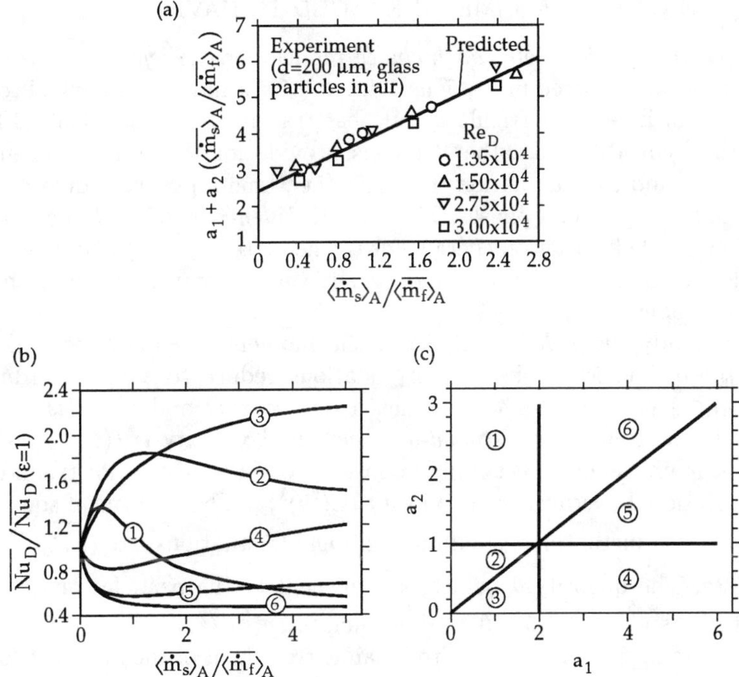

Figure 7.4. (a) Experimental evidence for constancy of a_1 and a_2 with respect to a solid loading ratio. (b) Variation of normalized mean Nusselt number with respect to a solid loading ratio and for a_1 and a_2 corresponding to the six different regimes shown in (c). (From Molodtsof and Muzyka, reproduced by permission ©1992 Pergamon Press.)

cantly simplified. Inclusion of radiation, in a simple form, is also discussed by Soo (1988).

7.1.3 Dense, Internal Particulate Flow

The time- and local phase volume-averaged momentum and thermal energy conservation equations for dense particle concentration flows, examined in Sections 5.4.2 and 5.4.3, have been applied to the internal particulate flow and heat transfer. The predictions have been compared with the *limited* experimental results for upflow through tubes with *dilute* concentrations, and the *time-averaged* behavior has been successfully predicted. These are reviewed below along with some measurements of the *instantaneous* heat transfer rate and the solid volume fraction which suggests that the hydrodynamics of the upflow are *not* yet well understood.

(A) Analysis of Time-Averaged Behavior

The *particle-particle collision included* models for the *hydrodynamics* and *heat transfer* reviewed in Sections 5.4.2 and 5.4.3 have been applied to the problem of internal particulate and heat transfer by Louge et al. (1993). For the hydrodynamics, these models provide for the *mean* momentum equations and the fluctuation turbulent (i.e., kinetic) energy equations for the fluid and particles phases. The *fluid turbulence* is *affected* and *coupled* to the particle fluctuation energy. For the heat transfer *turbulent* transport is allowed in the *fluid phase* and *collision conduction* is included in the particles phase.

For steady, *fully developed fields*, the momentum, fluctuation energy, and thermal energy conservation equations reduce to a set of ordinary differential equations. The *tube* has a diameter D and *particles* have a diameter d. The *cross-sectional-averaged* fluid velocity is $\langle\langle\overline{u}\rangle^f\rangle_A$, where $\langle\overline{u}\rangle^f$ is again the *local, mean* fluid-phase *volume*-averaged velocity, and the cross-sectional-averaged solid velocity is $\langle\langle\overline{u}\rangle^s\rangle_A$. The root-mean-square of the velocity fluctuations, assuming *isotropic* fluctuations, are $(\overline{{u'_f}^2})^{1/2}$ and $(\overline{{u'_s}^2})^{1/2}$. The quantity θ (which is equivalent to the particles phase temperature) is given by (5.149) and is equal to $(\overline{{u'_s}^2})^{1/2}/3$.

In the following, first the hydrodynamic treatment of the fully developed particulate flow in tubes, including the *effects* of the *presence* of the *tube surface* on the solid and fluid velocity fluctuations, is examined. Then its heat transfer counterpart is analyzed, and finally the numerical results of Louge et al. (1993) are discussed. The results are for both a *constant heat flux* and a *constant temperature* condition on the tube surface.

(i) Hydrodynamics

Based on the analysis and the results presented by Louge et al. (1993), the mean momentum equations and the equations for the fluctuation energy of the particles and the fluid phases were reviewed in Section 5.4.2. The gravity term in the particles-phase momentum equation is significant and the viscosity of the particles phase is derived from the kinetic theory. For a hydrodynamically fully developed flow, the particles-phase continuity equation becomes trivial and the *mean* momentum equation (5.147) becomes

$$0 = \frac{1}{r}\frac{d}{dr}r\overline{\tau}_s + \frac{\rho_s}{t_d}(1-\overline{\epsilon})(\langle\overline{u}\rangle^f - \langle\overline{u}\rangle^s) - \rho_s(1-\overline{\epsilon})g\,. \tag{7.26}$$

The particles-phase *mean viscous shear stress* given by (5.148) becomes

$$\overline{\tau}_s = \frac{5\pi^{1/2}}{96}\rho_s d\theta^{1/2}\frac{d\langle\overline{u}\rangle^s}{dr}\frac{1}{1+2\lambda/D}\,. \tag{7.27}$$

The particles relaxation time t_d is related to the drag coefficient. The drag coefficient (5.9) can be used, and for low Reynolds numbers Louge et al. (1993) recommend

$$\frac{\rho_s}{t_d} = c_d|\langle\overline{u}\rangle^f - \langle\overline{u}\rangle^s|\frac{3\rho_f}{4d} \ ,$$
$$c_d = \frac{24}{Re_d}(1+0.15Re_d^{0.687}) \quad Re_d \le 800 \ , \tag{7.28}$$

where

$$Re_d = \frac{|\langle\overline{u}\rangle^f - \langle\overline{u}\rangle^s|d}{\nu_f} \ . \tag{7.29}$$

The particles-phase fluctuations given by θ is obtained from the particles-phase fluctuation energy equation (5.150) which for the fully developed flow becomes

$$0 = \frac{1}{r}\frac{d}{dr}(r\frac{25\pi^{1/2}}{128}\rho_s d\frac{1}{1+2\lambda/D}\theta^{1/2}\frac{d\theta}{dr} + \overline{\tau}_s\frac{d\langle\overline{u}\rangle^s}{dr} -$$
$$\frac{24(1-e)\rho_s}{\pi^{1/2}d}(1-\overline{\epsilon})\theta^{3/2} + \frac{\rho_s}{t_d}(1-\overline{\epsilon})(\langle\overline{u'_f u'_s}\rangle - 3\theta) \ , \tag{7.30}$$

where e is the *coefficient of restitution* ($e \simeq 0.9$ is taken for polystyrene particles). The particles-fluid *cross* fluctuation correlation is given by (5.153) and for *isotropic* fluctuations become

$$\langle\overline{u'_f u'_s}\rangle = \frac{4}{\pi^{1/2}}\frac{d(\langle\overline{u}\rangle^f - \langle\overline{u}\rangle^s)}{t'_d\theta^{1/2}} \ , \tag{7.31}$$

where t'_d is the particles relaxation time based on the root-mean-square of the fluctuation slip velocity, i.e., $(\overline{u'^2_f})^{1/2} - (\overline{u'^2_s})^{1/2}$. As was mentioned in connection with (5.153), t_{d_o} (i.e., t_d for $Re_d \to 0$) is used for t'_d.

The particles-phase *mean* momentum equation (7.26) is *coupled* with the *mean* fluid-phase momentum equation through $\langle\overline{u}\rangle^f$. The fluid-phase continuity equation also becomes trivial and the mean fluid-phase momentum equation (5.155) becomes

$$0 = -\frac{\partial\overline{\epsilon}\overline{p}_f}{\partial x} + \frac{1}{r}\frac{d}{dr}r\overline{\epsilon}\overline{\tau}_f - \frac{\rho_s}{t_d}(1-\overline{\epsilon})(\langle\overline{u}\rangle^f - \langle\overline{u}\rangle^s) \ . \tag{7.32}$$

The *mean* fluid-phase *viscous shear stress* is given by (5.150) and becomes

$$\overline{\tau}_f = \rho_f(\nu+\nu_t)\frac{d\langle\overline{u}\rangle^f}{dr} = \rho_f[\nu + c_\mu\ell(\frac{1}{2}\langle\overline{u'_f u'_f}\rangle)^{1/2}]\frac{d\langle\overline{u}\rangle^f}{dr} \ , \tag{7.33}$$

where the turbulence is modeled using

$$c_\mu = 0.49 \ , \quad \kappa = 0.41 \ , \quad \frac{2\ell}{D} = \begin{cases} \kappa(1-2r/D) & 2r/D \ge 0.7 \\ 0.3\kappa & 2r/D < 0.7 \ . \end{cases} \tag{7.34}$$

The equation for the kinetic energy of the *isotropic* turbulence of the fluid phase is given by (5.159) and becomes (Louge et al., 1993)

$$0 = \frac{1}{r}\frac{d}{dr} r\overline{\epsilon}(\nu_f + \nu_t)\frac{d}{dr}\frac{1}{2}\langle\overline{u'_f u'_f}\rangle + \nu_t\overline{\epsilon}(\frac{d\langle\overline{u}\rangle^f}{dr})^2 -$$
$$\overline{\epsilon} c_\mu^3 \frac{\frac{1}{2}\langle\overline{u'_f u'_f}\rangle}{\ell} - \frac{\rho_s}{\rho_f t_d}(1-\overline{\epsilon})(\langle\overline{u'_f u'_f}\rangle - \langle\overline{u'_f u'_s}\rangle) \; , \tag{7.35}$$

where $\langle\overline{u'_f u'_s}\rangle$ is given by (7.31).

(ii) Boundary Conditions

The presence of the tube wall (i.e., the solid surface) modifies the particles motion and collision by using the kinetic theory. The particles-phase shear stress is modified to (Louge et al., 1993)

$$\overline{\tau}_s = -a_u\rho_s(1-\overline{\epsilon})\theta \quad \text{on a solid surface,} \tag{7.36}$$

where a_u is the *coefficient of dynamic friction* ($a_u \simeq 0.2$ is taken for polystyrene particles). The diffusion flux of θ defined by (5.151) for the *bulk* particulate flow, used in (7.30), is modified at the solid surface and is given by

$$q_\theta = \frac{3}{8}\rho_s(1-\overline{\epsilon})\theta(3\theta)^{1/2}[\frac{7}{2}(1-e_o)a_u^2 - (1-e_o)] \text{ on a solid surface,} \tag{7.37}$$

where e_o is the *surface* coefficient of restitution for a particle colliding with the solid *surface* ($e_o \simeq 0.7$ is taken for polystyrene particles).

For the fluid phase, the distribution of the mean velocity up to the scaled distance $y^* = 30$ is *prescribed*. The scaling of the distance and velocity are made similar to those used for the turbulent flows in Section 4.4, i.e., (4.74), and here becomes

$$y^* = \frac{y(\overline{\epsilon\tau}_{f_o}/\rho_f)^{1/2}}{\nu_f} \; , \quad \overline{u}_f^* = \frac{\langle\overline{u}\rangle^f}{(\overline{\epsilon\tau}_{f_o}/\rho_f)^{1/2}} \; , \tag{7.38}$$

where surface, fluid mean shear stress $\overline{\tau}_{f_o}$, or the mean shear velocity $u_\tau^2 = \overline{\epsilon\tau}_{f_o}/\rho_f$, is found from the tube cross-section-averaged, mean momentum equations (7.26) and (7.32), i.e.,

$$\frac{\partial\overline{\epsilon}\overline{p}_f}{\partial x} = \frac{4\rho_f u_\tau^{*2}}{D} + \frac{4\overline{\tau}_{s_o}}{D} - \rho_s(1-\langle\overline{\epsilon}\rangle_A)g \; , \tag{7.39}$$

where $\overline{p}_f$ is *uniform* across the tube (obtained from the radial component of the mean momentum equation) and $\overline{\tau}_{s_o}$ is the surface, particles mean shear stress.

Similar to the analysis based on the mixing length theory discussed in Section 4.4.1(B), the scaled fluid-phase mean velocity is prescribed by

$$\overline{u}_f^* = 5\ln y^* - 3 \quad 5 \le y^* \le 30 \; . \tag{7.40}$$

By assuming that the production and dissipation rates of the fluid-phase turbulent kinetic energy are in *balance*, the derivative of the turbulent kinetic energy at the surface is taken as zero. *Symmetry* conditions are used at the centerline.

(iii) Heat Transfer

The thermal energy equations for the particles and fluid phases were discussed in Section 5.4.3 and are given by (5.161) and (5.164). For the thermally fully developed fields, (5.161) becomes

$$(1-\overline{\epsilon})(\rho c_p)_s \langle \overline{u} \rangle^s \frac{\partial \langle \overline{T} \rangle^s}{\partial x} = \frac{1}{r}\frac{\partial}{\partial r} r (1-\overline{\epsilon}) k_{e_s} \frac{\partial \langle \overline{T} \rangle^s}{\partial r} + \frac{A_o k_f}{d} \langle Nu_d \rangle^{\rm ext}_{A_{sf}} (\langle \overline{T} \rangle^f - \langle \overline{T} \rangle^s) , \qquad (7.41)$$

where A_o is the specific surface area and the particle phase-effective conductivity given by (5.162) becomes

$$\langle k_e \rangle^s = k_{e_s} = \frac{\pi^{1/2}}{16} (\rho c_p)_s d \theta^{1/2} \frac{1}{1+2\lambda/D} \qquad (7.42)$$

and λ is defined by (5.163). The fluid-phase equation (5.164) becomes

$$\overline{\epsilon}(\rho c_p)_f \langle \overline{u} \rangle^f \frac{\partial \langle \overline{T} \rangle^f}{\partial x} = \frac{1}{r}\frac{\partial}{\partial r} r \overline{\epsilon} [k_f + (\rho c_p)_f \frac{\nu_t}{Pr_t}] \frac{\partial \langle \overline{T} \rangle^f}{\partial r} + \frac{A_o k_f}{d} \langle Nu_d \rangle^{\rm ext}_{A_{sf}} (\langle \overline{T} \rangle^s - \langle \overline{T} \rangle^f) , \qquad (7.43)$$

where ν_t is defined in (7.33) and $Pr_t = 0.9$. The particles-fluid interfacial Nusselt number is given through correlations such as (5.29). Louge et al. (1993) suggest

$$\langle Nu_d \rangle^{\rm ext}_{A_{sf}} = 2 + (0.4 {Re_d}^{1/2} + 0.06 {Re_d}^{2/3}) {Pr_f}^{0.4} . \qquad (7.44)$$

For the relative velocity in the Reynolds number Re_d, in *addition* to the *mean* slip velocity given above , they use the root-mean-square of the fluctuating components. However, their results show that the contribution from the mean slip velocity *dominates*.

Two different thermal boundary conditions (at the bounding surface) are used. These are the *constant heat flux* q_{s_b} and the *constant surface temperature* T_{s_b}. At the surface they *impose* a zero gradient on the particles-phase temperature, i.e., similar to (7.8),

$$\frac{d \langle \overline{T} \rangle^s}{dr} = 0 \ , \quad \overline{\epsilon} k_f \frac{d \langle \overline{T} \rangle^f}{dr} = q_{s_b} \qquad r = D/2 . \qquad (7.45)$$

At the centerline ($r = 0$), symmetry conditions are used. The *bulk mixture temperature* is defined as in (7.23), i.e.,

$$\langle \overline{T} \rangle_A^b = \frac{\int_0^{D/2} [(\rho c_p)_f \langle \overline{u} \rangle^f \overline{\epsilon} \langle \overline{T} \rangle^f + (\rho c_p)_s \langle \overline{u} \rangle^s (1 - \overline{\epsilon}) \langle \overline{T} \rangle^s] 2\pi r dr}{c_{p_f} \langle \overline{\dot{m}_f} \rangle_A \frac{\pi D^2}{4} (1 + \frac{\langle \overline{\dot{m}_s} \rangle_A}{\langle \overline{\dot{m}_f} \rangle_A} \frac{c_{p_s}}{c_{p_f}})} . \tag{7.46}$$

As with the analysis of Section (A) above, for the *constant surface heat flux* condition we have

$$\frac{dT_{s_b}}{dx} = \frac{d\langle \overline{T} \rangle^f}{dx} = \frac{d\langle \overline{T} \rangle^s}{dx} = \frac{d\langle \overline{T} \rangle_A^b}{dx}$$
$$= \frac{\pi D q_{s_b}}{c_{p_f} \langle \overline{\dot{m}_f} \rangle_A \frac{\pi D^2}{4} (1 + \frac{\langle \overline{\dot{m}_s} \rangle_A}{\langle \overline{\dot{m}_f} \rangle_A} \frac{c_{p_s}}{c_{p_f}})} . \tag{7.47}$$

The Nusselt number is defined as in (7.23).

For the *constant surface temperature* condition, q_{s_b} is an unknown and the self-similar form of the dimensionless temperatures, i.e.,

$$T_f^* = \frac{\langle \overline{T} \rangle^f - T_{s_b}}{\langle \overline{T} \rangle_A^b - T_{s_b}} , \quad T_s^* = \frac{\langle \overline{T} \rangle^s - T_{s_b}}{\langle \overline{T} \rangle_A^b - T_{s_b}} , \tag{7.48}$$

are used and the details are given by Louge et al. (1993).

(iv) Numerical Results

The coupled momentum equations (7.21) and (7.32) and the coupled energy equations (7.41) and (7.43) are solved numerically by Louge et al. (1993), and some of their results are shown in Figures 7.5(a) to (d). The parameters describing the problem are the *loading ratio* $\langle \overline{\dot{m}_s} \rangle_A / \langle \overline{\dot{m}_f} \rangle_A$, the *tube Reynolds number* $Re_D = \langle \langle \overline{u_f} \rangle \rangle_A D / \nu_f$, the *Archimedes number* $Ar_d = g \rho_s \rho_f d^3 / \mu_f^2$, the *density ratio* ρ_s / ρ_f, the *diameter ratio* D/d, the *coefficients* e, e_o, and a_u, the fluid-phase *Prandtl number* $Pr_f = \nu_f / \alpha_f$, and the *specific heat capacity ratio* c_{p_s} / c_{p_f}.

Figure 7.5(a) shows the predicted, normalized velocity distribution for both phases, as well as the available experimental results. The results are for $\langle \overline{u} \rangle_{r=0}^f = 9.65$ m/s, $d = 500$ μm, polystyrene particles of $\rho_s = 1020$ kg/m^3 in air, and $\langle \overline{\dot{m}_s} \rangle_A / \langle \overline{\dot{m}_f} \rangle_A = 1.1$ which in the *dilute* regime. The predicted results are in good agreement with the measurements. The dashed line is the prediction based on assuming a *zero* viscous shear stress in the particles phase and as evident does *not* agree with the experimental results (note that with $\overline{\tau}_s = 0$, the particles-phase velocity distribution nearly follows the fluid-phase velocity distribution).

Figure 7.5(b) shows the distribution of the normalized, fluid-phase root-mean-square of the velocity fluctuations. The predictions are for $\overline{\epsilon} = 1$ (i.e., no particles), and for $\overline{\epsilon} = 0.9984$ and $\overline{\epsilon} = 0.9982$. These are *very dilute* concentrations. The available experimental results are for $\langle \overline{\dot{m}_s} \rangle_A / \langle \overline{\dot{m}_f} \rangle_A = 1.3$

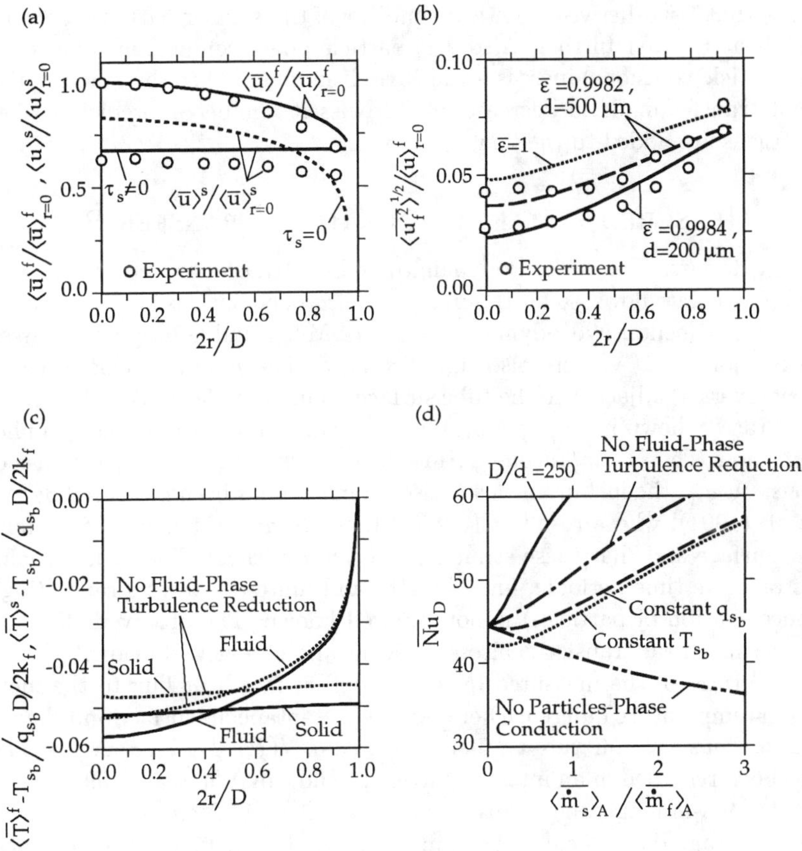

Figure 7.5. (a) Radial distribution of the normalized, axial phasic velocities. (b) Radial distribution of the normalized axial fluid velocity fluctuation. (c) Radial distribution of normalized mean phasic temperatures. (d) Variation of the mean Nusselt number with respect to the loading ratio. (From Louge et al., reproduced by permission ©1993 Pergamon Press.)

and $\langle\overline{u}\rangle^f_{r=0} \simeq 13$ m/s. The results show that the particles *reduce* the fluid-phase turbulence and that this is more significant for the *smaller* particles.

Figure 7.5(c) presents the predicted, normalized mean temperature distribution for the fluid and the particles phases. The results are for $\langle\overline{\dot{m}_s}\rangle_A/\langle\overline{\dot{m}_f}\rangle_A = 0.4$, $Re_D = 1.5 \times 10^4$, $\rho_s/\rho_f = 2100$, $Ar_d = 814$, $D/d = 180$, $c_{p_s}/c_{p_f} = 0.8$, and $Pr_f = 0.7$, which are typical of glass particles-air flow. Figure 7.5(d) shows the prediction of the variation of the mean Nusselt number for the cases of constant q_{s_b} and constant T_{s_b}, with respect to the loading ratio $\langle\overline{\dot{m}_s}\rangle_A/\langle\overline{\dot{m}_f}\rangle_A$. The results are for the conditions used in Figure 7.5(c). An *increase* in Nu_D is predicted for both the constant q_{s_b} and T_{s_b} conditions, and the reduction in the fluid-phase turbulence caused by the presence of the particles *reduces* the Nusselt number. The results also show that for otherwise identical conditions, the smaller particles *enhance* the heat transfer further. Also, the particle-phase conduction (due to interparticle collisions) increases the Nusselt number. Then, by ignoring this heat transfer mode, a decrease in the Nusselt number is predicted when particles are added to the flow.

(B) Measured Instantaneous Heat Transfer Rate

In the discussion of the hydrodynamic and heat transfer aspects of the internal particulate flows, so far the *time-averaged* behavior has been addressed. The experimental hydrodynamic results of Miller and Gidaspow discussed in Section 7.1.1(A) were also time averaged. The *negative* velocity (solid flow reversal) adjacent to the tube surface, found in upflows at high particle flow rates, shown in Figure 7.3(c), was considered as a time-averaged phenomenon. The *instantaneous*, surface heat transfer rate (or, for the case of constant q_{s_b}, the instantaneous Nusselt number) has been measured by Wu et al. (1991), whose results also show that *strands* of particles *fall* along the surface and that these strands have a *finite length.* The hydrodynamics of these time-periodic, finite-length, and finite-thickness strands (high concentration of particles) is not yet well known. The observed effect on the surface heat transfer is also significant and is reviewed below.

The ratio of the measured instantaneous surface heat flux to the measured temperature difference between the instantaneous surface and the *instantaneous* bulk mixture temperature, i.e., $q_{s_b}/(T_{s_b} - \langle T\rangle^b) = Nu_D k_f/D$, has been recorded in an internal particulate flow by Wu et al. (1991). They used a 9.3-m-long tube of inside diameter D = 15.2 cm, sand particles with average diameter d = 171 μm, with air flow of $\langle \overline{u_f}\rangle_A$ = 7 m/s, and several values of $\langle \overline{\dot{m}_s}\rangle_A$. Their results for $\langle \overline{\dot{m}_s}\rangle_A$ = 62.8 kg/m^2-s is shown in Figure 7.6(a). Also shown is the instantaneous solid volume fraction $1 - \epsilon$ measured 1 mm from the surface and adjacent to the location of the heat transfer measurement. The variations of $Nu_D k_f/D$ and $1 - \epsilon(r = D/2)$, with respect to time, show fluctuations with a large range of frequencies. The *time constant* for the solid-volume fraction probe is much *smaller* than that of the heat-transfer probe, and therefore, the fluctuations are more accurately demonstrated by the solid-volume probe. Note that the 50% variation in the heat transfer rate is *not* expected to be the result of the *flow turbulence* and *particle collisions*. Wu et al. (1991) also perform flow visualizations and observe time- and space-*periodic* variations in the *concentration* of the particles adjacent to the surface in the form of particle strands. Note that ϵ changes by nearly 0.4 as these strands pass by the voidage probe, indicating very *high* concentration of the particles (compared to $\langle\overline{\epsilon}\rangle_A$ which is about 0.75).

By examining the results of Figure 7.6(a) and those for other values of $\langle \overline{\dot{m}_s}\rangle_A$, Wu et al. (1991) obtain the time interval Δt_ϵ that a given *porosity* is

(a)

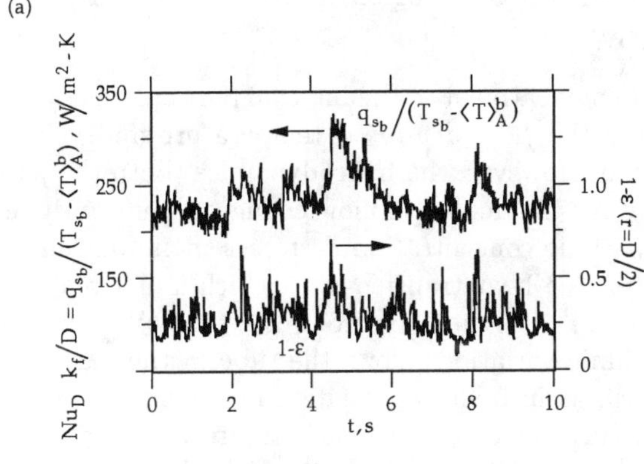

(b)

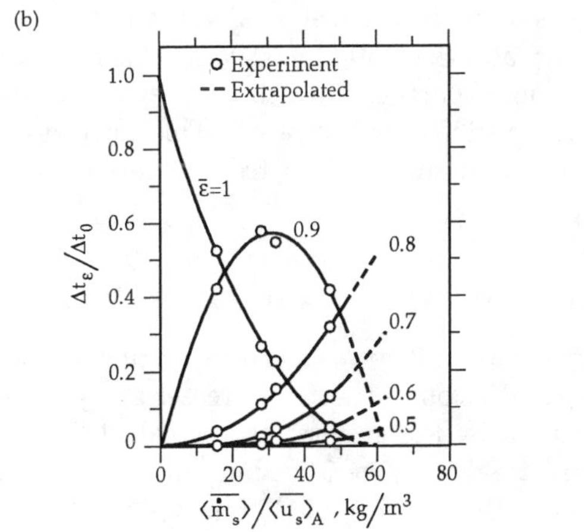

Figure 7.6. (a) Measured ratio of the instantaneous surface heat flux to the measured temperature difference. (b) Fraction of the total sampling time where a given mean porosity is observed, as a function of the particles-phase density. (From Wu et al., reproduced by permission ©1991 Pergamon Press.)

observed (recorded) adjacent to the surface. Then each porosity ϵ will occur during a *fraction* of the total sampling time Δt_o, and this ratio $\Delta t_\epsilon/\Delta t_o$ for *increments* of ϵ is plotted as a function of $\langle \dot{m}_s\rangle_A \langle \overline{\epsilon}\rangle_A / \langle \overline{u_s}\rangle_A$, which is defined as the time- and area-averaged solid density $\langle \overline{\rho_s}\rangle_A$. Their results are shown in Figure 7.6(b). For $\langle \overline{\rho_s}\rangle_A = 0$, $\epsilon = 1$ at all times. As $\langle \overline{\rho_s}\rangle_A$ increases the fraction of time that $\epsilon = 1$ occurs *decreases* monotonically. The fraction of time that $\epsilon = 0.9$ occurs *initially* increases and as $\langle \overline{\epsilon}\rangle_A$ increases with $\langle \overline{\rho_s}\rangle_A$, then after reaching a *peak* this fraction of time begins to *decrease*. The detailed *hydrodynamic* aspects of the internal particulate flows is *not* yet studied using the time- and local phase volume-averaged momentum equations which were discussed in Section 5.4.2, or simpler models.

7.1.4 Surfaces Submerged in Internal Particulate Flow

In the theoretical treatment of the internal particulate flow and heat transfer, so far only the *fully developed* temperature and velocity fields have been considered. However, the hydrodynamics (both the particles and the fluid phase) of the particulate upflow changes significantly along the tube, because the particle concentration *decreases* monotonically due to the action of *gravity*. The heat transfer rate, which is noticeably dependent on this *concentration*, also changes along the tube. We begin by examining the hydrodynamic variations along the tube in the particulate upflow. In some cases such as in boiler application involving *combustion* in the particulate flow, surfaces enclosing coolants are *submerged* in the *internal, exothermically reacting particulate flow*. The heat transfer rate to these submerged surfaces is of interest and similar to the heat exchange between the particles-fluid flow and the *confining surface*, this rate also depends on the local concentration of particles (which can be significantly influenced by the presence of these submerged surfaces). This correlation between the *local* particle concentration and the heat transfer rate is examined next.

(A) Axial Hydrodynamic Variations Along a Tube

In the internal particulate upflow (*concurrent* particulate and fluid flow), along the tube the particle concentration decreases and with it the pressure gradient. This is caused by the gravity (this is called the *gravity-affected stratification*), i.e., phase-density *negative* buoyancy, and is observed in the experimental results of Dou et al. (1991), Andersson and Leckner (1992), and Miller and Gidaspow (1992). For a large tube ($D \to \infty$), in the *axial components* of the momentum equations, (5.147) and (5.155), the contributions due to *inertia* and *boundary* (i.e., viscous shear stress) may become small. Then for the particles phase, the *volumetric drag* and the *gravity* force *dominate* and for the fluid phase the *pressure gradient* and the *volumetric drag* dominate. The pressure *decreases* along the tube, i.e., the x-axis, varies with a nearly logarithmic distribution, i.e., $\ln x$, and the mean, area-averaged particles-phase density $\langle\overline{\rho_s}\rangle_A \equiv \langle\overline{\dot{m}_s}\rangle_A\langle\overline{\epsilon}\rangle_A/\langle\overline{u_s}\rangle_A$ *decreases* with a nearly $1/x$ distribution (Andersson and Leckner, 1992). Then for a given $\langle\overline{\dot{m}_s}\rangle_A$ entering the tube, the velocity of the particles phase $\langle\overline{u_s}\rangle_A/\langle\overline{\epsilon}\rangle_A$ *increases* along x, i.e., the particles phase *accelerates*.

Also, the results of Miller and Gidaspow (1992), reviewed in Section 7.1.1 (A), show that $\langle\overline{\rho_s}\rangle$ varies *radially*. For example, if correlation (7.2) is used, it can be rewritten as

$$\frac{\langle\overline{\rho_s}\rangle}{\langle\overline{\rho_s}\rangle_A} = 2(\frac{2r}{D})^2 . \tag{7.49}$$

Then, both the *axial* and the *radial* variation of $\langle\overline{\rho_s}\rangle$ are significant and we expect these to influence the heat transfer rate when submerged surfaces are placed at various axial and radial locations.

(B) Local Nusselt Number for a Submerged Tube

In order to examine the *axial variation* of the local heat transfer rate to surfaces submerged in an internal particulate flow, Dou et al. (1991) place (i.e., submerge) a tube of diameter $D_1 = 9.5$ mm and of length $L_1 = 2.64$ m inside and on the centerline of an internal particulate flow with diameter $D = 150$ mm and length $L = 11$ m. Only a *portion* of submerged tube is heated. They use sand particles with a mean diameter $d = 185$ μm and air flowing at $\langle\overline{u_f}\rangle_A = 5.2$ m/s, and place the heated tube at two different axial locations along the flow. The heated tube allows for the *local* measurement of the *local heat flux* and the *local surface* temperature, i.e., $\overline{q}_{s_s}$ and $\overline{T}_{s_s}$, and $\langle\overline{T}\rangle_A^b$ is determined from other measurements. We define the *submerged surface mean* Nusselt number as

$$\overline{Nu}_{D_1} = \frac{\overline{q}_{s_s} D_1}{(\overline{T}_{s_s} - \langle\overline{T}\rangle_A^b)k_f} . \tag{7.50}$$

Figure 7.7(a) shows the measured variations of the local, mean particles-phase density (in absence of the heated submerged tube) and the local value of $\overline{Nu}_{D_1} k_f/D_1$ (i.e., internal heat transfer coefficient) as a function of the axial location x. Note that only a portion of the tube (1.35 m) is heated. The results are for $\langle\overline{\dot{m}_s}\rangle_A = 0$, 30, and 50 kg/m^2-s. The first case corresponds to no particle suspension and shows the expected axial *decrease* characteristic of external flow and heat transfer along heated surfaces. With the *addition* of *particles*, the local heat transfer *increases*, and this increase is *larger* where $\langle\overline{\rho_s}\rangle_A$ is larger. For the placement of the tube further downstream, a *constant* (asymptotic) $\overline{Nu}_{D_1}$ is observed.

They also allow for the heating the *confining* tube and measure the surface heat transfer rate $\overline{q}_{s_b}$ and the surface temperature $\overline{T}_{s_b}$ and also determine $\langle\overline{T}\rangle_A^b$ through other measurements. The *confining surface mean* Nusselt number is defined as in (7.23), i.e.,

$$\overline{Nu}_D = \frac{\overline{q}_{s_b} D}{(\overline{T}_{s_b} - \langle\overline{T}\rangle_A^b)k_f} . \tag{7.51}$$

Figure 7.7(b) shows the variations of $\overline{Nu}_{D_1} k_f/D_1$ and $\overline{Nu}_D k_f/D_1$ (i.e., the inner and outer surface heat transfer coefficients) with respect to the *local, mean* particle-phase density $\langle\overline{\rho_s}\rangle$. This local density is found by using a correlation such as (7.49), which shows higher concentrations *adjacent* to the tube surface. Considering the *complexity* of the hydrodynamics adjacent to the *outer* and *inner* radii, the *dominance* of the local, mean particles-phase density $\langle\overline{\rho_s}\rangle$ in determining the heat transfer rate is rather *interesting*. A linear curve fit suggests a *direct* proportionality.

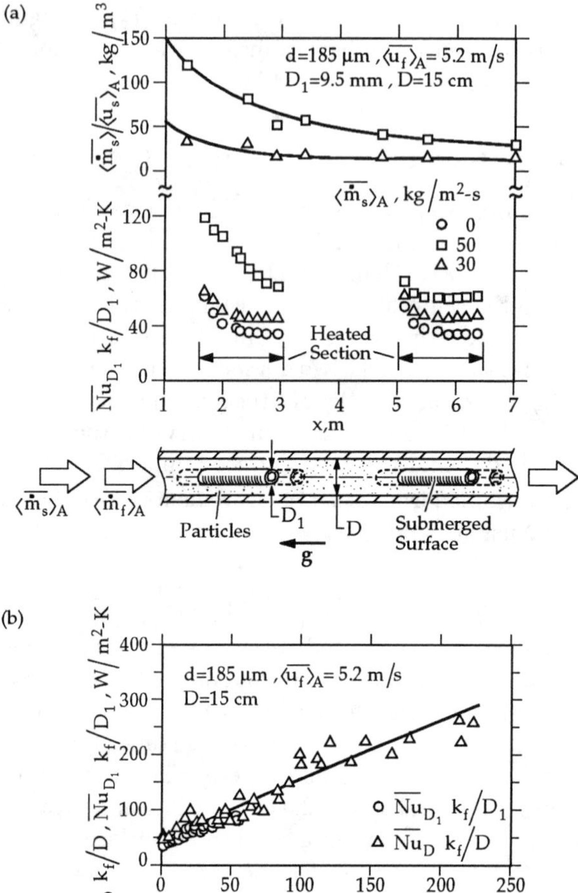

Figure 7.7. (a) Variations of the mean particles-phase density and the mean heat transfer coefficient for the immersed tube, with respect to the axial location. (b) Variation of the mean heat transfer coefficients (inner and outer surfaces) with respect to the mean particles-phase density. A best curve fit is also shown. (From Lou et al., reproduced by permission ©1991 Taylor and Francis.)

7.2 Fluidized Bed Bounding Solid Surfaces

Fluidization occurs when the vertical (i.e., upflow) velocity of the fluid through a bed of solid particles becomes larger than a *threshold velocity*. This threshold velocity is called the *minimum fluidization* velocity $\langle u_f \rangle_{mf}$. For $\langle u_f \rangle$ larger than $\langle u_f \rangle_{mf}$ the particles are no longer resting on each other and on the distributor, but are suspended, as shown in Figure 7.1(b). When $\langle u_f \rangle$ increases above $\langle u_f \rangle_{mf}$, the particles *remain in* the bed (i.e., *no net* particles flow across the bed) up to a *second critical velocity* beyond which the particles are *conveyed* out of the bed (or column), and this is known as the *hydraulic* or *pneumatic transport* (Couderc, 1985). This hydraulic transport is the phenomenon with the internal particulate flow discussed in Section 7.1 as being one of its special cases. In this section we consider fluidized beds with *no net* transport of particles. As shown in Figure 7.1 (b), the gas-fluidized beds generally operate at gas velocities $\langle u_f \rangle$ where *rising low-particle* (i.e., *dilute*) *concentration* regions are *dispersed* in a *high-particle* (i.e., *dense*) *concentration* surrounding. The *dense phase* is also called the *emulsion phase*, and the low-concentration regions are called *bubbles* (when their diameter is less than the bed diameter D) or *slags* (when their diameter approaches D; Clift and Grace, 1985).

In this section, the hydrodynamics of the fluidized beds is first reviewed, followed by the heat transfer to the *confining* (or *bounding*) surfaces of the fluidized bed. Then the heat transfer to surfaces *submerged* in fluidized beds is discussed. The principles governing the hydrodynamics and heat transfer are similar to those discussed for the internal particulate flow, except for the *recirculating* flow of the particles and the rather *abrupt* change in the particle concentration across the top of bed, as shown in Figures 7.1 (a) and (b).

7.2.1 Hydrodynamics

The *mechanics* of the fluidized beds addressing the *continuum models*, *rheological behavior, instabilities* and *waves, bubbling*, and other aspects, are in part known, but many aspects remain unknown. A recent appraisal is given by Homsy et al. (1992). In Sections 5.4.2 and 7.1.1, some aspects of the solid-fluid flow were discussed, and here features specific to fluidized beds, namely, the *minimum fluidization velocity*, the relationship between the superficial fluid velocity and the *average bed porosity*, the *stability* of the fluidized bed, the *structure* of *bubbles*, and the *regimes* of *fluidization*, are addressed.

(A) Minimum Fluidization Velocity

At the onset of fluidization, the fluid pressure gradient (which is uniform across a bed of length L of uniform porosity ϵ) is *equal* to the buoyant weight (i.e., phase-density buoyancy) of the solid particles per unit area (Coudrec, 1985), i.e.,

$$\frac{\Delta \langle p \rangle^f}{L} = -(1 - \epsilon_{m_f})(\rho_s - \rho_f) g \, . \tag{7.52}$$

This pressure drop is related to the superficial (or Darcean) fluid velocity $\langle u_f \rangle = u_D$, through the *extended Darcy law* discussed in Section 3.2.4. For a steady, one-dimensional flow along the x-axis and for no inertial and boundary effects, this momentum equation becomes the simplified form of (3.75), i.e.,

$$\frac{d\langle p\rangle^f}{dx} = -\frac{\mu_f\langle u_f\rangle}{K} - \frac{C_E}{K^{1/2}}\rho_f\langle u\rangle^2 \,. \tag{7.53}$$

The permeability K is given by the *Carman-Kozeny* empirical relation and along with the empirical *Ergun coefficient* C_E for spherical particles, are (Kaviany, 1991)

$$K = \frac{\epsilon^3}{150(1-\epsilon^2)}d^2 \,, \quad \frac{C_E}{K^{1/2}} = \frac{1.75(1-\epsilon)}{\epsilon^3}\frac{1}{d} \,. \tag{7.54}$$

Then at the minimum fluidization velocity the momentum equation (7.53) can be written as

$$\frac{\Delta\langle p\rangle^f}{L} = -\frac{150(1-\epsilon_{mf})^2}{\epsilon_{mf}^3}\frac{\mu_f\langle u_f\rangle}{d^2} - \frac{1.75(1-\epsilon_{mf})}{\epsilon_{mf}^3}\frac{\rho_f\langle u_f\rangle^2}{d} \,. \tag{7.55}$$

Now equating this pressure drop, which is due to the solid-fluid interfacial drag force exerted to the particles, to the negative phase-density buoyancy expression given by (7.52), we have the *equation* for the minimum fluidization velocity as

$$\frac{1.75\rho_f}{\epsilon_{mf}^3 d}\langle u_f\rangle_{mf}^2 + \frac{150(1-\epsilon_{mf})\mu_f}{\epsilon_{mf}^3 d^2}\langle u_f\rangle_{mf} - (\rho_s - \rho_f)g = 0 \,. \tag{7.56}$$

For spherical particles, the porosity for minimum fluidized is taken to be slightly larger than that for the random arrangement of monosize particles (which is between 0.39 and 0.41), i.e., $\epsilon_{mf} = 0.415$ to 0.420. Equation (7.56) is very sensitive to the prescribed ϵ_{mf} and as *alternative* correlations are developed for the determination of $\langle u_f\rangle_{mf}$. The parameters in (7.56) can be grouped into the *Galileo number* based on the particle diameter d, Ga_d, and the *density ratio*, ρ_s/ρ_f. Then one of these correlations (Coudrec, 1985) gives the predictive equation for $\langle u_f\rangle_{mf}$ for *liquid-solid* beds as

$$(Re_d)_{mf} = \frac{\langle u_f\rangle_{mf} d}{\nu_f} = 1.54\times 10^{-2}{Ga_d}^{0.66}(\frac{\rho_s-\rho_f}{\rho_f})^{0.70}$$

$$10 < (Re_d)_{mf} < 10^3 \quad \text{liquids} \,, \tag{7.57}$$

where

$$Ga_d = \frac{g\rho_f^2 d^3}{\mu_f^2} \,. \tag{7.58}$$

For *gas-solid beds*, where the Reynolds number (for the same $\langle u_f\rangle$ and d) is generally *smaller* for gases, different correlations are used and among these is (Coudrec, 1985)

$$(Re_d)_{mf} = (31.6^2 + 0.0425Ga_d\frac{\rho_s-\rho_f}{\rho_f})^{1/2} - 31.6 \quad \text{gases} \,. \tag{7.59}$$

(B) Average Bed Porosity

For *monosize* spherical particles with *liquid* as the fluid, *expansion* of the bed occurs as $\langle u_f \rangle$ increases above the minimum fluidization velocity $\langle u_f \rangle_{mf}$. The average bed porosity $\langle \epsilon \rangle^b$ has been correlated with three parameters, the *Reynolds* number, the *Galileo* number, and the *density ratio*. One of these correlations for liquid-solid beds is (Coudrec, 1985)

$$\langle \epsilon \rangle^b = \begin{cases} 1.58 Re_d^{0.33} Ga_d^{-0.21} (\frac{\rho_s - \rho_f}{\rho_f})^{-0.22} & \epsilon \le 0.85 \\ 1.20 Re_d^{0.17} Ga_d^{-0.11} (\frac{\rho_s - \rho_f}{\rho_f})^{-0.12} & \epsilon > 0.85 \quad \text{liquids} , \end{cases} \tag{7.60}$$

where

$$Re_d = \frac{\langle u_f \rangle d}{\nu_f} . \tag{7.61}$$

For *gas-solid* fluidized beds, between $\langle u_f \rangle_{mf}$ and the *minimum bubbling velocity* $\langle u_f \rangle_{mb}$, which will be discussed in Section (D), the bed also *expands* (i.e., $\langle \epsilon \rangle^b$ *increases*), but more significantly than for liquids. In general, even in this range the porosity is *not* uniform. A correlation for the variation of $\langle \epsilon \rangle^b$ in this range for *gas-solid* beds is given by (Coudrec, 1985)

$$\frac{(\langle \epsilon \rangle^b)^3}{1 - \langle \epsilon \rangle^b} \frac{g(\rho_s - \rho_f) d^2}{\mu_f} = 210(\langle u_f \rangle - \langle u_f \rangle_{mf}) + \frac{\epsilon_{mf}}{1 - \epsilon_{mf}} \frac{g(\rho_s - \rho_f) d^2}{\mu_f}$$

$$\langle u_f \rangle < \langle u_f \rangle_{mb} \quad \text{gases} . \tag{7.62}$$

(C) Equations of Motion for Particulate Flow

The *continuum* equations of motion (i.e., momentum conservation equations) for particluate flows has been under development with increased recent activities (Satake and Jenkins, 1988; Roco, 1993). Although presently a *unified* description (or a *set* of models applicable to various regimes) is *not* yet available, the studies dealing with the *granular flows* (i.e., the particles motion *not* influenced by the fluid motion), on the one hand, and the *dilute* particle concentration in the *particulate flows*, on the other, are addressing similar *rheological* and *kinetic* behavior of the moving particles.

For granular flows, under applied stresses the transition will occur from a *solidlike* behavior to a *fluidlike* behavior. A *quasi-static transitional* behavior is also found between these two asymptotic behaviors (Zhang and Campbell, 1992) when *soft* particles are used to allow for long-duration contacts between particles in the stagnant regions of the bed.

For particulate flows, where both the flow of the particles and the fluid phase are considered, the *sliding* and *rolling* of the particles and the *turbulence* in the fluid phase must be considered. Chart 7.3 gives a description

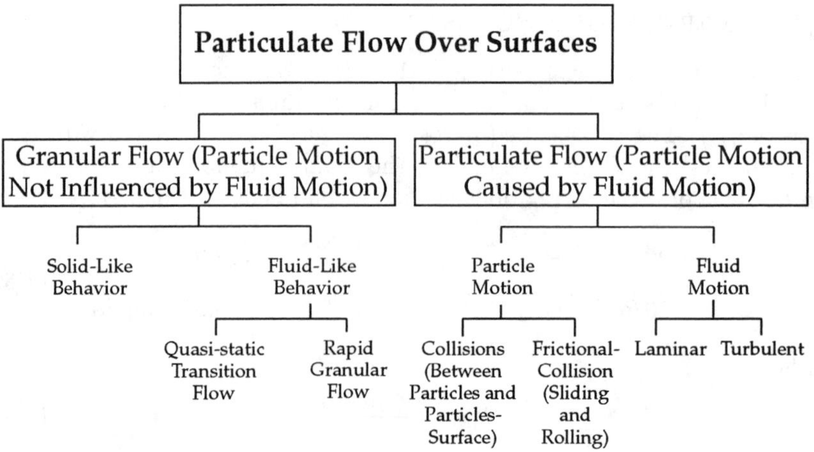

Chart 7.3. A classification of particulate flows over surfaces.

of particulate flows over surfaces with the first division made between the *presence* or *lack* of a fluid *affecting* the motion of the particles.

The *two-medium* modeling of the equation of motion for two-phase systems was discussed in Section 3.3.2, for two moving and *deforming continuous phases*. In Section 5.4.2 the collision-based two-medium modeling of the particulate flow was discussed. As expected, future inclusion of the interparticle sliding and the particle rolling, and other realistic and significant mechanisms, will allow for more accurate predictions of the particulate flow. Meantime, the flow instability and the heat transfer analyses use *various simplified* forms of the equation of motion for the particles phase. Then the results are useful and should be examined in the *limited range* for which they have been developed (e.g., particle *concentration range*, particle *size range*, particle *elasticity range*, fluid-phase *velocity fluctuations range*).

(D) Stability

The *hydrodynamic stability* of a *unidirectional* flow of a fluid with velocity $\langle u_f \rangle_o$ through a *uniform* particle concentration distribution $1-\epsilon_o$ with the particles being *initially stationary* (i.e., initially $\langle u_s \rangle_o = 0$) can be analyzed using the standard linear stability theory applied to the fluid flow. This is done by introducing a *disturbance* to a *base* (i.e., initial state) field, and an example was given in Section 4.12.1(C) for the thermobuoyant flow. The continuum equation of motion for the particles phase is needed and various (and continuing to be improved) forms are used. For an *isothermal* particulate flow, a *laminar* fluid flow is assumed and a rheological description of the particles phase is given. Then the growth or decay of the disturbances in the fields (i.e., velocities, porosity, and fluid pressure) are examined. A review of the hydrodynamic stability of fluid-particles systems is given by Jackson (1985), and some of the available results are briefly discussed below.

Assuming constant densities ρ_f and ρ_s, the continuity and momentum equations for the fluid and particle phases are those given by (5.154), (5.155), (5.146), and (5.147). For a *laminar* flow and by defining the stress tensors according to the model used by Jackson (1985), these equations become

$$\frac{\partial \epsilon}{\partial t} + \nabla \cdot \epsilon \langle \mathbf{u} \rangle^f = 0 \,, \tag{7.63}$$

$$\frac{\partial}{\partial t}(1-\epsilon) + \nabla \cdot (1-\epsilon) \langle \mathbf{u} \rangle^s = 0 \,, \tag{7.64}$$

$$\rho_f \epsilon (\frac{\partial \langle \mathbf{u} \rangle^f}{\partial t} + \langle \mathbf{u} \rangle^f \cdot \nabla \langle \mathbf{u} \rangle^f) = -\epsilon \nabla \langle p \rangle^f + \epsilon \nabla \cdot \mathbf{S}_f - \epsilon \mathbf{f}_d + \epsilon \rho_f \mathbf{g} \,, \tag{7.65}$$

$$\rho_s (1-\epsilon) (\frac{\partial \langle \mathbf{u} \rangle^s}{\partial t} + \langle \mathbf{u} \rangle^s \cdot \nabla \langle \mathbf{u} \rangle^s) = -(1-\epsilon) \nabla \langle p \rangle^s + (1-\epsilon) \nabla \cdot \mathbf{S}_f - \nabla \cdot \mathbf{S}_s + \epsilon \, \mathbf{f}_d + (1-\epsilon) \rho_s \mathbf{g} \,. \tag{7.66}$$

The volumetric drag force includes the effect of the *added mass* (Section 5.4.1) and is written as

$$\mathbf{f}_d = \beta (\langle \mathbf{u} \rangle^f - \langle \mathbf{u} \rangle^s) + \frac{1-\epsilon}{\epsilon} C \rho_f \frac{d}{dt} (\langle \mathbf{u} \rangle^f - \langle \mathbf{u} \rangle^s) \,, \tag{7.67}$$

where β is related to the drag coefficient through (5.144) and the particle drag coefficient in the presence of neighboring particles changes with the porosity as discussed in Section 5.2.1. The *virtual mass coefficient* C is also a function of the porosity. Then we have

$$\beta = \beta(\epsilon, |\langle \mathbf{u} \rangle^f - \langle \mathbf{u} \rangle^s|) \,, \quad C = C(\epsilon) \,. \tag{7.68}$$

For $Re_d \to 0$, the coefficient β becomes *only* a function of the porosity.

The components of the viscous stress tensors are modeled as

$$\tau_{f_{ij}} = \langle \lambda \rangle^f (\epsilon) \frac{\partial \langle u_k \rangle^f}{\partial x_k} \delta_{ij} + \langle \mu \rangle^f (\epsilon) (\frac{\partial \langle u_i \rangle^f}{\partial x_j} + \frac{\partial \langle u_j \rangle^f}{\partial x_i} - \frac{2}{3} \frac{\partial \langle u_k \rangle^f}{\partial x_k} \delta_{ij}) \,, \tag{7.69}$$

$$\tau_{s_{ij}} = \langle \lambda \rangle^s (\epsilon) \frac{\partial \langle u_k \rangle^s}{\partial x_k} \delta_{ij} + \langle \mu \rangle^s (\epsilon) (\frac{\partial \langle u_i \rangle^s}{\partial x_j} + \frac{\partial \langle u_j \rangle}{\partial x_i} - \frac{2}{3} \frac{\partial \langle u_k \rangle^s}{\partial x_k} \delta_{ij}) \,, \tag{7.70}$$

where $\langle \lambda \rangle^f = \langle \lambda \rangle^f (\epsilon)$ and $\langle \lambda \rangle^s = \langle \lambda \rangle^s (\epsilon)$ are the *effective bulk* (or *second*) viscosities and $\langle \mu \rangle^f = \langle \mu \rangle^f (\epsilon)$ and $\langle \mu \rangle^s = \langle \mu \rangle^s (\epsilon)$ are the *effective shear* viscosities. All viscosities are expected to *decrease* as ϵ increases. Similar forms of equation of motion are used by Homsy et al. (1980) in their stability analysis.

The *stability* of *unbounded, uniform suspension* unidirectional *base* flow (flowing against the gravity vector) given by (subscript o designates the *base flow condition*)

$$\langle \mathbf{u} \rangle^f = \langle \mathbf{u} \rangle^f_o = \mathbf{s}_x \langle u \rangle^f_o \ , \quad \langle \mathbf{u} \rangle^s = \langle \mathbf{u} \rangle^s_o = 0 \ , \quad \epsilon = \epsilon_o \ , \tag{7.71}$$

where $\mathbf{s}_x$ is the unit vector in the x direction, is examined by introducing disturbance components such that

$$\langle \mathbf{u} \rangle^f = \mathbf{s}_x \langle u \rangle^f_o + \mathbf{u}'_f \ , \quad \langle \mathbf{u} \rangle^s = \mathbf{u}'_s \ , \quad \epsilon = \epsilon_o + \epsilon' \ , \quad \langle p \rangle^f = \langle p \rangle^f_o + p'_f \ . \tag{7.72}$$

The base pressures are determined from (7.63) to (7.66) which for the base flow reduce to

$$\nabla \langle p \rangle^f_o + \mathbf{s}_x \beta(\epsilon_o) \langle u \rangle^f_o + \mathbf{s}_x \rho_f g = 0 \quad Re_d \to 0 \ , \tag{7.73}$$

$$\beta(\epsilon_o) \langle u \rangle^f_o - (1 - \epsilon_o)(\rho_s - \rho_f) g = 0 \quad Re_d \to 0 \ . \tag{7.74}$$

Substituting the perturbed conditions (7.72) into (7.63) to (7.66) and using (7.73) and (7.74), after *linearizing* we have

$$\frac{\partial \epsilon'}{\partial t} + \langle u \rangle^f_o \frac{\partial \epsilon'}{\partial x} + \epsilon_o \nabla \cdot \mathbf{u}'_f = 0 \ , \tag{7.75}$$

$$-\frac{\partial \epsilon'}{\partial t} + (1 - \epsilon_o) \nabla \cdot \mathbf{u}'_s = 0 \ , \tag{7.76}$$

$$\rho_f [1 + \frac{1 - \epsilon_o}{\epsilon_o} C(\epsilon_o)] (\frac{\partial \mathbf{u}'_f}{\partial t} + \langle u \rangle^f_o \frac{\partial \mathbf{u}'_f}{\partial x}) - \rho_f \frac{1 - \epsilon_o}{\epsilon_o} C(\epsilon_o) \frac{\partial \mathbf{u}'_s}{\partial t} =$$
$$-\nabla p'_f - \beta(\epsilon_o)(\mathbf{u}'_f - \mathbf{u}'_s) - \epsilon' \mathbf{s}_x \langle u \rangle^f_o \frac{\partial \beta(\epsilon_o)}{\partial \epsilon} +$$
$$[\langle \lambda \rangle^f (\epsilon_o) + \frac{1}{3} \langle u \rangle^f (\epsilon_o)] \nabla (\nabla \cdot \mathbf{u}'_f) + \langle \mu \rangle^f (\epsilon_o) \nabla^2 \mathbf{u}'_f \ , \tag{7.77}$$

$$[\rho_s (1 - \epsilon_o) + \rho_f \frac{1 - \epsilon_o}{\epsilon_o} C(\epsilon_o)] \frac{\partial \mathbf{u}'_s}{\partial t} - \rho_f (1 - \epsilon_o) \, [1 +$$
$$\frac{C(\epsilon_o)}{\epsilon_o}] (\frac{\partial \mathbf{u}'_f}{\partial t} + \langle u \rangle^f_o \frac{\partial \mathbf{u}'_f}{\partial x}) = \mathbf{s}_x (\rho_s - \rho_f) g \epsilon' +$$
$$\beta(\epsilon_o)(\mathbf{u}'_f - \mathbf{u}'_s) + \epsilon' \mathbf{s}_x \langle u \rangle^f_o \frac{\partial \beta(\epsilon_o)}{\partial \epsilon} - \frac{\pi d^3}{6} \frac{\partial \langle p \rangle^s}{\partial \epsilon} \nabla \epsilon' +$$
$$[\langle \lambda \rangle^s (\epsilon_o) + \frac{1}{3} \langle \mu \rangle^s (\epsilon_o)] \nabla (\nabla \cdot \mathbf{u}'_s) + \langle \mu \rangle^s (\epsilon_o) \nabla^2 \mathbf{u}'_s \ . \tag{7.78}$$

Now plane-wave solutions with periodic time variations, similar to (4.369) and (4.370), are assumed for the disturbances using a *wave number vector* $\boldsymbol{\eta}$ and the *complex* growth rate ω, i.e.,

$$\mathbf{u}'_f = \phi_{u_f} e^{\omega t} e^{i \boldsymbol{\eta} \cdot \mathbf{x}} \ , \tag{7.79}$$

$$\mathbf{u}'_s = \phi_{u_s} e^{\omega t} e^{i \boldsymbol{\eta} \cdot \mathbf{x}} \ , \tag{7.80}$$

$$\epsilon' = \phi_\epsilon e^{\omega t} e^{i \boldsymbol{\eta} \cdot \mathbf{x}} \ , \tag{7.81}$$

$$p'_f = \phi_{p_f} e^{\omega t} e^{i \boldsymbol{\eta} \cdot \mathbf{x}} \ , \tag{7.82}$$

$$\omega = \omega_r - i \omega_i \ , \quad \boldsymbol{\eta} = \mathbf{s}_x \eta_x + \mathbf{s}_y \eta_y + \mathbf{s}_z \eta_z \ .$$

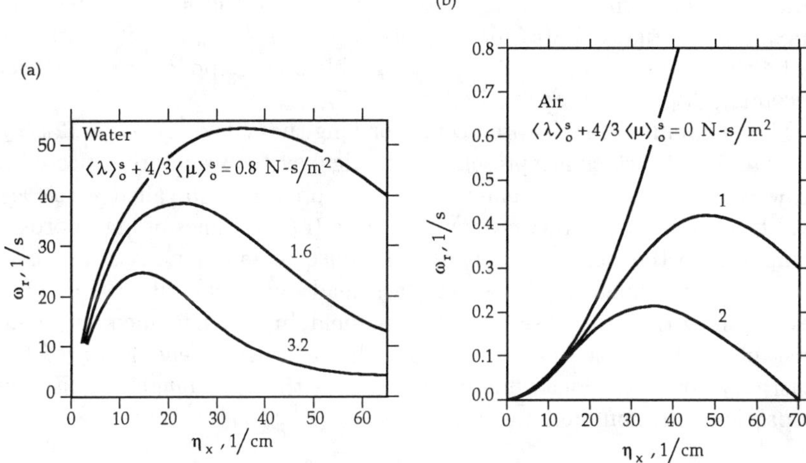

Figure 7.8. Variation of the growth-rate factor with respect to the longitudinal wave number. (a) Water, and (b) air. (From Jackson, reproduced by permission ©1985 Academic Press.)

The *disturbance wave speed* is given by $\omega_i/|\eta|$ and ω_r is the *growth-rate factor* such that for $\omega_r < 0$ the disturbances *decay*, while for $\omega_r > 0$ they grow and lead to an instability. There are six velocity amplitudes ϕ_u and along the porosity and pressure amplitudes ϕ_ϵ and ϕ_p. When (7.79) to (7.82) are substituted in (7.75) to (7.78), eight linear algebraic homogeneous equations will emerge which have nontrivial solutions only if the determinant of their coefficients vanishes. This determinant gives the *characteristic* equation which is used to determine ω as a function of η. Details of this are given by Jackson (1985), and here some of the results for $\omega_r = \omega_r(\eta_x)$ are discussed.

Figures 7.8(a) and (b) show the variation of the growth rate factor with respect to the longitudinal wave number η_x for *water* and *air*. The parameters influencing the stability are ϵ_o, ρ_s/ρ_f, $C(\epsilon_o)$, $\beta(\epsilon_o)$, $\partial\beta(\epsilon_o)/\partial\epsilon$, $|\partial p_s(\epsilon_o)/\partial\epsilon|$, d, and $\langle\lambda\rangle_o^s + (4/3)\langle\mu\rangle_o^s$. A relationship between $\langle\lambda\rangle_o^f$ and $\langle\mu\rangle_o^f$ is also prescribed. The results of Figure 7.8(a) are for *water* at room temperature and for $\epsilon_o = 0.46$, $d = 860$ μm, *glass beads*, $C(\epsilon_o) = 0.5$, $[\partial\beta(\epsilon_o)/\partial\epsilon)\epsilon_o/\beta_o = -3.47$, and $|\partial p_s(\epsilon_o)/\partial\epsilon| = 1.8$ Pa. The results of Figure 7.8(b) are for *air* with $\epsilon_o = 0.42$, $d = 860$ μm, *glass beads*, $C(\epsilon_o) = 0.5$, $[\partial\beta(\epsilon_o)/\partial\epsilon]\epsilon_o/\beta_o = -3.91$, and $|\partial p_s(\epsilon_o)/\partial\epsilon| = 2$ Pa. The results show that the disturbance plane waves, for the conditions considered here, grow rapidly for *small* wave numbers (i.e., *all* the small wave number disturbances grow rapidly). Very *large* wave numbers do *not* grow as fast (although all the results shown in the figures show a growth, i.e., $\omega_r \geq 0$, indicating that the particulate flow is *always* unstable). Comparison of Figures 7.8(a) and (b) shows that the waves grow much faster when the fluid is *air* (ω_r is higher by about two orders of magnitude). The lack of a maximum in Figure 7.8(a) for the case of $\langle\lambda\rangle_o^s + \frac{4}{3}\langle\mu\rangle_o^s = 0$ is considered to be the results of the mathematical problem corresponding to this case becoming ill-posed (Jackson, 1985).

The maximum growth rate occurs for longitudinal *wavelengths* $(2\pi/\eta_x)$ of 10 to 30 mm which are present in practical systems and are expected to grow *very* fast. The disturbance waves in porosity (also called *parwaves*) lead to the *formation* (i.e.,*onset*) of *bubbles* (i.e., volumes of high porosity compared to the bulk of the bed). The onset of bubbling has also been examined experimentally for water-glass beads by Ham et al. (1990). Their results show that the base uniform-flow field $\langle \mathbf{u}\rangle_o^f$ is *not* unconditionally unstable and that the bed can be stabilized by the phenomenon of *bulk elasticity* from the velocity fluctuations and by the *hydrodynamic dispersion* (a particle diffusion mechanism).

(E) Propagation of an Elastic Wave Through Bed

The onset of bubbling, also referred to as the *transition* from a *uniform* particulate flow to a *nonuniform* or *aggregate* particulate flow, has also been analyzed using a semiheuristic treatment based on the propagation of a porosity (or *medium* density) wave through the bed. When the *porosity propagation speed* u_ϵ is *greater* than (or *equal* to, for the case of the *marginal stability* or the onset of bubbling) the *speed of the elastic wave propagation* $u_e = (\partial\langle p\rangle/\partial\langle\rho\rangle)^{1/2}$, then *bubbling* will occur. Foscolo and Gibilaro (1984) relate the pressure variations to the combined interfacial drag and phase-density buoyancy force acting on particles which in turn depend on the porosity. The *porosity propagation speed* u_ϵ is the rate to the change of the superficial fluid velocity $\langle u\rangle^f$ with respect to the porosity. The analysis of Foscolo and Gibilaro (1984), leading to the determination of u_e and u_ϵ, is reviewed below.

In the presence of the surrounding particles, the drag force on a particle is modified and this is correlated as

$$\frac{c_d(\epsilon)}{c_{d_t}(\epsilon=1)} = (\frac{\langle u\rangle^f}{u_t})^{4.8/n}\epsilon^{-3.8}\ , \qquad (7.83)$$

where u_t is the single-particle *terminal velocity* (i.e., independent of porosity) and for highly turbulent flows (i.e., $Re_d = u_t d/\nu_f > 500$) becomes independent of the fluid viscosity. In (7.83), c_{d_t} is drag coefficient corresponding to u_t. The coefficient n relates the bed porosity ϵ to the superficial velocity. The correlation relating $\langle u\rangle^f$, u_t, and ϵ is called the *Richardson-Zaki equation* and is a variation in the relations given by (7.62). The Richardson-Zaki relation gives

$$\frac{\langle u\rangle^f}{u_t} = \epsilon^n \ , \quad n = n(Re_d = \frac{u_t d}{\nu_f}) \ , \tag{7.84}$$

and n is taken as 4.8 for *laminar* flows and 2.4 for *turbulent* flows. The single-particle drag coefficient (based on the terminal velocity), $c_{d_t}(\epsilon = 1)$, was examined in Section 5.1.1 for different flow regimes.

The sum of all forces acting on the particle (drag and buoyancy) per unit area, which is the product of the force *per particle* F and the number of particles per unit area N_p/A, is equal to the pressure. Then a *change* in the force results in a change in the pressure according to

$$\delta p = \frac{N_p}{A}\delta F = \frac{4(1-\epsilon)}{\pi d^2}\delta F \ , \tag{7.85}$$

where

$$F = F_d + F_b \tag{7.86}$$

$$= [c_{d_t}(\epsilon = 1)\frac{1}{2}\rho_f(\langle u\rangle^f)^2\frac{1}{4}\pi d^2](\frac{\langle u\rangle^f}{u_t})^{4.8/n}\epsilon^{-3.8} - g\epsilon\frac{\pi d^3}{6}(\rho_s - \rho_f) \ . \tag{7.87}$$

Then the speed of the elastic wave propagation becomes

$$u_e = (\frac{\partial\langle p\rangle}{\partial\langle\rho\rangle})^{1/2} \simeq [\frac{\partial\langle p\rangle}{\partial(1-\epsilon)\rho_s}]^{1/2} = [\frac{4(1-\epsilon)}{\pi d^2\rho_s}]^{1/2}(-\frac{\partial F_d}{\partial\epsilon})^{1/2} \ . \tag{7.88}$$

Now using (7.87) and noting that under the condition of stationary particles the drag and buoyancy force balance, we have

$$\frac{\partial F_d}{\partial\epsilon} = -3.8[c_{d_t}(\epsilon = 1)\frac{\pi d^2}{8}\rho_f(\langle u\rangle^f)^2](\frac{u}{u_t})^{4.8/n}\epsilon^{-4.8} - \frac{\pi d^3}{6}(\rho_s - \rho_f)g$$

$$= -4.8\frac{\pi d^3}{6}(\rho_s - \rho_f)g \ . \tag{7.89}$$

Then the expression for u_e, (7.82), becomes

$$u_e = [\frac{3.2gd(1-\epsilon)(\rho_s - \rho_f)}{\rho_s}]^{1/2} \ . \tag{7.90}$$

For an infinitesimal perturbation from a uniform porosity state, the porosity propagation speed u_ϵ is given by (Foscolo and Gibilaro, 1984)

$$u_\epsilon = (1-\epsilon)\frac{d\langle u\rangle^f}{d\epsilon} \ , \tag{7.91}$$

which upon using (7.84) becomes

$$u_\epsilon = n u_t(1-\epsilon)\epsilon^{n-1} \ . \tag{7.92}$$

The *onset of bubbling* (also called the *bubbling point*) is marked by the condition

$$u_e = u_\epsilon \qquad \text{onset of bubbling} \ . \tag{7.93}$$

For $u_\epsilon < u_e$ the porosity disturbances travel and result in a *uniform* change in the bed porosity and for $u_\epsilon > u_e$ they grow as bubbles.

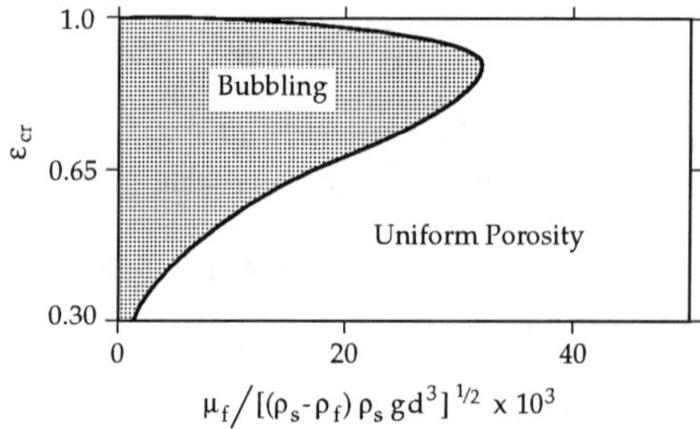

Figure 7.9. Variation of the critical porosity with respect of the product of the Galileo number and the density ratio. The unstable, or region of bubbling, is marked. (From Foscolo and Gibilaro, reproduced by permission ©1984 Pergamon Press.)

Substituting for u_e and u_ϵ in (7.93), the resulting equation for the *critical porosity* ϵ_{cr} becomes

$$(\frac{gd}{u_t^2})^{1/2}(\frac{\rho_s-\rho_f}{\rho_s})^{1/2} = 0.56n(1-\epsilon_{cr})^{1/2}\epsilon_{cr}^{n-1} \ . \tag{7.94}$$

Then ϵ_{cr} depends on the *Froude number* $Fr_d = (u_t^2/gd)^1/2$, $n = n(Re_d = u_t d/\nu_f)$, and the density ratio ρ_s/ρ_f. In (7.94) ϵ_{cr} is generally multivalued and in the range of $0.4 < \epsilon < 1$, there may be *two* solutions. The smaller range of ϵ_{cr} in these solutions is the *minimum bubbling porosity* ϵ_{mb}. Although in most fluidized beds the fluid flow is turbulent, for illustration Foscolo and Gibilaro (1984) show that for the case of a laminar flow, i.e., $Re_d = u_t d/\nu_f < 0.2$, the terminal velocity becomes

$$u_t = \frac{(\rho_s-\rho_f)gd^2}{18\mu} \tag{7.95}$$

and $n = 4.8$. Then (7.94) becomes

$$\frac{\mu}{[(\rho_s-\rho_f)\rho_s g d^3]^{1/2}} = 0.149(1-\epsilon_{cr})^{0.5}\epsilon_{cr}^{3.8} \ . \tag{7.96}$$

The term on the left-hand side is a product of the Galileo number Ga_d and the density ratio. The results of (7.96) are plotted in Figure 7.9, where a maximum in ϵ_{cr} is found below which ϵ_{cr} is double-valued. For porosities larger than the smaller of these values, the bubbles are formed, and for porosities larger than the larger of these values the bed porosity becomes uniform. The *bubbling region* is shaded.

(F) Bubble Structure and Interaction

Bubbles can form *inside* the bed due to the instabilities discussed in Sections (D) and (E) above, or can be formed at the *distributor*. For bubbles growing *at* the *distributor* due to a nonuniformity of fluid flow (e.g., *isolated* orifices, or velocity nonuniformity caused by the small flow resistance in the permeable distributor plate), a *critical* bubble diameter is reached before the bubbles *depart*. Unlike the gas bubbles in liquids, the fluidized bed bubbles *leak* through their surfaces, and in the analysis of the bubble dynamics and departure this leakage is included. Whitehead (1985) gives an account of the bubble formation at the distributor, the bubble dynamics and interaction after the departure, and the particle *recirculation*. This includes a classification of the distributors and the beds. The growth of bubbles at orifices has been analyzed using the existing lumped formulation applied to bubble growth and by solving the continuity and momentum equations for the fluid and particles phases. Kuipers et al. (1991) show that both approaches can predict the growth and the departure diameter and that the leakage at the bubble surface, which is a *prescribed* parameter in the lumped analysis, very *significantly* influences the predictions.

The bubble structure and interactions, i.e., the properties of isolated bubbles, the bubble coalescence, and the flow patterns, have been reviewed by Clift and Grace (1985). A bubble in a fluidized bed has a *spherical shape* over the entire *front* surface and is indented in the rear surface (this is called the wake *region*). This is rendered in Figure 7.1(b) and is similar to the gas bubbles in liquids which were shown in Figure 1.7, and for large Bond numbers, i.e., $Bo = g\rho_\ell d_b^2/\sigma > 40$, will have a *spherical cap* shape. The wake volume behind the bubble depends on the particle size and the bubble size. The *rise velocity* of the bubble is also treated similar to that of the gas bubbles in liquids. When reaching the surface, the bubble creates a *domelike* protuberance in the surface, and similar to the gas bubbles in liquids, the *mantle* of solids separating the top of dome from the roof of the bubble *thins* until the bubble *breaks through*. The fluid and particles motion around a pair of bubbles has been analyzed, and their attraction has been examined by Clift and Grace (1985).

(G) Regimes of Fluidization

The combination of the particle size and mechanical properties, superficial fluid velocity, density ratio, fluid viscosity, bed size and shape, and distributor characteristics, make for a very *large* range of fluid and particles flow behavior. A general regime map with sufficient details is *not* yet available (e.g., Davidson et al., 1985). However, for a given particle and fluid pair, a particle and bed size, and a distributor characteristic, the regimes of fluidization can be mapped as a function of the *superficial fluid velocity*. The regimes can be represented by the bed porosity. Yerushalmi and

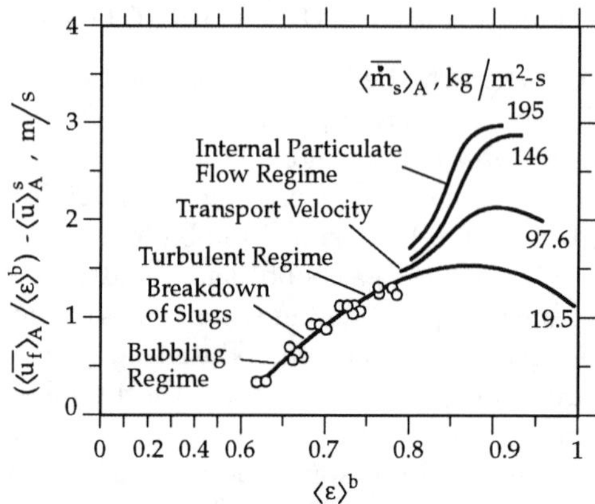

Figure 7.10. Regime diagram for a particulate flow showing the various fluidized regimes and the transition to the internal particulate flow regime. (From Yerushalmi and Avidan, reproduced by permission ©1985 Academic Press.)

Avidan (1985) have compiled the results for spherical particles with an average diameter of 49 μm and density of 1070 kg/m^3 (these are typical *fluid-cracking catalysts*) in *air* with a *terminal* velocity u_t of 7.8 cm/s at a pressure of one atmosphere. The bed is a 15.2-cm-diameter column with a *uniform* superficial fluid velocity through the distributor. Their results for the variation of the bed porosity $\langle \epsilon \rangle^b$ with respect to the *cross-sectional averaged slip* velocity defined as $(\langle \overline{u}_f \rangle_A / \langle \epsilon \rangle^b) - \langle \overline{u} \rangle_A^s$ are given in Figure 7.10. The mean, cross-sectional-averaged particles-phase velocity $\langle \overline{u} \rangle_A^s$ is equal to $\langle \overline{\dot{m}_s} \rangle_A / (1 - \langle \epsilon \rangle^b) \rho_s$. The case of $\langle \overline{\dot{m}_s} \rangle_A = 0$ is the *fluidized bed regime* and the case of $\langle \overline{\dot{m}_s} \rangle_A > 0$ corresponds to the *hydraulic transport* (e.g., *internal particulate flow regime*) discussed in Section 7.1. Figure 7.10 shows that for $\langle \overline{\dot{m}_s} \rangle_A = 0$, i.e., for the fluidized bed regime, the bed becomes *fluidized* at $\langle \epsilon \rangle^b = \epsilon_{mf}$ and as the superficial air velocity increases, the bubbling regime occurs for $\langle \epsilon \rangle^b$ between about 0.62 and 0.68. The velocity is still less than 0.5 m/s, but *many* times the terminal velocity. Then larger bubbles (slugs) are formed as the velocity is further increased, and then the *turbulent* regime is reached for $\langle \overline{u_f} \rangle / \langle \epsilon \rangle^b$ of about 1 m/s. As $\langle \overline{u_f} \rangle / \langle \epsilon \rangle^b$ is further increased the *net* solid flow begins, i.e., $\langle \overline{\dot{m}_s} \rangle_A > 0$, and here for a steady flow, particles must be *added* (i.e., *recirculating* particles) and the regime of hydraulic transport or internal particulate flow is encountered. Then depending on the magnitude of the particles flow rate $\langle \overline{\dot{m}_s} \rangle_A$, and the air flow rate $\langle \overline{\dot{m}_f} \rangle_A$, the cross-sectional-averaged porosity $\langle \epsilon \rangle_A$ changes.

7.2.2 Confining Surfaces

Convective heat transfer between the confining surfaces (i.e., the walls of the fluidized bed which generally are *parallel* to the gravity vector) and the fluid-particles flow in fluidized beds, is *influenced* by the *hydrodynamics* of the bed as well as the thermophysical properties of the fluid and the particles. As was mentioned, because of the presence of the bubbles and the confining surfaces, the fluid-particles flow in the bed is *nonuniform* and *unsteady*. Then the heat transfer through the confining surfaces is also unsteady and nonuniform along the surface. If the two-medium heat transfer treatment of the fluid and particles phases is adapted, then at the bounding surface the heat transfer occurs between the confining surface (i.e., the *third medium*) and the fluid and particles phases. Using this continuum treatments at a location on the *interface* of the confining surface and the fluid and particles media, the temperatures of the confining solid, fluid and particles are *equal*. Adjacent to the confining surface, in the fluid and particles phases, *boundary layers* (hydrodynamics and thermal) are formed. As will be shown, other than in these boundary layers, the bed is rather well mixed and a *nearly uniform* temperature exists in the *core* of the bed. Earlier analyses of the heat transfer to the confining surfaces used *semiheuristic* models for the fluid and particles phases hydrodynamic (including the transient effects of the bubble motion adjacent to the surface) and thermal boundary layers (Gutfinger and Abuaf, 1974; Xavier and Davidson, 1985). Recent treatments use the local, phase-volume-averaged treatments such as those given in Sections 5.4.2 and 5.4.3 which were used in Section 7.1.3 for the internal particulate flows. As an example, we discuss the prediction of the *local* wall to bed heat transfer in a *two-dimensional* bed, using rather simple fluid-phase and particles-phase conservation equations. The hydrodynamic simplifications include the assumption of a *laminar* fluid (gas) flow with a *negligible fluid-phase shear stress* and a simple expression for the *modulus of elasticity* of the particles phase. The heat transfer treatment simplifications include a simple expression for the *particles-phase effective conductivity* with a negligible dispersion in both phases. Numerical integration of the relevant conservation equations have been performed by Syamlal and Gidaspow (1985), Ding and Gidaspow (1990), and Kuipers et al. (1992). Their formulations are similar and allow for the bubbles formation at the distributor caused by a nonuniform flow. These formation along with their results are reviewed below.

(A) Formulation

The hydrodynamics of the fluidized bed has been studied more rigorously, compared to the heat transfer. The existing energy equations used for the predictions are not yet critically examined. The effective medium properties $\langle \mathbf{K} \rangle^f$, $\langle \mathbf{K} \rangle^s$, $\mathbf{D}^d_f$, and $\mathbf{D}^d_s$ are not well known. The *empirical* treatments of these coefficients are discussed by Gunn (1978) and Xavier and Davidson (1985). The equations for the conservation of mass, momentum, and energy, for each phase, are modeled similar to the hydrodynamic treatment of Section 7.2.1 (D), and are given by (Kuipers et al., 1992)

$$\frac{\partial}{\partial t}\epsilon\rho_f + \nabla\cdot\epsilon\rho_f\langle\mathbf{u}\rangle^f = 0 , \tag{7.97}$$

$$\frac{\partial}{\partial t}(1-\epsilon)\rho_f + \nabla\cdot(1-\epsilon)\rho_s\langle\mathbf{u}\rangle^s = 0 , \tag{7.98}$$

$$\frac{\partial}{\partial t}\epsilon\rho_f\langle\mathbf{u}\rangle^f + \nabla\cdot\epsilon\rho_f\langle\mathbf{u}\rangle^f\langle\mathbf{u}\rangle^f = -\epsilon\nabla\langle p\rangle^f - \frac{\rho_s}{t_d}(1-\epsilon)(\langle\mathbf{u}\rangle^f - \langle\mathbf{u}\rangle^s) + \epsilon\rho_f\mathbf{g} , \tag{7.99}$$

$$\frac{\partial}{\partial t}(1-\epsilon)\rho_s\langle\mathbf{u}\rangle^s + \nabla\cdot(1-\epsilon)\rho_s\langle\mathbf{u}\rangle^s\langle\mathbf{u}\rangle^s = -[(1-\epsilon)\nabla\langle p\rangle^s + \frac{\rho_s}{t_d}(1-\epsilon)(\langle\mathbf{u}\rangle^f - \langle\mathbf{u}\rangle^s) - E_\mu(\epsilon)\nabla\epsilon + (1-\epsilon)\rho_s\mathbf{g} , \tag{7.100}$$

$$\frac{\partial}{\partial t}\epsilon(\rho c_v)_f\langle T\rangle^f + \nabla\cdot\epsilon(\rho c_v)_f\langle\mathbf{u}\rangle^f\langle T\rangle^f = -\langle p\rangle^f(\frac{\partial\epsilon}{\partial t} + \nabla\cdot\epsilon\langle\mathbf{u}\rangle^f) + \nabla\cdot\epsilon\langle k\rangle^f\nabla\langle T\rangle^f + \frac{6(1-\epsilon)k_f}{d^2}\langle Nu_d\rangle^{\text{ext}}_{A_{sf}}(\langle T\rangle^s - \langle T\rangle^f) , \tag{7.101}$$

$$\frac{\partial}{\partial t}(1-\epsilon)(\rho c_p)_s\langle T\rangle^s + \nabla\cdot(1-\epsilon)(\rho c_p)_s\langle\mathbf{u}\rangle^s\langle T\rangle^s = -\langle p\rangle^s[\frac{\partial}{\partial t}(1-\epsilon) + \nabla\cdot(1-\epsilon)\langle\mathbf{u}\rangle^s] + \nabla\cdot(1-\epsilon)\langle k\rangle^s\nabla\langle T\rangle^s + \frac{6(1-\epsilon)k_f}{d^2}\langle Nu_d\rangle^{\text{ext}}_{A_{sf}}(\langle T\rangle^f - \langle T\rangle^s) . \tag{7.102}$$

The particle relaxation time t_d is defined by (5.145) and the drag coefficient is treated similar to that given by (7.83). The fluid- and particles-phase pressures are assumed to be *equal*.

The *modulus of elasticity* E_μ is given through a correlation (Syamlal and Gidaspow, 1985), i.e.,

$$-E_\mu(\epsilon) = 10^{-8.76\epsilon+5.43} . \tag{7.103}$$

A critical examination of the magnitude and physical interpretation of E_μ is given by Massoudi et al. (1992). The thermophysical properties are assumed constant, and $\langle k\rangle^f$ and $\langle k\rangle^s$ are modeled as

$$\frac{\langle k\rangle^f}{k_f} = \frac{1-(1-\epsilon)^{1/2}}{\epsilon} , \tag{7.104}$$

$$\frac{\langle k\rangle^s}{k_f} = \frac{k_e}{k_f}\frac{1}{1-\epsilon} , \tag{7.105}$$

where k_e is a modification to the available results for the effective conductivity under the condition of local thermal equilibrium, e.g., (3.127). These models are further discussed by Xavier and Davidson (1985), where the fluid-phase dispersion is also addressed.

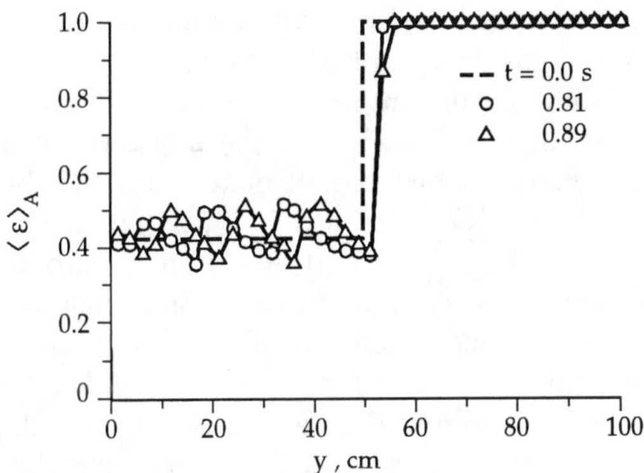

Figure 7.11. Computed propogation of a plane porosity disturbance introduced at the distributor. The results are for three different elapsed times. (From Kuipers et al., reproduced by permission ©1992 American Institute of Chemical Engineers.)

The Nusselt number $\langle Nu_d \rangle^{\rm ext}_{A_{sf}}$ is also a prescribed function of the particle Reynolds number and the fluid Prandtl number, using correlations similar to (7.44).

(B) Numerical Results

The numerical results of Kuipers et al. (1992) are reviewed in Figures 7.11 and 7.12, and are for a *two-dimensional* bed of *width* $W = 28.5$ cm and *height* $L = 100$ cm with an *initial* fluidized portion of height 50 cm. The initial fluidization is at a minimum fluidization velocity $\langle u_f \rangle_{mf} = 0.25$ m/s and a minimum fluidization porosity $\epsilon_{mf} = 0.402$. Air is used as the fluid, and glass particles of $d = 500$ μm are used as the particles ($\rho_s = 2000$ kg/m^3, $k_s = 1$ W/m-K, and $c_p = 737$ J/kg-K). The bed is heated on the right-hand-side surface by maintaining the surface temperature at $T_{s_b} = 100°$C, the left-hand side is *insulated*, and the inlet fluid is at 20°C. Bubbles are generated by allowing higher velocities through a slot next to the right-hand-side surface.

In order to examine the hydrodynamic stability of the bed, a time-periodic, plane porosity *disturbance* is introduced on top of the distributor and its propagation along the bed is examined. Figure 7.11 shows the growth and propagation of this porosity disturbance. The initial porosity distribution in the fluidized portion and the *free-board* portion of the bed are also shown. The distribution of the *instantaneous*, area-averaged porosity $\langle \epsilon \rangle_A$ is shown for two elapsed times of 0.81 and 0.89 s. The fluidized bed *expands* and some particles are also *present* in the free-board region. This *amplification* of plane waves shows that the fluidized bed is *unstable* even for a velocity approaching the minimum fluidization velocity, a general characteristic of the gas fluidized beds, as discussed in Section 7.2.1(D).

Figure 7.12 shows the instantaneous variation of the *local*, computed heat transfer rate (divided by $T_{s_b} - \langle T \rangle_A^b$) along the heated, right-hand side surface. The fluidized bed is well mixed and the thermal boundary layer is *small* such that $\langle T \rangle_A^b$ defined by (7.24) is *nearly* the inlet fluid temperature. The bubble is initiated at the distributor and next to the right-hand-side surface by specifying a *large* fluid inlet velocity (about 20 times $\langle u_f \rangle_{mf}$) over a *small* region (i.e., slot). The bubble grows and then departs and disturbes the bed such that later *multiple* bubbles are present in the bed. The results given in Figure 7.12 are for four elapsed times (the time after the simultaneous minimum fluidization and the introduction of the high velocity through the slot). At $t = 0.20$ s the bubble covers the lower portion of the right-hand side surface and *reduces* the heat transfer rate to that corresponding to the case of gas flow only (which is *negligibly small* compared to that due to the two-phase flow). The two-phase region of the flow near the heated surface has a relatively large heat transfer rate. Note that a local, instantaneous Nusselt number can be defined as

$$Nu_W = \frac{q_{s_b}}{T_{s_b} - \langle T \rangle_A^b} \frac{W}{k_f} \, . \tag{7.106}$$

Then because $\langle T \rangle_A^b$ is uniform along the bed, the Nusselt number distribution follows the distribution of the heat transfer rate.

The *detached* bubble travels along the surface, and with this bubble *sweeping* the surface the local Nusselt number changes with time. For $t =$ 0.37 s there are two peaks in the distribution, and the one in the *wake* of the bubble is *larger* in magnitude. This is due to the rather intense mixing in the *wake region*. For $t = 0.55$ s the first bubble breaks through the fluidized bed surface while a second and a third bubble, which are smaller, have also been produced by the distributor. The heat transfer rate is the largest where these smaller bubbles are sweeping the surface. For $t = 0.75$ s the larger bubble is reduced to an indentation in the top of the bed and the smaller bubbles move up. The peak in the instantaneous value of $q_{s_b}/(T_{s_b} - \langle T \rangle_A^b)$ reaches about 2500 W/m^2-K which is *very* large.

7.2.3 Submerged Surfaces

In practice, heat removal from the fluidized bed is done by placing tubes carrying a coolant inside the bed. The tubes are generally placed *horizontally* and at various locations above the distributor, including the *free-board region* and in the *uppermost* part of the *fluidized-bed region*. The *hydrodynamics* of the particulate flow around tubes and in the presence of bubbles,

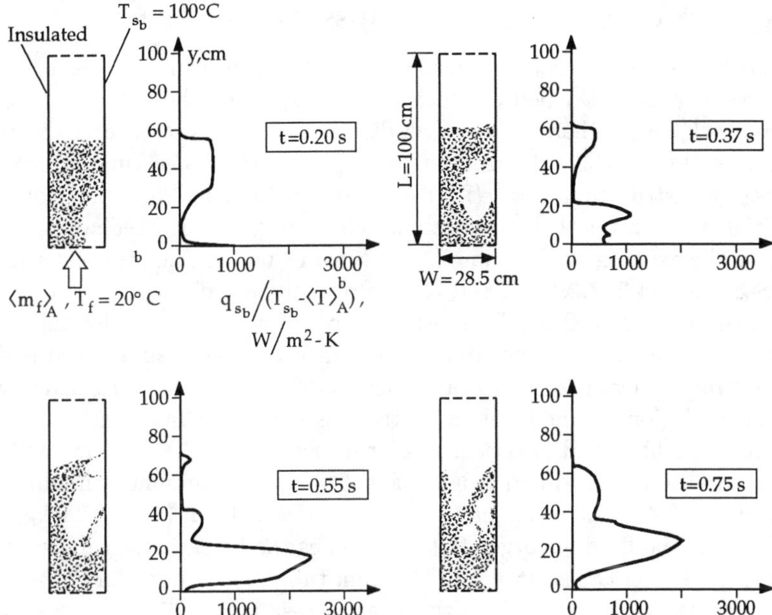

Figure 7.12. Computed distribution of the instantaneous heat transfer coefficient along the heated, right-hand side of a two-dimensional fluidized bed. (From Kuipers et al., reproduced by permission ©1992 American Institute of Chemical Engineers.)

which periodically sweep the tube, are rather complex and have not yet been treated in detail. A review of the available *empirical* treatments of the heat transfer to the submerged surfaces is given by both Xavier and Davidson (1985) and Saxena (1989). The empirical treatments include the parametric correlations of the effects of the superficial fluid-phase velocity, the bubble size and frequency, the tube location, and the radiation (e.g., Gabor, 1970; Alavizadeh et al., 1990; Dyrness et al., 1992).

In the following, the measured *local, instantaneous* heat transfer rate from a submerged tube is examined for insight into the *transient* particulate flow hydrodynamics and heat transfer. The existing experimental results suggest that as the fluid superficial velocity *increases*, the time-averaged heat transfer rate at the submerged surface first *increases* and then *decreases* after a maximum. This maximum heat transfer rate depends only on the Archimedes number $gd^3(\rho_s - \rho_f)\rho_f/\mu_f^2$, which is similar to the Galileo number. This is discussed and then the *time-averaged* local heat transfer rate is examined and an empirical treatment, which uses the *local porosity* for a correlation with the heat transfer rate, is reviewed.

(A) Instantaneous Local Nusselt Number

We consider the particulate flow around a horizontal tube of diameter D_1 placed in a fluidized bed of much larger diameter. The tube diameter is also much larger than the particle diameter d. The presence of the tube changes the porosity distribution in the vicinity of the tube. Then the porosity distribution $\epsilon = \epsilon(\mathbf{x}, t)$ around the tube has a rather complex spatial and temporal distribution which in principle is predictable using one of the existing two-medium hydrodynamic models, i.e., those discussed in Sections 5.4.2, 7.2.1(D), and 7.1.2. Presently, no solution is available for such submerged surfaces. The available experimental results for the local, instantaneous heat transfer rate on the surface of these submerged bodies are examined here in order to gain insight into this porosity variation as well as the variations in the local, instantaneous solid- and gas-phase velocities.

For a fluidized bed of rectangular cross-section (0.30 $\times$ 0.60 m) with an approximate fluidized-bed height of 0.46 m, at the minimum fluidization velocity of 0.47 m/s, using glass particles of $d = 1.0$ mm ($\rho_s = 2700$ kg/m^3) and air as the fluid, George (1993) examines the local, instantaneous heat transfer rate on the surface of a 50.8-mm tube placed horizontally at 0.36 m above the distributor. His experimental results for $\langle \overline{u_f} \rangle_A = 1$ m/s, and an average bed temperature $\langle T \rangle^b$ =56°C, are shown in Figure 7.13. The results are given in terms of $q_{s_s}/(\langle \overline{T} \rangle^b - \overline{T}_{s_s})$ and are for a 2-s time intervals and at various polar angles (measured from the front stagnation point).

Based on the local porosity and the particle dynamics, George (1993) defines different *regimes* of particles and fluid *contacts* with the surface. In the *bubble-phase contact regime*, the particle concentration adjacent to the surface is very *low* such that the convective heat transfer to the fluid phase *dominates* (and this heat transfer rate is rather low and does not change with time while this regime lasts). In the *nearly stationary emulsion-phase regime*, high concentration of particles exists on and adjacent to the surface, but it does not instantly *slide* on the surface, and therefore, the heat transfer rate is initially high and then begins to decrease as the particles cool. In the *moving enulsion-phase regime*, the particle concentration is high and the heat transfer rate and the frequency of its variations are also very high.

An examination of the results of Figure 7.13 shows that the lower stagnation point ($\theta = 0°$) is characterized by the presence of *long* periods of the bubble-phase contact regime (i.e., lowest value of $q_{s_s}/(\langle \overline{T} \rangle^b - \overline{T}_{s_s})$ and *short* periods of the nearly stationary and the moving emulsion-phase regimes. At $\theta = 45°$, the moving emulsion-phase regime is encountered more often and for $\theta = 90°$ this regime lasts even longer. At $\theta = 135°$ the bubble-phase contact regime is nearly absent and at $\theta = 180°$ (the near stagnation point) a stack of *defluidized* particles cover the surface and are moved by the passing bubbles.

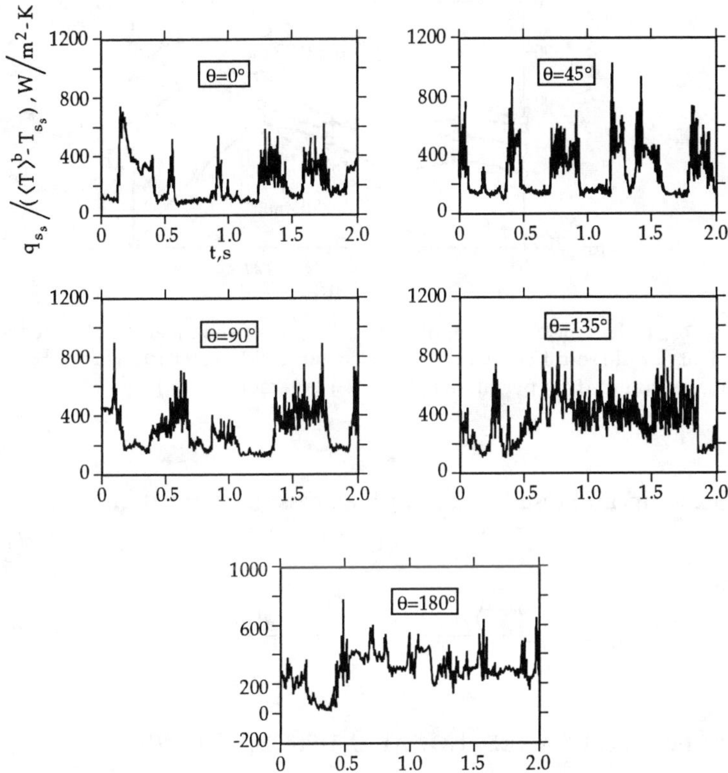

Figure 7.13. Measured variation of the local heat transfer coefficient for various polar angles around a tube submerged in a fluidized bed. (From George, reproduced by permission ©1993 Pergamon Press.)

(B) Maximum, Time-Averaged Local Nusselt Number

The time-averaged, local heat transfer rate is found similar to transient results discussed above. The results of Khan and Turton (1992) are shown in Figure 7.14 for $d = 0.73$ mm and for four different normalized fluid (air) mean superficial velocities (i.e., $\langle\overline{u}_f\rangle/\langle\overline{u}_f\rangle_{mf}$). The results show that the *angular distribution* of the mean local heat transfer rate changes with this velocity. Compared to the other locations, the mean heat transfer rate in the *rear stagnation point region* becomes the largest, for the larger values of $\langle\overline{u}_f\rangle/\langle\overline{u}_f\rangle_{mf}$. The magnitude of the local heat transfer rate does *not* increase monotonically with $\langle\overline{u_f}\rangle$, and for this particular experiment, there is a *maximum* in the area- and time-averaged heat transfer rate $\langle\overline{q_{s_s}}\rangle_A$. This is in part due to the bubble dynamics and the effect of the submerged tube on the local porosity distribution. Presently no detailed analysis is available; however, a correlation relating the maximum in $\langle\overline{q_{s_s}}\rangle_A$ to the Archimedes number Ar_d is available. Khan and Turton (1992) show that

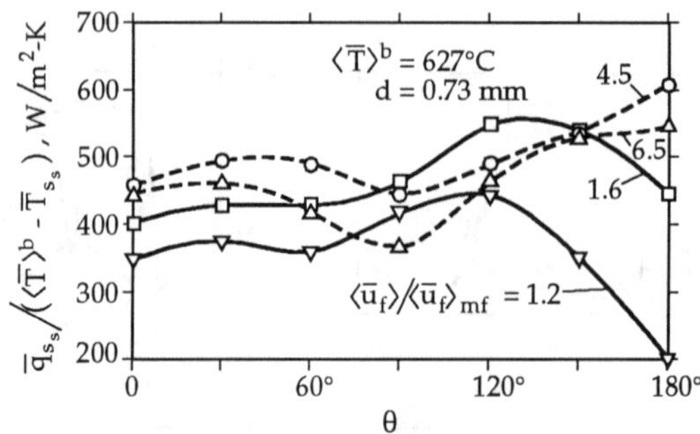

Figure 7.14. The measured angular distribution of the mean local heat transfer coefficient for three different normalized mean fluid velocities. (From Khan and Turton, reproduced by permission ©1992 Pergamon Press.)

their experimental results agree with an existing correlation

$$\langle \overline{Nu_d} \rangle_A \Big|_{max} = 0.88 Ar_d{}^{0.213} , \tag{7.107}$$

where

$$\langle \overline{Nu_d} \rangle_A = \frac{\langle \overline{q_{s_s}} \rangle_A d}{\langle \overline{T_{s_s}} \rangle_A - \langle \overline{T} \rangle_A^s} , \quad Ar_d = \frac{g(\rho_s - \rho_f)\rho_f d^3}{\mu_f^2} . \tag{7.108}$$

(C) Time-Averaged Local Nusselt Number

For practical usage, time-averaged local heat transfer rate is of interest. Using *simple hydrodynamic models*, the variation of the heat transfer with respect to the polar angle θ has been examined by both Adams and Welty (1982) and Adams (1984), among others. In these models the *fraction* of time that any of these three regimes discussed above are in contact with the surface must be known. The correlations which include these local variations can become fairly complex (Adams, 1984). Also, as was discussed above, the local heat transfer rate does *not* increase monotonically with the increase in $\langle \overline{u_f} \rangle$ and a maximum is observed in the heat transfer due to the fluidization at a velocity called the *optimum* velocity $\langle \overline{u}_f \rangle_{op}$. Then to simplify these correlations, Visser and Valk (1993) suggest the use of the *local porosity* ϵ, the Archimedes number Ar_d, the particle Reynolds number Re_d, the fluid Prandtl number Pr_f, the minimum fluidization $\langle \overline{u}_f \rangle_{mf}$, and the optimum velocity $\langle \overline{u}_f \rangle_{op}$. Their correlation for the local, time-averaged Nusselt number $\overline{Nu_d}$ is

$$\begin{aligned} \overline{Nu_d} &= \frac{\overline{q}_{s_s} d}{(\langle \overline{T} \rangle_A^b - \overline{T}_{s_s}) k_f} \\ &= 2.62 f_u Ar_d{}^{0.1} (1-\epsilon)^{2/3} + 0.033 Re_d Pr \frac{(1-\epsilon)^{2/3}}{\epsilon} , \end{aligned} \tag{7.109}$$

where

$$f_u = \begin{cases} 0 & \langle \overline{u_f} \rangle \le \langle \overline{u_f} \rangle_{mf} \\ (\dfrac{\langle \overline{u_f} \rangle - \langle \overline{u_f} \rangle_{mf}}{\langle \overline{u_f} \rangle_{op} - \langle \overline{u_f} \rangle_{mf}})^{0.3} & \langle \overline{u_f} \rangle_{mf} \le \langle \overline{u_f} \rangle \le \langle \overline{u_f} \rangle_{op} \\ 1 & \langle \overline{u_f} \rangle \ge \langle \overline{u_f} \rangle_{op} \,. \end{cases} \quad (7.110)$$

The local porosity used in (7.109) must be known with *high* accuracy in order to make any accurate predictions, and this is discussed by Visser and Valk (1993).

7.3 Phase Change

Convective heat transfer in solid-solid-fluid systems, with one of the solid phases being the bounding or the submerged solid phase, can involve a *solid-fluid phase change*. This can be *solidification/melting* for the solid-solid-liquid systems, or *frosting/sublimation* for the solid-solid-gas systems. The rate of phase change depends on the fluid motion, i.e., the convective heat and mass transfer. In this section we consider *solidification* of a *single-component* liquid as a *larger growth* on a cooled bounding solid surface. Then the *bulk crystal nucleation* and growth in a *multicomponent* liquid is considered. Next, layer *melting* of single- and multicomponent solids is examined. This section ends with a review of layer *frost* growth and densification.

7.3.1 Layer Growth in Single-Component Solidification

As an example of solid-liquid phase change influenced by the convective heat transfer, the solidification of a *pure, subcooled* liquid over a subcooled solid surface is examined. The heat removal is through the solidified layer (which grows as a *layer*, i.e., *directional growth*, as compared to the *bulk* growth which follows a bulk nucleation in a *supercooled*, i.e., metastable, liquid). The convection in the melt influences the rate of this layer growth. In the following, first an *ideal, one-dimensional* solidification growth is discussed and the role of the molten convection is examined. Then the layer (or film) solidification in *forced turbulent* flow through a *tube* is examined and the effect of the cross-sectional area reduction due to the solidification, and the *laminarization* and the *transition* again to a *turbulent* flow, on the solidified layer thickness is examined. Next, the *thermobuoyancy*-induced motion in solidification of a melt in a *cavity* is addressed. The more involved problem of crystal growth at the otherwise *free surface* of a melt in a cavity is discussed next. This involves the rotation of the cavity and or the solidified melt, and therefore, the *combined forced* and *thermobuoyant* flows influence the crystal growth rate and its stability.

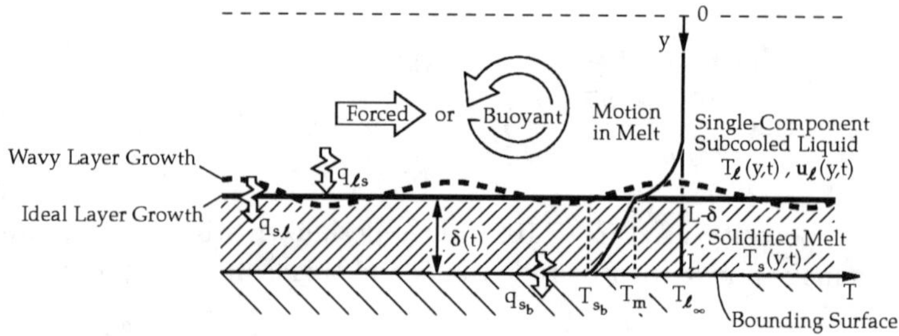

Figure 7.15. Layer (ideal and wavy) growth of a solidified layer in a single-component solidification with a motion within the melt.

As will become clear in the selected examples given below, there is a *strong interaction* between the growth rate and the molten motion. In solidification technology, these interactions (and their control) influence the quality of the solid formed. Pure *metals* as well as pure *nonmetals* are produced using various *continuous* and *batch* processes (Flemings, 1974; Vere, 1987; Szekely et al., 1988; Liebermann, 1993).

(A) Ideal, One-Dimensional Layer Growth

Consider the *ideal* problem of a *planar* solidification growth of a liquid initially at a *subcooled* temperature T_{ℓ_∞} and then suddenly cooled below its freezing temperature $T_{\ell s} = T_m$ at a plane solid surface located at $y = L$. The temperature of the plane-bounding surface is T_{s_b}. This ideal layer growth is rendered in Figure 7.15, where allowance has been made for the motion in the liquid (*forced* or *buoyant* flow). Heat is *withdrawn* from the liquid and solid phases and through the phase change at the solid–liquid interface.

The one-dimensional energy equation for the solidified layer is

$$(\rho c_p)_s \frac{\partial T_s}{\partial t} = k_s \frac{\partial^2 T_s}{\partial y^2} \quad L - \delta \le x \le L \, . \tag{7.111}$$

The solid–liquid *interface velocity* $u_{s\ell}$ is equal to the *growth rate* $d\delta/dt$. The heat balance at the liquid-solid interface gives

$$k_s \frac{\partial T_s}{\partial y} - k_\ell \frac{\partial T_\ell}{\partial y} = (\rho_s i_s - \rho_\ell i_\ell) u_{s\ell} + \rho_\ell i_\ell u_\ell \quad x = L - \delta \, . \tag{7.112}$$

The interfacial velocity $u_{s\ell}$ and the liquid velocity at $x = L - \delta$ are related through the interfacial mass balance

$$(\rho_s - \rho_\ell) u_{s\ell} + \rho_\ell u_\ell = 0 \, . \tag{7.113}$$

Using this in (7.113) and introducing the specific heat of solidification $\Delta i_{\ell s} = i_\ell - i_s$, (7.112) becomes

$$k_s \frac{\partial T_s}{\partial y} - k_\ell \frac{\partial T_\ell}{\partial y} = \rho_s \Delta i_{\ell s} \frac{d\delta}{dt} \quad x = L - \delta \,. \tag{7.114}$$

The Nusselt number for the liquid-side interfacial heat transfer is introduced as

$$k_\ell \frac{\partial T_\ell}{\partial y}|_{L-\delta} = Nu_{L-\delta} \frac{k_\ell}{L-\delta}(T_{\ell_\infty} - T_m) \,, \tag{7.115}$$

$$Nu_{L-\delta} = \frac{q_{\ell s}(L-\delta)}{(T_{\ell_\infty} - T_m)k_\ell} \,. \tag{7.116}$$

The energy balance at the interface (7.114) shows that when the rate of heat convected to the interface is *equal* to that conducted through the solidified layer, the layer thickness *ceases* to grow. Now, scaling the variables with the temperature difference across the solidified layer, the initial liquid-side length scale L, and the solid properties we have

$$T^* = \frac{T - T_m}{T_{\ell_\infty} - T_m} \,, \quad y^* = \frac{y}{L} \,, \quad \delta^* = \frac{\delta}{L} \,, \quad t^* = Fo_s Ste_s \,, \tag{7.117}$$

$$Fo_s = \frac{t\alpha_s}{L^2} \,, \quad Ste_s = \frac{c_{p_s}(T_{\ell_\infty} - T_m)}{\Delta i_{\ell s}} \,, \quad \theta_c = \frac{T_m - T_{s_b}}{T_{\ell_\infty} - T_m} \,. \tag{7.118}$$

The dimensionless time t^* is the product of the solid *Fourier* and *Stefan numbers*. The ratio of *solid subcooling* $T_m - T_{s_b}$ to the *liquid subcooling* $T_{\ell_\infty} - T_m$ is called the *cooling ratio* θ_c. By using these definitions, (7.111) and (7.114) become

$$Ste_s \frac{\partial T_s^*}{\partial t^*} = \frac{\partial^2 T_s^*}{\partial {y^*}^2} \quad 1 - \delta^* \le y^* \le 1 \,, \tag{7.119}$$

$$\frac{d\delta^*}{dt^*} = \frac{\partial T_s^*}{\partial y^*} - \frac{k_\ell}{k_s}\frac{1}{1-\delta^*} Nu_{L-\delta} \quad y^* = 1 - \delta^* \,. \tag{7.120}$$

The other boundary conditions and the initial conditions are

$$T_s^* = 1 \quad y^* = 0 \,, \tag{7.121}$$

$$T_s^* = 0 \quad y^* = 1 - \delta^* \,, \tag{7.122}$$

$$T_s^* = -\theta_c \quad y^* = 1 \,, \tag{7.123}$$

$$\delta^* = 0 \,, T_s^* = 1 \quad t^* = 0 \,. \tag{7.124}$$

In general, the Nusselt number *changes* with time $Nu_{L-\delta} = Nu_{L-\delta}(t^*)$, and using a *prescribed* temperature *distribution*, i.e., a second-order polynomial, (7.119) and (7.120) along with (7.121) to (7.124) can be simplified (i.e., the spatial derivative is eliminated) and then integrated numerically to solve for the variation of δ^* with respect to t^*. Gau and Viskanta (1984) find solutions for the case of *constant* $Nu_{L-\delta}$. The parameters are Ste_s, k_ℓ/k_s, θ_c, and $Nu_{L-\delta}$. They show that the growth is *not* sensitive to the Stefan number Ste_s for $Ste_s \le 0.1$ and that as $Nu_{L-\delta}$ *increases* the growth stops *earlier*. When the liquid temperature is *not* much *larger* than $T_{\ell s}$ (i.e., liquid is close to freezing/melting temperature), then δ is proportional to $t^{1/2}$. The increase in the liquid subcooling $T_{\ell_\infty} - T_m$, which decreases θ_c, plays the same role as an increase in $Nu_{L-\delta}$. Then as Nu_{L-s}, or convection, is increased, this growth rate is hindered.

In general, the solid-liquid interface is not stable and this instability is caused by a *deviation* from the equilibrium *thermodynamic* state (i.e., *metastability* and then a sudden nucleation and the following anisotropic growth in the *supercooled* liquid layer) or by the *hydrodynamic* effects such as the *nonuniformity* of the Nusselt number along the interface. The metastability can influence the crystal morphology and can result in anisotropic transport properties in the solid (i.e., anisotropic thermal conductivity). This is observed in the experiments of Gau and Viskanta (1985) with the solidification of gallium. The variation in the Nusselt number along the interface can *magnify* the small amplitude interfacial waves (Glicksman et al., 1986).

(B) Steady, Turbulent Forced Flow Through Channels

When a subcooled, single-component liquid flows through a channel with an entrance liquid temperature of $T_\ell(x = 0)$ and the surface of the channel is held at T_{s_b} which is below the freezing temperature of the liquid T_m, solidification of the liquid occurs. A rendering of this *turbulent* internal flow and solidification is given in Figure 7.16(a).

Depending on the area-averaged *liquid entrance velocity* $\langle \overline{u_\ell} \rangle_A$, the channel width $2L$, the channel length L_c, the *liquid* and *solid thermophysical properties*, the extent of the *liquid subcooling* $T_\ell(x = 0) - T_m$, and the *solidified layer subcooling* $T_m - T_{s_b}$, the *steady-state* solidification can either completely *close* the channel or allow for the *continuous* flow of the liquid. Also, the axial distribution of the thickness of the solidified layer $\delta(x)$ can take on many different forms. These can be *smooth, wavy* distributions with a nearly *regular* wavelength along the channel, or *nonregular* waves with wave forms which may *not* be axially symmetric. The complete mapping of these internal flow solidification *regimes* for circular tubes and parallel-plate channels is not yet available. However, some of the features have been classified by Epstein and Cheung (1983), Hirata and Matsuzawa (1987), and Weigand and Beer (1993), among others. For a given liquid, the Reynolds number based on the hydraulic diameter $Re_{d_h} = \langle \overline{u_\ell} \rangle_A d_h / \nu_\ell$, with $d_h = 4L$ for a channel flow, and the cooling ratio θ_c, defined by (7.118), have been used for this classification. When at or slightly downstream of the entrance the flow is *turbulent*, the formation of the solidified layer downstream causes an *acceleration* which can *laminarize* the flow. Since the interfacial liquid Nusselt number changes (here decreases) when transition from turbulent to laminar flow occurs, then the interface of the solidified layer undergoes *morphological* changes (here the thickness increases). The magnitude of the

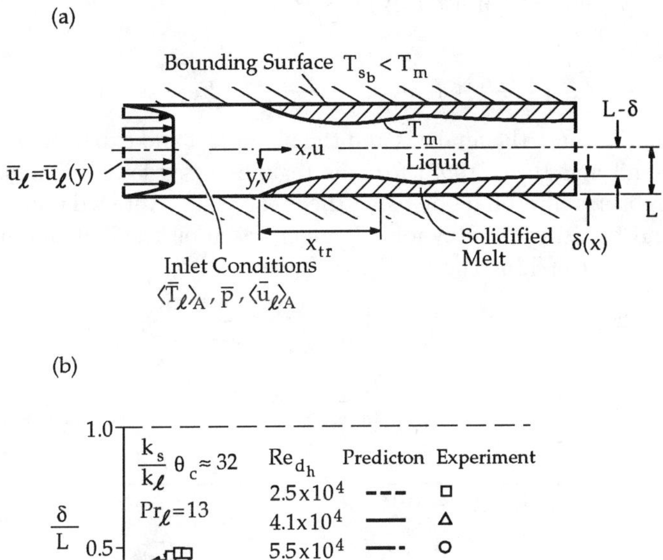

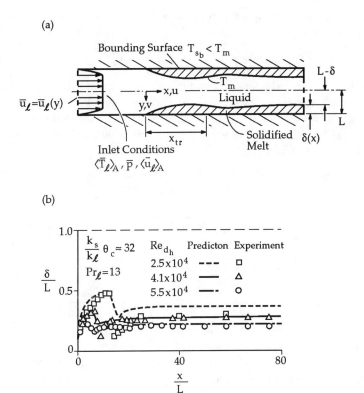

Figure 7.16. (a) Single-component solidification in a turbulent liquid flow through a two-dimensional channel with the channel walls cooled below the melting temperature. (b) Predicted and measured solidified-layer thickness distribution for three different Reynolds numbers. (From Weigand and Beer, reproduced by permission ©1993 Pergamon Press.)

acceleration needed for this transition is given by the critical value of the *acceleration parameter* (Weigand and Beer, 1993)

$$\frac{\nu_\ell}{\langle \overline{u_\ell} \rangle_A^2} \frac{d\langle \overline{u_\ell} \rangle_A}{dx} = \frac{1}{Re_L} \frac{d\delta}{dx} = 2 \times 10^{-6} \text{ to } 3 \times 10^{-6}. \tag{7.125}$$

The *increase* in δ along x is *hindered* due to the increased *conduction resistance* in the solidified layer, and an *asymptotic growth* for the laminarized flow is reached. Then flow acceleration *decreases* and the flow becomes *turbulent again*. When the value of the acceleration parameter decreases below the critical value given above, the flow is assumed to be turbulent and this axial location is marked with x_{tr} in Figure 7.16(a). For $x > x_{tr}$, the liquid Nusselt number *increases* and the solidified layer thickness *decreases*, as depicted in Figure 7.16(a). For *water*, this class of *single*, *asymmetric wave regime* depicted in Figure 7.16(a) has been observed to occur for (Weigand and Beer, 1993)

$$-0.41 + \frac{Re_{d_h}}{7077} \leq \theta_c \leq 2.4 + \frac{Re_{d_h}}{4421} \,, \quad Re_{d_h} > 6000 \,. \tag{7.126}$$

In the following, the analysis and the numerical and experimental results of Weigand and Beer (1993) for this regime are reviewed. The scaling of the variables is now made by using the mean, area-averaged liquid velocity $\langle \overline{u_\ell} \rangle_A$ and by using the Reynolds number based on the half-channel-width L. Then, by stretching the y-axis, we have

$$x^* = \frac{x}{L} \,, \quad y^* = \frac{y}{L} {Re_L}^{1/2} \,, \quad u^* = \frac{\overline{u}_\ell}{\langle \overline{u}_\ell \rangle_A (x=0)} \,, \tag{7.127}$$

$$v^* = \frac{\overline{v}_\ell}{\langle \overline{u}_\ell \rangle_A (x=0)} {Re_L}^{1/2} \,, \tag{7.128}$$

$$p^* = \frac{\overline{p}}{\rho_\ell [\langle \overline{u}_\ell \rangle_A (x=0)]^2} \,, \quad T^* = \frac{T - T_m}{\overline{T}_\ell (x=0) - T_m} \,, \tag{7.129}$$

$$Re_L = \frac{\langle u_\ell \rangle_A (x=0) L}{\nu_\ell} \,, \quad \theta_c = \frac{T_m - T_{sb}}{\overline{T}_\ell (x=0) - T_m} \,. \tag{7.130}$$

The equations of conservation of mass, momentum, and energy for the liquid phase are written for a steady, two-dimensional, *turbulent boundary-layer* flow and heat transfer, and the conduction in the solidified layer is assumed to be one-dimensional (normed to the liquid flow). These are

$$\frac{\partial u^*}{\partial x^*} + \frac{\partial v^*}{\partial y^*} = 0 \,, \tag{7.131}$$

$$u^* \frac{\partial u^*}{\partial x^*} + v^* \frac{\partial u^*}{\partial y^*} = -\frac{\partial p^*}{\partial x^*} + \frac{\partial}{\partial y^*} (1 + \frac{\nu_t}{\nu_\ell}) \frac{\partial u^*}{\partial y^*} \,, \tag{7.132}$$

$$u^* \frac{\partial T_\ell^*}{\partial x^*} + v^* \frac{\partial T_\ell^*}{\partial y^*} = \frac{1}{Pr_\ell} \frac{\partial}{\partial y^*} (1 + \frac{Pr_\ell}{Pr_t} \frac{\nu_t}{\nu_\ell}) \frac{\partial T_\ell^*}{\partial y^*} \,, \tag{7.133}$$

$$\frac{d^2 T_s^*}{d{y^*}^2} = 0 \,. \tag{7.134}$$

The boundary conditions are

$$u^*(x^* = 0, y^*) = f_u(y^*) \,, \; p^* = p^*(x=0) \,, \; T_\ell^* = 1 \quad x^* = 0 \,, \tag{7.135}$$

$$\frac{\partial u^*}{\partial y^*} = \frac{\partial T_\ell^*}{\partial y^*} = v^* = 0 \quad y^* = 0 \,, \tag{7.136}$$

$$u^* = v^* = T_\ell^* = T_s^* = 0 \,, \; \frac{\partial T_s^*}{\partial y^*} = \frac{k_\ell}{k_s} \frac{\partial T_\ell^*}{\partial y^*} \quad y^* = (1 - \delta^*) {Re_L}^{1/2} \,, \tag{7.137}$$

$$T_s^* = -\theta_c \quad y^* = {Re_L}^{1/2} \,. \tag{7.138}$$

The turbulence modeling, i.e., description of ν_t, is done using an algebraic model involving the mixing length and the wall function, i.e., a variation of the van Driest model discussed in Section 4.4.3. Some typical results of Weigand and Beer (1993) are shown in Figure 7.16(b). The numerical predictions and the experimental results are for water with $Pr_\ell = 13$ and the product of the ratio k_s/k_ℓ and θ_c (this product is called the *freezing parameter*) equal to 32 and for three different Re_{d_h}. As Re_{d_h} *increases*, the ice-layer thickness *decreases*. This is due to the increase in the liquid Nusselt number, defined by (7.115), caused by the increase in $\partial T_\ell/\partial y$ at the interface as Re_{d_h} and ν_t increase. Then the increase in Re_{d_h} tends to make the axial distribution of δ more *uniform* and *reduce* the magnitude of δ. The distance (measured from the entrance) for the transition to a turbulenct flow x_{tr} also *decreases* as Re_{d_h} *increases*.

(C) Transient, Thermobuoyant Flow

Although in mold casting the flow *induced* during the filling *persists* and is augmented by the thermobuoyant motion, the *ideal* case of a purely thermobuoyant flow in cavities and its effect on the solidification rate has been studied. This corresponds to a mold which is initially at the melt temperature (the melt can be subcooled) and is *suddenly* cooled. Depending on the pure substance used, the crystal formed can be *isotropic* or *nonisotropic* and the interface can be nonsmooth (with roughness of the order of the crystal size or larger). For example, for the case of *gallium*, the experimental results of Gau and Viskanta (1986) for a two-dimensional cavity cooled from one of the vertical surfaces, show an *irregular* crystal structure (solid anisotropy) and irreproducible interfacial geometry. For the case of solidification of *silicon oil*, Liu et al. (1993) find solid isotropy and reproducibility and regularity of the interfacial geometry.

When the solid formed is isotropic and the interface is stable, and when the flow is *laminar*, then the transient thermobuoyant motion in the melt can be predicted and the *local* growth of the interface can be determined. An example is the prediction of Liu et al. (1993), although their example is for a top free-surface and the thermocapillary effect is included. The incompressible liquid flow is described by the momentum conservation equation with the Boussinesq approximations, i.e., (1.120). The energy equation is written assuming no radiation, no viscous dissipation, no pressure and gravity work, and becomes the simplified form given by (1.121). These equations assume isotropic properties, and for the solid phase the convective term is dropped.

(D) Steady, Thermobuoyancy-Influenced Flow

One liquid-solid phase change application in which the thermobuoyancy influences the flow and the rate of phase change is in the *single-crystal growth* from melts. In one of the methods used, the melt is contained in a heated top-open *cavity* called the *crucible*, and a cooled *seed crystal* is dipped into the melt from the top and is *pulled* at a prescribed rate (which is

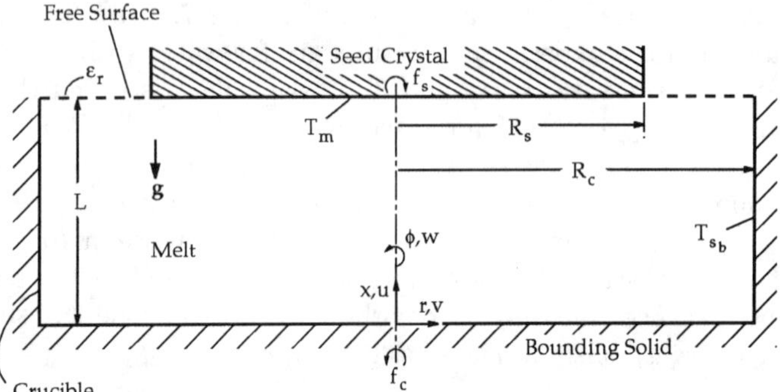

Figure 7.17. Single-crystal growth in a crucible holding the liquid with the seed crystal cooled to the freezing temperature of the single-component liquid. The direction of gravity and the system parameters are also shown.

the growth rate). The seed crystal and/or the crucible are also *rotated*. This is called the *Czochralski* method of single crystal growth and a review of the analysis of the fluid flow and heat transfer aspects of this phase change problem is given by Langlois (1985). Figure 7.17 depicts the geometry of the system and shows the coordinate system and the geometric parameters. The rotations of the seed crystal and the crucible are given by the magnitude of the *swirl* $\Omega = rw$, where w is the azimuthal component of the velocity. Then $\Omega = r^2 2\pi f_s$ on the *crystal surface* $A_{s\ell}$, and $\Omega = R_c^2 2\pi f_c$ on the *crucible vertical surface*, and $\Omega = r^2 2\pi f_c$ on the crucible lower surface, where f_s and f_c are the frequencies of the seed and crucible rotations (rotating in *opposite* direction). In the following, the analysis of the thermocapillarity- and thermobuoyancy-affected motion in the melt given by Langlois (1985), is reviewed.

Because of the presence of the free surface and a large temperature difference along the surface $T_{s_b} - T_m$, the *thermocapillarity* effect needs to be included. The *curvature* effect of this surface, i.e., the *curvature-capillarity* is assumed *negligible*. The *radiative heat loss* through the free surface is included as a surface radiation. Since *azimuthal symmetry* is assumed, the space variables are r and x and the stream function ψ is used for the velocities in the r and x directions. Because of this symmetry (i.e., $\partial/\partial\phi = 0$), the vorticity also takes on a simple form

$$\omega = \frac{\partial u}{\partial r} - \frac{\partial v}{\partial x} \, , \quad u = -\frac{1}{r}\frac{\partial \psi}{\partial r} \, , \quad v = \frac{1}{r}\frac{\partial \psi}{\partial x} \, . \tag{7.139}$$

The variables are *scaled* using R_c, $1/2\pi f_s$, and $T_{s_b} - T_m$, and the dimensionless equations for the distributions of the stream function, vorticity, swirl, and temperature are (Langlois, 1985)

$$\frac{\partial}{\partial r^*}\frac{1}{r^*}\frac{\partial \psi}{\partial r^*} + \frac{1}{r^*}\frac{\partial^2 \psi^*}{\partial x^{*2}} = -\omega^* \, , \tag{7.140}$$

$$\frac{1}{r^*}\frac{\partial}{\partial r^*}v^*\omega^* + \frac{\partial}{\partial x^*}\frac{u^*v^*}{r^*} + \frac{\partial}{\partial x^*}\frac{\Omega^{*2}}{r^{*4}} = \frac{Gr_{R_c}}{(R_c/R_s)^3 Re_{R_s}^2}\frac{1}{r^*}\frac{\partial T^*}{\partial r^*} +$$
$$\frac{1}{Re_{R_s}}\left(\frac{1}{r^*}\frac{\partial}{\partial r^*}\frac{1}{r^*}\frac{\partial}{\partial r^*}r^*\omega^* + \frac{\partial}{\partial x^{*2}}r^{*2}\omega^{*2}\right), \tag{7.141}$$

$$\frac{1}{r^*}\frac{\partial}{\partial r^*}r^*v^*\Omega^* + \frac{\partial}{\partial x^*}u^*\Omega^* = \frac{1}{Re_{R_s}}\left(\frac{1}{r^*}\frac{\partial}{\partial r^*}\frac{\Omega^*}{r^{*2}} + \frac{\partial^2\Omega^*}{\partial x^{*2}}\right), \tag{7.142}$$

$$\frac{1}{r^*}\frac{\partial}{\partial r^*}r^*u^*T^* + \frac{\partial}{\partial x^*}u^*T^* = \frac{1}{Re_{R_s}Pr}\left(\frac{1}{r^*}\frac{\partial}{\partial r^*}r\frac{\partial T^*}{\partial r^*} + \frac{\partial^2 T^*}{\partial x^{*2}}\right). \tag{7.143}$$

The boundary conditions for the stream function are $\psi = 0$ on all surfaces, and for the swirl and the temperature are the prescribed values on all surfaces except for the free surface where $\partial\Omega/\partial r = 0$ and the heat loss by surface radiation is balanced by surface conduction and convection. For the vorticity, the stream function equation is used for all surfaces, except for the free surface where

$$\frac{\omega^*}{r^*} = \frac{Ma_{R_s}}{Re_{R_s}}\frac{f_s}{f_c}\frac{1}{r^*}\frac{\partial T^*}{\partial x^*} \quad \text{on free surface.} \tag{7.144}$$

Including the surface radiation, the *nine* dimensionless parameters are

$$Gr_{R_c} = \frac{g\beta(T_{s_b} - T_m)R_c^3}{\nu^2}, \quad Re_{R_s} = \frac{2\pi f_s R_s}{\nu}, \quad Pr = \frac{\nu}{\alpha}, \tag{7.145}$$

$$Ma_{R_s} = -\left.\frac{\partial\sigma}{\partial T}\right|_p \frac{(T_{s_b} - T_m)R_s}{\mu\alpha}, \quad \frac{f_c}{f_s}, \quad \frac{R_c}{R_s}, \quad \frac{L}{R_c}, \tag{7.146}$$

$$\frac{2(T_{s_b} - T_m)}{T_{s_b} + T_m}, \quad \text{and } \epsilon_r^* = \frac{\epsilon_r\sigma_{SB}(T_{sb} + T_m)^4}{\rho c_v R_s}. \tag{7.147}$$

The numerical results for $Re_{R_s} = 8 \times 10^3$, $Pr = 0.035$, $Ma_{R_s} = 0$, $f_c/f_s = 0$, $R_c/R_s = 1.5$, $L/R_c = 0.6$, $\epsilon_r^* = 3.6$, $2(T_{s_b} - T_m)/(T_{s_b} + T_m) = 0.08$, and for four different values of Gr_{R_c} are shown in Figure 7.18. The crucible is *not* rotating and for the case of $Gr_{R_c} = 0$ the flow is *only* due to the rotation of the seed crystal. For this case a single *tip vortex* fills the cavity. For the largest value of Gr_{R_c} the *thermobuoyancy dominates* and a strong *outer vortex* (thermobuoyant) and two weaker *inner vortices* are present, one at the bottom and one at the top (the remnant of the tip vortex). The *transition* from the forced to thermobuoyant flow occurs around $Gr_{R_c} = 8 \times 10^6$ where the tip vortex is yet strong, and for $Gr_{R_c} = 3.2 \times 10^7$ this vortex splits into two small vortices. The equation governing the conservation of vorticity (7.141), suggests that the *dimensionless group* that governs this transition should be $Gr_{R_c}/Re_{R_s}^2(R_c/R_s)^3$, and this has been confirmed by Langlois (1985).

These vortices influence the local growth rate along the seed crystal surface and can lead to flow *instabilities* that can affect the crystal morphology. These aspects are discussed by Langlois (1985).

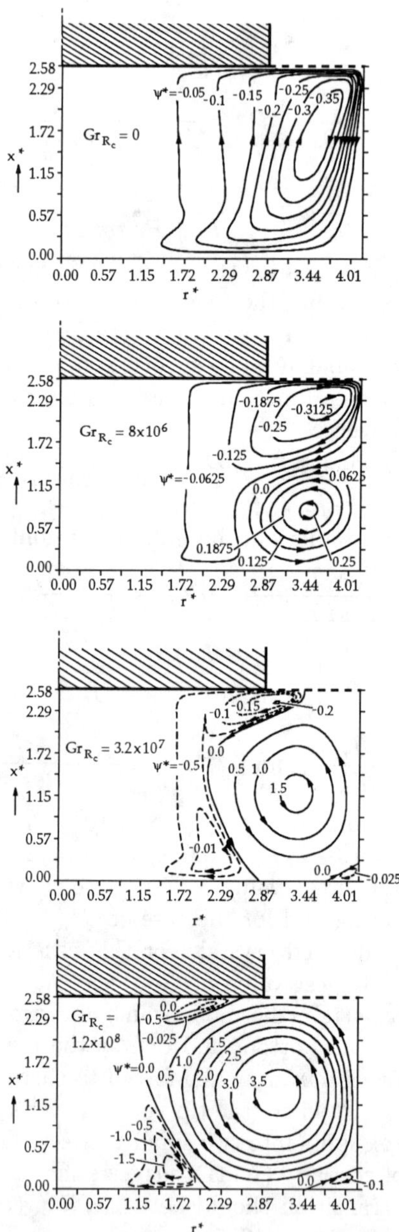

Figure 7.18. Predicted motion within the melt for four different Grashof numbers. (From Langlois, reproduced by permission ©1985 Annual Review Inc.)

7.3.2 Suspension Dendritic Growth

Solidification of *multicomponent* melts by heat removal from the confining surfaces (i.e., mold) of the melt is of practical interest in metallurgy, crystal growth, and geology (Hupport, 1990). *Motion* and convection in the melt may be caused by a *combination* of *volume change*, *thermo-* and *diffusobuoyancy*, *thermocapillarity*, and *forced flow*. The *binary* systems, being the simplest of the multicomponent systems and of some practical interest, have been studied extensively for insight into the *solidification growth rate* and the *crystal characteristics*. In Section 3.2.7 we examined the treatment of the solidification of multicomponent melts by assuming local *thermal* and *chemical* equilibria and using the specific enthalpy of the mixture of the liquid (melt) and solid phases. This allows for the prediction of the two-phase (or mushy zone) separating the solid and liquid phases. In Section 4.16 this formulation was applied to the solidification of a melt in a mold by assuming that the *solidification* (i.e., crystal *nucleation* and *growth*) *begins* at the *cooled* surface.

Figure 7.19 renders some features of the multicomponent solidification and shows the crystal growth on a cooled surface (which can be the side, top, or bottom surface of the mold). The crystals growing on the cooled side surfaces are *columnar* and their characteristics are discussed by Kurz and Fisher (1992). The columnar crystals can grow while having a nearly constant cross-sectional area (called the *faceted growth*) or can branch out as in *dendritic growth* (called *nonfaceted*). Nucleation can *also* occur in a *supercooled* melt and these crystals have some *symmetries* and are called *equiaxed*. These *dispersed* (or suspended) crystals and the liquid in its interstices can be treated as *dispersed elements*. Other than bulk nucleation, dispersed crystals can originate as a *fractured* and *fragmented* columnar crystal or be formed from *columns of liquid* with a very highly *dissimilar* temperature and concentration compared to the ambient liquid (such as the chimneys discussed in Section 4.16). The *bulk* solidification associated with these columnar (or channel) flows results in *freckles*. Chart 7.4 gives a classification of the morphology of the solid-liquid interface in multicomponent solidification with the first distinction made between the growth over a *subcooled solid* sruface and the growth in a *supercooled liquid*.

For a binary system (such as a metal alloy) such as the one depicted in Figures 1.6(d) and 3.6, in solidification of *pure* species B, *planar* or *dendritic* growth can be found. For concentrations *between* the *eutectic* concentration and the *pure* species B, a *dendritic plus dendritic-eutectic* solidification occurs and this is most common in applications.

The *bulk nucleation* in the liquid-vapor phase change discussed in Chapter 6 can also occur in the liquid-solid phase change, and then the nucleated phase can *grow* or *decay*. As an illustrative example of the effect of the convective heat transfer on solidification, in this section we examine the bulk nucleation and growth of crystals in a binary liquid supercooled by heat

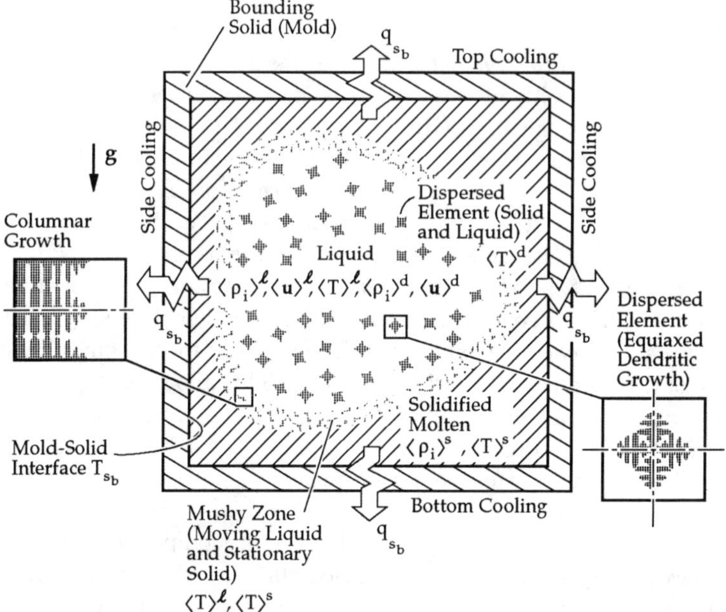

Figure 7.19. Columnar and dendritic growth in a multicomponent solidification. The growth on the solidified molten (with the mushy zone), as well as growth as dispersed elements, is depicted.

removal from the mold surface. We begin by reviewing the available results on bulk nucleation of crystals in multicomponent systems. Then the growth of these crystals will be examined. In general, this interface is unstable and the growth is *not* uniform *along* the surface. In order to gain some insight, the solid-liquid interfacial *instabilities* associated with phase growth in a binary system is examined. Then using a *dispersed-element model*, the growth of the *dendritic equiaxed crystal* is analyzed for a *diffusion-controlled* growth with a *prescribed* growth velocity for the *dendritic tip*. This is followed by the inclusion of the effect of liquid and crystal *motions* (caused by the thermo- and diffusobuoyancy). The continuum, hydrodynamic, and heat transfer treatments of the liquid and solid-particles phases and the effective media properties are discussed. Then the available numerical results for the solidification of a binary liquid in a two-dimensional *cavity* are reviewed.

(A) Bulk Nucleation

The metastable states and bulk and surface nucleations were briefly discussed in Section 1.5.1(C). As with the bulk bubble and droplet nucleations discussed in Section 6.2.2, a crystal grain can nucleate in a supercooled

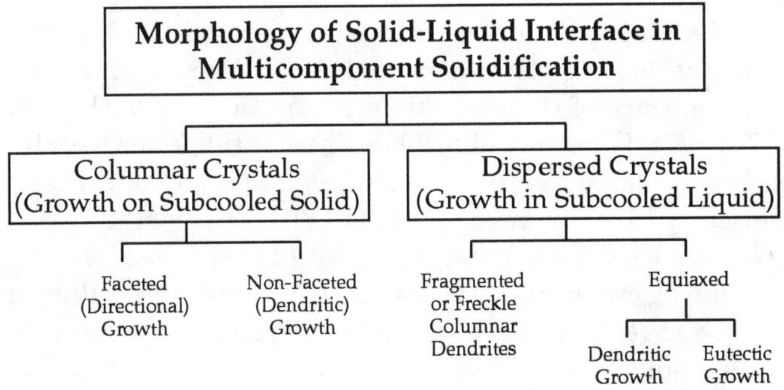

Chart 7.4. A classification of morphology of solid-liquid interface in the multicomponent solidification.

melt. Unlike bubble and droplet nucleations, the *maximum supercooling* for single- and multicomponent melts have not been examined to the same extent. The *thermodynamic* treatment of the volumetric nucleation number density and the nucleation rate has been reviewed by Kurz and Fisher (1992). The grains are formed from crystalline clusters where each cluster (or grain) contains a number of atoms (or molecules). The *equilibrium* solubility distribution of the these clusters in the melt is given by

$$\frac{n_s}{n_\ell} = \exp(-\frac{\Delta g_{\ell s}}{k_B T}) \,, \tag{7.148}$$

where n_s and n_ℓ are the volumetric number densities of the *clusters* and the *liquid* molecules, and $\Delta g_{s\ell}$ is the *specific Gibbs free energy* difference between the liquid and solid phases. The energy is given in terms of the surface tension $\sigma_{\ell s}$, the specific entropy of the phase trasnformation $\Delta s_{\ell s}$, and the grain radius R_s, as (Kurz and Fisher, 1992)

$$\Delta g_{\ell s} = \sigma_{\ell s} 4\pi R_s^2 - \Delta s_{\ell s} \Delta T_{sc} \,, \tag{7.149}$$

where ΔT_{sc} is the *supercooling*, i.e., the difference between the *equilibrium* melting temperature of the mixture and the actual melting temperature $T_m - T_\ell$. There is a maximum for $\Delta g_{\ell s}$ and it occurs at a crystal radius, i.e., $\partial \Delta g_{\ell s} / \partial R_s = 0$, given by

$$R_s = \frac{2\sigma_{\ell s}}{\Delta s_{\ell s} \Delta T_{sc}} \,, \tag{7.150}$$

which depends on the extent of supercooling.

The volumetric rate of the crystal formation is given by (Kurz and Fisher, 1992)

$$\dot{n}_s = \dot{n}_{s_o} \exp(-\frac{16\pi}{3 k_B T} \frac{\sigma_{\ell s}^3}{\Delta g_{\ell s}^2} - \frac{\Delta E_c}{k_B T}) \tag{7.151}$$

where $\dot{n}_{s_o}$ is a pre-exponential factor and ΔE_c is the activation energy for the formation of a cluster. In multicomponent systems, the magnitudes of $\sigma_{\ell s}$, $\Delta g_{\ell s}$, and ΔE_c would depend on the species concentrations.

As with the nucleation rates discussed in Section 6.2.2, (7.151) is *very* sensitive to the magnitude of ΔE_c and ΔT_{sc}. Experimental results have shown a dependence of the nucleation rate on the *rate* of the temperature change $\partial T/\partial t$ (Stefanescu et al., 1990). These correlations have been developed (with two constants) to express $\dot{n}_s$ as a function of ΔT_{sc} and $\partial T/\partial t$ and are discussed by Stefanescu et al. (1990).

An alternative, semi-empirical treatment of the nucleation number density has been suggested by both Thévoz et al. (1989) and Rappaz (1989), and uses a *prescribed*, empirical *maximum supercooling*. This model *limits* the maximum number of nuclei formed to that at the prescribed supercooling.

Other than bulk nucleation, suspended crystals are found in supercooled liquids and originating from fracture and fragmentation of columnar crystals formed on the solid surface. In extremely nonuniform concentration and temperature fields, such as around channels formed in the *thermo-solutal* convection discussed in Section 4.16, (bulk) solidification can occur leading to formation of freckles (Felicelli et al., 1991; Nielson and Incropera, 1993). Because of the presence of large gradients, the treatment of the crystal formation and growth in these columns should be based on thermal and chemical *nonequilibrium* models.

The crystal grain conservation equation can be written similar to (6.54), i.e.,

$$\frac{\partial n_s}{\partial t} + \nabla \cdot \mathbf{u}_s n_s = \dot{n}_s + \dot{n}_{s,f} - \dot{n}_{s,r} \ , \tag{7.152}$$

where $\dot{n}_s$ is the volumetric grain nucleation rate given by (7.151), $\dot{n}_{s,f}$ designates the volumetric grain *formation rate* due to the *columnar crystal fragmentation* (and *sources* other than bulk nucleation) and $\dot{n}_{s,r}$ designates the volumetric grain *removal rate*.

(B) Instabilities of Growing Crystal-Melt Interface

The growing solid-liquid interface may undergo a *morphological instability* and in pure liquids this is caused only by the interfacial *heat transfer* (i.e., *temperature distribution*). For multicomponent systems both the *temperature* and the *concentration* distributions influence this instability. As with other stability analyses, a disturbance, here in the interfacial morphology, i.e., an *introduced interface displacement wave*, grows in amplitude when the system is unstable. Starting with a *spherical* crystal growing in a *supercooled* (*metastable*) liquid, the temperature *decreases* with the distance from the interface. The *heat released* during solidification flows from the interface towards the liquid. We begin by assuming that an axisymmetric radial temperature gradient is present and that the interfacial temperature is *uniform*. When the interface is disturbed with a spatially sinusoidal variation in the radius R, the portion of the interface with an *increased* R will experience a *larger* gradient (interfacial temperature remains constant). Then this portion will *grow further* while the portion with a decreased R will grow less. Therefore, the growth of a spherical, pure crystal in a supercooled liquid is morphologically *unstable*, i.e., *dendritic*. A detailed analysis of this instability, but for the *isothermal, concentration-supersaturation* growth, is given by Mullins and Sekerka (1963). A two-dimensional, direct simulation of the dendritic growth in subcooled melts has been made by Murray et al. (1993). They use a *phase-field model* which introduces a phase-field variable which is continuous across the finite but thin interface. They include the capillarity and the interfacial kinetic in the phase-field conservation equation.

For a binary system with a phase diagram similar to that shown in Figure 3.6, as the crystal grows the concentration of species A *piles up* near the interface. Then in addition to a temperature gradient, a concentration gradient is also present in the liquid adjacent to the crystal. The effect of a local *increase* in R would still be an *increase* in the local heat transfer rate and an increase in the local growth rate. Note that for the growth of the bulk nucleated crystals considered, the liquid is in a *metastable* supercooled state.

Extensive studies of the morphological instability of the interface have been made for the case of *equilibrium liquid* and *subcooled solid* (Mullins and Sekerka, 1964; Hurle et al., 1983; Coriell et al., 1985; a review is given by Davis, 1990; Forth and Wheeler, 1992). In these cases the temperature *increases* with the distance from the interface; however, if the *actual* temperature is *below* the *liquidus* temperature (which also increases with the distance because the liquid concentration decreases), then *dendritic* solidification (mushy zone) occurs over the region confined by the interface and the location where the actual and liquidus temperatures become equal. This is further discussed by Kurz and Fisher (1992).

(C) A Dispersed-Element Model for Diffusion-Controlled Growth

Assuming that a *total* number n_s of *spherical* crystals are nucleated per unit volume at a supercooling of $\Delta T_{sc} = T_m - T_\ell$, then these crystals can grow to *final* grain radius of R_c

$$R_c = (\frac{3}{4\pi n_s})^{1/3} . \tag{7.153}$$

The initially spherical crystal will have a *dendritic* growth and between the nucleation at time $t = 0$ and the complete growth (end of the solidification) at $t = t_f$, the *grain envelope* radius R will grow from a radius R_{s_o} (i.e., the initial radius of the nuclei) to R_c. The grain envelop is initially around the crystal and during the dendritic growth the envelop contains solid and liquid phases, and finally at the end of the solidification will again contain *only* solid. The content of the grain envelope is treated as a *dispersed element* and the *liquid fraction* (or *porosity*) ϵ_s of this *dispersed element* is initially zero and then *increases* rapidly followed by an eventual *decrease*, finally becoming zero again. This dispersed element and its surrounding liquid makes a *unit-cell model* and has been introduced and analyzed by Rappaz and Thévoz (1987). A schematic of this unit-cell model is shown in Figure 7.20(a). The growth of the dispersed element subject to the *cooling rate* of the unit cell given by the heat transfer rate $q_c 4\pi R_c^2$ with *no liquid motion* has been analyzed.

The elemental-liquid fraction ϵ_e and the *cell porosity* ϵ_c are defined as

$$\epsilon_e = 1 - \frac{V_s}{V_e} \qquad \epsilon_c = 1 - \frac{V_e}{V_c} = 1 - \frac{R^3}{R_c^3} \,. \tag{7.154}$$

The solid volume is

$$V_s = (1-\epsilon_e)V_e = (1-\epsilon_e)\frac{4}{3}\pi R^3 \,. \tag{7.155}$$

The quantities ϵ_e and ϵ_c are determined from the heat and mass transfer analyses. An *apparent* solid radius can be defined as

$$R_s^3 = \frac{V_s}{\frac{4}{3}\pi} = (1-\epsilon_e)R^3 \ \text{ or } \ R_s = (1-\epsilon_e)^{1/3}R \,. \tag{7.156}$$

For convenience, the *volume fraction* of the *solid* in the *unit cell* f_s and the *volume fraction* of the *dispersed element* in the *unit cell* f_c, which are related to ϵ_e and ϵ_c, are used in the analysis. These are

$$f_s = \frac{V_s}{V_c} = \frac{R_s^3}{R_c^3} = (1-\epsilon_e)(1-\epsilon_c) \ , \ f_c = \frac{V_e}{V_c} = \frac{R^3}{R_c^3} \,. \tag{7.157}$$

Assuming an equilibrium phase diagram, such as that shown in Figure 7.20(b), the definition of the equilibrium partition ratio k_p was given by (3.96) and is repeated here

$$k_p = \frac{\frac{\rho_A}{\rho}|_{\ell s}}{\frac{\rho_A}{\rho}|_{s\ell}}|_{T,p} \,, \tag{7.158}$$

where the solid-liquid interfacial concentrations are $(\rho_A/\rho)_{s\ell}$ and $(\rho_A/\rho)_{\ell s}$ on the *solidus* and *liquidus* lines, respectively. The liquidus line mass fraction is related to the temperature using

$$T_{\ell s} - T_{s\ell} = -\gamma(\frac{\rho_A}{\rho}|_{\ell s} - \frac{\rho_A}{\rho}|_{s\ell}) \ \text{ or } \ T_e - T_{m,B} = \gamma\frac{\rho_A}{\rho}|_e \,, \tag{7.159}$$

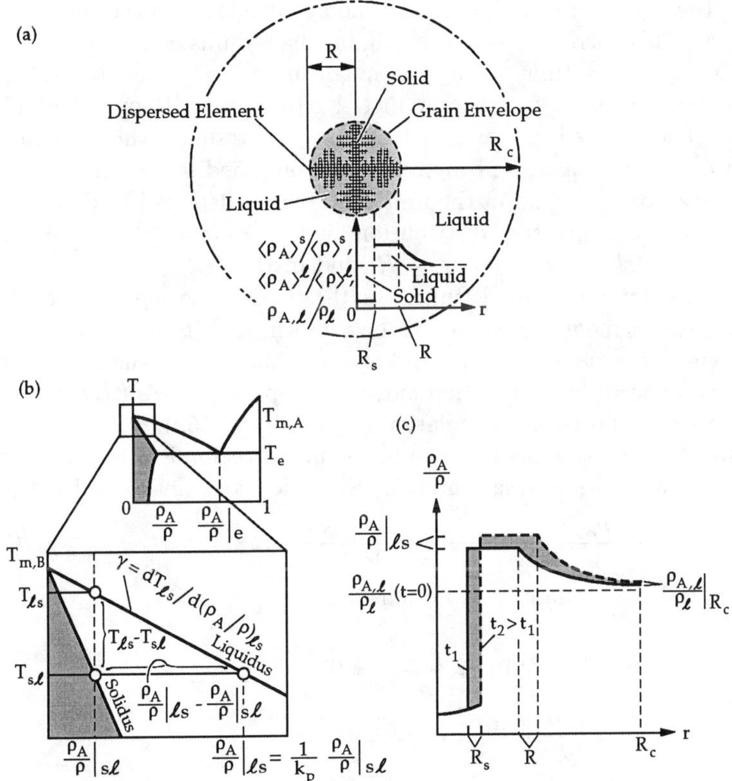

Figure 7.20. (a) Unit-cell model of the equiaxed dendritic growth of a crystal. The liquids within the grain envelope and within the element are shown along with the mass fraction distribution of the species A. (b) The idealized phase diagram. (c) The mass fraction distribution of the species A for two different elapsed times. (From Rappaz and Thévos, reproduced by permission ©1987 Pergamon Press.)

where

$$\gamma = \frac{dT_{\ell s}}{d(\rho_A/\rho)_{\ell s}}|_p = \frac{T_e - T_{m,B}}{(\rho_A/\rho)_e} . \tag{7.160}$$

The *mass fraction difference* can also be written as

$$(\rho_A/\rho)_{\ell s} - (\rho_A/\rho)_{s\ell} = (\rho_A/\rho)_{s\ell}\frac{1-k_p}{k_p} . \tag{7.161}$$

As the crystal grows the concentraion of species A in the solid increases *slightly*. This corresponds to a negligible mass diffusion in the solid ($D_s \rightarrow 0$). Other assumptions about the magnitude of D_s (i.e., $D_s \rightarrow \infty$ or a finite D_s) do *not* change the predicted growth rate (Rappaz and Thévoz, 1987). The rejected species A (i.e., *solute*) will result in the *increase* of the concentration of species A in the liquid contained in the envelope and in the remainder of liquid in the unit cell. This is depicted in Figure 7.20(c) where the concentration distributions in the *solid* ($0 \leq r \leq R_s$) in the *interelemental liquid* ($R_s \leq r \leq R$) and in the *cell liquid* ($R < r \leq R_c$) are shown for two elapsed times. In the model, the *liquid* concentration within the element is assumed to be uniform and its magnitude $(\rho_A/\rho)_{\ell s}$ is given by the phase diagram and as a function of T_e. Then in this model the *dendritic tip* concentration and the temperature are $(\rho_A/\rho)_{\ell s}$ and T_e, respectively, and their interrelation is given by (7.159).

The mass fraction distribution in the liquid region is determined for the species conservation equation (1.72) which for constant ρ_ℓ and D_ℓ gives

$$\frac{\partial}{\partial t}\frac{\rho_{A,\ell}}{\rho_\ell} = D_\ell\left(\frac{\partial}{\partial r^2} + \frac{2}{r}\frac{\partial}{\partial r}\right)\frac{\rho_{A,\ell}}{\rho_\ell} \quad R \leq r \leq R_c \, . \tag{7.162}$$

The initial and boundary conditions are

$$\frac{\rho_{A,\ell}}{\rho_\ell}(r,0) = \frac{\rho_{A,\ell}}{\rho_\ell}(t=0) \quad R_s \leq r \leq R_c \, , \tag{7.163}$$

$$\frac{\partial}{\partial r}\frac{\rho_{A,\ell}}{\rho_\ell} = 0 \quad r = R_c \, . \tag{7.164}$$

The concentration at $r = R$ is found from the species balance made over $0 \leq r \leq R_c$. This is done by the integration of the distributions shown in Figure 7.20(c), i.e.,

$$\int_0^{R_s} k_p \frac{\rho_A}{\rho}|_{\ell s} 4\pi r dr + \frac{\rho_A}{\rho}|_{\ell s}\frac{4}{3}\pi(R^3 - R_s^3) + \int_R^{R_c} \frac{\rho_{A,\ell}}{\rho_\ell} 4\pi r dr = \frac{\rho_{A,\ell}}{\rho}(t=0)\frac{4}{3}\pi R_c^3 \, . \tag{7.165}$$

By differentiating this with respect to time and using (7.162), we have

$$-4\pi D_\ell R^2 \frac{\partial}{\partial r}\frac{\rho_A}{\rho_\ell}|_R = -\frac{4}{3}\pi\frac{d}{dt}\frac{\rho_A}{\rho}|_{\ell s}(R^3 - R_s^3) + 4\pi R^2 \frac{dR}{dt}(1 - k_p)\frac{\rho_A}{\rho}|_{\ell s} \, . \tag{7.166}$$

This states that the outward solute flow rate through $r = R$ is determined by the solute *ejected* due to the solidification and the *temperature-caused change* in the mass fraction of the interdendritic liquid.

For *large* solid and liquid conductivities, within the cell the solid, the interdendritic liquid, and the cell liquid can all be in a *near-thermal equilibrium*. For a temperature change occurring on the boundaries of the cell, the assumption of a uniform temperature within the cell requires that the cell Biot number, i.e.,

$$Bi = \frac{Nu_{dc}^{\rm ext}}{6}\frac{k_\infty}{\langle k\rangle_{V_c}}\ , \quad Nu_{dc}^{\rm ext} = \frac{q_c 2R_c}{(T_c - T_\infty)k_\infty} \qquad (7.167)$$

be less than 0.1. This condition is assumed to hold (for metals) and a single temperature T_c is used for the unit cell.

The thermal energy balance on the unit cell gives

$$q_c 4\pi R_c^2 = [\rho_\ell \Delta i_{\ell s}\frac{df_s}{dt} + \langle\rho c_p\rangle_{V_c}\frac{dT_c}{dt}]\frac{4}{3}\pi R_c^3\ , \qquad (7.168)$$

where $\Delta i_{\ell s}$ is the heat of solidification and

$$\langle\rho c_p\rangle_{V_c} = (1-\epsilon_e)(1-\epsilon_c)(\rho c_p)_s + [\epsilon_e(1-\epsilon_c)+\epsilon_c](\rho c_p)_\ell\ . \qquad (7.169)$$

The variation of temperature can be replaced by that of the concentration by assuming that the cell temperature is the equilibrium temperature at the liquidus line. Then using (7.159), (7.168) becomes

$$\frac{3q_c}{R_c} = \Delta i_{\ell s}\frac{df_s}{dt} + \langle\rho c_p\rangle_{V_c}\gamma\frac{d}{dt}\frac{\rho_A}{\rho}|_{\ell s}\ . \qquad (7.170)$$

The growth rate of the *dendritic tip* is modeled using the available results for low Péclet number growth in *unbounded* liquid. The model used by Rappaz and Thévoz (1987) is

$$\frac{dR}{dt} = \frac{\gamma D_\ell}{\pi^2(k_p-1)\dfrac{\sigma_{\ell s}}{\Delta s_{\ell s}}\dfrac{\rho_{A,\ell}}{\rho_\ell}(t=0)}\left(\frac{\rho_A}{\rho}|_{\ell s} - \frac{\rho_{A,\ell}}{\rho_\ell}|_R\right)^2 . \qquad (7.171)$$

This kinetic condition is discussed by Hills and Roberts (1993). The distribution of $\rho_{A,\ell}/\rho_\ell$ in $R \le r \le R_c$ as well as the variations of f_s, f_c, $(\rho_A/\rho)_{\ell s}$, and $(\rho_{A,\ell}/\rho_\ell)_R$ with respect to time are determined by solving (7.156), (7.165), (7.166), (7.170), and (7.171), simultaneously. This is done numerically and some of their results are reviewed below.

The results are for the solidification of Al-Si with an initial silicon concentration $(\rho_{A,\ell}/\rho_\ell)(t=0) = 0.05$, an eutectic concentration of 0.108, $D_\ell = 3\times10^{-9}$ m^2/s, $k_p = 0.117$, $\gamma = -7.0$, $\rho_\ell\Delta i_{\ell s} = 9.5\times10^8$ J/m^3, $T_{m,B}$(aluminum) $= 660$°C, $T_e = 577$°C, $\sigma_{\ell s}/\Delta s_{s\ell} = 9\times10^{-8}$ m-K, and $\langle\rho c_p\rangle_{V_c} = 2.35\times10^6$ J/m^3-K. Figure 7.21 (a) shows the results for $\Delta T_{sc} = 0$, $R_c = 100$ μm, $R_{s_o} = 1$ μm, and a cooling rate that results in the *total* solidification of the cell in 10 s.

The variation of temperature T_c and solid and element volume fractions f_s and f_c, respectively, with respect to time, up to the time of the complete solidification, are shown. Since the mass fraction $(\rho_A/\rho)_{\ell s}$ has to *increase* in order for the growth of the dendrite to begin, as stated by (7.171), then no change in dT/dt occurs until the point of *undershoot* is reached, and growth begins with a *large* and sudden increase in f_c. Since the growth rate of the dendritic tip is larger than the rate

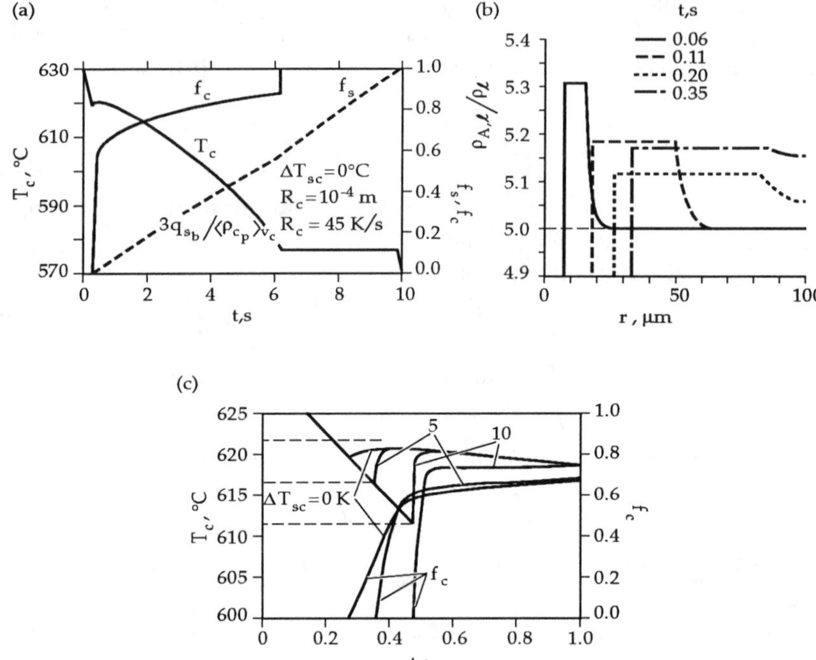

Figure 7.21. (a) Computed time variation of temperatures and solid and element volume fractions. (b) Radial distribution of the species A in the liquid for four different elapsed times. (c) Same as (a) but for three different supercooling conditions. (From Rappaz and Thévoz, reproduced by permission ©1987 Pergamon Press.)

of the solidification allowed by the heat removal, the volume fraction of the interdendritic liquid ϵ_s increases substantially. Therefore, f_s increases much slower than f_c. This increase in f_s *decreases* $(\rho_A/\rho)_{\ell s}$ and *increases* T_c, as stated by (7.168) and (7.170). The increase in T_c is called *recalescence*. The concentration distributions in the interdendritic liquid and in the cell liquid are shown in Figure 7.21 (b) for several elapsed times. The elapsed times correspond to the time of undershoot, shortly after the rise in T_c begins, when T_c reaches a maximum, and shortly after the decrease in T_c begins. For $t > 0.20$ s, the temperature *decreases* while $(\rho_A/\rho)_{\ell s}$ *increases*. For an elapsed time slightly larger than 6 s, the element grows to the maximum radius R_c and only the solidification of the interdendritic liquid occurs. This ends when all this liquid is solidified. The undershoot temperature predicted for no supercooling, shown in Figure 7.21(a), will correspond to the supercooling ΔT_{sc} when substantial liquid supercooling exists. Figure 7.21(c) shows the results for the same conditions as in Figures 7.21(a) and (b), except 5 and 10°C supercooling are allowed. The results show that the larger the supercooling, the faster the temperature rises after the initial growth (i.e., *accelerated recalescence*).

(D) Inclusion of Buoyant Liquid and Crystal Motions

The unit-cell-based, diffusion-controlled dendritic growth discussed above has been extended to thermo- and diffusobuoyant *convection* by using local phase-volume-averaged conservation equations and thermal and chemical *nonequilibrium* among the liquid phase, the solid-particles (equiaxed dendritics) phase, and the confining surfaces of the mold. As before, the crystals are assumed to be formed by the bulk nucleation in a supercooled liquid. The liquid temperature $\langle T\rangle^\ell$, the liquid volume fraction ϵ and its solid particle counterparts $\langle T\rangle^s$ and $1-\epsilon$, the velocities $\langle \mathbf{u}\rangle^\ell$ and $\langle \mathbf{u}\rangle^s$, the volumetric solidification rate $\langle \dot{n}\rangle^s$, and the concentration of species A in each phase $\langle \rho_A\rangle^\ell/\rho_\ell$, $\langle \rho_A\rangle^s/\rho_s$ are all determined from the solution of the local phase-volume-averaged conservation equations. The *thermodynamic* conditions are applied similar to those in the above diffusion treatment, but the growth rate of the dendritic tip is *not* prescribed. Instead, the interfacial heat and mass transfer is modeled using interfacial Nusselt and Sherwood numbers, which as used in Sections 7.1.3 and 7.2.2 for the particulate flow and heat transfer, are prescribed as functions of the Reynolds number based on the relative velocity and the solid particles diameter d.

The local, phase-volume-averaged treatment of flow and heat and mass transfer has been addressed by Voller et al. (1989), Prakash (1990), Beckermann and Ni (1992), Prescott et al. (1992), and Wang and Beckermann (1992), and a review is given by Beckermann and Viskanta (1993). In the following, the conservations and the thermodynamic conditions are examined, and then some of available results on the growth and the motion of the bulk-nucleated crystals are reviewed.

Laminar flow is assumed and the modeling of $\langle \mathbf{S}\rangle^\ell$, $\langle \mathbf{S}\rangle^s$, and τ_d are pursued similar to the hydrodynamics of the particulate flow discussed in Sections 5.4.2, 7.1.1, and 7.2.1. Since the solid particles are *not* spherical, the dendritic arms and other geometric parameters should be included in the models. Ahuja et al. (1992) develop a drag coefficients for equiaxed dendrites.

The effective media properties $\mathbf{D}_m^\ell$, $\mathbf{D}^\ell$, $\mathbf{D}_m^s$, and $\mathbf{D}^s$, which include both the *molecular* (i.e., *conductive*) and the hydrodynamic dispersion components, are also modeled. Due to the *lack* of any predictive correlations for the *nonequilibrium* transport, correlations similar to those discussed in Section 3.3.4, which are valid under local thermal equilibrium conditions are used. For the interfacial convective transport, the local Nusselt and Sherwood numbers are prescribed using correlations such as those discussed in Section 5.2.2. The effect of the solid particles geometry must also be addressed (Dash and Gill, 1984).

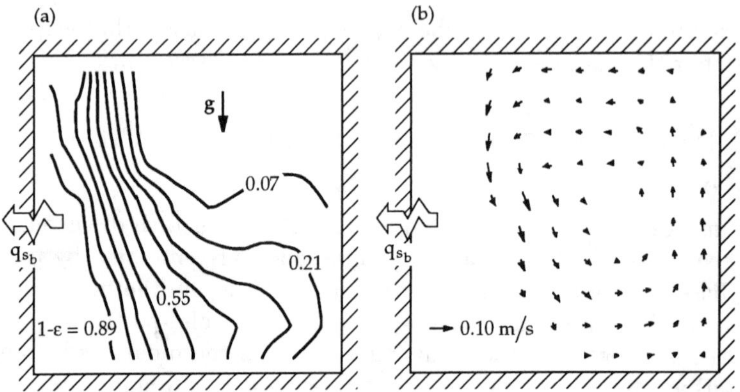

Figure 7.22. Typical solid-volume fraction and motion in a mold cooled on the left wall. (From Beckermann and Ni, reproduced by permission ©1992 Technomic Publishing Company.)

The numerical results of Beckermann and Ni (1992) for the solidification of a AlOCu mixture are shown in Figures 7.22(a) and (b). The results are for an initial copper concentration of 0.05 cooled in a two-dimensional square mold by extracting heat from the *left* vertical surface (all other surfaces are ideally *insulated*) at a constant and uniform rate q_{s_b}. An initial grain diameter (spherical crystal) of 1 μm is assumed along with a nucleation model. Their results for the solid-volume fraction $1-\epsilon$ and the solid-phase velocity $\langle \mathbf{u} \rangle^s$ for an elapsed time of 40 s are shown. The solid concentration is largest near the cooled surface and on the left lower corner, and the solid velocity vanishes in this areas. Thermo- and diffusobuoyant motion remains intensive in the areas of *low* solid-volume fraction.

The phasic, local-volume-averaged energy conservation equations are variations of those discussed in Sections 3.3.1 to 3.3.3. As was mentioned, the liquid- and solid-phase flows are assumed to be *laminar*. Then time averaging and the modeling of the turbulent transport are not needed. The interfacial heat and species fluxes are modeled using the *interfacial-averaged* Nusselt and Sherwood numbers. These interfacial convections resulting from the thermal and chemical nonequilibrium are modeled for *both* the liquid- and the solid-side of the interface. Then conservation equations similar to those used in Section 7.2.2 are found, except now the solid-side interfacial convection is added.

The phasic, local-volume-averaged overall mass, species, momentum, and energy equations for the liquid phase are

$$\frac{\partial}{\partial t}\epsilon\rho_\ell + \nabla \cdot \epsilon\rho_\ell \langle \mathbf{u} \rangle^\ell = \langle \dot{n} \rangle^\ell , \tag{7.172}$$

$$\frac{\partial}{\partial t}\epsilon\langle\rho_A\rangle^\ell + \nabla \cdot \epsilon\langle \mathbf{u} \rangle^\ell \langle\rho_A\rangle^\ell = \nabla \cdot \epsilon\rho_\ell \mathbf{D}_m^\ell \cdot \nabla \frac{\langle\rho_A\rangle^\ell}{\rho_\ell} + \langle \dot{n} \rangle^\ell \left. \frac{\langle\rho_A\rangle}{\rho} \right|_{\ell s} +$$

$$\frac{A_{\ell s}}{V}\langle Sh_d\rangle^{\ell}_{A_{\ell s}}\frac{\rho_\ell D_\ell}{d}\left(\frac{\rho_A}{\rho}\bigg|_{\ell s}-\frac{\langle\rho_A\rangle^\ell}{\rho_\ell}\right), \tag{7.173}$$

$$\frac{\partial}{\partial t}\epsilon\rho_\ell\langle\mathbf{u}\rangle^\ell+\nabla\cdot\epsilon\rho_\ell\langle\mathbf{u}\rangle^\ell\langle\mathbf{u}\rangle^\ell=-\epsilon\nabla\langle p\rangle^\ell+\nabla\cdot\epsilon\langle\mathbf{S}\rangle^\ell+\epsilon\rho_\ell\mathbf{g}+$$

$$\langle\dot n\rangle^\ell\langle\mathbf{u}_\ell\rangle_{A_{\ell s}}-\frac{\rho_s}{\tau_d}(1-\epsilon)(\langle\mathbf{u}\rangle^\ell-\langle\mathbf{u}\rangle^s), \tag{7.174}$$

$$\frac{\partial}{\partial t}\epsilon\rho_\ell\langle i\rangle^\ell+\nabla\cdot\rho_\ell\langle\mathbf{u}\rangle^\ell\langle i\rangle^\ell=\nabla\cdot\epsilon(\rho c_p)_\ell\mathbf{D}_\ell\cdot\nabla\langle T\rangle^\ell+$$

$$\langle\dot n\rangle^\ell\langle i_\ell\rangle_{A_{\ell s}}+\frac{A_{\ell s}}{V}\langle Nu_d\rangle^\ell_{A_{\ell s}}\frac{k_\ell}{d}(T_\ell-\langle T\rangle^\ell). \tag{7.175}$$

Similarly, for the solid phase we have

$$\frac{\partial}{\partial t}(1-\epsilon)\rho_s+\nabla\cdot\epsilon\rho_s\langle\mathbf{u}\rangle^s=\langle\dot n\rangle^s, \tag{7.176}$$

$$\frac{\partial}{\partial t}(1-\epsilon)\langle\rho_A\rangle^s+\nabla\cdot(1-\epsilon)\langle\mathbf{u}\rangle^s\langle\rho_A\rangle^s=\nabla\cdot(1-\epsilon)\rho_s\mathbf{D}^s_m\cdot\nabla\frac{\langle\rho_A\rangle^s}{\rho_s}+$$

$$\langle\dot n\rangle^s\frac{\rho_A}{\rho}\bigg|_{s\ell}+\frac{A_{\ell s}}{V}\langle Sh_d\rangle^s_{A_{\ell s}}\frac{\rho_s D_s}{d}\left(\frac{\rho_A}{\rho}\bigg|_{s\ell}-\frac{\langle\rho_A\rangle^s}{\rho_s}\right), \tag{7.177}$$

$$\frac{\partial}{\partial t}(1-\epsilon)\rho_s\langle\mathbf{u}\rangle^s+\nabla\cdot(1-\epsilon)\rho_s\langle\mathbf{u}\rangle^s\langle\mathbf{u}\rangle^s=-(1-\epsilon)\nabla\langle p\rangle^s+\nabla\cdot(1-\epsilon)\langle\mathbf{S}\rangle^s+$$

$$(1-\epsilon)\rho_s\mathbf{g}+\langle\dot n\rangle^s\langle\mathbf{u}_s\rangle_{A_{\ell s}}+\frac{\rho_s}{\tau_d}(1-\epsilon)(\langle\mathbf{u}\rangle^\ell-\langle\mathbf{u}\rangle^s), \tag{7.178}$$

$$\frac{\partial}{\partial t}(1-\epsilon)\rho_s\langle i\rangle^s+\nabla\cdot(1-\epsilon)\rho_s\langle\mathbf{u}\rangle^s\langle i\rangle^s=\nabla\cdot(1-\epsilon)(\rho c_p)_s\mathbf{D}_s\cdot\nabla\langle T\rangle^s+$$

$$\langle\dot n\rangle^s\langle i_s\rangle_{A_{\ell s}}+\frac{A_{\ell s}}{V}\langle Nu_d\rangle^s_{A_{\ell s}}\frac{k_s}{d}(T_\ell-\langle T\rangle^s). \tag{7.179}$$

The interfacial balance of the overall mass, species and energy yields

$$\langle\dot n\rangle^\ell=-\langle\dot n\rangle^s, \tag{7.180}$$

$$\frac{A_{\ell s}}{V}\langle Sh_d\rangle^\ell_{A_{\ell s}}\frac{\rho_\ell D_\ell}{d}\left(\frac{\rho_A}{\rho}\bigg|_{\ell s}-\frac{\langle\rho_A\rangle^\ell}{\rho_\ell}\right)+$$

$$\frac{A_{\ell s}}{V}\langle Sh_d\rangle^s_{A_{\ell s}}\frac{\rho_s D_s}{d}\left(\frac{\rho_A}{\rho}\bigg|_{s\ell}-\frac{\langle\rho_A\rangle^s}{\rho_s}\right),+\langle\dot n\rangle^s\left(\frac{\rho_A}{\rho}\bigg|_{s\ell}-\frac{\rho_A}{\rho}\bigg|_{\ell s}\right)=0, \tag{7.181}$$

$$\frac{A_{\ell s}}{V}\langle Nu_d\rangle^\ell_{A_{sf}}(T_\ell-\langle T\rangle^\ell)+\frac{A_{\ell s}}{V}\langle Nu_d\rangle^s_{A_{sf}}(T_\ell-\langle T\rangle^s)+$$

$$\langle\dot n\rangle^s(\langle i_s\rangle_{A_{\ell s}}-\langle i_\ell\rangle_{A_{\ell s}})=0. \tag{7.182}$$

The specific enthalpies are related to temperature using

$$\langle i_\ell\rangle_{A_{\ell s}}=c_{p_\ell}T_\ell+\Delta i_{\ell s},\quad\langle i_s\rangle_{A_{\ell s}}=c_{p_s}T_s,$$

$$\langle i \rangle^\ell = c_{p_\ell} \langle T \rangle_\ell + \Delta i_{\ell s} \,, \quad \langle i \rangle^s = c_{p_s} \langle T \rangle_s \,. \tag{7.183}$$

The interfacial temperature T_ℓ is related to the interfacial mass fraction $(\rho_A/\rho)_{\ell s}$ through (7.159) which can be written as

$$\left.\frac{\rho_A}{\rho}\right|_{\ell s} = \left.\frac{\rho_A}{\rho}\right|_e \frac{T_{m,B} - T_\ell}{T_{m,B} - T_e} \,. \tag{7.184}$$

A phase diagram similar to Figure 7.20(b) is assumed, and for a given temperature T_ℓ the mass fraction on the solidus line is related to that on the liquidus line using (7.159).

7.3.3 Layer Melting

Liquid formation and motion resulting from heating a solid to its melting temperature, where the local melting temperature depends on the local concentrations for *multicomponent* solids, occur in many applications, and the liquid motion in turn influences the rate of melting. In this section melting of solids by heat transfer through its *contacting melt* is examined. This begins by the examination of the *ideal, one-dimensional single-component melt layer growth* where the melt is contained between the *melting solid* and a *confining solid* which provides the heat for the phase change. This allows for the examination of the role of the liquid motion (i.e., convection) on the rate of melting (i.e., the melt layer growth rate). In practice, depending on the flow (i.e., *forced* or *buoyant*, *laminar* or *turbulent*), the liquid-solid interface may or may *not* remain *planar* and *stable*. This is examined next and a simple *stability analysis* is reviewed. The *thermobuoyant*- and *thermocapillary*-driven cavity motions occurring during melting of a single-component solid are examined, and the role of motion on the geometry of solidified layer is discussed. The *close-contact* melting which results when the melt is continuously *removed* and *increases* the rate of melting is discussed next. Instead of imposing an external heat flux (or a temperature difference) to *cause* melting of ice, a bounding surface *pressed* against an ice surface causes *lowering* of the *melting point*, and if the ice is just slightly above its melting point at the ambient pressure, melting of the ice occurs. This is called the *pressure melting* and is examined. Finally, the melting of binary solids is addressed and the role of the interfacial thermodynamics on the melting rate of a solid in contact with its melt, which is at a higher temperature, is examined.

(A) Ideal, One-Dimensional Layer Growth

When the surface temperature of a *pure* substance in the solid phase is raised to its melting temperature, under the assumption of equilibrium phase change, the surface melts. Figure 7.23 depicts the problem of a subcooled solid with an initial temperature of T_{s_∞} and *initially* in contact

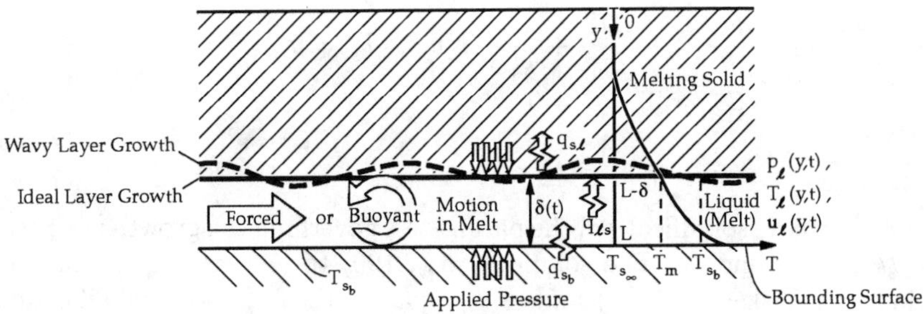

Figure 7.23. Layer (ideal and wavy) growth of a molten layer in a single-component melting with a motion within the melt and a lateral pressure applied across the melt causing its removal.

with a bounding solid surface. The bounding surface is maintained at a temperature T_{s_b} which is above the melting temperature $T_{s\ell} = T_m$. In this ideal, one-dimensional rendering, the melted layer of thickness δ *grows* and is assumed to have a planar, stable interface. The heat flux through the bounding surface towards the liquid is q_{s_b} and supplies the heat for the phase change as well as the sensible heat to the liquid and the solid phase.

Similar to the treatment of one-dimensional, layer solidification of Section 7.3.1(A), the energy equation for the solid phase is

$$(\rho c_p)_s \frac{\partial T_s}{\partial t} = k_s \frac{\partial^2 T_s}{\partial y^2} \quad 0 \le y \le L - \delta \,. \tag{7.185}$$

The bounding conditions are

$$T = T_{s_\infty} \quad y = 0 \,, \tag{7.186}$$

$$k_\ell \frac{\partial T_\ell}{\partial y} - k_s \frac{\partial T_s}{\partial y} = \rho_s \Delta i_{\ell s} \frac{\partial \delta}{\partial t} \,, \quad T = T_m \quad y = L - \delta \,. \tag{7.187}$$

The interfacial heat flux from the liquid to the solid can be given in terms of a Nusselt number defined, respectively, as

$$k_\ell \frac{\partial T_\ell}{\partial y} = Nu_\delta \frac{k_\ell}{\delta}(T_{s_b} - T_m) \,, \quad Nu_\delta = \frac{q_{\ell s}\delta}{(T_{s_b} - T_m)k_\ell} \,. \tag{7.188}$$

The initial condition is $\delta(t = 0) = 0$.

The variables are scaled similar to (7.117) and (7.118), except here the dimensionless temperature and the *heating ratio* θ_h are defined, respectively, as

$$T^* = \frac{T - T_m}{T_{s_\infty} - T_m} \,, \quad \theta_h = \frac{T_{s_b} - T_m}{T_m - T_{s_\infty}} \,. \tag{7.189}$$

The inverse of θ_h, $\theta_h^{-1} = (T_m - T_{s_\infty})/(T_{s_b} - T_m)$ is called the *subcooling parameter*. Then the dimensionless solid-phase energy equation (7.185) and the boundary and the initial conditions become

$$Ste_s \frac{\partial T_s^*}{\partial t^*} = \frac{\partial^2 T_s^*}{\partial {y^*}^2} \quad 1 - \delta^* \le y^* \le 1 \,, \tag{7.190}$$

$$\frac{d\delta^*}{dt} = -\frac{\partial T_s^*}{\partial y^*} + \frac{k_\ell}{k_s}\frac{1}{\delta^*} Nu_\delta \;, \quad T_s^* = 0 \quad y^* = 1 - \delta^* \;, \tag{7.191}$$

$$T_s^* = 1 \quad y^* = 0 \;, \tag{7.192}$$

$$T_s^* = -\theta_h \quad y^* = 1 \;, \tag{7.193}$$

$$\delta^* = 0 \;, \; T_s^* = 1 \quad t^* = 0 \;. \tag{7.194}$$

As with the solidification, the parameters governing the growth are Ste_{ℓ_s}, k_ℓ/k_s, Nu_δ, and θ_h. The solutions to (7.190) to (7.194) are obtained by Gau and Viskanta (1984) and are similar to those for the solidification. In general, the flow in the melt can be *forced, buoyant*, or a *combination* of them. The flow can also be *laminar, transitional*, or *turbulent*. Then for a nonuniform Nusselt numbers, the phase-change interface can become *unstable*. This interfacial instability is discussed below in Section (B).

In many applications the formed melt is partly *removed* by exerting an *external* force on the liquid. This can be the *passive weight* of the melting solid or the weight of the bounding solid or an *active* external force. When this melt removal causes a *close contact* (i.e., *only* a *thin liquid film* is present) between the heated bounding surface and the melting solid, the problem is called the *close-contact melting*. In addition to causing the flow of the melt, when the local liquid pressure rises because of this external force, the melting temperature *decreases*. This subclass of close-contact melting is referred to as the *pressure melting*.

(B) Morphological Instability of Melting Surfaces

The phase-change interface can become unstable when the Nusselt number distribution along with the interface is nonuniform. By assigning the x direction to be along the ideal planar interface, then the local variation of Nu_g along x leads to the *local* variations in the heat flow from the liquid to the interface $q_{\ell s}$, the heat flow from the interface to the solid $q_{\ell s}$, and the liquid-layer thickness δ. We begin by rewriting (7.187) as

$$q_{\ell s}(x,t) - q_{s\ell}(x,t) = \rho_s \Delta i_{\ell s} \frac{d\delta}{dt} \;. \tag{7.195}$$

For simplicity, assume that $q_{s\ell} \ll q_{\ell s}$. Then

$$\frac{q_{\ell s}}{\Delta i_{\ell_s}} \simeq \rho_s \frac{d\delta}{dt} \;. \tag{7.196}$$

Now a sinusoidal variation in the interface position (or liquid layer thickness) can be assumed to be of the form

$$\delta(x,t) = \langle\delta\rangle_x(t) + a_\delta(t)\cos(\eta x - \omega t) \;, \quad \eta = 2\pi/\lambda \;, \tag{7.197}$$

where a_δ is the amplitude of the interface displacement, η is the wave number, λ is the wavelength, and ω is the angular frequency of the disturbance. When the local change in $q_{\ell s}$ is *larger* in that part of the interface wave where δ is *larger* (i.e., in the *crest* of the wave observed from the liquid phase), then the amplitude of a_δ *increases* further and the surface waves are *amplified*.

Hanratty (1981) summarizes the analyses of the liquid-solid phase change interfacial instabilities and considers the general case of $q_{\ell s}$ *also* varying sinusoidally but *out* of phase with $\delta(t)$. This phase-lag angle is designated by θ, and using $a_q(t)$ as the amplitude of the Nusselt number (or $q_{\ell s}$) variation we have

$$\begin{aligned}\frac{q_{\ell s}(x,t)}{\Delta i_{\ell s}} &= \frac{\langle q_{\ell s}\rangle_x(t)}{\Delta i_{\ell s}} + a_q(t)\cos(\eta x + \theta - \omega t)\\ &= \frac{\langle q_{\ell s}\rangle_x(t)}{\Delta i_{\ell s}} + a_q(t)\cos\theta\cos(\eta x - \omega t) - a_q(t)\sin\theta\sin(\eta x - \omega t)\,.\end{aligned} \tag{7.198}$$

Now, depending on the driving force (i.e., external pressure, buoyancy, etc.) and the flow regime (i.e., laminar, transitional, or turbulent), various relations can exist between a_δ and a_q. Hanratty (1981) suggests a linear relation, with a proportionality constant $|b_{\delta q}|$, i.e.,

$$a_q(t) = a_\delta(t)|b_{\delta q}|\,. \tag{7.199}$$

Whether an increase in δ results in an increase in $q_{\ell s}$, would depend on the phase-lag angle. Inserting (7.196) and (7.198), with (7.199), in (7.195) we have

$$\rho_s\frac{d\langle\delta\rangle_x}{dt} = \frac{\langle q_{\ell s}\rangle_x}{\Delta i_{\ell s}}\,, \tag{7.200}$$

$$\rho_s\frac{da_\delta}{dt} = a_\delta|b_{\delta q}|\cos\theta\,, \tag{7.201}$$

$$\omega = \frac{|b_{\delta q}|\sin\theta}{\rho_s}\,. \tag{7.202}$$

The solution to (7.201) gives

$$a_\delta = a_{\delta_o}e^{|b_{\delta q}|\frac{\cos\theta}{\rho_s}}\,. \tag{7.203}$$

Then for the disturbances to grow, $\cos\theta$ needs to be *positive*, and therefore $-\pi/2 < \theta < \pi/2$. This implies that for the melting-front waves to grow, the *maximum* in the Nusselt number must occur in the *crest*. For forced flow over wavy surfaces, Hanratty (1981) shows that for *laminar* flows this condition is *not* met since the Nusselt number is *larger* in the *trough* (observed from the liquid phase). However, for *turbulent* flows the turbulent intensity is *expected* (Hanratty, 1981) to be *larger* in the crest region (or, the diverging region of a periodic diverging-converging channel). Based on this, Hanratty (1981) explains the experimental results showing interfacial instabilities in turbulent liquid flows and interfacial stability for laminar flows.

Epstein and Cheung (1983), Fang et al. (1985), and Glicksman et al. (1986) describe other morphological interfacial instabilities. These include melting in forced or buoyant, confined or semi-confined, and laminar or turbulent flows.

(C) Thermobuoyant and Thermocapillary Flows

The melt motion caused by thermobuoyancy can influence the phase change interfacial morphology and the rate of melting. Depending on the *orientation* of the heated, bounding surface *relative* to the *gravity*, the *circulating* melt motion can be *unicellular* or *multicellular*. Because of the *density difference* between the liquid and the solid $\rho_\ell - \rho_s$, the *volume change* excludes completely enclosed solid melting in rigid cavities. For a rectangular cavity, the heat can be added from the horizontal and/or the vertical surfaces. For the case of melting from the top, the liquid density stratification is stable and no motion occurs. When one of the surfaces of the finite-volume solid is maintained below the melting temperature, this *solid subcooling* $T_m - T_{s_{b2}}$ results in a *steady-state* phase distribution in which some solid *remains*. That is, at the phase-change interface the heat flux arriving from the liquid phase will equal that flowing into the solid phase.

One of the illustrative examples used has been the melting in *two-dimensional* cavities (and in some cases open-top cavities) with heat supplied from one *vertical* surface mentioned at $T_{s_{b_1}}$. When solid subcooling is maintained, the other vertical surface is cooled and maintained at $T_{s_{b2}}$. For this rather simple geometry and under the assumption of a *laminar* flow, the solution to the melt motion can be found numerically. This has been done by Sparrow et al. (1977), Beckermann and Viskanta (1989), Kim and Kaviany (1992), Lacroix and Arsenault (1993), and Liu et al. (1993), among others. Initially in these studies no solid subcooling was specified, but later the solid subcooling, the volume change, and the free surface *thermocapillary* effects were included. For melting of ice, because of the density extremum near 4°C, the thermobuoyant flow (and the diffusobuoyant flow for the presence of a solute in water) has some special features. These are discussed by Rieger and Beer (1986) and Gebhart et al. (1988).

In relation to the plasma-arc welding, the three-dimensional, combined plasma jet shear-thermocapillary-thermobouyant motion of the melt has been analyzed by Keanini and Rubinsky (1993). The less computationally intensive, combined thermobuoyant and thermocapillary melt motion in an open, two-dimensional cavity, depicted in Figure 7.24(a), has been analyzed by Liu et al. (1993), and their formulation and numerical results are reviewed below. The melting of the solid begins by raising the temperature of the left bounding surface to $T_{s_{b1}}$ and the interface moves to the right with the interface location given by $x_{\ell_s} = x_{\ell_s}(y, t)$. The length scale used is the horizontal witdth of the cavity L_h and the time scale uses the liquid thermal diffusivity α_ℓ, i.e., L_h^2/α_ℓ, and the temperature for the solid and

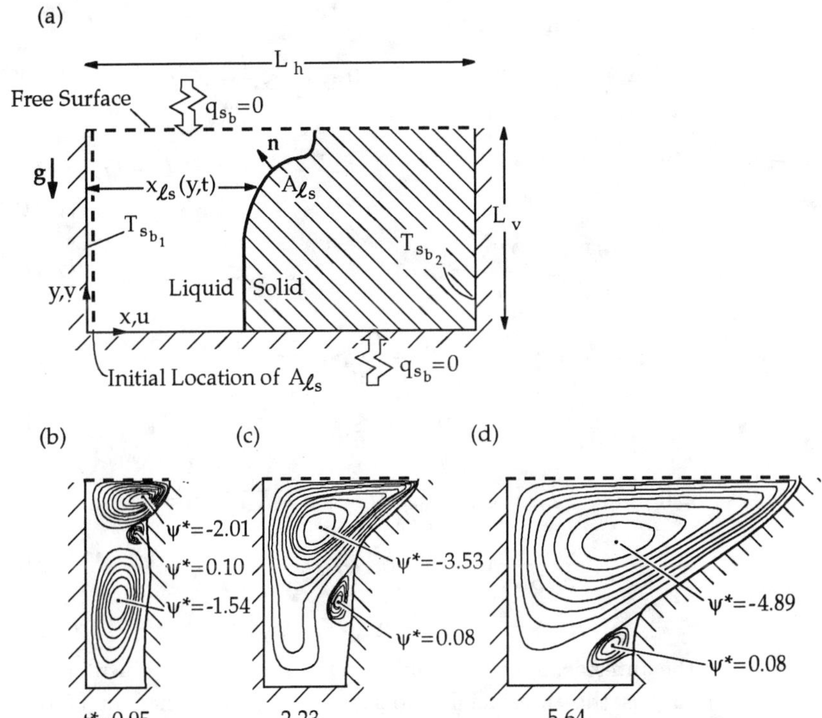

Figure 7.24. (a) A two-dimensional thermo-buoyancy and -capillarity motion during melting of a single-component liquid computed for the geometry and boundary conditions shown. (b)–(d) The motion at three different elapsed times is shown by the streamline contours. (From Liu et al., reproduced by permission ©1993 Pergamon Press.)

liquid phases are scaled using the solid and the liquid subcooling, i.e.,

$$T_s^* = \frac{T_s - T_m}{T_m - T_{s_{b2}}} \, , \quad T_\ell^* = \frac{T_\ell - T_m}{T_{s_{b1}} - T_m} \, . \tag{7.204}$$

The equations of conservation of mass, momentum, and energy (liquid and solid phases) in *vectorial, dimensionless* form are the simplified forms of (1.70), (1.74), and (1.84), i.e.,

$$\nabla \cdot u^* = 0 \, , \tag{7.205}$$

$$\frac{\partial \mathbf{u}^*}{\partial t^*} + (\mathbf{u}^* \cdot \nabla)\mathbf{u}^* = -\nabla p^* + Pr_\ell \nabla^2 \mathbf{u}^* + Ra_{L_h} Pr_\ell T_\ell^* \frac{\mathbf{g}}{g} \, , \tag{7.206}$$

$$\frac{\partial T_\ell^*}{\partial t^*} + \mathbf{u}^* \cdot \nabla T_\ell^* = \nabla^2 T_\ell^* \, , \tag{7.207}$$

$$\frac{\partial T_s^*}{\partial t^*} = \frac{\alpha_s}{\alpha_\ell} \nabla^2 T_s^* \, . \tag{7.208}$$

The boundary and initial conditions are

$$u^* = 0\,,\ \ T^*_\ell = 1 \quad x^* = 0\,,\ \ 0 \le y^* \le \frac{L_v}{L_h}\,, \tag{7.209}$$

$$u^* = 0\,,\ \ T^* = 1 \ \text{(no liquid subcooling)} \quad x^* = 1\,,\ \ 0 \le y^* \le \frac{L_v}{L_h}\,, \tag{7.210}$$

$$\mathbf{u}^* = 0\,,\ \ \frac{\partial T^*_\ell}{\partial y^*} = \frac{\partial T^*_s}{\partial y^*} = 0 \quad y^* = 0\,,\ \ 0 \le x^* \le 1\,, \tag{7.211}$$

$$\frac{\partial u^*}{\partial y^*} = Ma_{L_h}\frac{\partial T^*_\ell}{\partial x^*}\,,\ \ \mathbf{v}^* = \frac{\partial T^*_\ell}{\partial y^*} = \frac{\partial T^*_s}{\partial y^*} = 0$$
$$y^* = \frac{L_v}{L_h}\,,\ \ 0 \le x^* \le 1\,; \tag{7.212}$$

$$\mathbf{u}^* = 0\,,\ \ T^*_s = T^*_\ell = 0\,,\ \ \frac{k_s}{k_\ell}\frac{1}{\theta_h}\nabla T^*_s\cdot\mathbf{n} - \nabla T^*_\ell\cdot\mathbf{n} =$$
$$\frac{\rho_s}{\rho_\ell}\frac{1}{Ste_\ell}\mathbf{u}^*_{\ell s}\cdot\mathbf{n} \quad \text{on } A_{\ell s}\,, \tag{7.213}$$

$$\mathbf{u}^* = T^*_s = 0\,,\ \ T^*_\ell = 1 \quad t^* = 0\,, \tag{7.214}$$

where $\mathbf{n}$ is the interfacial normal unit vector on $A_{\ell s}$ as shown in Figure 7.24(a), $\mathbf{u}_{\ell_s}$ is the interfacial velocity, and θ_h is defined in (7.189). In addition to the *ratios* of the lengths, the solid and the liquid thermophysical properties, and θ_h, the dimensionless parameters are the Rayleigh, Prandtl, Marangoni, and Stefan numbers, i.e.,

$$Ra_{L_h} = \frac{g\beta(T_{s_{b1}} - T_m)L_h^3}{\nu_\ell\alpha_\ell}\,, \tag{7.215}$$

$$Pr_\ell = \frac{\nu_\ell}{\alpha_\ell}\,, \tag{7.216}$$

$$Ma_{L_h} = -\left.\frac{\partial\sigma}{\partial T}\right|_\rho \frac{(T_{s_{b1}} - T_m)L_h}{\mu_\ell\alpha_\ell}\,, \tag{7.217}$$

$$Ste_\ell = \frac{c_{p\ell}(T_{s_{b1}} - T_m)}{\Delta i_{\ell s}}\,. \tag{7.218}$$

The numerical results of Liu et al. (1993) for melting of *silicon* with $\theta_h^{-1} = 0$, $Pr = 0.031$, $Ra_{L_h} = 6.75\times10^5$, $Ma_{L_h} = 1.32\times10^4$, $L_v/L_h = 1/3$, $Ste_\ell = 5.14\times10^{-3}$, $\rho_s/\rho_\ell = 0.914$, $k_s/k_\ell = 0.688$, and $\alpha_s/\alpha_\ell = 0.747$, are shown in Figures 7.24(b) to (d). The elapsed times are $t^* = 0.95$, 2.23, and 5.64 and the values of the stream function ψ are also shown. Noting that the time is scaled with the diffusion time across the horizontal length L_h, the melting occurs rapidly. For $t^* = 0.95$ *three* vortices are present with the strongest being on top and assisted by the thermocapillary motion, the surface tension is lower at $A_{\ell s}$ and liquid is *pulled* towards $A_{\ell s}$. For $t^* = 5.64$ the thermobuoyancy dominates and the strongest vortex moves towards the center of the melt. The local Nusselt number is largest at the top portion of the interface, where the interfacial velocity $\mathbf{u}_{\ell s}$ is largest.

(D) CLOSE-CONTACT MELTING

The continuous removal of the melt, resulting in the *maintenance* of a *thin melt* layer between the heated bounding surface and the melting solid, occurs in many applications. The melt removal (i.e., squeezing of the forming melt out of the space) can be due to the weight of the melting solid when this solid is placed on top of the heated surface, as shown in Figure 7.25. In this close-contact melting, the *liquid film* thickness changes *radially* and with *time* (for a finite, initial melting-solid volume). The liquid flow in the thin liquid between the melting and the bounding solid surfaces is similar to that occurring in the *lubrication flows*, except for the phase change at the melting solid surface. This negligible inertia, liquid flow (i.e., Stokes flow) with melting has been examined by Moore and Bayazitoglu (1982), Emerman and Turcotte (1983), Moallemi and Viskanta (1986), Moallemi et al. (1986), Hong and Saito (1993), among others. Here a quasi-steady treatment of the melting of a cylindrical solid with its base placed on top of a heated bounding surface, and the determination of the melting rate, i.e., the reduction in the height of the cylinder, is considered. This problem has been analyzed by Moallemi et al. (1986) using simplifying assumptions and leading to a closed-form solution. Their formulations and results are discussed below. The numerical treatment of Hong and Saito (1993) confirms the validity of the quasi-steady fields and other assumptions made by Moallemi et al. (1986), except for a very short initial period.

A rendering of the *axisymmetric* liquid flow and temperature fields are also shown in Figure 7.25. The melt-layer thickness δ varies radially and its magnitude is assumed to be much *smaller* than the cylinder radius R (i.e., $\delta/R \ll 1$. The melt Reynolds number $Re_\ell = \langle v_\ell \rangle_{A_\ell} \delta / \nu_\ell$, where $\langle v_\ell \rangle_{A_\ell}$ is the area-averaged radial liquid velocity, is also assumed to be very small. Then the inertial terms in the liquid momentum equation are negligible when $(\delta/R) Re_\ell \ll 1$. The viscous dissipation in the liquid energy conservation equation, the radial diffusions, and the variation of the thermophysical properties are also neglected. The *quasi-steady* formulation allows for the variation of the cylinder height L with respect to time, and in turn the downward velocity of the solid u_s and the liquid film thickness will be functions of time. The liquid pressure p_ℓ and velocity u_ℓ, which depend on u_s and δ, are also time dependent.

Subject to the above-mentioned assumptions, the liquid mass, momentum, and thermal energy equations become

$$\frac{\partial u_\ell}{\partial x} + \frac{1}{r}\frac{\partial}{\partial r} r v_\ell = 0 \ , \qquad (7.219)$$

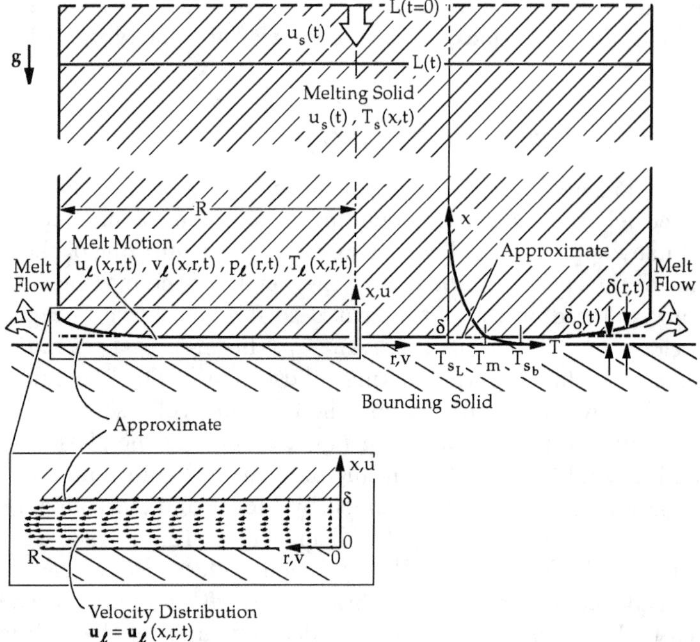

Figure 7.25. A rendering of the close-contact melting when a single-component solid is melting due to the heat transfer from a bounding solid maintained above the melting temperature of the first solid. The liquid motion in an idealized melt space is also shown.

$$0 = -\frac{dp_\ell}{dr} + \mu_\ell \frac{\partial^2 v_\ell}{\partial x^2} , \tag{7.220}$$

$$u_\ell \frac{\partial T_\ell}{\partial x} + v_\ell \frac{\partial T_\ell}{\partial r} = \alpha_\ell \frac{\partial^2 T_\ell}{\partial x^2} . \tag{7.221}$$

The boundary conditions are

$$v_\ell = \frac{\partial u_\ell}{\partial x} = \frac{\partial T_\ell}{\partial x} = 0 \quad r = 0\,, \;\; 0 \le x \le \delta\,, \tag{7.222}$$

$$u_\ell = v_\ell = 0\,, \;\; T = T_{s_b} \quad x = 0\,, \;\; 0 \le r \le R\,, \tag{7.223}$$

$$u_\ell = -|u_s|\,, \;\; v_\ell = 0\,, \;\; -k_\ell[1 + (\frac{\partial \delta}{\partial r})^2]\frac{\partial T_\ell}{\partial x} = \rho_s |u_s| [\Delta i_{\ell s} +$$

$$c_{p_s}(T_m - T_{s_L})]\,, \;\; T = T_m \quad x = \delta\,, \;\; 0 \le r \le R\,. \tag{7.224}$$

The radial velocity is found by the integration of (7.220) subject to the boundary condition on v_ℓ given by (7.222) to (7.224). This solution is

$$v_\ell = \frac{1}{2\mu_\ell}\frac{dp}{dr} x(x-\delta)\,. \tag{7.225}$$

The pressure distribution is found by inserting (7.225) in (7.219) and then integrating the result with respect to x and using the boundary conditions on u_ℓ. The result is

$$\frac{dp}{dr} = -\frac{6\mu_\ell}{\delta^3}|u_s|r \ , \tag{7.226}$$

i.e., the magnitude of the pressure gradient *increases* linearly with r because the flow rate of the liquid *increases* with r as the melt accumulates. The increase in flow rate can be noted by inserting (7.226) into (7.225) which gives

$$v_\ell = -\frac{6x(x-\delta)}{2\delta^3}|u_s|r \ . \tag{7.227}$$

This *radial* and *lateral* distribution of velocity is rendered in the inset of Figure 7.25.

The temperature distribution in the liquid is found by assuming a distribution along x. Moallemi et al. (1986) use a quadratic polynomial. The following distribution of T_ℓ satisfied the boundary condition (7.224):

$$T_\ell = T_{s_b} + x\{\frac{-2(T_{s_b}-T_m)}{\delta} + \frac{\rho_s[\Delta i_{\ell s} + c_{p_s}(T_m - T_{s_L})]|u_s|}{k_\ell[1+(\frac{\partial\delta}{\partial r})^2]}\} +$$

$$x^2\{\frac{T_{s_b}-T_m}{\delta^2} - \frac{\rho_s[\Delta i_{\ell s} + c_{p_s}(T_m - T_{s_L})]|u_s|}{\delta k_\ell[1+(\frac{\partial\delta}{\partial r})^2]}\} \ . \tag{7.228}$$

The equation for the melt-layer thickness is found by using this distribution in the x direction, the integrated form of the energy equation (7.221), and the application of the interfacial energy balance. The solution for v_ℓ is given by (7.227), and the result is the dimensionless equation

$$\frac{1}{r^*}\frac{\partial}{\partial r^*}\frac{r^{*2}\delta^*}{1+(\frac{\partial\delta^*}{\partial r^*})^2} + \frac{3Ste_\ell}{|u_s^*|} + \frac{20}{|u_s^*|[1+(\frac{\partial\delta^*}{\partial r^*})^2]} - \frac{20Ste_\ell}{|u_s^*|\delta^*} = 0 \ , \tag{7.229}$$

where the dimensionless variables and parameters are

$$r^* = \frac{r}{R} \ , \quad \delta^* = \frac{\delta}{R} \ ,$$

$$u_s^* = \frac{u_s R}{\alpha_\ell} \ , \quad Ste_\ell = \frac{c_{p_\ell}(T_{s_b}-T_m)}{\Delta i_{\ell s} + c_{p_s}(T_m - T_{s_L})} \ . \tag{7.230}$$

The first-order solution for δ^* in (7.229) gives δ_o^* which is *independent* of r^*. This solution is (Moallemi et al., 1986)

$$\delta_o^* \simeq \frac{Ste_\ell}{|u_s^*|} \qquad Ste_\ell \le 0.1 \ . \tag{7.231}$$

The solid velocity u_s is found from a force balance on the cylinder which gives

$$\rho_s \pi R^2 L(t) g = \int_{A_{\ell s}} p_\ell dA \ . \tag{7.232}$$

Using (7.226) and the ambient pressure condition $p_\ell(R) = p_{\ell_R}$, gives

$$p_\ell - p_{\ell_R} = \frac{6\mu_\ell(R^2 - r^2)}{2\delta^3}|u_s| \,. \qquad (7.233)$$

Using this in (7.232) with the substitution for δ made using (7.231) and integrating gives the relation between the solid velocity and the solid height as

$$|u_s^*|(t) = |\frac{u_s R}{\alpha_\ell}| = (\frac{\rho_s}{\rho_\ell}\frac{gR^3}{\alpha_\ell^2}\frac{Ste_\ell^3}{\frac{3}{2}Pr_\ell})^{1/4}(\frac{L}{R})^{1/4}\,, \quad Pr_\ell = \frac{\nu_\ell}{\alpha_\ell}\,. \qquad (7.234)$$

Also, by assuming that the time rate of change of δ is *negligible* compared to that of L, then $|u_s| = |dL/dt|$. This relation, when used in the above and when the *time* integration is performed, gives

$$\frac{L(t)}{R} = \{[\frac{L(t=0)}{R}]^{3/4} - \frac{3}{4}\frac{t\alpha_\ell}{R^2}(\frac{\rho_s}{\rho_\ell}\frac{gR^3}{\alpha_\ell^2}\frac{Ste_\ell^3}{\frac{3}{2}Pr_\ell})^{1/4}\}^{4/3}\,. \qquad (7.235)$$

The measured values of $|u_s|$ and $L(t)$, have been found to be in good agreement with those predicted by (7.234) and (7.235), and as was mentioned, the numerical treatment of Hong and Saito (1993) which relaxes some of the assumptions of Moallemi et al. (1986), are also in good agreement with the predictions of (7.234) and (7.235).

(E) Pressure Melting

For water, the melting temperature *decreases* with an increase in the pressure, and when a bounding solid surface rests and exerts pressure on the surface of a slightly subcooled volume of ice, melting of the ice can occur. This is called the *pressure melting*, and the formation of the melt and its flow has been analyzed by Bejan and Tyvand (1992). Figure 7.26 depicts the *axisymmetric, steady* melt formation and motion under a cylindrical, bounding solid of radius R, which exerts a force f_{s_b} on the melt, causing a pressure distribution $p = p(r)$ and velocity $\mathbf{u}_\ell = \mathbf{u}_\ell(x, r)$. The interface of the melting solid and the melt is at the melting temperature $T_m = T_m(p)$ which is *lower* than the temperature of the bounding and melting solid far from the interface $T_{s_b} = T_{s_\infty} = T_o$. The Clausius equilibrium relation (1.51) gives

$$\frac{\partial p}{\partial T} = \frac{\Delta i_{\ell s}}{T\Delta v_{\ell s}} < 0 \quad \text{liquid-solid (ice I) equilibrium for water.} \qquad (7.236)$$

This can be linearized as

$$T_m = T_o + \frac{p - p_o}{\Delta i_{\ell s}/T_o\Delta v_{\ell s}}\,. \qquad (7.237)$$

For water, $\Delta i_{\ell s}/T_o\Delta v_{\ell s} = -1.36 \times 10^7$ Pa/°C, indicating that for every nearly 100 atmosphere of increase in pressure, the melting point *decreases* by 1°C (Bejan and Tyvand, 1992).

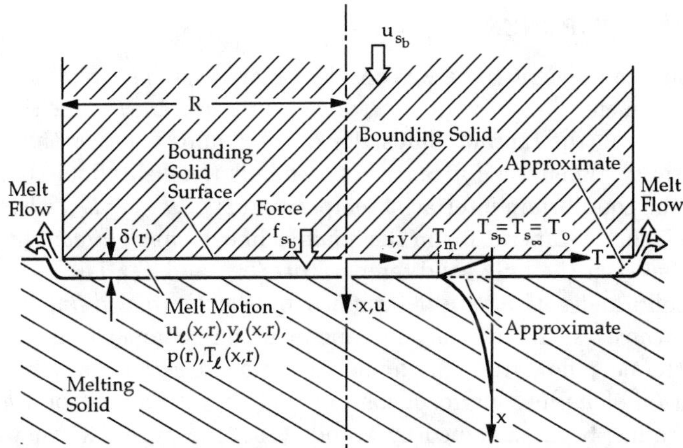

Figure 7.26. A rendering of the pressure melting where the bounding and melting solids are brought into close contact by a large external force.

Following an analysis similar to the one given in Section (D) above for the close-contact melting under *imposed temperature*, Bejan and Tyvand (1992) show that the interfacial energy balance can be approximated as (assuming *conduction* heat transfer across the melt)

$$\frac{k_\ell}{\delta}[T_o - T_m(p)] = \rho_s[\Delta i_{\ell s} + c_p(T_m - T_{s_\infty})]u_{s_b} \tag{7.238}$$

or

$$-\frac{k_\ell}{(\Delta i_{\ell s}/T_o \Delta v_{\ell s})\delta}(p - p_o) = \rho_s[\Delta i_{\ell s} + c_p(T_m - T_{s_\infty})]u_{s_b} \; . \tag{7.239}$$

Then they show that *steady-state* pressure distribution is given by

$$p(r) - p_o = \frac{[12\mu_\ell u_{s_b} R^2(1 - \frac{r^2}{R^2})]^{1/4}}{\{-\frac{\rho_s[\Delta i_{\ell s} + c_{p_s}(T_m - T_o)]\Delta i_{\ell s} u_{s_b}}{T_o \Delta v_{\ell s}}\}^{3/4}} \; . \tag{7.240}$$

The relation between the applied force f_{s_L} and the velocity of the bounding solid u_{s_b} is

$$f_{s_b} = \frac{4\pi}{5}(12)^{1/4} u_{s_b} R^{5/2}\{-\frac{\mu_\ell^{1/3}\rho_s[\Delta i_{\ell s} + c_{p_s}(T_m - T_o)]\Delta i_{\ell s}}{k_\ell T_o \Delta v_{\ell s}}\}^{3/4} . \tag{7.241}$$

The radial variation of the melt-layer thickness is given by

$$\frac{\delta(r)}{R} = \{-\frac{12\mu_\ell k_\ell T_o \Delta v_{\ell s}(1 - \frac{r^2}{R^2})}{\rho[\Delta i_{\ell s} + c_{p_s}(T_m - T_o)]\Delta i_{\ell s} R^2}\}^{1/4} \; , \tag{7.242}$$

which shows a *sharp* increase of the thickness at the edge ($r = R$), shown by the dashed line in Figure 7.26, followed by nearly *uniform* distribution in the area away from the edge.

(F) Multicomponent Systems

When a multicomponent (here we consider binary systems) solid is in contact with its melt, *dissolving* (i.e., *species transport*) and/or melting can occur. In general, the thermal diffusivity of the liquids is larger than their mass diffusivity. When the *difference* between bulk (far-field) liquid temperature and the equilibrium interfacial temperature, i.e., $T_{\ell_\infty} - T_{\ell s}$, where $T_{\ell s}$ is the equilibrium temperature on the liquidus line, is *much larger* than the difference between the bulk solid temperature T_{s_∞} and $T_{\ell s}$, i.e.,$T_{\ell s} - T_{s_\infty}$, then the solid melts at a rate determined by the *thermal diffusivity*. As $T_{\ell_\infty} - T_{\ell s}$ *decreases*, the interfacial temperature and composition may *just* allow for the heat flow from the liquid to the solid *without* any melting. The condition of *no* melting for a solid in contact with a melt of a *higher* temperature has been examined by Woods (1992). Figure 7.27 shows the conditions T_{ℓ_∞} and $(\rho_{A,\ell}/\rho_\ell)_\infty$ far from the interface in semi-infinite liquid layer and T_{s_∞} and $(\rho_{A,s}/\rho_s)_\infty$ far from the interface in a semi-infinite solid in contact with the liquid at $x_{\ell s}$. For $T_{\ell_\infty} > T_{s_\infty}$, heat flows from the liquid (i.e., melt) to the solid. Then beginning with a point in the liquid approaching the interface, passing through the interface, and reaching far from the interface into the solid, a path is taken with the temperatures and the concentrations marked in the figure. The path assumes a liquid mass fraction $(\rho_{A,\ell}/\rho_\ell)_\infty$ which is to the *left* of the eutectic point, and a solidus line with an *infinite* slope is assumed (typical of aqueous salts), i.e., the solidus is *independent* of concentration as shown in Figure 1.6(e).

The analysis of Woods (1992) determines the required *minimum* liquid temperature for melting for a given set of T_{s_∞} and $(\rho_{A,\ell}/\rho_\ell)_\infty$, $(\rho_{A,s}/\rho_s)_\infty$, and a given phase diagram. The results of his analysis are summarized below.

Assuming $\rho_\ell = \rho_s = \rho$ and taking this density and the diffusivity to be constant, the one-dimesional transient conservation equations for the species and thermal energy (for each phase), and the interfacial species and heat balances are

$$\frac{\partial}{\partial t}\frac{\rho_{A,\ell}}{\rho} = D_\ell \frac{\partial^2}{\partial x^2}\frac{\rho_{A,\ell}}{\rho} \quad x < x_{\ell s}(t) , \tag{7.243}$$

$$\frac{\partial T_\ell}{\partial t} = \alpha_\ell \frac{\partial^2 T_\ell}{\partial x^2} \quad x < x_{\ell s}(t) , \tag{7.244}$$

$$\frac{\partial T_s}{\partial t} = \alpha_s \frac{\partial^2 T_s}{\partial x^2} \quad x > x_{\ell s}(t) , \tag{7.245}$$

$$D_\ell \frac{\partial^2}{\partial x^2}\frac{\rho_{A,\ell}}{\rho} = -[(\frac{\rho_A}{\rho})_{\ell s} - (\frac{\rho_{A,s}}{\rho_s})_s]\frac{dx_{\ell s}}{dt} \quad x = x_{\ell s}(t) , \tag{7.246}$$

$$T_\ell = T_s = T_{\ell s} , \quad k_\ell \frac{\partial T_\ell}{\partial x} - k_s \frac{\partial T_s}{\partial x} = -\rho \Delta i_{\ell s} \frac{dx_{\ell s}}{dt} \quad x = x_{\ell s}(t) . \tag{7.247}$$

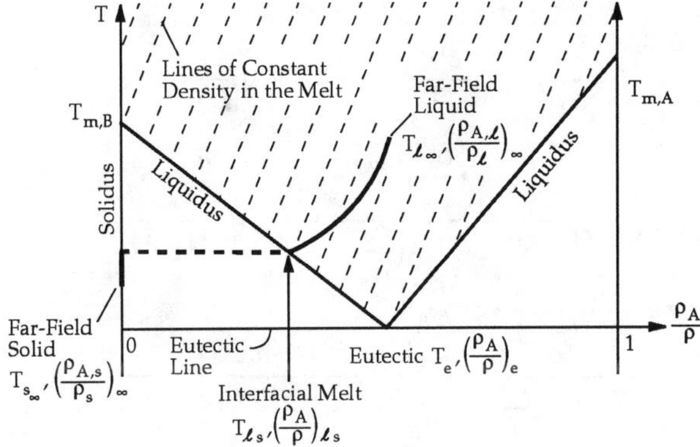

Figure 7.27. Melting of a multicomponent solid shown in a phase diagram. The far-field liquid and solid conditions, as well as the conditions within the solid, mushy zone, and liquid, are also shown.

The interfacial thermodynamic constraint (7.159) can be written in terms of the eutectic conditions as

$$T_\ell - T_e = -\gamma[(\frac{\rho_A}{\rho})_{\ell s} - (\frac{\rho_A}{\rho})_e]x \quad x = x_{\ell s}(t) \,. \tag{7.248}$$

A *similarity* solution exists for the semi-infinite liquid and solid domains, and the similarity variable and the general form of the solution for the interfacial location $x_{\ell s}$ are

$$\eta = \frac{x}{2(\alpha_\ell t)^{1/2}} \,, \tag{7.249}$$

$$x_{\ell s}(t) = 2\lambda(\alpha_\ell t)^{1/2} \,, \tag{7.250}$$

where λ is the *dimensionless* rate of phase change which is obtained as part of the solution.

The solution to (7.243) to (7.248), when the temperatures and concentrations are normalized using

$$T^* = \frac{T - T_{\ell s}[(\frac{\rho_A}{\rho})_{\ell s}]}{\gamma(\frac{\rho_A}{\rho})_{\ell s}} \,, \quad \rho^* = \frac{\frac{\rho_{A,\ell}}{\rho}}{(\frac{\rho_A}{\rho})_{\ell s}} \tag{7.251}$$

and for the case of $\rho^*_{s_\infty} = 0$, becomes

$$\rho^* = 1 + \frac{\rho^*_{\ell s} - 1}{\mathrm{erfc}[-\lambda(\frac{\alpha_\ell}{D_\ell})^{1/2}]}\mathrm{erfc}[-\eta(\frac{\alpha_\ell}{D_\ell})^{1/2}] \quad \eta < \lambda \,, \tag{7.252}$$

$$T^*_\ell = T^*_{\ell_\infty} + \frac{T^*_{\ell s} - T^*_{\ell_\infty}}{\mathrm{erfc}(-\lambda)}\mathrm{erfc}(-\eta) \quad \eta < \lambda \,, \tag{7.253}$$

$$T^*_s = T^*_s + \frac{T^*_{\ell s} - T^*_{s_\infty}}{\mathrm{erfc}[\lambda(\frac{\alpha_\ell}{\alpha_s})^{1/2}]}\mathrm{erfc}[\eta(\frac{\alpha_\ell}{\alpha_s})^{1/2}] \quad \eta > \lambda \,. \tag{7.254}$$

The interfacial boundary conditions give

$$\rho_{\ell s}^* = (\frac{D_\ell}{\alpha_\ell})^{1/2}\{(\frac{D_\ell}{\alpha_\ell})^{1/2} + \lambda f[-\lambda(\frac{\alpha_\ell}{D_\ell})^{1/2}]\}^{-1} , \tag{7.255}$$

$$T_{\ell s}^*\{\frac{1}{f(-\lambda)} + \frac{(\frac{\alpha_s}{\alpha_\ell})^{1/2}\frac{(\rho c_p)_s}{(\rho c_p)_\ell}}{f[\lambda(\frac{\alpha_\ell}{\alpha_s})^{1/2}]}\} = \frac{T_{\ell_\infty}^*}{f(-\lambda)} +$$

$$\frac{T_{s_\infty}(\frac{\alpha_s}{\alpha_\ell})^{1/2}\frac{(\rho c_p)_s}{(\rho c_p)_\ell}}{f[\lambda(\frac{\alpha_\ell}{\alpha_s})^{1/2}]} - \frac{\rho_s \Delta i_{\ell s}}{(\rho_{A,\ell})_\infty \gamma(\rho_A/\rho)_{\ell s}} , \tag{7.256}$$

where

$$f(z) = \pi^{1/2} e^{z^2} \mathrm{erfc}(z) . \tag{7.257}$$

Combining these two boundary conditions, i.e., using $\rho_{\ell s}^* + T_{\ell s}^* = 1$, gives the *eigenvalue relation* for λ as

$$T_{\ell_\infty}^* = f(-\lambda)\{\frac{\rho_s \Delta i_{\ell s}}{(\rho_{A,\ell})_\infty \gamma(\rho_A/\rho)_{\ell s}} + T_{s_\infty}^* \frac{\frac{(\rho c_p)_s}{(\rho c_p)_\ell}}{f[\lambda(\frac{\alpha_\ell}{\alpha_s})^{1/2}]} + g\} , \tag{7.258}$$

$$g = \{\frac{1}{f(-\lambda)} + \frac{\frac{(\rho c_p)_s}{(\rho c_p)_\ell}(\frac{\alpha_\ell}{\alpha_s})^{1/2}}{f[\lambda(\frac{\alpha_\ell}{\alpha_s})^{1/2}]}\}\{\frac{\lambda f[-\lambda(\frac{\alpha_\ell}{D_\ell})^{1/2}]}{(\frac{D_\ell}{\alpha_\ell})^{1/2} + \lambda f[-\lambda(\frac{\alpha_\ell}{D_\ell})^{1/2}]}\} . \tag{7.259}$$

For no melting, i.e., $\lambda = 0$, the interfacial mass fraction and temperature are $\rho_{\ell s}^* = 1$ and $T_{\ell s}^* = 0$. Then from (7.256), i.e., the energy balance across the interface, we have

$$T_{\ell_\infty}^* = \frac{T_{\ell_\infty} - T_{\ell s}[(\frac{\rho_A}{\rho})_{\ell s}]}{\gamma(\frac{\rho_A}{\rho})_{\ell s}}$$

$$= -(\frac{\alpha_s}{\alpha_\ell})^{1/2}\frac{(\rho c_p)_s}{(\rho c_p)_\ell}\frac{T_{s_\infty} - T_{\ell s}[(\frac{\rho_A}{\rho})_{\ell s}]}{\gamma(\frac{\rho_A}{\rho})_{\ell s}} \qquad (\frac{\rho_{A,s}}{\rho})_\infty = 0 . \tag{7.260}$$

At the condition of *equality*, the liquid temperature T_{ℓ_∞} is at the *threshold* value for which the heat supplied through the liquid into the interface $-k_\ell \partial T_\ell/\partial x|_{x_{\ell s}}$ is *equal* to that flowing into the solid $-k_s \partial T_s/\partial x|_{x_{\ell s}}$ and *no* phase change (melting) occurs. For the melting to occur T_{ℓ_∞} should be *larger* than this threshold value.

For solidus lines with a *finite* slope, which is characteristic of metals, similar phenomena occur and are discussed by Woods (1992). The melt motion and its effect on the transport in multicomponent melting have also been discussed by Woods (1991, 1992).

7.3.4 Layer-Frost Growth and Densification

As an example of a solid-solid-fluid system with thermal *nonequilibrium* and *phase change*, consider the *fluid* being a *multicomponent gas* and *flowing* near a bounding surface which is at a temperature *much* below the melting temperature of the *condensable species* in the gas. Then the condensable species (i.e., *vapor*) form a *frost layer* on the bounding surface. The frost is made of *dendritic crystals* which are *anisotropic*. The *formation* and *growth* of these crystals, and the various *mechnisms* and *regimes* associated with these phenomena, are discussed below. The *modeling* of the *hydrodynamics* and *heat* and *mass transfer* for these regimes is also reviewed. Because of its natural as well as engineering occurrences, *ice* frost formations, in a flow of moist air (i.e., a mixture of water vapor as the condensable, and nitrogen, oxygen, and trace of argon as noncondensables) with temperatures above the sublimation temperature of water T_{vs}, over solid surfaces with temperatures below T_{vs}, has been studied extensively.

(A) Growth Mechanisms and Regimes

The microscopic observations of Tao et al. (1993b) on *transient* frost formation on a flat surface cooled below T_{vs} with the moist air flow over and parallel to the surface has led to the identification of *three regimes* (or *periods*). These are depicted in Figure 7.28. In the initial period, *water droplet nucleation* occurs on the surface cavity sites (i.e., surface or heterogeneous droplet nucleation). This is referred to as the *dropwise condensation period* (surface droplet nucleation will be discussed in Section 8.6), and during this period the droplet-coalescence and formation of larger drops occur. At the end of this period, i.e., $t = t_c$, it is assumed that surface unit cells of linear dimensions ℓ are present with the liquid droplets having spherical cap shape with a base diameter d. The *solidification and tip-growth period* follows and lasts up to $t = t_s$. During this period the liquid droplet solidifies and a columnar crystal growth occurs with the tip of the column growing while its base diameter and the solidified liquid base diameter remain near constant. During this period solid phase growth is influenced by the convective mass transfer to the columnar crystal. The last period is the *densification and bulk-growth period*, when *dendritic*, *lateral* (i.e., *filling*) growth occurs. The height (or thickness) of the frost layer δ increases as the elapsed time increases and the various periods are encountered.

The first period is generally *short* compared to the rest, and the second period is also short compared to the third. In some existing analyses, only the third period is considered. For example, the analysis of White and Cremers (1981) allows for the simultaneous densification (i.e., *decrease* in

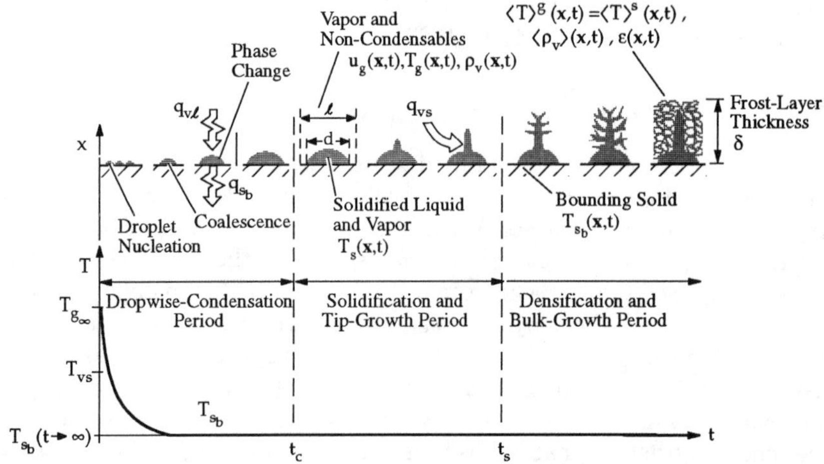

Figure 7.28. A microscopic rendering of the frost growth on a surface cooled below the triple point of the condensing component of the gas. The three different regimes are also shown. (From Tao et al., reproduced by permission ©1993 ASHRAE.)

porosity) and bulk growth (i.e., increase in δ), with time. There are an *asymptotic average* porosity and thickness which are reached when the local heat flowing into the layer from the bounding surface is *equal* to that flowing from the layer to the gas. The asymptotic frost-layer averaged porosity $\langle \epsilon \rangle_\delta$ and frost-layer thickness δ have been predicted using various models.

Since the details of the crystal growth during the second and third periods are not yet well known, some tentative models have been suggested with large adjustable constants which are determined empirically. A *unit-cell* model for the second period, $t_c \le t \le t_s$ and a *local volume-averaged* model for the third period are mentioned below.

(B) Solidification and Tip-Growth Period

A unit-cell model for this period has been proposed by Tao et al. (1993a). A *variable* cross-sectional area solid cylinder in this unit cell grows in length δ and local diameter. This *cylinder-in-cell* growth is governed by interfacial *convection* (heat and mass transfer) described by the local Nusselt and Sherwood numbers, Nu_{vs} and Sh_{vs}, interfacial *phase change* (solid-phase growth), and solid-phase *conduction*. The length of the periods $t_s - t_c$, and Nu_{vs} and Sh_{vs} are *prescribed*. The heat and mass transfer are assumed to be associated with a *cross flow* with ρ_{v_∞} and T_{g_∞} evaluated as an average of the bulk gas and the bounding surface values, respectively.

(C) Densification Bulk-Growth Period

For this period, the frost layer is treated as a *porous medium* and the *local volume*-averaged description has been proposed by both Tao et al. (1993a) and Tao and Besant (1993). Local thermal and chemical equilibria are assumed between the solid and the fluid phases in the frost layer. In the porous medium diffusion of species and heat is allowed, as described in Section 3.2, and *effect medium properties* $\langle k \rangle$ and $\langle D \rangle$ have been assumed and prescribed. At the interface of the porous medium and the flowing moist air, the interfacial convective coefficients for heat and mass transfer, i.e., $\langle Nu_\delta \rangle_{A_i}$ and $\langle Sh_\delta \rangle_{A_i}$ are prescribed. The *growth* rates of the frost layer are determined from the interfacial heat and mass balances, i.e.,

$$-\langle Nu_\delta \rangle_{A_i} \frac{k_g}{\delta}(T_{g\infty} - \langle T \rangle) + \langle k \rangle \frac{\partial \langle T \rangle}{\partial x} = \langle \rho \rangle \Delta i_{vs} \frac{d\delta}{dt} \,,$$

$$\langle Sh_\delta \rangle_{A_i} \frac{D_g}{\delta}(\rho_{v\infty} - \langle \rho_v \rangle) - \langle D \rangle \frac{\partial \langle \rho_v \rangle}{\partial x} = \langle \rho \rangle \frac{d\delta}{dt} \,,$$

$$\text{on} \quad x = \delta \,, \tag{7.261}$$

where

$$\langle \rho \rangle = \rho_s(1-\epsilon) + \rho_g \epsilon \,. \tag{7.262}$$

The interfacial gradients of $\langle T \rangle$ and $\langle \rho_v \rangle$, and the distributions of $\langle T \rangle$, $\langle \rho_v \rangle$, and ϵ in the porous medium are determined by solving (simultaneously) these and other boundary conditions and the *transient diffusion* equations.

7.4 References

Adams, R.L., 1984, "Heat Transfer in Large-Particle Bubbling Fluidized Beds," *ASME J. Heat Transfer*, 106, 85–90.

Adams, R.L., and Welty, J.R., 1982, "An Analytical Study of Bubble and Adjacent Tube Influence on Heat Transfer to a Horizontal Tube in a Gas Fluidized Bed," *ASME J. Heat Transfer*, 104, 206–209.

Ahuja, S., Beckermann, C., Zakhem, R., Weidman, P.D., and de Groh III, H.C., 1992, "Drag Coefficient of an Equiaxed Dendrite Settling in an Infinite Medium," Beckermann, C., et al., Editors, ASME HTD-Vol. 218, 85–91, American Society of Mechanical Engineers, New York.

Alavizadeh, N., Adams, R.L., Welty, J.R., and Goshayeshi, A., 1990, "An Instrument for Local Radiative Heat Transfer Measurement Around a Horizontal Tube Immersed in a Fluidized Bed," *ASME J. Heat Transfer*, 112, 486–491.

Andersson, B.-Å., and Leckner, B., 1992, "Experimental Methods of Estimating Heat Transfer in Circulating Fluidized Bed Boilers," *Int. J. Heat Mass Transfer*, 35, 3353–3362.

Anderson, K.G., and Jackson, R., 1992, "A Comparison of Solutions of Some Proposed Equations of Motion of Granular Materials for Fully Developed Flow Down Inclined Planes," *J. Fluid Mech.*, 241, 145–168.

Beckermann, C., and Ni, J., 1992, "Modeling of Equiaxed Solidification with Convection," in *Proceedings, First International Conference on Transport Phenomena in Processing*, Güceri, S.J., Editor, 308–317, Technomic Publishing Company, Lancaster, PA.

Beckermann, C., and Viskanta, R., 1989, "Effect of Solid Subcooling on Natural Convection Melting of a Pure Metal," *ASME J. Heat Transfer*, 111, 416–424.

Beckermann, C., and Viskanta, R., 1993, "Mathematical Modeling of Transport Phenomena During Alloy Solidification," *Appl. Mech. Rev.*, 46, 1–27.

Bejan, A., and Tyvand, P.A., 1992, "The Pressure Melting of Ice Under a Body with a Flat Base,", *ASME J. Heat Transfer*, 114, 529–531.

Ben-Ammar, F., Kaviany, M., and Barber, J.R., 1992, "Heat Tranfer During Impact," *Int. J. Heat Mass Transfer*, 35, 1495–1506.

Clift, R., and Grace, J.R., 1985, "Continuous Bubbling and Slugging," in *Fluidization*, Davidson, J.E., et al., Editors, Second Edition, 73–132, Academic Press, London.

Coriell, S.R., McFadden, G.B., and Sekerka, R.F., 1985, "Cellular Growth During Directional Solidification," *Ann. Rev. Mater. Sci.*, 15, 119–145.

Couderc, J.-P., 1985, "Incipient Fluidization and Particulate Systems," in *Fluidization*, Davidson, J.E., et al., Editors, Second Edition, 1–46, Academic Press, London.

Dash, S.K., and Gill, N.M., 1984, "Forced Convection Heat and Momentum Transfer to Dendritic Structures (Parabolic Cylinder and Paraboloids of Revolution)," *Int. J. Heat Mass Transfer*, 27, 1345–1356.

Davidson, J.F., Clift, R., and Harrison, D., Editors, 1985, *Fluidization*, Second Edition, Academic Press, London.

Davis, S.H., 1990, "Hydrodynamic Interactions in Directional Solidification," *J. Fluid Mech.*, 212, 241–262.

Ding, J., and Gidaspow, D., 1990, "A Bubbling Fluidzation Model Using the Kinetic Theory of Granular Flow," *AIChE J.*, 36, 523–538.

Dou, S., Herb, B., Tuzla, K., and Chen, J.C., 1991, "Heat Transfer Coefficients for Tubes Submerged in a Circulating Fluidized Bed," *Exper. Heat Transfer*, 4, 343–353.

Dyrness, A., Glicksman, L.R., and Yule, T., 1992, "Heat Transfer in the Splash Zone of a Bubbling Fluidized Bed," *Int. J. Heat Mass Transfer*, 35, 847–860.

Emerman, S.H., and Turcotte, D.L., 1983, "Stokes' Problem with Melting," *Int. J. Heat Mass Transfer*, 26, 1625–1630.

Epstein, M., and Cheung, F.B., 1983, "Complex Freezing-Melting Interfaces in Fluid Flow," *Ann. Rev. Fluid Mech.*, 15, 293–319.

Fang, Q.T., Glicksman, M.E., Coriell, S.R., McFadden, G.B., and Boisvert, R.F., 1985, "Convective Influence on the Stability of a Cylinderical Solid-Liquid Interface," *J. Fluid Mech.*, 151, 121–140.

Felicelli, S.D., Heinrich, J.C., and Poirier, D.R., 1991, "Simulation of Freckles During Vertical Solidification of Binary Alloys," *Metall. Trans.*, 22B, 847–859.

Flemings, M.C., 1974, "Solidification Processing," McGraw-Hill, New York.

Forth, S.A., and Wheeler, A.A., 1992, "Coupled Convective and Morphological Instability in a Simple Model of Solidification of a Binary Alloy, Including a Shear Flow," *J. Fluid Mech.*, 236, 61–94.

Foscolo, P.U., and Gibilaro, L.G., 1984, "A Fully Predictive Criterion for the Transition Between Particulate and Aggregate Fluidization," *Chem. Engng.*

Sci., 39, 1667–1675.

Gabor, J.D., 1970, "Heat Transfer to Particle Beds with Gas Flows Less Than or Equal to That Required for Incipient Fluidization," *Chem. Engng. Sci.*, 25, 979–984.

Gau, C., and Viskanta, R., 1984, "Melting and Solidification of a Metal System in a Rectangular Cavity," *Int. J. Heat Mass Transfer*, 27, 113–123.

Gau, C., and Viskanta, R., 1985, "Effect of Natural Convection on Solidification from Above and Melting from Below of a Pure Metal," *Int. J. Heat Mass Transfer*, 28, 573–587.

Gau, C., and Viskanta, R., 1986, "Melting and Solidification of a Pure Metal on a Vertical Wall," *J. Heat Transfer*, 108, 174–181.

Gebhart, B., Jaluria, Y., Mahajan, R.L., and Sammakia, B., 1988, *Buoyancy-Induced Flows and Transport*, Hemisphere Publishing Corporation, Washington, D.C.

George, A.H., 1993, "Instantaneous Local Heat Transfer Coefficients and Related Frequency Spectra for a Horizontal Cylinder in a High-Temperature Fluidized Bed," *Int. J. Heat Mass Transfer*, 36, 337–345.

Gidaspow, D., 1993, "Hydrodynamic Modeling of Circulating and Bubbling Fluidized Beds," in *Particulate Two Phase Flow*, Roco, M.C., Editor, Butterworth-Heinemann, Boston.

Glicksman, M.E., Coriell, S.R., and McFadden, G.B., 1986, "Interaction of Flows with Crystals-Melt Interface," *Ann. Rev. Fluid Mech.*, 18, 307–335.

Grace, J.R., 1986, "Contacting Modes and Behavior Classification of Gas-Solid and Other Two-Phase Suspensions," *Can. J. Chem. Eng.*, 64, 353–363.

Gunn, D.J., 1978, "Transfer of Heat or Mass to Particles in Fixed and Fluidized Beds," *Int. J. Heat Mass Transfer*, 21, 467–476.

Gutfinger, C., and Abuaf, N., 1974, "Heat Transfer in Fluidized Beds," *Andvan. Heat Transfer*, 10, 167–218.

Ham, J.M., Thomas, S., Guazzelli, G., Homsy, G.M., and Anselmet, M.-C., 1990, "An Experimental Study of the Stability of Liquid-Fluidized Beds," *Int. J. Multiphase Flow*, 16, 171–185.

Hanratty, T.J., 1981, "Stability of Surfaces That Are Dissolving or Being Formed by Convective Diffusion," *Ann. Rev. Fluid Mech.*, 13, 231–252.

Hasegawa, S., Echigo, R., Kanemura, K., Ichimiya, K., and Sanui, M., 1983, "Experimental Study on Forced Convective Heat Transfer of Flowing Gaseous Solid Suspension at High Temperatures," *Int. J. Multiphase Flow*, 9, 131–145.

Hills, R.N., and Roberts, P.H., 1993, "A Note on the Kinetic Conditions at a Supercooled Interface," *Int. Comm. Heat Mass Transfer*, 20, 407–416.

Hirata, T., and Matsuzawa, H., 1987, "A Study of Ice-Formation Phenomena on Freezing of Flowing Water in a Pipe," *ASME J. Heat Transfer*, 109, 965–970.

Homsy, G.M., El-Kaissy, M.M., and Didwania, A., 1980, "Instability Waves and the Origin of Bubbles in Fluidized Beds. II: Comparison with Theory," *Int. J. Multiphase Flow*, 6, 305–318.

Homsy, G.M., Jackson, R., and Grace, J.R., 1992, "Report of a Symposium on Mechanics of Fluidized Beds," *J. Fluid Mech.*, 236, 477–495.

Hong, H., and Saito, A., 1993, "Numerical Method for Direct Contact Melting in a Transient Process," *Int. J. Heat Mass Transfer*, 36, 2093–2103.

Hupport, H.E., 1990, "The Fluid Mechanics of Solidification," *J. Fluid Mech.*, 212, 209–240.

Hurle, D.T.J., Jakeman, E., and Wheeler, A.A., 1983, "Hydrodynamic Stability of the Melt During Solidification of a Binary Alloy," *Phys. Fluids*, 26, 624–626.

Jackson, R., 1985, "Hydrodynamic Stability of Fluid-Particle Systems," in *Fluidization*, Davidson, J.F., et al., Editors, Second Edition, Academic Press, London.

Kaviany, M., 1988, "Effect of a Moving Particle on Wall Heat Transfer in a Channel Flow," *Num. Heat Transfer*, 13, 119–124.

Kaviany, M., 1991, *Principles of Heat Transfer in Porous Media*, Springer-Verlag, New York.

Keanini, R.C., and Rubinsky, B., 1993, "Three-Dimensional Simulation of Plasma-Arc Welding Process," *Int. J. Heat Mass Transfer*, 36, 3283–3298.

Khan, T., and Turton, R., 1992, "The Measurement of Instantaneous Heat Transfer Coefficient Around the Circumference of a Tube Immersed in a High-Temperature Fluidized Bed," *Int. J. Heat Mass Transfer*, 35, 3397–3406.

Kim, C.-J., and Kaviany, M., 1992, "A Numerical Method for Phase Change Problems with Convection and Diffusion," *Int. J. Heat Mass Transfer*, 35, 457–467.

Kim, J.M., and Seader, J.D., 1983, "Heat Transfer to Gas-Solids Suspensions Flowing Concurrently Downward in a Circular Tube," *AIChE J.*, 29, 306–312.

Kuipers, J.A.M., Prins, W., and van Swaaij, W.P.M., 1991, "Theoretical and Experimental Bubble Formation at a Single Orifice in a Two-Dimensional Gas-Fluidized Beds," *Chem. Engng. Sci.*, 46, 2881–2894.

Kuipers, J.A.M., Prins, W., and van Swaaij, W.P.M., 1992, "Numerical Calculation of Wall-to-Bed Heat-Transfer Coefficients in Gas-Fluidized Beds," *AIChE J.*, 38, 1079–1091.

Kurz, W., and Fisher, D.J., 1992, *Fundamentals of Solidification*, Third Edition, Trans Tech Publications, Switzerland.

Langlois, W.E., 1985, "Buoyancy-Driven Flows in Crystal-Growth Melts," *Ann. Rev. Fluid Mech.*, 17, 191–215.

Lacroix, M., and Arsenault, A., 1993, "Analysis of Natural Convective Melting of a Subcooled Pure Metal," *Num. Heat Transfer*, 23A, 21–34.

Liebermann, H.H., Editor, 1993, *Rapidly Solidified Alloys*, Marcel Dekker, New York.

Liu, A., Voth, J.E., and Bergman, T.L., 1993, "Pure Material Melting and Solidification with Liquid Phase Buoyancy and Surface Tension Force," *Int. J. Heat Mass Transfer*, 36, 411–422.

Louge, M., Yusof, J.M., and Jenkins, J.T., 1993, "Heat Transfer in the Pneumatic Transport of Massive Particles," *Int. J. Heat Mass Transfer*, 36, 265–275.

Massoudi, M., Rajagopal, K.R., Ekman, J.M., and Mathur, M.P., 1992, "Remarks of the Modeling of Fluidized Beds," *AIChE J.*, 38, 471–472.

Michaelides, E.E., 1986, "Heat Transfer in Particulate Flows," *Int. J. Heat Mass Transfer*, 29, 265–273.

Michaelides, E.E., and Lasek, A., 1987, "Fluid-Solids Flow with Thermal and Hydrodynamic Nonequilibrium," *Int. J. Heat Mass Transfer*, 30, 2663–2669.

Miller, A., and Gidaspow, D., 1992, "Dense, Vertical Gas-Solid Flow in a Pipe," *AIChE J.*, 38, 1801–1815.

Moallemi, M.K., and Viskanta, R., 1986, "Analysis of Close-Contact Melting Heat Transfer," *Int. J. Heat Mass Transfer*, 29, 855–867.

Moallemi, M.K., Webb, B.W., and Viskanta, R., 1986, "An Experimental and Analytical Study of Close-Contact Melting," *ASME J. Heat Transfer*, 108, 894–899.

Molodtsof, Y., and Muzyka, D.W., 1989, "General Probablistic Multiphase Flow Equations for Analyzing Gas-Solid Mixtures," *Int. J. Eng. Fluid Mech.*, 2, 1-24.

Molodtsof, Y., and Muzyka, D.W., 1991, "A Similar Profile Regime in the Vertical Fully Developed Flow of Gas-Solids Suspensions," *Int. J. Multiphase Flow*, 17, 573-583.

Molodtsof, Y., and Muzyka, D.W., 1992, "Wall to Suspension Heat Tranfer in the Similar Profile Regime," *Int. J. Heat Mass Transfer*, 35, 2665–2673.

Moore, F.E., and Bayazitoglu, Y., 1982, "Melting Within a Spherical Enclosure," *ASME J. Heat Transfer*, 104, 19–23.

Mullins, W.W., and Sekerka, R.F., 1963, "Morphological Stability of a Particle Growing by Diffusion on Heat Flow," *J. Appl. Phys.*, 34, 323–329.

Mullins, W.W., and Sekerka, R.F., 1964, "Stability of a Planar Interface During Solidification of a Dilute Binary Alloy," *J. Appl. Phys.*, 35, 444–451.

Murray, B.T., Boettlinger, W.J., McFadden, G.B., and Wheeler, A.A., 1993, "Computation of Dendritic Solidification Using a Phase-Field Model," in *Heat Transfer in Melting, Solidification and Crystal Growth*, Habib, I.S., and Thynell, S., Editors, ASME HTD Vol. 234, 67–76, American Society of Mechanical Engineers, New York.

Nielson, D.G., and Incropera, F.P., 1993, "Effect of Rotation on Fluid Motion and Channel Formation During Unidirectional Solidification of a Binary Alloy," *Int. J. Heat Mass Transfer*, 36, 489–505.

Ocone, R., Sundraresan, S., and Jackson, R., 1993, "Gas-Particle Flow in a Duct of Arbitrary Inclination with Particle-Particle Interactions," *AIChE J.*, 39, 1261–1271.

Prakash, C., 1990, "Two-Phase Model for Binary Solid-Liquid Phase Change, Part I: Governing Equations, Part II:Some Illustration Examples," *Num. Heat Transfer*, 18B, 131–167.

Prescott, P.J., Incropera, F.P., and Gaskell, D.R., 1992, "The Effects of Undercooling, Recalescence and Solid Transport on the Solidification of Binary Metal Alloys," in *Transport Phenomena in Materials Processing and Manufacturing*, ASME HTD-Vol. 196, American Society of Mechanical Engineers, New York.

Rappaz, M., 1989, "Modeling of Microstructure Formation in Solidification Processes," *Int. Mat. Rev.*, 34, 93–123.

Rappaz, M., and Thévoz, Ph., 1987, "Solute Model for Equiaxed Dendritic Growth," *Acta Metall.*, 35, 1487–1497.

Rhodes, M., Zhou, S., and Benkreira, H., 1992, "Flow of Dilute Gas-Particle Suspensions," *AIChE J.*, 38, 1913–1915.

Rieger, H., and Beer, H., 1986, "The Melting Process of Ice Inside a Horizontal Cylinder: Effects of Density Anomaly," *ASME J. Heat Transfer*, 108, 167–173.

Roco, M.C., Editor, 1993, *Particulate Two-Phase Flows*, Butterworth-Heinemann, Boston.

Satake, M., and Jenkins, J.T., Editors, 1988, *Micromechanics of Granular Materials*, Elsevier, Amsterdam.

Saxena, S.C., 1989, "Heat Transfer Between Immersed Surfaces and Gas-Fluidized Beds," *Andvan. Heat Transfer*, 19, 97–190.

Soo, S.-L., 1989, *Particulates and Continuum*, Hemisphere Publishing Corporation, Washington, D.C.

Sparrow, E.M., Patankar, S.V., and Ramadhyani, S., 1977, "Analysis of Melting in the Presence of Natural Convection in the Melt Region," *ASME J. Heat Transfer*, 99, 520–526.

Stefanescu, D.M., Upadhya, G., and Bandyopadhyay, D., 1990, "Heat Transfer-Solidification Kinetics Modeling of Solidification of Castings, *Metall. Trans.*, 21A, 997–1005.

Syamlal, M., and Gidaspow, D., 1985, "Hydrodynamics of Fluidization: Prediction of Wall to Bed Heat Transfer Coefficients," *AIChE J.*, 31, 127–135.

Szekely, J., Evans, J.W., and Brimacombe, J.K., 1988, *The Mathematical and Physical Modeling of Primary Metals Processing Operations*, John Wiley and Sons, New York.

Tamehiro, H., Echigo, R., and Hasegawa, S., 1973, "Radiative Heat Transfer by Flowing Multiphase Medium-Part III. An Analysis on Heat Transfer of Turbulent Flow in a Circular Tube," *Int. J. Heat Mass Transfer*, 16, 1199–1213.

Tao, Y.-X., and Besant, R.W., 1993a, "Prediction of Spatial and Temporal Distribution of Frost Growth on a Flat Plate Under Forced Convection," *ASME J. Heat Transfer*, 115, 278–281.

Tao, Y.-X., Besant, R.W., and Rezkallah, K.S., 1993b, "A Mathematical Model for Predicting the Densification and Growth of Frost on a Flat Plate," *Int. J. Heat Mass Transfer*, 36, 353–363.

Tao, Y.-X., Besant, R.W., and Mao, Y. 1993, "Characteristics of Frost Growth on Flat Plate During the Early Growth Period," *ASHRAE Trans. Symposia*, CH-93-2-2, 746-753.

Thévoz, Ph., Desbiolles, J.L., and Rappaz, M., 1989, "Modeling of Equiaxed Microstructure Formation in Casting," *Metall. Trans.*, 20A. 311–322.

Tien, C.L., 1961, "Heat Transfer to a Turbulently Flowing-Solids Mixture in a Pipe," *ASME J. Heat Transfer*, 83, 183–188.

Vere, A.W., 1987, *Crystal Growth: Principles and Progress*, Plenum Press, New York.

Visser, G., and Valk, M., 1993, "The Porosity in a Fluidized Bed Heat Transfer Model," *Int. J. Heat Mass Transfer*, 36, 627–632.

Voller, V.R., Brent, A.D., and Prakash, C., 1989, "The Modeling of Heat, Mass and Solute Transport in Solidification Systems," *Int. J. Heat Mass Transfer*, 32, 1719–1731.

Wang, C.Y., and Beckermann, C., 1992, "A Multiphase Micro-Macroscopic Model of Solute Diffusion in Dendritic Alloy Solidification," in *Micro/Macro Scale Phenomena in Solidification*, Beckermann, C., et al., Editors, ASME HTD-Vol. 218, 43–57, American Society of Mechanical Engineers, New York.

Weigand, B., and Beer, H., 1993, "Ice Formation Phenomena for Water Flow Inside a Cooled Parallel Plate Channel: An Experimental and Theoretical Investigation of Wavy Ice Layers," *Int. J. Heat Mass Transfer*, 35, 685–693.

White, J.E., and Cremers, C.J., 1981, "Prediction of Growth Parameters of Frost Deposits in Forced Convection," *ASME J. Heat Transfer*, 103, 3–6.

Whitehead, A.B., 1985, "Distributor Characteristics and Bed Properties," in *Fluidization*, Davidson, J.F., et al., Editors, Second Edition, 173–199, Academic Press, London.

Woods, A.W., 1991, "Fluid Mixing During Melting,", *Phys. Fluids*, A3, 1393–1404.

Woods, A.W., 1992, "Melting and Dissolving," *J. Fluid Mech.*, 239, 429–448.

Wu, R.L., Lim, C.J., Grace, J.R., and Brereton, C.M.H., 1991, "Instantaneous Local Heat Transfer and Hydrodynamics in a Circulating Fluidized Bed," *Int. J. Heat Mass Transfer*, 34, 2019–2027.

Xavier, A.M., and Davidson, J.F., 1985, "Heat Transfer in Fluidized Beds," in *Fluidization*, Davidson, J.F., et al., Editors, Second Edition, Academic Press, London.

Yerushalmi, J., and Avidan, A., 1985, "High-Velocity Fluidization," in *Fluidization*, Davidson, J.F., et al., Editors, Second Edition, Academic Press, London.

Zhang, Y., and Campbell, C.S., 1992, "The Interface Between Fluidlike and Solid-like Behavior in Two-Dimensional Granular Flows," *J. Fluid Mech.*, 237, 541–568.

8

Solid–Liquid–Gas Systems

In this chapter, the *liquid-gas, two-phase flow*, and *heat transfer*—including the *evaporation/condensation phase change*—around *solid* surfaces are considered. *Temperature* and *velocity nonuniformities* are allowed in the liquid and gas phases, and all *three phases* are in thermal *nonequilibrium*. The heat flows across the *phase interfaces*, and *mass transfer* (evaporation/condensation) occurs across the *liquid-gas interface*. The liquid and the gas can be *multicomponent* mixtures. Chart 8.1 lists the various *fluid, solid surface,* and *flow* variables affecting the evaporation and condensation rate and the solid surface temperature, in heat transfer across a solid surface with liquid/vapor phase change. Throughout this chapter these variables will be discussed. We begin with a more general classification. Chart 8.2 gives a classification of the *nonisothermal* liquid-gas two-phase flow and heat transfer (including phase change) around solid surfaces. In some applications *only* one fluid phase is in *contact* with the solid and the liquid-gas interface is *continuous*, as in the *vapor* or *liquid films*. In others, both liquid and gas are in contact with the surface, as in *surface bubble* and *droplet nucleation* and in *droplet impingement* on heated surfaces. The solid–liquid–gas *contact line* is also called the *common line* or *interline*, as discussed in Section 1.4.4 (C). In the following the flow and heat transfer in the vapor and liquid films covering solid surfaces are considered first. Then the *nonisothermal* common line and its dynamics are examined. The surface bubble and droplet nucleation and growth are examined next, and finally the impinging droplets are examined.

8.1 Vapor Films

In liquid flow over heated bounding solid surfaces, in many applications either by design or accident, the surface temperature of the heated surface *sufficiently exceeds* the *saturation* temperature of the liquid such that a vapor film is formed at the surface. In steady heat and fluid flow, this vapor film is maintained by a steady vaporization and the liquid is presented from *wetting* the solid surface (*intermittent* surface wetting has been observed in thin vapor films). This class of evaporation problems, i.e., superheated vapor existing between the liquid and the heated solid surface, is considered here. The liquid phase can be *continuous* or the liquid can be *dispersed* in the vapor as *droplets*. As illustrative examples, in Section 8.1.1 the vapor-film flow with a continuous liquid phase is considered for the case where the vapor-film flow is due to *phase-density buoyancy* only. In Section 8.1.2 the

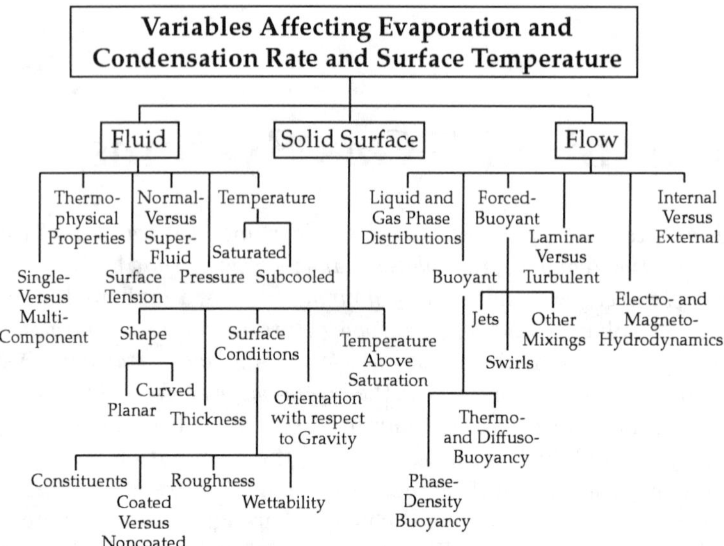

Chart 8.1. A list of fluid, solid surface, and flow variables affecting the evaporation and condensation rate and the surface temperature in heat transfer across a surface with a liquid/vapor phase change.

combined phase-density buoyant and *forced* vapor-film flow, with a continuous liquid phase, will be discussed. Finally, in Section 8.1.3 the *dispersed droplet*-superheated vapor flow and heat transfer will be examined.

8.1.1 Phase-Density Buoyant Flow

A problem of interest in single-component, vapor-film flow and heat transfer is that of *evaporation* of a liquid bounded by a heated solid surface with the vapor and liquid *motions* caused primarily by the *density difference* between the vapor and the liquid phase. We refer to this as the vapor-film, *phase-density buoyant flow* with *evaporation* (this is also called the *pool film-boiling*). The thermobuoyant flow within the vapor is generally negligible and in the liquid is only *secondary*. *Normal gravity* (at the earth surface) or a *significant* gravity is assumed to be present.

When a *subcooled liquid* (i.e., the liquid temperature T_{ℓ_∞} being *lower* than the *liquid-vapor equilibrium* or the *saturation* temperature $T_{\ell g} = T_s$) is placed adjacent to a heated bounding surface and the temperature of this bounding surface T_{s_b} is *above* the saturation temperature, the evaporation phase change *may* occur. For a *semi-infinite liquid body* with the bounding surface making an angle θ with the horizontal axis, this evaporation process is rendered in Figure 8.1. As the *surface superheat* $T_{s_b} - T_s$ is increased, the evaporation does not occur until a *threshold superheat* for the *surface bubble*

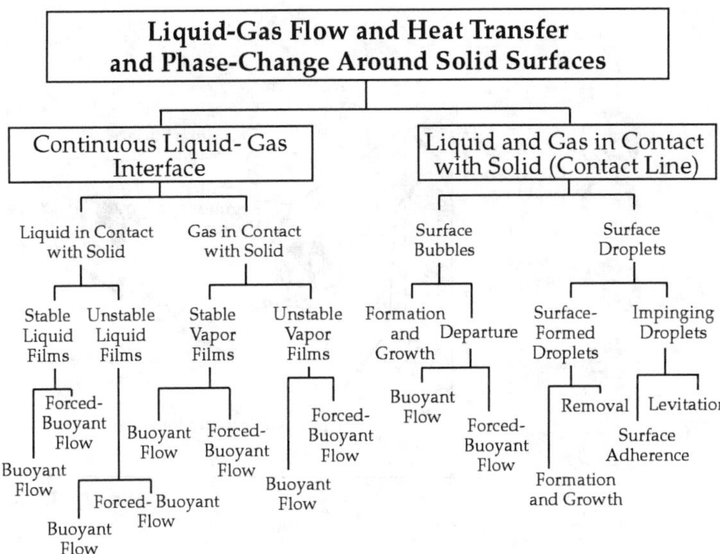

Chart 8.2. A classification of two-phase (liquid-gas) flow and heat transfer, including evaporation and condensation, around solid surfaces.

nucleation inception $(T_{s_b} - T_s)_{inc}$ is reached. This *single-phase* (liquid) heat transfer regime is shown in the figure.

This threshold or incipient superheat depends on the *surface conditions* (*roughness, cavities, coating, gas trapped* in the cavities, etc.), the *fluid* (*dissolved gas* content, *wettability, miscible liquid mixture, pressure, subcooling*, etc.), the *hydrodynamics* (*forced liquid flow, orientation* of the surface with respect to the gravity vector, etc.), and *surface geometry*. The dependency on the trapped gas and other conditions results in the *incipient nucleation hysteresis* which will be discussed in Section 8.5. The maximum for this superheat is expected to be the homogeneous nucleation limit discussed in Sections 1.5.1(C) and 6.1.1 (A) with the superheat given by (1.57) or (6.1).

When the solid surface is heated very *rapidly* (thin solid surfaces), the surface nucleation may be delayed and a higher incipient superheat is found. Also, for the *ideally nonwetting* liquids the surface bubble nucleation may *not* occur. This will be discussed in Sections 8.4 and 8.5.

As the surface superheat is further increased, the surface bubble nucleation regime is encountered which corresponds to a sharp increase in the surface heat flux q_{s_b}. This is called the *surface nucleation regime* (or *nucleate-boiling regime*) and has several subregimes. As shown in the figure, at low surface heat flux, isolated bubbles are formed at active nucleation sites and this is called the *isolated bubbles regime*. At higher q_{s_b} (or $T_{s_b} - T_s$), the frequency of the bubble departure increases and the number of active sites also increases. This is referred to as the *combined isolated bubbles* and *jets regime*. Next, these neighboring jets coalesce and the *vapor-mushroom*

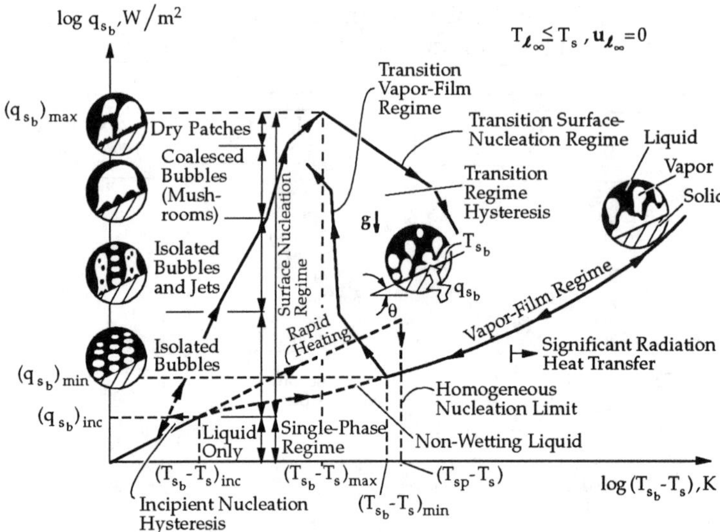

Figure 8.1. Heat transfer and evaporation caused by raising the bounding solid surface to a temperature T_{s_b} above the liquid saturation temperature T_s. The various regimes and subregimes are shown along with the incipient nucleation and transition regime hysteresis. The variation of the surface heat flux is shown versus the superheat.

regime is found. As $T_{s_b} - T_s$ (or q_{s_b}) is further increased, the *dry patches regime* appears. The surface nucleation regime will be discussed in Section 8.5. At a surface superheat $(T_{s_b} - T_s)_{max}$, a maximum in q_{s_b} is found. This $(q_{s_b})_{max}$ is called the *critical heat flux*, and any further increase in $T_{s_b} - T_s$ *decreases* q_{s_b}. In the *transition regime*, bubble nucleation occurs, but the vapor removal mechanism is hindered.

In the *vapor-film regime* which is discussed here, it is *assumed* that the liquid is *not* in contact with the bounding surface and that the heat is transferred by conduction and radiation across the vapor film to the vapor-liquid interface where the *vapor production* occurs.

Now, starting from the vapor-film regime and *decreasing* the superheat, it is possible to reduce the *superheat* to a *minimum* value $(T_{s_b} - T_s)_{min}$, while maintaining a vapor-film flow. There is a corresponding surface heat flux which is called the *minimum heat flux* $(q_{s_b})_{min}$. The two *branches* of the transition regime, i.e., the *transition vapor-film regime* and the transition surface-*nucleation regime*, have been discussed by Witte and Lienhard (1982). This *transition-regime hysteresis* will be discussed in Section 8.5. The mechanisms of the surface-nucleation regimes have been reviewed by Bergles (1992) and Unal et al. (1992), and will also be discussed in Section 8.5.

In this section, the phase-density buoyant, vapor-film flow regime will be examined first for the case where the vapor film is *nonwavy* and then where it becomes *wavy*. We consider the case of a *vertical*, bounding surface, which is more amenable to analysis because the vapor generated at the liquid-vapor interface moves *through* the vapor film (instead of moving *into* the liquid phase as bubbles for other inclination angles). Then a discussion of the minimum surface superheat encountered in the *decreasing* superheat branch will be given. The surface bubble nucleation will be discussed in Section 8.5.

(A) Nonwavy Films

The *two-dimensional*, boundary-layer *buoyant* vapor-film flow with liquid *subcooling* is rendered in Figure 8.2. The temperature of the bounding solid surface is T_{s_b} and is higher than the saturation temperature of the *single-component* fluid T_s by a superheat amount $T_{s_b} - T_s$ required for a stable buoyant vapor-film flow. The vapor flow rate is low enough such that the flow is *laminar* and *wave free*. As will be discussed in Section (B), the liquid-vapor interface becomes wavy a short distance from the leading edge. This *wavy film regime* is followed by a *turbulent film regime* further downstream, where the velocities become rather large. The transition to turbulent flow has been experimentally observed by Hsu and Westwater (1960). They use the *local* vapor-liquid *interface* velocity $u_{\ell v}$ and the local vapor-film *thickness* δ_v and give the laminar-to-turbulent vapor-film flow *transition Reynolds* number as

$$(Re_{\delta_v})_{tr} = \frac{u_{\ell v}\delta_v}{\nu_v} = 100 \,. \tag{8.1}$$

Since in the laminar regime $u_{\ell v}$ and δ_v monotonically increase with the axial location x, the vapor-film flow becomes turbulent at a downstream axial location x_{tr}. The waves appear *before* the transition to turbulence and in this section a *wave-free, laminar vapor-film* flow is considered.

The vapor is formed at the liquid-vapor interface located at $y = \delta_v$ and the heat is supplied to this interface by conduction (the convection contribution is negligible) and radiation across the vapor film which is considered to be *transparent*. As T_{s_b} becomes very large, the radiation contribution becomes significant and asymptotically dominates over conduction. When the liquid is *subcooled*, i.e., $T_{\ell_\infty} < T_s$, some of the heat flowing from the bounding surface to the interface flows into the *liquid* boundary layer, thus *reducing* the evaporation rate. In general, the thermal and the viscous liquid boundary-layer thicknesses are not equal (for liquid Prandtl numbers other than unity); however, in the rendering of Figure 8.2, they are assumed equal.

Similarity solution to the steady, two-dimensional laminar boundary-layer buoyant vapor film flow, *without* liquid subcooling, has been obtained by Koh (1962). A review of the analysis of the buoyant vapor-film flow and heat transfer is given by Sakurai and Shiotsu (1992). This includes bound-

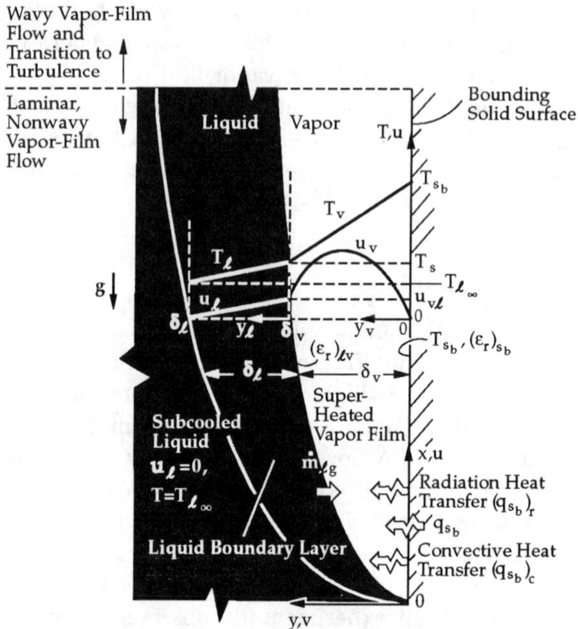

Figure 8.2. Nonwavy film evaporation adjacent to an isothermal, vertical bounding surface. The vapor film and the liquid boundary layer, and their respective, approximate temperature and velocity distributions, are also shown.

ing surface geometries other than the vertical, semi-infinite planar surface considered here (e.g., *horizontal* cylinders and planar surfaces and spheres). The liquid subcooling is *included* with boundary-layer integral analysis of Sakurai et al. (1990), which assumes that the inertial and *convective terms* are *negligible*, and their results are reviewed below.

Assuming negligible inertial and convective terms, the continuity, momentum, and energy equations become (Figure 8.2 shows the coordinate system used)

$$\frac{\partial u_v}{\partial x} + \frac{\partial v_v}{\partial y} = 0 \qquad 0 \le y \le \delta_v \ , \tag{8.2}$$

$$\frac{\partial u_\ell}{\partial x} + \frac{\partial v_\ell}{\partial y} = 0 \qquad 0 \le y_\ell \le \delta_\ell \ , \tag{8.3}$$

$$0 = \mu_v \frac{\partial^2 u_v}{\partial y^2} + g(\rho_\ell - \rho_v) \qquad 0 \le y \le \delta_v \ , \tag{8.4}$$

$$0 = \mu_\ell \frac{\partial^2 u_\ell}{\partial {y_\ell}^2} \qquad 0 \le y_\ell \le \delta_\ell \ , \tag{8.5}$$

$$0 = k_v \frac{\partial^2 T_v}{\partial y^2} \qquad 0 \le y \le \delta_v \ , \tag{8.6}$$

$$0 = k_\ell \frac{\partial^2 T_\ell}{\partial {y_\ell}^2} \qquad 0 \le y_\ell \le \delta_\ell \, . \tag{8.7}$$

The boundary conditions, except for the interfacial energy balance, are

$$u_v = u_\ell = 0 \, , \quad T_\ell = T_{\ell_\infty} \qquad x \le 0 \, , \tag{8.8}$$

$$u_v = v_v = 0 \, , \quad T_v = T_{s_b} \qquad y = 0 \, , \tag{8.9}$$

$$u_v = u_\ell = u_{\ell_v} \, , \quad \mu_v \frac{\partial u_v}{\partial y} = \mu_\ell \frac{\partial u_\ell}{\partial y_\ell} \, , \quad T_v = T_\ell = T_{\ell_v} = T_s \quad y = \delta_v \, , \tag{8.10}$$

$$u_\ell = v_\ell = 0 \, , \quad T_\ell = T_{\ell_\infty} \qquad y = \delta_\ell \, . \tag{8.11}$$

Since the velocity component v does not appear in the momentum and energy equations, the continuity equations become trivial.

The distributions of the velocity components u_v and u_ℓ and temperatures T_v and T_ℓ are obtained by the integration of (8.4) to (8.7) using the y-direction boundary conditions. The results are (Sakurai et al., 1990)

$$u_v = \frac{g(\rho_\ell - \rho_v)}{2\mu_v} \left(\frac{\dfrac{\mu_\ell}{\mu_v} + 2\dfrac{\delta_\ell}{\delta_v}}{\dfrac{\mu_\ell}{\mu_v} + \dfrac{\delta_\ell}{\delta_v}} \delta_v y - y^2 \right) \quad 0 \le y \le \delta_v \, , \tag{8.12}$$

$$u_\ell = -\frac{g(\rho_\ell - \rho_v)\delta_v}{2\mu_\ell} \frac{\dfrac{\mu_\ell}{\mu_v}}{\dfrac{\mu_\ell}{\mu_v} + \dfrac{\delta_\ell}{\delta_v}} (\delta_v \delta_\ell - \delta_v y_\ell) \quad 0 \le y_\ell \le \delta_\ell \, , \tag{8.13}$$

$$T_v = T_{s_b} - (T_{s_b} - T_s) \frac{y}{\delta_v} \, , \tag{8.14}$$

$$T_\ell = T_s - (T_s - T_{\ell_\infty}) \frac{y_\ell}{\delta_\ell} \, . \tag{8.15}$$

The temperature profile for the liquid does *not* result in a zero heat flux at $y_\ell = \delta_\ell$. The interfacial energy balances written in terms of the lateral conduction and axial convection (the radiation heat flux will be addressed later) are

$$\rho_v [\Delta i_{\ell g} + \frac{1}{2} c_{p_v} (T_{s_b} - T_s)] \frac{d}{dx} \int_0^\delta u_v dy =$$
$$k_v \frac{T_{s_b} - T_s}{\delta_v} - k_\ell \frac{T_s - T_{\ell_\infty}}{\delta_\ell} \, , \tag{8.16}$$

$$(\rho c_p)_\ell \frac{d}{dx} \int_0^{\delta_\ell} u_\ell (T_\ell - T_{\ell_\infty}) dy_\ell = k_\ell \frac{T_s - T_{\ell_\infty}}{\delta_\ell} \, . \tag{8.17}$$

The vapor-film thickness and the ratio of film thicknesses

$$\xi = \frac{\delta_\ell}{\delta_v} \tag{8.18}$$

are determined from the interfacial energy balances. Using the distributions

(8.12) to (8.15) in the interfacial energy balances (8.16) and (8.17), the results are (Sakurai et al., 1990)

$$\frac{d}{dx} f_v \delta_v{}^3 = \frac{a_2}{\delta_v} - \frac{a_3}{\delta_v \xi}\,, \quad \delta_v(x=0) = 0\,, \tag{8.19}$$

$$\frac{d}{dx} f_\ell \delta_\ell{}^3 = \frac{a_4}{\delta_v \xi}\,, \tag{8.20}$$

where f_v and f_ℓ and constants a_i are

$$f_v = \frac{a_1 + 4\xi}{a_1 + \xi}\,, \quad f_\ell = \frac{a_1 \xi^2}{a_1 + \xi}\,, \quad a_1 = \frac{\mu_\ell}{\mu_v}\,,$$

$$a_2 = \frac{12\mu_v k_v (T_{s_b} - T_s)}{[\Delta i_{\ell g} + \frac{1}{2} c_{p_v}(T_{s_b} - T_s)] g \rho_v (\rho_\ell - \rho_v)}\,,$$

$$a_3 = a_1 \frac{T_s - T_{\ell_\infty}}{T_{s_b} - T_s}\,, \quad a_4 = \frac{6\mu_\ell k_\ell}{g(\rho c_p)_\ell (\rho_\ell - \rho_v)}\,. \tag{8.21}$$

Equations (8.19) and (8.20) admit solutions with ξ independent of x, i.e., ξ being a *constant* designated by ξ_o. Then the solutions to (8.19) and (8.20) are

$$\frac{\delta_v{}^4}{x} = \frac{4(a_1 + \xi_o)(a_2 \xi_o - a_3)}{3(a_1 + 4\xi_o)\xi_o}\,, \tag{8.22}$$

$$\frac{\delta_v{}^4}{x} = \frac{4a_4(a_1 + \xi_o)}{3a_1 \xi_o{}^3}\,. \tag{8.23}$$

Then ξ_o is found from equating these two equation and the acceptable roots of the resultant equation gives

$$\xi_o = a_5 + a_6 + \frac{a_3}{3a_2}\,, \tag{8.24}$$

where the constants a_5 and a_6 are functions of the *primary* constants a_1 to a_4 and are given by Sakurai et al. (1990).

The growth of the vapor-film thickness follows a $x^{1/4}$ relation characteristic of the laminar buoyant flow along vertical surfaces discussed in Section 4.12.1(A). The *local* Nusselt number (*excluding* the radiation contribution which will be discussed below) is

$$(Nu_x)_c = \frac{(q_{s_b})_c}{(T_{s_b} - T_s)k_v} = \frac{-k_v \left.\frac{\partial T_v}{\partial y}\right|_o x}{(T_{s_b} - T_s)k_v} = \frac{x}{\delta_v}\,. \tag{8.25}$$

The *x-direction-averaged* Nusselt number (which is the more useful and also the more readily measurable) is given in terms of x, ξ_o, and other constants, as

$$\langle (Nu_x)_c \rangle_x = \frac{1}{x}\int_0^x (Nu_x)_c dx = \frac{4}{3}\left[\frac{3a_1 \xi_o^3 x^3}{4a_4(a_1 + \xi_o)}\right]^{1/4}\,. \tag{8.26}$$

The *primary* constants a_1 to a_4 can be given in terms of the *liquid* and *vapor* Prandtl *numbers*, Pr_ℓ and Pr_v, respectively; the *vapor superheat* and *liquid subcooling Jakob numbers*, Ja_{sh} and Ja_{sc}, respectively; the ratio $(\rho\mu)_v/(\rho\mu)_\ell$; and the *Grashof number* Gr_x. The new parameters are

$$Pr_\ell = \frac{\nu_\ell}{\alpha_\ell}, \quad Pr_v = \frac{\nu_v}{\alpha_v},$$

$$Ja_{sh} = \frac{c_{p_v}(T_{s_b} - T_s)}{[\Delta i_{\ell g} + \frac{1}{2}c_{p_v}(T_{s_b} - T_s)]Pr_v}, \quad Ja_{sc} = \frac{c_{p_\ell}(T_s - T_{\ell_\infty})}{\Delta i_{\ell g} + \frac{1}{2}c_{p_v}(T_{s_b} - T_s)},$$

$$(\rho\mu)^* = \frac{(\rho\mu)_v}{(\rho\mu)_\ell}, \quad Gr_x = \frac{g(\rho_\ell - \rho_v)x^3}{\rho_v {\nu_v}^2}. \tag{8.27}$$

The expression for the Nusselt number as a function of these parameters is rather lengthy and is given by Sakurai et al. (1990).

Inclusion of the radiation heat flux into the interfacial energy balance adds $(q_{s_b})_r$ into (8.16) with $(q_{s_b})_r$ given by a surface radiation exchange approximation between two *parallel* plate as

$$(q_{s_b})_r = \frac{\sigma_{SB}(T_{s_b}^4 - T_s^4)}{\dfrac{1}{(\epsilon_r)_{s_b}} + \dfrac{1}{(\epsilon_r)_{\ell v}} - 1}. \tag{8.28}$$

This would *change* the solution for $(Nu_x)_c$. Sakurai et al. (1990) suggest a *correction* to the convection $\langle (Nu_x)_c \rangle_x$ to include the radiation. This is to represent the total heat flux q_{s_b}, or the total Nusselt number $\langle Nu_x \rangle_x$, as

$$\langle Nu_x \rangle_x = \frac{\langle q_{s_b} \rangle_x}{(T_{s_b} - T_s)k_v} = \langle (Nu_x)_c \rangle_x +$$

$$f_J[Ja_{sh}, Ja_{sc}, (\rho\mu)^*] \frac{\sigma_{SB}(T_{s_b}^4 - T_s^4)x}{[\dfrac{1}{(\epsilon_r)_{s_b}} + \dfrac{1}{(\epsilon_r)_{\ell v}} - 1](T_{s_b} - T_s)k_v}. \tag{8.29}$$

A correlation for f_J is given by Sakurai et al. (1990), where the Jakob numbers used in this correlation do *not* include the average, specific sensible heat of the vapor $c_{p_v}(T_{s_b} - T_s)/2$.

The similarity solution for the case where the *variation* in densities is allowed and the inertial and convection terms are *included* in the momentum and energy equations is also considered by Sakurai et al. (1990) using the stream function and a space similarity variable. The solution for the *horizontal cylinders* is also obtained by them, including radiation. For *horizontal cylinders*, the vapor formed at the liquid-vapor interface is collected on top of the cylinder where it departs as *bubbles*. For small-diameter bubbles, the liquid-vapor interface is *nonwavy* except at the top, while for large bubble diameters the interface is wavy except for a small region around the front stagnation point. These are discussed by Jordan (1968), Carey (1992), and Nishio and Ohtake (1992). We will examine the evaporation and vapor-film flow around horizontal cylinders, for the case of combined buoyant-forced flows in Section 8.1.2.

(B) Wavy Films

The laminar vapor film thickness δ_v discussed above is generally *small* and of the order of 1 nm, and the vapor-liquid interface is potentially *unstable*. This instability is due to the difference in the vapor and liquid *velocities* away from the interface and also due to the difference in the vapor and liquid densities. Then a short distance from the leading edge ($x = 0$), the interface becomes *wavy*. Experimental results show that the Nusselt number predicted by (8.26), or $\langle(q_{s_b})_c\rangle_x/(T_{s_b} - T_s)$, does *not* occur beyond this short distance from the leading edge. The experimental results show that instead of the $x^{3/4}$ relations for $\langle(Nu_x)_c\rangle_x$, or $x^{-1/4}$ relation for $\langle(q_{s_b})_c\rangle_x/(T_{s_b} - T_s)$, in *the wavy-film regime* this relation becomes x^1, or x^0 (i.e., independent of x). This observed *uniformity* of $\langle(q_{s_b})_c\rangle_x$ has been used to develop models for the flow and heat transfer and for the prediction of the magnitude of $\langle(q_{s_b})_c\rangle_x$. These models assume *periodic unit cells* made of a *vapor film* and an *attached vapor bubble*. The linear dimension of the unit cell along the surface is related to the wavelength of the *fastest* growing interfacial wave, as predicted by the linear stability theory governing growth of infinitesimal interface displacements.

In the following, the results of the linear stability theory for the interfacial displacement waves, the unit-cell models used, and the treatments of the vapor and liquid boundary layers in the unit cell are reviewed and discussed, leading to the prediction of $\langle(q_{s_b})_c\rangle_x$ or $\langle(Nu_x)_c\rangle_x$ for the wavy-film regime.

(i) Critical Wavelength of Interfacial Displacement

As with the single-phase, thermobuoyant boundary-layer flow adjacent to an isothermal vertical surface discussed in Section 4.12.1(c), the phase-density buoyant vapor-film can also become unstable. However, the *mechanism* of the instability are different for this two-phase flow problem. The instability occurs as the growth of *infinitesimal* displacements on the liquid-vapor interface. The growth can be due to an unstable density stratification, i.e., the *discontinuity* in the density across the interface which can place the heavier fluid on top of the lighter fluid (this is called the *Rayleigh-Taylor instability*), or due to the difference in the *bulk* vapor and liquid velocities (this is called the *Kelvin-Helmholtz instability*). These instabilities occur also in the *isothermal*, liquid-gas two-phase flow and have been examined for simple liquid and gas velocity fields (mostly *uniform velocities* with *zero viscosties* such that a *jump* in the *velocity* also occurs across the interface). These *inviscid, isothermal* liquid-gas interfacial instabilities have been reviewed by Lamb (1945) and Chandrasekhar (1961). The applications in the vapor-film flows with phase change have been reviewed Zuber (1958), Berenson (1961), Jordan (1968), Anderson (1976), and van Stralen and Cole (1979).

The *linear* stability analysis is made by introducing infinitesimal displacement disturbances to the *base* flows and *interface* location and by determination the growth or decay of these disturbances. For *unconditionally* unstable flows, the growth of the *fastest* growing disturbance wavelength is followed. This *critical* wavelength is then assumed to prevail even after the displacement amplitudes are *not* infinitesimal. However, the *finite amplitude* disturbances (or waves) are governed by nonlinearities which can make the dominant wavelength (i.e., the largest amplitude wave) different from the one which grows the fastest at the initial stages, where the linearization is valid because of the small amplitudes. For two-dimensional, liquid-vapor *base* flow discussed in Section (A), no detailed stability analysis is available. Experiments have shown that despite many idealizations, such as the uniform (one-dimensional) velocity and the zero viscosities, using the *simple* base flows can relatively accurately predict the *dominant* wavelength (Nishio et al., 1991).

The solution for the *critical* wavelength for an *isothermal* vapor film of *uniform* velocity $\langle u_v \rangle_\delta$ and thickness δ_v flowing between and *parallel* to a *vertical* bounding solid surface and an *initially* quiescent liquid has been examined by Nishio et al. (1991). The two-dimensional, *inviscid* infinitesimal amplitude liquid and vapor *disturbance* flow fields are described by the *velocity potential* ϕ. The continuity equations (8.2) and (8.3) become (referring to Figure 8.2)

$$\frac{\partial^2 \phi_v}{\partial x^2} + \frac{\partial^2 \phi_v}{\partial y^2} = 0 \qquad \delta_v \geq y \geq 0\,, \tag{8.30}$$

$$\frac{\partial^2 \phi_\ell}{\partial x^2} + \frac{\partial^2 \phi_\ell}{\partial y^2} = 0 \qquad \delta_\ell \geq y \geq \delta_v\,, \tag{8.31}$$

where the *disturbed* components of the velocity are

$$u' = \frac{\partial \phi}{\partial x}\,, \quad v' = \frac{\partial \phi}{\partial y}\,. \tag{8.32}$$

The inviscid momentum equations (i.e., *Bernoulli equations*) along the y direction can be written as (e.g., Valentine, 1967)

$$\frac{p'_v}{\rho_v} = \frac{\partial \phi_v}{\partial t} + \langle u_v \rangle_{\delta_v} \frac{\partial \phi_v}{\partial x}\,, \tag{8.33}$$

$$\frac{p'_\ell}{\rho_\ell} = \frac{\partial \phi_\ell}{\partial t}\,. \tag{8.34}$$

Choosing a coordinate system centered at the interface, i.e., $y_\ell = y - \delta_v$, the boundary conditions are

$$\frac{\partial \phi_v}{\partial y_\ell} = 0 \qquad y_\ell = -\delta_v\,, \tag{8.35}$$

$$\frac{\partial \phi_\ell}{\partial y_\ell} = 0 \qquad y_\ell = \delta_\ell\,, \tag{8.36}$$

$$-\frac{\partial \phi_v}{\partial y_\ell} = \langle u_v \rangle_{\delta_v} \frac{\partial \delta_v'}{\partial t} + \frac{\partial \delta_v'}{\partial t}\,, \quad -\frac{\partial \phi_\ell}{\partial y_\ell} = \frac{\partial \delta_v'}{\partial t} \qquad y_\ell = \delta_v'\,, \tag{8.37}$$

$$p_v' - p_\ell' = -\sigma \frac{\partial^2 \delta_v'}{\partial x^2}\,, \tag{8.38}$$

where δ_v' is the amplitude of the interfacial displacement and the linearization is made by dropping the second- and higher-order terms and one of the principal radii of the interface curvature is zero.

An interfacial displacement disturbance of *angular frequency* ω and wave number η, where $\eta = 2\pi/\lambda$ and λ is the wavelength, i.e.,

$$\delta_v' = a_\delta e^{i(\omega t + \eta x)} \tag{8.39}$$

is introduced.

The solution to the *Laplace* equations (8.30) and (8.31), subject to the boundary conditions (8.35) to (8.37) and the interface displacement (8.39), is (Nishio et al, 1990)

$$\phi_v = -i a_\delta (\omega + \eta \langle u_v \rangle_{\delta_v}) \frac{\cosh \eta (y_\ell + \delta_v)}{\eta \sinh \eta \delta_v} e^{i(\omega t + \eta x)}\,, \tag{8.40}$$

$$\phi_\ell = i a_\delta \omega \frac{\cosh \eta (y_\ell - \delta_\ell)}{\eta \sinh \eta \delta_\ell} e^{i(\omega t + \eta x)}\,. \tag{8.41}$$

Using these solutions in (8.33) and (8.34) and the capillary pressure $p_c' = p_v' - p_\ell'$, expression (8.38), the result for the frequency, and the growth rate defined as $(i\omega)^2$, is

$$(\rho_v \cosh \eta \delta_v + \rho_\ell)\omega^2 + 2(\rho_v \eta \langle u_v \rangle_{\delta_v} \cosh \eta \delta_v)\omega + \rho_v \eta^2 \langle u_v \rangle_{\delta_v}^2 \cosh \eta \delta_v - \sigma \eta^3 = 0 \quad \delta_\ell \to \infty\,. \tag{8.42}$$

When the vapor density is much smaller than the liquid density, the wave number that would result in the largest ω, i.e., the *critical* wave number η_{cr} is (Nishio et al., 1991)

$$\eta_{cr} = \frac{2\pi}{\lambda_{cr}} = \left(\frac{\rho_v \langle u_v \rangle_{\delta_v}^2}{3\sigma \delta_v}\right)^{1/2} \quad \frac{\rho_v}{\rho_\ell} \ll 1\,. \tag{8.43}$$

The critical wavelength λ_{cr} increases with an increase in liquid-vapor interfacial tension σ and also with an increase in δ_v, and decreases with the vapor axial momentum flux $\rho_v \langle u_v \rangle_{\delta v}^2$.

(ii) Space-Periodic, Moving Unit-Cell Model

Continuous efforts have been made to predict $(\langle q_{s_b} \rangle_x)_c$ using the predicted wavelength for the fastest growing interfacial wave and the experimental observations such as the uniformity of $(\langle q_{s_b} \rangle_x)_c$ for the *uniform* T_{s_b} condition. Among these models is the *space-periodic, moving* unit-cell model depicted in Figure 8.3 which allows for a laminar *vapor film* which grows in thickness in the unit and empties into an *attached bubble*. The unit cell

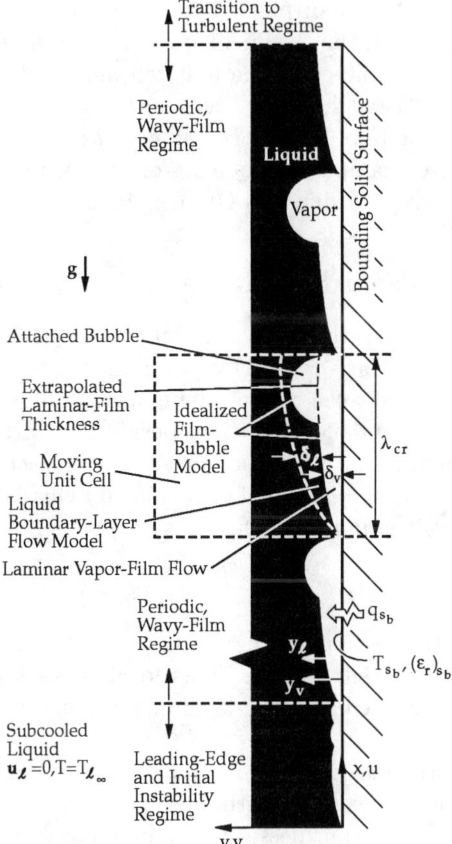

Figure 8.3. The unit-cell model of a wavy vapor film adjacent to an isothermal, vertical bounding solid surface. The unit-cell variables are also shown.

moves upward on the bounding surface and has an axial unit-cell length λ_{cr}. Observations of Nishio et al. (1991) show that *after* a distance of λ_{cr} from the *leading edge* this periodic structure is maintained. The size, geometry, and heat transfer in the attached bubble have been modeled with various extent of rigor (e.g., Anderson, 1976; Bui and Dhir, 1985; Nishio and Ohtake, 1993). The models assume that the vapor-film thickness is zero at the *upstream* edge of the unit cell and that the film is *laminar* and predictable by a boundary-layer analysis similar to that discussed in Section (A) above. Observations show that the attached bubble *enlarges* with distance from the leading edge as increasingly more vapor flows into it. In the model of Bui and Dhir (1985), heat transfer is also allowed in the attached-bubble region (about 30 percent of the unit cell heat transfer), while in the model of Nishio and Ohtake (1993) the bubble heat transfer is *not* considered and instead it is assumed that the laminar vapor film extends to the *end* of the unit cell (as shown in Figure 8.3).

The liquid-phase and vapor-phase hydrodynamics of the wavy interface are not well known, and the flow fields are *not* two-dimensional, periodic, and quasi-steady, as depicted in the unit-cell model. Bui and Dhir (1985) indicate that *recirculating* flow may occur in the *liquid between* the attached bubbles, and that the bubble can become *detached*.

As an illustrative example in modeling of this two-phase flow and heat transfer, the analysis of Nishio and Ohtake (1993) is reviewed below.

(iii) Liquid and Vapor Boundary Layers

In the wavy-film regime, the Nusselt number, or $(\langle q_{s_b}\rangle_x)_c$, is *larger* than that predicted by an *ever-growing* vapor film boundary layer discussed in Section (A). The unit-cell model of Figure 8.3 allows for the *periodic destruction* of the boundary layer (i.e., periodic restart of the boundary layer). This is similar to the other *boundary-layer renewal models* commonly used. In the *two-dimensional* model of Nishio and Ohtake (1993), the unit-cell averaged Nusselt number is defined as

$$\langle (Nu_x)_c\rangle_x = \frac{1}{\lambda_{cr}}\int_0^{\lambda_{cr}} (Nu_x)_c dx \,, \tag{8.44}$$

where $(Nu_x)_c$ is the *local* nonradiative (i.e., conduction and convection) component of the Nusselt number. This local Nusselt number is generally determined using the boundary-layer integral method discussed in Section (A).

Some of the approximations made in the determination of $(Nu_x)_c$, which in turn depends on the vapor-film thickness δ_v and the ratio of the liquid to vapor boundary-layer thicknesses ξ, can be *relaxed*. These are discussed by Nishio and Ohtake (1992) and some of them are briefly mentioned.

The *nonzero* heat flux at $y_\ell = \delta_\ell$, which was allowed by the *linear* liquid temperature profile given by (8.15), is made to be zero by choosing the profile

$$T_\ell = T_{\ell_\infty} + (T_s - T_{\ell_\infty})(\frac{y_\ell}{\delta_\ell} - 1) \,. \tag{8.45}$$

The inertial and the thermobuoyancy term in the liquid momentum equation, which were neglected in (8.5), are added and we have

$$\frac{\partial}{\partial x}\rho_{\ell_o} u_\ell^2 = \mu_\ell \frac{\partial^2 u_\ell}{\partial {y_\ell}^2} + \rho_{\ell_o} g\beta (T_\ell - T_{\ell_\infty}) \,. \tag{8.46}$$

This will allow for motion caused by both the *liquid thermobuoyant* force (Boussinesq approximations) and the *liquid-vapor interfacial viscous drag*, in addition to the phase-density buoyancy. The integration of (8.46) across the liquid boundary layer gives

$$\rho_{\ell_o}\frac{d}{dx}\int_0^{\delta_\ell} u_\ell^2 dy_\ell = \rho_{\ell_o} g\beta \int_0^{\delta_\ell} (T_\ell - T_{\ell_\infty}) dy_\ell - \mu_\ell \left.\frac{\partial u_\ell}{\partial y_\ell}\right|_{y_\ell = 0} \,. \tag{8.47}$$

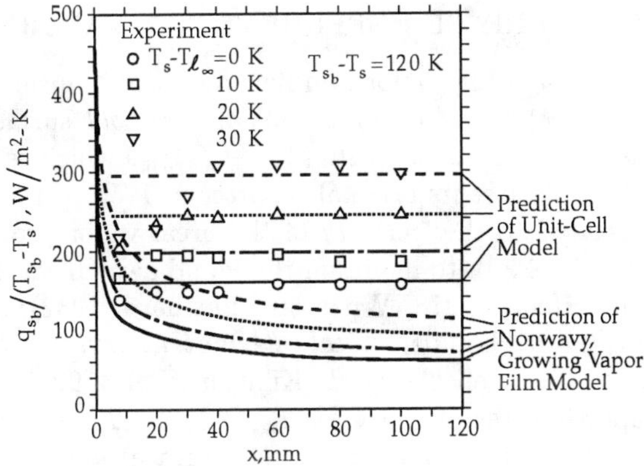

Figure 8.4. Comparison of the experimental results for the axial distribution of the heat transfer coefficient, for film evaporation adjacent to an isothermal vertical surface, with the predictions of the laminar wavy and nonwavy vapor-film models. (From Nishio and Ohtake, reproduced by permission ©1993 Pergamon Press.)

A liquid velocity profile which satisfies the zero velocity and shear stress at $y_\ell = \delta_\ell$ has a form

$$u_\ell = a_{\ell_1}(1 - \frac{y_\ell}{\delta_\ell})^2(\frac{y_\ell}{\delta_\ell} + a_{\ell_2}) \,, \tag{8.48}$$

where the constant a_{ℓ_1} and a_{ℓ_2} are determined by matching the velocity and the shear stress at $y_\ell = 0$ (or $y = \delta_v$) as given by (8.10).

The local Nusselt number is given by (8.25); however, δ_v (and δ_ℓ, or ξ) which are determined from (8.22) and (8.23), will have different constants in these equations. The x-dependence of $(Nu_x)_c$ remains the same, i.e., $x^{3/4}$ as given by (8.26), but the constants appearing in (8.26) are also altered. These new constants are given by Nishio and Ohtake (1992).

The results of the moving unit-cell model, are compared with the available experimental results by Nishio and Ohtake (1993), and some of their typical results are shown in Figure 8.4. The results for the *nonwavy, growing* vapor-film model, i.e., similar to those discussed in Section (A), which predict a *decrease* in $(q_{s_b})_c$ along the surface, i.e., a $x^{-1/4}$ dependency, are also shown. The radiation heat transfer is *not* included in this unit-cell model, but the predictions are in good agreement with the experimental results. Noting the approximations made in the determination of λ_{cr} and in the film-bubble unit-cell models, this shows how approximate models assist identifying some of the key mechanisms in this rather *complex* and *coupled thermal-hydrodynamic* problem.

(C) Minimum Surface Superheat for Vapor Films

The phase-density buoyant, vapor-film flow and heat transfer discussed in Sections (A) and (B) above occur *beyond* a *threshold* surface superheat $T_{s_b} - T_s$. This *minimum superheat*, $(T_{s_b} - T_s)_{\rm min}$, which is also named the *Leidenfrost limit* after its original reporter in 1756 (Carey, 1992), had been initially related to the *inability* of the areal vapor production rate, i.e., $k_v(T_{s_b} - T_s)/(\delta_v \Delta_{i_{\ell g}})$, to maintain the liquid away from the bounding surface (Yao and Henry, 1978; Witte and Lienhard, 1982). However, for vapor films in *saturated* liquids, a *local intermittent liquid contact* with the bounding surface has been observed (Kikuchi et al., 1992) in the otherwise stable vapor-film regime, i.e., for $T_{s_b} - T_s > (T_{s_b} - T_s)_{\rm min}$. Then it is hypothesized that the *heterogeneous* (i.e., surface) *nucleation* of the intermittently and locally contacting liquid prevents the vapor film collapse (Ramilison and Lienhard, 1987; Sakurai and Shiotsu, 1992). The *minimum superheat* is then related to the *heterogeneous nucleation superheat limit* of the liquid. This limit is related *not* only to the fluid properties (as in the homogeneous, or bulk, bubble nucleation superheat limit discussed in Section 6.2.2), but also to the solid surface properties and flow conditions. This will be discussed in Section 8.5.1. The *wettability* of the solid surface is combined into the *contact angle* θ_c. Since this liquid contact is intermittent and dynamic, the *dynamic* contact angle, discussed in Section 1.4.4(c), should be used. For bounding surfaces with less than ideal conductivity, the contacting liquid will *decrease* the surface temperature below its initial and far-field value. Then this *transient interfacial temperature* is expected to be *above* the heterogeneous superheat limit for the stable vapor-film regime. For wetting liquids, the heterogeneous superheat is *lower* than the homogeneous superheat limit (because the threshold nucleation energy is lower when the surface nonhomogeneities and foreign species are present).

This *thermodynamic, phenomenological* model of the minimum surface superheat makes it *nearly hydrodynamic independent*—e.g., the orientation of the bounding surface with respect to gravity, the forced versus buoyant flows, the magnitude of the velocity all do not influence this superheat—and this has not been extensively verified. This minimum surface superheat is then given by

$$(T_{s_b} - T_s)_{\rm min} = f(T_{sp} - T_s, \theta_c) \,. \tag{8.49}$$

Ramilison and Lienhard (1987) find a correlation for *saturated* liquids (Carey, 1992) as

$$\frac{(T_{s_b} - T_s)_{\rm min}}{T_{sp} - T_s} = 0.97 e^{-6\times 10^{-4} \theta_c^{1.8}} \quad T_{\ell_\infty} = T_s \,, \tag{8.50}$$

where T_{sp} is given by (1.57) and θ_c can be taken as the *advancing* contact angle. In practice, rather than using θ_c in terms of the variables given is Section 1.4.4(C), it is taken as a constant. Also, the liquid *subcooling* substantially *suppresses* the intermittent local liquid-solid contact (Kikuchi et al., 1992). The phenomenological models for the *minimum heat flux* $(q_{s_b})_{\min}$ are *hydrodynamic dependent* and involve modeling of the *contact frequency* and the *contact area* (Jordan, 1968; van Stralen and Cole, 1979; Ramilison and Lienhard, 1987). Semi-empirical treatments use the results of the hydrodynamic instability theory and the transient heat conduction in the vapor and solid. Witte and Lienhard (1982) discuss the need for a *controlled* experiment in which $(q_{s_b})_{\min}$ is reached from the stable vapor-film regime by *decreasing* $T_{s_b} - T_s$ and *avoiding* the *transition* surface-nucleation regime *branch* of the transition regime in Figure 8.4. Otherwise, a $(q_{s_b})_{\min}$ *higher* than the actual value will be obtained, i.e., the reported *multivalueness* of $(q_{s_b})_{\min}$.

8.1.2 Buoyant-Forced Flow

For solid surfaces *submerged* in a flowing liquid, the vapor flow due to the phase-density buoyancy (discussed in Section 8.1.1) can be *assisted*, *opposed*, or *deflected* by this *forced* flow depending on the *direction* of the far-field liquid flow *relative* to the gravity vector. When the forced flow is against the gravity vector, the structure of this forced-flow-*assisted* buoyant vapor flow is *not* significantly different from that of the *purely* buoyant flow. For *opposing* and *cross* forced-liquid flows, the vapor flow is significantly altered, and when the forced flow dominates, the vapor-film flow field is determined only by the liquid-flow direction. One of the solid surface geometries most studied because of the practical interest is the *horizontally* placed *cylinder*. The vapor film flow around this cylinder was originally examined by Bromley, and the details of these contributions and of some others immediately following him are given by Jordan (1968). Based on the experimental results for assisting forced flow, the *buoyancy-flow dominated* regime is marked by the *Froude number* $Fr_d = v_{\ell_\infty}/(g_d)^{1/2} < 1$, where u_{ℓ_∞} is the far-field velocity of the forced liquid flow and d is the cylinder diameter. The *forced-flow dominated* regime is marked by $Fr_d > 2$.

Recent analyses of vapor-film flow and heat transfer with *assisting* liquid flow have been made by Chou and Witte (1992) and Liu et al. (1992). A rendering of the idealized *two-dimensional* flow and heat transfer is given in Figure 8.5. The surface of the solid cylinder is at T_{s_b} which is much larger than the saturation temperature T_s such that the vapor-film regime is realized. The *minimum superheat* $(T_{s_b} - T_s)$ for *forced*-film flow is examined by Chang and Witte (1990), and as was discussed in Section 8.1.1(c) this minimum nearly corresponds to the temperature of the cylinder surface approaching the *heterogeneous nucleation superheat limit*. The vapor-film has its smallest thickness at the angle $\theta = 0$, i.e., the *front* stagnation point. The vapor flow *separates* at a *separation* angle $\theta = \theta_s$ and then a *wake* is formed for $\theta > \theta_s$, where the vapor wake is *three-dimensional*. Liquid thermal and viscous boundary layers are present around both the vapor

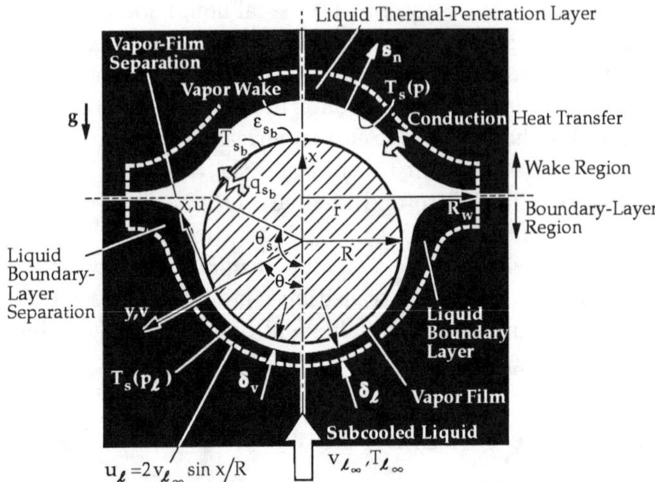

Figure 8.5. Film evaporation around an isothermal, horizontal cylinder with combined upward-forced and phase-density buoyant flows. The vapor film and wake regions, the liquid boundary layer, and the thermal penetration layer are also shown.

film and the wake. Since a significant pressure drop can occur across the cylinder, the saturation temperature occurring at the liquid-vapor interface is allowed to change with the pressure. The liquid flow outside the thin viscous boundary layer is approximated by a *potential flow*, and at the edge of the liquid boundary layer the velocity is given as a function of the location on the x-axis shown in Figure 8.5, with $x = 0$ at the front stagnation point. Some of the hydrodynamic aspects of the purely forced vapor-film flow around the cylinder have been discussed by Witte and Orozco (1984). The wake-region flow and heat transfer is complex, and two-dimensional approximations have been used. In the following, the analysis of Chou and Witte (1992) is reviewed for the prediction of the *local* and the *surface area-averaged* heat transfer rate from the cylinder.

The radiation heat transfer is included by assuming a perfect emitter-absorber liquid-vapor interface, i.e., $(\epsilon_r)_{\ell v} = 1$, and then (8.28) gives

$$(q_{s_b})_r = (\epsilon_r)_{s_b} \sigma_{SB} [T_{s_b}^4 - T_s^4(p_\ell)] \,. \tag{8.51}$$

The liquid velocity at the edge of the liquid boundary layer u_{ℓ_e} is given by the *inviscid liquid flow* solution

$$u_{\ell_e} = 2v_{\ell_\infty} \sin \frac{x}{R} \,. \tag{8.52}$$

Then the pressure gradient at the edge of the liquid boundary layer is given by

$$\rho_{\ell_e} u_{\ell_e} \frac{\partial u_{\ell_e}}{\partial x} = -\frac{\partial p_\ell}{\partial x} \,. \tag{8.53}$$

The interfacial curvature-capillarity effect is neglected, i.e., the capillary pressure is taken as zero, $p_{\ell v} = p_\ell = p_v$ and this gives

$$\frac{\partial p_\ell}{\partial x} = \frac{\partial p_v}{\partial x} = -\rho_{\ell_\infty} v_{\ell_\infty}^2 \frac{2}{R} \sin\frac{2x}{R} . \tag{8.54}$$

The laminar, steady two-dimensional boundary-layer continuity, momentum, and energy equations for the vapor and liquid phases are

$$\frac{\partial}{\partial x}\rho_v u_v + \frac{\partial}{\partial y}\rho_v v_v = 0 , \tag{8.55}$$

$$\frac{\partial}{\partial x}\rho_\ell u_\ell + \frac{\partial}{\partial y}\rho_\ell v_\ell = 0 , \tag{8.56}$$

$$\rho_v u_v \frac{\partial u_v}{\partial x} + \rho_v v_v \frac{\partial u_v}{\partial y} = \rho_{\ell_\infty} v_{\ell_\infty}^2 \frac{2}{R} \sin\frac{2x}{R} + g(\rho_{\ell_\infty} - \rho_v)\sin\frac{x}{R} + \frac{\partial}{\partial y}\mu_v \frac{\partial u_v}{\partial y} , \tag{8.57}$$

$$\rho_\ell u_\ell \frac{\partial u_\ell}{\partial x} + \rho_\ell v_\ell \frac{\partial u_\ell}{\partial y} = \rho_{\ell_\infty} v_{\ell_\infty}^2 \frac{2}{R} \sin\frac{2x}{R} + g(\rho_{\ell_\infty} - \rho_\ell)\sin\frac{x}{R} + \frac{\partial}{\partial y}\mu_\ell \frac{\partial u_\ell}{\partial y} , \tag{8.58}$$

$$(\rho c_p)_v (u_v \frac{\partial T_v}{\partial x} + v_v \frac{\partial T_v}{\partial y}) = -\rho_{\ell_\infty} v_{\ell_\infty}^2 \frac{2}{R} \sin(\frac{2x}{R}) u_v + \frac{\partial}{\partial y} k_v \frac{\partial T_v}{\partial y} , \tag{8.59}$$

$$(\rho c_p)_\ell (u_\ell \frac{\partial T_\ell}{\partial x} + v_\ell \frac{\partial T_\ell}{\partial y}) = \frac{\partial}{\partial y} k_\ell \frac{\partial T_\ell}{\partial y} . \tag{8.60}$$

The term $(\partial p/\partial x)u_v$, i.e., the *compressibility work* in (1.83), has been *retained* and a perfect gas behavior is used which gives $\beta_v = 1/T_v$.

The boundary conditions are

$$u_v = v_v = 0 , \quad T = T_{s_b} \quad y = 0 , \tag{8.61}$$

$$u_v = u_\ell , \ \mu_v \frac{\partial u_v}{\partial y} = \mu_\ell \frac{\partial u_\ell}{\partial y} ,$$
$$\dot{m}_{\ell g} = \rho_v (v_v - u_v \frac{d\delta_v}{dx}) = \rho_\ell (v_\ell - u_\ell \frac{d\delta_v}{dx}) , \ T_v = T_\ell = T_s(p_\ell) ,$$
$$-k_v \frac{\partial T_v}{\partial y} + (q_{s_b})_r = -k_\ell \frac{\partial T_\ell}{\partial y} - \rho_v (v_v - u_v \frac{d\delta_v}{dx}) \Delta i_{\ell g} \quad y = \delta_v , \tag{8.62}$$

$$u_\ell = 2 v_{\ell_\infty} \sin\frac{x}{R} , \quad T_\ell = T_{\ell_\infty} \quad y = \delta_\ell . \tag{8.63}$$

The interfacial mass balance equation given in (8.62) is derived by noting that for the vapor-film flow the *increase* in the vapor flow rate is

$$\begin{aligned} \rho_v \frac{d}{dx}\int_0^{\delta_v} u_v dy &= \rho_v u_v|_{\delta_v} \frac{\partial \delta_v}{\partial x} + \int_0^{\delta_v} \frac{\partial u_v}{\partial x} dy \\ &= \rho_v u_v|_{\delta_v} \frac{d\delta_v}{dx} - \rho_v v_v|_{\delta_v} . \end{aligned} \tag{8.64}$$

Similarly, for the liquid boundary-layer flow

$$\rho_\ell \frac{d}{dx}\int_0^{\delta_\ell} u_\ell dy = -\ \rho_\ell u_\ell|_{\delta_v}\frac{d\delta_v}{dx} - \rho_\ell(v_\ell|_{\delta_\ell} - v_\ell|_{\delta_v}) \ . \tag{8.65}$$

The mass balance over the vapor film *and* the liquid boundary layer gives

$$\rho_v \frac{d}{dx}(\int_0^{\delta_v} u_v dy)dx + \rho_\ell\frac{d}{dx}(\int_0^{\delta_\ell} u_\ell dy)dx + \rho_\ell v_\ell|_{\delta_\ell}dx = 0 \ . \tag{8.66}$$

Now substituting (8.64) and (8.65) into this equation, the interfacial mass balance in (8.62) is obtained.

The *wake region*, i.e., $\theta > \theta_s$, has been *modeled* as a *two-dimensional cap* with an effective radius R_w. This radius is chosen with hydrodynamic instability and heat transfer considerations and is given by (Chou and Witte, 1992)

$$R_w^2 = \frac{2(\pi - \theta_s)R(2\pi 3^{1/2})}{\pi[\dfrac{g(\rho_L - \rho_v)}{\sigma} + \dfrac{1}{2R^2}]^{1/2}} \ . \tag{8.67}$$

Only *conduction* heat transfer is allowed in the wake region (both vapor and liquid phases) and the thermal conductivities are assumed constant.

For the wake region, we have the energy equations

$$0 = \frac{1}{r}\frac{\partial}{\partial r}r\frac{\partial T_v}{\partial r} + \frac{\partial^2 T_v}{\partial x^2} \ , \tag{8.68}$$

$$0 = \frac{1}{r}\frac{\partial}{\partial r}r\frac{\partial T_\ell}{\partial r} + \frac{\partial^2 T_\ell}{\partial x^2} \ . \tag{8.69}$$

The boundary conditions are

$$\frac{d\delta_v}{dr} = \frac{\partial T_v}{\partial r} = \frac{\partial T_\ell}{\partial r} = 0 \ , \qquad r = 0 \ , \tag{8.70}$$

$$\delta_v = \delta_v(\theta_s) \ , \ T_v = T_v(\theta_s) \ , \ T_\ell = T_\ell(\theta_s) \qquad r = R_w \ , \tag{8.71}$$

$$T_v = T_{s_b} \qquad x = 0 \ , \tag{8.72}$$

$$T_v = T_\ell = T_s(p_\ell) \ , \ k_v\frac{\partial T_v}{\partial n} = k_\ell\frac{\partial T_\ell}{\partial n} \qquad x = \delta_v(r) \ , \tag{8.73}$$

$$T_\ell = T_{\ell_\infty} \ , \ \frac{\partial T_\ell}{\partial x} = \frac{\partial T_\ell}{\partial r} = 0 \qquad x = \delta_\ell(r) \ , \tag{8.74}$$

where n indicates the normal direction.

Chou and Witte (1992) show that similarity solutions exist for the velocity and temperature fields of the vapor and the liquid boundary-layer equations (8.55) to (8.60). These boundary-layer solutions predict the vapor and liquid boundary-layer *separations* (i.e., the liquid separates *before* the vapor). For the wake region, a numerical integration scheme is used, and depending on the liquid subcooling, $T_{\ell_\infty} - T_s(p_\ell)$, compared to the vapor superheat, $T_{s_b} - T_s$, the *maxima* in the *thermal* buoyancy-layer thicknesses,

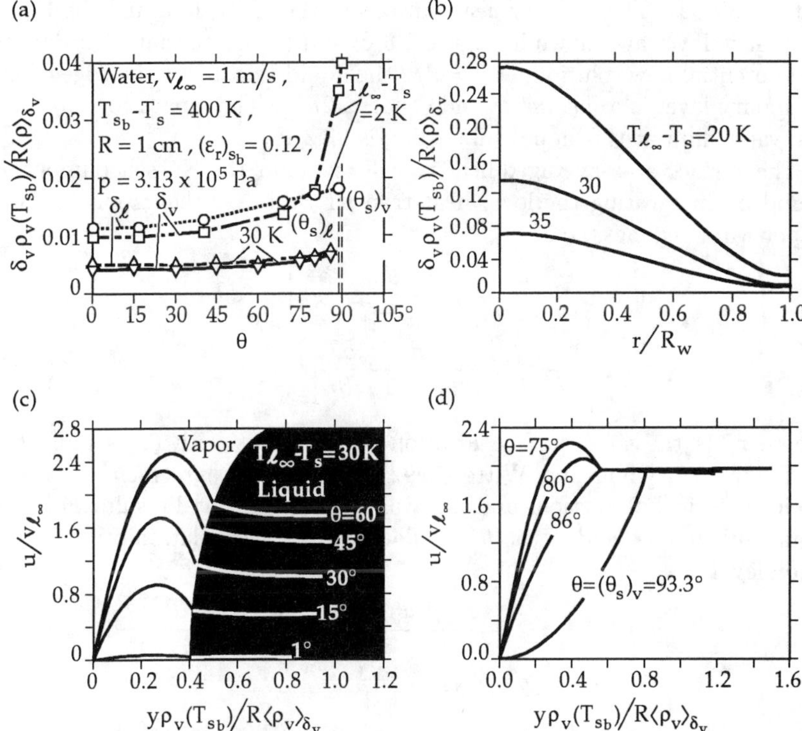

Figure 8.6. (a) Computed angular distribution of the thickness of the vapor film and liquid boundary layer around the heated horizontal cylinder. The separation angles are also shown. (b) Radial distribution of the vapor-film thickness in the wake region. (c) Distribution of the normalized tangential velocity in the vapor film and liquid boundary layer for various polar angles. (d) Same as (c), except near the vapor-film separation angle. (From Chou and Witte, reproduced by permission ©1992 American Society of Mechanical Engineers.)

δ_v and δ_ℓ, change while the profile remains the same. Some of their typical results are shown in Figures 8.6(a) to (d).

The vapor-film and the liquid boundary-layer separations are shown in Figure 8.6(a) for water, $R = 1$ cm, and a large vapor superheat with small liquid subcooling. The Froude number $Fr_d = v_{\ell_\infty}/(g_d)^{1/2} = 2.26$ is in the *forced-flow dominated regime*. The increase in the vapor-film and the liquid boundary layer thickness is initially nearly linear and the liquid separates *before* the vapor (but the difference in the separation angles is small). The results show that as the subcooling *increases*, the vapor-film thickness *decreases*. The distribution of the vapor-film thickness in the wake region is shown in Figure 8.6(b) for three different values of the liquid subcooling. Similar to the vapor-film regime, as the liquid subcooling increases, the vapor-film thickness *decreases*, and therefore, the local heat transfer rate *increases*. The velocity distributions in the vapor film and the liquid boundary layer are shown in Figure 8.6(c) and (d). Note that according to the potential flow solution, at $\theta = \pi/2$ the liquid velocity at the edge of the boundary layer is a *maximum* and equal to $2u_{\ell_\infty}$. The separation angle for the vapor film at this liquid subcooling is 93.3°.

The surface area-averaged heat transfer rate, or the Nusselt number, is found by integrating the local heat transfer rate over the vapor-film and vapor-wake regimes, i.e.,

$$\langle Nu_d \rangle_{A_{sv}} = \frac{1}{\theta_s} \int_0^{\theta} -\frac{d}{T_{s_b} - T_s} \left. \frac{\partial T}{\partial y} \right|_{y=0} d\theta + \frac{1}{\pi - \theta_s} \int_0^{\pi} -\frac{d}{T_{s_b} - T_s} \left. \frac{\partial T}{\partial r} \right|_{r_s} d\theta \,, \tag{8.75}$$

where r_s is the solid surface location which varies with θ, as indicated in Figure 8.5. Chou and Witte (1992) report good agreements between their numerical predictions and the available experimental results. For *zero* liquid subcooling, and in the *forced-flow regime*, a correlation proposed by Bromley, i.e.,

$$\begin{aligned} \langle (Nu_d)_c \rangle_{A_{sv}} &= 2.7 \{ \frac{v_{\ell_\infty} \rho_v d[\Delta i_{\ell g} + \frac{1}{2} c_{p_v} (T_{s_b} - T_s)]}{k_v (T_{s_b} - T_s)} \}^{\frac{1}{2}} \\ &= 2.7 Pe_d^{\frac{1}{2}} Ja_{sh}^{-\frac{1}{2}} \,, \end{aligned}$$

$$Pe_d = \frac{v_{\ell_\infty} d}{\alpha_v} \,, \quad Ja_{sh} = \frac{c_{p_v} (T_{s_b} - T_s)}{[\Delta i_{\ell g} + \frac{1}{2} c_{p_v} (T_{s_b} - T_s)]} \,,$$

$$(\epsilon_r)_{s_b} = 0 \,, \quad T_{\ell_\infty} = T_s \,, \quad Fr_d = \frac{v_{\ell_\infty}}{(gd)^{1/2}} > 2 \,, \tag{8.76}$$

predicts the experimental results accurately. The heat of vaporization *includes* the vapor superheat to the average vapor-film temperature $(T_{s_b} - T_s)/2$. For the *buoyancy-forced flow regime* with subcooling and radiation, a *large* number of parameters are specified (Chou and Witte, 199; Liu et al., 1992). These include the liquid and vapor *Prandtl numbers* Pr_ℓ and Pr_v, respectively, the *Froude number* Fr, the liquid and vapor *Jakob numbers*, Ja_{sc} and Ja_{sh}, respectively, the *ratio* of the product of the density and the viscosity $(\rho\mu)_v/(\rho\mu)_\ell$, the vapor *Grashof number* Gr_d, and the surface emissivity $(\epsilon_r)_{s_b}$. Other than Fr_d, which signifies the forced flow, the remainder of these parameters also appeared in the analysis of the buoyant flow, e.g., in (8.27).

8.1.3 Dispersed Droplets

In Sections 8.1.1 and 8.1.2, we considered vapor flow adjacent to a solid surface *submerged* in an *unbounded* liquid (with the solid temperature above the minimum vapor-film regime limit). Outside these vapor films, the far-field, *single-phase* liquid was saturated or subcooled and quiescent or moving. In *internal* (i.e., *bounded*) *flows*, these far-field conditions become the *entrance* conditions. As an initially subcooled, single-phase liquid flows in a *heated* tube and undergoes phase change (evaporation), the generated vapor (if not collapsed) flows *downstream*. Then the liquid-vapor interface will *not* be continuous and simple. Figure 8.7 depicts an *internal upward* flow of an *initially* (entrance condition) *subcooled liquid* through a vertical tube heated with a *constant* heat flux q_{s_b}. Right at the entrance to the tube, the solid surface temperature T_{s_b} is nearly the inlet subcooled liquid temperature T_{ℓ_i}, because no thermal boundary layer is yet developed. With the increase in the distance from the entrance x, the surface temperature *increases* and when the surface bubble-nucleation inception temperature is reached, bubble nucleation begins on the surface. These bubbles collapse in the subcooled liquid and heat the liquid until some distance downstream where the liquid subcooling diminishes and the bubbles no longer collapse and begin to coalesce. As the amount of vapor in the cross-section of the tube *increases* with the distance x, i.e., the *void* (or *vapor*) volume *fraction increases*, the liquid will flow as a liquid film and as dispersed droplets. Finally, the liquid film disappears and only dispersed droplets flow, which then completely evaporate. Then the superheated vapor flow occurs.

There are many regimes of two-phase flow and heat transfer in this upward liquid flow and evaporation. A simple classification identifying *seven regimes* is shown in Figure 8.7 (Collier, 1972; Carey, 1992). Starting from the entrance to the tube, after the *regime* of the *subcooled-liquid* flow with *solid-surface bubble nucleation*, the *regime* of the *bubbly flow*, which was discussed in Section 1.6.1(A)(ii) and shown in Figure 1.8, with continued solid-surface bubble nucleation, is encountered. As the void (or vapor) volume fraction *increases*, the *plug- and churn-flow regimes* with the continued solid-surface bubble nucleation, are observed. Further downstream, with the liquid fraction *decreasing*, the *thin liquid film* (in the *annular-flow regime*) evaporation occurs and the solid-surface bubble nucleation is *suppressed*. As a result of this change in the mechanism of vapor formation, the surface temperature decreases. The surface temperature *increases* once the film totally evaporates. In the *dispersed droplet-superheated vapor* flow regime the heat of evaporation of the droplets prevents the occurrence of a large temperature gradient in the solid-surface temperature which develops in the *superheated-vapor flow regime*. The anticipated surface temperature distribution along the tube is shown in Figure 8.7. Due to the decrease in the local pressure along the tube, the saturation temperature also decreases. The beginning of the dispersed-droplet regime is generally marked by the minimum vapor-film regime superheat $(T_{s_b} - T_s)_{\min}$.

In relation to the vapor-film flow, the flow and heat transfer in the superheated vapor-dispersed droplet flow regime is considered below. This allows for the introduction of *dispersed-liquid phase* as an extension of the continuous-liquid phase treatment discussed in Sections 8.1.1 and 8.1.2,

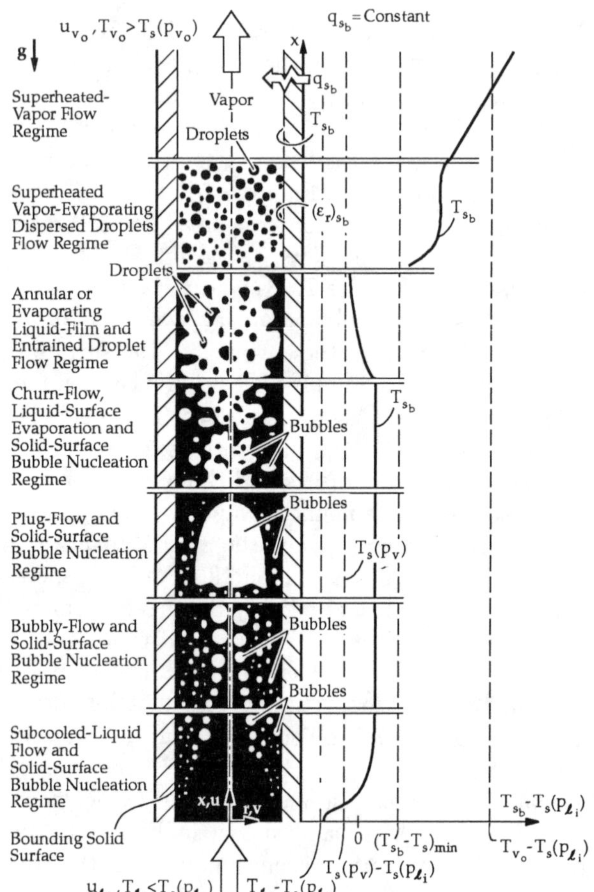

Figure 8.7. A rendering of the upward liquid and vapor flow in a tube with a constant heat flux and a surface temperature above the saturation temperature. The inlet liquid is subcooled. The tube surface temperature distribution, as well as the various regimes encountered along the tube axis, is also shown.

and allows for the review of the *thermal nonequilibrium* treatment of this particular solid–liquid–gas system.

First, the physical characteristics of the dispersed droplets-superheated vapor regime are examined. Then the models used in the three-medium treatment of the heat transfer are reviewed. The treatments of the droplets as a *continuum* is discussed first, followed by the *discrete*, i.e., *droplet tracking, Lagrangian* treatments.

(A) Dispersed Droplets-Superheated Vapor Regime

In this regime of internal, two-phase flow and heat transfer with evaporation, the droplets are generally spherical and dispersed in the vapor phase.

The droplet size and concentration are generally small such that the *void fraction* ϵ (which is the *volume* fraction of the *continuous phase*) is greater than 0.95. In the *thermodynamic* consideration, the *thermodynamic quality* x_t, which is *mass flow fraction* of the vapor, is used. The relationship between ϵ and x_t for a one-dimensional flow is

$$x_t = \frac{\dot{m}_v}{\dot{m}} = \frac{\dot{m}_v}{\dot{m}_v + \dot{m}_d} = \frac{\rho_v \langle u \rangle^v \epsilon}{\rho_v \langle u \rangle^v \epsilon + \rho_d \langle u \rangle^d (1-\epsilon)} = \frac{1}{1 + \dfrac{\rho_d}{\rho_v} \dfrac{\langle u \rangle^d}{\langle u \rangle^v} \dfrac{1-\epsilon}{\epsilon}} \tag{8.77}$$

or

$$\epsilon = \frac{1}{\dfrac{1 - x_t}{x_t \dfrac{\rho_d}{\rho_v} \dfrac{\langle u \rangle^d}{\langle u \rangle^v}} + 1} . \tag{8.78}$$

Then for large values of ρ_d/ρ_v, which is the case when the pressure is much less than that at the thermodynamic critical state, x_t is much *smaller* than ϵ. In the dispersed-droplet regime considered, typically $x_t \geq 0.20$ (Webb and Chen, 1982) or even lower (Andreari and Yadigaroglu, 1992). Since the droplet and the vapor are *not* in thermal equilibrium, the *actual* (i.e., nonequilibrium) quality (i.e., the actual vapor mass flow fraction) is *lower* than the *equilibrium* quality calculated based on local thermal equilibrium. This is because some of the heat transferred from the solid surface results in the vapor superheating instead of the droplet evaporation.

The droplet and vapor flow in a circular tube of diameter D placed vertically with the vapor-phase flow against the gravity vector is depicted in Figure 8.8. The droplets are formed in the upstream churn- and annular-flow regimes and are nearly at the saturation temperature $T_d = T_s(p_v)$, where p_v is the *local* vapor pressure (however, the pressure drop in the dispersed droplet regime is not significant). These droplets of average diameter d can *break up* or *coalesce* as they *travel* and can *impinge* on the solid bounding surface (i.e., the tube surface). When droplets collide with the surface, depending on the *inertial force*, the *angle of impact*, and the *surface superheat* $T_{s_b} - T_s$, they can *wet* the surface or remain *separated* by a vapor film. The droplet impingement on heated surfaces will be discussed in Section 8.6. The specification of the average droplet diameter (or a diameter distribution when the *variations* are significant) and their *number concentration* n_d, at the entrance (i.e., onset) to the dispersed-droplet regime, are made based on the *hydrodynamic instability* analysis and also based on *empiricism* (Yoder and Rohsenow, 1983). The droplet diameter at the onset of the regime is of the order of 100 μm.

The heat transfer from the *bounding surface* is by *convection* to the superheat vapor, *radiation* to saturated droplets, and by a combination of modes in the droplet-surface impact. There is also a *convective* heat

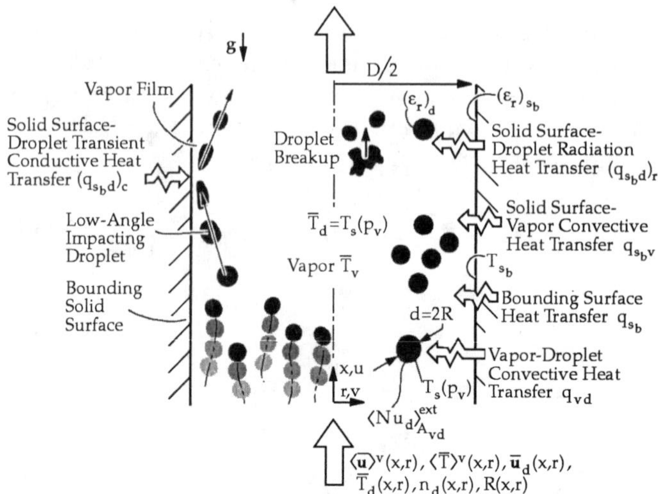

Figure 8.8. The droplet dynamics and the various heat transfer mechanisms in the dispersed droplets-superheated vapor regime in a tube subject to a constant heat flux.

transfer between the superheated vapor and the droplets. These are shown in Figure 8.8. The droplets evaporate at a rate determined by the *direct* heat transfer from the surface and that *indirectly* through the vapor. The vapor flow is *turbulent* and this turbulence is affected by the droplets, similar to the dispersed *solid particles* flow and heat transfer discussed in Sections 5.4, 7.1, and 7.2.

The analysis leading to the determination of the axial distributions of the cross-sectional area-averaged evaporation rate $\langle \dot{n}_v \rangle_A(x)$ and the surface temperature $T_{s_b}(x)$, can be made with various extents of vigor. Most of the existing analyses assume that the direct droplet-surface heat transfer, i.e., two of the three components of the surface heat transfer, are *negligible*. However, as indicated by Chen and Costigan (1992), the droplet impact heat transfer may *not* be negligible in the *entrance* (i.e., onset) region of the dispersed-droplet regime, where the droplets have a significant *radial* velocity component. The presence of the gradient of the axial velocity adjacent to the surface also causes a droplet *migration* towards the surface. This is caused by a radial force proportional to the square root of the absolute value of this gradient, and is called the *Saffman force* (Osiptsov and Shapiro, 1993). Also in most of these analyses the vapor-droplet convective heat transfer rate q_{vd} is treated with a *prescribed* Nusselt number $\langle Nu_d \rangle_{A_{vd}}$. This is done using the available results such as those reviewed in Section 6.4, including the interdroplet convective interactions. Then the surface-vapor heat transfer, including the effect of the presence of the droplets and their distribution on the *turbulence* in the vapor flow, is the *determining* heat transfer mode. In the existing *one-dimensional* (*aver-*

aged in the *radial* direction) treatments (e.g., Yoder and Rohsenow, 1983; Varone and Rohsenow, 1986), this bounding surface-vapor heat transfer rate $q_{s_\ell v}$ is also prescribed with a Nusselt number Nu_D. As expected, this Nusselt number depends on the *local* droplet diameter and concentration distribution, and because of the significant radial and axial variations of the vapor superheat, the *variations* of the thermophysical properties should also be included. These resulting correlations are complex and not of general use (Yadigaroglu and Bensalem, 1987).

In the *two-dimensional, axisymmetric* treatments of the vapor flow and heat transfer, the droplet flow and heat transfer can be described using the *continuum* models discussed in Sections 3.3 and 5.4, or by using the *Lagrangian* (or *discrete*) models with droplet tracking and *ensemble averaging* (Andreani and Yadigaroglu, 1992). The droplet tracking facilitates the inclusion of the droplet-breakup and the surface droplet impingement. In the following, first the continuum and then the discrete (i.e., Lagrangian) models, and some of their results, are discussed.

(B) Continuum Models for Droplets

Detailed *hydrodynamic, thermodynamic*, and *heat transfer* treatments of the dispersed droplet-superheat vapor flow is expected to be complex. The presence of the solid bounding surface influences the droplet concentration and the transport in general. As a three-medium, two-phase flow, a significant number of experimental and analytical studies of this problem have been made. As an illustrative example, some features of the models used for the analysis of this three-medium problem are examined below. The improvement in these models continues, and the discrete-droplet model described in Section (D) has also been used.

The continuum (or Eulerian) description of the droplets assumes that a droplet is *present* at *any* point in the spatial domain of interest. As compared to the Lagrangian (or discrete) description which tracks and determines the varying position of the individual droplets. When ensemble averages are taken over many such trajectories, and in the limit of the number of these realizations tending to infinity, the continuum and the discrete descriptions will give similar results, if both models contain the same description of the various mechanisms. However, the continuum treatment of the droplet breakup and surface impingement and reflection requires specification of some *probability* distributions which are not readily available.

The vapor flow is *turbulent* and at the entrance to the dispersed-droplet regime the droplet size is of the order of 100 μm. Therefore, the droplets are generally considered to be *massive* particles. As was mentioned in Section 5.4.2, for massive droplets the hydrodynamic relaxation time τ_d, defined by (5.145), is large, i.e., the characteristic period of droplet fluctuations is larger than the roll-over time of a typical turbulent eddy in the vapor phase. Then the droplet will not follow the vapor velocity fluctuation.

As droplets evaporate and their diameter *decreases*, they can follow the turbulent fluctuations. The droplet concentration is generally considered *dilute*, because the vapor-volume fraction ϵ is nearly unity (i.e., $\epsilon \geq 0.95$).

The general, *time-dependent, mean-component three-dimensional* conservation equations for the *vapor phase* (local, *representative* vapor-phase volume averaged) and for the *droplets* (local, finite droplet-volume averaged) are written by considering turbulent flows. The *mean* mass, momentum, and energy equations are the appropriate and approximate forms of (3.117), (3.118), and (3.112), respectively. These are written using the vaporside Nusselt number for the vapor-droplet interfacial heat transfer $\langle Nu_d\rangle^{\rm ext}_{A_{vd}}$ discussed in Section 6.4, and the interfacial drag force for *fluid particles* discussed in Section 6.1.2(B) along with the definition of the hydrodynamic relaxation time. The radiation heat transfer from the bounding surface to the droplet is given as a source term in the droplet energy equation, where $(q_{s_b d})_r$ is in terms of the *droplet* unit surface area. The conservation equations are

$$\frac{\partial}{\partial t}\overline{\epsilon}\rho_v + \nabla\cdot\rho_v\langle\overline{\mathbf{u}}\rangle^v = \langle\dot{n}_v\rangle^v\,, \tag{8.79}$$

$$\frac{\partial}{\partial t}(1-\overline{\epsilon})\rho_d + \nabla\cdot(1-\overline{\epsilon})\overline{\mathbf{u}}_d = -\langle\dot{n}_v\rangle^v\,, \tag{8.80}$$

$$\begin{aligned}\frac{D}{Dt}\overline{\epsilon}\rho_v\langle\overline{\mathbf{u}}\rangle^v &= -\nabla\overline{\epsilon}\langle p\rangle^v + \nabla\cdot\epsilon(\langle\overline{\mathbf{S}}\rangle^v + \langle\overline{\mathbf{S}_t}\rangle^v) + \\ &\quad \overline{\epsilon}\rho_v\mathbf{g} + \frac{\rho_d}{\tau_d}(1-\overline{\epsilon})(\langle\overline{\mathbf{u}}\rangle^v - \overline{\mathbf{u}}_d)\,,\end{aligned} \tag{8.81}$$

$$\frac{D}{Dt}(1-\overline{\epsilon})\rho_d\overline{\mathbf{u}}_d = (1-\overline{\epsilon})\rho_d\mathbf{g} - \frac{\rho_d}{\tau_d}(1-\overline{\epsilon})(\langle\overline{\mathbf{u}}\rangle^v - \overline{\mathbf{u}}_d)\,, \tag{8.82}$$

$$\begin{aligned}\frac{D}{Dt}\overline{\epsilon}(\rho c_p)_v\langle\overline{T}\rangle^v &= \nabla\cdot[\mathbf{K}_e + \epsilon(\rho c_p)_v\mathbf{D}^d_v]\cdot\nabla\langle\overline{T}\rangle^v + \\ &\quad \frac{6(1-\overline{\epsilon})}{d}\langle Nu_d\rangle^{\rm ext}_{vd}\frac{k_v}{d}(\overline{T}_d - \langle\overline{T}\rangle^v) + \langle\dot{n}_v\rangle^v\overline{i}_v\,,\end{aligned} \tag{8.83}$$

$$\frac{D}{Dt}(1-\overline{\epsilon})(\rho c_p)_d\overline{T}_d = -\langle\dot{n}_v\rangle^v\overline{i}_d + \frac{6(1-\overline{\epsilon})}{d}(q_{s_b d})_r\,. \tag{8.84}$$

Various other models, or modifications of the above models, have been proposed (e.g., Rane and Yao, 1981; Kirillov et al., 1987; Chen and Costigan, 1992; Osiptsov and Shapiro, 1993). The vapor-droplet interfacial mass and force balance have been used in the continuity and momentum equations. The interfacial heat balance gives

$$\langle\dot{n}_v\rangle^v\Delta i_{\ell g} = \frac{6(1-\overline{\epsilon})}{d}\langle Nu_d\rangle^{\rm ext}_{A_{vd}}\frac{k_v}{d}(\langle\overline{T}\rangle^v - \overline{T}_d)\,. \tag{8.85}$$

For *locally monosize* droplets of diameter d, the number concentration n_d is related to the vapor-volume fraction ϵ and d through

$$n_d = \frac{6(1-\epsilon)}{\pi d^3} \quad \text{or} \quad \epsilon = 1 - \frac{\dot{n}_d \pi d^3}{6} . \tag{8.86}$$

The *droplet number* concentration *conservation* equation is given by (6.54) and is

$$\frac{\partial n_d}{\partial t} + \nabla \cdot \overline{\mathbf{u}}_d n_d = \dot{n}_d . \tag{8.87}$$

The source/sink term $\dot{n}_d$ can be due to breakup/coalescence of droplets. When breakup/coalescence occurs, the *polysize* version of (8.87), which allows for the integration (i.e., *ensemble average*) over the a known *continuous* or *discrete* diameter distribution, is used. A more general treatment of the vapor-droplet interfacial geometry and transport is given by Ishii and Mishima (1984).

As was mentioned, the droplet temperature is taken as the saturation temperature, i.e., $\overline{T}_d = T_s(p_v)$, and since the pressure drop in the dispersed-droplet regime is negligibly small, the droplet temperature is taken as *uniform* in this regime. Because of its relevance to the heat dissipation from radiative bounding solid surfaces, the *constant heat flux* boundary condition is generally used in the analyses. The solid surface emissivity is $(\epsilon_r)_{s_b}$ and the surface temperature which changes throughout the regime is T_{s_b}. The distribution $T_{s_b} = T_{s_b}(x)$, where $x = 0$ marks the *axial entrance* or *inlet* (to the dispersed-droplet regime) location, depends on the *entrance* conditions, $\langle u_v \rangle_i^v$, $\langle \overline{T} \rangle_i^v$, $\langle \overline{u} \rangle_{d_i}$, $\langle \overline{T} \rangle_{d_i}, n_{d_i}, d_i$, and $T_{s_{b_i}}$, and the surface heat flux q_{s_b}, as well as on the thermophysical properties.

The vapor-phase boundary conditions at $r = D/2$ and $r = 0$ are

$$\langle \overline{\mathbf{u}} \rangle^v = 0 \, , \quad -k \frac{\partial \langle \overline{T} \rangle^v}{\partial r} = (q_{s_b v})_c \qquad r = D/2 \, , \tag{8.88}$$

$$\frac{\partial \langle \overline{u} \rangle^v}{\partial r} = \langle \overline{v} \rangle^v = \frac{\partial \langle \overline{T} \rangle^v}{\partial r} = 0 \qquad r = 0 \, . \tag{8.89}$$

The vapor-phase turbulent shear stress tensor $\langle \mathbf{S}_t \rangle^v$, which in principle is influenced by the presence of the droplets (Varone and Rohsenow, 1986), the vapor-phase effective molecular conductivity tensor $\mathbf{K}_e$, and the turbulent diffusivity (or dispersion) tensor $\mathbf{D}_v^d$ need to be modeled. As was discussed in connection with the *solid* particles in Section 5.4, with no solid surface present, and in Section 7.1, with a bounding solid surface present, the presence of the particles does influence the vapor-phase transport, including the turbulent transport. However, presently this is not thoroughly analyzed, and generally turbulent vapor transport *without* any droplet effects is used. This turbulent flow is assumed to be *hydrodynamically* fully developed, i.e., the radial distribution of the mean, axial velocity $\langle \overline{u} \rangle^v$ is assumed to remain unchanged as x increases. The turbulence models used are similar to those discussed in convection with the single-phase channel flows in Section 4.4.3(B), where the effect of the presence of the surface is modeled using the van Driest mixing-length model. A turbulent Prandtl number Pr_t is also used for the evaluation of the eddy diffusivity (here referred to as dispersion). The temperature distribution is assumed to be turbulent, *two-dimensional*, and not fully developed. Rane and Yao (1981), Webb and Chen (1982), and Kirillov et al. (1987) use different eddy viscosity and diffusivity models.

The entrance droplet diameter d_i and number density n_{d_i} are determined from the hydrodynamic instabilities of the annular liquid-film flow which results in the *entrained* droplets, and these are added to the droplets arriving directly from the churn-flow regime. Yoder and Rohsenow (1983) review the available correlations, and Pilch and Erdman (1987) give a criterion for the maximum stable droplet diameter, which is

$$d = We_{cr}\frac{\sigma}{\rho(\langle\overline{u}\rangle^v - \overline{u}_d)^2}(1 - \frac{|u_c|}{|\langle\overline{u}\rangle^v - \overline{u}_d|}) , \tag{8.90}$$

where the critical *Weber number* is correlated as

$$We_{cr} = \frac{\rho(\langle\overline{u}\rangle^v - \overline{u}_d)^2 d}{\sigma} = 12\{1 + 1.077[\frac{\mu_d}{(\rho_d d\sigma)^{1/2}}]^{1.6}\} \tag{8.91}$$

and u_c is the *relative velocity* of the droplet-fragment cloud, which is *smaller* than $|\langle u\rangle^v - \overline{u}^d|$ and is determined experimentally. The ratio of the viscous force to the surface tension force $\mu_d/(\rho_d d\sigma)^{1/2}$, appearing in (8.91), is called the *Ohnesorge number*.

(C) Results of Continuum Models

The prediction of the axial surface temperature distribution $T_{s_b} = T_{s_b}(x)$ has been made using various models. Because of a *large* number of phenomena that require modeling—e.g., vapor-phase turbulence, droplet diameter and diameter distribution, radial forces acting on the droplets—many droplet continuum models have been able to predict this surface temperature distribution with good accuracy (Rane and Yao, 1981; Webb and Chen, 1982; Kirillov et al. 1987; Osiptsov and Shapiro, 1993).

A typical prediction of the surface temperature and the vapor temperature distributions along the dispersed-droplet regime from Webb and Chenn (1982) is shown in Figure 8.9. Their experimental results are also shown. The solid surface T_{s_b} and the area-averaged mean vapor temperature $\langle\langle\overline{T}\rangle^v\rangle_A$ are both assumed to be equal to T_s at the entrance. The entrance quality x_{t_i} for the water experiment used for comparison is 0.52. Note that the surface temperature *rises* very rapidly, and further downstream, as the vapor temperature increases and the evaporation of the droplets becomes significant, the surface temperature does not increase as rapidly. The radial distribution of the vapor velocity and droplet concentration are examined by Kirillov et al. (1987).

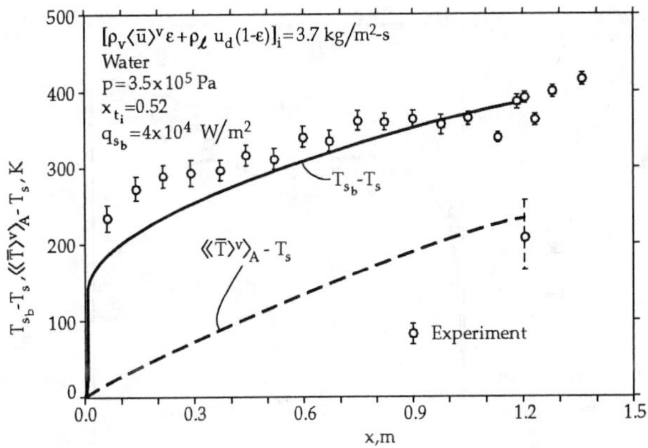

Figure 8.9. Typical predictions and measurements of the surface and vapor temperature distributions along the axis of a heated tube at constant heat flux. The results are for the dispersed droplets-superheated vapor regime. (From Webb and Chen, reproduced by permission ©1982 Pergamon Press.)

(D) Droplet-Tracking Discrete Models

The droplet *breakup* and surface *impingement* can be readily implemented by using a *Lagrangian* description of the droplet motion and heat transfer (including evaporation) in a Eulerian mesh. This is called the particle-mesh (or Lagrangian-Eulerian) formulation and is a method of computer simulation using particles, and a general discussion is given by Hockney and Eastwood (1989). By *tracking* many droplets and by *ensemble averaging* over these droplet trajectories (or *realizations*), a *statistical* description of the droplet transport can be made. This *discrete simulation* and *statistical-averaging* (i.e., a Monte Carlo method) has been used by both Andreani and Yadigaroglu (1992) and by Michaelides et al. (1992).

The entrance velocity of the droplets is specified with axial and radial components, i.e., u_{d_i} and v_{d_i}, and in the simulations of Andreani and Yadigaroglu (1992) a *radial droplet* thrust force $F_{t,r}$ is specified in addition to the radial drag force $F_{d,r}$. Then the radial position of the droplet $\overline{r}_d$ is determined by the time integration of the radial velocity. The droplet can *impinge* on the wall and *reflect*. The *breakup* of the droplets is also modeled as discussed in Section (B). The droplet radial position, radial velocity, and the rate of change of the droplet radius are given by

$$\frac{d}{dt}\overline{r}_d = \overline{v}_d \,, \tag{8.92}$$

$$\rho_d \frac{4}{3}\pi R^3 \frac{d}{dt}\overline{v}_d = F_{d,r} + F_{t,r} \,, \tag{8.93}$$

$$\rho_d \Delta i_{\ell g} \frac{d}{dt} R = (q_{s_b d})_r + (q_{s_b d})_c + q_{vd} \,. \tag{8.94}$$

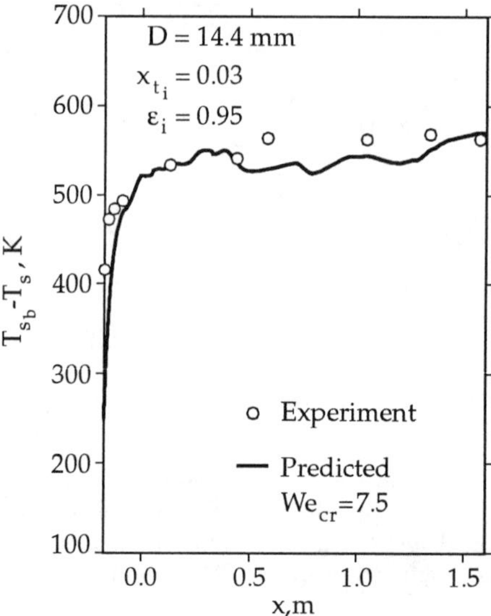

Figure 8.10. Axial distribution of the tube surface temperature in the dispersed droplets-superheated vapor regime with a constant heat flux. (From Andreani and Yadigaroglu, reproduced by permission ©1992 Taylor and Francis.)

The axial droplet velocity can be written in this Lagrangian form and the vapor-phase conservation equations are written in the Eulerian forms given in Section (B). In the simulations of Michaelides et al. (1992), the vapor-phase turbulence is also modeled with *finite*-size eddies, and as the droplet size *decreases* these eddies can fluctuate the radial and axial position of the droplets.

Some typical results of Andreani and Yadigaroglu (1992) are shown in Figure 8.10. The distribution of the surface temperature is shown in the dispersed-droplet regime. The results are for water, entrance quality of 0.03, and entrance void fraction of 0.95. The ensemble-averaged results, with the assumptions made about the droplet-surface impingement and heat transfer given by Andreani and Yadigaroglu (1992) are in good agreement with the available experiment results.

8.2 Liquid Films

Consider a *binary* mixture of species A and B with a liquid-gas *equilibrium* phase diagram similar to that shown in Figure 1.6(b), i.e., *miscible* liquid phases of species A and B with the boiling point of pure species A, $T_{b,A}$ being *lower* than that for species B, $T_{b,B}$. As an example of a three-medium treatment with a liquid film flow, we consider a bounding solid surface with a temperature lower than $T_{b,B}$, but higher than $T_{b,A}$ (and also higher than the *freezing* temperature of B, $T_{m,B}$. Then *condensation* of species B on this solid surface may occur and here it is assumed that a *condensate* film is already *formed* and *flows* under the action of *gravity* and/or an external (i.e., imposed) *pressure* gradient. The phenomenon of *surface* droplet nucleation in condensation on cooled solid surfaces will be discussed in Section 8.6.

The liquid (condensate) *film* flowing under the *negative* phase-density buoyancy or an external pressure gradient, with the *vapor* flowing *concurrent* or *counter* to this flow, will become *wavy* and unstable. The *ideal, laminar nonwavy film condensation* has been analyzed extensively and reviews are given by Sparrow and Gregg (1959), Shekriladze and Gomelauri (1966), Merte (1973), Rose (1988), Fujii (1991), and Carey (1992). The formation and evolution of the *liquid-gas interfacial waves*, in film condensation, have also been examined (including their effects on the local condensation rate), and reviews are given by Kutateladze (1982) and Stuhlträger et al. (1993). Because of the *strong coupling* between *thermodynamics, hydrodynamics*, and *heat* and *mass transfer* in the film condensation, many interesting and practical combinations of the *bounding surface geometry* (and position with respect to the gravity vector), *gas mixture*, and *external pressure gradient* have been studied. Here as an illustrative example the, film condensation adjacent to a *planar, vertical surface* (extension to small tilt angles can be made by choosing the component of the gravity vector along the surface) is considered in Section 8.2.1. The *similarity solutions* for the phase-density *buoyant* (*negative*) flow of the *nonwavy* film is reviewed. Then the interfacial waves are discussed in Section 8.2.2.

8.2.1 Negatively Buoyant, Nonwavy Film Flow

The ideal, laminar film condensation on a planar surface, with a small tilt angle θ from the direction of the gravity vector, is rendered in Figure 8.11. The temperature of the bounding surface T_{s_b} is assumed to be uniform, and therefore, the solid–liquid interfacial heat flux q_{s_b} is *nonuniform* because the liquid film thickness δ_ℓ is nonuniform and the heat transfer *across* the liquid film is dominated by conduction. Due to the much larger liquid density assumed $\rho_\ell \gg \rho_g$, which is valid when the pressure and temperature are much below the mixture critical pressure shown in Figure 1.6(a), the condensate flows due to the *negative phase-density buoyancy*. The gas phase can have a far-field velocity u_{g_∞} which can be concurrent or counter to the condensate flow. Since the *gas* temperature is lower at the gas-liquid interface, compared to that far from the interface T_{g_∞}, then the *thermobuoyancy* of the gas-phase is also *negative*. The gas-phase diffusobuoyancy depends on the species distribution and the ratio of their molecular weights. For a binary system with the species B having a higher boiling temperature, the concentration of species A is *lower* in the conden-

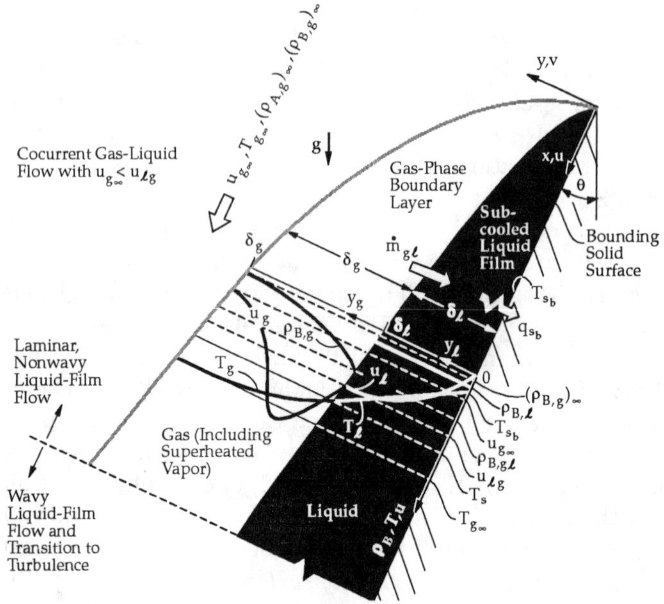

Figure 8.11. Nearly downward liquid-film (condensate) flow in a nearly downward flowing ambient gas. The temperature, velocity, and species concentration distributions across the liquid-film layer and the gas boudary layer are also shown.

sate, and therefore, there is an *accumulation* of species A at the gas-liquid interface, i.e., $\rho_{A,g}(y = \delta_\ell) > (\rho_{A,g})_\infty$.

(A) Formulation

The combined thermo- and diffusobuoyancy of the gas phase can be written assuming an ideal-gas behavior. The density of the gas is

$$\frac{\rho_g}{\rho_{g\infty}} = \frac{T_{g\infty}}{T_g} \frac{M_A - (M_A - M_B)\dfrac{\rho_{A,g}}{\rho_g}}{M_A - (M_A - M_B)(\dfrac{\rho_{A,g}}{\rho_g})_\infty} . \tag{8.95}$$

Then the thermo- and diffusobuoyancy source term in the gas momentum equation becomes

$$1 - \frac{\rho_{g\infty}}{\rho_g} = a_A[\frac{\rho_{A,g}}{\rho_g} - (\frac{\rho_{A,g}}{\rho_g})_\infty] + \frac{T_{g\infty} - T_g}{T_{g\infty}} - \frac{a_A}{T_{g\infty}}[\frac{\rho_{A,g}}{\rho_g} - (\frac{\rho_{A,g}}{\rho_g})_\infty](T_{g\infty} - T_g) , \tag{8.96}$$

where

$$a_A = \frac{M_A - M_B}{M_A - (M_A - M_B)\dfrac{(\rho_{A,g})_\infty}{\rho_{g\infty}}} . \tag{8.97}$$

Since the $T_g \le T_{g\infty}$ and $\rho_{A,g} \ge (\rho_{A,g})_\infty$ in the gas-phase boundary layer, when $M_A > M_B$ the diffusobuoyancy also results in a *negative* buoyancy. This concurrent, downward flow of the gas and liquid is depicted in Figure 8.11. The gas-phase free-stream velocity $u_{g\infty}$ *can* be *larger* than the gas-liquid phase interfacial velocity u_{ℓ_g}, and therefore, the gas exerts an interfacial shear stress which *pulls* the film downward. When $u_{g\infty} < u_{\ell g}$, the downward motion of the liquid film is *hindered* and this is also depicted in Figure 8.11.

A rather complete review and description of the laminar film condensation adjacent to a vertical surface is given by Fujii (1991), and as an illustrative example, the *similarity* solution for the case of buoyant flow of (phase-density buoyancy of the liquid phase and thermo- and diffuso-buoyancy of the gas phase) film condensation is reviewed below.

(B) Similarity Solution

The two-dimensional, steady boundary-layer species (binary mixture), overall mass, momentum, and energy conservation equations for the liquid and gas phases and for $\theta = 0$ are (Fujii, 1991)

$$u_\ell \frac{\partial}{\partial x}\frac{\rho_{A,\ell}}{\rho_\ell} + v_\ell \frac{\partial}{\partial y}\frac{\rho_{A,\ell}}{\rho_\ell} = D_\ell \frac{\partial^2}{\partial y^2}\frac{\rho_{A,\ell}}{\rho_\ell} \quad 0 \le y \le \delta_\ell , \tag{8.98}$$

$$u_g \frac{\partial}{\partial x}\frac{\rho_{A,g}}{\rho_g} + v_g \frac{\partial}{\partial y_g}\frac{\rho_{A,g}}{\rho_g} = D_g \frac{\partial^2}{\partial {y_g}^2}\frac{\rho_{A,g}}{\rho_g} \quad 0 \le y_g \le \delta_g , \tag{8.99}$$

$$\frac{\partial u_\ell}{\partial x} + \frac{\partial v_\ell}{\partial y} = 0 \quad 0 \le y \le \delta_\ell , \tag{8.100}$$

$$\frac{\partial u_g}{\partial x} + \frac{\partial v_g}{\partial y_g} = 0 \quad 0 \le y_g \le \delta_g , \tag{8.101}$$

$$u_\ell \frac{\partial u_\ell}{\partial x} + v_\ell \frac{\partial u_\ell}{\partial y} = -\frac{1}{\rho_\ell}\frac{\partial p_\ell}{\partial x} + \nu_\ell \frac{\partial^2 u_\ell}{\partial y^2} + g \quad 0 \le y \le \delta_\ell , \tag{8.102}$$

$$u_g \frac{\partial u_g}{\partial x} + v_g \frac{\partial u_g}{\partial y_g} = -\frac{1}{\rho_g}\frac{\partial p_g}{\partial x} + \nu_g \frac{\partial^2 u_g}{\partial {y_g}^2} + g\left(1 - \frac{\rho_{g,\infty}}{\rho_g}\right) \quad 0 \le y_g \le \delta_g , \tag{8.103}$$

$$u_\ell \frac{\partial T_\ell}{\partial x} + v_\ell \frac{\partial T_\ell}{\partial y} = \alpha_\ell \frac{\partial^2 T_\ell}{\partial y^2} \quad 0 \le y \le \delta_\ell , \tag{8.104}$$

$$u_g \frac{\partial T_g}{\partial x} + v_g \frac{\partial T_g}{\partial y_g} = \alpha_g \frac{\partial^2 T_g}{\partial {y_g}^2} +$$
$$\frac{D_g(c_{p_{A,g}} - c_{p_{B,g}})}{c_{p_g}} \frac{\partial T_g}{\partial y_g} \frac{\partial}{\partial y_g} \frac{\rho_{A,g}}{\rho_g} \quad 0 \le y_g \le \delta_g \, . \tag{8.105}$$

The boundary conditions are

$$u_\ell = 0 \, , \;\; u_g = u_{g\infty} \, , \;\; T_g = T_{g\infty} \quad x \le 0 \, , \tag{8.106}$$

$$u_\ell = v_\ell = 0 \, , \;\; T_\ell = T_{s_b} \quad y = 0 \, , \tag{8.107}$$

$$u_\ell = u_g = u_{\ell g} \, , \;\; \mu_\ell \frac{\partial u_\ell}{\partial y} = \mu_g \frac{\partial u_g}{\partial y_g} \, ,$$
$$\dot{m}_{g\ell} = \dot{m}_{A,g\ell} + \dot{m}_{B,g\ell} = \rho_\ell \frac{d}{dx} \int_0^{\delta_\ell} u_\ell dy = \rho_\ell (u_\ell \frac{d\delta_g}{dx} - v_\ell) =$$
$$\rho_g (u_g \frac{d\delta_g}{dx} - v_g) \, , \;\; T_\ell = T_g = T_{\ell g} = T_s \, ,$$
$$k_\ell \frac{\partial T_\ell}{\partial y} = \Delta i_{\ell g} \rho_\ell (u_\ell \frac{d\delta_\ell}{dx} - v_\ell) + k_g \frac{\partial T_g}{\partial y_g} \, ,$$
$$\rho_{B,g} = \rho_{B,g}(T_s, p) \, , \;\; \rho_{B,\ell} = \rho_{B,\ell}(T_s, \rho_{B,g}) \, ,$$
$$\dot{m}_{A,g\ell} = (\rho_{A,g} u_g - \rho_g D_g \frac{\partial}{\partial y_g} \frac{\rho_{A,g}}{\rho_g}) \frac{d\delta_\ell}{dx} - \rho_{A,g} v_g +$$
$$\rho_g D_g \frac{\partial}{\partial y_g} \frac{\rho_{A,g}}{\rho_g} = \rho_{A,g}(u_g \frac{d\delta_\ell}{dx} - v_g) + \rho_g D_g \frac{\partial}{\partial y_g} \frac{\rho_{A,g}}{\rho_g} \, ,$$
$$\rho_g D_g \frac{\partial}{\partial y_g} \frac{\rho_{B,g}}{\rho_g} = \dot{m}_{B,g\ell} - \frac{\rho_{B,g}}{\rho_g} \dot{m}_{g\ell} \, , \quad y = \delta_\ell \, , \tag{8.108}$$

$$u_g = u_{g\infty} \, , \;\; T_g = T_{g\infty} = T_{s\infty} + \Delta T_{g\infty} \, ,$$
$$\rho_{A,g} = (\rho_{A,g})_\infty \, , \;\; \rho_{B,g} = (\rho_{B,g})_\infty \quad y_g = \delta_g \to \infty \, . \tag{8.109}$$

Only one of the species mass flow interfacial conditions in (8.108) needs to be used, because only one species concentration distribution is used along with the local summation of the species concentrations, i.e.,

$$\frac{\rho_{A,g}}{\rho_g} + \frac{\rho_{B,g}}{\rho_g} = 1 \, . \tag{8.110}$$

The partial pressure and concentration of species A, and the average gas-phase specific heat, are

$$\frac{p_{A,g}}{p_g} = (1 + \frac{M_A \frac{\rho_{B,g}}{\rho_g}}{M_B \frac{\rho_{A,g}}{\rho_g}})^{-1} \quad \text{or} \quad \frac{\rho_{A,g}}{\rho_g} = (1 + \frac{M_B p_{B,g}}{M_A p_{A,g}})^{-1} \, , \tag{8.111}$$

$$c_{p_g} = c_{p_{A,g}} \frac{\rho_{A,g}}{\rho_g} + c_{p_{B,g}} \frac{\rho_{B,g}}{\rho_g} \, . \tag{8.112}$$

Other mixture properties are determined as outlined by Fujii (1991).

The chemical equilibrium state, i.e., the equilibrium species concentrations in the gas and liquid phases, can be determined from the Raoult law for the *ideal solutions* of miscible liquids, Section 1.5.2, and the Henry law for the *ideal dilute solutions* of miscible liquids, Section 6.3.1(D), depending on the relative concentration of species A in the liquid phase. In practice, since it does influence the fluid flow and heat transfer (except for the very small variation of the thermophysical properties within the liquid mixture), the concentration of species B in the liquid phase is taken as *uniform* and *equal* to the equilibrium condition at the interfacial temperature T_s, i.e., $\rho_{A,\ell} = \rho_{A,\ell}(\delta_\ell)$. Then

$$\frac{\rho_{A,\ell}}{\rho_\ell} = \frac{\dot{m}_{A,g\ell}}{\dot{m}_{g\ell}} . \tag{8.113}$$

The *similarity solutions* to the boundary-layer equations (8.98) to (8.109) for $u_{g\infty} = 0$, $\partial p_\ell/\partial x = \partial p_g/\partial x = 0$, i.e., buoyant flow only, has been obtained and the various contributions are listed by Fujii (1991). The liquid-film, and the gas boundary-layer spatial similarity variables, η_ℓ and η_g, the dimensionless stream functions, f_ℓ and f_g, the dimensionless temperatures, T_ℓ^* and T_g^*, and the dimensionless mass concentration of species A in the gas phase $\rho_{A,g}^*$ are defined as

$$\eta_\ell = y_\ell(\frac{g}{4\nu_\ell^2 x})^{1/2} , \quad \eta_g = y_g(\frac{g}{4\nu_g^2 x})^{1/2} ,$$

$$f_\ell = \frac{\psi_\ell}{2^{3/2}(g\nu_\ell^2 x^3)^{1/4}} , \quad f_g = \frac{\psi_g}{2^{3/2}(g\nu_g^2 x^3)^{1/4}} ,$$

$$u_\ell = \frac{\partial \psi_\ell}{\partial y} , \quad v_\ell = -\frac{\partial \psi_\ell}{\partial x} , \quad u_g = \frac{\partial \psi_g}{\partial y} , \quad v_g = -\frac{\partial \psi_g}{\partial x} ,$$

$$T_\ell^* = \frac{T_s - T_\ell}{T_s - T_{s_b}} , \quad T_g^* = \frac{T_{g\infty} - T_g}{T_{g\infty} - T_s} , \quad \rho_{A,g}^* = \frac{\dfrac{\rho_{A,g}}{\rho_g} - (\dfrac{\rho_{A,g}}{\rho_g})_\infty}{(\dfrac{\rho_{A,g}}{\rho_g})_{\delta_\ell} - (\dfrac{\rho_{A,g}}{\rho_g})_\infty} . \tag{8.114}$$

Then (8.98) to (8.105), including the liquid-phase species equation which becomes trivial for the assumed uniform distribution of $\rho_{A,\ell}$ and the continuity equations, which become trivial with the use of the stream functions, become (Fujii, 1991)

$$\frac{d^2}{d\eta_g^{\ 2}}\rho_{A,g}^* + 3Sc_g f_g \frac{d}{d\eta_g}\rho_{A,g}^* = 0 , \tag{8.115}$$

$$\frac{d^3 f_\ell}{d\eta_\ell^{\ 3}} + 3f_\ell \frac{d^2 f_\ell}{d\eta_\ell^{\ 2}} - 2(\frac{df_\ell}{d\eta_\ell})^2 + 1 = 0 , \tag{8.116}$$

$$\frac{d^3 f_g}{d\eta_g^{\ 3}} + 3f_g \frac{d^2 f_g}{d\eta_g^{\ 2}} - 2(\frac{df_g}{d\eta_g})^2 + a_\rho \rho_{A,g}^* + a_T T_g^* - a_\rho a_T \rho_{A,g}^* T_g^* = 0 , \tag{8.117}$$

$$\frac{d^2 T_\ell^*}{d\eta_\ell^{\ 2}} + 3Pr_\ell f_\ell \frac{dT_\ell^*}{d\eta_\ell} = 0 , \tag{8.118}$$

$$\frac{d^2T_g^*}{d\eta_g{}^2} + 3Pr_g f_g \frac{dT_g^*}{d\eta_g} +$$

$$\frac{Pr_g}{Sc_g}\frac{c_{p_{A,g}} - c_{p_{B,g}}}{c_{p_g}}[(\frac{\rho_{A,g}}{\rho_g})_{\delta_\ell} - (\frac{\rho_{A,g}}{\rho_g})_\infty]\frac{dT_g^*}{d\eta_g}\frac{d}{d\eta_g}\rho_{A,g}^* = 0 \,, \quad (8.119)$$

where

$$a_\rho = \frac{M_A - M_B}{M_A - (M_A - M_B)(\frac{\rho_{A,g}}{\rho_g})_\infty}[(\frac{\rho_{A,g}}{\rho_g})_{\delta_\ell} - (\frac{\rho_{A,g}}{\rho_g})_\infty] \,,$$

$$a_T = \frac{T_{g\infty} - T_s}{T_{g\infty}} \,, \quad Sc_g = \frac{\nu_g}{D_g} \,, \quad Pr_g = \frac{\nu_g}{\alpha_g} \,, \quad Pr_\ell = \frac{\nu_\ell}{\alpha_\ell} \,. \quad (8.120)$$

The boundary conditions (8.106) to (8.109) for y, y_g, and x combined and give

$$\frac{df_\ell}{d\eta_\ell} = \frac{df_g}{d\eta_g} = 0 \,, \quad T_\ell^* = 1 \qquad \eta_\ell = 0 \,, \quad (8.121)$$

$$\frac{df_\ell}{d\eta_\ell} = \frac{df_g}{d\eta_g} \,, \quad \frac{d^2 f_g}{d\eta_g{}^2} = (\rho u)^{*1/2}\frac{d^2 f_\ell}{d\eta_\ell{}^2} \,,$$

$$f_g = (\rho\mu)^{*1/2} f_\ell = \frac{1}{3}(\rho u)^{*1/2}(\dot{m}_{A,g\ell}^* + \dot{m}_{B,g\ell}^*) = \frac{1}{3}(\rho u)^{*1/2}\dot{m}_{g\ell}^* \,,$$

$$T_\ell^* = 0 \,, \quad T_g^* = 1 \,, \quad -\frac{dT_\ell^*}{d\eta_\ell} = \frac{Pr_\ell \dot{m}_{g\ell}^*}{Ja_{sc}} - \frac{k_g}{k_\ell}(\frac{\nu_\ell}{\nu_g})^{1/2}\frac{T_{g\infty} - T_s}{T_s - T_{s_b}}\frac{dT_g^*}{d\eta_g} \,,$$

$$\rho_{A,g}^* = 1 \,, \quad -\frac{d}{d\eta}\rho_{A,g}^* =$$

$$(\rho\mu)^{*1/2} Sc_g \frac{(\frac{\rho_{A,g}}{\rho_g})_{\delta_\ell}\dot{m}_{B,g\ell}^* - [1 - (\frac{\rho_{A,g}}{\rho_g})_\infty]\dot{m}_{A,g\ell}^*}{(\frac{\rho_{A,g}}{\rho_g})_{\delta_\ell} - (\frac{\rho_{A,g}}{\rho_g})_\infty} \qquad \eta_\ell = \eta_{\delta_\ell} \,, \quad \eta_g = 0$$

(8.122)

$$\frac{df_g}{d\eta_g} = \frac{dT_g^*}{d\eta_g} = \rho_{A,g}^* = 0 \qquad \eta_g \to \infty \,, \quad (8.123)$$

where

$$(\rho\mu)^* = \frac{(\rho\mu)_\ell}{(\rho\mu)_g} \,, \quad \dot{m}_{g\ell}^* = \frac{\dot{m}_{g\ell} x}{\mu_\ell}(\frac{Gr_x}{4})^{-1/4} \,,$$

$$Gr_x = \frac{g x^3}{\nu_\ell^2} \,, \quad Ja_{sc} = \frac{c_{p\ell}(T_s - T_{s_b})}{\Delta i_{\ell g}} \,. \quad (8.124)$$

The local *Galileo* (or *Archimedes*) number Ga_x is occasionally used instead of the local *Grashof* number Gr_x to make a distinction between the phase-density and thermo- and diffusobuoyant flows. The concentration of

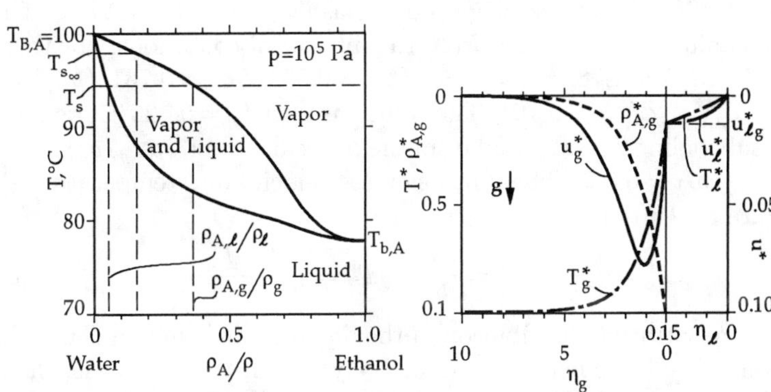

Figure 8.12. (a) Liquid-vapor phase diagram for the water-ethanol system, at atmospheric pressure. (b) Predicted distributions of the temperature, velocity, and species A concentration, across the liquid film layer and the gas boundary layer, for the case of an otherwise stagnant gas. (From Fujii, reproduced by permission ©1991 Springer-Verlag.)

species in the liquid phase given by (8.113) becomes

$$\frac{\rho_{A,\ell}}{\rho_\ell} = \frac{\dot{m}^*_{A,g\ell}}{\dot{m}^*_{g\ell}} . \tag{8.125}$$

Similarity solutions corresponding to constant $\dot{m}^*_{g\ell}$, i.e., $\dot{m}_{\ell g}$ *proportional* to $x^{-1/4}$, have been obtained by many investigators and are listed by Fujii (1991). The ordinary differential equations (8.115) to (8.119) are solved *numerically*.

The local Nusselt number is defined as

$$Nu_x = \frac{q_{s_b} x}{(T_s - T_{s_b})k_\ell} = -\frac{x}{(T_s - T_{s_b})} \left.\frac{\partial T_\ell}{\partial y}\right|_{y=0} =$$

$$-(\frac{Gr_x}{4})^{1/4} \left.\frac{dT^*_\ell}{d\eta_\ell}\right|_{\eta_\ell=0} = Nu_x[Pr_\ell, Pr_g, Sc_g, Ja_{sc},$$

$$(\rho\mu)^*, \frac{k_g}{k_\ell}, \frac{\nu_\ell}{\nu_g}, \frac{T_{g\infty} - T_s}{T_s - T_{s_b}}, (\frac{\rho_{A,g}}{\rho_g})_\infty, (\frac{\rho_{A,g}}{\rho_g})_{\delta_\ell}, a_T, a_\rho, \frac{M_A}{M_B}, \text{etc.}] . \tag{8.126}$$

Fujii (1991) *correlates* Nu_x and $\dot{m}^*_{g\ell}$ and finds a simple relation

$$Nu_x = (\frac{Gr_x}{4})^{1/4} \dot{m}^{*\,-1/3}_{g\ell} . \tag{8.127}$$

The interfacial temperature T_s and concentrations $(\rho_{A,g}/\rho_g)_{\delta_\ell}$ and $\rho_{A,\ell}/\rho_\ell$ are found as part of the solution. Various correlations are presented by Fujii (1991) for some *binary mixtures* of common use (including water-air). Some typical results for the distributions of the temperature, velocity, and concentration are shown in Figure 8.12(a) and (b) for *water* (species B)-*ethanol* (species A). Figure 8.12(a) shows the phase diagram for this binary mixture. For the results shown in Figure 8.12(b), the conditions are $p = 10^5$ Pa, $T_{s_\ell} = 98°\text{C}$, $(\rho_{A,g}/\rho_g)_\infty = 0.145$, $\Delta T_{g_\infty} = 100°\text{C}$, and $T_{s_b} = 93.76°\text{C}$. Note that the gas phase has *large* superheat ΔT_{g_∞}. The liquid subcooling is determined from the interfacial conditions. For this fluid pair $(M_A/M_B) = 1.78$, $Pr_\ell = 1.84$, $Pr_g = 0.992$, $Sc_g = 0.731$, $(\rho\mu)^* = 176$, $(c_{p_{A,g}} - c_{p_{B,g}})/c_{p_g} = -0.101$. The solutions give $T_s = 94.26°\text{C}$, i.e., a *small* liquid subcooling $T_s - T_{s_b}$, and from the phase diagram, $(\rho_{A,g}/\rho_g)_{\delta_\ell} = 0.382$ and $(\rho_{A,g}/\rho_\ell) = 0.055$. The dimensionless velocity and temperature shown in Figure 8.12(b) are

$$u_\ell^* = \frac{df_\ell}{d\eta_\ell}\ , \quad u_g^* = \frac{df_g}{d\eta_g}\ , \quad T^* = \frac{T - T_{s_b}}{T_{g_\infty} - T_{s_b}}\ . \tag{8.128}$$

The concentration distribution in the liquid is uniform and is not shown. Because of the large negative thermobouyancy in the gas phase, the maximum gas-phase velocity is *larger* than the interfacial velocity $u_{\ell g}^*$. The accumulation of the ethanol at the interface is evident and this *lowers* the saturation temperature of water compared to the far-field value, i.e., where $(\rho_{A,g}/\rho_g)_\infty = 0.145$, by $T_{s_\infty} - T_s = 3.74°\text{C}$. This results in a temperature difference *across* the *liquid film* of only $T_s - T_{s_b} = 0.5°\text{C}$. Therefore, the presence of a *less*-volatile species *reduces* the liquid subcooling and the heat withdrawal rate.

For completeness and for a later usage, the integral-boundary layer solution of *Nusselt* for the *asymptotic* case of $\Delta T_{g_\infty} = (\rho_{A,g})_\infty = \mu_g = 0$ is also mentioned (Carey, 1992), and the results are

$$\delta_\ell = [\frac{4k_\ell \nu_\ell (T_s - T_{s_b})x}{\rho_\ell g \Delta i_{\ell g}}]^{1/4}\ , \tag{8.129}$$

$$\dot{m}_{g\ell} = \frac{\rho_\ell g \delta_\ell^2}{\nu_\ell}\ , \tag{8.130}$$

$$\langle u_\ell \rangle_{\delta_\ell} = \frac{g\delta_\ell^2}{3\nu_\ell}\ ,$$

$$Nu_x = \frac{x}{\delta_\ell} = [\frac{\rho_\ell g \Delta i_{\ell g} x^3}{4k_\ell \nu_\ell (T_s - T_{s_b})}]^{1/4}$$

$$\Delta T_{g_\infty} = (\rho_{A,g})_\infty = \mu_g = 0\ , \quad Ja_{sc} \to 0\ , \quad Pr_\ell \to \infty\ . \tag{8.131}$$

8.2.2 Negatively Buoyant, Wavy Film Flow

For the downward condensate-film flow over a vertical surface, a short distance from the leading edge (i.e., $x = 0$), surface *waves* are formed on the liquid-gas interface. For negligible liquid-gas interfacial drag, the waves are the ambient disturbances magnified when the destabilizing forces of inertia and gravitational acceleration overcome the stabilizing forces of viscous shear and surface tension. This hydrodynamic instability is referred to as the *Kapitza instability* after his original simplified analysis in 1948 (Kutateladze, 1982).

The liquid Reynolds number based on the hydraulic diameter (which is equal to $4\delta_\ell$) is

$$Re_{\delta_\ell} = \frac{4\delta_\ell \langle u_\ell \rangle_{\delta_\ell}}{\nu_\ell} = \frac{4}{\nu_\ell} \int_0^{\delta_\ell} u_\ell dy \,. \tag{8.132}$$

Experimental results of Nakorykov et al. (1976) for downward-flowing liquid films with no phase change show that for a Re_{δ_ℓ} (i.e., corresponding to a downstream location in a *condensate* flow) of less than 20 these waves are *not* visible. For $Re_{\delta_t} > 20$, *small-amplitude* waves appear further downstream, and as Re_{δ_t} increases the amplitude of these waves also *increases*. Because of the *nonlinearity* and *asymmetry* of some of the forces, the *shape* of the wave changes from the *initially sinusoidal* form to *distorted sinusoidal* waves or *intermediate* waves, for $100 < Re_{\delta_\ell} < 1500$. For $Re_{\delta_\ell} > 1500$, the wave front becomes very *steep* and *large roll* waves are formed with *smaller bow* waves appearing in *front* of them.

The *hydrodynamic* aspects of these waves have been analyzed by Hirshburg and Florschuetz (1982), Karapantsios et al. (1989), Wasden and Duckler (1989), Bach and Villadsen (1984), Chang (1987), and Stuhlträger et al. (1993), among others. These include the *normal form* analysis of long (compared to the film thickness) waves, the *two-dimensional, periodic* intermediate wave analysis, and the *numerical, direct simulation* of two-dimensional large-amplitude short waves.

Here some aspects of the growing interfacial displacement disturbances in the condensate flow are examined. The local *undisturbed* (i.e., nonwavy) liquid film thickness δ_{ℓ_o} and the film-averaged axial velocity $\langle u_\ell \rangle_{\delta_{\ell_o}}$, such as those given by (8.129) and (8.131), are used to scale the variables. The *two-dimensional*, continuity and transient momentum equations for the liquid are similar to (8.100) and (8.102), but here because the wavy liquid flow is not a boundary-layer flow the boundary-layer approximations are not made. Then we have

$$\frac{\partial u_\ell^*}{\partial x^*} + \frac{\partial v_\ell^*}{\partial y^*} = 0 \,, \tag{8.133}$$

$$\frac{\partial u_\ell^*}{\partial t^*} + u_\ell^* \frac{\partial u_\ell^*}{\partial x^*} + v_\ell^* \frac{\partial u_\ell^*}{\partial y^*} = -\frac{\partial p_\ell^*}{\partial x^*} + \frac{4}{Re_{\delta_{\ell_o}}} (\frac{\partial^2 u_\ell^*}{\partial x^{*2}} + \frac{\partial^2 u_\ell^*}{\partial y^{*2}}) + \frac{1}{Fr^2_{\delta_{\ell_o}}} \,, \tag{8.134}$$

$$\frac{\partial v_\ell^*}{\partial t^*} + u_\ell^* \frac{\partial v_\ell^*}{\partial x^*} + v_\ell^* \frac{\partial v_\ell^*}{\partial y^*} = -\frac{\partial p_\ell^*}{\partial y^*} + \frac{4}{Re_{\delta_{\ell_o}}} (\frac{\partial^2 v_\ell^*}{\partial x^{*2}} + \frac{\partial^2 v_\ell^*}{\partial y^{*2}}) \,, \tag{8.135}$$

where the *Froude number* is defined as

$$Fr_{\delta_{\ell_o}} = \frac{\langle u_\ell \rangle_{\delta_{\ell_o}}}{(g\delta_{\ell_o})^{1/2}} \,. \tag{8.136}$$

The liquid-gas interface will be *curved* and the interfacial force balance (1.131) is applied. The effect of the gas phase is neglected by assuming that $\mu_g = 0$ and the two-dimensional surface curvature is given by

$$2H = \frac{1}{R_1} + \frac{1}{R_2} = \frac{\partial^2 \delta_\ell}{\partial x^2}[1 + (\frac{\partial \delta_\ell}{\partial x})^2] \simeq \frac{\partial^2 \delta_\ell}{\partial x^2} . \tag{8.137}$$

Then the dimensionless interfacial *normal* force balance gives

$$p_\ell^* - p_g^* + \frac{1}{We_{\delta_{\ell_o}}} \frac{\partial \delta_\ell^*}{\partial x^*} + \frac{8}{Re_{\delta_{\ell_o}}} \frac{\partial u_\ell^*}{\partial x^*} \quad y^* = \delta_\ell , \tag{8.138}$$

where the Weber number is defined as

$$We_{\delta_{\ell_o}} = \frac{\rho_\ell \langle u_\ell \rangle^2_{\delta_{\ell_o}} \delta_{\ell_o}}{\sigma} . \tag{8.139}$$

The wave speed (i.e., wave *celerity*) u_w is *larger* than the liquid-gas interfacial velocity and *depends* on $Re_{\delta_{\ell_o}}$, $Fr_{\delta_{\ell_o}}$, and $We_{\delta_{\ell_o}}$. The relation can be found by the numerical integration, and a correlation based on the available experimental results is given by Chang (1987). A typical, two-dimensional *instantaneous* surface wave located at a downstream location x^*, found by the numerical simulation of Stuhlträger et al. (1993), is shown in Figure 8.13(a). The results are for Freon-R11 with $Pr_\ell = 4.3$ and a liquid subcooling which results in $Ja_{sc} = 0.06$. The analysis of Hirshburg and Florschuetz (1982) leads to the introduction of the *dimensionless surface tension* which is called the *Kapitza number*

$$Ka = \frac{\sigma}{\rho_\ell (\nu_\ell^4 g)^{1/3}} . \tag{8.140}$$

For Freon-R11 used, $Ka = 2671$. The analysis of Stuhlträger et al. (1993) for heat transfer assumes conduction heat transfer *only*. Also shown in Figure 8.13(a) is the velocity distribution as the *relative* velocity $\mathbf{u} - \mathbf{u}_w$. These nonlinear waves have extensive *recirculation* under the wave crest. The *time-averaged* liquid-film thickness $\bar{\delta}_\ell$ distribution is shown in Figure 8.13(b). The distribution of the time-averaged axial velocity $\bar{u}_\ell$ is also shown. The maximum in $\bar{u}_\ell$ occurs at the interface and at the location of the wave *inception* (called the *inception point*). The time-averaged condensate film thickness δ_ℓ becomes *smaller* than the nonwavy thickness δ_{ℓ_o} (note that here the nonwavy film thickness δ_{ℓ_o} used for scaling applies to $x^* = 4500$). In this example, at $x^* = 4500$, $\bar{\delta}_\ell$ is less than δ_{ℓ_o} predicted for this location from the Nusselt analysis.

The *heat transfer* aspects of the wavy condensate flow have been examined by Kutateladze (1982), who also has considered the *turbulent* condensate flow regime, and by both Hirshburg and Florschretz (1982) and Faghri and Seban (1985) for *laminar* wavy films and Lyu and Mudawar (1991) for laminar and turbulent wavy films with no phase change. In the semi-empirical analysis of Hirshburg and Florschretz (1982), the intermediate waves are modeled as having a frequency *smaller*, by a ratio f^*, than the most dangerous frequency (the ratio of the wave celerity to wavelength)

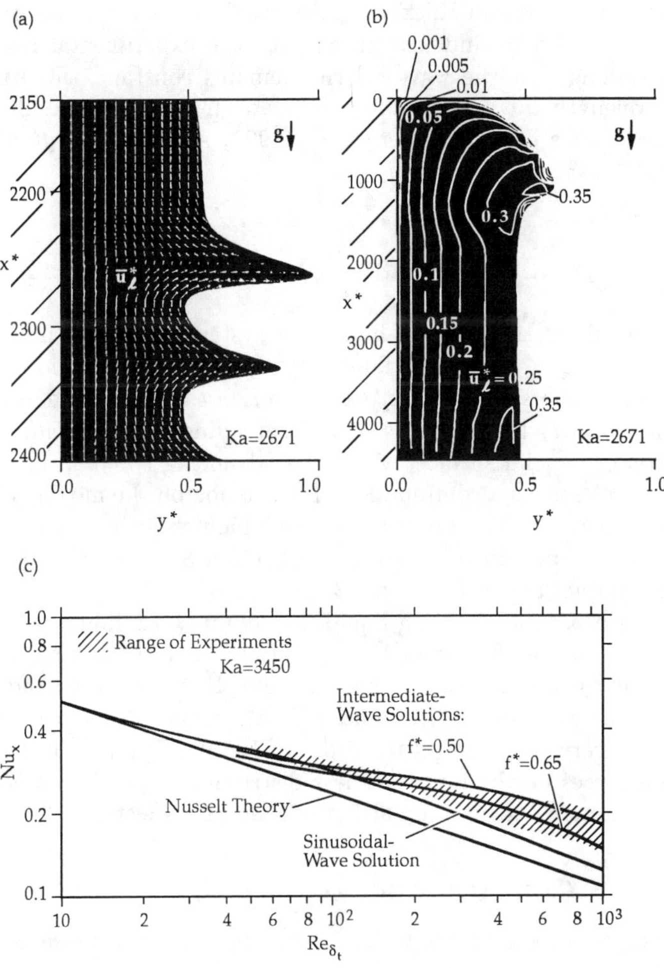

Figure 8.13. (a) Predicted, wavy downward liquid film flow showing the instantaneous local liquid velocity in the film (with arrows) in the region of wave amplification. (b) Predicted mean film thickness and local liquid velocity distributions. (c) Effect of condensate-film waveness on the local Nusselt number. (From Stuhlträger et al. and Hirshburg and Florschuetz, reproduced by permission ©1993 Pergamon Press, and ©1982 American Society of Mechanical Engineers.)

determined from their stability analysis. Their results for the local Nusselt number, for $f^* = 1$, 0.65, and 0.50, and $Ka = 3450$, are shown in Figure 8.13(c). The Nusselt solution for the nonwavy film and the available experimental results are also shown. The results show that for $Re_{\delta_\ell} > 20$, a deviation from the Nusselt solutions is predicted due to a *decrease* in the time-averaged film thickness. Also, as Re_{δ_ℓ} *increase* the frequency of the wave needed to find agreement with the experimental results *decreases*. Assuming that the wave celerity remains constant, this implies that the wavelength *increases* with an increase in Re_{δ_ℓ}. The results of the direct simulations of Stuhlträger et al. (1993) shows that *both* the wavelength and u_w *increase* with Re_{δ_ℓ}.

8.3 Nonisothermal Common- and Interline

In our discussion of solid–liquid–gas systems, so far we have assumed that a continuous and flowing fluid film is present on the solid surface. Here we consider the *simultaneous* presence of all these three phases at a common (or *contact*) line, such as in the spreading of a *partially* wetting liquid drop on a solid surface, where the advancing common line *slips* over the surface. This is a continuation of discussion on the moving common line in Section 1.4.4(C) and here the hydrodynamics of the *nonisothermal* moving common line will be discussed in Section 8.3.2 after a re-examination of the contact angles (Section 8.3.1).

For a *perfectly* wetting liquid, by *thinning* the liquid film over a heated surface, using gravity, high evaporation rates can occur with a small conduction resistance across this thin film. However, when the film thickness becomes very small, the van der Waals surface forces prevent evaporation of the very thin film (of the order of 100 Å). The role of these intermolecular forces on the film thickness distribution and the evaporation rate of the extended menisci are also discussed here (Section 8.3.3).

8.3.1 Contact Angles

The *static* equlibrium between the three surface tensions given by the Young equation (1.47) defines the *static contact angle*. The *penetration* of the solid surface forces into the liquid is *not directly* addressed by this equation and it is assumed that this penetration distance is small. Since in many very thin liquid films (of the order of 100 Å, i.e., 10 nm) with a common line these surface forces are important, a molecular treatment of the contact angle has also been made.

The van der Waals attracting forces between two liquid molecules ($\ell\ell$) a distance r apart, and that between a solid molecule and a liquid molecule ($s\ell$) (neglecting the liquid-gas molecular attractive force because of the small molecular number density for the gas compared to the liquid and solid), are modeled by the *intermolecular potential energy distributions* (Miller and Ruckenstein, 1974)

$$\phi_{\ell\ell} = -\frac{B_{\ell\ell}}{r^6}, \quad \phi_{\ell s} = -\frac{B_{s\ell}}{r^6}, \tag{8.141}$$

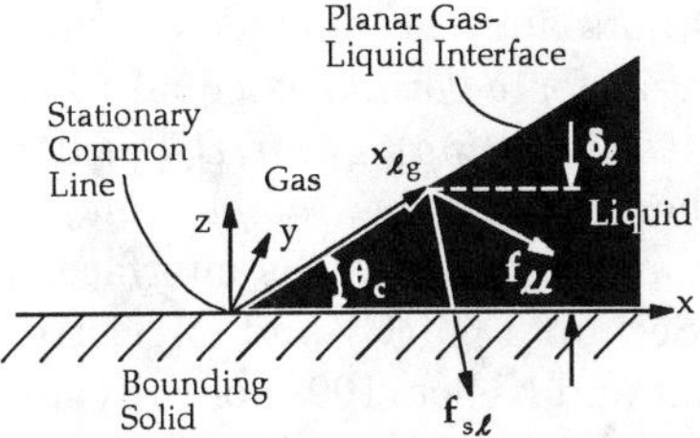

Figure 8.14. A liquid-gas-solid common line and the interfacial forces, acting on the liquid surface.

where $B_{\ell\ell}$ and $B_{s\ell}$ are constants that depend on the atomic properties of the liquid and solid constituents, respectively. As was mentioned in Section 1.4.2, a *repulsion* force is also present for *small* r, and this is neglected here by assuming that r is larger than r_o, i.e., the location of the minimum in $\phi(r)$ when the repulsion force is also included.

For a point along a *planar* (i.e., neglecting the *gravity* and *curvature* effects) liquid-gas interface given by the position vector $\mathbf{x}_{\ell s}$ in the Cartesian coordinate shown in Figure 8.14, the integrated potential energy over the solid phase (half-space $z < 0$) is

$$\Delta\phi_{s\ell} = -\int_0^\infty \int_{-\infty}^\infty \int_{-\infty}^0 n_s \frac{B_{s\ell}}{[(x-\delta_\ell \cot\theta_c)^2 + y^2 + (z-\delta_\ell)^2]^3} dxdydz \ , \tag{8.142}$$

where n_s is the molecular number density of the solid. The local liquid film thickness δ_ℓ and the contact angle θ_c are also used. The result of this integral is (Miller and Ruckenstein, 1974; Hocking, 1993)

$$\Delta\phi_{s\ell} = -\frac{\pi n_s B_{s\ell}}{6\delta_\ell^3} \ . \tag{8.143}$$

The integration over the liquid volume is made in two parts (Miller and Ruckenstein, 1974; Hocking, 1993) and when these are added to the above gives

$$\Delta\phi = -\frac{\pi n_s B_{s\ell}}{6\delta_\ell^3} - \frac{2\pi n_\ell B_{\ell\ell}}{3r_o^3} + \frac{\pi n_\ell B_{\ell\ell}}{6\delta_\ell^3}(\frac{1}{2} + \frac{3}{4}\cos\theta_c - \frac{1}{4}\cos^3\theta_c) \ , \tag{8.144}$$

where n_ℓ is the molecular number density of the liquid.

At the planar, liquid-gas interface the potential energy given by (8.144) must be constant. Then the coefficients of the terms containing δ_ℓ^{-3} should be zero giving

$$\frac{1}{2} + \frac{3}{4}\cos\theta_c - \frac{1}{4}\cos^3\theta_c = \frac{n_s B_{s\ell}}{n_\ell B_{\ell\ell}} \ . \tag{8.145}$$

This expression gives the *static contact angle* as a function of the intermolecular force constants and the number densities. For $n_s B_{s\ell} = n_\ell B_{\ell\ell}$ (i.e., when the potential for *adhesion* is equal to that for *cohesion*), this gives $\theta_c = 0$ (i.e., *complete wetting*). For $n_s B_{s\ell} \to 0$ (i.e., when the potential for *adhesion* diminishes), $\theta_c = \pi$ (i.e., *nonwetting*). The contact angle predicted by (8.145) for a *planar* liquid-gas interface is the *intrinsic* contact angle mentioned in Section 1.4.4(C) and is referred to as the *boundary-layer region* (Merchant and Keller, 1992) of the contact region. The Young equation, i.e., (1.47) has been referred to as the *outer region* contact angle. Note that in the derivation of (8.144) the contributions of the gas molecules are neglected (i.e., the liquid molecules on the liquid-gas interface and the solid molecules on the solid–gas interface only experience *internal* attraction forces). In order to discuss the outer region, an allowance has to be made for the liquid-gas interfacial *curvature*. For a *nonplanar* liquid-gas interface the effect of *curvature* and the potential energy due to *gravity* (assuming a *horizontal* solid surface) are added, and we then have

$$2\sigma H - \rho_\ell g \delta_\ell + \Delta\phi = \Delta\phi_o \, , \tag{8.146}$$

where $2H$ is the average liquid-gas interfacial curvature defined in Section 1.12.3. Equation (8.146) is the modified *Young-Laplace equation* (Rowlinson and Widom, 1989) with the van der Waals intermolecular forces included (note that $\Delta\phi \to 0$ as δ_ℓ increases). Since $\Delta\phi$ is only significant when the liquid-film thickness δ_ℓ is very small, while the gravity term is only significant when δ_ℓ is large, a *boundary-layer* exists over which $\Delta\phi$ *dominates*. Merchant and Keller (1992) use the matched asymptotic expansion with the expansion parameter taken as the *ratio* of the *range* of the van der Waals forces (of the order of 100 Å) to the meniscus length scale. The *other region* (where $\Delta\phi = 0$) solution is matched with the solution of the boundary-layer equation which is similar to (8.146). Then they show that when the leading term in the other solution is evaluated at the solid (i.e., by applying the outer solution to the surface), the Young equation is *recovered*. Therefore, the contact angle used in the Young equation (1.47) is an *apparent* (or *distant*) contact angle valid beyond the range of the intermolecular forces. Merchant and Keller (1992) show that the transition between the two static contact angles is *not monotonic* with the distance from the surface.

For a *moving* contact line, Hocking (1993) suggests that by including the intermolecular forces in a *dynamic* form of (8.146) the intrinsic contact angle condition can be satisfied while allowing for the apparent (i.e., outer region) contact angle to vary with the various parameters given in (1.49). This allows for the imposition of the liquid, *no-slip* velocity boundary condition, while without the inclusion of this intermolecular force in the liquid momentum equation, a *slip* in the liquid velocity at the surface must be allowed in order to obtain a solution (de Gennes et al., 1990). When the intermolecular forces are *not* included, a semi-empirical treatment of the

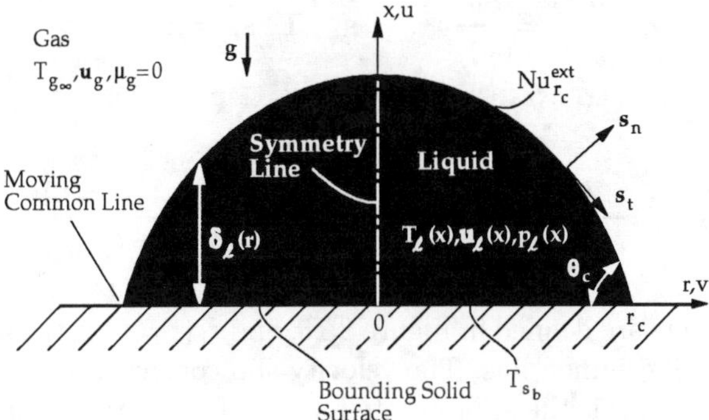

Figure 8.15. A schematic of a wetting droplet undergoing surface heating or cooling at the solid bounding surface and heat transfer to the ambient gas. The coordinate system is also shown.

dynamic contact angle is made (Dussan V, 1979; Chen and Wada, 1982; Dussan V et al., 1991; Hocking, 1992). The velocity slip boundary condition, and the models for the dynamic contact angle, are used below in an analysis of the nonisothermal, moving common line.

8.3.2 Nonisothermal, Moving Contact Line

The *hydrodynamics* of the isothermal, moving common line examines the role of various *dynamic interfacial* (three interfaces) and *bulk* (three phases) *forces* at the various *length scales* associated with the dominance of these forces. For *nonisothermal*, moving common lines, the variation of the interfacial forces with temperatures (i.e., *thermocapillarity*), and possible *phase changes* can modify the hydrodynamics significantly. Here, as an illustrative example, the *spreading* of a droplet *wetting* a solid surface with a *surface temperature* T_{s_b} *higher* or *lower* than the initial droplet temperature, is examined. The liquid is assumed to be *nonvolatile* (i.e., no phase change). the problem has been analyzed by Ehrhard and Davis (1991) using the thin liquid film approximations and a model for the moving common line.

The liquid continuity, momentum, and energy equations are

$$\nabla \cdot \mathbf{u}_\ell = 0 \,, \tag{8.147}$$

$$\rho_\ell(\frac{\partial \mathbf{u}_\ell}{\partial t} + \mathbf{u}_\ell \cdot \nabla \mathbf{u}_\ell) = -\nabla p_\ell + \mu_\ell \nabla^2 \mathbf{u}_\ell - \rho_\ell \mathbf{g} \,, \tag{8.148}$$

$$(\rho c_p)_\ell(\frac{\partial T_\ell}{\partial t} + \mathbf{u}_\ell \cdot \nabla T_\ell) = k_\ell \nabla^2 T_\ell \,. \tag{8.149}$$

Using the coordinate system in Figure 8.15 and for the assumed *axisymmetric* fields, the boundary conditions at the liquid-solid and the liquid-gas interface are

$$u_\ell = 0 \,, \;\; v_{\ell_i} = \alpha_u \frac{\partial v_\ell}{\partial x} \,, \;\; T_\ell = T_{SB} \quad x = 0 \,, \tag{8.150}$$

$$u_\ell - \frac{\partial \delta_\ell}{\partial t} = \frac{\partial \delta_\ell}{\partial r} v_\ell \ , \qquad \mathbf{s}_n \cdot \mathbf{T} \cdot \mathbf{s}_n = 2\sigma H \ ,$$

$$2H = \nabla \cdot (1 + |\nabla \delta_\ell|^2)^{-1/2} \nabla \delta_\ell \ , \qquad \mathbf{s}_t \cdot \mathbf{T} \cdot \mathbf{s}_t = \mathbf{s}_t \cdot \nabla \sigma \ ,$$

$$-k_\ell \frac{\partial T}{\partial n} = Nu_{r_c}^{\text{ext}} \frac{k_g}{r_c(t=0)} (T_\ell - T_{g_\infty}) \ , \ Nu_{r_c}^{\text{ext}} = \frac{q_{\ell g} r_c(t=0)}{(T_\ell - T_{g_\infty}) k_g} \qquad x = \delta_\ell . \tag{8.151}$$

The parameters are defined in Figure 8.15, and $r_c(t=0)$ is the initial radial position of the common line. The velocity slip coefficient α_u is prescribed. The first of (8.151) is the appropriate form of the *kinematic* condition for a moving interface. The normal and tangential interfacial force balances are the vectorial forms of (1.131) and the interfacial viscosity is assumed to be zero. The variation of the surface tension with temperature, generally shown through correlations such as (1.44), is *linearized* as

$$\sigma = \sigma(T_{s_b}) - \left. \frac{\partial \sigma}{\partial T} \right|_p (T_\ell - T_{s_b}) \tag{8.152}$$

with $\partial \sigma / \partial T|_p$ treated as a *constant* coefficient. Further boundary, smoothness, contact-angle, and volume-constraint conditions on δ_ℓ are

$$\delta_\ell(r = r_c, t) = \frac{\partial \delta_\ell}{\partial r}(r = 0, t) = 0 \ , \tag{8.153}$$

$$\lim_{r \to 0} r \frac{\partial^3 \delta_\ell}{\partial r^3}(r = 0, t) = 0 \ , \quad \frac{\partial \delta_\ell}{\partial r}(r = r_c, t) = -\tan \theta_c(t) \ ,$$

$$2\pi \int_0^{r_c(t)} r \delta_\ell(r, t) dr = V(t = 0) \ . \tag{8.154}$$

The relation at $r = r_c$ between the radial variation in δ_ℓ and the contact angle θ_c, is called the *compatibility* condition. The variation of θ_c with respect to $u_c = \partial r_c / \partial t$, i.e., the *velocity* of the *advancing common line*, can be modeled using correlations such as (1.49). Ehrhard and Davis (1991) use

$$u_c = a_1 [\theta_c - \theta_c(u_c = 0)]^{a_2} \ , \tag{8.155}$$

which can be written as

$$\theta_c = \theta_c(u_c = 0) + (\frac{u_c}{a_1})^{1/a_2} \qquad a_1 > 0 \ , \ a_2 \geq 1 \ . \tag{8.156}$$

This dynamic contact angle and the velocity slip on the solid surface given by (8.156) are among the models used for the moving common lines. Other models are the contact angle model of Hocking (1993) which includes the van der Waals forces (called the *microscopic model*), and models which *prescribe* the flow in the immediate vicinity of the moving common line (called the *excision* models).

The solution to the above conservation and constitutive equations and boundary conditions, and the initial conditions which include a prescribed initial common line velocity $u_c(t = 0) = a_1[\theta_c(t = 0) - \theta_c(u_c = 0)]^{a_2}$, are solved by Ehrhard and Davis (1991) subject to the *thin* liquid-film (i.e., $\delta_\ell \rightarrow 0$) or lubrication-theory approximations. This is based on $\theta_c \rightarrow 0$ and allows for neglecting the initial, convective, and radial diffusion terms. The variables are scaled using the initial position of the common line $r_c(t=0)$, the initial common velocity $u_c(t=0)$ or the initial contact angle $\theta_c(t=0)$, the liquid density, the initial volume, and temperatures T_{s_b} and T_{g_∞}. This gives

$$
\begin{aligned}
\rho_\ell^* &= \frac{\delta_\ell}{r_c(t=0)\theta_c(t=0)} , \quad x^* = \frac{x}{r_c(t=0)\theta_c(t=0)} ,\\
r^* &= \frac{r}{r_c(t=0)} , \quad t^* = \frac{a_1[\theta_c(t=0)]^{a_2} t}{r_c(t=0)} ,\\
v_\ell^* &= \frac{v_\ell}{a_1[\theta_c(t=0)]^{a_2}} , \quad u_\ell^* = \frac{u_\ell}{a_1[\theta_c(t=0)]^{1+a_2}} ,\\
p_\ell^* &= \frac{r_c(t=0)[\theta_c(t=0)]^{2-a_2} p_\ell}{\mu_\ell a_1} , T_\ell^* = \frac{T_\ell - T_{g_\infty}}{T_{s_b} - T_{g_\infty}} ,\\
\theta_c^* &= \frac{\theta_c}{\theta_c(t=0)} , \quad V_\ell^* = \frac{V_\ell}{V(t=0)} .
\end{aligned}
\tag{8.157}
$$

The conservation equations (8.147) to (8.149), subject to the lubrication theory approximations, become

$$\frac{\partial u_\ell^*}{\partial x^*} + \frac{1}{r^*}\frac{\partial r^* v_\ell^*}{\partial r^*} = 0 , \tag{8.158}$$

$$0 = -\frac{\partial p_\ell^*}{\partial x^*} + \frac{Bo_{r_c}}{Ca} , \tag{8.159}$$

$$0 = -\frac{\partial p_\ell^*}{\partial r^*} + \frac{\partial^2 v_\ell^*}{\partial x^{*2}} , \tag{8.160}$$

$$0 = \frac{\partial^2 T_\ell^*}{\partial x^{*2}} . \tag{8.161}$$

The boundary conditions (8.150) and (8.151) become

$$u_\ell^* = 0 , \quad v_{\ell_i}^* = \alpha_u^* \frac{\partial v_\ell^*}{\partial x^*} , \quad T^* = 1 \quad x^* = 0 , \tag{8.162}$$

$$
\begin{aligned}
&u_\ell^* - \frac{\partial \delta_\ell^*}{\partial t^*} = v_\ell^* \frac{\partial \delta_\ell^*}{\partial r^*} , \quad -Ca\, p_\ell^* = \frac{\partial^2 \delta_\ell^*}{\partial r^{*2}} + \frac{1}{r^*}\frac{\partial \delta_\ell^*}{\partial r^*} ,\\
&Ma_1 \frac{\partial v_\ell^*}{\partial x^*} = -\frac{\partial T^*}{\partial r^*} - \frac{\partial \delta_\ell^*}{\partial r^*}\frac{\partial T^*}{\partial x^*} , \quad \frac{\partial T^*}{\partial x^*} + Nu_{r_c}^{\text{ext}} \frac{k_g}{k_\ell} T^* = 0 \quad x^* = \delta_\ell^* .
\end{aligned}
\tag{8.163}
$$

The gas-phase pressure and density are neglected compared to the liquid pressure and density. The variation of the surface tension along the liquid-gas interface is related to the one-dimensional temperature variation using (1.132). The dimensionless parameter appearing above are the Bond Bo_{r_c}, Capillary Ca, Nusselt $Nu_{r_c}^{\text{ext}}$, and Marangoni Ma_1 numbers, the dimensionless slip coefficient α_u^*, and the ratio of gas to liquid conductivities k_g/k_ℓ. The gas-phase Nusselt number is defined in (8.151), and the other dimensionless parameters are

$$Bo_{r_c} = \frac{\rho_\ell g r_c^2(t=0)}{\sigma(T_{s_b})}, \quad Ca = \frac{\mu_\ell a_1}{\sigma(T_{s_b})[\theta_c(t=0)]^{3-a_2}},$$

$$Ma_1 = \frac{\mu_\ell a_1}{\dfrac{\partial \sigma}{\partial T}(T_{s_b} - T_\infty)[\theta_c(t=0)]^{1-a_2}}, \quad \alpha_u^* = \frac{\alpha_u}{r_c(t=0)\theta_c(t=0)}. \tag{8.164}$$

In the *limit* of $Bo_{r_c} = Nu_{r_c}^{\text{ext}} k_g/k_\ell \rightarrow 0$, Ehrhard and Davis (1991) show that the modified Marangoni number Ma_2 combines the effect of Ca, $Nu_{r_c}^{\text{ext}} k_g/k_\ell$, and Ma_1, i.e.,

$$Ma_2 = \frac{Ca\, Nu_{r_c}^{\text{ext}} \dfrac{k_g}{k_\ell}}{Ma_1}. \tag{8.165}$$

For $Ma_2 > 0$, the droplet is *heated* (with respect to the gas) by the bounding solid surface, and for $Ma_2 < 0$, it is *cooled*. Some typical results of Ehrhard and Davis (1991) are shown in Figures 8.16(a) and (b) for $Ma_2 = 0.2$ and $Ma_2 = -0.1$, respectively. The effect of *gravity* is *neglected*, i.e., $Bo_{r_c} = 0$. The results are for three different elapsed times, and in Figure 8.16(b) the asymptotic solution for $t^* \rightarrow \infty$ is also shown. Results of Figure 8.16(a) for *heating*, i.e., $T_{s_b} > T_{g_\infty}$, show that heating the droplet *retards* the spreading of the droplet by inducing a thermocapillary motion (i.e., liquid flow towards the higher surface tension created by a lower temperature) which *opposes* the flow of the liquid towards the common line. The *recirculation* caused by this thermocapillary effect persists to where the *spreading stops*. The center vortex *ring* moves outward as the elapsed time increases. In contrast, the results of Figure 8.16(b) for *cooling*, i.e., $T_{s_b} < T_{g_\infty}$, show that cooling the droplet *enhances* the spreading of the droplet, because the thermocapillary motion is towards the moving common line. After the droplet spreading stops, the thermocapillary motion causes a recirculatory flow, as was the case for the heated droplet.

8.3.3 Evaporation from Extended Menisci

For a liquid pool (i.e., a finite volume of liquid with a liquid motion caused *only* by *gravity*, *capillarity*, and short-range *surface* forces) of a *perfect* wetting liquid, i.e., $\theta_c = 0$, covering an *inclined*, heated bounding surface, the liquid film covering the solid surface decreases in thickness with an increase of the distance from the pool surface until a layer of *adsorbed* (*nonevaporating*) film is reached. This is rendered in Figure 8.17. The distribution of the liquid film thickness in the *region* between the pool and the adsorbed film is influenced by the gravity, capillarity, and the solid–liquid–gas inter-

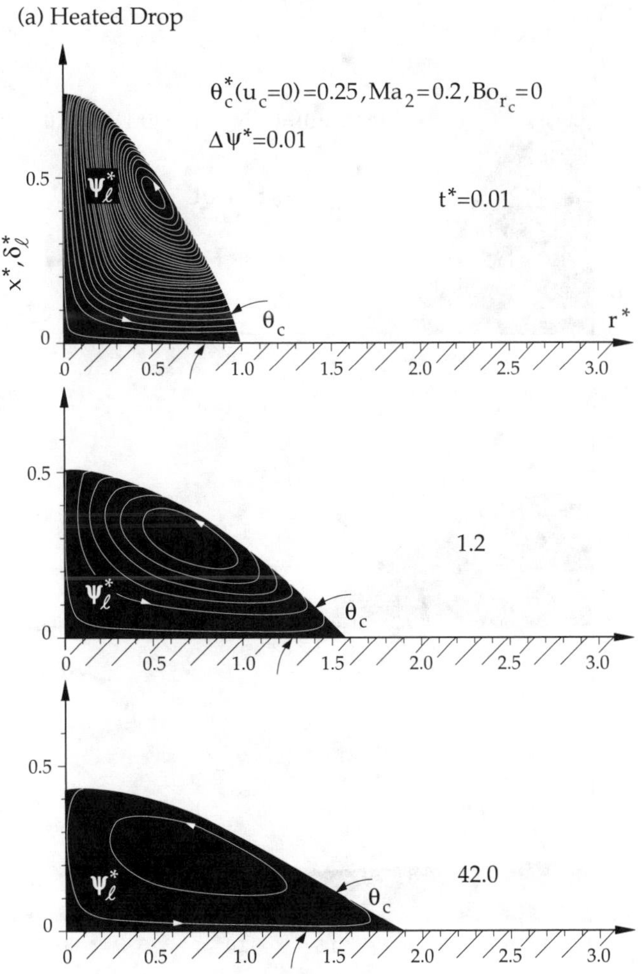

Figure 8.16. (a) Predicted motion inside a droplet heated on a solid surface. The streamlines are for constant increment in the stream function. The results are for three different dimensionless elapsed times. (b) A cooled droplet. (From Erhard and Davis, reproduced by permission ©1992 Cambridge University Press.)

molecular forces (i.e., the van der Waals forces) acting over this film. In this section, first for an *isothermal* system, this *static* equilibrium is examined. Then, because of the interest in *evaporation* from this liquid film (because of the small conductive resistance), an *estimate* of the evaporation rate for the *nonisothermal* extended menisci is given. This is followed by a

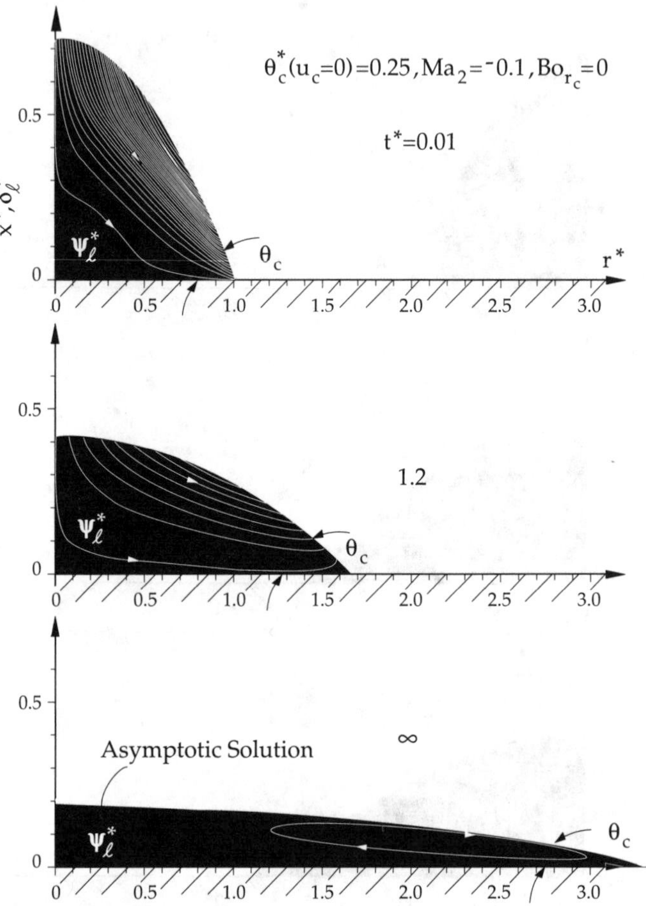

Figure 8.16 (Continued.)

more accurate prediction of the film thickness and the evaporation rate. Since *no* external pressure is used for the flow of the liquid, the liquid flow rate through the meniscus is *limited* by the curvature capillarity and the surface forces. This flow limitation as well as the heat conduction and the vaporization kinetics (including *nonequilibrium* interfacial conditions) determines the evaporation rate.

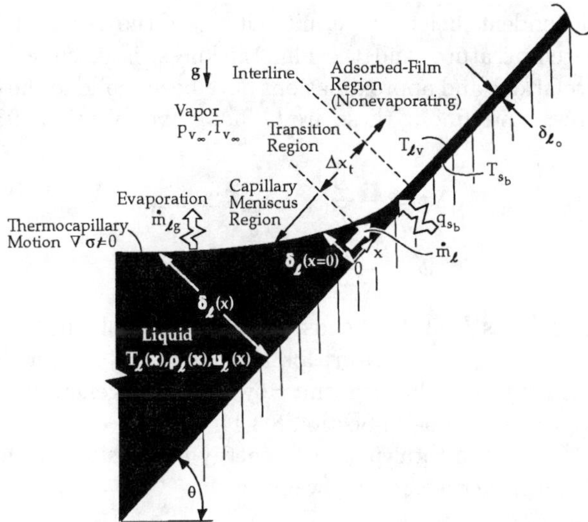

Figure 8.17. An extended meniscus heated through the bounding solid surface and with evaporation from the liquid-gas interface. The various regions and the coordinate system are also shown.

(A) Isothermal, Equilibrium Liquid Film Distribution on a Wedge

Consider a pool of a wetting fluid in *static* equilibrium with a bounding planar solid surface inclined at an angle of θ with respect to the horizontal axis. In this section an isothermal system is considered and nonisothermal systems and evaporation will be discussed in Sections (B) and (C). The liquid covers the surface and due to the curvature-*capillarity* and *van der Waals forces* (or *potentials*) the liquid rises, while the *gravitational potential* limits this rise. Figure 8.17 shows the pool, where far from the solid surface (i.e., the wedge) the liquid-gas interface is planar and horizontal. As the wedge is approached, this interface is curved, and before the liquid film layer thickness δ_ℓ becomes very small (i.e., for $\delta_\ell > 1$ μm), the meniscus curvature is governed by the curvature capillarity and by gravity. This region is called the *capillary-meniscus region*. For $\delta_\ell < 1$ μm, the van der Waals potentials (i.e., *dispersive forces*) become important and the gravity becomes *less* significant. Up to the point where the van der Waals forces dominate, the region is called the *transition region*. This is followed by a region of very thin *absorbed* (a *condensed* phase which is strongly *adhered* to the surface) layer with δ_ℓ of the order of 10 nm, and is called the *absorbed-film region*. The boundary between the transition and the adsorbed-film region is called the *interline* (as compared to the common line for the *partially* wetting liquids, i.e., $\theta_c > 0$). The theory of the van der Waals forces (i.e., potentials) in thin adsorbed films has been reviewed by Dzyaloshinskii et al. (1961) and Israelachvili (1989). The calculation of these forces which involves the frequency-dependent dielectric permittivity ϵ_e of the gas, liquid, and solid phases, the temperature, and the film thickness, is discussed by Truong (1987). Correlations and approximations have been made to the expressions for these forces, and among these are (Truong and Wayner, 1987)

$$\Delta\phi = -\frac{B}{\delta_\ell^3} \qquad \delta_\ell \leq 20 \text{ nm} , \tag{8.166}$$

$$\Delta\phi = -\frac{B}{\delta_\ell^4} \qquad \delta_\ell \geq 40 \text{ nm} . \tag{8.167}$$

In these correlations B also depends on δ_ℓ, for example, in (8.167), $0.32 \times 10^{-29} < -B < 4.36 \times 10^{-29}$ N-m^2 for $10^{-3} < \delta_\ell < 1$ μm for the air(gas)-octane (liquid)-silicon (solid) system. However, B is generally treated as a constant (as was also done in Section 8.3.1).

Using (8.167) in the total potential energy expression (8.146) (i.e., the *modified Young-Laplace equation*) we have

$$2\sigma H - \rho_\ell g x \sin\theta - \frac{B}{\delta_\ell^4} = \Delta\phi_o . \tag{8.168}$$

For $\delta_\ell > 1$ μm, the van der Waals potentials (or the dispersive forces) diminish. Then by choosing the origin of x such that $\delta_\ell(x=0) > 1$ μm, we have

$$\begin{aligned} 2\sigma H - \frac{B}{\delta_\ell^4} &= (2\sigma H - \frac{B}{\delta_\ell^4})_{x=0} + \rho_\ell g x \sin\theta + \rho_\ell g[\delta_\ell - \delta_\ell(x=0)]\cos\theta \\ &\simeq 2\sigma H(x=0) \quad \delta_\ell > 1\ \mu\text{m} , \quad x \geq 0 . \end{aligned} \tag{8.169}$$

The location $x = 0$, which is in the capillary meniscus region near the boundary with the transition region, is shown in Figure 8.17.

The average curvature is given by

$$2H = \frac{\dfrac{d^2\delta_\ell}{dx^2}}{[1 + (\dfrac{d\delta_\ell}{dx})^2]^{3/2}} \simeq \frac{d^2\delta_\ell}{dx^2} . \tag{8.170}$$

Using *two* length scales such that

$$\delta_\ell^* = \frac{\delta_\ell}{\delta_\ell(x=0)} , \quad x^* = \frac{x}{[\delta_\ell(x=0)/2H(x=0)]^{1/2}} ,$$
$$\gamma^4 = -\frac{B}{\sigma 2H(x=0)\delta_\ell^4(x=0)} , \tag{8.171}$$

then (8.169) becomes

$$\frac{d^2\delta_\ell^*}{dx^{*2}} + \frac{\gamma^4}{\delta_\ell^{*4}} = 1 . \tag{8.172}$$

The two boundary conditions are

$$\delta_\ell^*(x^* = 0) = 1 , \quad \frac{d^2\delta_\ell^*}{dx^{*2}}(x^* \to \infty) = 0 . \tag{8.173}$$

The second boundary condition, when used in (8.172), gives

$$\delta_\ell^* = \gamma \, , \quad \frac{d\delta_\ell^*}{dx^*} = 0 \qquad x^* \to \infty \, . \tag{8.174}$$

When this is used after the integration of (8.172), we have

$$\frac{d\delta_\ell^*}{dx^*} = -(2\delta_\ell^* + \frac{2\gamma^4}{3\delta_\ell^{*3}} - \frac{8}{3}\gamma)^{1/2} \, . \tag{8.175}$$

A perturbation solution for *small* γ is found for this by Truong and Wayner (1987) and is

$$\delta_\ell^* = 1 - 2^{1/2}x^* + \frac{x^{*2}}{2} + \frac{2^{3/2}}{3}\gamma\xi + O(\gamma^3) \, , \tag{8.176}$$

where ξ is defined by Truong and Wayner (1987). This shows that the adsorbed layer is reached, i.e., $d\delta_\ell^*/dx^* \to 0$ for $\Delta x_t^* \simeq 2^{1/2}$ (which is the approximate, dimensionless *length* of the transition region).

For the air-octane-silicon system, Truong and Wayner (1987) find $\delta_\ell(x = 0) = 1.078$ μm, $2H(x = 0) = 63$ m^{-1}, and $[\delta_\ell(x = 0)/2H(x = 0)]^{1/2} = 130.8$ μm, for $B = -5.8 \times 10^{-29}$ N-m^2, $\gamma = 0.0743$, and δ_{ℓ_o} (adsorbed film thickness) $= 19.5$ nm. They find that δ_{ℓ_o} is of the order of 10 nm for the fluids examined, and the *length* of the transition region Δx_t is of the order of 100 μm with the transition from 20% capillarity to 20% surface force occurring over a distance of about 20 μm. This illustrates that the transition region is rather *small* in extent (less than 1 mm) with the film thickness in this region being much smaller (less than 1 μm).

(B) Estimation of Evaporation Rate in Transition Region

One of the applications of the extended meniscus has been in liquid film evaporation (Moosman and Homsy, 1980; Mirzamoghadam and Catton, 1988; Kaviany, 1989; Tao and Kaviany, 1991; Swanson and Herdt, 1992). The *thin* liquid film in the transition region can result in a *large* heat transfer rate across the film, and therefore a large evaporation rate, even for a *small* temperature difference (i.e., small liquid superheat). Using the van der Waals potential, for driving the liquid flow (called the *negative* or *disjoining* pressure), the simple and approximate liquid and heat flow analysis of Wayner (1989) allows for the estimation of the evaporation rate across the transition region. This analysis assumes that van der Waals potentials do *not* influence the evaporation from the thin liquid film.

When the van der Waals potentials are added to the liquid momentum equation and for a horizontal solid surface, i.e., $\theta = 0$, and a *negligible* gravity effect, we have (Erneux and Davis, 1993)

$$\rho\frac{D\mathbf{u}_\ell}{Dt} = -\nabla(p_\ell + \Delta\phi) + \mu_\ell\nabla^2\mathbf{u}_\ell \, . \tag{8.177}$$

Now with no external pressure, the one-dimensional thin liquid flow, with no inertial force (i.e., the lubrication-theory approximations), is given by

$$0 = -\frac{d\phi}{dx} + \mu_\ell \frac{d^2 u_\ell}{dy^2} . \tag{8.178}$$

The boundary conditions are

$$u_\ell(y=0) = \frac{du_\ell}{dy}(y=\delta_\ell) = 0 . \tag{8.179}$$

Using the van der Waals potentials given by (8.166), we have

$$\frac{d\phi}{dx} = -\frac{d}{dx}\frac{B}{\delta_\ell^3(x)} = \frac{3B}{\delta_\ell^4}\frac{d\delta_\ell}{dx} \qquad \delta_\ell \le 20 \text{ nm} . \tag{8.180}$$

Using the *film-thickness* averaged velocity $\langle u_\ell \rangle_{\delta_\ell}$, the solutions (8.178) and (8.180) for a small variation of the liquid-film thickness along the x-axis and when written in terms of the liquid mass flow rate per unit depth become

$$\dot{m}_\ell \delta_\ell = \rho_\ell \langle u_\ell \rangle_{\delta_\ell} \delta_\ell = \frac{\delta_\ell^3}{3\nu_\ell}\frac{d\phi}{dx} = \frac{B}{\nu_\ell}\frac{1}{\delta_\ell}\frac{d\delta_\ell}{dx} . \tag{8.181}$$

The change in the film thickness along the x-axis is determined from the energy equation which allows for the evaporation. Also, using an average film thickness $\langle \delta_\ell \rangle_{\Delta x_t}$ across the transition region Δx_t, and assuming a *linear* variation of δ_ℓ across Δx_t (which is a poor approximation), the evaporation rate per unit area is

$$\dot{m}_{\ell g} = -\frac{d\dot{m}_\ell \delta_\ell}{dx} \simeq -\dot{m}_\ell \frac{1}{\langle \delta_\ell \rangle_{\Delta x_t}}\frac{d\delta_\ell}{dx} . \tag{8.182}$$

Then *heat flux* required for this evaporation mass flux is

$$\langle q_{s_b} \rangle_{\Delta x_t} = -\Delta i_{\ell g} \dot{m}_{\ell g} \simeq -\frac{B \Delta i_{\ell g}}{\nu_\ell \langle \delta_\ell \rangle_{\Delta x_t}} (\frac{d\delta_\ell}{dx})^2 . \tag{8.183}$$

Using $\langle \delta_\ell \rangle_{\Delta x_t} = 10$ nm, and $d\delta_\ell/dx = 0.017$ (for an air-octane-gold system with the gold surface heated *slightly above* the boiling point of octane), and typical values for B, discussed in Section (A) above, and the properties of octane, Wayner (1989) shows that $\langle q_{s_b} \rangle_{\Delta x_t} = 8 \times 10^3$ W/m^2. The liquid superheat needed is much *less* than 1°C, and therefore, this is an attractively large heat flux for a small superheat. Since Δx_t is small, then per extended meniscus this heat transfer rate (per unit depth), i.e., $\langle q_{s_b} \rangle_{\Delta x_t} \Delta x_t$, is *not* large. However, a large volumetric concentration of these menisci can be used for a larger heat transfer rate. In practice, the *control* of the small superheat (needed to prevent boiling) can be difficult (Xu and Carey, 1990).

(C) Evaporation Rate in Capillary and Transition Regions

A more accurate description of the evaporation rate (and film thickness distribution) in the transition region has been made by DasGupta et al.

(1993) by allowing for a nonequilibrium liquid-gas interface and using the *kinetically affected* evaporation rate. This model does *not* impose a saturation temperature (corresponding to p_{v_∞}) on the liquid-vapor interface, i.e., $T_{\ell v} \neq T_v(p_{v_\infty})$, so a *nonequilibrium* interface is assumed. The results of the kinetic theory for a *single-component* system is used for the gas mass flux at the liquid-vapor interface. A constant solid surface heat flux q_{s_b} is assumed. Consider evaporation from the extended meniscus shown in Figure 8.17.

For a *horizontal* solid surface, i.e., $\theta = 0$, and for thin liquid films considered, the gravitional potential is negligible compared to the capillarity and surface potentials. Using the thermodynamic relations (Wayner, 1991), the modified Young-Laplace equation can be written in terms of the pressure difference (or *jump*) across the liquid-vapor interface as

$$p_\ell - p_{v_\infty} + 2\sigma H + \Delta\phi = 0 \, . \tag{8.184}$$

Then using the potential given by (8.166) and approximating the expression for the average surface curvature, we have for the two-dimensional meniscus of Figure 8.17

$$p_\ell - p_{v_\infty} = \frac{B}{\delta_\ell^3} - \sigma \frac{d^2 \delta_\ell}{dx^2} \, . \tag{8.185}$$

Now using a nonequilibrium interface and the gas mass flux determined from the kinetic theory (Schrage, 1953; DasGupta et al., 1993), for $T_{v_\infty} = T_{v_\infty}(p_{v_\infty})$, i.e., saturated, *uniform* vapor temperature and pressure, and using the pressure jump across the interface $p_\ell - p_{v_\infty}$, and the *unknown* interface temperature $T_{\ell v}$, we have

$$\dot{m}_{\ell g} = a_1 (T_{\ell v} - T_{v_\infty}) + a_1 (p_\ell - p_{v_\infty}) \, , \tag{8.186}$$

where the kinetic variables are

$$a_1 = 2 (\frac{M}{2\pi R_g T_{\ell v}})^{1/2} \frac{p_{v_\infty} M \Delta i_{\ell g}}{R_g T_{v_\infty} T_{\ell_v}} \, , \tag{8.187}$$

$$a_2 = 2 (\frac{M}{2\pi R_g T_{\ell v}})^{1/2} \frac{v_{\ell_\infty} p_{v_\infty}}{R_g T_{v_\infty}} \, , \tag{8.188}$$

where v_{ℓ_∞} is the *molar* specific volume of the liquid and M is the molecular weight (a single-component system is considered). The far-field conditions for the vapor and the interfacial quantities are also shown in Figure 8.17.

Now assuming heat *conduction* across the thin liquid film in the transition region, we have

$$\frac{k_\ell}{\delta_\ell} (T_{s_b} - T_{\ell v}) = \dot{m}_{\ell g} \Delta i_{\ell g} \, . \tag{8.189}$$

Since a constant surface heat flux q_s is assumed for an *ideally* conducting solid, T_{s_b} remains constant while $T_{\ell v}$ varies as δ_ℓ varies. The evaporation rate is determined using the liquid momentum equation, as discussed in

Section (B), and (8.181) is rewritten as

$$\dot{m}_{\ell g} = -\frac{d\dot{m}_\ell \delta_\ell}{dx} = \frac{1}{3\nu_\ell}\frac{d}{dx}\delta_\ell^3\frac{dp_\ell}{dx}\,. \tag{8.190}$$

Now combining this with (8.186) and (8.189), by eliminating $T_{\ell v}$ between the latter two equations, we have (DasGupta et al., 1993)

$$\begin{aligned}\dot{m}_{\ell g} &= \frac{1}{3\nu_\ell}\frac{d}{dx}\delta_\ell^3\frac{dp_\ell}{dx}\\ &= \frac{1}{1+\dfrac{a_1\Delta i_{\ell g}\delta_\ell}{k_\ell}}[a_1(T_{s_b}-T_{v_\infty})+a_2(p_\ell-p_{v_\infty})]\,.\end{aligned} \tag{8.191}$$

Equations (8.185) and (8.191) describe the variations of the liquid pressure p_ℓ and the liquid film thickness δ_ℓ with respect to x. These equations are made dimensionless by using

$$\delta_\ell^* = \frac{\delta_\ell}{\delta_{\ell_o}}\,,\quad p_\ell^* = \frac{p_\ell - p_v}{\dfrac{a_1}{a_2}(T_{s_b}-T_{v_\infty})}\,,\quad x^* = \frac{x}{[\dfrac{-B}{\nu_\ell a_1(T_{s_b}-T_{v_\infty})}]^{1/2}}\,,$$
$$a_1^* = \frac{a_1\Delta i_{\ell g}\delta_{\ell_o}}{k_\ell}\,,\quad a_2^* = \frac{a_2\sigma\nu_\ell\delta_{\ell_o}}{-B}\,, \tag{8.192}$$

where δ_{ℓ_o} is the thickness of the adsorbed film. In these equations, a_1^* is the ratio of the *kinetic* and *conductive* resistances, and a_2^* is the ratio of the *surface tension* (and viscous force) to the *van der Waals* forces.

The dimensionless equations are

$$p_\ell^* = -\frac{1}{\delta_\ell^{*3}} - a_2^*\frac{d^2\delta_\ell^*}{dx^{*2}}\,, \tag{8.193}$$

$$\frac{1}{3}\frac{d}{dx^*}\delta_\ell^{*3}\frac{dp_\ell^*}{dx} = \frac{1}{1+a_1^*\delta_\ell^*}(1+p_\ell^*)\,. \tag{8.194}$$

The boundary conditions for the *absorbed-film region asymptote* and the *capillary-meniscus region asymptote* are

$$\begin{aligned} p_\ell^* &= -1\,,\quad \delta_\ell^* = 1 \quad x^* \to \infty\,,\\ p_\ell^* &= p_{\ell_m}^*\,,\quad \delta_\ell^* \to \infty \quad x^* \to -\infty\,,\end{aligned} \tag{8.195}$$

where $p_{\ell_m}^*$ is the prescribed normalized liquid pressure in the *capillary-meniscus region* and far from the boundary of this region and the transition region.

Equations (8.193) and (8.194) are solved numerically as described by DasGupta et al. (1993). Some of their results are shown in Figures 8.18(a) to (c) for the distributions of liquid film thickness δ_ℓ, the dimensionless liquid pressure p_ℓ^*, the dimensionless liquid-vapor interfacial curvature $d^2\delta_\ell^*/dx^{*2}$, and the dimensionless evaporation rate $\dot{m}_{\ell g}^* = \dot{m}_{\ell g}/a_1(T_{s_b} - T_{v_\infty})$. The results are for an air-heptane-silicon system with $\delta_{\ell_o} = 5.6$ nm, $a_1^* = 2.44\times10^{-2}$, $a_2^* = 1.05$, $a_1(T_{s_b}-T_{v_\infty})/a_2 = 6.32\times10^2$ N/m^2, $a_1(T_{s_b}-T_{v_\infty}) =$

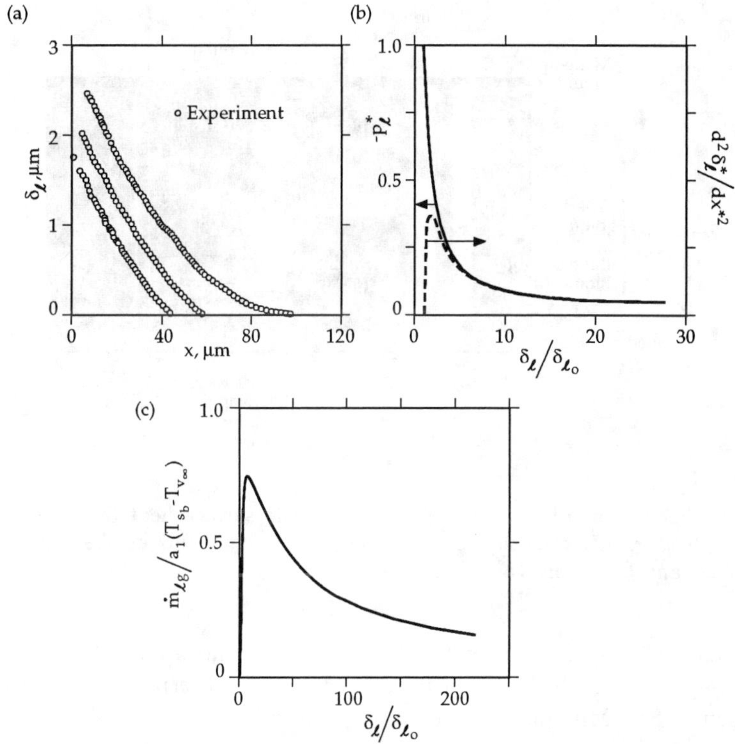

Figure 8.18. (a) Distribution of the liquid-film thickness. (b) Variation of the dimensionless liquid pressure and the interfacial curvature, with respect to the normalized liquid-film thickness. (c) Variation of the dimensionless evaporation rate with respect to the normalized liquid-film thickness. The results are for the air-heptane-silicon system. (From DasGupta et al., reproduced by permission ©1993 American Society of Mechanical Engineers.)

1.15×10^{-3} kg/m^2-s, $B = 1.14 \times 10^{-22}$ J, $T_{s_b} - T_{v_\infty} = 7.6 \times 10^{-4}$ K, $a_1(T_{s_b} - T_{v_\infty})\Delta i_{\ell g} = 4.17 \times 10^2$ W/m^2. Their measured film thickness distribution (using ellipsometry) is also shown in Figure 8.18(a) and good agreement is found compared with the predicted results. The applied liquid superheat, $T_{s_b} - T_{v_\infty}$, is *very small* (because of the low q_{s_b} applied). The evaporation rate has a *maximum* between the capillary-meniscus region, where because of the large δ_ℓ the evaporation rate is negligible (for *conduction-controlled* heat transfer), and the *near* adsorbed-film region, where the temperature drop across the thin film is very large (*kinetically controlled*). The *viscosity* of the liquid also limits the liquid flow caused by the curvature capillarity and the surface forces. This *viscosity effect* exists throughout the transition region. The maximum occurs where the liquid-film conduction resitance, viscous resistance, and interfacial evaporation

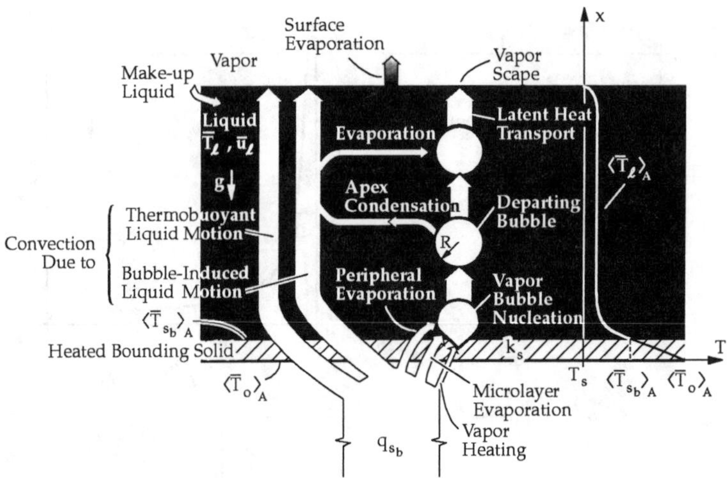

Figure 8.19. A rendering of the division of the surface heat flux, among the various heat transfer mechanisms in the liquid phase and in the nucleating-, growing-, and departing-vapor bubbles.

kinetic resistance are all significant. Here the thermocapillarity is *negligible* because of the small temperature variations. The thermocapillarity effect is discussed by Mirzamoghadam and Catton (1988).

8.4 Kinetic Upper Bound on Evaporation/Condensation Rate

The *continuous vapor* (or *condensate*) production rate is *controlled* by the *rate of heat* delivered (or removed) to the liquid-vapor interface and by the *rate of vapor* (or condensate) removal. The evaporation in a *pool* of liquid (i.e., *no* net liquid motion) was discussed in Section 8.1.1 and in Figure 8.2. It was indicated that a *maximum* in q_{s_b} is observed at a surface superheat $T_{s_b} - T_s$. This maximum (or the critical heat flux) is determined by the heat delivery and the vapor removal limitations.

In Figure 8.19, an approximate *division* of q_{s_b} among the various heat transfer mechanisms is shown (Leiner and Gorenflo, 1992) for a pool of liquid under steady-state heating from below. The flow of heat experiences resistances as it flows through the heated solid surface and *toward* the *vapor* bubble nucleated on the surface (this bubble nucleation will be discussed in Section 8.5.1) and *toward* the *liquid*. Because of the *cyclic* bubble formation and departure, the temperature distribution in the solid and the adjacent liquid changes rapidly, i.e., the *space* and *time* derivatives of the local solid and liquid temperatures are very large. The convection through the liquid

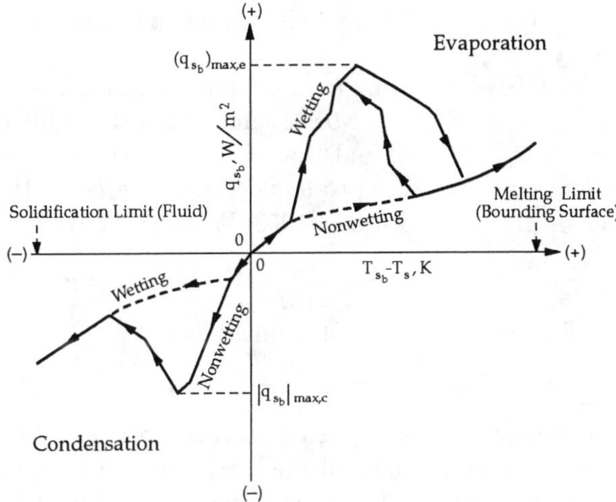

Figure 8.20. Variation of the solid surface temperature with respect to the surface heat flux for heat addition to a liquid ($T_{s_b} > T_s$), i.e., evaporation, and heat removal from a vapor ($T_{s_b} < T_s$), i.e., condensation. The maximum heat flux added for the nucleate boiling and the maximum heat flux withdrawn for the dropwise condensation are also shown.

occurs with the motions *influenced* by the thermobuoyancy (unstable liquid density distribution) and by the phase-density (or bubble) buoyant motion.

The contribution of the bubble-induced motion is expected to be significant. The heat flow towards vapor is through the solid-vapor interface (expected to be a small fraction of q_{s_b}) and through the liquid-vapor interface. The heat flow through a thin liquid layer underneath the attached bubble (called the *microlayer*) is expected to be large compared to that through the remainder of the bubble periphery. The heat transfer to the bubble is labeled as the *latent* heat transport, and for the example given in Figure 8.20 is only a fraction of q_{s_b}. The division between the *sensible* and *latent* heat transfer will be discussed in Section 8.5.2.

A generalization of Figure 8.1 that also addresses *condensation* is suggested by Gambill and Lienhard (1989) and is given in Figure 8.20. For evaporation, the solid surface *wetting* by the liquid is required for a *continuous* bubble formation. For an *ideally nonwetting* liquid, only a vapor film is formed (after a threshold surface superheat $T_{s_b} - T_s$, which can be *small*). For wetting liquids a maximum in q_{s_b} indicates the *smallest combined* resistance to the heat delivery (to the liquid-vapor interface) and vapor removal. A similar behavior is expected for condensation (*negative* q_{s_b} and $T_{s_b} - T_s$) and for a nonwetting liquid a *dropwise* condensation (similar to the surface bubble nucleation) occurs. This will be discussed in Section 8.6.

Gambill and Lienhard (1989) suggest that as the combined resistance to the heat delivery (or removal) and vapor (or condensate) removal is decreased, $(q_{s_b})_{\max,e}$ (or $|q_{s_b}|_{\max,c}$) can be increased up to the limit determined by kinetic theory.

The molecular flux evaporating from a liquid surface was given in terms of the liquid properties by (1.45). Now assuming that the liquid evaporation takes place at an *infinite* rate and that vapor molecules are *immediately* removed (*no* molecules returning to liquid-vapor interface), the mass flux of gas molecule is given by (Schrage, 1953; Walton, 1989)

$$\dot{m}_{\ell v} = \frac{mn_v}{4}\overline{u}_m \ , \quad \overline{u}_m = (\frac{8R_g T_v}{\pi M})^{1/2} \ , \quad mn_v = \rho_v \ . \tag{8.196}$$

Then the heat flux associated with this mass flux is

$$q_{\max} = \dot{m}_{\ell v}\Delta i_{\ell g} = \rho_v(\frac{R_g T_v}{2\pi M})^{1/2}\Delta i_{\ell g} \ . \tag{8.197}$$

As the temperature *increases*, $\rho_v \Delta i_{\ell g}$ *increases* and $q_{\max}$ increases until the *critical point* is approached and the $\Delta i_{\ell g}$ tends to zero. Gambill and Lienhard (1989) evaluate $q_{\max}$ for water, ethanol, and Freon-12, and their results are shown in Figure 8.21(a) for variation of $q_{\max}$ as a function of the *reduced pressure* $p_r^* = p/p_c$. As is shown, very large heat flux is possible by delivering the heat directly on the liquid-vapor interface and by removing the vapor once it is formed. This *ideal* heat transfer (i.e., no thermal resistance) would require $T_{s_b} = T_{\ell v} = T_v$.

In practice only a fraction of $q_{\max}$ is obtained and the dimensionless maximum *solid bounding surface* heat flux

$$(q^*_{s_b})_{\max,i} = \frac{|q_{s_b}|_{\max,i}}{q_{\max}} \qquad i = e, c \tag{8.198}$$

is less than unity. Gambill and Lienhard (1989) compile the experimental results for $(q^*_{s_b})_{\max,i}$, and these results are also given in Figure 8.21(b). They suggest that since $\rho_v \Delta i_{\ell g}$ increases with p_r^*, then *less* vapor (or liquid) is produced for a given q_{s_b}, as p_r^* increases. Then a large fraction of the heat flows through the liquid (or vapor). Numerous techniques introducing high velocities, swirls, jets, and subcooling have been used in the results compiled in Figure 8.21(b). Note that an experiment with water at $q_{\max} = 2 \times 10^9$ W/m^2 using a solid slab of silver (high k_s) with a 1-mm thickness, requires a $T_o - T_{s_b} = 4700$ K. This is *not* possible because it *melts* the solid. Other practical limitations preventing the achievement of $q_{\max}$ are discussed by Gambill and Lienhard (1989).

8.5 Surface Bubble Formation and Dynamics

Here we address the role of convective heat transfer in the evaporation rate of a liquid heated on a surface with a superheat in excess of that required for the inception of surface bubble nucleation, i.e., $(T_{s_b} - T_s)_{\rm inc}$. This role may be significant in the surface bubble-nucleation regimes shown in Figure 8.1,

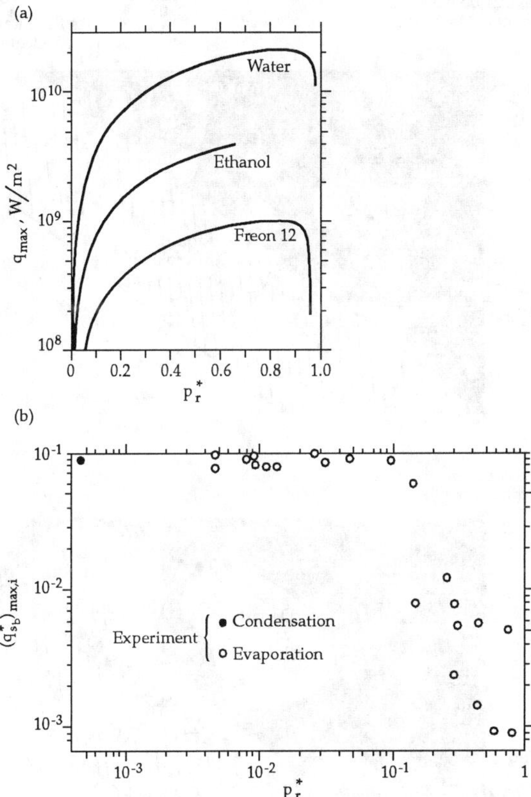

Figure 8.21. (a) Variation of the predicted, kinetic-limit heat flux for evaporation of water, ethanol, and Freon 12, with respect to the reduced pressure. (b) Variation of the ratio of the measured maximum heat flux for evaporation and condensation to the predicted limit, with respect to the reduced pressure. (From Gambill and Lienhard, reproduced by permission ©1989 American Society of Mechanical Engineers.)

where the bubbles are isolated or when jets and vapor columns are forming but *no* bubble *crowding* has yet occurred. In the bubble coalescence regime, the role of the convective heat transfer becomes *less* significant. Presently *separate* models are used for the isolated- and crowded-bubbles regimes. These models rely significantly on the *geometrical*, the *surface conditions*, and the *hydrodynamic* parameters. In the following, first a discussion of the incipient surface superheat is made. Then some of the models for the isolated-bubbles regime and the coalesced-bubbles regime are reviewed. These models are rendered in Figures 8.22(a) and (b). The role of *gravity* (and heated surface orientation) and the *liquid* free-stream condition, T_{ℓ_∞}, p_{ℓ_∞}, $\mathbf{u}_{\ell_\infty}$, as well as the surface conditions and the surface superheat, are included in some of these models. The complete analysis of these models is rather impossible because of the large number of variables.

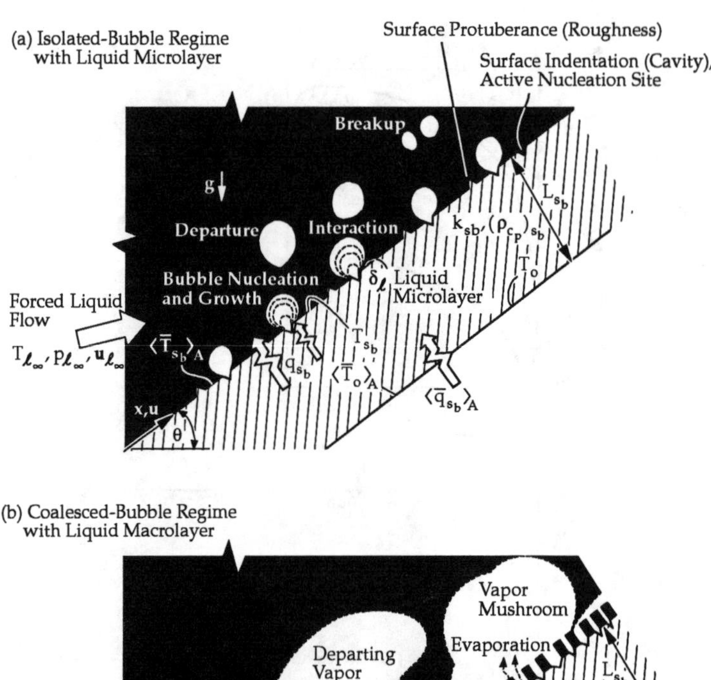

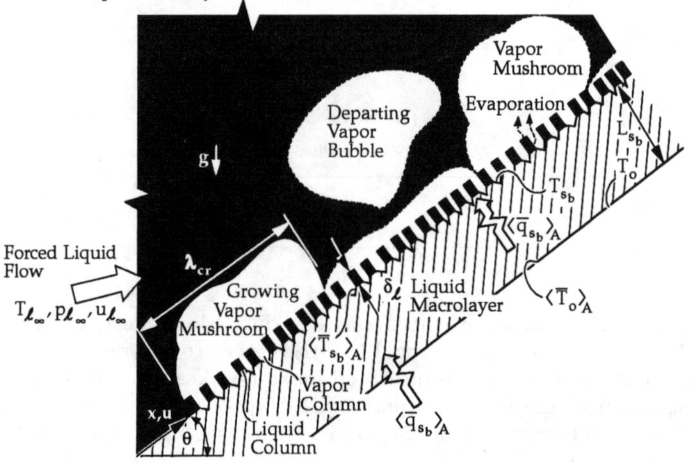

Figure 8.22. Surface bubble nucleation regime with (a) isolated bubbles and (b) coalesced bubbles. The heated bounding surface and the forced liquid flow are also shown.

8.5.1 Incipient Surface Bubble Nucleation

As the solid surface temperature T_{s_b} is *raised* above the saturation temperature T_s and beyond a *threshold superheat* $T_{s_b} - T_s$, surface vapor bubble nucleation occurs. The bubble nucleation occurs at fixed *sites* which are randomly distributed over conventionally prepared (e.g., polished) surfaces. This incipient solid surface superheat can be very small for rough surfaces and wetting liquids with a *slow* heating rate of the surface. However, for smooth surfaces and perfectly wetting liquids, and when the surface is *rapidly* heated, this superheat can reach the *homogeneous* superheat limit (e.g., Iida et al., 1993) as predicted by the thermodynamic considerations, i.e., (1.57), or the kinetic theory, i.e., (6.1). This range of superheat for the incipient surface vapor bubble nucleation $(T_{s_b} - T_s)_{\rm inc}$ was shown in Figure 8.1. The magnitude of $(T_{s_b} - T_s)_{\rm inc}$ depends on the *surface conditions*, such as the *surface roughness* (including the *statistics* of size, geometry, angle of orientation with respect to the normal surface, etc.), the *surface cavities* (size, geometry, orientation, connectivity), the *surface coating* (constituents, thickness, distribution along the surface), and noncondensable gas trapped in the cavities (amount in the cavities, fraction *adsorbed* to the cavity surface by the van der Waals forces). It also depends on the *fluid* and the *hydrodynamics conditions*, such as the *disolved gas* in the liquid, the *pressure*, the *wettability* (contact angle), the *liquid velocity* adjacent to the surface, the magnitude and orientation of the *gravitational acceleration*, the *surface geometry*, and the *liquid subcooling* $T_s - T_{\ell_\infty}$. The hydrodynamics and the subcooling influence the *temperature distribution* on the surface as well in the liquid adjacent to the surface.

This dependence on a large number of parameters has resulted in a lack of a general prediction for $(T_{s_b} - T_s)_{\rm min}$. However, predictions for specific and *limited* range of *conditions* have been made by introducing modifications to the homogeneous superheat limit and a summary is given by Cole (1974). He uses another approximate form of the kinetic theory prediction of the homogeneous superheat limit, i.e.,

$$T_{sp} - T_s = \frac{T_s}{\rho_v \Delta i_{\ell g}} \left(\frac{16\pi\sigma^3}{3k_B T_{sp} \ln \dfrac{n_\ell k_B T_{sp}}{h_P}} \right)^{1/2} , \qquad (8.199)$$

where in this *implicit* relation for T_{sp} the parameters other than those defined in connection with (6.1) are the Planck constant h_P which enters through the specification of the frequency of collisions, resulting in the formation of *vapor clusters*.

The earlier attempts have been along the correlations developed for the minimum superheat for the vapor-film evaporation given by (8.49) and (8.50). For example, one single-variable modification of (8.199) to accommodate the surface, fluid, and hydrodynamic effects has been (Cole, 1974)

$$\begin{aligned} T_{sp} - T_s &= \frac{T_s}{\rho_v \Delta i_{\ell g}} \left(\frac{16\pi\sigma^3 f_n}{3k_B T_{sp} \ln \dfrac{n_\ell k_B T_{sp}}{h_P}} \right)^{1/2} \quad 0 \le f_n \le 1 , \\ f_n &= f_n(\theta_c, \text{ surface cavity, trapped gas, etc.}) , \end{aligned} \qquad (8.200)$$

where the *surface nucleation factor* f_n has been determined for some simple cases. For the upper asymptote of $f_n \rightarrow 1$, the heterogeneous and homogeneous superheat limits are identical. This is expected to be the case for *perfectly* wetting liquids. For *ideally* nonwetting liquids, $f_n \to 0$ and no significant superheat is needed before a *vapor film* is formed on the surface.

The relation between f_n and the contact angle, surface cavity parameters, and the amount of noncondensable gas trapped in the cavities has been reviewed by Cole (1974) and van Stralen and Cole (1979).

Nghiem et al. (1981) use *two* variables f_n and $\dot{n}_b$ (the volumetric bubble nucleation rate) and observe that $\dot{n}_b$ varies with the *heating rate* (or for a solid film of finite thickness, the *impulsive*, i.e., step, heat flux q_{s_b}). Further discussions on the incipient surface bubble nucleation are given by Cornwell (1981), Bar-Cohen (1992), Bräuer and Mayinger (1992), Carey (1992), Jansen et al. (1992), Leiner (1992), and Zhukov and Barelko (1992). These include some *localization* phenomena (Kennig, 1992, 1993).

Application of the surface nucleation factor f_n to the prediction of the volumetric nucleation rate, has been suggested by Cole (1974) and Nghiem et al. (1981) and the volumetric homogeneous vapor bubble nucleation rate given by (6.55) is modified similar to (8.200). The surface area, heterogeneous bubble nucleation rate, which is the *product* of the *active* surface nucleation sites and the frequency of the bubble departure, has also been addressed using empirical relations (Judd, 1981; Kocamustafaogullari and Ishii, 1983; Barthau, 1992; Fujita, 1992). Up to 5000 nucleation sites per cm^2 have been measured by Barthau (1992) on a grinded (average roughness of about 1 μm) copper tube surface. The number of active nucleation sites *increases* with the heat flux q_{s_b} and *decreases* with pressure (and *decreases* as the wettability increases; Wang and Dhir, 1992).

8.5.2 Isolated-Bubbles Regime

A rendering of the surface bubble nucleation, growth, and departure on a heated solid surface is given in Figure 8.22(a). The surface bubble *nucleation* number per unit area N_b/A is influenced by the surface, fluid, and hydrodynamic parameters and, for example, decreases as the free-stream liquid velocity $\mathbf{u}_{\ell_\infty}$ increases (Zeng and Klausner, 1993). The *growth* of the *bubbles* on the surface is influenced by the liquid motion around the bubble as well as the heat transfer through the liquid and the solid surface. The *bubble departure* is caused by a combination of the *phase-density buoyancy force*, the *hydrostatic pressure gradient,* the *interfacial drag force,* the *liquid flow inertia*, and the *surface forces*, and other forces, the hydrodynamic consideration leads to a semiempirical prediction of the *departure diameter* (Klausner et al., 1993). The analysis of the convective heat transfer to the bubble is coupled to the hydrodynamics and heat transfer adjacent to the heated surface and the presence of surface bubble nucleation and growth. Because of the strong effect of the solid surface, the analyses of bubble dynamics, in the absence of a solid surface (given in Section 6.3), do *not* directly apply.

The hydrodynamics and heat transfer aspects of the *isolated-bubbles*

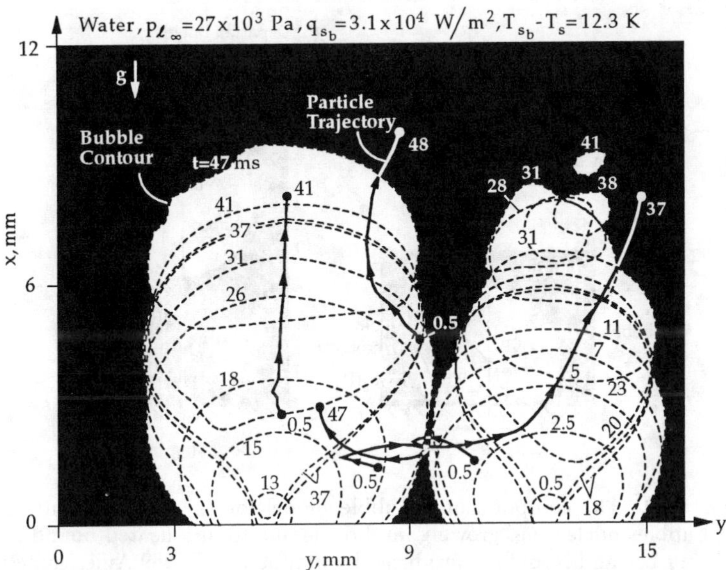

Figure 8.23. Controlled bubble nucleation, growth, and departure on a heated surface. The bubble contour at various elapsed times and the tracer-particle trajectory are shown. (From Nakayama and Kano, reproduced by permission ©1992 Scripta Technica.)

regime have been analyzed as an extension of the thermobuoyant boundary-layer flow and heat transfer, by Zuber (1963), where an analogy between the bubbles and the *thermals* (i.e., blubs of the fluid departing the surface due to thermobuoyancy, discussed in Section 2.7) has been used. Further refinements in these models have included the liquid motion induced by the bubble growth, the presence of turbulence, and more elaborate models for the thickness of the liquid *microlayer* (e.g., Ruckenstein, 1966; van Stralen and Cole, 1979).

As an illustrative example of the hydrodynamics around a bubble nucleation site on a horizontal surface ($\theta = 0$), the visual recording of Nakayama and Kano (1992) are shown in Figure 8.23. The *controlled* bubble nucleation is by *two* hydrogen-generating electrodes separated by 7.2 mm placed on a heated brass surface in a pool of water. The growth of the hydrogen-vapor bubble is controlled in part by the current provided by the electrodes (and mostly by the heat transfer to the liquid-vapor interface resulting in vaporization). The bubbles are photographed every 0.5 ms, and a composite drawing is given in Figure 8.23, along with the traces of *four* mutually buoyant particles suspended in the liquid boundary layer and around the bubble sites. With the spatial *resolution* shown, it *appears* that the bubbles grow as a nearly hemispherical cap and before their departure the contact (i.e., common) line *recedes*. In modeling of the surface bubble growth a liquid *microlayer* is prescribed underneath the bubble so that the bubble

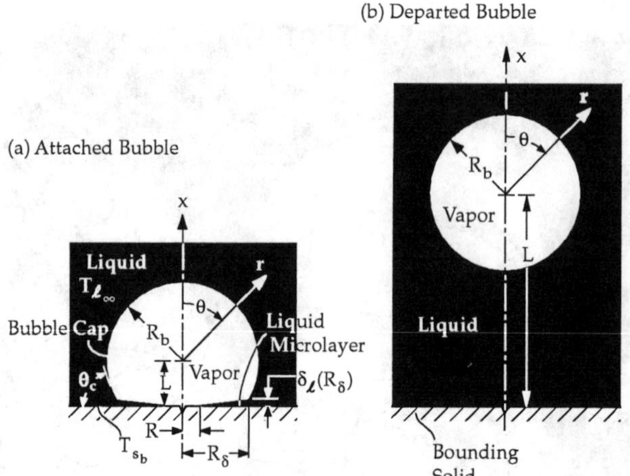

Figure 8.24. The computational bubble model for (a) attached and (b) detached bubbles nucleating, growing, and departing from a heated bounding surface. (From Lee and Nydahl, reproduced by permission ©1989 American Society of Mechanical Engineers.)

remains *attached* at the edge of the surface cavity from which it originates. The bubbles shown in Figure 8.23 grow about 6 mm before departure and the right-hand side bubble which is nucleated first breaks up into smaller bubbles a short time after departure. The lateral and upward motion of the liquid is shown through the traces of the suspended particles, and the lateral motion is more pronounced in a region near the surface. The thickness of this region is approximately equal to the *radius* of the departing bubbles.

The heat transfer analysis of the *attached* and *departed* bubbles is made by using a computational bubble *model* depicted in Figures 8.24(a) and (b). The specification of the *attached* bubble model includes the *dry patch* radius R_d, the apparent contact angle θ_c, and the thickness of the liquid microlayer at its outer edge $\delta_\ell(R_\delta)$. Then the radius of the microlayer region R_δ can be determined. The prescription of $\delta_\ell(R_\delta)$ is made using the *hydrodynamics* of a *growing* bubble (van Stralen and Cole, 1979) and has a $t^{1/2}$ time dependence. Numerical simulations of the axisymmetric bubble growth and departure have been made by Lee and Nydahl (1989). The laminar, transient continuity, momentum (without thermobuoyancy), and energy equations for the liquid phase are solved using a coordinate transformation which maps the physical domain into a two-dimensional domain that facilitates discretization. The variation of the liquid pressure with the height above the surface, and variation of the vapor pressure with the surface curvature, through the Laplace equation (6.42), and with the saturation temperature, through the Clapeyron equation (1.51), are also

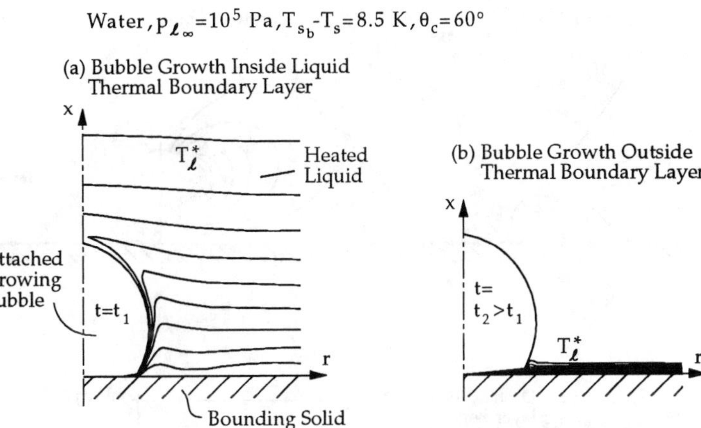

Figure 8.25. Predicted temperature distribution around a growing bubble during the early growth period. (a) Elapsed time corresponding to a small bubble growing inside the liquid thermal boundary layer. (b) Elapsed time corresponding to a bubble size larger than the liquid thermal boundary layer. (From Lee and Nydahl, reproduced by permission ©1989 American Society of Mechanical Engineers.)

allowed. Typical numerical predictions for the *early growth*, i.e., when the bubble is *within* the thermal buoyancy layer, and for the *later growth*, when the bubble becomes much larger than the thermal bound layer, are shown in Figures 8.25(a) and (b). Since this early growth takes place over a short time, the microlayer heat transfer which provides most of the heat transfer during the later growth dominates the cycle-averaged heat transfer. Then the heat transfer through the *bubble cap* is not a significant portion of the cycle-averaged heat transfer.

The liquid flow and temperature fields around the bubble for large elapsed times corresponding to the *before* and *after* departures are shown in Figures 8.26(a) and (b). The liquid flows underneath the departing bubble (called the *wall scavenging* effect), causing a temperature disturbance which is limited to a small region around the line passing through the bubble center. The heat transfer *due* to the growth and departure of the bubble has been determined by Lee and Nydahl (1989) by subtracting the heat transfer due to the *transient heat conduction* to the liquid $(q_{s_b})_o$, i.e.,

$$\Delta Q_{s_b} = 2\pi \int_0^\infty q_{s_b} - (q_{s_b})_o r dr \ . \tag{8.201}$$

The result for this instantaneous *excess* heat flow per bubble, for the conditions used in their numerical simulation, is shown in Figure 8.26(c). The *first* peak in ΔQ_{s_b} is at the elapsed time corresponding to the radius of the bubble becoming equal to the departure radius and the beginning of the *destruction* of the liquid microlayer to accommodate a spherical bubble shape. This destruction decreases the heat transfer rate. The *second* peak

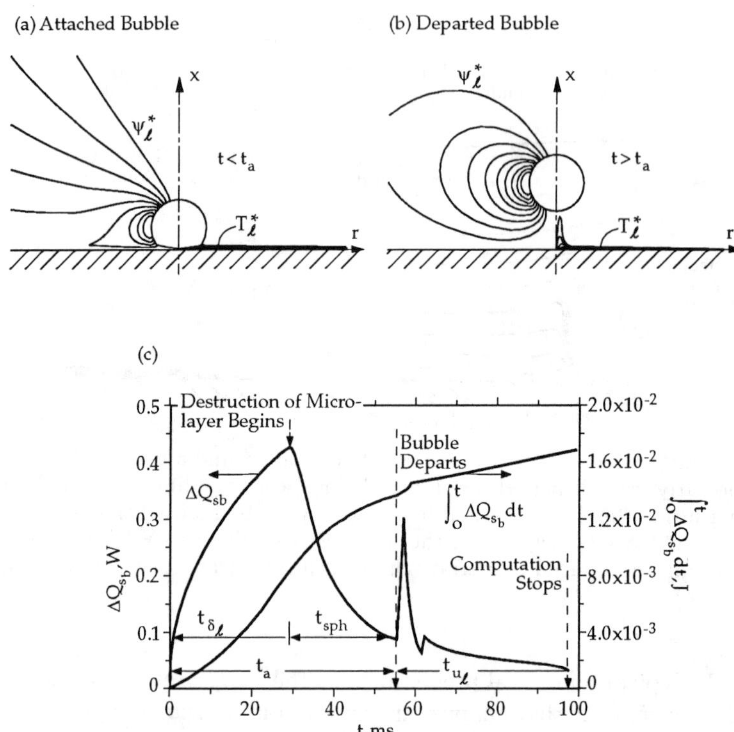

Figure 8.26. Predicted temperature and liquid flow field around (a) an attached and (b) a departed bubble. (c) The predicted temporal variation of the instantaneous and accumulated excess heat flow per bubble. (From Lee and Nydahl, reproduced by permission ©1989 American Society of Mechanical Engineers.)

is just after the departure time. The *microlayer period* t_{δ_ℓ}, the *shape-changeperiod* t_{sph}, the sum of these periods referred to as the *attached-bubble period* t_a ($t_a = t_{\delta_\ell} + t_{sph}$), and the *liquid-convection period* t_{u_ℓ} are marked in the figure. The required prescription of δ_ℓ and the required shape-change period, involve empiricism and computational integration constraints which influence the bubble *model*.

For *periodic* departures, a complete cycle will have an attached-bubble (i.e., *growth*) period t_a and a waiting (i.e., liquid convection) period (Judd, 1989) t_w. Then the *frequency* of the event involving the bubble nucleation, growth, and departure and nucleation of the next bubble, f_b, is $1/(t_a + t_w)$. The heat transfer *rate* per unit area q_{s_b} is found by multiplying the heat transferred by each bubble over the period $t_a + t_w$, by the number of active nucleation sites per unit area N_b/A, and the frequency f. The results of Lee and Nydahl (1989) are in general agreement with the available experimental results and show how the combined effect of the latent heat transport and the convection due to the phase-density buoyant motion increases the surface heat transfer rate. As was shown in Figure 8.19, this division of the heat transfer among the various modes depends on the liquid subcooling, the far-field liquid velocity $\mathbf{u}_{\ell_\infty}$, and other variables, and also on the particular surface bubble-nucleation *regime*. For example, Guo and El-Genk (1992) experimentally show the effect of the surface orientation, and Ulucakli and Merte (1990) show the effects of the gravitational constant and the subcooling on the heat transfer rate. The combined effects of the surface orientation and a small far-field liquid velocity are examined by Kirk et al. (1992). The experimental results on the effect of an *imposed local* liquid motion, i.e., a liquid jet, on the heat transfer are summarized by Wolf et al. (1993). These effects have not been included in any simulations, because of the anticipated extreme complexity of these simulations. *Imposition* of *time* and *space periodicity* and integration over these periods allows for a direct comparison with the experimental results which are generally time- and space-averaged.

8.5.3 Coalesced-Bubbles Regime

A rendering of the formation and departure of bubble mushrooms on a heated surface is given in Figure 8.22(b). As the surface heat flux q_{s_b} (or surface superheat $T_{s_b} - T_s$) increases, the bubbles coalesce and large mushroom-shape bubbles appear on the surface and periodically depart. Observations have shown that the *liquid* reaches the solid surface *inside* these vapor mushrooms and evaporates, causing their growth. When the phase-density buoyant and interfacial viscous drag force can overcome the surface tension forces, the bubble departs, as depicted in Figure 8.22(b). Many models have been proposed for the evaporation within the vapor mushrooms. Most of them assume a *discontinuous* liquid layer called the *macrolayer* which surrounds the vapor columns or chimneys which are envisioned as being on the nucleation sites. One of these models (Dhir and Liaw, 1989) assumes a cylindrical vapor column of diameter d and a *square* unit-cell *base* of linear dimension ℓ, as depicted in Figure 8.27. This linear dimension is taken as the average distance between two adjacent *active* nucleation sites. As was mentioned, this distance depends on p_{ℓ_∞}, u_{ℓ_∞}, θ_c, and q_{s_b}, and in the model of Dhir and Liaw (1989) is given in the form

$$\ell = a_1(p_{\ell_\infty}, u_{\ell_\infty}, \theta_c) q_{s_b}^{-0.75} , \qquad (8.202)$$

where the function $a_1 = a_1(p_{\ell_\infty}, u_{\ell_\infty}, \theta_c)$ is assumed known. The diameter of the vapor column is related to the *surface void fraction* ϵ_{s_b} which is found empirically as

$$\epsilon_{s_b} = \epsilon_{s_b}(T_{s_b} - T_s, \theta_c) . \qquad (8.203)$$

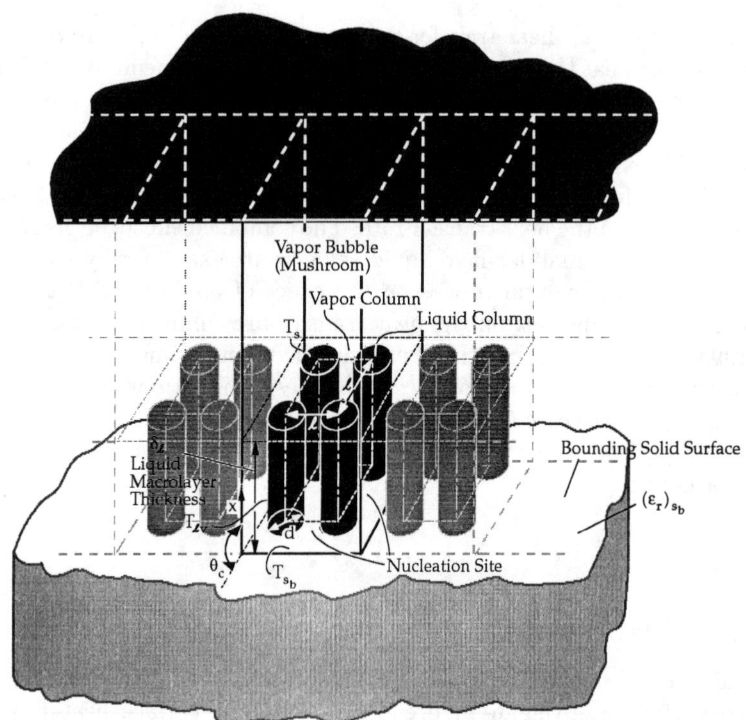

Figure 8.27. The unit-cell model for the surface bubble nucleation and growth in the coalesced-bubbles regime. The liquid macrolayer thickness and liquid-column spacing are also shown.

Then d is determined from the *two geometrical parameters* of the model, i.e., the prescribed ϵ_{s_b} and ℓ, using the geometric relation

$$\epsilon_{s_b} = \frac{\pi}{4}\frac{d^2}{\ell^2} . \tag{8.204}$$

The liquid filling the space between the vapor columns extends to a distance δ_ℓ from the surface (this is the liquid macrolayer thickness). Evaporation is allowed (but d is assumed constant along the column axis) from the liquid vapor interface $A_{\ell v}$ by assuming that this surface is *not* at equilibrium, i.e., $T_{\ell v} \neq T_s(p_v)$, and then the evaporation rate is prescribed using the result of the kinetic theory for the maximum evaporation rate, i.e., (8.196). Because of the rather insignificant subsequences in the model of Dhir and Liaw (1989) instead of the interfacial temperature $T_{\ell v}$, which is determined from the steady-state conduction equation in the liquid column, the saturation temperature is used in (8.196).

The liquid column *removes* heat from the surface as an *extended* surface (by conduction), and this heat is in turn removed from the liquid surface by evaporation. The temperature of the tip of the liquid column is assumed to be the saturation temperature T_s. Since the evaporative heat flux is very large (however, the liquid cross-section is *not* allowed to change along the column axis), the temperature in the liquid column decreases from T_{s_b} at the base to T_s over a short distance, and therefore, the length of the liquid column δ_ℓ is not a significant parameter and in the model of Dhir and Liaw (1989) $\delta_\ell \rightarrow \infty$. It is assumed that the liquid is constantly supplied to the surface, and therefore, a steady-state conduction-evaporation is used. The heat conduction through the liquid, with these boundary conditions, is solved to determine the heat transfer through the liquid at the solid–liquid interface. For contact angles other than 90°, Dhir and Liaw (1989) suggest different liquid column geometries. The effect of wettability is also discussed by Wang and Dhir (1993).

Other variations of this macrolayer evaporation model for the coalesced-bubbles (or bubble-growding) regime are discussed by Katto (1985), Lienhard (1988), Jairajpuri and Saini (1991), Unal et al. (1992), Bonetto et al. (1993), Shoji (1992), and Gallaway and Mudawar (1993). In some of these models, the *hydrodynamic aspects* are also included. Dhir and Liaw (1989) suggest that at high heat fluxes the temperature gradient are very large, and the phenomena occurring over a distance of the order 10 μm, *controls* the heat removal and evaporation rates. Then the effect of the liquid free-stream velocity and the surface orientation is expected to diminish, and this is in general agreement with the experimental observations. Based on this, Dhir and Liaw (1989) suggest that this model can *also* apply to the *dry-patch* and the *transition* regimes shown in Figure 8.1. The result of their predictions are compared with some experimental results, including those of Liaw and Dhir (1989) for a vertical plate, and some of their typical results are shown in Figure 8.28. Because of the use of empiricism and adjustable constants, the predictions of the model are in a very good agreement with the experimental results, including the magnitude of $(q_{s_b})_{\rm max}$. In this conduction-evaporation model this maximum corresponds to the maximum in the liquid-vapor interfacial area per unit solid surface area, or $(A_{\ell_v})_{\rm cell}/\ell^2$. Note that this *geometrical* maximum has been obtained by the selection of the two geometric parameters ϵ_{s_b} and ℓ.

In the surface bubble-nucleation regime, the contribution of the direct heat transfer to the vapor is negligible. Beyond the surface superheat for the maximum (i.e., critical) heat flux, in the transition regime, the superheating of the vapor becomes significant. Dhir and Liaw (1989) *combine* the results of the *vapor-film regime* heat transfer, discussed in Section 8.1, with the results of the bubble-nucleation regime to extend the results for the transition regime and also the surface superheat values where the vapor-film regime *dominates*. The intermittency of the liquid contact in the transition regime has been discussed by Bankoff and Mehra (1962) and Chen and Hsu (1992). This *combination* is made by a linear superposition of the results of the two regimes using the surface void fraction ϵ_{s_b}, i.e.,

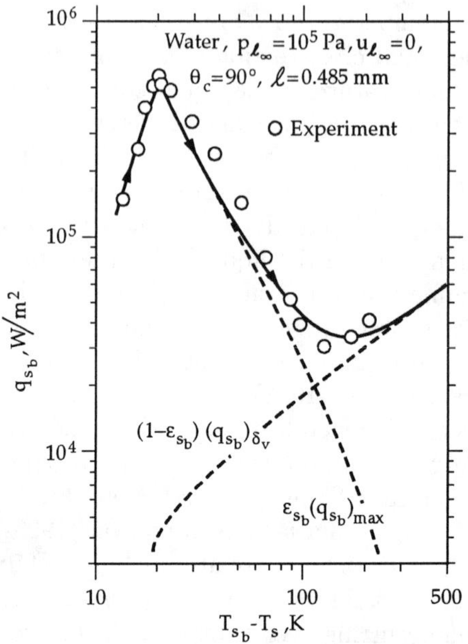

Figure 8.28. Comparison of the predicted and measured variation of the surface heat flux with respect to the surface superheat. The results are for both the surface bubble-nucleation and the transition surface-nucleation regimes. (From Dhir and Liaw, reproduced by permission ©1989 American Society of Mechanical Engineers.)

$$q_{s_b} = (1 - \epsilon_{s_b})(q_{s_b})_{\max} + \epsilon_{s_b}(q_{s_b})_{\delta_v} , \tag{8.205}$$

where $(q_{s_b})_{\delta_v}$ is found as mentioned in the discussion of the vapor-film flow (Section 8.1). Again, good agreement is found for the vertical, vapor-film flow asymptote ($\epsilon_{s_b} \rightarrow 1$), as shown in Figure 8.28.

8.6 Surface Droplet Formation and Dynamics

We now examine the surface droplet nucleation, growth, and departure on a cooled surface bounding a saturated vapor, as depicted in Figure 8.29. The role of convective heat transfer in condensation on surfaces with a partially wetting condensate resulting in surface droplet formation is *not* expected to be significant at relatively *low* vapor velocities. Most existing analyses of droplet formation and dynamics on cooled surfaces bounding a vapor are for an otherwise *quiescent* vapor. The interest in the droplet form of condensation, versus the liquid-film form discussed in Section 8.2, is that the surface droplets which have a *spherical cap* shape offer a small conductive resistance to the heat flow around their *peripheries*. This can increase the heat transfer and therefore the condensation rate. Various surface coatings

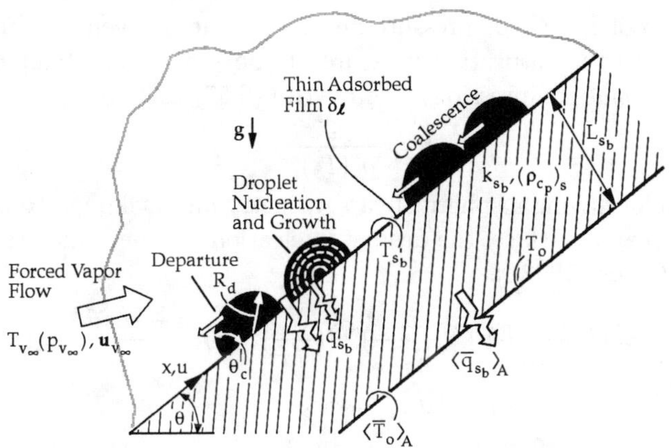

Figure 8.29. A rendering of the droplet nucleation, growth, departure, and coalescence on a cooled surface bounding a saturated vapor.

which increase the contact angle of the liquid, i.e., hinder wetting, have been examined (e.g., Griffith, 1985; Zhao, et al., 1991; Haraguchi, et al., 1991).

Similar to Section 8.5, in this section the *incipient surface subcooling* required for the surface droplet nucleation is addressed. Then the heat transfer aspects of the *isolated-droplets regime* are examined. Finally, we examine the *coalesced-droplets regime*, and the *transition* to a liquid-film flow, which occurs as the surface subcooling, is *increased*.

8.6.1 Incipient Surface Droplet Nucleation

The condensation of vapor on a cooled solid surface occurs when the surface temperature T_{s_b} becomes lower than a threshold (or incipient) temperature. This is designated by the *threshold supercooling* (or *supersaturation*), i.e., $T_s - T_{s_b}$, where T_s is the saturation temperature of the vapor. As with the incipient surface bubble nucleation discussed in Section 8.5.1, this supercooling would depend on the fluid and surface properties, and therefore would be different from the *bulk* or *homogeneous* supercooling. The *thermodynamic* considerations, i.e., the *vapor spinodal* temperature discussed in Section 1.5.1(C), and the results of the *kinetic theory*, i.e., the volumetric droplet nucleation rate discussed in Section 6.2.2(C), allow for the prediction of the *homogeneous* supercooling (or supersaturation). As with the vapor bubbles, the general practice is to *modify* the homogeneous nucleation theory to accommodate the surface energy, e.g., contact angle θ_c. This is done by introducing a *surface nucleation factor* f_n similar to that in (8.200).

In discussions of vapor condensation, in place of the surface subcooling $T_s - T_{s_b}$, the vapor *supersaturation ratio* S, defined in (6.57), is used. This is the ratio of the vapor pressure needed for the incipient surface droplet nucleation to the saturation pressure at the same vapor temperature T_s (i.e., *isothermal* compression). We rewrite (6.57) as

$$S = \frac{p_v}{p_v(T_s)} \,. \tag{8.206}$$

For a droplet having a spherical-cap geometry and making a contact angle θ_c with a planar surface, the droplet nucleation rate per unit *area* $\dot{N}_d/A$ is given by Carey (1992) as

$$\frac{\dot{N}_d}{A} = (\frac{2\sigma N_A f_n}{\pi M})^{1/2}(\frac{p_v}{R_g T_s})^{5/3}(\frac{N_A}{M})^{2/3} v_\ell (\frac{1-\cos\theta_c}{2}) f_n$$
$$\exp\{-\frac{16\pi(\frac{\sigma f_n}{R_g T_s})^3 v_\ell^3 N_A}{3M[\ln\frac{p_v}{p_v(T_s)}]^2}\} \,, \tag{8.207}$$

where the surface nucleation factor is given by

$$f_n = \frac{2 - 3\cos\theta_c + \cos^3\theta_c}{4} \,. \tag{8.208}$$

This is not valid for the limiting cases of a *perfectly* wetting liquid $\theta_c = 0°$, and an ideally nonwetting liquid $\theta_c = 180°$, because of the assumed geometrical model of the droplet meniscus. Carey (1992) computes $\dot{N}_d/A = (\dot{N}_d/A)(S, T_s, \theta_c)$ for water with $T_s = 100°\text{C}$. His results for several values of the contact angle are shown in Figure 8.30. This kinetic theory prediction of the supersaturation ratio requires a prescribed value of $\dot{N}_d/A$ in order to determine the supersaturation ratio for a given θ_c. Note that for $\dot{N}_d/A \rightarrow 0$, this supersaturation ratio becomes independent of $\dot{N}_d/A$. The supersaturation ratio *increases* with the *increase* in the contact angle and becomes nearly 2 for asymptotically nonwetting liquids ($\theta_c \rightarrow 180°$).

The calculation of the surface supercooling, $T_{s_b} - T_s$, is made in a similar manner but with the *specification* of the system vapor pressure p_v, and then the isobaric cooling gives the required pressure $p_v(T_{s_b})$ for the prescribed nucleation rate. In practice, the surface temperature is difficult to measure, and no extensive measurement of the surface supercooling for various *liquids*, *hydrodynamic* and *surface conditions*, and *surface geometry* are available. In general, the measured supercooling is *smaller* than the superheat.

8.6.2 Isolated-Droplet Regime

Beyond the threshold subcooling, the number of nucleations per unit area and the growth rate of the formed droplet *increase* as the subcooling increases. Then for small values of liquid subcooling an *isolated-droplets regime* exists and this is *followed* by the *interacting-droplets regime* which results in the droplet coalescence and eventually in the *liquid-film flow*.

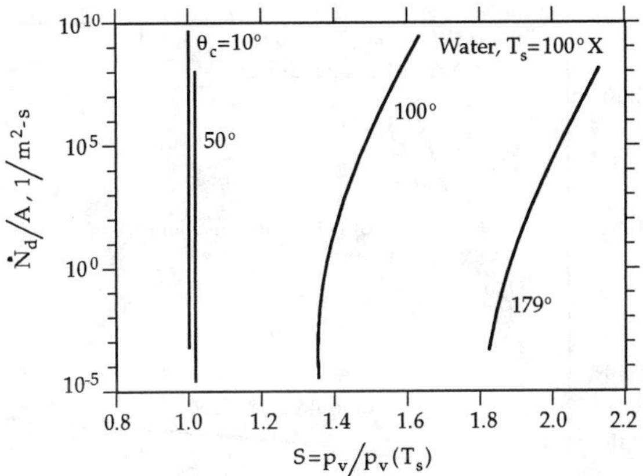

Figure 8.30. Predicted variation of the droplet nucleation rate per unit surface area, with respect to the vapor supersaturation ratio. The results are for water and for various contact angles. (From Carey, reproduced by permission ©1992 Taylor and Francis.)

The *growth* of a single, surface-nucleated droplet is by condensation on the droplet surface and is controlled by the heat transfer rate through the droplet and across the droplet-solid interface $A_{\ell s}$. Due to the gravity (i.e., negative phase-density buoyancy) or the drag force at the liquid-vapor interface $A_{\ell v}$ (for the case of a flowing vapor), the droplet *moves* when a *departure* size (or an equivalent diameter) is reached. This growth and departure is depicted in Figure 8.29. For vertical surfaces with no vapor motion, the departure diameter can be as large as 2 mm (Haraguchi et al., 1991).

The *hydrodynamic* theory of droplet condensation addresses the droplet shape and its sliding motion on the surface due to gravity and/or vapor drag. The solid surface around the droplets is covered by a thin, adsorbed film of thickness δ_ℓ (Song et al., 1991; Zhao et al., 1991). Merte and Son (1987) use the static Young-Laplace equation (8.146) and the droplet energy minimization to determine the departure volume. They consider a *two-dimensional, asymmetric* droplet and allow for different *advancing* (front of the droplet) and *receding* (back of the droplet) *angles*. In principle, the droplet spreading of Section 8.3.2 can be applied here by including phase change. However, in general, the droplets are treated as a *hemispherical* cap of radius R_d with a prescribed contact angle θ_c, as shown in Figure 8.29.

The *heat transfer* and growth rate of the droplet has been modeled by including the conduction heat transfer *in* the *solid surface*, allowing for a *nonequilibrium evaporation* at the liquid-vapor interface, and allowing for the suppression of the liquid-vapor saturation temperature due to *surface curvature capillarity*. The inclusion of the substrate (i.e., the solid) heat transfer results in a *nonuniform* solid surface temperature T_{s_l} and surface heat flux q_{s_l}. This is equivalent to the expected nonuniformity at the solid-

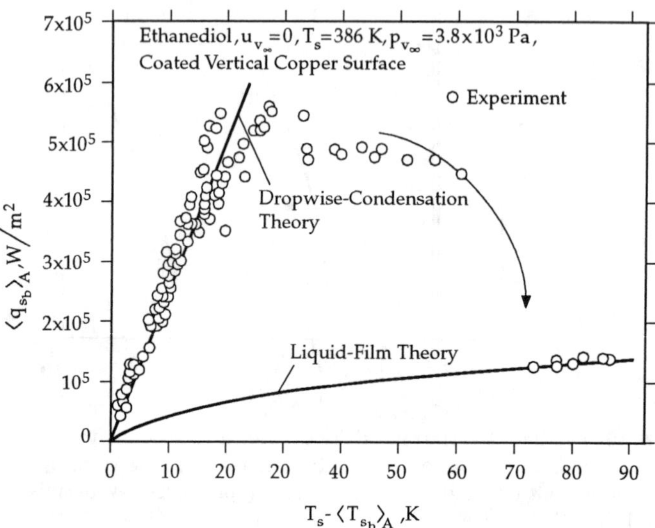

Figure 8.31. Measured and predicted variation of the surface heat flux with respect to the surface subcooling for dropwise and filmwise (and transition) condensation on a coated vertical surface. (From Stylianou and Rose, reproduced by permission ©1983 Pergamon Press.)

vapor interface for the case of bubble growth depicted in Figure 8.22. For $k_s \to \infty$ the *uniform* surface temperature assumption becomes valid. The heat transfer through the adsorbed film is generally *neglected* (Griffith, 1985; Merte 1973; Merte, et al., 1986; Carey, 1992).

When the liquid-vapor interfacial nonequilibrium and the curvature depression of the interfacial temperature are given as functions of R_d and T_s, then the *transient conduction* through the liquid droplet and the solid substrate determines the droplet growth rate. An *axisymmetric unit-cell*, transient conduction model of the droplet-substrate has been analyzed by Tsuruta and Tanaka (1991). The time- and space-averaged heat flux $\langle \overline{q_{s_b}} \rangle_A$ through the droplet-substrate is computed for the dropwise condensation of water (steam) on a vertical surface with $\theta_c = 90°$. Their results show that as the vapor pressure increases, the effect of the solid conductivity becomes more pronounced and for $p_v = 10^5$ Pa a nearly tenfold difference in the heat flux is predicted for $k_s \to 0$ and $k_s \to \infty$ (with q_{s_b} increasing with k_s for a given $T_s - T_{s_b}$). This has been found to be in good agreement with the experimental results (Tsuruta et al., 1991).

8.6.3 Coalescence and Transition to Liquid-Film Flow

At a given time, droplets of *different* diameters occupy the surface. This *diameter distribution* has been determined empirically (Stylianou and Rose,

1983; Griffith, 1985). As the surface subcooling, $T_s - T_{s_l}$, *increases*, the number of droplets per unit area increases and the departed droplets begin to *coalesce*. As the supercooling is further increased, the moving coalesced droplets join and form a *liquid-film flow*. Then the *hydrodynamics* of the moving droplets as influenced by the liquid viscosity μ_ℓ (assuming that the far-field vapor velocity is zero and neglecting the vapor viscosity), surface tension σ, and negative phase-density buoyancy $(\rho_\ell - \rho_v)g \simeq \rho_\ell g$, influence this *transition* to *droplet coalesence regime* and the later *transition* to *liquid-film* (or *filmwise*) *flow regime*. The increase in nucleation rate per unit surface area, as influenced by $T_s - T_{s_l}$, also influence these transitions. Figure 8.31 shows the experimental results of Stylianou and Rose (1983) for the condensation of ethanediol on a copper surface coated with various dropwise-condensation coatings (oleic acid and polytetrafluoroethylne which is also called ptfe, or Teflon). The results show that only a *small* supercooling is required for the incipient condensation. Then dropwise condensation continues until the *hydrodynamic* or *nucleation* (or *thermodynamic*) limits are reached and transitions to the coalescing-droplets regime and the liquid-film flow regime occurs. The theory of liquid-film flow (or filmwise) condensation was discussed in Section 8.2. This theory predicts the heat transfer rate accurately. The dropwise condensation theory, which is based on the *ensemble average* (over the droplet diameter distributions) of the isolated-droplet models discussed in Section 8.6.2, also predicts the results for this regime. Stylianou and Rose (1983) suggest the following criteria for *transition* to liquid-film flow regime for *vertical* surfaces (i.e., $\theta = 90^\circ$) with $u_{v_\infty} = 0$,

$$\frac{\langle q_{s_l}\rangle_A \mu_\ell}{\sigma \rho_\ell \Delta i_{\ell g}} \geq 2 \times 10^{-5} \quad \text{hydrodynamic transition,} \quad (8.209)$$

$$\frac{\sigma T_s}{\rho_\ell \Delta i_{\ell g}(\langle T_{s_l}\rangle_A - T_s)} < 10^{-9}\ \text{m} \quad \text{thermodynamic transition.} \quad (8.210)$$

8.7 Impinging Droplets

Impinging droplets can reduce the resistances preventing the realization of the upper limit on the evaporation rate. The resistance to the heat transfer to the liquid-vapor interface can be reduced by sweeping the heated solid surface by a rapidly moving liquid surface in close contact (due to droplet inertia). The resistance to vapor removal can be reduced by entraining the vapor in the rapidly moving gas surrounding the droplets. In order to understand and reduce these resistances, the *droplet impingement dynamics* (for multiple droplets)—which is influenced by the surface heat transfer—and the time- and space-averaged surface heat transfer rate have been studied extensively. However, because of the complexity of the *droplet impingement dynamics* (i.e., the droplet-surface interaction), the *interdroplet* interactions (i.e., interaction *before*, *during*, and *after* impingement), and the existence of various *evaporation regimes* (e.g., the surface bubble nucleation and the vapor film regimes), *no* comprehensive discussion is yet available (Yao, 1993).

Impinging droplets have many applications, for example, they are used for heat dissipation from a solid surface (i.e., spray cooling) or occur in the dispersed-droplet regime of the internal flow and evaporation (Section 8.1.3). Other applications are mentioned by Grissom and Wierum (1981), Shraiber and Khelemskiy (1986), Yao et al. (1988), Cokmez-Tuzla et al. (1992), Dawson and di Marzo (1993), and Yao (1993). In droplet impingement heat transfer, the heat is removed in *part* by the *surface evaporation* of the droplet through the *surface bubble nucleation* or the *vapor-film* evaporation. As with the *continuous* bulk liquid covering a surface discussed in Section 8.5, the rate of heat transfer is large when surface bubble nucleation occurs (as compared to the vapor-film regime discussed in Section 8.1). Therefore, when achievable, the surface bubble nucleation regime which results in the larger evaporation rate is the preferred regime of heat transfer. The evaporation rate is influenced by the *droplet parameters*, such as size (which generally is not uniformly distributed among droplets), velocity, thermophysical properties, and subcooling. It is also influenced by the *spray parameters*, such as the number of droplets per unit area (and their spatial distribution, i.e., the spray pattern, which changes with the distance from the nuzzle exit), and the velocity of the *gas surrounding* the droplets (which can have the same constituents as the droplets or be made of different species). It is further influenced by the *surface parameters*, such as its orientation with respect to the gravity vector, the surface superheat, and surface conditions (e.g., coating, roughness, entrapped gas). A rendering of the evaporating droplet impingement is shown in Figure 8.32. Some aspects of the *single-bubble* impingement and heat transfer are discussed below. This is followed by a discussion of the surface bubble-nucleation and vapor-film regimes in spray cooling. Finally, the *dryout* of the surface occurring in the surface bubble-nucleation regime, when the droplet flow rate is reduced below a threshold value, is discussed.

8.7.1 Isolated-Droplet Dynamics and Evaporation

The breakup of a droplet upon impingement on a surface is described by the *droplet Weber number* based on the droplet diameter d, i.e.,

$$We_d = \frac{\rho_\ell |v_d|^2 d}{\sigma}, \tag{8.211}$$

where $|v_d|$ is the magnitude of the droplet velocity *normal* to the surface (the coordinate axis is shown in Figure 8.32). The Weber number is the ratio of the inertial to surface tension forces, and for large droplet inertia

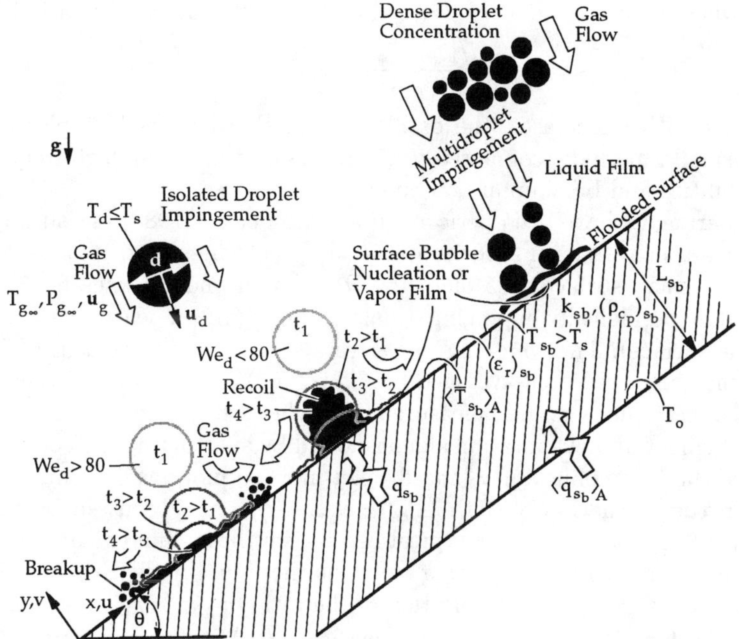

Figure 8.32. A rendering of the impinging and evaporating droplets showing the droplet impingement dynamics for the isolated-droplet (for two different droplet Weber numbers) and for the multidroplet impingement (with interdroplet interactions upon surface impingement).

the droplet breaks into many smaller droplets. The *isothermal* deformation and breakup of impinging droplets have been numerically simulated by Fukai et al. (1993). The surface temperature (compared to the saturation temperature of the liquid) influences this breakup (Bolle and Moureau, 1982; Yao et al., 1988; Chandra and Avedisian, 1992). For example, for a *water* droplet at room conditions with the surface temperature at 400°C (i.e., when a vapor film may be present between the droplet and surface), for $We_d < 30$ the droplet *bounces* on the surface without breakup. For $30 < We_d < 80$, the droplet *deforms* significantly but recovers its spherical shape (i.e., *recoils*) and in some cases a *few* smaller satellite droplets are formed. For $80 < We_d$, first the droplet spreads on the surface and forms a *liquid film* which becomes unstable and breaks up into many (of the order of 100) smaller droplets (Bolle and Moureau, 1982). At lower surface temperatures, a different behavior is expected and the surface conditions (including permeability) also influence this behavior (Senda et al., 1988; Chandra and Avedisian, 1992; Pias et al., 1992; Inada and Yang, 1993). The rendering given in Figure 8.32 for isolated droplets includes the case of surface bubble nucleation which influences recoiling. The *natural period*, or the *time constant* t_r, of the droplet is of the order of $\pi(16\rho_\ell d^3/\sigma)^{1/2}$

and when compared to a *residence* time t_d which is $d/|v_d|$ the ratio is

$$\frac{t_d}{t_r} = \frac{d/|v_d|}{\pi(16\rho_\ell d^3/\sigma)^{1/2}} = \frac{4}{\pi} We_d^{\ -1/2} . \tag{8.212}$$

Then, as We_d increases, the oscillations in the deformed droplet occur with smaller periods (compared to the residence time), and high-frequency instabilities and breakup are observed.

Experiments have been performed by Senda et al. (1988) measuring the heat transfer rate from a horizontal ($\theta = 0$), chrome-plated copper-heated surface (small surface area diameter $D = 8$ mm and thickness of 5 mm) cooled by periodically impinging droplets (impinging on nearly the location on the surface). The *time*- and *area*-averaged surface temperature $\langle \overline{T}_{s_b} \rangle_A$ and the time- and area-averaged heat flux $\langle \overline{q}_{s_b} \rangle_A$ are used, and there are some *uncertainties* in their measurements. The heat transfer due to droplet impingement and evaporation alone is found by subtracting the convection due to the thermobuoyancy- and droplet-induced motions. The uncertainties in the evaluation of both $\langle \overline{T}_{s_b} \rangle_A$ and $\langle \overline{q}_{s_b} \rangle_A$ are discussed by Senda et al. (1988). Some of their typical results for three different droplet impingement frequencies and using water as the liquid with $We_d = 116$ are shown in Figure 8.33. The variation of the ratio of $\langle \overline{q}_{s_b} \rangle_A / (\langle \overline{T}_{s_b} \rangle_A - T_s)$ is shown with respect to the surface superheat $\langle \overline{T}_{s_b} \rangle_A - T_s$. The results show a similarity with the distribution for the case of a continuous liquid phase, e.g., Figure 8.1.

The maximum heat flux occurs over the range of a superheat between 10 to 20°C. As the frequency of impaction f increases, the heat transfer rate increases (for a given superheat). This increase is *not* expected to be monotonic, because the surface becomes *flooded* when the frequency becomes very large. The concept of the heat transfer *effectiveness* η, the ratio of the actual heat transfer rate and the maximum possible when the droplet sensible and latent heat are *completely* used, i.e.,

$$\eta = \frac{\langle \overline{q}_{s_b} \rangle_A}{\langle \dot{m}_d \rangle_A [\Delta i_{\ell g} + c_{p_\ell}(T_s - T_d)]} , \tag{8.213}$$

has been used. As f becomes very large, η diminishes. Also, as the superheat increases and a *vapor film* is formed underneath the droplet, η decreases. These trends also exist in the droplet *sprays* (i.e., multiple, simultaneous impingement), which are discussed next.

8.7.2 Droplet Spray and Evaporation Regimes

Characteristics of sprays, i.e., droplet size distribution, spatial distribution of the droplets at various locations from the nozzle, the gas leaving the nozzle exit with the droplets (or the liquid *ligaments* which break into droplets), and the entrained gas are discussed by Clift et al. (1978), Bolle and Moureau (1982), Zhou and Yao (1992), and Yao (1993). A uniform

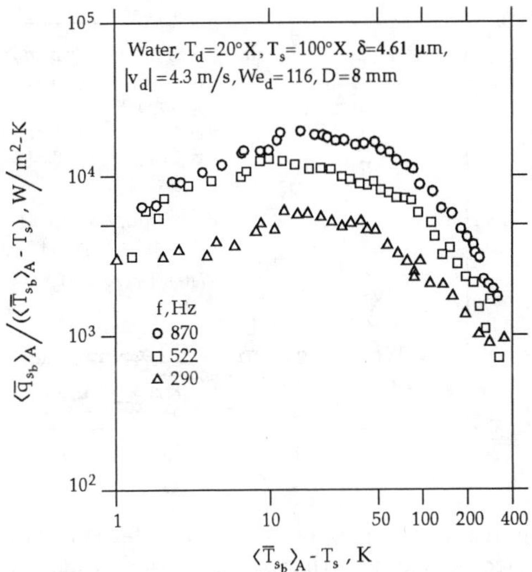

Figure 8.33. The measured variation of the surface heat transfer coefficient, for the impinging and evaporating droplets on a heated surface, with respect to the surface superheat. The results are for water and three different droplet-impingement frequencies. (From Senda et al., reproduced by permission ©1988 JSME International Journal.)

spatial distribution of the monosize droplets is the *ideal* description and is used to represent an average droplet mass flux $\langle \overline{\dot{m}_d} \rangle_A$ and diameter d.

As was mentioned, the evaporation of liquid supplied to the heated surface as droplets (i.e., discrete liquid phase) has some similarity with that for the continuous liquid phase, shown in Figure 8.1, and the surface bubble-nucleation regime or vapor-film regime is observed. The results of Choi and Yao (1987) for spray cooling (i.e., multiple and simultaneous droplet impingement) with water droplets impinging on a 14.4-cm-diameter chrome-coated horizontal copper plate are shown in Figure 8.34. The surface bubble-nucleation regime on the left-hand side, and the transition regime and vapor-film regimes on the right-hand side are evident. The results are for four different values of droplet time- and cross-sectional-averaged mass flux $\langle \overline{\dot{m}_d} \rangle_A$ with a slight variation of the droplet velocity perpendicular to the surface $|v_d|$. The gas velocity normal to the surface $|v_g|$ has nearly the same values as $|v_d|$. The Weber number for the 480-μm average diameter droplets is between 67 and 84. In the vapor-film regime, the heat transfer does not change significantly with the superheat. The total emissivity of the surface $(\epsilon_r)_{s_b}$ is estimated to be about 0.15. This makes the radiation contribution rather *small*. The experimental results of Choi and Yao (1987) show that an increase in d to 541 μm does *not* change the results noticeably.

An analysis of the vapor-film regime has been made by Ito et al. (1992),

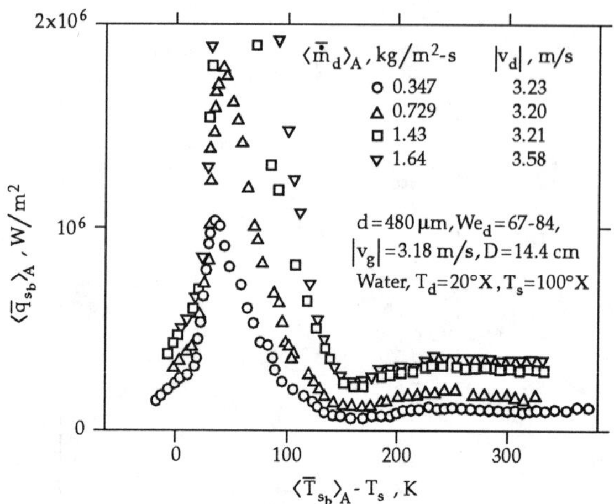

Figure 8.34. Liquid spray cooling of a heated surface showing the variation of the measured surface heat flux with respect to the surface superheat, for four different droplet mass fluxes. (From Choi and Yao, reproduced by permission ©1987 Pergamon Press.)

which allows for a *transient* (quasi-steady) droplet *deformation* and a vapor-film flow under the assumed isolated droplets. Time integration is made to allow for the heat transfer contribution of each droplet. Deb and Yao (1989) use a correlation to relate the vapor-film regime heat transfer rate to the various parameters, i.e.,

$$\eta = \eta[We_d, Ja, Pr_g, \frac{(\rho c_p k)_c}{(\rho c_p k)_{s_b}}] , \tag{8.214}$$

where the Jakob and Prandtl numbers are

$$Ja = \frac{c_{p_g}(\langle \overline{T_{s_b}} \rangle_A - T_s)}{\Delta i_{\ell g}} , \quad Pr_g = \frac{(\mu c_p)_g}{k_g} \tag{8.215}$$

and the last term in the square bracket is the ratio of the coating to substrate (chrome and copper for Figure 8.34) *thermal activity* or *effusivity*. The effects of liquid *subcooling* and surface orientation are not included in this correlation. The role of the gas-phase *convection* (called *air-assist* when the gas is made of air and the vapor of the liquid) is included through Pr_g and nearly equal gas-phase and droplet velocities (perpendicular to the surface) are assumed. This description does not address the droplet *crowding*, i.e., *flooding* of the surface, and does not predict the surface *dryout* at low droplet mass fluxes observed experimentally. These phenomena are discussed below.

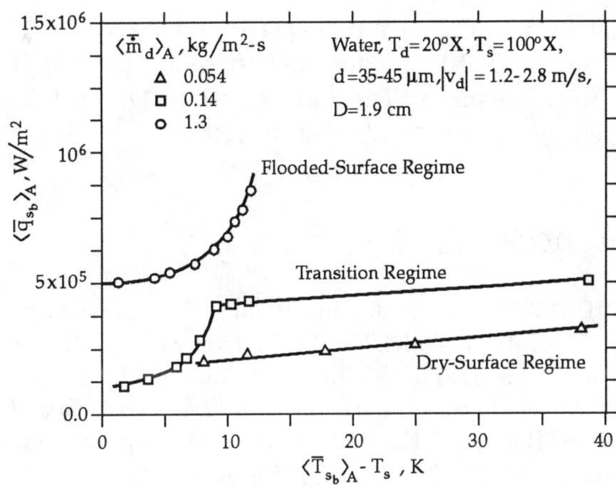

Figure 8.35. Liquid spray cooling of a heated surface showing the variation of the measured heat flux with respect to the surface superheat for three different droplet mass fluxes. The flooded, dry-surface and transitional regime are shown. (From Webb et al., reproduced by permission ©1992 Taylor and Francis.)

8.7.3 Dryout in Surface-Bubble Nucleation Regime

As the droplet mass flux is *decreased*, the surface dryout is encountered, where all the impinging liquids are evaporated. Any further decrease in the droplet mass flux increases the surface temperature *significantly*. This surface dryout occurs in the surface bubble-nucleation regime, where the surface superheat is *low* for the droplet mass flux $\langle \overline{\dot{m}_d} \rangle_A$ above the dryout threshold value. Figure 8.35 shows the experimental results of Webb et al. (1992) for water (with *no* assisting air flow through the nozzle) droplets impinging on an aluminum surface. Their results show that for a high droplet mass flux, the surface is *flooded* and the surface bubble nucleation occurs with a *small* surface superheat and a rather large surface heat flux. When the droplet mass flux is lowered below a threshold value, which is above 0.14 kg/m^2-s in Figure 8.35, the surface superheat is increased substantially with a slight increase in the surface heat flux. At the *transition* to dryout marked by a transitional liquid mass flux, *partial* or *total surface dryout* can occur. As the liquid mass flux is further reduced, the surface superheat becomes large even for small values of the surface heat flux and this is designated as the *surface dryout* in Figure 8.35. The surface bubble-nucleation regime is *not* observed for the dryout droplet mass flux. Webb et al. (1992) recommend that the ratio of the droplet mass flux to the droplet velocity $\langle \overline{\dot{m}_d} \rangle_A / |v_d|$ be used as the criterion for the *onset* of the dryout. The ratio is called the area- and time-averaged droplet density, i.e.,

$$\langle \overline{\rho_d} \rangle_A = (1 - \overline{\epsilon})\rho_d = \frac{\langle \overline{\dot{m}_d} \rangle_A}{|v_d|} . \tag{8.216}$$

They show that for water the transition criterion is given by $(1-\overline{\epsilon})\rho_d = 0.2$ kg/m^3, i.e., $1-\overline{\epsilon} \simeq 2 \times 10^{-4}$. They suggest that when the liquid fraction is below this limit, the surface encounters dryout. Other discussions of the dryout are given by Grissom and Wierum (1981).

8.8 References

Anderson, J.G.H., 1976, "Low-Flow Film Boiling Heat Transfer on Vertical Surfaces, Part I: Theoretical Model," *AIChE Symp. Ser.*, 73, 2–6.

Andreani, M., and Yadigaroglu, G., 1992, "A Mechanistic Eulerian-Lagrangian Model for Dispersed Flow Film Boiling," in *Phase-Interface Phenomena in Multiphase Flow*, Hewitt, G.F., et al., Editors, Hemisphere Publishing Corporation, New York.

Bach, P., and Villadsen, J., 1984, "Simulation of the Vertical Flow of Thin, Wavy Film Using a Finite-Element Methods," *Int. J. Heat Mass Transfer*, 27, 815-827.

Bankoff, S.G., and Mehra, V.S., 1962, "A Quenching Theory for Transition Boiling," *I & EC Fundamentals*, 1, 38–40.

Bar-Cohen, A., 1992, "Hysteresis Phenomena at the Onset of Nucleate Boiling," in *Pool and External Flow Boiling*, Dhir, V.K., and Bergles, A.E., Editors, American Society of Mechanical Engineers, New York.

Barthau, G., 1992, "Active Nucleation Site Density and Pool Boiling Heat Transfer—An Experimental Study," *Int. J. Heat Mass Transfer*, 35, 271-278.

Berenson, P.J., 1961, "Film-Boiling Heat Transfer from a Horizontal Surface," *ASME J. Heat Transfer*, 83, 351–358.

Bergles, A.E., 1992, "What Is the Real Mechanism of CHF in Pool Boiling?" in *Pool and External Flow Boiling*, Dhir, V.K., and Bergles, A.E., Editors, American Society of Mechanical Engineers, New York.

Bolle, L., and Moureau, J.C., 1982, "Spray Cooling of Hot Surfaces," in *Multiphase Science and Technology*, Hewitt, G.F., et al., Editors, Volume 1, 1–97, Hemisphere Publishing Corporation, Washington, D.C.

Bonetto, F., Clausse, A., and Converti, J., 1993, "Two-Phase Flow in the Localized Boiling Field Adjacent to Heated Wall," *Int. J. Heat Mass Transfer*, 36, 1367–1372.

Bräuer, H., and Mayinger, F., 1992, "Onset of Nucleate Boiling and Hysteresis Effects Under Forced Convection and Pool Boiling," in *Pool and External Flow Boiling*, Dhir, V.K., and Bergles, A.E., Editors, American Society of Mechanical Engineers, New York.

Bui, T.D., and Dhir, V.K., 1985, "Film Boiling Heat Transfer on an Isothermal Vertical Surface," *ASME J. Heat Transfer*, 107, 764–771.

Carey, V.P., 1992, *Liquid-Vapor Phase-Change Phenomena*, Hemisphere Publishing Corporation, Washington, D.C.

Chandra, S., and Avedisian, C.T., 1992, "Observations of Droplet Impingement on a Ceramic Porous Surface," *Int. J. Heat Mass Transfer*, 35, 2377–2388.

Chandrasekhar, S., 1961, *Hydrodynamic and Hydromagnetic Stability*, Dover, New York.

Chang, H.-C., 1987, "Evolution of Nonlinear Waves on Vertically Falling Films—A Normal Form Analysis," *Chem. Engng. Sci.* 42, 515–533.

Chang, K.H., and Witte, L.C., 1990, "Liquid-Solid Contact During Flow Film Boiling of Subcooled Freon-11," *ASME J. Heat Transfer*, 112, 465–471.

Chen J.C., and Costigan, G., 1992, "Review of Post-Dryout Heat Transfer in Dispered Two-Phase Flow," in *Post-Dryout Heat Transfer*, Hewitt, G.F., et al., Editors, CRC Press, Boca Raton, FL.

Chen, J.C., and Hsu, K.K., 1992, "Heat Transfer During Liquid Contact on Superheated Surface," in *Pool and External Flow Boiling*, Dhir, V.K.., and Bergles, A.E., Editors, American Society of Mechanical Engineers, New York.

Chen, J.D., and Wada, N., 1982, "Wetting Dynamics of the Edge of a Spreading Drop," *Phys. Rev. Lett.*, 62, 3050–3053.

Choi, K.J., and Yao, S.-C., 1987, "Mechanisms of Film Boiling Heat Transfer of Normally Impacting Spray," *Int. J. Heat Mass Transfer*, 30, 311–318.

Chou, X.S., and Witte, L.C., 1992, "A Theoretical Model For Flow Film Boiling Across Horizontal Cylinders," in *Pool and External Flow Boiling*, Dhir, V.K., and Bergles, A.E., Editors, American Society of Mechanical Engineers, New York.

Clift, R., Grace, J.R., and Weber, M.E., 1978, *Bubbles, Drops and Particles*, Academic Press, New York.

Cokmez-Tuzla, A.F., Tuzla, K., and Chen, J.C., 1992, "Experimental Assessment of Liquid-Wall Contacts in Post-CHF Convective Boiling," in *ANS Proceedings*, HTC-Vol. 6, 143–150, American Nuclear Society, La Grange Park, IL.

Cole, R., 1974, "Boiling Nucleation," *Advan. Heat Transfer*, 10, 85–166.

Collier, J.G., 1972, *Convective Boiling and Condensation*, McGraw-Hill, New York.

Cornwell, K., 1981, "On Boiling Incipience Due to Contact Angle Hysteresis," *Int. J. Heat Mass Transfer*, 25, 205–211.

DasGupta, S., Schonberg, J.A., and Wayner, P.C., Jr., 1993, "Investigation of an Evaporating Extended Meniscus Based on the Augmented Young-Laplace Equation," *ASME J. Heat Transfer*, 115, 201–208.

Dawson, H.F., and di Marzo, M., 1993, "Multidroplet Evaporative Cooling: Experimental Results," in *Heat Transfer-Atlanta, 1993, AIChE Symposium Series*, Volume 89, American Institute of Chemical Engineers, New York.

Deb, S., and Yao, S.-C., 1989, "Analysis on Film Boiling Heat Transfer of Impacting Sprays," *Int. J. Heat Mass Transfer*, 32, 2099–2122.

de Gennes, P.G., Hua, X., and Levinson, P., 1990, "Dynamics of Wetting: Local Contact Angles," *J. Fluid Mech.*, 212, 55–63.

Dhir, V.K., and Liaw, S.P., 1989, "Framework for a Unified Model for Nucleate and Transition Pool Boiling," *ASME J. Heat Transfer*, 111, 739–746.

Dussan V, E.B., 1979, "On the Spreading of Liquids on Solid Surfaces: Static and Dynamic Contact Lines," *Ann. Rev. Fluid Mech.*, 11, 371–400.

Dussan V, E.B., Ráme, E., and Garoff, S., 1991, "On Identifying the Appropriate Boundary Conditions at a Moving Contact Line: An Experimental Investigation," *J. Fluid Mech.*, 230, 97–116.

Dzyaloshinskii, I.E., Lifshitz, E.M., and Pitaevskii, L.P., 1961, "The General Theory of van der Waals Forces," *Adv. Phys.*, 10, 165–209.

Ehrhard, P., and Davis, S.H., 1991, "Nonisothermal Spreading of Liquid Drops on Horizontal Plates," *J. Fluid Mech.*, 229, 365–388.

Erneux, T., and Davis, S.H., 1993, "Nonlinear Rapture of Free Films," *Phys. Fluids*, A5, 1117–1122.

Faghri, M., and Seban, R.A., 1985, "Heat Transfer in Wavy Liquid Films," *Int. J. Heat Mass Transfer*, 28, 506–508.

Fujii, T., 1991, *Theory of Laminar Film Condensation*, Springer-Verlag, New York.

Fujita, Y., 1992, "The State-of-the-Art Nucleate Boiling Phenomena," in *Pool and External Flow Boiling*, Dhir, V.K., and Bergles, A.E., Editors, American Society of Mechanical Engineers, New York.

Fukai, J., Zhao, Z., Poulikakos, D., Mergaridis, C.M., and Miyataki, O., 1993, "Modeling of the Deformation of a Liquid Droplet Impinging upon a Flat Surface," *Phys. Fluids*, A5, 2588–2599.

Galloway, J.E., and Mudawar, I., 1993, "CHF Mechanism in Flow Boiling from a Short Heated Wall—II. Theoretical CHF Modal," *Int. J. Heat Mass Transfer*, 36, 2527–2540.

Gambill, W.R., and Lienhard, J.H., 1989, "An Upper Bound for Critical Boiling Heat Flux," *ASME J. Heat Transfer*, 111, 815–818.

Griffith, P., 1985, "Dropwise Condensation," in *Handbook of Heat Transfer*, Rohsenow, W.M., et al., Editors, Second Edition, McGraw-Hill, New York.

Grissom, W.M., and Wierum, F.A., 1981, "Liquid Spray Cooling of a Heated Surface," *Int. J. Heat Mass Transfer*, 24, 261-271.

Guo, Z., and El-Genk, M.S., 1992, "An Experimental Study of Saturated Pool Boiling from Downward Facing and Inclined Surfaces," *Int. J. Heat Mass Transfer*, 35, 2109–2117.

Haraguchi, T., Shimada, R., Humagai, S., and Takeyama, T., 1991, "The Effect of Polyvinylidene Chloride Coating Thickness on Promotion of Dropwise Steam Condensation," *Int. J. Heat Mass Transfer*, 34, 3047–3054.

Hirshburg, R.I., and Florschuetz, L.W., 1982, "Laminar Wavy-Film Flow: Part I, Hydrodynamics Analysis, Part II, Condensation and Evaporation," *ASME J. Heat Transfer*, 104, 452–464.

Hocking, L.M., 1992, "Rival Contact-Angle Models of the Spreading of Drops," *J. Fluid Mech.*, 239, 671–681.

Hocking, L.M., 1993, "The Influence of Intermolecular Forces on Thin Fluid Layers," *Phys. Fluids*, A5, 793–799.

Hockney, R.W., and Eastwood, J.W., 1989, *Computer Simulation Using Particles*, Adam Hilger, Bristol, England.

Hsu, Y.Y., and Westwater, J.W., 1960, "Approximate Theory for Film Boiling on Vertical Surfaces," *Chem. Eng. Prog. Symp. Ser.*, 56, 15–24.

Iida, Y., Okuyama, K., and Sakurai, K., 1993, "Peculiar Bubble Generation on a Film Heater Submerged in Ethyl Alcohol and Imposed a High Heating Rate over 10^7 Ks^{-1}," *Int. J. Heat Mass Transfer*, 36, 2699–2701.

Inada, S., and Yang, W.-J., 1993, "Effects of Heating Surface Materials on a Liquid-Solid Contact State in a Sessible Drop-Boiling System," *ASME J. Heat Transfer*, 115, 222-230.

Ishii, M., and Mishima, K., 1984, "Two-Fluid Model and Hydrodynamic Constititutive Relations," *Nucl. Eng. Design*, 82, 107–126.

Israelachvili, J.W., 1989, *Intermolecular and Surface Forces*, Academic Press, London.

Ito, T., Takata, Y., and Mousa, M.M., 1992, "Studies on the Water Cooling of Hot Surfaces (Analysis of Spray Cooling in the Region Associated with Film Boiling," *JSME Int. J.*, Series II, 35, 589–598.

Jairajpuri, A.M., and Saini, J.S., 1991, "A New Model for Heat Flow Through a Macrolayer in Pool Boiling at High Heat Flux," *Int. J. Heat Mass Transfer*, 34, 1579–1591.

Jansen, F., Minlar, M., and Schröder, J.J., 1992, "The Influence of Wetting Behavior on Pool Boiling Heat Transfer," in *Pool and External Flow Boiling*, Dhir, V.K., and Bergles, A.E., Editors, American Society of Mechanical Engineers, New York.

Jordan, D.P., 1968, "Film and Transition Boiling," *Advan. Heat Transfer*, 5, 55–128.

Judd, R.L., 1981, "Applicability of the Bubble Flux Density Concept," *ASME J. Heat Transfer*, 103, 175–177.

Judd, R.L., 1989, "The Influence of Subcooling on the Frequency of Bubble Emission in Nucleate Boiling," *ASME J. Heat Transfer*, 111, 747-751.

Karapantsios, T.D., Paras, S.V., and Karabelas, A.J., 1989, "Statistical Characteristics of Free-Falling Films at High Reynolds Numbers," *Int. J. Multiphase Flow*, 15, 1–21.

Katto, Y., 1985, "Critical Heat Flux," *Advan. Heat Transfer*, 17, 1–64.

Kaviany, M., 1989, "Forced Convection Heat and Mass Transfer from a Partially Liquid-Covered Surface," *Num. Heat Transfer*, A15, 445–469.

Kennig, D.B.R., 1992, "Wall Temperature Variations and the Modelling of Bubble Nucleation Sites," in *Pool and External Flow Boiling*, Dhir, V.K., and Bergles, A.E., Editors, American Society of Mechanical Engineers, New York.

Kennig, D.B.R., 1993, "Wau Temperature Pattern in Nucleate Boiling," *Int. J. Heat Mass Transfer*, 35, 73–86.

Kikuchi, Y., Ebisu, T., and Michiyoshi, I., 1992, "Measurement of Liquid-Solid Contact in Film Boiling," *Int. J. Heat Mass Transfer*, 35, 1589–1594.

Kirillov, P.L., Kashcheyev, V.M., Muranov, Yu.V., and Yuriev, Yu.S., 1987, "A Two-Dimensional Mathematical Model of Annular-Dispersed and Dispersed Flows—I and II," *Int. J. Heat Mass Transfer*, 30, 791-800.

Kirk, K.M., Merte, H., Jr., and Keller, R.B., 1992,"Low-Velocity Nucleate Flow Boiling at Various Orientations," in *Fluid Mechanics Phenomena in Microgravity*, AMD-Vol-154, 1-10, American Society of Mechanical Engineers, New York.

Klausner, J.F., Mei, R., Bernhard, D.M., and Zeng, L.Z., 1993, "Vapor Bubble Departure in Forced Convection Boiling," *Int. J Heat Mass Transfer*, 36, 651–662.

Kocamustafaogullari, G., and Ishii, M., 1983, "Interfacial Area and Nucleation Site Density in Boiling Systems," *Int. J. Heat Mass Transfer*, 26, 1377–1387.

Koh, J.C.Y., 1962,"Analysis of Film Boiling on Vertical Surfaces," *ASME J. Heat Transfer*, 84, 55–62.

Kutateladze, S.S., 1982, "Semi-Empirical Theory of Film Condensation of Pure Vapours," *Int. J. Heat Mass Transfer*, 25, 653–660.

Lamb, H., 1945, *Hydrodynamics*, Dover, New York.

Lee, R.C., and Nydahl, J.E., 1989, "Numerical Calculation of Bubble Growth in Nucleation Boiling from Inception Through Departure," *ASME J. Heat Transfer*, 111, 474–479.

Leiner, W., 1992, "Transition from Convective to Nucleate Pool Boiling and Effects of Nucleation Activities on Hysteresis," in *Pool and External Flow Boiling*, Dhir, V.K., and Bergles, A.E., Editors, American Society of Mechanical Engineers, New York.

Leiner, W., and Gorenflo, D., 1992, "Methods of Predicting the Boiling Curve and a New Equation Based on Thermodynamic Similarity," in *Pool and External Flow Boiling*, Dhir, V.K., and Bergles, A.E., Editors, American Society of Mechanical Engineers, New York.

Liaw, S.P., and Dhir, V.K., 1989, "Void Fraction Measurement During Saturated Pool Boiling of Water on Partially Wetted Vertical Surface," *ASME J. Heat Transfer*, 111, 731–738.

Lienhard, J.H., 1988, "Burnout on Cylinders," *ASME J. Heat Transfer*, 110, 1271–1286.

Liu, Q.S., Shiotsu, M., and Sakurai, A., 1992, "A Correlation for Forced Convection Film Boiling Heat Transfer from a Horizontal Cylinder," in *Two-Phase Flow and Heat Transfer*, Kim, J.H., et al., Editors, HTD-Vol. 197, American Society of Mechanical Engineers, New York.

Lyu, T.H., and Mudawar, I., 1991, "Determination of Wave-Induced Fluctuation of Wall Temperature and Convection Heat Transfer Coefficient in the Heating of a Turbulent Falling Liquid Film," *Int. J. Heat Mass Transfer*, 34, 2521–2534.

Merchant, G.J., and Keller, J.B., 1992, "Contact Angles," *Phys. Fluids*, A4, 477–485.

Merte, H., Jr., 1973, "Condensation Heat Transfer," *Advan. Heat Transfer*, 9, 181–272.

Merte, H., Jr., and Son, S., 1987, "Further Consideration of Two-Dimensional Condensation Drop Profiles and Departure Sizes," *Wärme- und Stoffübertragung*, 21, 163–168.

Merte, H., Jr., Yamali, C., and Son, S., 1986, "A Simple Model for Dropwise Condensation Heat Transfer Neglecting Sweeping," *Proceedings, Int. Heat Transfer Conference*, Volume 4, 1659–1664, 1986.

Michaelides, E.E., Liang, L., and Lasek, A., 1992, "The Effect of Turbulence on the Phase Change of Droplets and Particles Under Nonequilibrium Conditions," *Int. J. Heat Mass Transfer*, 35, 2069–2076.

Miller, C.A., and Ruckenstein, E., 1974, "The Origin of Flow During Wetting of Solids," *J. Colloid Interface Sci.*, 48, 368–373.

Mirzamoghadam, A., and Catton, F., 1988, "A Physical Model of the Evaporating Meniscus," *ASME J. Heat Transfer*, 110, 201–213.

Moosman, S., and Homsy, G.M., 1980, "Evaporating Menisci of Wetting Fluids," *J. Colloid Interface Sci.*, 14, 1835–1842.

Naghiem, L., Merte, H., Winter, E.R.F., and Beer, H., 1981, "Prediction of Transient Inception of Boiling in Terms of a Heterogeneous Nucleation Theory," *ASME J. Heat Transfer*, 103, 69–73.

Nakayama, A., and Kano, M., 1992, "Liquid Motion in Saturated-Pool Nucleate Boiling by Interfering Bubbles," *Heat Transfer-Jap. Res.*, 805–815.

Nakorykov, V.E., Pokusaev, B.G., and Alekseenkon, S.V., 1976, "Stationary Rolling Waves on a Vertical Film of Liquid," *J. Engr. Phys.*, 30, 517–521.

Nishio, S., Chandratilleke, G.R., and Ozu, T., 1991, "Natural-Convection Film-Boiling Heat Transfer, Saturated Film Boiling with Long Vapor-Film," *JSME Int. J. Series II*, 34, 202–212.

Nishio, S., and Ohtake, H., 1992, "Natural-Convection Film Boiling Heat Transfer," *JSME Int. J., Series II*, 35, 580–588.

Nishio, S., and Ohtake, H., 1993, "Vapor-Film-Unit Model and Heat Transfer Correlation for Natural-Convection Film Boiling with Wave Motion Under Subcooled Conditions," *Int. J. Heat Mass Transfer*, 36, 2541–2552.

Osiptsov, A.N., and Shapiro, Ye.G., 1993, "Heat Transfer in Boundary Layer of 'Gas-Evaporating Drops' Two-Phase Mixture," *Int. J. Heat Mass Transfer*, 36, 71–78.

Pias, M.R., Chow, L.C., and Mahefkey, E.T., 1992, "Surface Roughness and Its Effects on the Heat Transfer Mechanism in Spray Cooling," *ASME J. Heat Transfer*, 114, 211-219.

Pilch, M., and Erdman, C.A., 1987, "Use of Breakup Time Data and Velocity History Data to Predict the Maximum Size of Stable Fragments for Acceleration-Induced Breakup of a Liquid Drop," *Int. J. Multiphase Flow*, 13, 741–757.

Ramilison, J.M., and Lienhard, J.H., 1987, "Transition Boiling Heat Transfer and the Film Transition Regime," *ASME J. Heat Transfer*, 109, 746–752.

Rane, A.G., and Yao, S.-C., 1981, "Convective Heat Transfer to Turbulent Droplet Flows in Circular Tubes," *ASME J. Heat Transfer*, 103, 679–684.

Rose, J.W., 1988, "Fundamentals of Condensation Heat Transfer: Laminar Flow Condensation," *JSME Int. J., Series II*, 31 357–375.

Rowlinson, J.S., and Widom, B., 1989, *Molecular Theory of Capillarity*, Oxford University Press, Oxford.

Ruckenstein, E., 1966, "Remarks on Nucleate Boiling Heat Transfer from a Horizontal Surface," *Int. J. Heat Mass Transfer*, 9, 229–237.

Sakurai, A., and Shiotsu, M., 1992, "Pool Film Boiling Heat Transfer and Minimum Film Boiling Temperature," in *Pool and External Flow Boiling*, Dhir, V.J., and Bergles, A.E., Editors, American Society of Mechanical Engineers, New York.

Sakurai, A., Shiotsu, M., and Hata, K., 1990, "General Correlation for Pool Film Boiling Heat Transfer from a Horizontal Cylinder to Subcooled Liquid: Part I: A Theoretical Pool Film Boiling Heat Transfer Model Including Radiation Contributions and Its Analytical Solution," *ASME J. Heat Transfer*, 112, 430–440.

Schrage, R.W., 1953, *A Theoretical Study of Interface Mass Transfer*, Columbia University Press, New York.

Senda, J., Yamada, K., Fujimoto, H., and Miki, H., 1988, "The Heat-Transfer Characteristics of a Small Droplet Impinging Upon a Hot Surface," *JSME Int. J.*, Series II, 31, 105–111.

Shekriladze, I.G., and Gomelauri, V.I., 1966, "Theoretical Study of Laminar Film Condensation of Flowing Vapor," *Int. J. Heat Mass Transfer*, 9, 581–591.

Shoji, M., 1992, "A Study of Steady Transition Boiling of Water: Experimental Verification of Macrolayer Vaporation Model," in *Pool and External Flow Boiling*, Dhir, V.K., and Bergles, A.E., Editors, American Society of Mechanical Engineers, New York.

Shraiber, A.A., and Khelemskiy, S.L., 1986, "Heat Transfer of Rapidly Moving Droplets Interacting with a Wall," *Heat Transfer-Sov. Res.*, 18, 107–112.

Song, Y., Xu, D., and Lin, J., 1991, "A Study of the Mechanism of Dropwise Condensation," *Int. J. Heat Mass Transfer*, 34, 2827–2831.

Sparrow, E.M., and Gregg, J.L., 1959, "A Boundary-Layer Treatment of Film Condensation," *ASME J. Heat Transfer*, 81, 13–18.

Stuhlträger, E., Naridomi, Y., Miyara, A., and Uehara, H., 1993, "Flow Dynamics and Heat Transfer of a Condensate Film on a Vertical Wall—I. Numerical Analysis and Flow Dynamics," *Int. J. Heat Mass Transfer*, 36, 1677–1686.

Stylianou, S.A., and Rose, J.W., 1983, "Drop-to-Filmwise Condensation Transition: Heat Transfer Measurement for Ethanediol," *Int. J. Heat Mass Transfer*, 26, 747–760.

Swanson, L.W., and Herdt, G.C., 1992, "Model of the Evaporating Meniscus in a Capillary Tube," *ASME J. Heat Transfer*, 114, 434–441.

Tao, Y.-X., and Kaviany, M., 1991, "Simultaneous Heat and Mass Transfer from a Two-Dimensional, Partially Liquid Covered Surface," *ASME J. Heat Transfer*, 113, 875–882.

Truong, J.G., 1987, *Ellipsometric and Interferometric Studies of Thin Liquid Film Wetting on Isothermal and Nonisothermal Solid Surface*, Ph.D. thesis, Rensselaer Polytechnic Institute, Troy, NY.

Truong, J.G., and Wayner, P.C., Jr., 1987, "Effects of Capillary and van der Waals Dispersion Forces on the Equilibrium Profile of a Wetting Liquid: Theory and Experiment," *J. Chem. Phys.*, 87, 4180–4188.

Tsuruta, T., and Tanaka, H., 1991, "A Theoretical Study on the Constriction Resistance in Dropwise Condensation," *Int. J. Heat Mass Transfer*, 34, 2779–2786.

Tsuruta, T., Tanaka, H., and Togashi, S., 1991, "Experimental Verification of Constriction Resistance Theory in Dropwise Condensation Heat Transfer," *Int. J. Heat Mass Transfer*, 34, 2787–2796.

Ulucakli, M.E., and Merte, H., Jr., "Nucleate Boiling with High Gravity and Large Subcooling," *ASME J. Heat Transfer*, 112, 451–457.

Unal, C., Daw, V., and Nelson, R.A., 1992, "Unifying the Controlling Mechanisms for the Critical Heat Flux and Quenching: The Ability of Liquid to Contact the Hot Surface," *ASME J. Heat Transfer*, 114, 972–982.

Valentine, H.R., 1967, *Applied Hydrodynamics*, Plenum Press, New York.

van Stralen, S.J., and Cole, R., Editors, 1979, *Boiling Phenomena*, Volumes 1 and 2, Hemisphere Publishing Corporation, Washington, D.C.

Varone, A.F., and Rohsenow, W.H., 1986, "Post-Dryout Heat Transfer Predictions," *Nucl. Eng. Design*, 95, 315–327.

Walton, A.J., 1989, *Three Phases of Matter*, Second Edition, Oxford University Press, Oxford.

Wang, C.H., and Dhir, V.K., 1992, "On the Prediction of Active Nucleation Sites Including the Effect of Surface Wettability," in *Pool and External Flow Boiling*, Dhir, V.K., and Bergles, A.E., Editors, American Society of Mechanical Engineers, New York.

Wang, C.H., and Dhir,V.K., 1993,"Effect of Surface Wettability on Active Nucleation Site Density During Pool Boiling of Water on a Vertical Surface," *ASME J. Heat Transfer*, 115, 659–669.

Wasden, F.K., and Ducker, A.E., 1989, "Numerical Investigation of Large-Wave Interactions on Free Falling Films," *Int. J. Multiphase Flow*, 15, 357–370.

Wayner, P.C., Jr., 1989, "A Dimensionless Number for the Contact Line Evaporative Heat Sink," *ASME J. Heat Transfer*, 111, 813–814.

Wayner, P.C., Jr., 1991, "The Effect of Interfacial Mass Transfer on Flow in Thin Liquid Films," *Colloids Surfaces*, 52, 71–84.

Webb, S.W., and Chen, J.C., 1982, "A Numerical Model for Turbulent Nonequilibrium Dispersed Flow Heat Transfer," *Int. J. Heat Mass Transfer*, 25, 325–335.

Webb, B.W., Queiroz, M., Oliphant, K.N., and Bonin, M.P., 1992, "Onset of Dry-Wall Heat Transfer in Low-Mass-Flux Spray Cooling," *Exp. Heat Transfer*, 5, 33–50.

Witte, L.C., and Lienhard, J.H., 1982, "On the Existence of Two Transition Boiling Curves," *Int. J. Heat Mass Transfer*, 25, 771–779.

Witte, L.C., and Orozco, J., 1984, "The Effect of Vapor-Velocity Profile Shape on Flow Film Boiling from Submerged Bodies," *ASME J. Heat Transfer*, 106, 191–197.

Wolf, D.H., Incropera, F.P., and Viskanta, R., 1993, "Jet Impingement Boiling," *Advan. Heat Transfer*, 23, 1–132.

Xu, X., and Carey, V.P., 1990, "Film Evaporation from a Micro-Grooved Surface: An Approximate Heat Transfer Model and Its Comparison with Experimental Data," *J. Thermophysics*, 4, 512–520.

Yadigaroglu, G., and Bensalem, A., 1987, "Interfacial Mass Generation Rate Modeling in Nonequilibrium Two-Phase Flow," in *Multiphase Science and Technology*, Hewitt, G.F., et al., Editors, Volume 3, 85–127, Hemisphere Publishing Corporation, Washington, D.C.

Yao, S.-C., 1993, "Dynamics and Heat Transfer of Impacting Sprays," *Ann. Rev. Heat Transfer*, 5, 351–382.

Yao, S.-C., and Henry, R.E., 1978, "An Investigation of the Minimum Film Boiling Temperature on Horizontal Surfaces," *ASME J. Heat Transfer*, 100, 260–267.

Yao, S.-C., Hochreiter, L.E., and Cai, K.Y., 1988, "Dynamics of Droplets Impacting on Thin Heated Strips," *ASME J. Heat Transfer*, 110, 214–220.

Yoder, G.L., and Rohsenow, W.H., 1983, "A Solution for Dispersed Flow Heat Transfer Using Equilibrium Fluid Conditions," *ASME J. Heat Transfer*, 105, 11–17.

Zeng, L.Z., and Klausner, J.F., 1993, "Nucleation Site Density in Forced Convection Boiling," *ASME J. Heat Transfer*, 115, 215–221.

Zhao, Q., Zhang, D., and Lin, J., 1991, "Surface Materials with Dropwise Condensation Made by Ion Implantation Technology," *Int. J. Heat Mass Transfer*, 34, 2833–2835.

Zhou, Q., and Yao, S.-C., 1992, "Group Modeling of Impacting Spray Dynamics," *Int. J. Heat Mass Transfer*, 35, 121–129.

Zhukov, S.A., and Barelko, V.V., 1992, "Dynamic and Structural Aspects of the Processes of Single-Phase Convective Heat Transfer Metastable Regime Decay and Bubble Boiling Formation," *Int. J. Heat Mass Transfer*, 35, 759–775.

Zuber, N., 1958, "On the Stability of Boiling Heat Transfer," *Trans. ASME*, 80, 711-720.

Zuber, N., 1963, "Nucleation Boiling: The Region of Isolated Bubbles and the Similarity with Natural Convection," *Int. J. Heat Mass Transfer*, 6, 53–78.

Nomenclature

Uppercase bold letters indicate that the quantity is a second-order *tensor* and *lowercase* bold letters indicate that the quantity is a *vector* (or *spatial tensor*). Some symbols which are introduced briefly through derivations and otherwise are not referred to in the text are locally defined in the appropriate locations and are not listed here. The *mks* units are used throughout, and the magnitude of the fundamental physical constants used are from *Physics Today*, August 1993, BG9-13.

a	phase distribution function
$\mathbf{a}$	vector (magnetic) potential (V/s-m)
a_j	$j = 1, 2, \cdots$, constants
a_s	speed of sound (m/s)
A	area (m^2), pre-exponential factor
A_0	volumetric surface area (1/m)
Ar_x	interfacial area between i and j phases (m^2)
Arx	Archimedes or Galileo number $g\rho\Delta\rho x^3/\mu^2$
$\mathbf{b}$	magnetic induction (T or V-s/m^2 or N/A-m) or closure vector function (m)
b	half-width (m)
B	blowing parameter $\rho_s u_s/\rho_\infty u_\infty$
Bi_d	Biot number $h_{sf}d/k_s$
Bo_d	Bond or Eötvös number $\rho_\ell g d^2/\sigma$, where ℓ stands for the wetting phase
c_d	drag coefficient $8f_d/\pi d^2\rho u_\infty^2$
c_f	friction coefficient $2\tau_s/\rho u_\infty^2$
c_p	specific heat capacity at constant pressure (J/kg-K)
c_v	specific heat capacity at constant volume (J/kg-K)
C	mean clearance distance (m)
C_E	coefficient in the Ergun modification of the Darcy law
Ca	Capillary number where ℓ stands for the wetting phase $\mu_\ell u_{D\ell}/\sigma$
d	diameter (m) or pore-level linear length scale (m)
d_h	hydraulic diameter (m)
d_{ij}	components of deformation rate tensor (1/s)
$\mathbf{d}$	displacement current vector (A/m^2-s)
D	binary mass diffusion coefficient (m^2/s) or total thermal diffusion coefficient (m^2/s) or diameter (m)
D^d	dispersion coefficient (m^2/s)
D_K	Knudsen diffusivity (m^2/s)
$\mathbf{D}$	total thermal diffusivity tensor (m^2/s)
$\mathbf{D}^d$	dispersion tensor (m^2/s)

$\mathbf{D}_m$	total mass diffusivity tensor (m^2/s)
$\mathbf{D}_{m_e}$	effective mass diffusivity tensor (m^2/s)
e	specific internal energy (J/kg) or electric field intensity (V/m)
$\mathbf{e}$	electric field intensity vector (V/m)
e_c	electron charge 1.602×10^{-19} C or A-s
E_b	total blackbody radiation emissive power $n^2\sigma T^4$ (W/m^2)
Ec	Eckert number $u^2/c_p\Delta T$
E_μ	modulus of elasticity (Pa)
$E\ell$	normalized electric field force $\epsilon_e\epsilon_o e_\infty^2/\{\rho_o^{1/3}[\mu_o\beta_o g(T_s - T_\infty)]^{2/3}\}$
f	volumetric force (N/m^3) or frequency (Hz) or fugacity (Pa)
f_d	volumetric interfacial drag force (N/m^3)
f_e	volumetric electromagnetic force (N/m^3)
f_δ	volumetric van der Waals force (N/m^3)
$\mathbf{f}$	volumetric force vector (N/m^3)
F	radiation exchange factor
F_d	drag force (N)
$\mathbf{F}$	Force vector (N)
Fo_x	Fourier number $t\alpha/x^2$
Fr_d	Froude number $[u^2\rho/gd(\rho_\infty - \rho)]^{1/2}$ or $(u^2/gd)^{1/2}$
g	gravitational acceleration (m/s^2), standard (or normal) gravitational acceleration 9.8067 m/s^2 or specific Gibbs energy (J/kg)
$\mathbf{g}$	gravitational acceleration vector (m/s^2)
Ga_d	Galileo or Archimedes number $g\rho\Delta\rho d^3/\mu^2$
h	magnetic field intensity (A/m) or specific Helmholtz energy (J/kg)
h_{ij}	interfacial heat transfer coefficient between phases i and j (W/m^2-K)
h_P	Planck constant 6.626×10^{-34} (J-s)
$\mathbf{h}$	magnetic field intensity vector (A/m)
H	mean curvature of the meniscus $(1/2)(1/R_1 + 1/R_2)$ (m) where R_1 and R_2 are the two principal radii of curvature
Ha_x	Hartmann number $\sigma_e b_o^2 x^2/\mu$
i	specific enthalpy (J/kg) or $(-1)^{1/2}$
I	radiation intensity (W/m^2)
$\mathbf{I}$	second-order identity tensor
$\mathbf{j}_e$	current density (A/m^2)
Ja	Jakob number $c_p\Delta T/\Delta i_{\ell g}$, $(\rho c_p)_\ell\Delta T/\rho_v\Delta i_{\ell g}$
J_i	Bessel function
k	thermal conductivity (W/m-K), or wave number (1/m)
k_B	Boltzmann constant 1.381×10^{-23} (J/K), $k_B = R_g/N_A$
k_e	effective thermal conductivity (W/m-K)
K	permeability (m^2)
Ka	Kapitza number $\sigma/\rho_\ell({\nu_\ell}^4 g)^{1/3}$
K_A	Henry law constant (Pa)
Kn	Knudsen number, λ (mean free path)/C (average interparticle clearance)
$\mathbf{K}$	thermal conductivity tensor (W/m-K) or permeability tensor (m^2)
$\mathbf{K}_e$	effective thermal conductivity tensor (W/m-K)
ℓ	linear length scale for a representative elementary volume, unit-cell length (m), or mixing length (m)
L	system linear length scale (m)
Le	Lewis number $\alpha/D = k/\rho c_p D$
Ly_x	Lykoudis number $2Ha_x^2/{Gr_x}^{1/2}$
m	molecular mass (kg) or mass (kg)
m_e	electron mass 9.109×10^{-31} kg

$\dot{m}$	mass flux (kg/m^2-s)
$\dot{\mathbf{m}}$	mass flux vector (kg/m^2-s)
M	molecular weight (kg/kg·mole)
Ma	Mach number u/a_s
Ma_x	Marangoni number $(\partial\sigma/\partial T)\Delta Tx/\alpha_\ell\mu_\ell$
Mo	Morton number $g\mu^4\Delta p/\rho^2\sigma^3$
n	index of refraction or number of molecules per unit volume (molecule/m^3) or number density (number/m^3)
$\dot{n}$	volumetric nucleation rate (number/m^3-s) or volumetric production rate (kg/m^3-s)
$\dot{n}_i$	volumetric rate of production of species i (kg/m^3-s)
n_r	number of components in gas mixture
$\mathbf{n}$	normal unit vector
N	number of molecules
N_A	Avogadro number 6.0221×10^{23} (molecule/mole)
Na	Nahme number $\mu\Delta E_\mu u^2/kR_gT^2$
Nu_x	Nusselt number $qx/(T_s - T_\infty)k$
p	pressure (Pa)
p_c	capillary pressure (Pa)
Pe_x	Péclet number $Re_xPr = ux/\alpha$
P_k	production rate of turbulent kinetic energy (m^2/s^3)
Pr	Prandtl number $\nu/\alpha = (\mu c_p)_f/k_f$
Pr_e	effective Prandtl number $\nu/\alpha_e = (\mu c_p)_f/k_e$
Pr_t	turbulent Prandtl number ν_t/α_t
q	heat flux (W/m^2)
$\mathbf{q}$	heat flux vector (W/m^2)
Q	heat flow rate (W)
r	radial coordinate axis (m) or separation distance (m)
r_f	recovery factor $(T_s - T_\infty)/(u_\infty^2/2c_p)$
$\mathbf{r}$	radial position vector (m)
R	radius (m)
R_g	universal gas constant 8.3145×10^3 J/kg·mole-K, $R_g = k_BN_A$
Ra_x	Rayleigh number $g\beta\Delta Tx^3/\alpha\nu$
Re_x	Reynolds number ux/ν
Ri_x	Richardson number $(g\partial\rho/\partial x)/\rho(\partial u/\partial x)^2$ or $\Delta\rho gx/\rho u^2$ or $\beta\Delta gx/u^2$
$\dot{s}$	volumetric heat generation (W/m^3)
$\mathbf{s}_i$	unit vector in the i direction
S	path length (m) or supersaturation ratio
$\mathbf{S}$	viscous component of stress tensor (Pa)
Sc	Schmidt number $\nu/D = \mu/\rho D$
Sh_x	Sherwood number $\dot{m}x/(\rho_s - \rho_\infty)D$
Sr_d	Strouhal number fd/u_∞
St	Stanton number $q/(T_s - T_\infty)\rho c_pu_\infty$
Ste	Stefan number $c_{p_s}\Delta T/\Delta i_{\ell s}$
Sto	Stokes number $(R^2/\omega/\nu)^{1/2}$
t	time (s)
t_b	turbulent burst period (s)
t_d	particle relaxation time (s)
t_μ	relaxation time of non-Newtonian liquids (s)
T	temperature (K)
$\mathbf{T}$	stress tensor (Pa)
u, v, w	component of velocity vector in x, y, and z directions (m/s)
u_F	front or flame velocity (m/s)

$\overline{u}_m$	mean molecular speed (m/s)
$(\overline{u}_m^2)^{1/2}$	root-mean-square molecular speed (m/s)
$\mathbf{u}$	velocity vector (m/s)
$\mathbf{u}_D$	Darcean (or superficial) velocity vector (m/s)
v	specific volume (m^3/kg)
V	volume (m^3)
W	width (m)
We	Weber number $\rho_\ell u_{D\ell}^2 d/\sigma$
Ws	Weisenberg number $t_\mu u/x$
$\mathbf{x}$	position vector (m)
x, y, z	coordinate axes (m)
x_t	thermodynamic quality
Y_A	species A mass (or mole) fraction
Ze	Zel'dovich number $\Delta E_a \Delta T / R_g T^2$

Greek

α	thermal diffusivity (m^2/s)
α_t	turbulent eddy diffusivity (m^2/s)
α_T	temperature slip coefficient
α_u	velocity slip coefficient
β	volumetric thermal expansion coefficient (1/K)
β_m	volumetric species-concentration expansion coefficient (kg-mixture/kg-species)
γ	c_p/c_v
$\dot{\gamma}$	magnitude of deformation rate tensor (1/s)
Γ	vortex strength (1/s)
δ	boundary layer thickness (m), film thickness (m), or thickness (m)
ΔB	dimensionless transfer driving force (transfer number)
ΔE_D	diffusion activation energy (J/kg-mole)
ΔE_a	molar activation energy (J/kg-mole)
ΔE_μ	molar activation energy for the temperature dependence of viscosity (J/kg-mole)
$\Delta i_{\ell g}$	heat of evaporation (J/kg)
Δi_c	heat of combustion (J/kg or J/kg-mole)
ε	porosity or volume fraction
ε_e	dielectric constant (relative permittivity)
$\varepsilon_e \varepsilon_o$	permittivity (A^2-s^2/N-m^2)
ϵ_k	dissipation rate of turbulent kinetic energy (m^2/s^3)
ϵ_o	free space permittivity 8.8542×10^{-12} A^2-s^2/N-m^2
ε_r	surface emissivity
η	similarity variable, film cooling effectiveness, or wave number (1/m)
θ	polar (or callatitude) angle (rad)
θ_c	contact angle (rad) measured through the wetting fluid
θ_o	angle between incident and scattered beam (rad)
θ_s	separation angle (rad)
κ	isothermal compressibility (1/Pa) or index of extinction or von Kármán constant
λ	wavelength (m) or mean free path (m)
λ_μ	second viscosity (kg/m-s)
μ	dynamic viscosity (kg/m-s) or $\cos\theta$ (used in radiation) or chemical potential (J/mole)
μ_e	magnetic permeability (N/A^2)

ν	kinematic viscosity μ/ρ (m^2/s)
ν_k	eddy viscosity for turbulent kinetic energy (m^2/s)
ν_t	turbulent eddy viscosity (m^2/s)
ρ	density (kg/m^3), electrical resistivity (ohm/m), or reflectivity
ρ_e	electric charge density (Coulomb/m^3)
σ	surface tension (N/m)
σ_a	absorption coefficient (1/m)
σ_e	electric conductivity (mho/m) or extinction coefficient (1/m)
σ_s	scattering coefficient (1/m)
σ_{SB}	Stefan-Boltzmann constant 5.6705×10^{-8} W/m^2-K^4
τ	shear stress (Pa)
τ_r	radiation optical thickness (m)
τ_s	surface shear stress (Pa)
ϕ	azimuthal angle (rad), scalar
Φ_μ	viscous dissipation ($1/s^2$)
φ	molecular potential energy (J/molecule) or electrical (voltage) potential (V)
ψ	stream function (m^2/s)
ω	angular frequency (rad/s) or vorticity (1/s)
$\boldsymbol{\omega}$	vorticity vector (1/s)
Ω	solid angle (sr)

Superscripts

—	time average
*	dimensionless quantity
′	deviation from volume- or time-averaged value
b	bulk-mixed, or bed
d	dispersion component
ℓ	liquid
f	fluid
g	gas
s	solid or surface
t	transpose

Subscripts

o	reference
a	absorption or ambient
ad	adsorbed
b	boundary or blackbody radiation
c	condensible or convection
cr	critical
d	particle or downstream
D	Darcy
e	effective, extinction, electrical, or electromagnetic
f	fluid phase
fs	solid-fluid interface
g	gas phase
i	interfacial, incident, component i, or inside
j	component j
K	Knudsen
ℓ	liquid (or wetting) phase or representative elementary volume

L	system dimension
m	molecular or mass or mean
m_f	minimum fluidization
n	normal or nonreacted
o	reference or initial
p	pore or pressure
r	radiation, reflected, relative, or reacted
s	surface, solid phase, scattering, or saturation state
s_b	bounding surface
s_s	submerged surface
sc	subcooled
sf	solid-fluid interface
sg	solid-gas interface
sh	superheated
$s\ell$	solid-liquid interface
sp	spinodal limit
t	turbulent
tr	transition
T	temperature
u	velocity or upstream
v	vapor or volume
$v\ell$	vapor-liquid interface
x, y, z	x-, y-, and z-component
λ	wavelength dependent
μ	viscous

Others

[ ]	matrix
$\langle\ \rangle$	spatial (line, area, or volume) average
$O(\ )$	order of magnitude

Citation Index

Subject Index